STUDIES IN EARTH AND SPACE SCIENCES

Harry Hammond Hess
1906–1969

The Geological Society of America, Inc.
Memoir 132

Studies in Earth and Space Sciences

*A Memoir in Honor of
Harry Hammond Hess*

Edited by

R. Shagam   R. B. Hargraves

W. J. Morgan   F. B. Van Houten   C. A. Burk

H. D. Holland   L. C. Hollister

1972

Published by
THE GEOLOGICAL SOCIETY OF AMERICA, INC.
3300 Penrose Place
Boulder, Colorado 80302

Issued January, 1973

Printed in the United States of America

The publication of this volume
has been made possible through the bequest of
Richard Alexander Fullerton Penrose, Jr.

Contents

Caribbean Geology

Petrology

Mineralogy

Deformation of Materials

Environmental Geology

Space Science

Preface

Commenting on an essay by Macaulay, an editor once noted that the famous historian was not ". . . altogether the man to measure the full sweep of Byron's wing." How more precarious my predicament in writing this! The task is doubly compounded by the fact that tributes to Harry Hess have already appeared or are in press from the able pens of A. F. Buddington (1972, Geological Society of America Memorials Volume for 1969), H. L. James (National Academy of Sciences, in press), R. Girdler (*Nature*, 1969, v. 224, no. 5217, p. 393), T. W. Donnelly (1971, Geological Society of America, Introduction to *Memoir 130*), and Scott McVay (October 28, 1969, Princeton Alumni Weekly, v. 70, no. 6, p. 10–17) among others. If I am to avoid extensive repetition, there is little alternative but to seek the topics for discussion here within the psyche of one no longer living, with all the accompanying risks and pitfalls.

Common responses to the question "What was Hess's outstanding forte as a scientist?" are "remarkable imagination," "superb intuition," "brilliant thinker," and "creative mind." The implication is that his was a mind which appears fortuitously every once in a great while, rather than one which was shaped by educational experience. McVay (*ibid.*, p. 10, last line) quotes Hess's statement: "Scientific discoveries and ideas are produced by the intuition, creativeness and genius of a man . . . ," suggesting that his own view of scientific ability embraced, at least in part, a similar mystical quality. A metaphysical discussion of this subject is not indicated here; let us, however, take a closer look at the character of his special genius.

It appears to me that "remarkable imagination" and the other eulogistic phrases are largely synonymous with *impartial mind*. He brought to bear on geological problems an enquiring mind unfettered by preconceived notions. It is standard practice in geological research first to attempt to set broad limits and then to search, empirically or theoretically, for the truth between those limits. In some cases the limits are set carelessly and prove not to be as limiting as was at first supposed, or at least their validity is questionable. A modern example concerns the lunar rilles which have the appearance of river systems or comparable erosional agencies. Inasmuch as there is no significant lunar atmosphere one could apparently set one limit so as to eliminate explanation of the origin of the rilles by earthly fluids and processes of erosion, and turn to other fields (tectonics, volcanism) in search of an answer. Whether ultimately right or wrong, Hess, by interpreting the rilles in terms of the melting of subsurface lunar ice (*see* Helsley, this volume), showed that the limit was not as inescapable as at first appeared. Another example with far greater impact concerns the development of the science of oceanography. Prior to his famous paper on guyots (1946, Drowned ancient islands of the Pacific basin: *Am. Jour., Sci.,* v. 244, no. 11, p. 772–791), one unstated but understood limit *known* to all was set by the presence of one or more kilometers of sea water. Clearly one could not effect and interpret geological surveys as on land. Thus a limit was set and oceanography took the form of endless static compilations of data from dredge hauls with few attempts at dynamic synthesis. Hess in effect ignored the unstated limit and was able to integrate deep-sea data collected from the surface into a cogent dynamic hypothesis which, even though he had later to modify it markedly, paved the way for the present science of oceanography. Menard (1964 *Marine Geology of the Pacific:* New York, McGraw-Hill Book Co., p. 4) has pointed out his enormous contribution in this regard.

The importance of his impartial intellect is brought into sharper focus by the fact that many of his most stimulating syntheses can be traced, in part, to observations so elemental that they are standard fare in undergraduate geology courses. The wave-cut explanation for the beveled edges of guyots and the stream-course interpretation of the lunar rilles are good examples. Tackling the same or related problems, most workers quite simply would not believe the evidence of their own eyes, whereas Hess did. A common corollary to the finding of such standard features in apparently anomalous environments (after all, wave-cut platforms are supposed to occur along coastlines, not under 2 km of water in mid-ocean) inevitably lead him to *dynamic* solutions; assuming a wave cut for guyots, then either oceanic sea level had risen with time or the guyots had subsided (or both).

These facets of his scientific method were not restricted to the newly expanding fields of oceanography and lunar studies but characterized his every work. They emerged strikingly in the course of the graduate research project he supervised in the Caribbean Mountains of north-central Venezuela. There, extensive work by several investigators had failed to solve the problem of the stratigraphic relations between two major rock groups, one meta-sedimentary, the other metavolcanic, separated by a major high angle fault. The simple observation that the fault maintained a roughly constant stratigraphic position relative to reference horizons in each group suggested to him that it was originally a subhorizontal surface of décollement. This in turn led him to formulate the startling hypothesis that the volcanic terrain, some 300 km long and 20 to 30 km wide, was a vast allochthonous block which had reached its present location by sliding from a root zone located far to the north in the Caribbean Sea. Thus he overcame the unstated but implied limiting case that both rock groups were autochthonous. The dynamic solutions became his trademark; he did not merely rearrange the nomenclature and definition of pyroxene species but related them to a reasonable picture of the crystallization of basalt; he not only brought out the elegant order of island-arc geology but proposed a dynamic mechanism to explain it; he not only codified the characteristic features of ocean basins but, in what became his climactic work, integrated them into an active geotectonic mechanism.

That impartiality was a fundamental aspect of his brilliance is suggested to some degree by the fact that when some of his dynamic solutions proved wanting he was the first to renounce them. The role he envisaged for hypotheses of that kind is discussed below. For the moment it may be said that they served both as guideposts for extension of the concepts and as targets for those in disagreement with his views. If his "History of Ocean Basins" (1962, *in Petrologic Studies: A Volume to Honor A. F. Buddington:* Geological Society of America, 600 p.; *see also* Menard, this volume) set off a flood of papers by others caught up with the beauty of the concept, it also led to a marked focussing of papers in rebuttal. In this volume, Bellizzia and Bell extend Hess's hypothesis that the volcanic terrain in Venezuela is allochthonous, whereas Harvey returns to the originally accepted autochthonous relation.

In addition to his impartial intellectual faculties there were two other complementary facets basic to his outstanding contributions to the science: he combined professional mastery of the basic tools of geology with extensive, world-wide field experience. In the late 1950s he described himself as a mineralogist. His contributions to our knowledge of the pyroxenes have become classic papers and are invariably the source for recent studies (*see,* for example, Brown, this volume). Though he himself did no experimental work, he maintained close touch with workers and the literature in that field. Many forget that he was scarcely less of a geophysicist than he was a mineralogist. Not only did he work with a master of that field, Vening Meinesz, but there too he kept in close touch with new developments. His basic grasp of that branch of the science was evident in many papers, for example, his explanation for attenuated seismic velocities under oceanic ridges (*see also* MacKenzie, this volume). He was adamant on the importance of field work as the starting basis for geological research. One of his oft-repeated dicta was "If you can't hit it with a hammer, it probably is not a very good thesis." Unhappily, many of his modern imitators do not combine the same degree of professional mastery and field experience.

The reader may perhaps agree with my selection, if not the order of importance, of the talents he combined. In my opinion the extensive field experience served to locate and define the essence of the problem, and the command of basic science provided the tools to dissect and examine the problem, but it took the impartial mind to weave elicited data into a cogent, unified, rational, and dynamic whole. In the light of the foregoing, the exclamation "what a brilliant mind," however true, does not adequately convey the character of his genius. If my thesis is correct, his principal advantage over his contemporaries lay in the realm of psychology. Although his co-workers and graduate students were constantly kicking themselves for failing to register the very simple observations that were key elements in his dynamic solutions, it should be remembered that his solutions were seldom instantaneous affairs. In the Venezuelan study there were some 10 years of frustration before he came through with the hypothesis outlined above. The "geo-poetry" of his "History of Ocean Basins" was also not the product of a momentary flash of insight. In almost all his research projects he observed, examined, tested, and rejected, and returned to test and test again before finally arriving at a suitable explanation. The image conveyed to me is that of mental order and discipline, especially discipline. There is an echo here of a comment by the late Robert Oppenheimer to the effect that a good deal of the aura of genius surrounding Einstein for his Theory of Relativity stemmed not so much from his brilliance as from the laziness of many to buckle down and read the work! Though it is negative evidence I am further convinced of the legitimacy of my view on the basis of the intellectual development of some of Hess's students and research associates. They strove to emulate similar qualities of intellect and some of their dynamic solutions would have been a credit to their teacher. Clearly, abundance of gray matter did not in itself suffice.

If the opinions expressed are correct, there are important implications for our science. Fundamental training in science alone may not be enough. The matter is especially pertinent to the rapidly growing field of geological education. The excellent training manuals with their ingenious experimental exercises may serve only to prolong and exacerbate lives of biased science if the basic psychological problem is left unattended.

Where Hess will ultimately rank among the great men of our profession will be decided naturally and in due course by his peers. Whether or not in rank, certainly in *kind,* the order he brought to geology parallels that of the Bohr atom or of Darwinian evolution. It is interesting to speculate whether he came to his integrated geotectonic syntheses by chance or whether, as a young scientist, he deliberately set his sights on major accomplishments. Over the fireplace in the Hess living room there hung a portrait in oils of a cherubic, curly headed Harry in sailor suit at 5 years of age, entitled "The Future Admiral." I will not strain your credulity by suggesting that at that tender age he had set his sights on the rank he ultimately attained in the U.S. Naval Reserves. In science, however, there are indications that he worked according to at least a loose master plan, beginning as early as his graduate student years (1929 to 1932) at Princeton. Both James and Donnelly (*ibid.*) have noted that interest in ultramafic rocks runs as a continuous thread through his career, from his thesis study on serpentinite at Schuyler, Virginia, to his last field trip in South Africa. Apparently from the outset he foresaw the direct relevance of mantle petrology to the solution of geotectonic problems. But on closer inspection it appears that in the initial 5- to 6-year period (ca. 1931 to 1936) he had already begun to grapple with the other fields (gravity anomalies and island arcs, pyroxenes, plagioclases, layered complexes and the crystallization of basalt, and physical oceanography) in which he was to maintain an interest to the end. The bulk of this period was devoted to mastering the fundamental tools (mineralogy, petrology, geophysics) he subsequently employed in the geotectonic syntheses. In subsequent years these tools were further honed, and variation in the master plan was wrought largely by additional field experience. The intense concentration on island arcs in the 1930s and early 1940s gradually broadened via the guyots to mid-ocean ridges. In that sense only was it a loose master plan.

Signs of a rough master plan are also evidenced by his prescient appreciation for the importance of certain fields in which he was not a major contributor; for example, the measurement of radio-isotope content of rocks which might serve as possible mantle models. Another indication was the way he treasured his time. For a limited period prior to and following World War II he toyed with amphiboles but decided not to make them the subject of major research because, I suspect, he foresaw that their complexity demanded more time than he was prepared to devote to them. While he was always available to students for consultation on theses and went to great lengths to promote affairs of intellectual import, he spent little time on matters of mundane administration. Yet another pointer was his striking single-mindedness of purpose which could be quite unnerving. More than once he approached me with such opening gambits as: "No, I don't think so." I was invariably taken aback and only after several exchanges would it emerge that he was referring to casual opinions made many months previously in the field in

Venezuela. It is said he once picked up in mid-stream a discussion interrupted only by World War II!

Despite the charge of provincial partisanship which may be levelled, I suggest that he was significantly influenced toward his remarkable paper on the history of ocean basins by his contacts, spanning 40 years, with the geology and geologists of southern Africa. That region constituted one of the few outposts of vehement "driftism" during the lean years between the early efforts of Wegener and Taylor, and those of Hess himself and others (see Vine; Morgan; and others, this volume) in the 1960s. As reported by James (ibid.), Hess's initial field experience was 2 years in the bush of Zambia (then Northern Rhodesia) and his last, only a month before his death, saw him in the Barberton Mountain Land, Eastern Transvaal. He repeatedly expressed admiration for du Toit's "Wandering Continents" (Edinburgh, Oliver and Boyd, 1937, 366 p.) and was a member of the Geological Society of South Africa and a devoted reader of its publications.

His attitude over the years toward the concept of "drift" constitutes an interesting microcosm of the implicit master plan unfolding. In the early 1950s, in reply to my question, he commented that he found "drift" an intriguing idea and that some of the geological evidence was impressive, but that as long as there was no known source of energy adequate to power ships of continental dimensions he could not go along with the hypothesis. Not only was his reaction a rational one but it revealed the importance he assigned to developing an integrated dynamic solution that would reasonably explain the facts. When in due course studies on the physical state of the mantle revealed the feasibility of solid-state convection, it was again a rational matter not merely to jump on the bandwagon but to come to hold the reins. This he accomplished against the tide of popular opinion. Drift was anathema to most of the American school, and reactions varied from the ribald to the earnest but illogical. There were those who argued, for example, that if the former joining of South America and Africa was predicated on the correlation of geologic features, one could by analogy also stipulate drift between Arkansas and South Africa on the basis of the occurrence of kimberlite. If reactions in such prodrift centers as South Africa were philosophically sounder, they were still little more than observational empiricism: if one could show a marked geological continuity in space and time, statistically beyond the possibility of simple coincidence, then drift between South America and Africa was virtually proven. On the subject of energy source, with rare exceptions, the attitude was almost pragmatic ("I don't know how it happened, but it did happen"). Hess's logical and impartial approach to the problem was largely instrumental in achieving a volte-face of American workers toward the concept with the ironic result that workers from the New World are among the leaders in the study of sea-floor spreading and the concept of plate tectonics which stemmed from his work.

There is an important lesson to be learned here, one lucidly stated by Hess in the introduction to his paper "Mid-Oceanic Ridges and Tectonics of the Sea Floor" (1965, Colston Papers, Vol. XVII: London, Butterworth and Co.). Remarking on the rapid accumulation of data in the field of oceanography he noted "To bring the problem into focus and guide continued exploration, co-ordinating hypotheses are needed and necessary. Even incorrect or partly incorrect speculations serve to identify the crucial observations needed for progress. Blind, usually called objective or unprejudiced, collection of data without a framework of hypothesis by which it can be tested is wasteful and commonly unproductive, and leads to an accumulation of an indigestible mass of data of minor significance." Too many detractors of sea-floor spreading appear to believe that he developed the concept as a rigid, do-or-die theory of geotectonism and report gleefully every little fact which appears to contradict the hypothesis. There is no special fount of wisdom I can appeal to that would enable me to pass on to you the ultimate truth concerning sea-floor spreading. All that can be said, from the viewpoint of those interested in the process of orogeny, is that during the long dark night of static tectonics that dominated geologic thinking until Hess published his paper on the origin of ocean basins, just such an indigestible mass of data was accumulated and led only to complete puzzlement. Since the development of the sea-floor spreading hypothesis, a distinct ray of light has come to illuminate the enigma of orogeny and to provide rational explanations for the essential energy source and feasible mechanical models that have so long eluded us. This is what it is all about; this is what Hess accomplished with those remarkable dynamic solutions. Should it prove necessary to introduce radical changes to the hypothesis, even completely negate it, his fundamental contribution in terms of scientific stimulus will scarcely be affected. It should be recalled that he himself had to withdraw his hypothesis of rising sea level with time, contained in the "Drowned Ancient Islands" paper. I know of no thoughtful scientist who is now disdainful of that work; indeed it effectively ushered in the modern era of oceanography. It would be too much to expect that one man in his professional life could devise final solutions to the immense and complex geotectonic enigma of our planet. Still, I should be surprised if final solutions to such problems as those of orogeny did not show a direct line of descent from the paper on the history of ocean basins.

By the same token, those who now match faults on opposite sides of this or that ocean and thereby attempt to support the sea-floor spreading hypothesis are returning to the old observational empiricism and will not significantly contribute to the solution of the overall problem. Hess implied the need for continuous development of fresh hypotheses; hopefully the reader will detect some in this volume.

Inasmuch as the contributors to this volume have elected to promote the memory of Harry Hess the scientist through this collection of geological studies, it is appropriate that the preface should be largely concerned with the work and mind of the man. Yet his accomplishments as a scientist, remarkable as they were, in no way outshone his qualities as a human being. The traits of careful observation, rational thought, and impartiality he brought to bear on geological problems were but the extension of his basic attitude to his fellow man.

We shall not easily find another like him. One hopes that his successors will bring with them the same qualities of mind and heart. It will take a very human and intellectual giant to fill his shoes; such was the whole man.

REGINALD SHAGAM
Department of Geology
University of the Negev
Beer-Sheva, Israel

Acknowledgments

A remarkable degree of cooperation was obtained from all those connected with the preparation of this volume. It is a pleasure to acknowledge my indebtedness to my fellow editors who in reality bore the brunt of the editorial load. An inordinate share of the burden fell on the shoulders of R. B. Hargraves to whom I am particularly grateful. The editors, in turn, are beholden to critical readers who attended to their thankless task with outstanding alarcrity and considerable beneficial effect. All the editors participated in the critical review of manuscripts. In addition reviews were rendered by:

D. L. Anderson, California Institute of Technology; T. Atwater, Stanford University; H. G. Avé Lallement, Rice University; D. R. Baker, Rice University; S. K. Banerjee, University of Minnesota; M. N. Bass, NASA, Houston; W. E. Bonini, Princeton University; S. Bonis; M. G. Bown, Cambridge University, England; J. C. Briden, Leeds University, England; P. Butler Jr., NASA, Houston; A. R. Byers, University of Saskatchewan; N. Carter, Yale University; J. M. Christie, University of California, Los Angeles; S. P. Das Gupta, Geological Survey of India, Hyderabad, India; K. S. Deffeyes, Princeton University; J. Dewey, Cambridge University, England; R. S. Dietz, NOAA, Miami; T. W. Donnelly, State University of New York, Binghamton; C. L. Drake, Dartmouth College; K. O. Emery, Woods Hole Oceanographic Institute; A. G. Fischer, Princeton University; R. S. Fiske, U. S. Geological Survey, Washington, D. C.; F. Frey, Massachusetts Institute of Technology; R. F. Fudali, U. S. National Museum, Washington, D. C.; A. M. Gaines, University of Pennsylvania; R. F. Giegengack, University of Pennsylvania; J. Goguel, Service de la Carte Géologique, Paris, France; J. Handin, Texas A & M University; M. C. Hironaka, Naval Civil Engineering Laboratory, Port Hueneme, Cal.; C. A. Hopson, University of California, Santa Barbara; K. A. Howard, U. S. Geological Survey (Astrogeologic Studies), Menlo Park, Cal.; A. L. Inderbitzen, Lockheed Ocean Laboratory, San Diego, Cal.; T. N. Irvine, Geological Survey of Canada, Ottawa; Y. Isachsen, New York State Museum and Science Service, Albany; E. D. Jackson, U. S. Geological Survey, Menlo Park, Cal.; S. Judson, Princeton University; J. O. Kalliokoski, Michigan Technological University; G. deV. Klein, University of Illinois, Urbana; W. D. MacDonald, State University of New York, Binghamton; R. J. Malloy, Naval Civil Engineering Laboratory, Port Hueneme, Cal.; W. Maresch, Princeton University; J. Martignole, Université de Montreal, Canada; J. C. Maxwell, University of Texas, Austin; G. E. McGill, University of Massachusetts; A. A. Meyerhoff, American Association of Petroleum Geologists, Tulsa; C. B. Moore, Arizona State University; G. W. Moore, U. S. Geological Survey, La Jolla, Cal.; J. C. Moore, University of California, Santa Cruz; E. M. Moores, University of California, Davis; C. G. Murray, Princeton University; B. E. Nordlie, University of Arizona; L. Ogniben, Università di Catania, Sicily, Italy; N. Opdyke, Lamont-Doherty Geological Observatory, Columbia University; E. R. Oxburgh, Oxford University, England; R. A. Phinney, Princeton University; G. Plafker, U. S. Geological Survey, Menlo Park, Cal.; C. B. Raleigh, U. S. Geological Survey, Menlo Park, Cal.; J. J. W. Rogers, Rice University; W. W. Rubey, University of California, Los Angeles; W. W. Rutley, University of California, Los Angeles; K. N. Sachs Jr., U. S. National Museum, Washington, D. C.; A. Shapiro, University of the Negev, Beer Sheva, Israel; H. R. Shaw, U. S. Geological Survey, Washington, D. C.;

F. G. Stehli, Case Western Reserve University; C. H. Stockwell, Geological Survey of Canada, Ottawa; T. P. Thayer, U. S. Geological Survey, Washington, D. C.; M. J. Viljoen, Johannesburg Consolidated Investment, Johannesburg, South Africa; R. P. Viljoen, Johannesburg Consolidated Investment, Johannesburg, South Africa; F. J. Vine, University of East Anglia, England; K. M. Waage, Yale University; D. R. Waldbaum, Princeton University; L. Weiss, University of California, Berkeley; H. S. Yoder, Carnegie Geophysical Laboratory, Washington, D. C.; E-an Zen, U. S. Geological Survey, Washington, D. C.; and J. Zussman, The University of Manchester, England.

The Department of Geological and Geophysical Sciences, Princeton University, bore by far the major portion of the secretarial expenses involved. I am obliged to the chairman, Professor S. Judson, for his generous assistance in these and related matters. At the University of the Negev, Mr. I. Ben Amitai, Director General, and Professor E. Azmon, chairman of the Geology Department, were instrumental in arranging liberal secretarial and financial assistance in the closing phases of the project. Some initial secretarial costs were born by the Geology Department, University of Pennsylvania.

At every step sage advice and support was provided by the editorial staff of the Geological Society of America.

Mrs. H. Black, Princeton University, Mrs. M. R. Fanok and Mr. B. Blanchard, University of Pennsylvania, and Mrs. H. Weinberg and Mr. E. Z. Shertok, University of the Negev, rendered outstanding secretarial aid.

REGINALD SHAGAM

Geotectonics

THE GEOLOGICAL SOCIETY OF AMERICA, INC.
MEMOIR 132
© 1972

Citations in a Scientific Revolution*

H. W. MENARD

*Institute of Marine Resources and Scripps Institution of Oceanography,
University of California, San Diego, California 92037*

CITATIONS IN A SCIENTIFIC REVOLUTION

The earth sciences are in a revolution because of the demonstration a few years ago that continents drift, the sea floor spreads, and the earth has had a lively history. A few details of the demonstration are needed in order to understand the significance of a citation analysis of the subject.

Harry Hammond Hess published what he called an "essay in geopoetry" in 1962. He suggested that the mantle rises under mid-ocean ridges in the center of the ocean basins and is converted by a reversible reaction to oceanic crust. It then spreads laterally, rafting continents with it, to oceanic trenches, where it is reconverted to mantle. His reasons were multitudinous but included the facts that the ocean basins contain no old rocks, that the sediment is too thin for the basins to be old, that various criteria suggest mantle convection under the oceanic crust, and so on. In the following year Fred Vine and Drummond Matthews took note of the fact that the earth's magnetic field reverses itself episodically and that the orientation of the field is recorded in the magnetic minerals in cooling lava flows. Thus, they reasoned that the cooling and spreading of lavas in the center of ocean basins should produce magnetic stripes parallel to the central ridge. They accepted Hess's hypothesis but did not cite his 1962 paper.

The stripes were soon found and by overwhelming evidence were shown to be caused by the suggested reversals of the magnetic field. The stripes are bilaterally symmetrical for thousands of kilometers on each side of a mid-ocean ridge—which has been explained only by Hess's spreading mechanism. The stripes have been dated and the distance from the center is proportional to the age. Thus, it is known that the spreading has had a constant speed for a long time. Most geological evidence is relatively vague because of the many uncontrolled variables involved. It is important to emphasize that this evidence is quite adequate to convince mathema-

---

* Modified from "Science: growth and change," H. W. Menard, 1971, by permission of Harvard University Press.

ticians and physicists. These measurements plus countless confirmations showed that the fundamental elements of Hess's argument were correct and provided the foundation for the present scientific revolution.

By tracing the history of citations to Hess's paper we can examine how well they reflect the impact of a truly important contribution. Regrettably, the picture is clouded. Hess originally expected to publish his geopoetry in volume three of a multi-author work called *The Sea: Ideas and observations.* He cannot have felt any urgency in publication because it is well known that such volumes are invariably delayed for unconscionable periods while the most indolent authors try to meet their commitments. He did the standard thing, issuing a mimeographed preprint which most of us received in 1960. No one got very excited, but confusion resulted when Robert Dietz published very similar concepts in *Nature* in 1961. Hess, meanwhile, gave up *The Sea* and published his paper, "History of ocean basins," in yet another symposium volume called *Petrologic Studies: A Volume in Honor of A. F. Buddington,* published by the Geological Society of America in 1962. Dietz tidied things up in 1963 by giving priority to Hess for the basic ideas of interest here. Considering the prevalence of multiple discovery, this was a most unusual act. Hess then created some bibliographic complications by writing a further elaboration of his hypothesis in yet another symposium volume, the *Colston Papers,* in 1965. Thus, in 1960–1962, anyone wishing to refer to these ideas could cite either Dietz or the preprint by Hess. Then for two years either of two papers might be cited and after 1965 any one of three. Nonetheless, the 1962 paper by Hess was by far the most frequently referenced.

Citations to the paper increased fairly regularly from 2 in 1963 to 36 in 1968 and then declined to 18 in 1969.[1] This gave an exponential doubling in about 17 months from 1964 to 1968 and then a marked retardation of growth (Fig. 1). The books and journals studied can be grouped according to the publication delay in order to see if the citing authors had any sense of urgency about priorities. Apparently they were concerned, because in every year but 1965 most of the citations appeared in journals with the fastest available publication. In 1965, three symposium volumes swamped the citations to the Hess paper.

We identify 76 authors who cited "History of ocean basins" in the literature studied. Their number increased almost as rapidly as the citations. The distribution is 58 authors who cited once; 12, twice; 3, three times; 2, four times; and 1, seven times. In only eight years few of them had much time to cite the paper more than once. Even so, the prevalence of single citers suggests some factor at work. To understand this we must first look at what the authors said about the paper when they cited it.

As we have already observed, one of the principal reasons for the dormancy of the earth sciences was a surfeit of geopoetry without the means to distinguish the work of a Keats from that of an Edgar Guest. All the facts that Hess explained had already been explained by other hypotheses. Consequently, the real novelty in Hess's paper was that he *called* it geopoetry or that he explained so many different

---

[1] We have examined the following journals from 1962 through 1969: *American Journal of Science, Journal of Geology, Marine Geology, Bulletin of the Geological Society of America, Science, Nature, Journal of Geophysical Research, Earth and Planetary Science Letters, Tectonophysics,* and the *Journal of Petrology.* We have also studied citations in seventeen symposium volumes resulting from meetings during the period.

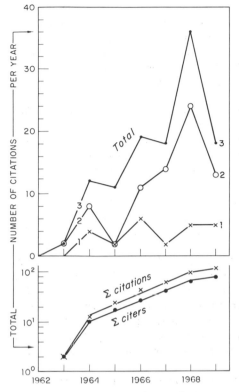

Figure 1. Citations to the revolutionary paper, "History of ocean basins," published by Harry Hess. (1) Slower journals: Geol. Soc. America Bull., Jour. Geology, Am. Jour. Sci., Jour. Petrology, Tectonophysics, Marine Geology; (2) Faster journals: Nature, Science, Earth and Planetary Sci. Letters, Jour. Geophys. Research; (3) total.

things. Eleven citations were given in 1963 and 1964. One was by Hess, who said "if it is correct," and one by Wilson, who had very similar ideas anyway. These two plus others by Orowan and Schuiling, who are applied physicists, gave a total of four citations which accepted the idea of sea-floor spreading. The seven remaining citations were made by geologists and geophysicists who responded to the idea with a hardly disguised "ho-hum." One paper merely cited some data from Hess and never referred to his ideas at all.

A paper by Maurice Ewing and others in 1964 set the tone for many in 1965 and even 1966. Ewing and his colleagues stated that the object of their study of sediment distribution and thickness in the North Atlantic was to test the hypotheses of continental drift and sea-floor spreading. If Hess had been right, they reasoned, the sediment would be thin in the center of the basin where the crust is newly created and it would thicken on the increasingly older crust away from the center. What they found is that the sediment is indeed thin or absent in a central band 75 miles wide but that in the surrounding region it is uniformly thick. This was baffling and remains so. Consequently, in the summary of conclusions they say nothing at all about the avowed object of the study except that (p. 33) "the sediment distribution also denies the possibility that any part of the ridge crust, except the crest, is appreciably older than another."

Given apparently conflicting data, attention was focused on similarities rather than differences. The time was not ripe for a revolution.

Precisely the same thing occurred in a paper by van Andel and others published in 1965. In a comparatively detailed geological survey they found that by every criterion available to them the crest of the Mid-Altantic Ridge is young and the flanks are older. However, the conclusion they reached was (p. 1216) "We do not, however, believe that our data either require or materially support such an interpretation" of sea-floor spreading. Their conclusion was absolutely correct because the data were subject to other interpretations and the time was not ripe.

I went through this phase at about the same time. After offering a hypothesis to explain many aspects of marine geology, I pointed out that the "bolder hypothesis" by Hess would also explain them but that the facts did not require acceptance of his ideas. Dale Krause in 1965 was in about the same position when he found his data "in harmony" with sea-floor spreading but also with other ideas. By the end of 1965, the count was 12 citations by unconcerned or skeptical authors, 8 citations by those who had faith, and finally a crucial citation by Vine and Wilson (1965), who gave proof of sea-floor spreading. (Proof does not mean instantaneous enlightenment for all scientists; the overwhelming power of the proof was only gradually apparent to most of us.)

In 1966 and 1967 the members of the marine geology establishment looked at their data and went through the same experience. Most said "My observations are not compatible with sea-floor spreading and I shall prepare a critical demonstration that this is so and thus demolish this nutty idea and we can all get back to work." One by one they found (for they were honest scientists) that in fact their data, regardless of the subject, were compatible with sea-floor spreading. An elaborate network of confirmations appeared and most marine geologists got back to work, but with a new paradigm. During this period several people reexamined data on which they had already published interpretations not favorable to Hess's hypothesis. The time was ripe and gradually the various uncertainties in interpretation were found to balance in favor of the hypothesis. Several ad hoc explanations were still available for most facts and relations. What was changed was the paradigm, because of a few crucial observations and field experiments. The citers in 1968 and 1969 were not the established marine geologists who were on to other things by then. They were mostly very young marine geologists and members of the establishment of continental geologists who were examining *their* data in a new light. Almost all accepted the proof of sea-floor spreading and began to reinterpret what they had already done.

We can now return to the question of why so many authors cite "History of ocean basins" only once. A scientific revolution means that a large number of apparently unrelated facts are seen to be in an unsuspected relation. Many established workers have available the means to apply critical tests. Each type of test is important but there is little occasion for many repetitions; thus, one author, one subject, one test, one citation.[2]

The graph suggests that the 36 citations of 1968 may not again be matched. The number probably will drop off to some much lower steady-state value. It appears that the extremely important papers that trigger a revolution may not receive a proportionately large number of citations. The normal procedures of referencing are not used for folklore. A real scientific revolution, like any other revolution, is news. *The Origin of Species* sold out as fast as it could be printed and was denounced from the pulpit almost immediately. Sea-floor spreading has been explained, perhaps not well, in leading newspapers, magazines, books, and most recently in a color motion picture. When your elementary school children talk about something at dinner, you rarely continue to cite it.

We have attempted to verify some of the citation patterns of a revolutionary paper by a brief study of the very famous one in which Watson and Crick (1953)

---

[2] I have cited the paper seven times. The fact came as a surprise to me and certainly was not deliberate. As it happens, we had at Scripps some brilliant students, an enormous stock of unprocessed magnetic data, and a data processing capacity which my associates and I had developed to analyze echograms. The instruments and techniques could also analyze the magnetic data just when it was demonstrated that they provide the best information about sea-floor spreading. (There was a windfall!)

announced the structure of DNA. Citations in the *Journal of Molecular Biology* have remained relatively constant from 1959, when the journal was founded, to the present. The only major perturbation was in 1964–1965, not long after the authors and M. H. F. Wilkins won the Nobel Prize for this work. In the *Science Citation Index* the annual number of citations from 1964 through 1968 was 48, 49, 45, 40, and 36, respectively, and is estimated at about 30 for 1969.[3] This suggests a declining trend but also suggests that it is a paper that will be cited extensively for some time to come. By 1965, it was being cited in the *Journal of Dairy Science* and the *Journal of the American Oil Chemists Society,* among others.

To a geologist, the number of journals in which citations occur is astonishing, and it would be interesting to follow the expanding recognition of the importance of a paper and its broad applicability.

REFERENCES CITED

Ewing, M., Ewing, J., and Talwani, M., 1964, Sediment distribution in the oceans: The Mid-Atlantic Ridge: Geol. Soc. America Bull., v. 75, p. 17–36.

van Andel, Tj. H., Bowen, V. T., Sachs, P. L., and Siever, R., 1965, Morphology and sediments of a portion of the Mid-Atlantic Ridge: Science, v. 148, p. 1214–1216.

Vine, F. J., and Wilson, J. Tuzo, 1965, Magnetic anomalies over a young oceanic ridge off Vancouver Island: Science, v. 150, p. 485–489.

Watson, J. D., and Crick, F.H.C., 1953, Molecular structure of nucleic acids—A structure for deoxyribose nucleic acid: Nature, v. 171, p. 737.

MANUSCRIPT RECEIVED BY THE SOCIETY MARCH 29, 1971

[3] This paper gives a neat example of the difficulties of using these very useful volumes. Sometimes the journal is listed as *Nature* and sometimes it is separately listed as *Nature London.* We have identified the cited years as 1953, 1958, and 1963; the volume numbers as 171 and 177; the page numbers as 737, 373, 727, 137, and 780.

THE GEOLOGICAL SOCIETY OF AMERICA, INC.
MEMOIR 132
© 1972

Plate Motions and
Deep Mantle Convection

W. JASON MORGAN

Princeton University, Princeton, New Jersey 08540

ABSTRACT

A scheme of deep mantle convection is proposed in which narrow plumes of deep material rise and then spread out radially in the asthenosphere. These vertical plumes spreading outward in the asthenosphere produce stresses on the bottoms of the lithospheric plates, causing them to move and thus providing the driving mechanism for continental drift. One such plume is beneath Iceland, and the out-pouring of unusual lava at this spot produced the submarine ridge between Green-land and Great Britain as the Atlantic opened up. It is concluded that all the aseismic ridges, for example, the Walvis Ridge, the Ninetyeast Ridge, the Tuamotu Archipelago, and so on, were produced in this manner, and thus their strikes show the direction the plates were moving as they were formed. Another plume is beneath Hawaii (perhaps of lesser strength, as it has not torn the Pacific plate apart), and the Hawaiian Islands and Emperor Seamount Chain were formed as the Pacific plate passed over this "hot spot."

Three studies are presented to support the above conclusion. (1) The Hawaiian-Emperor, Tuamotu-Line, and Austral-Gilbert-Marshall island chains show a re-markable parallelism and all three can be generated by the same motion of the Pacific plate over three fixed hot spots. The accuracy of the fit shows that the hot spots have remained practically fixed relative to one another in this 100 m.y. period, thus implying a deep source below the asthenosphere. (2) The above motion of the Pacific plate agrees with the paleo-reconstruction based on magnetic studies of Pacific seamounts. The paleomotion of the African plate was deduced from the Walvis Ridge and trends from Bouvet, Reunion, and Ascension Islands. This motion did not agree well with the paleomagnetic studies of the orientation of Africa since the Cretaceous; however, better agreement with the paleomagnetic studies of Africa and of seamounts in the Pacific can be made if some polar wan-dering is permitted in addition to the motion of the plates. (3) A system of absolute plate motions was found which agrees with the present day relative plate motions (deduced from fault strikes and spreading rates) and with the present trends of island chains–aseismic ridges away from hot-spots. This shows that the hot spots form a fixed reference frame and that, within allowable errors, the hot spots do not move about in this frame.

BASIC MODEL

Let us suppose there is convection deep in the mantle. The arguments presented here do not depend on the depth of such convection—any depth from just beneath the asthenosphere to the core-mantle boundary would suffice—but, for present purposes, let us say such convection extends to a 2,000 km depth. It is common knowledge that such deep convection is improbable due to the efficiency of heat transport by radiation at this depth, but let us explore the possibility of such convection and then come back to the heat flow "proofs" of impossibility. Suppose there are several (approximately 20) plumes of deep mantle rising upward and the rest of the mantle is slowly sinking downward in a pattern analogous to a thunderhead or a coffee percolator. To add concreteness, suppose there are several "pipes" in the rigid middle mantle and that very hot lower mantle is coming upward in these pipes and being added to the asthenosphere. The more rigid middle mantle, including the "walls" of the pipes, is slowly moving downward to fill the void created below in the more fluid lower mantle, and this rigid middle mantle is being added to at its top as the asthenosphere cools and welds itself to the mesosphere. The 400 or 600 km discontinuity may mark this boundary between mesosphere and asthenosphere.

Such a model has the following features. There are about 20 pipes to the deep mantle bringing up heat and relatively primordial material to the asthenosphere (Fig. 1). Within the asthenosphere, there will be horizontal flow radially away from each of these pipes. These points of upwelling will have unique petrologic and kinematic properties, but there will be no corresponding unique downwelling points, as the return flow is assumed to be uniformly distributed throughout the remainder of the mantle. The pattern of localized upwelling without localized downwelling was suggested by the gravity map (Fig. 8, to be discussed later). How will such a flow pattern interact with the crustal plates above? A plate will respond to the net sum of all stresses acting on it (the shear stress acting on its bottom due to currents in the asthenosphere plus the stresses on its sides due to its motion relative to adjacent plates). It appears that the plate-to-plate interactions are very important in determining the net forces on a plate, that is, the existing rises, faults, and trenches have a self-perpetuating tendency. This claim is based on two observations: (1) rise crests do not commonly die out and jump to new locations (Labrador and Rockall are the only places for which the evidence strongly suggests extinct rise crests), and (2) points of deep upwelling do not always coincide with ridge crests (for example, the Galápagos and Reunion upwellings are near triple junctions in the Pacific and Indian Oceans; asthenosphere motion radially away from these points would help drive the plates away from the triple junction, but there is considerable displacement between these pipes to the deep mantle and the lines of weakness in the lithosphere which enable the surface plates to move apart). Also note the toughness of the plates as exemplified by the fact that the upwelling beneath Hawaii has not torn apart the Pacific plate.

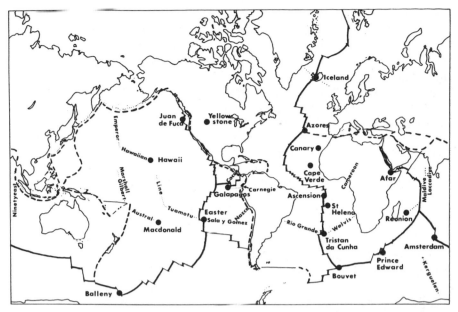

Figure 1. Map showing the locations of the probable hot spots and the names of some features cited in the text.

This model is compatible with the observation that oceanic island basalts are different from oceanic ridge basalts (Gast, 1968). Island type basalts, as on Iceland or Hawaii, would have access to relatively primordial material from deep in the mantle. In contrast, the ridge crests tap only the asthenosphere—the asthenosphere passively rising up to fill the void created as the plates are pulled apart by the stresses acting on them. The oceanic ridge basalts are known to be relatively low in potassium and in some trace elements. We may account for this by claiming that the asthenosphere source has been reworked and cleaned out of lighter elements in previous sea-floor spreading episodes, or that the lighter elements have had sufficient time to migrate upward (to the bottom of the lithosphere) and are not present to rise to the ridge crest where the plates are pulled apart. The oceanic island basalts are rich in potassium and have a rare earth distribution implying more fractionation, in accord with their deep primordial source. If we relate the observed island basalt fractionation to the composition of the parent rock, we should have a new picture of the composition of the deep mantle. Such an estimate will undoubtedly be higher in potassium than those estimates based on ridge basalts. The implied increased estimate of radiogenic heat production is desirable in this scheme, in that the deep convection model requires more heat production at depth than radiative transport alone can cope with.

As the Pacific plate moves over the upwelling beneath Hawaii, the continuous outpouring of basalt from this point produces a linear basaltic ridge on the sea floor—the Hawaiian Islands. Likewise, the excessive flow from Tristan de Cunha has produced the Walvis and Rio Grande Ridges in the South Atlantic. Here we require Africa to drift to the northeast (parallel to the Walvis Ridge) and South America to drift roughly northwest (parallel to the Rio Grande Ridge). Note that the transform faults between Africa and South America trend east-west; the transform faults show the *relative* motion of the African and South American plates, the

Walvis and Rio Grande Ridges show the *absolute* motion of Africa and South America (plus the effects of the migration of the hot spot, to be discussed later).

We assume that all such aseismic ridges are produced by plate motion over hot spots fixed in the mantle. Thus the aseismic ridges indicate the trajectories of the plates over fixed points and we may reconstruct continental positions with both latitude and longitude control, an important addition to paleomagnetic reconstructions. This interpretation of the aseismic ridges and island chains is identical to that presented by Wilson (1963, 1965) except that here we attribute a more fundamental nature to the hot spots—we associate the hot spots with major convection deep in the mantle, providing the motive force for sea-floor spreading.

We shall now examine three aspects of the worldwide pattern of island chains and aseismic ridges consistent with the concept of plate motions over fixed hot spots.

ISLAND CHAINS IN THE PACIFIC

There are only two presently active volcanos in the interior of the Pacific plate, Hawaii and Macdonald Seamount (Johnson, 1970). It has long been noted that the active Hawaiian volcano is at the southeast extreme of the Hawaiian chain and that there is a linear progression of the age of these islands as they become farther from Hawaii (Fig. 2). Johnson has noted that Macdonald Seamount (29.0° S.,

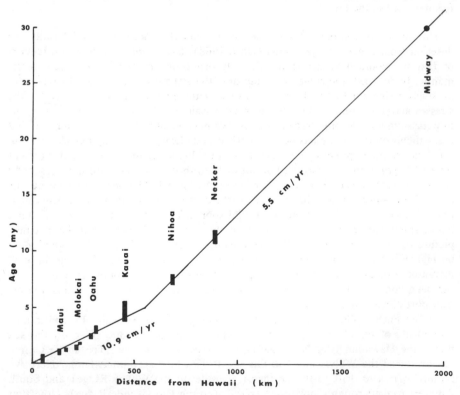

Figure 2. The ages of Hawaiian volcanos versus distance from the presently active volcano at Hawaii. The point at Midway is based on Miocene fossils; the other ages are K-Ar results reported by Funkhouser and others (1968).

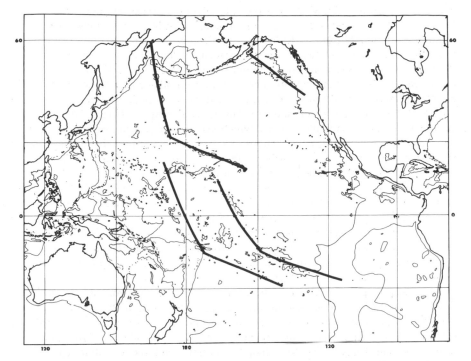

Figure 3. Hot spot trajectories constructed by rotating the Pacific plate 34 degrees about a pole at 67° N., 73° W., and then 45 degrees about a pole at 23° N., 110° W.

140.3° W.) is likewise situated at the southeastern extreme of the Austral Islands chain. Could both of these island chains have been generated by a single motion of the Pacific plate over these two hot spots? The Hawaiian chain is terminated by the Emperor Seamounts; is there an analogous feature for the Austral chain? The answer to these questions is shown in Figure 3. We have assumed three fixed hot spots located at 19° N., 155° W. (Hawaii), at 29° S., 140° W. (Macdonald), and at 27° S., 114° W. (where the East Pacific Rise intersects the Tuamotu and Sala y Gomez Ridges). The solid lines show the points which would pass over these hot spots if the Pacific plate were rotated 34 degrees about a pole at 67° N., 73° W. and then 45 degrees about a pole at 23° N., 110° W. (The fourth solid line in Figure 3, extending from the Juan de Fuca Ridge to Kodiak Island, will be discussed later.) We thus note the similarity of the Hawaiian-Emperor, Tuamotu-Line, and Austral-Gilbert-Marshall chains with the lines generated from present-day active hot spots. In particular, the Marshall-Gilbert Islands do not coincide with the proposed locus of the Pacific over the fixed hot spot. We shall use this to measure the constancy of the fixed spots, but first let us estimate ages along these chains.

We have the rate of recent motion of the Pacific plate past the fixed points from the K-Ar ages of the Hawaiian Islands shown in Figure 2, and we see that the rate may be variable with about 10 cm/yr motion for the past 5 m.y. and about 5 cm/yr before that time. (Note that all three of these island chains are nearly 90 degrees from the pole at 67° N., 73° W. and so all have essentially the same velocities.) Another point on this curve is the age of Midway Island, which has been dated pre-Miocene from drill holes through the coral cap. We estimate the age of the

Hawaiian-Emperor "elbow" two ways. (1) From a linear extrapolation of the Hawaii to Midway distance and age difference, we estimate the elbow to have an age of 43 m.y. (2) The Nazca Ridge—Sala y Gomez Ridge intersection presumably represents the equivalent feature in the eastern Pacific. The hot spot which made the Sala y Gomez-Nazca and the Tuamotu-Line Ridges is directly on the crest of the spreading rise, so the magnetic anomaly pattern adjacent to these features will directly give the age during which each feature was made. From Morgan and others (1969) we see that anomaly 13 (38 m.y.) is near the Nazca-Sala y Gomez conjunction. A third line of evidence which could have bearing on the age of this change in trend is the study by Menard and Atwater (1968) of the changes in the fracture zones pattern in the northeast Pacific. They do not find a major change at about anomaly 13 (although they note a change near the coast of California at the time of anomaly 11); the major change in this pattern occurred at anomalies 21 to 24 (55 m.y.). We shall assume that the bend in the Hawaiian-Emperor chain was made 40 m.y. ago, while keeping in mind that a 55 m.y. age for this feature cannot be ruled out. What is the age of the northern end of the Emperor or Line features? Again using the assumption that the Tuamotu-Line chain was generated at a ridge crest, we infer from the nearness of anomaly 32 that 100 m.y. is a good estimate for the age of the northernmost features.

These age assignments will now be compared to ages determined by drilling or dredging on atolls and guyots in the Pacific. The Mid-Pacific Mountains and Magellan Seamounts will be featured in this discussion, so we first present our interpretation of these in terms of hot spots. The Mid-Pacific Mountains (or Marcus-Necker Ridge), the Magellan Seamounts, and the Caroline Islands are here regarded as east-west island chains formed from 100 to 150 m.y. ago by a rotation of the Pacific plate about a pole near the present North Pole. This motion was not displayed in Figure 3 because all of the island chains are close together and do not have a geometry to accurately determine the pole, and also because of the complications introduced by the "wandering" hot spot, which will be discussed later. The Mid-Pacific Mountains and the Magellan Seamounts are regarded as continuations of the Tuamotu-Line and the Austral-Marshall-Gilbert chains and the Caroline Islands as the continuation of another hot spot chain not present today. Hamilton (1956) reports the following ages (here converted from his age classification name to millions of years) for dredge and core samples obtained from five guyots in the Mid-Pacific Mountains: Hess Guyot (18° N., 174° W.), 120 m.y.; Cape Johnson Guyot (17° N., 177° W.), 120 m.y.; Guyot 20171 (21° N., 171° W.), 80 m.y. On two of the guyots sampled, Horizon (19° N., 169° W.) and Guyot 19171 (19° N., 171° W.), no age older than about 55 m.y. was obtained. A younger age does not contradict these guyots being formed at about 100 m.y., as only surface samples were obtained and much older sediments may not be exposed. Hamilton and Rex (1959) summarize the fossil ages found in the Marshall Islands. Bikini (12° N., 165° E.) and Eniwetok (12° N., 162° E.), just west of the Marshall chain, have been drilled and dated. The age at the bottom of the Bikini hole (about halfway to the basalt basement) is 35 m.y., and the oldest age sample dredged from the adjacent Sylvania Guyot is 55 m.y. More important, two drill holes on Eniwetok penetrated the coral cap and reached the basalt basement. The fossils at the bottom of these holes are about 55 m.y. old. This implies that Eniwetok had a different history than that suggested by its position in the supposed hot spot trajectory. We also note that two of the Japanese Seamounts discussed in the following section on paleomagnetics (near 28° N., 148° E.) have been dredged and dated by the K-Ar method. Their ages, 80 m.y., also contradict their position on the western end of the Mid-Pacific Mountains. We thus have conflicting evidence in the western Pacific and some additional factors must be found if we are to reconcile this with the simple hotspot pattern that we observe farther east.

The solid lines generated by rotating a rigid Pacific plate over the hot spots do not exactly follow the island chains. We may use this systematic departure to estimate the rate of migration of the hot spots relative to one another. The trajectories follow the Hawaiian-Emperor and Tuamotu-Line chains fairly exactly, so we may use the departures of the Austral-Marshall-Gilbert chain to measure mobility. The measured distance from Macdonald Seamount to the turning point is 10 percent longer than the predicted distance; the distance from the turning point to the northernmost of the Marshall Islands is 25 percent less than the corresponding prediction. The rate of plate motion over the "fixed" hot spots is about 7 cm/yr (30 degrees in 40 m.y. and 40 degrees in the following 60 m.y.). Thus this hot spot moving at about 1 cm/yr relative to the others is a good measure of its mobility in the deep mantle. If we had chosen trajectories based on a more compromise set of rotations which did not agree so well with the Hawaiian-Emperor or Tuamotu-Line chains, then each of the hot spots migrating at about .5 cm/yr in this reference frame would match the observations.

Why are the Hawaiian Islands islands? In the simple model presented above the continuous eruption of deep material should make a smooth continuous ridge—what geological complexities must we introduce to get isolated episodic volcanos? We adapt Menard's (1969) model of growing volcanos to this problem. We suppose that the light fractionation from the deep plume continuously flows up but is trapped by the asthenosphere-lithosphere "interface" (not a sharp boundary but a gradual transition in rigidity). This trapped island-type basalt accumulates in pockets, analogous to oil trapped by certain formations, and its unstable situation causes vents to the surface to form, which tap the reservoir and cause volcanos at the surface. This complexity has the possibility of answering a number of questions. (1) The plume may plaster the asthenosphere-lithosphere boundary over an area 100 mi square, but a single vent to the surface can tap this reservoir and concentrate this into a single volcano (as opposed to a continuous ridge). (2) The motion of the lithospheric plate eventually displaces the vent from the area above the deep plume sufficiently far so that a new vent forms. The old vent then taps only the remains in the reservoir in its immediate vicinity and soon dies out. We thus might expect to find a simple relation among the spacing of volcanos, the rate of plate motion, and the magnitude of the hot spot (as measured by the volume of the volcanic chain). (3) The activity of each island ends with alkali-rich eruptions. This different chemistry may result from the remains in the old vent after it has migrated and has been cut off from the hot spot. Does the volume of alkali-rich basalt agree with this? Is the chemistry compatible with this less than 100 km origin? Does the start of the alkali eruptions on an old volcano coincide with the start of a new volcano next in line? (4) This model allows a volcano to continue to grow for millions of years even after it has left its source area, in agreement with the model described by Menard (1969). The bulk of Menard's data supporting this growing seamount model comes from the Juan de Fuca Ridge region. We claim this region has two minor hot spots creating the line of seamounts and guyots between Cobb Seamount and Kodiak Island and the Explorer Seamount—

Pratt-Welker guyot string farther north. (We thus limit the applicability of the growing seamount model to regions of hot spots.) The question as to whether all off-ridge seamounts are produced by minor hot spots raises interesting possibilities, but we shall sidestep this generality.

It is said, based on dredge samples, that seamounts such as Cobb Seamount are capped with an alkali-rich basalt. We claim that such a seamount is not primarily made of ocean ridge–type basalt capped in its last stages of growth with a more alkaline skin, but that it is made of the island-type basalt throughout. Dredging cannot answer this question; only deep drilling can distinguish between an alkali-rich coating or island-type throughout basalt.

Having minor hot spots at the Juan de Fuca Ridge offers an explanation as to why this ridge exists in the first place. The North American and Pacific plates could quite logically have their present motions without there even being an oblique Juan de Fuca Ridge; but placing one of the world's driving mechanisms here assures the continuing existence of a spreading ridge at this location—a ridge that may change its orientation but must pivot about this hot spot.

PALEOMAGNETISM

Francheteau and others (1970) have presented a polar wandering diagram for the Pacific plate based primarily on studies of the magnetic field around seamounts. Figure 4a is a reproduction of their Figure 17 with the following changes. First, we assign definite, though of course possibly inaccurate, ages to each pole position based on our understanding of their discussion of the possible ages of each seamount. The number in the name of each pole position shows our estimate of its age in millions of years. Second, we have greatly enlarged the error circle of the Midway data point. Francheteau and others used the Midway determination of Vine (1968), and Vine (personal commun.) states their estimate of the error is

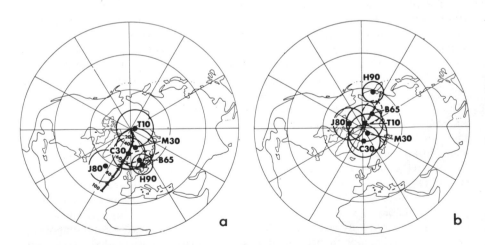

Figure 4. (a) The Pacific paleomagnetic pole positions of Francheteau and others (1970) and the polar wander curve predicted by the motion shown in Figure 3. (b) The paleopoles are "corrected" for the presumed motion of the Pacific plate; the paleopoles of all ages should now coincide at the north pole.

too small. The measured drill core samples were all from the same flow and the scatter reported represents differences in the single flow, not the scatter that might result from polar migration about the average dipole if many flows had been sampled. In addition to the paleomagnetic data points, Figure 4a also shows a predicted polar wander curve for the Pacific plate made from the rotations needed to make the hot spot trajectories shown in Figure 2 (0.85°/m.y. for 40 m.y. about a pole at 67° N., 73° W.; 0.75°/m.y. for 60 m.y. about a pole at 23° N., 110° W.). Each dot on the polar wander curve shows the predicted position for successive 10 m.y. ages. The observed paleomagnetic pole positions should coincide with the dot for its age if (1) the motion of the Pacific plate is as described above, and (2) the magnetic pole does not migrate relative to the fixed hot spots. Figure 4b shows the paleomagnetic data corrected for the predicted motion of the Pacific plate. In principle we have taken the inclination and declination of the original measurement, rotated the Pacific plate back to its orientation at the time the feature was magnetized, and computed the position of the paleopole at that time. Ideally all data points would form a tight cluster about the "north" pole. We see that, except for the Hawaiian Seamounts of presumed 90 m.y. age, there is excellent agreement between the predicted polar wandering and the observed paleomagnetic positions.

The African plate offers another test of a paleoreconstruction based on plate motion over hot spots versus paleomagnetic data. The Walvis Ridge is the most conspicuous aseismic ridge in this region, but there are also submarine ridges trending northeast away from present-day active volcanos at Reunion Island, Bouvet Island, Ascension Island, and the Cape Verde and Canary Islands. A similar trend exists for St. Helena Island and the Cameroon trend, but here we have the peculiar situation of an active volcano at both ends of the trend. Perhaps the trapped basalt at the asthenosphere-lithosphere boundary has taken nearly 100 m.y. to find a vent to the surface at Mt. Cameroon—if so, we have a mechanism to account for the anomalously young ages (Eocene) of much of the activity in the Marshall and Gilbert island chains as discussed above. A rotation of 27 degrees about a pole at 25° N., 55° W. was found to best fit this data, and the trajectories of the hot spots on the African plate based on this are shown in Figure 5. The time at which a hot spot was beneath points on these trajectories was computed by assuming linear interpolation between 110 m.y. and the present—the uniformity of this motion is based on the JOIDES results in the South Atlantic.

Figure 6a shows the paleomagnetic pole determination of Africa as tabulated by McElhinny and others (1968). The pole positions B14, B15, B16, B17, B18, and B19 in McElhinny and others' classification were used. Numbers representing the age of the site are used to identify each pole. The solid line shows the polar wandering curve predicted for the motion of Africa shown in Figure 5. Figure 6b shows the paleomagnetic data corrected for the presumed motion of Africa, analogous to how Figure 4b was obtained from 4a. The clustering at the "north" pole is not as good as the Pacific data. We may claim this is due in part to the slower motion of Africa, hence the migration of the hot spots would be more noticeable than in the Pacific. The African data does not lend support to the hypothesis presented here, but it could be reconciled with this model if there was an episode of rapid polar wandering between 90 and 110 m.y. ago. Such polar wandering would be a rapid shift of the whole mantle (in which all the hot spots would move in unison) to a new axis of rotation, as envisaged by Goldreich and Toomre (1969).

No other plate contains a variety of aseismic ridges so that a direct test of the

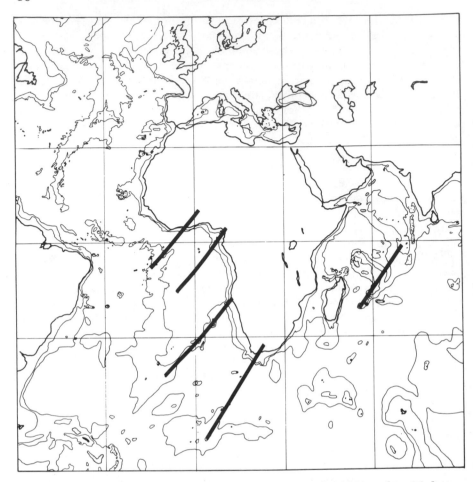

Figure 5. Hot spot trajectories constructed by rotating the African plate 27 degrees about a pole at 25° N., 55° W.

motions inferred from the ridge pattern and from paleomagnetism may be made. However, we may use the deduced motion of Africa and the Pacific together with the known relative motions of the plates to infer the motion over the mantle for the other plates. A tentative inference of the motion of North America, based on a counterclockwise rotation of Africa over the mantle about 30° N., 60° W. and the clockwise rotation of North America relative to Africa about 60° N., 30° W., shows that North America has rotated 30 degrees clockwise about the present north pole since mid-Cretaceous time. North American Tertiary paleomagnetics cluster near the present North Pole but the Cretaceous paleopoles are distinctly different, in the Bering Sea, in agreement with the pattern shown here for Africa.

PRESENT MOTION OF THE PLATES

Table 1 lists the components of an angular velocity vector for each crustal plate. The relative motions of adjacent plates have been determined from fracture zone

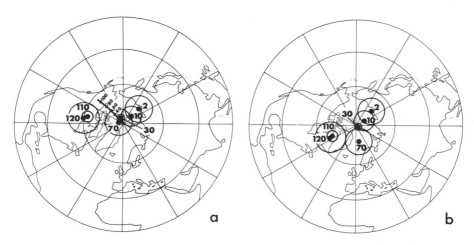

Figure 6. (a) African paleomagnetic pole positions of McElhinny and others (1968) and the polar wander curve predicted by the motion shown in Figure 5. (b) The paleopoles are "corrected" for the presumed motion of the African plate; the poles of all ages should now coincide at the North Pole. The discrepancy of the 110 and 120 m.y. poles may be due to a shift of the entire mantle shell relative to the core.

strikes and spreading rates for most ridge systems and Table 1 attempts a synthesis of this data into a worldwide self-consistent model. Table 2 shows relative motions computed from the vectors in Table 1 for most of the spreading pairs. The reader may compare these relative motion poles and rates with his own favorite data to judge accuracy of this synthesis. (A paper discussing the data used to arrive at Table 1 is in preparation.)

The relative motions are found by subtracting vectors in Table 1, so any constant vector may be added to all the vectors in Table 1 without affecting the relative motions. We have added that constant vector so that, in addition to satisfying the relative motion data, it also satisfies the hot spot data. The heavy vectors in Figure 7 show the motion of the crustal plates over each hot spot; in each case the vector is closely parallel to an aseismic ridge or island chain. Thus the hot spots form a reference frame fixed in the mantle, and Table 1 and Figure 7 show the absolute motion of each plate over the mantle. The good agreement between the arrows in Figure 7 and the trends of the island chains–aseismic ridges leads to two conclusions: (1) there has not been a major reorganization of plate motion in the past 40 m.y. or so, and (2) the hot spots have remained relatively fixed in the mantle. The slight disagreement is most pronounced in the Atlantic region where the slowly spreading plates are most vulnerable to "noise"—the magnitude of this divergence suggests that each hot spot wanders at less than ∼½ centimeter per year.

Kaula's (1970) recent gravity map of the earth is shown in Figure 8. This is an isostatic anomaly map computed for spherical harmonics of order 6 through 16, so the features of 1,000 to 10,000 kilometers length are displayed. Note that there are gravity highs over Iceland, Hawaii, and most of the other hot spots (Galápagos is a conspicuous exception). Such gravity highs are symptomatic of rising currents in the mantle; the less dense material in the rising current produces a negative gravity anomaly, but the satellite passes closer to the elevated surface pushed up by this current and the net gravity field is positive. From formulas in Morgan

TABLE 1. ABSOLUTE MOTIONS OF THE CRUSTAL PLATES IN DEGREES/M.Y.

Plate Name	W_x	W_y	W_z
AM American	.023	−.022	−.140
PA Pacific	−.173	.334	−.702
AN Antarctic	−.117	−.033	.268
IN Indian	.459	.315	.350
AF African	.149	−.112	.147
EU Eurasian	−.050	.052	.039
CH Chinese	−.145	.176	.223
NZ Nazca	−.118	−.314	.616
CO Cocos	−.693	−.921	.624
CR Caribbean	−.180	.430	.090
JF Juan de Fuca	.907	1.234	−1.512
PH Philippine	1.295	−.694	−.859
SM Somalian	.113	−.143	.181
AR Arabian	.412	−.025	.352
PR Persian	−.367	.023	−.141

(1965) we can estimate the size of the rising current. Take 10 mgal excess and 1,000 km diameter as typical for the hot spots; such a geoid high could be produced by a mass deficiency of 10^{20} gm centered at about 300 km depth, or roughly a cylindrical plug 100 km in diameter extending from the surface to 600 km depth with a density deficiency of 1 percent.

Note the paradox: both rising currents and oceanic trenches are associated with positive gravity anomalies. This behavior has been explained by Kaula and others as due to flow in a nonuniform viscous material. At the rises, the light ascending current buoys up the surface for a net positive gravity effect (and descending currents of the same pattern would pull down the surface for a net gravity minimum). However, a deep lithospheric plate may push down onto a hard bottom surface. If the lower surface supports part of the weight of the sinking plate, the top surface will not be depressed by the flow pattern and the satellite will sense only the excess mass of the plunging lithosphere for a positive gravity effect.

TABLE 2. RELATIVE PLATE MOTIONS DEDUCED FROM TABLE 1

	Latitude (°N.)	Longitude (°E.)	Spreading Rate (cm/yr)
EU-AM	60	135	1.2
AF-AM	62	−36	1.8
AM-PA	54	−61	3.9
PA-AN	−69	99	5.8
AF-AN	−24	−16	1.7
IN-AN	7	31	3.7
AR-AF	36	18	1.9
AR-SM	28	22	2.0
IN-SM	16	53	3.3
SM-AN	−19	−26	1.5
CO-PA	44	−113	10.5
CO-NZ	1	−133	4.6
NZ-PA	64	−85	8.2
NZ-AN	51	−90	2.6

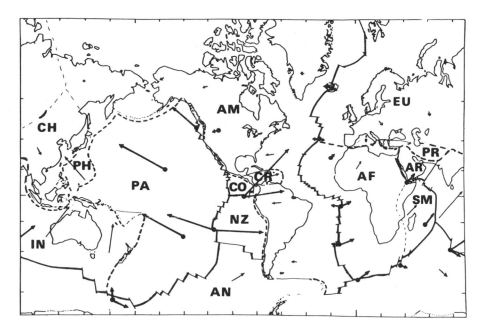

Figure 7. The present motion of the plates over the hot spots. This motion is computed from Table 1 and agrees with the relative motion data, as well as with the trends of the aseismic ridges–island chains. The length of each arrow is proportional to the plate speed.

The gravity measurements appear to offer the best method to assess the strength of the different plumes. More measurements are needed as the geoid maps change dramatically each year (compare Kaula's 1970 statements with those of earlier years). In choosing the possible plumes shown in Figures 1 and 7; the gravity measurements were augmented by what is known locally as the "Hess Gravity Theorem," namely that one does not need a gravimeter to measure gravity, one needs only to look at the topography. This is a corollary of the statements made above; the flow patterns associated with positive gravity anomalies raise the surface, thus high topography means positive gravity and vice versa. We thus look for those abnormally shallow places in the oceans, such as the areas near the Galapagos, the Juan de Fuca Ridge, and Prince Edward Island. The National Geographic Society globe has contours at particularly apt intervals and spiderlike fingers radiate away from many of these topographic highs. Whether the unusually high Tibetan Plateau or southern Africa should be considered symptomatic of a subcontinental hot spot is an open question; the more uniform oceans are more amenable to this type of analysis. The best case for a present day subcontinental hot spot could be made for the Snake River flood basalts in analogy to the Deccan Traps of the early Reunion hot spot.

The data presented in this paper, the parallelism of the Pacific island chains, the agreement of this motion of the Pacific with the paleomagnetic results, and the agreement of the present relative motions of the plates with the trends of the island chains–aseismic ridges all substantiate that plate motion over mantle hot spots is a valid and useful concept but this data contributes little to the hypothesis that these hot spots provide the motive force for continental drift. The case for this

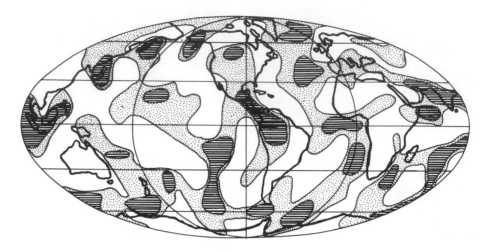

Figure 8. Isostatic gravity map of the earth, redrafted from Kaula (1970), to emphasize positive anomalies. The shaded areas show regions of positive anomalies; the
heavier shaded areas show regions where the anomalies are greater than +20 mgal.
Note the correlations of the gravity highs with Iceland, Hawaii, and most of the other
hot spots.

association rests on three facts: (1) Most of the hot spots are near a ridge, and a
hot spot is near each of the triple ridge junctions; (2) the gravity and regional
high topography suggests that more than just surface volcanism is involved at
each hot spot; and (3) neither rises nor trenches appear capable of driving the
plates, implying that asthenospheric currents acting on the plate bottoms must
exist.

The symmetric magnetic patterns and the mid-ocean position of the rises suggest
that the ridges are passive. The first deduction of plate tectonics was that if two
plates are pulled apart, they split along some line of weakness and *in response,*
asthenosphere rises to fill the void. With further pulling of the plates, the laws of
heat conduction and the temperature dependence of strength dictate that future
cracks appear right down the center of the previous "dike" injection. If the two
plates are displaced equally in opposite direction or if only one plate is moved and
the other held fixed, perfect symmetry of the magnetic pattern will be generated.
The axis of the ridge must be free to migrate (as shown by the near closure of rises
around Africa and Antarctica). If the "dikes" on the ridge axis are required to
push the plates apart, it is not clear how the symmetric character of the rises is to
be maintained.

The best argument against the sinking lithospheric plates providing the main
motive force is that small trench-bounded plates such as the Cocos do not move
faster than the large Pacific plate. Also, the slow compressive systems, as in Iran,
would not appear to have the ability to pull other plates, such as the Arabian plate,
away from other units. The pull of the sinking plate is needed to explain the gravity
minimum and topographic deep locally associated with the trench system (see
Morgan, 1965), but we do not wish to invoke this pull as the main tectonic stress.

We are left with sublithospheric currents in the mantle. The question now is
whether these currents are great rolls—mirrors of the rise and trench systems—or
whether they are localized upwellings, that is, hot spots. Also, how deep do such

currents extend? The circumstantial evidence seems to favor the hot spot mode, but there are several tests which could answer this question. (1) The most dramatic proof would be to seismically detect the shadow cast by a deep plume (the large time delay of teleseismic events in Iceland may be a plume effect). (2) Assumptions as to the magnitude of each plume and of the stresses at rise, fault, and trench plate-to-plate boundaries could be made and, the directions of the resulting plate motions could be deduced from these simplified dynamics. (3) A re-evaluation of the heat flow problem may show that convection deep in the mantle is necessary to remove heat from the lower mantle. The near equality of the oceanic and continental heat flux may be explained in terms of hotter than normal asthenosphere flowing away from each hot spot. (4) Finally, a continuing study of the Cenozoic and Cretaceous sea-floor spreading may show that major reorganizations of the spreading pattern coincide with the disappearance or emergence of new hot spots.

ACKNOWLEDGMENTS

Many of the points presented here arose during discussions with K. S. Deffeyes and F. J. Vine, and I thank them for their many suggestions. I also thank the organizers of the Birch Symposium at Harvard in 1970, as the near simultaneous presentation of papers by P. W. Gast, W. M. Kaula, and W. H. Menard directly lead to this model. This work was partially supported by the National Science Foundation and the Office of Naval Research.

REFERENCES CITED

Francheteau, J., Harrison, C. G. A., Sclater, J. G., and Richards, M. L., 1970, Magnetization of Pacific seamounts, a preliminary polar curve for the northeastern Pacific: Jour. Geophys. Research, v. 75, p. 2035–2061.

Funkhouser, J. G., Barnes, I. L., and Naughton, J. J., 1968, Determination of ages of Hawaiian volcanoes by K-Ar method: Pacific Sci., v. 22, p. 369–372.

Gast, P. W., 1968, Trace element fractionation and the origin of tholeiitic and alkaline magma types: Geochim. et Cosmochim. Acta, v. 32, p. 1057–1086.

Goldreich, P., and Toomre, A., 1969, Some remarks on polar wandering: Jour. Geophys. Research, v. 74, p. 2555–2569.

Hamilton, E. L., 1956, Sunken islands of the Mid-Pacific Mountains: Geol. Soc. America Mem. 64, 97 p.

Hamilton, E. L., and Rex, R. W., 1959, Lower Eocene phosphatized globigerina ooze from Sylvania Guyot: U.S. Geol. Survey Prof. Paper 260-W, p. 785–797.

Johnson, R. H., 1970, Active submarine volcanism in the Austral Islands: Science, v. 167, p. 977–979.

Kaula, W. M., 1970, Earth's gravity field: Relation to global tectonics: Science, v. 169, p. 982–985.

McElhinny, M. W., Briden, J. C., Jones, D. L., and Brock, A., 1968, Geological and geophysical implications of paleomagnetic results from Africa: Rev. Geophysics, v. 6, p. 201–238.

Menard, H. W., 1969, Growth of drifting volcanoes: Jour. Geophys. Research, v. 74, p. 4827–4837.

Menard, H. W., and Atwater, T. M., 1968, Changes in direction of sea-floor spreading: Nature, v. 219, p. 463–467.

Morgan, W. J., 1965, Gravity anomalies and convection currents: Jour. Geophys. Research, v. 70, p. 6175–6204.

Morgan, W. J., Vogt, P. R., and Falls, D. F., 1969, Magnetic anomalies and sea-floor spreading on the Chile Rise: Nature, v. 222, p. 137–142.

Vine, F. J., 1968, Paleomagnetic evidence for the northward movement of the North Pacific basin during the past 100 m.y. [abs.]: Am. Geophys. Union Trans., v. 49, 156 p.

Wilson, J. T., 1963, Continental drift: Sci. American, v. 208, p. 86–100.

—— 1965, Evidence from ocean islands suggesting movement in the earth: Royal Soc. London, Philos. Trans., v. 258 (Symposium on Continental Drift), p. 145–165.

MANUSCRIPT RECEIVED BY THE SOCIETY MARCH 29, 1971

THE GEOLOGICAL SOCIETY OF AMERICA, INC.
MEMOIR 132
© 1972

Estimation of Tectonic Rotation Poles of Inactive Structures

A. GILBERT SMITH
Department of Geology, Sedgwick Museum, Cambridge, England

ABSTRACT

Some inexact methods for estimating tectonic rotation poles are described. It is shown geometrically that each compressive plate margin has a "forbidden sector" in which the pole cannot lie. This may help to delimit the pole position, particularly for large orogenic structures. Five present-day compressive zones are examined, and their respective poles lie generally in line with the zone and about 20° to 40° from one end of it. A possible Atlantic-Mediterranean reconstruction is consistent with these findings.

INTRODUCTION

The notion of a tectonic rotation pole arises from realizing that continental drift is the net result of the relative motion among rigid parts of the earth's crust. In writing about this topic, I should point out that during my stay at Princeton (1959–1963), I was a complete skeptic about continental drift, despite the fact that this was a period when Hess wrote his most important papers on the subject. My attitude stemmed from accepting those physical arguments that "proved" drift could not happen, rather than examining the evidence for it. Later at Cambridge, still unconvinced of the reality of Mesozoic and Tertiary drift, I took part in an attempt to reassemble continents by fitting their edges together with the aid of a computer, and since then have maintained an active interest in the problems involved.

Needless to say, I have subsequently been strongly influenced by Hess's ideas, which have since been strikingly confirmed and have played a major role in the recent revolutionary developments in the earth sciences. I will always be indebted to Harry Hess for many things, but am particularly grateful to him for opening my mind to new ways of looking at geology.

REASSEMBLING CONTINENTS IN PAST TIME

In Mesozoic and Tertiary time the rotations needed to move the major continents to their former relative positions may be found by matching pairs of sea-floor spreading anomalies in oceans with aseismic margins (Le Pichon, 1968). These rotations may also be found from the fit of the continental margins on the opposite sides of such oceans (Bullard and others, 1965; McKenzie and others, 1970; Smith and Hallam, 1970). Similar results have been obtained by best-fit methods in which the pole and angle are not explicitly given, though they are implicit in the result (Sproll and Dietz, 1969; Dietz and Sproll, 1970). A further rotation may be used to orient the reassembly parallel to the paleomagnetic axis, currently the best estimate of the spin or geographic axis.

Providing the continents have not moved relative to one another for a time interval during which substantial polar wandering occurs, it is possible to position them uniquely from paleomagnetic data (Creer and others, 1958; Irving, 1958). The method has been used to reconstruct Gondwanaland (McElhinny and Luck, 1970). As in any reconstruction, the rotation poles and angles are implicit in the result.

All the above methods fail for much of Paleozoic and earlier time. Continents cannot be uniquely positioned from paleomagnetic data when their relative motion masks polar wandering or when no polar wandering occurs. Matching of pre-Mesozoic sea-floor spreading anomalies is not possible because sea floor of this age has mostly vanished into the mantle. The best-fit of pre-Mesozoic aseismic continental margins, supposing such margins could be recognized, might be a misleading indication of the original positions of the continents concerned because of unknown but possibly significant changes to their original shapes.

Nevertheless, plate theory may be used to delineate the pre-Mesozoic fragments, and paleomagnetic data may be used to orient them into their former latitudes, but the assessment of relative longitudes cannot yet be made. Similar uncertainties in relative longitude exist for the smaller continental fragments in parts of the Alpine-Himalayan and Caribbean regions during Mesozoic and Tertiary time.

The relative positions of the major continents in Mesozoic and Tertiary time are therefore readily determined by winding up the present sea floor in the Atlantic and Indian Oceans. In principle, their pre-Mesozoic positions could be found if we could discover a method for unwinding the sea floor lost in intracontinental orogenic belts, such as the Urals. The simplest possible case would be that in which the rotation pole is fixed and only one plate margin lay between the approaching continents. If this pole could be found from tectonic or other data and differs from the paleomagnetic pole, the rotation angle consistent with turning one or the other continent about the rotation pole could be estimated from paleomagnetic data, and the two continents could be repositioned. This method assumes the magnetic field has always resembled a centered axisymmetric dipole, which may not always have been true, at least not in Permo-Triassic time (Briden and others, 1970).

Some simple geometrical methods are suggested for limiting the positions of rotation poles of compressional plate margins. The methods are examined with reference to present-day margins and to part of a possible Permo-Triassic reassembly of the world, where the relevant poles are known or may be found independently.

ROTATION POLES AND PLATE MARGINS

The sense and direction of the slip vector at a plate margin determines the behavior of the margin (Morgan, 1968). The slip vector is tangent to the small circle centered on the rotation pole. If the pole position is unknown, its locus is the great circle through the point on the margin that is perpendicular to the slip vector. Currently, transform faults are the only structures whose slip vectors can be unambiguously determined from their shapes. The rotation pole may be found from the intersections of great circles drawn perpendicular to different parts of a sufficiently long transform fault, or a set of related transform faults (Le Pichon, 1968; Morgan, 1968). "Fault-plane solutions" from earthquakes give the slip vector for all types of present-day margins (Isacks and others, 1968), but obviously cannot be used if the margin is inactive.

The rotation poles of inactive oceanic ridges or of old orogenic belts cannot be directly inferred from their shapes. In the case of an orogenic belt, it may eventually be possible to determine the slip vector at a particular point from the internal structure, and hence the great-circle locus for the pole. The intersections of two or more such loci drawn on a belt several thousand kilometers long would give the rotation pole. The ability to make such inferences will mark an important step forward in tectonics.

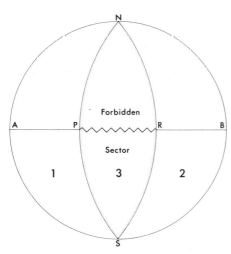

Figure 1. PR is a compressive margin and also part of a great circle. Sectors 1 and 2 contain all rotation poles that would make PR compressive along its entire length. No pole lying in sector 3 can make PR compressive along its entire length. This is the so-called "forbidden sector." Poles lying on AP and RB make PR a purely compressional margin, with no translational components along it. Stereographic equal angle projection.

The remainder of this section is restricted to a discussion of orogenic belts. In the absence of fault-plane solutions, all that we can say with certainty about the slip vector at a trench or near any other active orogenic belt is that it always contains some compressional component, and an unknown, but possibly zero, translational component. Despite these uncertainties, we may limit the positions of the rotation poles of orogenic belts by idealizing their shapes. For example, parts of present-day orogenic belts resemble great circle arcs. Were they exactly great circle arcs, the rotation pole could not lie within the sector of the globe for which the arc is an equator (Fig. 1). This result is of little use for small structures, but may have application to large orogenic belts, such as the Caledonian-Appalachian system, where the forbidden sector may occupy more than half the globe. A pole in this sector, hereafter referred to as "the forbidden sector," causes nontranslational slip components along the equator to change from compressional to extensional orientations (Fig. 2). Stated another way, if the structures on a margin that is a great circle arc change from compressional to extensional, the pole must lie on the great circle perpendicular to and passing through the change-over point (NQS in Fig. 2). More exact conditions follow from considering the detailed shape of the belt.

A more questionable approximation would be that insignificant translational components are present in the slip vectors of large orogenic belts. Such structures would be meridian arcs with respect to the rotation pole, which would lie on some part of the arc outside of the forbidden sector (Fig. 1). The extent to which present-day orogenic belts approximate meridian arcs may be found by fitting great circle arcs to them and noting where the rotation pole lies.

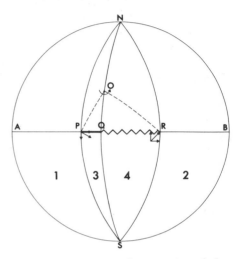

ROTATION POLES OF PRESENT-DAY COMPRESSIONAL PLATE MARGINS

Fault-plane solutions for the Aleutian part of the Pacific/America plate margin give slip vectors having an almost constant geographical trend, as required by plate theory. Because the margin itself is curved, there is a wide range in the angle between it and the slip vector (Isacks and others, 1968). This relation might be expected because the margin neither resembles a great circle, nor is it very long.

Five other longer and (or) less sinuous compressional margins have been selected for investigation, whose extent and rotation poles have been taken from Le Pichon (1968). Great circle arcs have been drawn through their ends and the relation of the pole to the arc has been found (Fig. 3). The method is obviously imprecise. For example, only Le Pichon's poles and margins are used. He assigned present-day plate margins to six plates and computed the rotation poles of the compressional margins so that they were consistent with spreading rates and transform fault directions on the major ocean ridges. At least twice as many plates are known to exist and some of his spreading poles differ considerably from other estimates (*see* Morgan, 1968). No attempt has been

Figure 2. PQR is a plate margin and also part of a great circle. It has extensional components along PQ and compressional components along QR. The pole for PQ must lie outside or on the boundaries of sector 3; the pole for QR must lie outside or on the boundaries of sector 4. The southern plate is assumed to be rotating counterclockwise with respect to the northern plate. Sectors 1 and 2 contain all the rotation poles producing either extensional or compressional slip components on PQR. Arcs AP and RB are loci of all rotation poles producing either purely extensional or purely compressional components on PQR. No pole in sectors 1 and 2 can simultaneously produce compression and extension on PQR. Thus the only possible locus for the pole is NQS—the common boundary of the two forbidden sectors. O is such a pole. For O the slip vector at P has extensional and translational components; that at Q has entirely translational components; that at R has compressional and translational components. Q is the only pole that can produce purely extensional and purely compressional components on margin PQR. Stereographic equal angle projection.

made to evaluate here the significance of these differences or to assess the probable errors in his compressional poles; no entirely objective methods have been used to decide on the limits of the compressional margins discussed; and finally, rather than fitting arcs to the whole margin, arcs have been drawn instead between the ends of the margins.

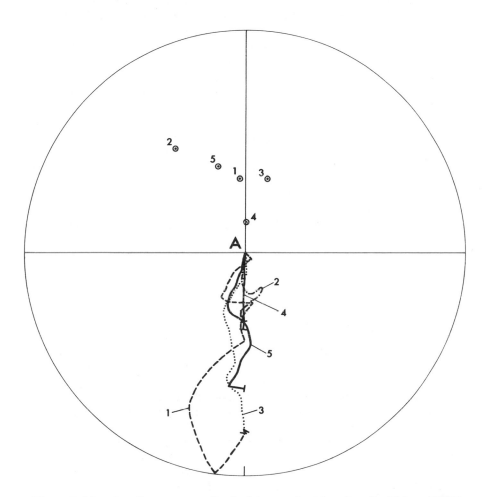

Figure 3. Lines 1 to 5 are compressional plate margins taken from Le Pichon (1968): (1) India/Eurasia; (2) Africa/Eurasia from Gibraltar to Syria; (3) America/Antarctica from Central America to Cape Horn; (4) India/Pacific from the Tonga Islands to central New Zealand; (5) Eurasia/Pacific. Great circle arcs have been fitted to the ends of these margins. The arcs have all been oriented so that they are north-south, with the end closest to the rotation pole at A. The positions of the corresponding rotation poles relative to the margins have been plotted as numbered open circles. Stereographic equal angle projection. The arc that fits the India/Eurasia margin is longer than 90° and continues into the back hemisphere as shown. The distribution of rotation poles shows that none of them lie in forbidden sectors (below the equator) and it also suggests that all margins except number 2 approximate meridian arcs with respect to their poles.

All five poles lie outside their forbidden sectors and four of them resemble meridian arcs. Thus the basic assumption of classical tectonics—that an orogenic belt automatically implies that the principal compressional stress is at right angles to its trend—is a reasonable approximation for straight and (or) long present-day belts. Therefore, providing we can trace the limits of a similar ancient orogenic belt and fit a great circle arc to its ends, we could, in the absence of any other information, assume that all the rotation poles describing its behavior in time lay on the arc within 40° of one end. This assumption does not mean that the pole was always in the same position, nor that it always lay at the same end of the arc.

In the case of intracontinental orogenic belts, two or more plate margins may have existed while the intervening ocean vanished. If more than one of these margins was simultaneously active, the relative positions of two approaching continents cannot be uniquely determined from a knowledge of the rotation poles and paleomagnetic data alone. In theory, one of the two margins could have been a spreading ridge. If creation exceeded consumption, a compressive plate margin on one of the continents may appear to be bringing them closer together, while they are in fact moving farther apart. Presently available techniques cannot be used to recognize this possibility in pre-Mesozoic intracontinental orogenic belts. But, when beginning to unravel their history, we may have to assume—perhaps incorrectly—that this case was rare or unimportant. Obviously, the tectonics of pre-Mesozoic orogenic belts cannot yield more precise information about rotation poles until the number, position, and duration of the major plate margins in them has been gauged from field data.

POSSIBLE SIGNIFICANCE OF SOME BEST-FIT ROTATION POLES BETWEEN ASEISMIC CONTINENTAL MARGINS

When the best-fit join of the Atlantic continents is rotated about the three rotation poles used to bring the Atlantic continents together, then the trace of part of it sweeps out a pattern resembling the sea-floor spreading pattern (Funnell and Smith, 1968). Published data suggest the same relation may hold for part of the join between Africa and Antarctica, and also between Australia and Antarctica. Such areas of the oceans have been formed by spreading from median ridges in the sense of Menard (1958).

The best-fit poles describing these patterns are points having more than a merely geometrical significance. They are also close to the average spreading pole of oceans created between parts of aseismic continental margins. If fixed pole opening models prove to be adequate descriptions of the spreading history of some Mesozoic and Tertiary oceans, then we may be able to use the models to describe a possible history of some pre-Mesozoic oceans, bordered by formerly aseismic continental margins, before their floors began to vanish into the mantle.

ROTATION POLES OF CONTINENTS THAT HAVE COLLIDED

The only known rotation poles of continents that have collided are those describing the net movement of Africa, Arabia, and India relative to Eurasia from the time of the best-fit reassembly—probably most of Permo-Triassic time—to the present day. A large oceanic area, the Tethys, vanished as the result of this movement, and the Alpine-Himalayan chain was formed. In the case of Africa, the single rotation pole and angle needed to bring Africa to Eurasia may be found by combining the two rotations needed to bring Africa to North America and then to Eurasia. The pole and angle merely give the net effect of all the relative movements between Africa and Eurasia from the time of breakup of the reassembly to the present day. There is no necessary connection between this pole and angle and the tectonic rotation pole(s) of any branch of the Alpine-Himalayan system.

ESTIMATE OF A SPREADING POLE FROM STRUCTURES PRESERVED ON CONTINENTS

A provisional best-fit has been made of the continental fragments around the Mediterranean based entirely on geometrical matching of the 1,000-m contour, much of it deformed (Smith, 1971, Fig. 4, p. 2045). It has been assumed that the reassembly began to break up in Lower Jurassic time. Inferred plate margins of this age have been plotted on Figure 4. All these margins may be joined together to form a single, irregularly shaped plate margin separating the northern from the southern continents. Triple junctions might have existed at the northern and southern ends of the central Atlantic (A and B on Fig. 4), and in southern Yugoslavia (C). A and B would imply that the South Atlantic and North Atlantic were also opening during Lower Jurassic time, but this is not supported by the available data (Hsü and Andrews, 1970; Avery and others, 1969; Talwani and others, 1969). C implies that the spreading age of the ophiolite-chert sequences in Greece, here regarded as old ocean floor, is partly of Lower Jurassic age, a supposition not yet supported by field evidence.

If the irregular margin shown existed as a single continuous margin without any intervening triple points, then all the structures formed on it should be consistent with a single rotation pole, and this pole should also be the initial opening pole of the central Atlantic. The area that could be occupied by the rotation pole is very small. It is limited mostly by the field of the supposed transform fault marking the northern boundary of Africa (Fig. 4, no. 2). The fields of all the other structures (1 and 4 to 7) except 3 are consistent with the field marked out by this fault. Why 3 should be inconsistent is readily accounted for by the geometry. The fit between Greece-Turkey and Italy-Sicily-Africa causes northern Yugoslavia to overlap onto Italy. This overlap arises from treating Yugoslavia-Greece-Turkey as a rigid fragment whose shape has not changed since the supposed breakup of the area. It is highly probable that the overlap never existed, and that in Lower Jurassic time Yugoslavia paralleled Italy as it does today. If so, the sector traced out by the structures in Yugoslavia and Albania (3') corresponds with all the other fields. This may seem to be a good instance of "fudging" the results, but exactly analogous overlaps occur in the Caribbean region, this time arising from treating the region as one that has not moved relative to North America (*see* Fig. 5). What is perhaps most surprising is that the field of the pole consistent with all the structures (P') is then remarkably close to the average spreading pole (P) of the central Atlantic found from the best-fit of the continental margins.

What this shows is that under favorable circumstances, the initial spreading pole of the ocean between two separating continents may be estimated by plotting tec-

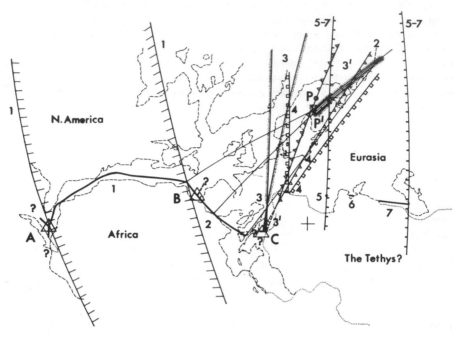

Figure 4. Graphically determined limitations on the initial spreading pole of central Atlantic opening. Opening is assumed to be of Lower Jurassic age and the Mediterranean region is assumed to have been as shown (Smith, 1971, Fig. 4, p. 2045). Possible Lower Jurassic plate margins (*ibid.*, Fig. 14, p. 2061): (1) assumed spreading of central Atlantic; (2) inferred transform fault; (3) inferred ridge and transform system represented by ophiolite-chert sequence in Yugoslavia and Albania; (4) transform in Eastern Alps; (5) small granite plutons in Romania; (6) acid and intermediate volcanism and folding in Crimea, and (7) Great Caucasus. Margins idealized as: great circle arcs (1), (5 to 7), poles outside sectors shown; (2) transform, pole limited by intersection of orthogonal great circles drawn on ends and middle of transform; (3) orthogonal ridge and transform system, great circles drawn parallel to supposed range of ridge azimuths (3′ is 3 rotated so that Yugoslavia parallels Italy instead of overlapping it); (4) transform with great circles drawn orthogonal to range of present orientations. P′ is field consistent with structures 1, 2, 3′, and 4 to 7. P is the best-fit pole of Africa and North America plotted on the reassembly. A, B, and C are possible but improbable Lower Jurassic triple junctions. Equal angle stereographic projection centered on cross near C.

I have recently learned of two alternative views of parts of the postulated plate margin: (1) the transform zone in southern Yugoslavia (the Scutari-Pec line) may not cut the coast as it is shown to on the 1:2,500,000 Tectonic Map of Europe (Aubouin and Ndojaj, 1964, Fig. 7, p. 620); the ophiolites in southern Yugoslavia may be of Upper Jurassic or Lower Cretaceous age (Rampnoux, 1967). If either view is correct, the interpretation of Figure 4 will need to be revised, though the methods for limiting the pole will still apply.

tonic data on a reassembly. Moreover, in the case considered, this initial pole is close to the average spreading pole. From the point of view of reassembling continents, we would wish to invert this procedure in some way, but knowing the answer beforehand suggests no method whereby this can be done. All that can be concluded is that a reassembly may be tested for consistency by this method, but at present it cannot be constructed by inverting it.

ESTIMATION OF A CLOSING POLE AND ANGLE

As stated above, the closing pole and angle between Africa and Eurasia is merely the geometrical pole and angle required to bring Africa from its position on the reassembly to its present-day position. The net effect of this change in position has been to eliminate most, if not all, of the Tethys. It is quite clear that ocean elimination is a highly efficient process, for no trace of unclosed but infilled and undeformed oceanic areas has been recognized in the Caledonian, Appalachian, Hercynian, or Ural Mountains, even though the Appalachian-Caledonian and Ural systems, like the Alpine-Himalayan belt, are most reasonably interpreted as resulting from the coalescence of formerly separated continents (*see,* for example, Wilson, 1966; Bird and Dewey, 1970; Hamilton, 1970). The initial condition of the Alpine-Himalayan belt is known from the reassembly, and the final condition is simply

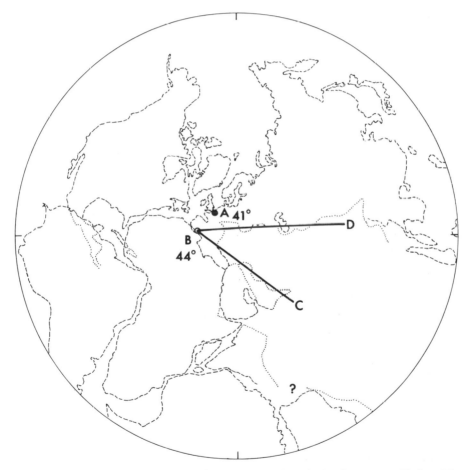

Figure 5. Graphically determined closing pole and angle for the western Tethys. The exact shape of the Tethys is not known but was roughly as shown. BC and BD are best-fitting great circle arcs, intersecting at B at about 44°. A single rotation of Africa about A through 41° brings it from its best-fitting position relative to Eurasia to its present-day position.

that of no ocean. We might hazard a guess that the pole and angle for this over-all change could be found from the simplest geometrical manner of closing the ocean.

The western part of this ocean is roughly wedge-shaped. The simplest way to close it is to turn Africa about a point near the apex of the wedge. The point and angle may be visually estimated by fitting great circles to the opposite margins and finding their intersection angle and point. On inspection, these are close to the pole and angle determined from the best-fit geometry (Fig. 4). The fact that the angle estimated is greater than the best-fit angle may reflect the fact that there is still some ocean to be eliminated in the region. Presumably, future compressional movements will eventually consume all the remaining patches of oceanic crust there.

CONCLUSIONS

Opening and closing poles may be estimated from continental reassemblies, but the methods used cannot currently be inverted to make the reassemblies in the first place. In the absence of such methods, the problem of repositioning continents in Paleozoic or earlier time is that of unwinding the ocean floor lost in intracontinental orogenic belts. This in turn depends on being able to infer the position, number, and duration of the major plate margins in these belts and in being able to assign a rotation pole (or poles) to each margin. Provided only one margin has been active at any given time, the rotation angles needed for a reassembly may, in most cases, be found from paleomagnetic data.

A rotation pole could be found from the intersection of great circles drawn perpendicular to two or more slip vectors. At present, we do not even know how to find the slip vector in an orogenic belt from the observable structures. Therefore it might be worth calculating the Tertiary and Mesozoic slip vector history of compressional plate margins from the sea-floor spreading data. They could be re-examined in the field to learn what features are significant for slip vector determination and what are not. We could apply these findings to pre-Mesozoic belts and perhaps find the rotation poles. If this can be done it may be possible to unravel the Paleozoic and older drift history of the earth with some confidence.

ACKNOWLEDGMENTS

I thank C. P. Hughes, W. J. Morgan, E. R. Oxburgh, and F. J. Vine for reading and suggesting several improvements of the manuscript, and J. Dercourt for drawing attention to some relevant work.

REFERENCES CITED

Aubouin, J., and Ndojaj, I., 1964, Regard sur la géologie de l'Albanie: Bull. Soc. Géol. France, sér. 7, v. 6, p. 593–625.

Avery, O. E., Vogt, P. R., and Higgs, R. H., 1969, Morphology, magnetic anomalies and evolution of the northeast Atlantic and Labrador Sea—Part II—Magnetic anomalies [abs. 023]: Am. Geophys. Union Trans., v. 50, p. 184.

Bird, J., and Dewey, J., 1970, Lithosphere plate-continental margin tectonics and the evolution of the Appalachian orogen: Geol. Soc. America Bull., v. 81, p. 1031–1059.

Briden, J. C., Smith, A. G., and Sallomy, J. T., 1970, The geomagnetic field in Permo-Triassic time: Royal Astron. Soc. Geophys. Jour., v. 23, p. 101–117.

Bullard, E. C., Everett, J. F., and Smith, A. G., 1965, The fit of the continents around the Atlantic: Royal Soc. London, Philos. Trans., v. 258A, p. 41–51.

Creer, K. M., Irving, E., Nairn, A.E.M., and Runcorn, S. K., 1958, Palaeomagnetic results from different continents and their relation to the problem of continental drift: Annales Géophysique, v. 14, p. 492–501.

Dietz, R. S., and Sproll, R. P., 1970, Fit between Africa and Antarctica: A continental drift reconstruction: Science, v. 167, p. 1612–1614.

Funnell, B. M., and Smith, A. G., 1968, Opening of the Atlantic Ocean: Nature, v. 219, p. 1328–1333.

Hamilton, W., 1970, The Uralides and the motion of the Russian and Siberian platforms: Geol. Soc. America Bull., v. 81, p. 2553–2576.

Hsü, K. J., and Andrews, J. E., 1970, History of south Atlantic basin, leg 3, deep sea drilling project, in Maxwell, A. E., and others, Initial reports of the deep sea drilling project, v. 3: Washington, U.S. Govt. Printing Office, p. 464–467.

Irving, E., 1958, Palaeogeographic reconstructions from palaeomagnetism: Royal Astron. Soc. Geophys. Jour., v. 1, p. 224–237.

Isacks, B., Oliver, J., and Sykes, L. R., 1968, Seismology and the new global tectonics: Jour. Geophys. Research, v. 73, p. 5855–5899.

Le Pichon, X., 1968, Sea-floor spreading and continental drift: Jour. Geophys. Research, v. 73, p. 3661–3697.

McElhinny, M. W., and Luck, G. R., 1970, Paleomagnetism and Gondwanaland: Science, v. 168, p. 830–832.

McKenzie, D. P., Davies, D., and Molnar, P., 1970, Plate tectonics of the Red Sea and East Africa: Nature, v. 226, p. 243–248.

Menard, H. W., 1958, Development of median elevations in ocean basins: Geol. Soc. America Bull., v. 69, p. 1179–1185.

Morgan, W. J., 1968, Rises, trenches, great faults and crustal blocks: Jour. Geophys. Research, v. 73, p. 3661–3697.

Rampnoux, J. P., 1967, Contribution à l'étude des séries ophiolitoques: Titograd Geological Institute of Montenegro, Bull. Géologique, v. 5, p. 431–434.

Smith, A. G., 1971, Alpine deformation and the oceanic areas of the Tethys, Mediterranean and Atlantic: Geol. Soc. America Bull., v. 82, p. 2039–2070.

Smith, A. G., and Hallam, A., 1970, The fit of the southern continents: Nature, v. 219, p. 1328–1333.

Sproll, W. P., and Dietz, R. S., 1969, Morphological continental drift fit of Australia and Antarctica: Nature, v. 222, p. 345–348.

Talwani, M., Pitman, W., and Heirtzler, J. R., 1969, Magnetic anomalies in the north Atlantic [abs. 046]: Am. Geophys. Union Trans., v. 50, p. 189.

Wilson, J. T., 1966, Did the Atlantic close and then re-open? Nature, v. 211, p. 676–681.

MANUSCRIPT RECEIVED BY THE SOCIETY MARCH 29, 1971

The Geological Society of America, Inc.
MEMOIR 132
© 1972

Origin of Lithosphere behind Island Arcs, with Reference to the Western Pacific

RALPH MOBERLY

Hawaii Institute of Geophysics, Honolulu, Hawaii 96822

ABSTRACT

The sea floor inside island arcs characteristically is less deep and has higher heat flow than the ocean floor outside the arc and trench system. Direct evidence from drilling and indirect evidence based on thin sediment cover, interrupted geologic trends, paleomagnetic studies, and fitting of pre-drift continental margins show that the lithosphere behind island arcs is young and commonly did not form on the mid-oceanic ridge system. The slab of dense lithosphere that flexes and sinks spontaneously through the asthenosphere under arcs is shown to sink at an angle that is steeper than the plane of the earthquakes. As a consequence, the trench and arc migrate seaward against the retreating line of flexure of the suboceanic lithosphere. Part of the warm asthenosphere pushed aside by the plunging slab migrates up by creep and as magma, then cools and forms new lithosphere in the extensional region behind the advancing island arc. Extension is favored where the lithospheric plate behind the arc is moving tangentially or away from the plate outside the arc.

A series of maps shows the tectonic development of the western Pacific from mid-Eocene to the present. The maps are based on concepts developed from sea-floor spreading and the new global tectonics, and incorporate the postulate that new lithosphere can form behind advancing island arcs. The origin and later deformation of arcs and basins are shown as resulting mainly from the great shear between the northward-moving Australian plate and the northwestward-moving Pacific plate.

INTRODUCTION

The radio message reporting the death of Harry Hess was sent to us on board the drilling vessel *Glomar Challenger* by Professor A. G. Fischer, while we were working in the Caroline Basin in August 1969. The shock of the news was intensified by the fact that his name had been on our lips repeatedly. The bathymetry and tectonic elements of that part of the western Pacific were described in a classic paper by Hess (1948). The concept that the ocean basins are formed at, and spread from, the ridge system (Hess, 1962, 1965) had been amply supported by cores obtained during earlier legs of the *Glomar Challenger,* but Leg 6 in the western Pacific (Fischer and others, 1970) and our own Leg 7 (Winterer and others, 1969) had been finding evidence of young oceanic crust that apparently had not formed on part of the mid-ocean ridge system.

During the ensuing months any reminder of Harry's passing brought back to mind that evidence, the topic of this presentation. As a point of departure, it is assumed that the reader is familiar with the main points in the growth of global tectonic theory during the 1960s, as summarized for example by Phinney (1968), or by Isacks and others (1968). I will especially hold to the following two premises. (1) The lithosphere moves as great blocks (McKenzie and Parker, 1967; Morgan, 1968), whose rate and direction of movement can be determined from studies of seismology, geomagnetism, and structural geometry (*see,* for example, Isacks and others, 1968; Vine, 1966; Heirtzler and others, 1968; Menard and Atwater, 1968), so that past positions of the lithospheric blocks or plates can be approximated (Le Pichon, 1968). (2) The lithosphere is denser than the asthenosphere, a suggestion going back at least to Daly (1925) and supported today by interpretations of mechanical properties of the lithosphere (Elsasser, 1968), distribution of satellite-determined gravity anomalies (Moberly and Khan, 1969), stress orientation from earthquake-mechanism solutions (Isacks and Molnar, 1969), and randomly generated earth models (Press, 1969; for a contrary view, *see* Wang, 1970). Bass (1969, unpub. data) and Laubscher (1969) have commented on the general tectonic effects of a dense, spontaneously sinking lithosphere.

YOUTHFUL CRUST BEHIND ISLAND ARCS

Location

In recent years interpretations of the geomagnetic stripe patterns have shown that the greatest part of the deep-sea floor has formed along the oceanic rise or mid-ocean ridge system (Vine and Matthews, 1963; Wilson, 1965; Heirtzler and others, 1968). There are exceptions (Fig. 1).

The computer-fitting of the edges of continental masses now separated by ocean has been a powerful means of reconstructing paleogeography before a particular episode of sea-floor spreading. Often the reconstructions are aided by restoring a smaller block that presumably has rotated (Spain: Bullard and others, 1965) or split away (east Canary Islands: Dietz and Sproll, 1970) from the main continental block. For the Atlantic fit, however, North and South America are close together. Even if the Gulf of Mexico was formed by fragmentation and drifting of small blocks that became Honduras, Cuba, and so on, as proposed by Bass (1968, unpub. data), there is still the embarrassment of accounting for the sea floor of the Caribbean, which must have formed after the initiation of drift (Funnell and Smith, 1968).

Karig (1970) concluded from its thin sediment cover that the Lau-Havre Basin west of the Tonga-Kermadec Ridge was a youthful feature, where new sea floor formed by extensional rifting very late in the Cenozoic. He also suggested that

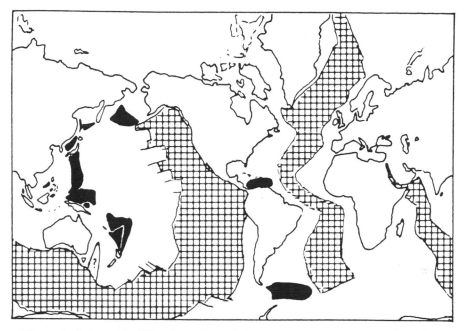

Figure 1. Suboceanic lithosphere formed in Cenozoic times. Cross-ruling: formed on mid-ocean ridge system (after Vine, 1968); black: known or suspected of having formed behind advancing island arcs.

several marginal basins on the concave side of island arcs, characterized by higher heat flow (Vacquier and others, 1966) and thinner sediments than might otherwise be expected, could be other regions of new oceanic crust. What is known of the attitude, amplitude, and nonsymmetry of magnetic anomalies of the sea floor behind the Scotia arc (Kroenke and Woollard, 1968) suggests that it is not forming from a part of the oceanic rise system.

It seems, therefore, that besides the places in the western Pacific where drilling has shown young oceanic crust to be present, there are other places where it is reasonable to conclude that young crust has also formed. These areas are behind island arc systems at the present day, or as will be shown below, almost certainly were behind arcs in mid-Cenozoic time.

Origin

Oceanic crust behind island arcs has been attributed by some to foundering of continental crust (Belousov, 1967), a concept that many others consider difficult to support on petrogenic and isostatic grounds. Several workers have advanced theories that the arcs are shaped around rising convection currents (shallow but powerful: Holmes, 1965; subsidiary circulation: Wright, 1966; the Melanesian Rise: Menard, 1964). In the present-day context of global tectonics, it would follow that the crust (or more properly, the lithosphere, or tectosphere, composed of crust and uppermost mantle) would have formed at the crest of those small rises. The average bathymetry and heat flow certainly do support the concept that the lithosphere is new, but actual patterns of bathymetric and magnetic trends are difficult to reconcile with plate theory and orthogonal spreading.

The post-Miocene Andaman Basin is explained by Rodolfo (1969) as a rhombochasm that resulted from a south-southwestward movement of southeast Asia.

Oxburgh and Turcotte (1970) believe that the trench and front part of an arc migrate oceanward as a consequence of the addition of sedimentary material to the front part of the island arc system, but that the migration is rather small.

McKenzie and Sclater (1968) discussed two possible sources of the high heat flow inside the island arcs of the northwestern Pacific. One was the upward movement of magmas and volcanism, but they argued that an enormous volume of volcanic material would be required to maintain the heat flow anomaly and the rate of intrusion and extrusion would be very high. They calculated, for instance, that the crust in the Sea of Japan would have formed in about the last 10 m.y. Because they believed that the trench and island arc features of the entire northwestern Pacific are unlikely to be so recent, McKenzie and Sclater hesitated to assign the unexpectedly high heat flow to a source from the upward movement of magma. Their text is not entirely clear, but apparently they were considering only andesitic volcanism (McKenzie and Sclater, 1968, p. 3176–3177), and they did point out the great difficulties of dating deformation in island arcs, where old deformed rocks may have either been formed by the present arc or have been carried in from elsewhere by moving lithosphere.

Of special interest is Karig's (1970) paper in which he has demonstrated the strong probability that only a particular segment ("interarc basin" or the Lau-Havre Trough) of the region behind the Tonga-Kermadec arc has been newly formed. The generation and extension of new oceanic lithosphere without mid-ocean ridges follows behind the "frontal" (volcanic and coralline) island arc and trench, which are visualized as migrating eastward. Karig postulates the extension as having been related to the initiation of the present episode of island arc volcanism, which he places in the late Tertiary at no more than 5 m.y. ago. Other marginal basins behind island arcs of the western Pacific are also held to be active or previously active extensional areas. Karig does not, however, offer a mechanism by which the island arc is able to advance against the oceanic plate that is spreading toward it. Specifically for his area of major work (Karig, 1970), he does not explain how the Tonga-Kermadec Ridge can migrate eastward from the Australian plate against the Pacific plate that is spreading west-northwestward from the East Pacific Rise and into the Tonga-Kermadec Trench.

In order to integrate these observations of Karig, and those of Fischer, and of others, into the model of global tectonics, two explanations are necessary. One of these is the development of a generalized account that would explain how new lithosphere might form behind island arcs. The other is a specific application to the varied geometry of plate boundaries and spreading vectors and to the geologic and geophysical patterns that actually exist.

Spontaneously Sinking Lithosphere

Figure 2, showing successive stages in the sinking of a plate of dense lithosphere, is after an illustration by Isacks and Molnar (1969), and represents a typical and current model of sinking at plate boundaries. Compare it with Figure 3, which shows the trajectory of the dense slab as being not parallel to the top and bottom of the slab, but actually inclined at a steeper angle. The initial position is at the margin of an area of the upper mantle that is warmer than adjacent areas. The lithosphere would thereby be thinner over the warm area, and as pointed out by Bass (unpub. data, 1969), that place would be the most likely site for initial failure of the lithosphere. As the edge of the thicker slab sags, new lithosphere forms above it, probably by basaltic volcanism at the sea floor and with cumulate and refractory bodies below. When the top of the slab, with its basalt and sediment that formerly were at the sea floor, sinks to the depth at which it can partially melt, calc-alkaline, typically andesitic volcanism commences (Fig. 3b). As the slab continues to descend, the belt of calc-alkaline volcanism maintains about the same position relative to its source at the top surface of the slab, and therefore also relative to the trench (Raleigh and Lee, 1969). Whereas the motion of the leading edge of the slab was initially vertical, it soon gains a lateral component toward the arc because the ratio of the cross-sectional area of the end of the slab to that of the bottom of the slab decreases as more and more slab descends (and also as the slab thins by warming); those are the areas that must force aside the asthenosphere to make room for the intruding lithosphere. The down-dip extension caused by the falling slab provides the gravity anomalies and earthquake patterns characteristic of arc-trench systems. It also provides the setting for structural features observed at trenches, whose tensional tectonics have been interpreted by such writers as Elsasser (1968), Lliboutry (1969), and Malahoff (1970). As the motion of the slab past the trench accelerates (Fig. 3c), the thermal regime of the area is probably like that described by Minear and Toksoz (1970) or Oxburgh and Turcotte (1970), with the effects of heating by adiabatic compression and radioactivity, phase changes, and—at least on the upper surface—strain heating. In addition, there must be an accounting for the mass of asthenosphere pushed aside. Almost certainly it would be the least viscous and thereby the warmest and lightest fraction (Moberly and Khan, 1969). Inside the arc new lithosphere would form from igneous rocks whose parent magmas were the warm asthenosphere. The dikes and flows would permeate the existing lithosphere so that the apparent area of new lithosphere might seem greater than that caused by the new material alone.

The slab's descent would slow as it entered the increased density and rigidity of the mesosphere (Fig. 3e). Additional movement could take place in several ways (Fig. 4). The slab might buckle at, or be thrust along, the top of the mesosphere. It might be thinned by heating at a rate equivalent to its sinking. In these instances, new lithosphere might be added behind the island arc for as long as the trench and arc continued to advance toward the falling slab. Or new edges of slabs might break and start to sink where especially thinned.

To this point the discussion has been illustrated with generalized vertical cross sections across island arc systems (Figs. 3 and 4). The two lithospheric plates (inside the arc and outside the arc) are postulated as initially having no motion relative to one another. The motion of the thicker outside plate toward the trench is initially nil (Fig. 3a), then accelerates (Fig. 3b to 3c), next slows (Fig. 3c to 3d), and finally reaches a steady state or ceases (Fig. 3e). Relative motion of the two plates leads to possible modifications (Fig. 5). Where the plate inside the arc is moving away from the trench, pulled perhaps by sinking into another trench nearby, the swath of new lithosphere may be wider. On the other hand, if both plates move toward the trench there may be little if any new lithosphere formed behind the arc, and the greatest proportion of the asthenosphere shoved aside from the sinking plate would migrate toward the mid-oceanic ridge system. A third possibility exists where the motion of one block is tangential to the other. Finally, where continental crust does lie over a sinking slab it may, without significant basaltic volcanism, be deformed into the immense horsts and even larger basins characteristic of germanotectonics (Malahoff and Moberly, 1968).

MIGRATING ISLAND ARCS OF THE WESTERN PACIFIC

Tectonic Patterns

If the assumptions are allowed that the lithosphere moves as blocks and that new lithosphere can form behind advancing island arcs, reconstruction of the tectonic pattern of the western Pacific can be attempted.

The areas of the western Pacific known or suspected to have youthful crust are shown on Figure 1. The spreading direction of the Indian-Australian block from the Mid-Indian Ocean Ridge is tangential to the direction of the Pacific block moving from the East Pacific Rise. The tectonics of the Melanesian region, commented on by Hess and Maxwell (1953) and many subsequent investigators, have in large measure resulted from that great shear. Several island arcs lie between the two blocks. Tonga and Kermadec arcs face toward the Pacific block; others face the Australian block. It seems virtually certain that a mid-Cenozoic festoon of south-facing arcs extended from the Sunda arc to the South New Hebrides arc, and that the arcs east of Java had advanced southward, forming the new lithosphere in their wake, perhaps by the method proposed above. Part of the Australian lithospheric block has continental crust,

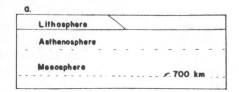
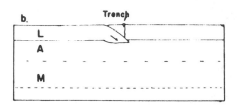
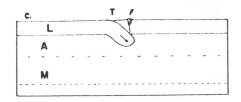
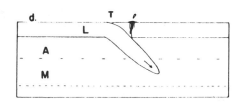
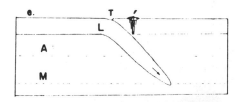

Figure 2. Possible model of sinking edge of a plate of lithosphere, with trajectory of plate parallel to surfaces of plate. Initial fracture (a) and origin of island arc system (b) added to concept of Isacks and Molnar (1969) of edge of slab sinking into asthenosphere (c), into increasing strength of upper part of mesosphere (d), and hitting high-strength part of mesosphere (e). The sequence (c), (d), and (e) are after Isacks and Molnar (1969, Fig. 3). Arrows show direction of movement of sinking plate edge. Except for growth of island arc, there is no new lithosphere formed above the sinking slab by this model.

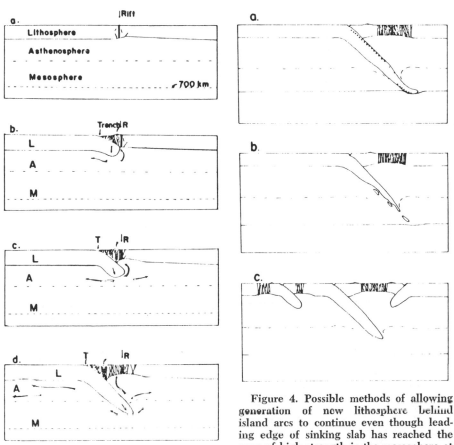

Figure 3. Preferred model of sinking edge of a plate of lithosphere, with trajectory of plate steeper than the angle of the plate. After rupture of lithosphere (a), the thicker plate sinks more rapidly than the thinner plate (b). Part of asthenosphere pushed aside from leading edge of plate, which initially sinks vertically, becomes new lithosphere by creep and igneous activity. The zone between trench and initial position of rift fills first. After island arc forms (b), new lithosphere grows mainly behind arc (c) and (d). As displaced asthenosphere warms and thins base of old lithosphere, later intrusions may cover a wider region (e). Arrows show

Figure 4. Possible methods of allowing generation of new lithosphere behind island arcs to continue even though leading edge of sinking slab has reached the zone of high strength in the mesosphere at about 600 to 700 km. Slab may bend and be thrust along top of layer of high strength (a), so that more of the plate outside the arc may fall away from the trench. Dots indicate a probable pattern of earthquake hypocenters (see Sykes and others, 1969, Fig. 2). Slab may be thinned by heating in the asthenosphere or upper mesosphere, so that little or no lithosphere collides with the zone of high-strength mesosphere (b). Where lithosphere is thinned by heating, rifting may allow additional slab edges to form and sink into the asthenosphere (c).

direction of movement of sinking plate edge as changing progressively from vertical to nearly parallel to plate surfaces. As lithosphere falls away from position of initial rupture, the distance from trench to rift increases, and new lithosphere forms behind advancing island arcs by this model.

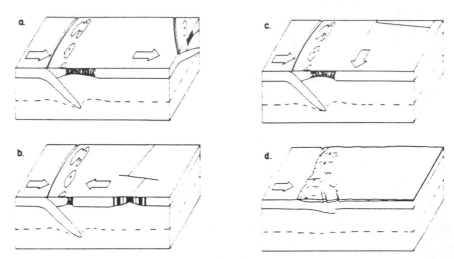

Figure 5. Modifications of theory of formation of new lithosphere behind advancing island arcs. Where plate behind an advancing arc is being pulled away by falling below an adjacent trench-arc system, (a) the formation of new lithosphere behind the first island arc may be more extensive than if there were no relative motion between the two plates. If the two plates move toward one another (b), formation of new lithosphere may be less extensive. Where the movement of the plates on opposite sides of the trench-arc system are tangential to one another (c), formation of new lithosphere may be more extensive. Where a trench-arc system strikes across the boundary between oceanic and continental crust (d), the continental crust may not sink sufficiently to allow new lithosphere to form. Possible examples might be (a) view to south across Bonin arc, (b) view to north across Macquarie arc, (c) view to south across Tonga arc, (d) view to north across Middle Rocky Mountains in Late Cretaceous times.

and during its northward drift the leading edge of continental crust did not sink into the trench. Rather, it has smothered a part of the trench festoon.

Several tectonic elements are present in New Guinea. The main mountain system of central and northern New Guinea has been termed the Papuan geosyncline (David, 1950). It bordered the edge of the Australian block, and is farther northeast and younger than the Tasman geosyncline. The Papuan geosyncline is composite. The northernmost ranges of New Guinea are of ultramafic and metamorphosed mafic rocks overlain by latest Cretaceous and Paleogene basalts and deepsea deposits, in turn overlain by shallower water limestones and detrital rocks. In western New Guinea (Irian) these rocks make up the north New Guinea province described by Visser and Hermes (1962) and equivalent provinces along the coast of Australian New Guinea and into New Britain. The more southerly belt of ultramafics in Papua (Davies, 1968) may extend via New Caledonia to New Zealand (Hess and Maxwell, 1953). The central and southern part of New Guinea is an extension of the Australian shield and bordering mountains. Visser and Hermes (1962) and Krause (1965) emphasized the geological differences between the regions.

The tectonic complexities west and east of New Guinea are results of the differing effects of oceanic crust and continental crust along the leading edge of the Australian block reaching the trench. The Sula Spur and adjoining Banda and Halmahera arcs west of New Guinea (Fitch, 1970) have a pattern unlike the one of the New Britain arc and Bismarck Archipelago east of New Guinea. The asymmetry is due to the shear between the major plates.

The paleogeographic maps that follow (Figs. 6 through 9) were constructed backwards in time from the present-day configuration (Fig. 10), using existing compilations (*see* Visser and Hermes, 1962, Fig. IV-2) as modified by more recently published data on plate boundaries and their directions of relative movement from earthquake studies (*see* Katsuma and Sykes, 1969) and spreading rates from the mid-ocean ridge system (*see* Isacks and others, 1968). Adjustments between plates, which at times must have been necessary (McKenzie and Morgan, 1969), are also allowed by small changes of the poles of relative rotation and by lateral, probably transform, faulting along median tectonic lines of the island arcs and along sharp bathymetric features that certainly were faults. New lithosphere is added behind advancing arcs by the process suggested above.

Figures 6 through 10 are palinspastic and paleotectonic. Although the maps were constructed by hindcasting, they are presented in historical order so that the reader can follow sequentially the development of the western Pacific from the Eocene to the present.

Middle Eocene

The paleogeography of the western Pacific in the middle Eocene, about 45 m.y. ago, is shown diagrammatically in Figure 6. Immediately prior to that time, peninsular India apparently was spreading away from Malaysia and Australia in the Indian Ocean, perhaps centered on the Ninetyeast Ridge. The time shown is of the initial spreading along the Southeast Indian Ridge, marking the beginning of the separation of the Australian and Antarctic plates. The Pacific plate was being formed along two ridges. One was the Pacific-Antarctic Ridge portion of the East Pacific Rise (orthogonal to Eltanin and similar fracture zones). The other was the "fossil" ridge in the northeast Pacific (orthogonal to Mendocino and similar fracture zones), from a center of rotation at about 77° N., 175° W. (Francheteau and others, 1970). Sinking of the Pacific plate under the tectonic belts from Kamchatka to Indonesia, and perhaps also from New Guinea to New Zealand, accommodated the relative convergence of the plates in the western Pacific.

The Japanese arc is shown straighter than at present (Takeuchi and others, 1967, p. 182), with new lithosphere not yet formed, that now underlies part of the Sea of Japan. The Philippine arc is shown as partly formed from earlier extension of the South China, Sulu, and Celebes Seas, although the timing of that extension is not yet well known. The Philippine Islands (Gervasio, 1967), like Japan, have a long complex history of orogenesis. The separate Philippine plate and its enclosing rim of the ancestral Bonin, Mariana, Yap, Palau, and Banda arcs are hypothetical. The northern part, from Japan to Palau, may have formed in the early Cenozoic (Fig. 6), but the Banda Islands east of Java, including Timor, Ceram (Seram), Buru, and also the central to southeastern part of Celebes (Sulawesi), are much older. The southern part of the Philippine plate may have been formed by the earlier advance of the Banda arc from near Borneo (Kalimantan). However, the northern part, a remnant of which is the present-day Philippine Sea west of 135° E., is deep and probably was the old western part of the Pacific plate, separated from it by the Mariana arc system newly formed in early Cenozoic times. The South Fiji Basin may also have started to form by advancement of the Lau-with-Tonga ridge

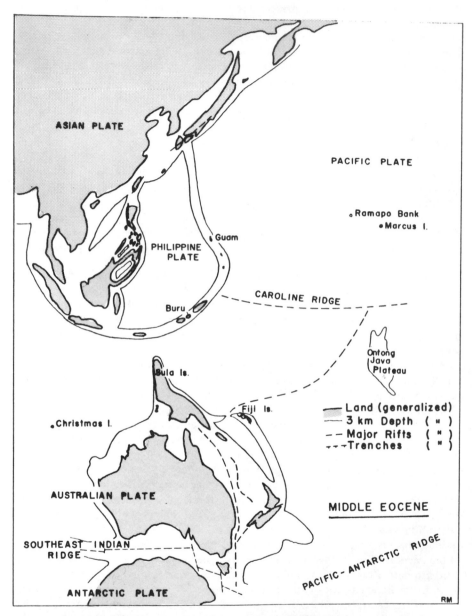

Figure 6. Hypothetical restoration of the western Pacific during the middle Eocene (at about 45 m.y.) at the time of initiation of trench development south of the Caroline Ridge and of spreading on the Southeast Indian Ridge. "Land" is generalized outline of the more important present-day features at this time in the past. Not all continental areas and islands marked as "Land" were actually above water in mid-Eocene time.

in the early Cenozoic (Karig, 1970). New Zealand, then close to Australia, soon separated (Le Pichon, 1968).

Relative to the Pacific plate, the Australian plate had been approaching from west to east. With mid-Eocene initiation of spreading on the Southeast Indian Ridge, the approach shifted more to the north. In order to accommodate the shift, or perhaps actually causing it, the lithosphere ruptured, and an edge sank along what was to become the present-day Caroline Ridge. If the trend of Molokai fracture zone were extended, it would bound the northern Caroline Islands. It would have provided a younger, higher, thinner, and warmer northern block against a thicker southern block. Though more complex, it clearly does extend some distance west of Hawaii (Malahoff and others, 1966) and, like other fracture zones, probably well into the central Pacific (Menard, 1967). Irrespective of whether or not the locus of rupture specifically was along an old fracture zone, the band of lithosphere that failed by rupture was the origin of the south-facing arc that today is mainly welded to the north coast of New Guinea. The ends of the band remain as the badly deformed arcs west of New Guinea to Halmahera and Morotai, and east of New Guinea through New Britain and New Ireland. The northeastern limit of the Australian plate presumably was another rupture that extended from the eastern Carolines to Fiji, but its initial position is less well known. A part of that plate boundary that later became the Solomon Islands was block-faulted and sufficiently sheared and compressed so that mild metamorphism resulted during the Eocene (Coleman, 1970). Those events marked the birth of the Solomons as a tectonic entity, even though the crust there had formed earlier.

Middle Oligocene

By middle or late Oligocene time, about 30 m.y. ago, the tectonic pattern of the western Pacific was about as shown on Figure 7. During the 15 m.y. between Figures 6 and 7, new lithosphere had been formed behind advancing island arcs and along the mid-oceanic ridge system in the areas shown by the stipple pattern. Existing lithosphere had been consumed by arc systems from areas equivalent to the lined pattern. Figure 7 therefore differs from Figure 6 mainly by the addition of the new areas and the removal of the old ones. Minor adjustments between plates are accommodated along fault systems and by folding.

The largest areas of new lithosphere not formed on mid-oceanic ridges are the eastern part of the Philippine Sea between the Palau-Kyushu and the South Honshu ridges, the Caroline Basins between the Caroline and New Britain ridges, and the South Fiji Basin between the Loyalty and Lau ridges. The transform fault north of the nascent South Fiji Basin may have been ancestral to the fracture zones trending west from Espiritu Santo and adjacent areas today. The proposed origin in Paleogene times for the South Fiji Basin and the Shikoku Basin of the northeastern Philippine Sea has not yet been tested by drilling.

Of possible origins for the Caroline Basins, the one shown on Figure 7 is of new lithosphere formed behind the New Britain arc that advanced to the south. A second possibility, considered by the Leg 7 party on the *Glomar Challenger,* was that Eauripik Rise may be the remnant of a small north-trending segment of the oceanic rise system. Lithosphere of the East and West Caroline Basins was thought to have formed on the now extinct Eauripik Rise and spread west toward the Yap-

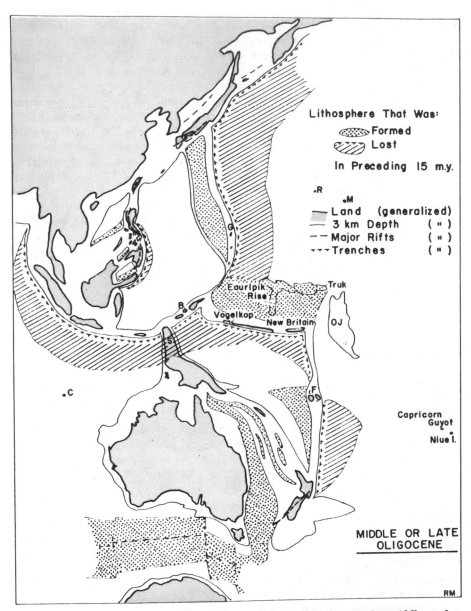

Figure 7. Hypothetical restoration of the western Pacific during the middle to late Oligocene (at about 30 m.y. ago). Letters locate features named on Figure 6. After the time shown by Figure 6, new lithosphere is known or assumed to have formed in the stippled areas. Lithosphere was lost into trenches from plates outside the trenches equivalent to the lined areas. Much of the new lithosphere is thought to have formed behind advancing island arcs by the method proposed in this paper. Individual areas are discussed in the text.

Palau trenches and east toward the Mussau Trough, an extinct trench. However, the magnetic anomalies of the southern parts of Eauripik Rise and East Caroline Basin, where Leg 7 holes were drilled, trend east and so could not have formed parallel to a north-trending center of spreading. Still another possibility might have been that of spreading from an extinct rise that trended east, but the bathymetric patterns and the position of pre-Oligocene crust surrounding the Caroline Basins rule strongly against that possibility.

Young lithosphere is also shown in areas east of Australia. Seismic refraction and continuous reflection work demonstrate that no episode of spreading centered on Lord Howe Rise has occurred that would have separated Australia from New Zealand and the ridges north of New Zealand—as Taylor and Brennan (1969) have also concluded from Project Magnet data. Their illustrations of magnetic profiles projected to even parallels of latitude do not, however, allow one to decide whether some other center of spreading might have existed in that region. The actual magnetic profiles do not rule out spreading from within the Tasman Sea (A. Malahoff, personal commun.). The opening of the Tasman Sea resulted from a faster drift of Australia northward relative to New Zealand from the Southeast Indian Ridge. More data are needed to decide whether there was a separate New Zealand plate and how long Macquarie Ridge may have been an island arc before one can reconstruct accurately the origin of the Tasman Sea and the splitting of the "Tasmantis Complex" (Cullen, 1970) into the Lord Howe and Norfolk ridges, and other ridges. Those ridges are intermediate in thickness between typical oceanic and typical continental crust. Cullen (1970) describes them as rifted apart "comparable with the spreading of fingers of a hand" by northeastward-directed convection currents. However, the mid-Cenozoic was also the time of glaucophane-facies metamorphism and the emplacement of the great ultramafic masses of New Caledonia, probably by powerful thrusting from the northeast (Lillie and Brothers, 1970). With evidence for both extension and compression, perhaps the ridges can be attributed to rifting, as well as to rotation of a separate New Zealand plate and differential spreading in the bight between the Southeast Indian and the Pacific-Antarctic ridges.

Middle Miocene

Shortly after the time shown on Figure 7, at about the time of anomaly 7 or 8, spreading on the fossil ridge of the northeast Pacific ceased, and the East Pacific Rise extended northward as the Gorda Rise and other rises. Relative to some fixed point on the earth, that event would have been accompanied by slowing of the eastward drift of the rise system (note the problem of two rises meeting at an angle on a nonexpanding earth) and consequent acceleration of the Pacific plate in a more northerly direction. That motion of the Pacific plate has continued to the present day.

By the middle Miocene, about 15 m.y. ago, the western Pacific may have appeared about as shown on Figure 8. Past the northwest rim of the Pacific plate, the initial opening of the basins of the Sea of Japan and those of the southern Sea of Okhotsk are shown. These regions of new lithosphere, and probably also the Aleutian Basin, are behind the arcs toward which the Pacific plate was accelerated. A corollary of the theory advanced in this paper would be that accelerated spontaneous sinking of the lithosphere at the arcs of the northern rim of the Pacific allowed the spreading center of the northeastern Pacific to change from the fossil rises to the present ones.

The northward drift of the Australian plate against the northwestward drift of the Pacific plate resulted in structural complexities along the plate margins. The less-dense leading edge of the continent, the present-day Sula Islands and part of northwestern New Guinea, could not plunge down the Banda and New Britain Trenches with the suboceanic lithosphere. The arcs were bent by the north-drifting spur. Ultramafic and mafic ocean floor pushed up as a ridge ahead of the Sula Spur ultimately became the eastern arms of Celebes, now wrapped around the Sula Islands and pushed against central and southeastern Celebes. On the east side, pushed-up ocean floor became the main part of present-day Halmahera. The continental projection, shoved westward by the relative motion between the plates, began to split and rotate at present-day Geelvink (Sarera) Bay. The leading edge of the Australian lithospheric plate split where it could not sink. The western part ruptured at Buru, and continued to sink under the Banda arc while that arc was pressed relatively westward. The eastern part sank under the New Britain arc, but as the triangular remnant of suboceanic lithosphere ahead of New Guinea sank, the seam between the two blocks closed to the southeast.

Sinking into the Mindanao Trench pulled the Philippine plate into clockwise rotation. The western edge of the Pacific block near the equator was broken by lateral faults. Some, in the Caroline Islands and along the southwest part of the Ontong Java Plateau, were sites of volcanism in the Neogene. The larger faults continue to have bathymetric expression today as sharp ridges and trenches (Fig. 8). Sorol, Lyra, Malaita, and Vitiaz faults, as well as linear structure mapped on the Ontong Java Plateau (Kroenke, 1970, unpub. data), trend northwesterly. They are left-lateral faults that allowed the main part of the Pacific plate to be pulled northwest into the Mariana Trench and more northerly trench systems. The Mussau fault could accommodate differential transverse motion ahead of the thick Ontong Java block. The Mussau, Cape Johnson, and Malaita faults may not have been active until after the time shown on Figure 8.

Continued sundering of the Tasmantis Complex is shown northwest of New Zealand. The New Hebrides and Fiji volcanic and volcanogenic sediments had continued to accumulate, unaffected by cratonic influences (Dickinson, 1967).

Middle Pliocene

By mid-Pliocene time (Fig. 9), about 5 m.y. ago, the western Pacific appeared much as it does today. The northern arcs continued to advance to the southeast, enlarging the new lithosphere of the Sea of Japan and Sea of Okhotsk. Rotation of the Philippine Sea plate (Katsuma and Sykes, 1969) was accomplished mainly by sinking at the Mindanao and Nansi Shoto Trenches, by sinking at Luzon Trench, and by left-lateral faulting along the great dislocation zones of the Philippines and Taiwan.

The northern part of the leading edge of the Pacific plate continued to be pulled below the Kurile, Japan, Bonin, and Mariana Trenches. The central or tropical part of the edge remained a great shear zone between the main Pacific plate and the Philippine Sea plate and the arcs and basins of Melanesia. At the northwest

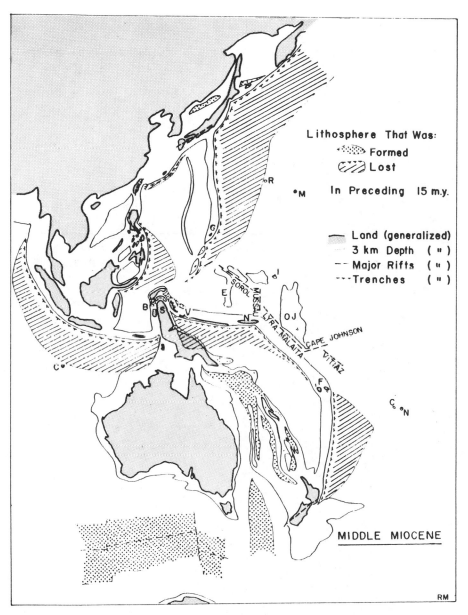

Figure 8. Hypothetical restoration of the western Pacific during the middle Miocene (about 15 m.y. ago), shortly before spreading from the fossil northeastern Pacific ridge ceased. Names in capital letters are of important faults or rifts, other symbols as on previous figures. Individual areas are discussed in the text.

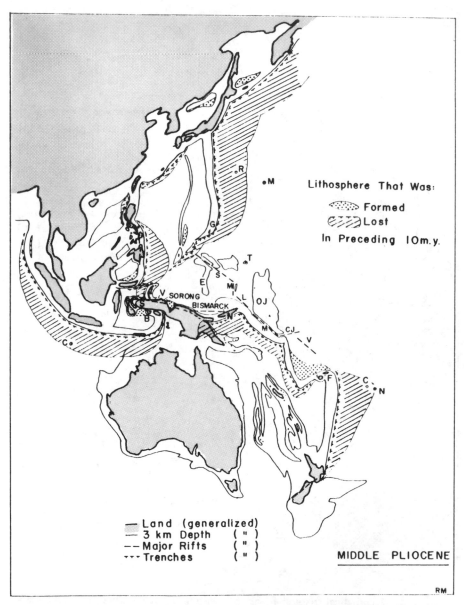

Figure 9. Hypothetical restoration of the western Pacific during the middle Pliocene (about 5 m.y. ago), showing tectonic changes of previous 10 m.y. Symbols as on previous figures. Individual areas are discussed in the text.

end, the Yap and Palau festoons and the Sorol and other faults in the western Carolines accommodated part of the differential movement. Part, however, continued at the southern edge of the Caroline Basins with additional bending westward of the Sula Spur and Vogelkop, and further tightening of the Banda arc. After the Sula Spur was bent sufficiently far, the Halmahera arc advanced westward by left-lateral faulting on the Sorong fracture zone.

The Ontong Java Plateau, intermediate in thickness between typical oceanic and typical continental crust (Furumoto and others, 1970), impinged on what was to be the Solomons arc shortly before the time shown on Figure 9. The Cape Johnson Trench, aligned with the vector of relative movement between the Australian and Pacific plates, acted as a strike-slip fault bounding crustal blocks, as suggested by Fisher and Hess (1963). If one removes the present-day offset of the southern Solomons arc (San Cristobal Island) from the northern New Hebrides arc (Santa Cruz Islands) by restoring the plates at their present-day rate of convergence, the impingement would have been about 7 m.y. ago. The same late Miocene age was determined independently (Kroenke, 1970, unpub. data) by correlating reflection profiles between the Malaita Trough, where flat-lying beds in the trough overlap warped beds at the southwestern edge of the plateau, and the JOIDES site cored on Leg 7 of the Deep-Sea Drilling Project. Moreover, late Miocene was one of two major phases of deformation on the island of Malaita (the other, as noted above, was in the Eocene); Coleman (1970) postulated a westward movement of Malaita toward Guadalcanal as having started in the late Miocene.

The west edge of the Ontong Java Plateau block pushed the New Ireland end of the New Britain arc west against the Caroline Basins, and probably also initiated movement along the rupture in the middle of the Bismarck Sea. That fault zone, called Bismarck on Figure 9, is an intense seismic belt today (U.S. Coast and Geodetic Survey, 1970).

Southeast of the Ontong Java crustal block the oblique movement of the Pacific plate behind the New Hebrides arc (Fig. 5c) allowed the New Hebrides arc to advance southward and new lithosphere formed near Fiji.

Present Day

The last of this series of maps shows the present-day appearance of the western Pacific (Fig. 10). Ramapo Bank, Capricorn Guyot, and Christmas Island have reached the trench edges, soon to be pulled down. Areas of highest heat flow in the Sea of Japan and elsewhere are assumed to be the youngest areas of new lithosphere behind advancing arcs. The Sorol, Lyra, Bismarck, and Sorong remain as the more active lateral faults (U.S. Coast and Geodetic Survey, 1970). The westward thrust of the Sula Spur has pushed together the formerly separate peninsulas of Celebes. Limits of the sinking lithospheric plates of the Banda, Halmahera, and Mindanao arcs are well outlined by their gravity fields (Vening Meinesz, 1948) and seismicity (Fitch, 1970; Fitch and Molnar, 1970).

Continuing the hypothesis of Karig (1970), the advance of the Tonga-Kermadec arc from the Lau-Colville Ridge occurred very recently, and is so shown in Figure 10. The Fiji Plateau is now made of small plates bounded by belts of small shallow earthquakes (Sykes and others, 1969) and is characterized by high heat flow (Sclater and Menard, 1967) and zones of *en echelon* faults marking extensional

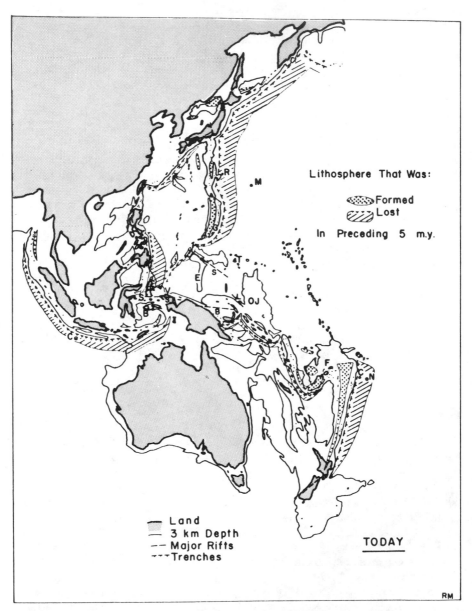

Figure 10. Present-day map of the western Pacific showing tectonic changes of previous 5 m.y. Symbols as on previous figures. Individual areas are discussed in the text.

areas (Malahoff, 1970, unpub. data). Fiji Plateau has grown by accretion in a spiral fashion, crudely analogous to a snowball garnet in metamorphic rock.

ACKNOWLEDGMENTS

My interest in marine geology was dominated by the influence of Harry Hess. It began in 1947 when he was the first faculty member I encountered in Guyot Hall when trying to decide whether to enroll in introductory geology. I was a sophomore, newly back from an NROTC cruise; I was flattered that he seemed as interested in my comments about ships as I certainly was in his comments about geology. My interest heightened in his part of our senior course, Advanced General Geology, and continued in graduate school. Later, the single strongest influence on my decision to apply for an opening in Hawaii was his letter to me predicting a bright future for marine geology here.

I appreciate the opportunity to study unpublished maps and manuscripts of Manuel N. Bass, Loren W. Kroenke, and Alexander Malahoff. I also appreciate their criticism of this manuscript.

REFERENCES CITED

Belousov, V. V., 1967, Some problems concerning the oceanic earth's crust and upper mantle evolution: Geotectonics, v. 1, p. 1–6.

Bullard, E. C., Everett, J. E., and Smith, A. G., 1965, The fit of the continents around the Atlantic: Royal Soc. London Philos. Trans., Ser. A, v. 258, p. 41–51.

Coleman, P. J., 1970, Geology of the Solomon and New Hebrides Islands, as part of the Melanesian re-entrant, Southwest Pacific: Pacific Sci., v. 24, p. 289–314.

Cullen, D. J., 1970, "Two-way stretch" of sialic crust and plate tectonics in the Southwest Pacific: Nature, v. 226, p. 741–742.

Daly, R. A., 1925, Relation of mountain building to igneous action: Am. Philos. Soc. Trans., v. 64, p. 283–307.

David, T.W.E., 1950, The geology of the Commonwealth of Australia: London, Edward Arnold Co., v. 1, 747 p.

Davies, H. L., 1968, Papuan ultramafic belt: Internat. Geol. Cong., 23rd, Prague 1952, v. 1, p. 209–220.

Dickinson, W. R., 1967, Tectonic development of Fiji: Tectonophysics, v. 4, p. 543–553.

Dietz, R. S., and Sproll, W. P., 1970, East Canary Islands as a microcontinent within the Africa-North America continental drift fit: Nature, v. 226, p. 1043–1045.

Elsasser, W. M., 1968, Submarine trenches and deformation: Science, v. 160, p. 1024.

Fischer, A. G., Heezen, B. C., Boyce, R. E., Bukry, D., Douglas, R. G., Garrison, R. E., Kling, S. A., Krasheninnikov, V., Lisitzin, A. P., and Pimm, A. C., 1970, Geological history of the western North Pacific: Science, v. 168, p. 1210–1214.

Fisher, R. L., and Hess, H. H., 1963. Trenches, in Hill, M. N., ed., The sea, v. 3: New York, Interscience, p. 411–436.

Fitch, T. J., 1970, Earthquake mechanisms and island arc tectonics in the Indonesian-Philippine region: Seismol. Soc. America Bull., v. 60, p. 565–591.

Fitch, T. J., and Molnar, P., 1970, Focal mechanisms along inclined earthquake zones in the Indonesia-Philippine region: Jour. Geophys. Research, v. 75, p. 1431–1444.

Francheteau, J., Sclater, J. G., and Menard, H. W., 1970, Pattern of relative motion from fracture zone and spreading rate data in the Northeastern Pacific: Nature, v. 226, p. 746–748.

Funnell, B. M., and Smith, A. G., 1968, Opening of the Atlantic Ocean: Nature, v. 219, p. 1328–1333.

Furumoto, A. S., Hussong, D. M., Campbell, J. F., Sutton, G. H., Malahoff, A., Rose, J. C., and Woollard, G. P., 1970, Crustal and upper mantle structure of the Solomon

Islands as revealed by seismic refraction survey of November-December 1966: Pacific Sci., v. 24, p. 315–332.

Gervasio, F. C., 1967, Age and nature of orogenesis of the Philippines: Tectonophysics, v. 4, p. 379–402.

Heirtzler, J. R., Dickson, G. O., Herron, E. M., Pitman, W. C., III, and Le Pichon, X., 1968, Marine magnetic anomalies, geomagnetic field reversals, and motions of the ocean floor and continents: Jour. Geophys. Research, v. 73, p. 2119–2136.

Hess, H. H., 1948, Major structural features of the western North Pacific, an interpretation of H. O. 5485, bathymetric chart, Korea to New Guinea: Geol. Soc. America Bull., v. 59, p. 417–466.

—— 1962, History of ocean basins, in Engel, A.E.J., James, H. L., and Leonard, B. F., eds., Petrologic studies (Buddington volume): Geol. Soc. America, p. 559–620.

—— 1965, Mid-oceanic ridges and tectonics of the sea-floor, in Whittard, W. F., and Brawshaw, R., eds., Submarine geology and geophysics: London, Butterworths, p. 317–333.

Hess, H. H., and Maxwell, J. C., 1953, Major structural features of the Southwest Pacific: A preliminary interpretation of H. O. 5484, bathymetric chart, New Guinea to New Zealand: 7th Pacific Sci. Congr. Proc., v. 2, p. 14–17.

Holmes, A., 1965, Principles of physical geology (2d ed.): New York, Roland Press Co., 1288 p.

Isacks, B., and Molnar, P., 1969, Mantle earthquake mechanisms and the sinking of the lithosphere: Nature, v. 223, p. 1121–1124.

Isacks, B., Oliver, J., and Sykes, L. R., 1968, Seismology and the new global tectonics: Jour. Geophys. Research, v. 73, p. 5855–5899.

Karig, D. E., 1970, Ridges and basins of the Tonga-Kermadec island-arc system: Jour. Geophys. Research, v. 75, p. 239–259.

Katsuma, M., and Sykes, L. R., 1969, Seismicity and tectonics of the Western Pacific: Izu-Mariana-Caroline and Ryukyu-Taiwan regions: Jour. Geophys. Research, v. 75, p. 5923–5948.

Krause, D. C., 1965, Submarine geology north of New Guinea: Geol. Soc. America Bull., v. 76, p. 27–42.

Kroenke, L. W., and Woollard, G. P., 1968, Magnetic investigations in the Labrador and Scotia Seas, USNS Eltanin cruises 1–10, 1962–1963: Hawaii Inst. Geophysics Rept. 68–4, 59 p.

Laubscher, H., 1969, Mountain building: Tectonophysics, v. 7, p. 551–563.

Le Pichon, Xavier, 1968, Sea-floor spreading and continental drift: Jour. Geophys. Research, v. 73, p. 3661–3697.

Lillie, A. R., and Brothers, R. N., 1970, The geology of New Caledonia: New Zealand Jour. Geology and Geophysics, v. 13, p. 145–183.

Lliboutry, L., 1969, Sea-floor spreading, continental drift, and lithosphere sinking with an asthenosphere at melting point: Jour. Geophys. Research, v. 74, p. 6525–6540.

Malahoff, A., 1970a, Some possible mechanisms for gravity and thrust faults under oceanic trenches: Jour. Geophys. Research, v. 75, p. 1992–2001.

Malahoff, A., and Moberly, R., Jr., 1968, Effects of structure on the gravity field of Wyoming: Geophysics, v. 33, p. 781–804.

Malahoff, A., Strange, W. E., and Woollard, G. P., 1966, Molokai fracture zone: Continuation west of the Hawaiian Ridge: Science, v. 153, p. 521–522.

McKenzie, D. P., and Morgan, W. J., 1969, Evolution of triple junctions: Nature, v. 224, p. 125–133.

McKenzie, D. P., and Parker, R. L., 1967, The North Pacific: An example of tectonics on a sphere: Nature, v. 216, p. 1276–1280.

McKenzie, D. P., and Sclater, J. G., 1968, Heat flow inside the island arcs of the northwestern Pacific: Jour. Geophys. Research, v. 73, p. 3173–3179.

Menard, H. W., 1964, Marine geology of the Pacific: New York, McGraw-Hill Book Co., 271 p.

—— 1967, Extension of northeastern-Pacific fracture zones: Science, v. 155, p. 72–74.

Menard, H. W., and Atwater, T., 1968, Changes in direction of sea-floor spreading: Nature, v. 219, p. 463–467.

Minear, J. W., and Toksoz, M. N., 1970, Thermal regime of a downgoing slab and new global tectonics: Jour. Geophys. Research, v. 75, p. 1397–1419.

Moberly, R., Jr., and Khan, M. A., 1969, Interpretation of the sources of the satellite-determined gravity field: Nature, v. 223, p. 263–267.

Morgan, W. J., 1968, Rises, trenches, great faults, and crustal blocks: Jour. Geophys. Research, v. 73, p. 1959–1982.

Oxburgh, E. R., and Turcotte, D. L., 1970, Thermal structure of island arcs: Geol. Soc. America Bull., v. 81, p. 1665–1688.

Phinney, R. A., 1968, Introduction, in Phinney, R. A., ed., The history of the Earth's crust: Princeton, Princeton Univ. Press, p. 3–12.

Press, F., 1969, The sub-oceanic mantle: Science, v. 165, p. 174–176.

Raleigh, C. B., and Lee, W.H.K., 1969, Sea-floor spreading and island-arc tectonics, in McBirney, A. R., ed., Proceedings of the andesite conference: Oregon Dept. Geology and Mineral Industries Bull. 65, 193 p.

Rodolfo, K. S., 1969, Bathymetry and marine geology of the Andaman Basin, and tectonic implications for southeast Asia: Geol. Soc. America Bull., v. 80, p. 1203–1230.

Sclater, J. G., and Menard, H. W., 1967, Topography and heat flow of the Fiji Plateau: Nature, v. 216, p. 991–993.

Sykes, L. R., Isacks, B. L., and Oliver, J., 1969, Spatial distribution of deep and shallow earthquakes of small magnitude in the Fiji-Tonga region: Seismol. Soc. America Bull., v. 59, p. 1093–1113.

Takeuchi, H., Uyeda, S., and Kanamori, H., 1967, Debate about the Earth: San Francisco, Freeman, Cooper and Co., 253 p.

Taylor, P. T., and Brennan, J. A., 1969, Airborne magnetic data across the Tasman Sea: Nature, v. 224, p. 1100–1102.

U.S. Coast and Geodetic Survey, 1970, World seismicity 1961–1969: Map NEIC 3005.

Vacquier, V., Uyeda, S., Yasui, M., Sclater, J., Corrie, C., and Watanabe, T., 1966. Heat flow measurements in the northern Pacific: Tokyo Univ. Earthquake Research Inst. Bull., v. 44, p. 1519–1535.

Vening Meinesz, F. A., 1948, Gravity expeditions at sea: Netherlands Geodetic Comm. Publ. 233 p.

Vine, F. J., 1966, Spreading of the ocean floor; new evidence: Science, v. 154, p. 1405–1415.

―― 1968, Evidence from submarine geology: Am. Philos. Soc. Proc., v. 112, p. 325–334.

Vine, F. J., and Matthews, D. H., 1963, Magnetic anomalies over oceanic ridges: Nature, v. 199, p. 947–949.

Visser, W. A., and Hermes, J. J., 1962, Geological results of the exploration for oil in Netherlands New Guinea: Verh. Konin, Nederlands Geol.-Mijnb. Genoot., Geol. Ser., deel XX, Spec. Num., 265 p.

Wang, C. Y., 1970, Density and constitution of the mantle: Jour. Geophys. Research, v. 75, p. 3264–3284.

Wilson, J. T., 1965, A new class of faults and their bearing on continental drift: Nature, v. 207, p. 343–347.

Winterer, E. L., Riedel, W. R., Moberly, R., Jr., Resig, J. M., Kroenke, L. W., Gealy, E. L., Heath, G. R., Bronnimann, P., Martini, E., and Worsley, T. R., 1969, Deep Sea Drilling Prospect: Leg 7: Geotimes, v. 14, no. 10, p. 14–15.

Wright, J. B., 1966, Convection and continental drift in the Southwest Pacific: Tectonophysics, v. 3, p. 69–81.

Manuscript Received by the Society March 29, 1971

Contribution No. 423, Hawaii Institute of Geophysics, Honolulu, Hawaii

THE GEOLOGICAL SOCIETY OF AMERICA, INC.
MEMOIR 132
© 1972

Outcropping Layer A and A"
Correlatives in the Greater Antilles

PETER H. MATTSON
Queens College, City University of New York, Flushing, New York 11367

E. A. PESSAGNO, JR.

AND

C. E. HELSLEY
University of Texas at Dallas, Dallas, Texas 75230

ABSTRACT

Paleocene and Eocene green cherty tuffs and pelagic sediments, in a generally volcanic rock sequence, crop out in Puerto Rico and the Dominican Republic. These rocks arc here correlated by lithology and age with oceanic horizons found by reflection profiling and coring: Layer A in the Atlantic Ocean, and Layer A" and the Carib beds in the Caribbean Sea. The generally thin pelagic or airborne sediments in the ocean basins flank thicker clastic, pelagic, and volcanic deposits on the Antillean Ridge. Trenches north and south of the ridge apparently trapped most of the coarser volcanic material, preventing its reaching the ocean basins. Lithification (except that forming chert) and deformation of these sediments was also restricted to the Antillean Ridge. The zone of deformation is about 180 km wide, and the transition between deformed and undeformed rocks takes place over a distance in some places as short as 10 km.

INTRODUCTION

Seismic reflection profiles from the world oceans have recorded several prominent reflecting horizons. The most widespread of these have been designated Layers or Horizons A and B in the Atlantic Ocean (J. Ewing and others, 1966; Saito and others, 1966) and A" and B" in the Caribbean Sea (J. Ewing and others, 1967; Fig. 1). A and A" plunge beneath undeformed turbidite deposits in the Puerto Rico and Dominican Trenches north and south of the eastern Greater Antilles, respectively (Fig. 2). This report suggests a correlation of A and A" with rock of early Tertiary age exposed in Puerto Rico and the Dominican Repubic. Although generally undeformed on the ocean floor, these beds are intensely deformed and uplifted where they crop out in the Greater Antilles.

LAYER A IN THE NORTH ATLANTIC

Layer A in the north Atlantic Ocean consists of a set of closely spaced, conformable reflectors. On the Bermuda Rise, it apparently splits into two parts. At sites 6 and 7 of JOIDES Leg 1, on the Bermuda Rise, the sediments at the reflecting levels are cherty turbidites, lutites, and siliceous mudstones or cherts (M. Ewing and others, 1969). Both reflecting and nonreflecting layers are rich in *Radiolaria,* diatoms, and silicic sponge spicules; many layers are pale green in color. The lutites contain zeolites and abundant montmorillonite. The upper reflector at sites 6 and 7 is of early-middle Eocene age; the age of the lower horizon is not yet known. Red clays overlie and underlie the reflecting zone.

Core samples from the region north of San Salvador, Bahamas (25° N., 73° W.), are composed of turbidites and interbedded red pelagic clays of late Maestrichtian age. These may represent horizons between Layers A and B. The vertical distance between the topmost Layer A reflector and the top of the next set of prominent reflectors (called Layer B) is variable but is about 500 m north of the Bahamas.

LAYER A″ IN THE CARIBBEAN; THE CARIB BEDS

Prominent reflecting horizons in the Caribbean area plunge beneath well-layered coarse turbidite deposits in the Dominican Trench south of Puerto Rico (J. Ewing and others, 1967). The reflecting horizons are designated A″ and B″. The acoustically transparent layers above and between them are called the Carib beds. Layer A″ is the uppermost of a series of closely spaced reflecting horizons. The depth corresponds to layers of Eocene chert cored during JOIDES Legs 3 and 4 (Bader and others, 1969). The Carib beds are thought to be uniformly deposited pelagic sediments because of their transparent acoustical character and their uniform thickness. Where sampled, they consist of pelagic radiolarian ooze, but a fine-grained turbidite composition or a mixture of pelagic and turbidite deposits has also been suggested (Edgar and others, 1971).

EXTENT OF LAYERS A AND A″ OUTSIDE THE NORTH ATLANTIC AND CARIBBEAN

J. Ewing and others (1970) suggest that Layer A does not occur extensively on the mid-Atlantic Ridge, but that it may reappear as Eocene chert forming a good reflector about 160 to 200 m below the sea floor at JOIDES site 12, northwest of the Cape Verde Islands, and also at 180 m below the sea floor at JOIDES site 13 in the Sierra Leone Basin west of Africa. A few other sites in the south Atlantic

---

Figure 1. A, Exposure areas of the Maestrichtian to Eocene rock sequences in Puerto Rico. Letters A through E correspond to geologic columns shown in Figure 3. The west end of area B and the northeast part of the island are not yet mapped in detail, so boundaries in those areas are probably inaccurate. B, Layer A and A″ in the western Atlantic Ocean and the Caribbean Sea. Dashed lines outline areas where Layers A or A″ have been recorded by coring or profiling; the solid line labeled *outcrop* marks the outcrop of Layer A. Crosses mark core holes from the JOIDES project, Leg 1, and lines 1–1′, 2–2′, and 3–3′ mark the locations of cross sections shown in Figure 2. Data are from M. Ewing and others (1969) and J. Ewing and others (1966, 1967). Distribution of Layer A″ in the western Caribbean and Gulf of Mexico is not shown.

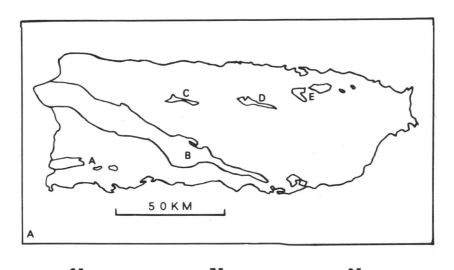

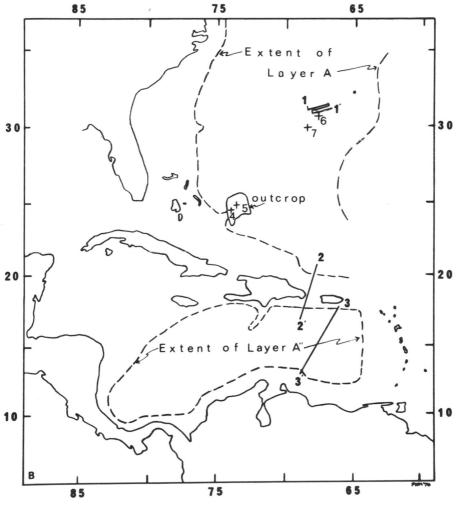

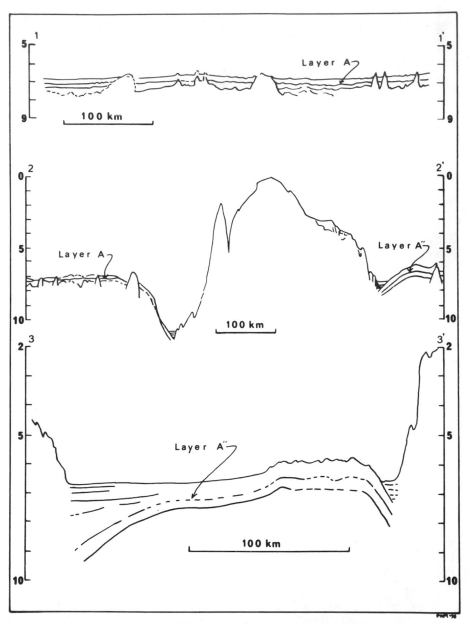

Figure 2. Layer A and A″ in traced seismic profiler sections. Horizontal scale is different in each section and is shown by a 100-km scale bar; vertical scale is reflection time in seconds, where 1 second is approximately 400 fathoms. Vertical exaggeration is approximately ×25. Section 1–1′ from J. Ewing and others (1966); 2–2′ and 3–3′ from J. Ewing and others (1967). Sections are located on Figure 1.

and in the Pacific report reflecting Eocene chert horizons, but other sites in these same regions contain reflectors of different ages and not of an Eocene age; so if present, the Eocene horizon is not uniformly distributed or has been missed by the sampling procedure. The relation of these distant localities to the possible origin of the chert is discussed in a later section.

MAESTRICHTIAN TO EOCENE ROCK SEQUENCE IN PUERTO RICO

Typical stratigraphic columns of Maestrichtian to Eocene rocks in Puerto Rico are shown in Figure 3. The Cretaceous portion of the sequence is predominantly shallow marine and nonmarine volcanic and volcaniclastic rock containing Late Cretaceous (probably Maestrichtian) rudists and *Foraminifera* (Mattson, 1960). Overlying these rocks is the Jacaguas Group (Pessagno, 1961), an extensive series of rocks generally containing early-middle Eocene *Foraminifera* and *Radiolaria,* but also fossils as old as Paleocene in a few places (Glover and Mattson, 1967). The predominant sediment types within the Jacaguas Group are well-bedded tuffaceous graywackes, calcareous and noncalcareous lutites and cherts, and rare bedded and reefoid limestones. The noncalcareous lutites are largely lithified volcanic ash deposits. They are brown, gray, purple, and green in color. In some exposures the color is due to green montmorillonite (Fig. 4).

Exposures of these rocks occur within fault-bounded areas throughout Puerto Rico (Fig. 1). Where the lower boundary is exposed, the Maestrichtian or early Tertiary rocks rest unconformably upon deformed rocks of Maestrichtian or older ages (Mattson, 1960, 1968b). The sequence is covered unconformably by Oligocene and younger clastic and carbonate rocks. In several places unconformities of apparently local extent divide the Maestrichtian–lower Tertiary sequence. Structures in the sequence vary from broad folds to tight disharmonic slump-folds (Mattson, 1972).

PALEOCENE TO LOWER EOCENE IMBERT FORMATION IN DOMINICAN REPUBLIC

Nagle (1971) has defined the Imbert Formation as "a 1,000-m-thick succession of fine-grained, graded bedded, calcareous tuffs, which grade upward to vitric andesite and dacite tuffs, with rare interbedded green radiolarian cherts and thin white aphanitic limestones." The formation rests unconformably upon a volcanic unit of unknown age, and supplied clasts for early-late Eocene conglomerates and sandstones. Nagle suggests that the formation is of deep-water origin because of the presence of radiolarian cherts and pelagic *Foraminifera,* and the absence of coarse clastic material. He believes that uplift of the area in middle Eocene time caused part of the Imbert Formation to form a chaotic gravity-slide deposit, the San Marco Olistostrome.

CORRELATION WITH LAYER A AND A″

Layer A and A″, the Jacaguas Group, and the Imbert Formation are all of Maestrichtian to Eocene age. Moreover, they contain many similar rock types. The Puerto Rican sequence contains thin chert beds interbedded with lutites, carbonates, sandstones, and dacitic volcanic rocks. Some of the cherts are green; they are the only consistently green layers in the Albian to Recent volcanic and sedimentary sequences that form Puerto Rico. It is here suggested that the green chert strata

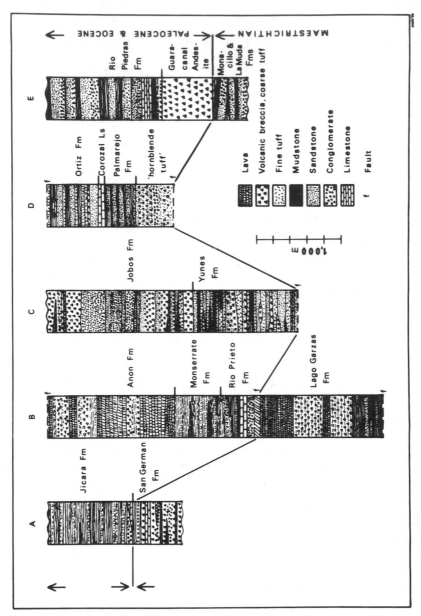

Figure 3. Diagrammatic geologic columns for Maestrichtian to Eocene rocks in Puerto Rico. Columns are located in Figure 1; rock types are shown in correct proportions as known, but vertical positions within formations are of varying accuracy. Data for A from Mattson (1960); B, Mattson (1967, 1968a, 1968b); C, Nelson (1967b); D, Nelson (1966, 1967a); E, Pease (1968a, 1968b, 1968c).

found in Layer A and A'' correlate with the outcropping green cherts of the Greater Antilles. In the Greater Antilles, green cherts and dacite are common only in the Imbert Formation and the Jacaguas Group. Accordingly, the Imbert Formation and the Jacaguas Group may be correlated by rock type and age with the A and A'' layers. McIntyre and Aaron (1971) have independently made the same correlation, based on mapping in the northwestern part of the longest exposure area on Figure 1.

Differences between the land exposures and Layer A and A'' are the relative amounts of coarse volcanic material, thicknesses, tectonic patterns, and the depths of deposition. No pelagic red clays or claystones have been mapped in the Jacaguas sequence in Puerto Rico. Shallow-water reefoid limestones, marine and nonmarine lavas and volcanic sediments may have been localized on the volcanic ridge that now forms the Greater Antilles. Finer grained carbonates, lutites, and fine-grained clastic rocks of the land sequence may represent the more acoustically transparent parts of the Carib beds and Layer A.

If this correlation is correct, one must assume that the Greater Antilles contain the nonpelagic sediments deposited during the Maestrichtian-Eocene interval, and that little or none of this coarse clastic or pyroclastic material reached the ocean basins to the north or south. Trench systems analogous to the present-day Puerto Rico Trench and Dominican Trench presumably trapped these coarse deposits before they reached the basins. No thickening of Layer A or A'' is seen on profiles across either trench (Fig. 2), suggesting that no significant thickness of coarse-grained sediment was contributed by the Antillean Ridge to the ocean basin deposits.

Some of the finer grained material in the oceanic layers may originate in the Greater Antilles, however. The green montmorillonite- and zeolite-bearing sediment present in the Eocene part of Layer A (Peterson and others, 1970) may be derived from very fine volcanic ash spread by wind and currents over large parts of the North Atlantic. One source for this volcanic material could have been the Eocene volcanism common in the Greater Antilles. The fine wind-blown material would be able to cross the trench systems that trap the coarser material. The lack of a variation in thickness of the oceanic reflecting beds may be due to their pelagic or airborne character, being composed of grains too small for noticeable further sorting by water or wind.

It is not the intention of this paper to suggest that all Eocene oceanic cherts are due to volcanism in the Greater Antilles. Fantastic ocean current or wind patterns would be necessary to distribute the silica about the world from such a local source. Early Tertiary volcanism is common in many parts of the world, however, and perhaps winds and currents at that time were adequate to uniformly distribute silica from many sources in the world oceans in the amounts necessary for the formation of chert.

ORIGIN OF THE CHERT

The formation of oceanic chert is usually thought to be due to either the proportionally greater preservation of siliceous skeletal material, or an increase in the silica content of sea water by increased volcanism. J. Ewing and others (1970) have summarized the possible origins of the Eocene cherts found in the deep oceans, suggesting a correlation with the abundance of radiolarian remains, perhaps influenced by other factors such as climate, volcanism, or changes in deep-ocean water circulation. It is tempting to emphasize volcanic activity, as shown by the rock association in the Greater Antilles and by the presence of montmorillonite (presumably volcanically derived) in both the exposed and submarine Eocene strata. However, montmorillonite occurs in the JOIDES cores at levels above the reflecting horizons and is less abundant in the cherty horizons than in the clays, so the mere presence of volcanism is not a sufficient condition for the formation of chert. But certainly explosive or submarine volcanism will increase the silica concentration in the oceans, perhaps to levels sufficient to permit the formation of chert by an organic or inorganic mechanism.

TRANSITION BETWEEN DEEP-OCEAN AND LAND EXPOSURES OF EOCENE CHERTS

Layer A and A″ are not visible in the profiles between the two trench axes, a distance of about 180 km near Puerto Rico. Folding into steep or vertical orientations has occurred in the correlatives of these layers in the Greater Antilles, and gravitational slumping has also affected the beds. Such intense deformation would probably render layered sediments acoustically incoherent and no simple layering could be seen. This may be the explanation for the lack of reflections recorded in the landward slopes of the two trenches.

Deformation was confined to the Antillean Ridge, as was deposition of the coarse clastic and volcanic facies. A relatively sharp facies and tectonic boundary apparently existed near the trench axes north and south of the ridge. The maximum width of the transition between deformed and undeformed rocks is the width of

Figure 4. Jicara Formation, Puerto Rico. Exposures on Route 327, 100 m south of intersection with Route 117, Sabana Grande quadrangle. Pessagno found calcareous beds in this exposure with an early Eocene foraminiferal assemblage from the *Globorotalia rex* zone.

A, Silicified light green fine-grained tuff. Thick-bedded and laminated on the left,

the slopes, about 90 km south of Puerto Rico. As stated above, deformed layers rather than undeformed layers probably underlie the landward slopes, so that the transition might be across a narrow zone of no more than 10 km near the trench axis.

SUMMARY

Layer A and A″, widespread lithologic units discovered by geophysical exploration in the deep ocean, apparently can be recognized in land exposures in the Greater Antilles. Greater thickness and more clastic lithologic facies in the land exposures are due to the restriction of volcanic centers and deformation to a narrow band at the junction of the two oceanic plates. Restriction of volcaniclastic sediments to the narrow zone may have been facilitated by two parallel oceanic trenches flanking the Antillean ridge in early Tertiary time.

ACKNOWLEDGMENTS

Much of the basic geological work in the Greater Antilles was sponsored by Princeton University in the research program of H. H. Hess. Additional support was provided by the U.S. Geologic Survey and the Economic Development Administration of Puerto Rico, under the directorships of R. Fernandez Garcia and Dr. C. Vincenty, respectively. Office and laboratory facilities were provided by Queens College of City University of New York, and the University of Leeds (United Kingdom). We are grateful for discussions with N. T. Edgar, Lynn Glover

medium-bedded and alternating with dark green (glass-rich?) highly altered rock on the right. B, Closeup view of thick tuff beds, showing joints and solution features where calcite has been leached from the siliceous rock. C, Closeup view of alternating beds of dark- and light-colored green tuffs. Laminae show a suggestion of graded bedding with tops toward the left (south).

III, and F. Nagle, and for the helpful comments on the manuscript by J. C. Briden and A. G. Fischer.

REFERENCES CITED

Bader, R. G., Gerard, R. D., Benson, W. E., Bolli, H. M., Hay, W. W., Riedel, W. R., Rothwell, W. T., Jr., Reuf, M. H., and Sayles, F. L., 1969, Deep-sea drilling project: Leg 3 and leg 4: Geotimes, v. 14, no. 6, p. 13–16.

Edgar, N. T., Ewing, J. I., and Hennion, J., 1971, Seismic refraction and reflection in Caribbean Sea: Am. Assoc. Petroleum Geologists Bull., v. 55, p. 833–870.

Ewing, J., Worzel, J. L., Ewing, M., and Windisch, C., 1966, Ages of Horizon A and the oldest Atlantic sediments: Science, v. 154, no. 3753, p. 1125–1132.

Ewing, J., Talwani, M., Ewing, M., and Edgar, T., 1967, Sediments of the Caribbean: Univ. of Miami, Studies in Tropical Oceanography, v. 5, p. 88–102.

Ewing, J., Windisch, C., and Ewing, M., 1970, Correlation of horizon A with JOIDES bore-hole results: Jour. Geophys. Research, v. 75, no. 29, p. 5645–5653.

Ewing, M., Worzel, J. L., Beall, A. O., Jr., Bukry, D., Berggren, W. A., Fischer, A. G., Burk, C. A., and Pessagno, E. A., 1969, Initial reports of the deep-sea drilling project: Washington, D.C., Natl. Sci. Foundation, v. 1, 672 p.

Glover, Lynn, III, and Mattson, P. H., 1967, The Jacaguas Group in central-southern Puerto Rico: U.S. Geol. Survey Bull. 1254A, p. 29–39.

Mattson, P. H., 1960, Geology of the Mayaguez area, Puerto Rico: Geol. Soc. America Bull., v. 71, p. 319–362.

—— 1967, Cretaceous and lower Tertiary stratigraphy in west-central Puerto Rico: U.S. Geol. Survey Bull. 1254B, p. 1–35.

—— 1968a, Geologic map of the Adjuntas quadrangle, Puerto Rico: U.S. Geol. Survey Misc. Geol. Inv. Map I–519.

—— 1968b, Geologic map of the Jayuya quadrangle, Puerto Rico: U.S. Geol. Survey Misc. Geol. Inv. Map I–520.

—— 1972, Puerto Rico-Virgin Islands, in Spencer, A. M. ed., Data for orogenic studies: Geol. Soc. London Spec. Pub. no. 4.

McIntyre, D. H., and Aaron, J. M., 1971, Possible subaerial outcropping of Horizon A, northwestern Puerto Rico (abs.): Sixth Caribbean Geological Conf., Margarita, Venezuela, p. 30–31.

Nagle, F., 1971, Geology of the Punta Plata area, Dominican Republic, relative to the Puerto Rico Trench, p. 79–84 in Mattson, P. H., ed., Transaction of the Fifth Caribbean Geological Conference, St. Thomas, Virgin Islands: Queens Coll. Geol. Bull. no. 5, 256 p.

Nelson, A. E., 1966, Cretaceous and Tertiary rocks in the Corozal quadrangle, northern Puerto Rico: U.S. Geol. Survey Bull. 1244C, p. 1–20.

—— 1967a, Geologic map of the Corozal quadrangle, Puerto Rico: U.S. Geol. Survey Misc. Geol. Inv. Map I–473.

—— 1967b, Geologic map of the Utuado quadrangle, Puerto Rico: U.S. Geol. Survey Misc. Geol. Inv. Map I–480.

Pease, M. H., Jr., 1968a, Cretaceous and lower Tertiary stratigraphy of the Naranjito and Aguas Buenas quadrangles and adjacent areas, Puerto Rico: U.S. Geol. Survey Bull. 1253, 57 p.

—— 1968b, Geologic map of the Aguas Buenas quadrangle, Puerto Rico: U.S. Geol. Survey Misc. Geol. Inv. Map I–479.

—— 1968c, Geologic map of the Naranjito quadrangle, Puerto Rico: U.S. Geol. Survey Misc. Geol. Inv. Map I–508.

Pessagno, E. A., Jr., 1961, The micropaleontology and biostratigraphy of the middle Eocene Jacaguas group, Puerto Rico: Micropaleontology, v. 7, p. 351–358.

Peterson, M.N.A., Edgar, N. T., von der Borch, C. C., Gartner, S., Goll, R. M., Nigrini, C., Cita, M., and Josephson, P., 1970, Initial reports of the deep-sea drilling project: Washington, D.C., Natl. Sci. Foundation, v. 2, 490 p.

Saito, T., Burckle, L. H., and Ewing, M., 1966, Lithology and paleontology of the reflective layer Horizon A: Science, v. 154, no. 3753, p. 1173–1176.

MANUSCRIPT RECEIVED BY THE SOCIETY MARCH 29, 1971   PRINTED IN U.S.A.

THE GEOLOGICAL SOCIETY OF AMERICA, INC.
MEMOIR 132
© 1972

Pulsations, Interpulsations, and Sea-Floor Spreading

EMILE A. PESSAGNO, JR.
The University of Texas at Dallas, Geosciences Division, Dallas, Texas 75230

ABSTRACT

It is postulated that worldwide transgressions (pulsations) and regressions (interpulsations) throughout the course of geologic time are related to the elevation and subsidence of oceanic ridge systems and to sea-floor spreading. During the Mesozoic-Cenozoic interval, for example, the Cretaceous represents a period of worldwide transgression of the seas over the continents. Such a transgression may have been caused by the elevation of the old Mid-Pacific Ridge system, which in turn displaced a considerable amount of sea water from the ocean basins to the continents.

Two multiple working hypotheses are proposed to explain major transgressions and regressions and the elevation and subsidence of oceanic ridge systems. One hypothesis interrelates the sea-floor spreading hypothesis to the hypothesis of sub-Mohorovičić serpentinization. The second hypothesis relates the sea-floor spreading hypothesis to a hypothesis involving thermal expansion and contraction.

INTRODUCTION

Amadeus W. Grabau (1940, p. vii) stated that one of the two most important phenomena influencing the development of the earth is ". . . the rhythmic movement or pulsation of the sea, which, unlike tidal phenomena, is of simultaneous occurrence in all the oceans of the earth. This is shown by the record of transgressions and regressions everywhere in the strata of all continents, and I have embodied it in the law of slow pulsatory rise and fall of the sea-level in each geological period, or more briefly: *The Pulsation Theory*."

Grabau referred to major transgressions of the sea as *pulsations* and to major regressions as *interpulsations*. He believed that major transgressions and regressions of the sea during the course of geologic time were caused by the expansion and contraction of the oceanic crust. Expansion and, accordingly, transgression were

67

said to be due to heat generated in the earth's crust resulting from radioactivity. Contraction and regression were attributed to loss of heat to the "cold ocean floor" through conductivity.

In his book entitled *The Rhythm of the Ages,* Grabau went so far as to propose a classification of geologic systems on the basis of pulsations and interpulsations. This classification was never widely accepted, partly because it departed from traditional stratigraphic terminology. It is also apparent that Grabau's pulsations and interpulsations were clouded in part by his focusing on transgressive and regressive cycles in mobile belts rather than those affecting the continental cratons. In spite of such deficiencies, Grabau's ideas on transgressions and regressions as well as his ideas concerning continental drifting and polar wandering were about thirty years ahead of their time.

GLOBAL TECTONICS AND THE STRATIGRAPHIC RECORD

With the important data resulting particularly from Legs 2 and 3 of the Atlantic phase of the JOIDES Deep-Sea Drilling Project, the Hess-Dietz hypothesis (Hess, 1962, p. 599–620; Dietz, 1961, p. 854–857) of sea-floor spreading has become a well-documented theory. In an age of "global tectonics"—spreading sea floors and drifting continents—a stratigrapher should consider whether such events are reflected in the stratigraphic record. If the sea-floor spreading theory is valid, the older portions of the oceanic crust are being consumed periodically at sites of downward convection. Hence, the completest stratigraphic records are to be found on the continents.

In this paper a major transgression or pulsation is defined as one in which large areas of all of the continental cratons were invaded by the sea during a period or epoch of earth history. A major regression or interpulsation is defined as one in which the cratons of all continents were emergent during a period or epoch of earth history. These definitions are meant to exclude local orogenic or epeirogenic events that affect a portion of a given continental craton (for example, the formation of the Mississippi Embayment during Mesozoic and Cenozoic times) and to include global tectonic events that influenced all continents similarly during a given interval of geologic time.

One of the greatest transgressive cycles or pulsations during all of earth history occurred during the Cretaceous Period. During the Mesozoic-Cenozoic interval the Triassic, Tertiary, and Quaternary represent major regressive cycles or interpulsations. The case for the Jurassic is less clear. In the present paper the Jurassic is tentatively treated as regressive, although this may be an oversimplification. For instance, in western North America there was a transgression of the craton by the so-called "Sundance Sea" during Callovian and Oxfordian times. In Eurasia an apparent transgression took place over the Russian Platform during the same time interval. However, in the latter case the transgression may be due to local epeirogenic movement. Definitive evidence from other continents of a Callovian-Oxfordian transgression is lacking. Exclusive of the Callovian and Oxfordian, the remainder of the Jurassic record appears to be regressive.

It is apparent that if a major transgression or pulsation was to occur, a sea-level change had to be brought about either (1) by lowering all continental cratons in unison, or (2) by raising the floor of the ocean basins (or an ocean basin) over large areas. Like Grabau, the writer considers the second of these two alternatives to be more likely. It is difficult to imagine, particularly during the Mesozoic-Cenozoic interval, that all continental cratons were lowered or raised in unison.

HYPOTHESES EXPLAINING MAJOR SEA-LEVEL CHANGES— ELEVATION AND SUBSIDENCE OF OCEANIC RIDGES

The Hypothesis of Sub-Mohorovičić Serpentinization

In various papers, Harry H. Hess (1954a, 1954b, 1955) proposed that epeiro-genic movement in the ocean basins or, for that matter, under parts of the continents could be a direct result of a phase change in the upper mantle. Assuming the mantle to be composed of peridotite and utilizing the well-known Bowen and Tuttle (1949) equation (below), Hess (1954a, p. 404) proposed what he called the "hypothesis of sub-Mohorovičić serpentinization."

$$\text{Olivine} + H_2O \underset{500° C}{\overset{\longleftarrow}{}} \text{serpentine} + \text{heat}$$

$$\overleftarrow{\text{25\% volume decrease}} \qquad \overrightarrow{\text{25\% volume increase}}$$

He suggested that water from the mantle which leaked across the 500° C isotherm would result in the serpentinization of the upper mantle and a 25 percent volume increase. The net result would be a rise in the sea floor over the serpentinized area. According to Hess's hypothesis, deserpentinization and a resulting 25 percent volume decrease would occur when the 500° C isotherm was caused to rise due to convective overturn, or perhaps basaltic intrusion. Hess (1954a, p. 404) stated that ". . . deserpentinization would occur at the base of the zone and serpentiniza tion higher up without any necessary effect on the surface topography. But once the top of the zone of serpentinization reached the Mohorovičić, then any deserpentinization at the base would cause the overlying topography to subside. Thus, we have here a possible mechanism for epeirogenic movements either upward or downward. . . ."

For the Hess hypothesis to be implemented, it is necessary that a great volume of water be available from the mantle. The source of such a great volume of water is not apparent. Yet, when one considers that no one satisfactory theory has been proposed for the origin of the oceans themselves, this problem does not seem so great.

Expansion and Contraction Hypothesis

This hypothesis was supported by Holmes (1965), Hess (1962), and in part by Grabau (1940). These workers generally maintained that an oceanic ridge could be elevated by a marked temperature increase in the upper mantle and a subsequent density decrease–volume increase produced by an ascending convection cell. Subsidence of an oceanic ridge would result from the waning of the rising convection cell, a decrease in temperature, and subsequent density increase–volume decrease in the upper mantle. Hallam (1963), Russell (1969), and Valentine and Moores (1970) have pointed to connections between the development of oceanic ridges and major transgressions and regressions.

THE PULSATION HYPOTHESIS OF GRABAU, EMENDED

Hess (1954a, p. 405) originally believed that the hypothesis of sub-Mohorovičić serpentinization could explain the elevation of oceanic ridges and formation of guyots. At a later date, however, he discarded this hypothesis because it interfered with his ideas on continuous sea-floor spreading and continuous upward convection (1962, p. 605). He preferred to attribute the rise of oceanic ridges to thermal expansion in the upper mantle at sites of ascending convection and to a resulting density decrease–volume increase (Hess, 1962, p. 611).

Either the hypothesis of sub-Mohorovičić serpentinization or the expansion-contraction hypothesis can be used to explain the elevation and subsidence of major oceanic ridge systems. Furthermore, either hypothesis can be integrated with the sea-floor spreading hypothesis and an emended version of the pulsation hypothesis of Grabau (1940) to explain major trangressions and regressions throughout the course of geologic time.

Two multiple working hypotheses are thus proposed herein.

Hypothesis One

This hypothesis has the following components.

(1) *Major transgressions or pulsations during the course of geologic time are due to serpentinization along oceanic ridge systems.* The 25 percent volume increase that occurs along a major oceanic ridge system such as the old Mesozoic Mid-Pacific Ridge or the present Mid-Atlantic Ridge could displace a considerable amount of sea water onto the continents, thus causing a major transgressive cycle like that of the Cretaceous.

Corollary. Sea-floor spreading during a pulsation cycle would essentially cease. According to the hypothesis of sub-Mohorovičić serpentinization, serpentinization occurs at a time when convective upturn ceases. However, my colleague, C. E. Helsley (personal commun., 1970), suggests that serpentinization might actually occur at the initiation of a convection cycle at a time when spreading is still relatively slow and temperatures were not yet great enough to effect a rise of the 500° C isotherm. At such an early stage of a convection cycle, slow ascending convection might actually serve as a mechanism for concentrating water in the upper mantle, which in turn would facilitate serpentinization.

(2) *Major regressions or interpulsations during the course of geologic time are due to deserpentinization along oceanic ridge systems.* The 25 percent volume decrease that occurs with deserpentinization would result in the withdrawal of the seas from the continents.

Corollary. Sea-floor spreading during an interpulsation would be active. Marked convective overturn would cause both deserpentinization and the generation of basaltic crust along oceanic ridge systems.

Hypothesis Two

The second hypothesis has the following components.

(1) *Major transgressions or pulsations during the course of geologic time are*

due to a marked temperature increase in the upper mantle and subsequent density decrease–volume increase along major oceanic systems. Holmes (1965, p. 1041) states in discussing the old Mesozoic Mid-Pacific Ridge: "That the subcrustal material of the Pacific was formerly hotter than now is indicated by the enormous numbers of seamounts and guyots discovered in recent years. Some of these date from the middle Cretaceous. At that time and certainly during several later epochs there must have been many more volcanic islands than remain today. The suboceanic upper mantle would then be correspondingly hotter and its thermal expansion would have the effect of raising the ocean floor and so spilling the displaced sea water over the low-lying parts of the continents. . . ."

Should a continent override part of an oceanic ridge system or an active ascending convection cell as a result of drifting, then this continent would most likely stand higher than other continents and be less affected by a pulsation cycle.

Corollary. Sea-floor spreading during a pulsation cycle would be more rapid as a result of a marked temperature increase in the upper mantle.

(2) *Major regressions or interpulsations during the course of geologic time are due to a temperature decrease in the upper mantle and subsequent density increase–volume decrease along major oceanic ridge systems.*

Corollary. Sea-floor spreading during an interpulsation would be slower as a result of a temperature decrease in the upper mantle or a more uniform temperature distribution in the upper mantle.

DISCUSSION

Figure 1 attempts to apply multiple working hypotheses one and two to the Mesozoic-Cenozoic interval. The stratigraphic record indicates that the Triassic, Jurassic(?), and Quaternary represent times of worldwide regression or interpulsation. The Cretaceous (particularly the Late Cretaceous) represents one of the greatest transgressions or pulsations that has occurred during the course of geologic time. By hypothesis one (Fig. 1), the Triassic, Jurassic(?), Tertiary, and Quaternary represent times of active sea-floor spreading and deserpentinization along one or more oceanic ridge systems. On the other hand, the Cretaceous represents a time of inactive sea-floor spreading and a time of serpentinization along one or more oceanic ridge systems.

By hypothesis two (Fig. 1), the Triassic, Jurassic(?), Tertiary, and Quaternary represent times of slower sea-floor spreading and thermal contraction in the upper mantle at the sites of one or more oceanic ridges. The Cretaceous, in contrast, represents a time of rapid sea-floor spreading and thermal expansion in the upper mantle at the site of one or more oceanic ridges.

The test of the two multiple working hypotheses proposed above largely awaits data from the JOIDES Deep-Sea Drilling Project and the calculation of sea-floor spreading rates—particularly for the Cretaceous. If the Cretaceous records show more rapid spreading rates than either the Jurassic or the Tertiary, the second hypothesis would seem the more plausible one.

Circumstantial evidence may indicate that interpulsations or major regressive cycles are related to times of mixed polarity. Pulsations or transgressive cycles appear to be related to times of constant polarity. The work of Burek (1967) and Helsley (1969) demonstrates that the Triassic—particularly the Early Triassic—

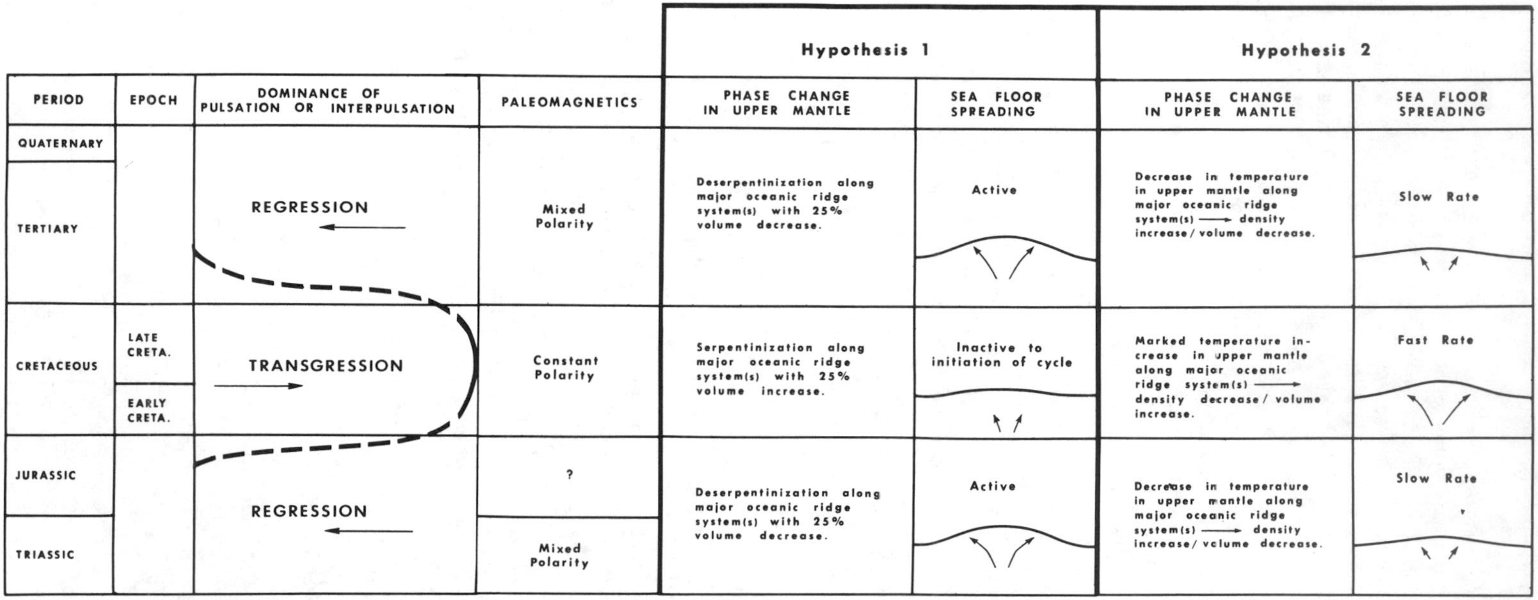

Figure 1. Multiple working hypotheses one and two.

was a time of mixed polarity. Unfortunately, few data are available for the Jurassic. Extensive work by Helsley and Steiner (1969) and Helsley (in prep.) indicates that the Cretaceous was dominantly a period of constant polarity. Apparently only three magnetic reversals have been recorded from the Cretaceous. Paleomagnetic data presented by Vine (1966), Heirtzler and Hayes (1967), and others clearly show that the Cenozoic was a time of mixed polarity.

ACKNOWLEDGMENTS

The writer wishes to thank his assistant, Sheila Martin, for preparing the illustrations and his colleague, Charles E. Helsley, for his constructive criticism. This work was supported in part by the general NASA grant (NGL–44–004–001) to the Southwest Center for Advanced Studies.

REFERENCES CITED

Bowen, N. L., and Tuttle, O. F., 1949, The system Mgo–SiO$_2$–H$_2$O: Geol. Soc. America Bull., v. 60, p. 439–460.
Burek, P. J., 1967, Korrelation revers magnetisierter Gesteinsfolgen im Oberen Buntsandstein SW Deutschlands [Correlation of reversely magnetized rock sequences in the Upper Buntsandstein, SW-Germany]: Geol. Jahrb., v. 84, p. 591–616.
Dietz, R. S., 1961, Continent and ocean basin evolution by spreading of the sea floor: Nature, v. 190, no. 4779, p. 854–857.
Grabau, A. W., 1940, The rhythm of the ages; earth history in the light of the pulsation and polar control theories: Henri Vetch, Peking, p. 1–561.
Hallam, A., 1963, Major epeirogenic and eustatic changes since the Cretaceous, and their possible relationship to crustal structure: Am. Jour. Sci., v. 261, p. 397–423.
Helsley, C. H., 1969, Magnetic reversal stratigraphy of the Lower Triassic Moenkopi Formation of western Colorado: Geol. Soc. America Bull., v. 80, p. 2431–2450.
Helsley, C. E., and Steiner, M. B., 1969, Evidence for long intervals of normal polarity during the Cretaceous period: Earth and Planetary Sci. Letters, v. 5, p. 325–332.
Heirtzler, J. R., and Hayes, D. E., 1967, Magnetic boundaries in the North Atlantic Ocean: Science, v. 157, p. 185–187.
Hess, H. H., 1954a, Serpentines, orogeny, and epeirogeny, in Poldervaart, A., ed., Crust of the earth, Geol. Soc. America Spec. Paper 62, p. 391–407.
——1954b, Geological hypotheses and the earth's crust under the oceans: Royal Soc. [London] Proc., p. 341–348.
——1955, The oceanic crust: Jour. Marine Research, v. 14, p. 423–439.
——1962, History of ocean basins, in Engel, A.E.J., James, H. D., and Leonard, B. F., eds., Petrologic studies: A volume to honor A. F. Buddington: New York, Geol. Soc. America, p. 599–620.
Holmes, A., 1965, Principles of physical geology: New York, Ronald Press Co., 1288 p.
Russell, K., 1969, Oceanic ridges and eustatic changes in sea level: Nature, v. 218, p. 861–862.
Valentine, J. W., and Moores, E. M., 1970, Plate-tectonic regulation of faunal diversity and sea level: A model: Nature, v. 228, p. 657–659.
Vine, F. J., 1966, Spreading of the ocean floor; new evidence: Science, v. 154, p. 1405–1415.

MANUSCRIPT RECEIVED BY THE SOCIETY MARCH 29, 1971
CONTRIBUTION NO. 157, GEOSCIENCES DIVISION,
UNIVERSITY OF TEXAS AT DALLAS, DALLAS, TEXAS

THE GEOLOGICAL SOCIETY OF AMERICA, INC.
MEMOIR 132
© 1972

Uplifted Eugeosynclines and Continental Margins

C. A. BURK

Mobil Oil Corporation, Princeton, New Jersey 08540

ABSTRACT

Mesozoic eugeosynclinal sequences of turbidites with radiolarian cherts, pillowed basalts, and ultramafic rocks, appear to characterize much of the exposed Pacific continental margins and much of the Tethyan tectonic belts. Extremely great stratigraphic thicknesses have been reported for many of these deep-ocean sequences.

These eugeosynclinal rocks have had a complex history of penecontemporaneous deformation and subsequent tectonic displacements, and have been uplifted and added to the margins of the continents. Based on studies of the southern continental margin of Alaska, the apparently great thicknesses and the subsequent uplift of these eugeosynclinal sequences seem best explained by deposition in oceanward-migrating trenches and the repeated landward uplift of the sedimentary fill in these successive trenches.

INTRODUCTION

The uplifted margins of many continents consist of thick sequences of eugeosynclinal sediments, presumed to have been deposited at great oceanic depths. This is particularly true of Mesozoic rocks around the periphery of the Pacific Ocean. Since these deep-water sediments are now widely exposed at continental margins, they must have been uplifted by an amount at least equal to their stratigraphic thickness plus the original oceanic depth of the crust on which they were deposited.

According to present concepts, based on sea-floor spreading (Hess, 1962), much of the Pacific margin is the site of convergent and downward-directed forces, which may have provided both the depressed oceanic trenches in which the eugeosynclinal deposits accumulated and the stresses to account for their severe deformation. Within such a tectonic framework, the nature of any large-scale and repeated uplift becomes particularly important.

75

EUGEOSYNCLINAL ENVIRONMENT

Eugeosynclinal sequences are considered here to consist primarily of thick accumulations of graywacke and argillite flysch, commonly containing radiolarian cherts, submarine flows of basic lavas, and masses of ultramafic rocks (particularly in the lower parts of these sequences). The presence of radiolarian cherts and the great scarcity of calcareous shell material suggest that these sequences were deposited in very deep water, well below the oceanic depth of carbonate compensation. The abundance of deep-ocean *lebensspuren* and the great scarcity of benthonic fossils also indicate a similarly deep environment. These sequences are considered here as having accumulated largely at typical oceanic depths and largely on oceanic crust.

There has been much discussion in recent years as to whether eugeosynclines might be compared to modern continental rises or whether they represent the filling of present deep-ocean trenches, or both (for example, Drake and others, 1959). One distinguishing feature of these sequences is the dominance of turbidites; it is difficult to imagine such deposits accumulating to any significant thickness on the inclined surface of continental rises. It is possible that rise deposits may constitute part of the younger beds, but these eugeosynclinal accumulations are overwhelmingly dominated by turbidites. For the purpose of this discussion we will consider the exposed eugeosynclinal sequences bordering the Pacific Ocean as having accumulated largely in ancient oceanic trenches, rather than on a broad continental rise.

The minimum uplift of the youngest exposed rocks of these sequences is probably equivalent at least to normal oceanic depths (4 to 5 km), and the total uplift would exceed this amount by at least the original stratigraphic thickness between the youngest and oldest rocks exposed—representing the added depression of the underlying oceanic crust during accumulation of the deposits.

THICKNESS OF EUGEOSYNCLINAL SEQUENCES

Although thick sedimentary accumulations have been reported in a great variety of geological settings, it seems to be almost characteristic of eugeosynclinal sequences that they are credited with truly amazing stratigraphic thicknesses.

Bailey and others (1964) reported that the Franciscan Formation of California probably exceeds 15 km in thickness. A reportedly comparable eugeosynclinal sequence may be of even greater thickness in the coastal areas of the Koryak Mountains and Kamchatka Peninsula in eastern Siberia (N. A. Bogdanov, 1971, personal commun.). Campbell (*in* Stoneley, 1969, p. 224) estimates the Cretaceous eugeosynclinal deposits of Colombia to be on the order of 15 km thick. Murray (1969) reports nearly 20 km in the Paleozoic eugeosyncline of eastern Australia. The Southland geosyncline of New Zealand is reported by Mutch (1957) to have accumulated more than 30 km of rocks. The Mesozoic and early Tertiary eugeosyncline of eastern New Guinea is credited with a sequence as much as 35 km thick. Moore (1969) has suggested from field studies that the eugeosynclinal sediments at Kodiak Island along the southern continental margin of Alaska may be at least 40 km thick.

It should be emphasized that meaningful estimates of true depositional thickness in such sequences are extremely difficult. They are commonly tightly folded and faulted on a small scale, and lithologic or paleontologic markers are extremely rare in these monotonous, barren sequences. However, these factors have been recognized in all field studies by experienced geologists and, as noted above, the reported stratigraphic thicknesses are still remarkably great.

Excessive and erroneous thicknesses in many instances can be accounted for by assuming lateral progradation of the strata; however, it is difficult to allow any significant depositional dip in turbidites (Kuenen, 1964). It seems extremely unlikely that these great thicknesses of deep-ocean deposits could have accumulated continuously at any one locality. It is more likely that the reported thickness is actually cumulative and includes several cycles of eugeosynclinal subsidence, deposition, and uplift. This appears to be true along the southern margin of Alaska.

Stoneley (1969) has recently discussed many of the perplexing physical problems related to the apparent accumulation of great sedimentary thicknesses, and the reader is referred to that report for a broader view of these problems. The deepest oceanic trench in the world is 11 km below sea level, and a more typical figure would be 8 to 9 km. Trenches are normally depressed 4 to 5 km below the level of the adjacent sea floor. Filling of the trench with sediments to the level of overflow could provide isostatic adjustment to accommodate eventually another 2 to 3 km of deposits. Thus, accumulation of 6 to 8 km of deep-ocean turbidites is not unreasonable.

The presence of deep-ocean sediments at continental margins obviously also requires significant uplift of these sequences, many of which are exposed adjacent to presently existing oceanic trenches. The apparently great cumulative thicknesses of these eugeosynclinal rocks would seem to require repeated accumulation and uplift of such sequences at continental margins.

EUGEOSYNCLINAL SEQUENCE OF SOUTHERN ALASKA

Pertinent aspects of the history and structure of southern Alaska are taken largely from recent field studies by Burk (1965, 1966), Plafker and MacNeil (1966), Plafker (1969), G. Moore (1967, 1969) and J. C. Moore (1972). The geological history of the Alaska Peninsula (Fig. 1) is characterized by the continued and thick accumulation of shallow-water sediments as old as Permian, which were intruded by granitic plutons in the Early Jurassic and deformed moderately in the mid-Cretaceous. Granodioritic plutons were intruded in mid-Tertiary, and the peninsula underwent its most widespread tectonic deformation in the Pliocene. Since that time, the area has been rising continuously to form a prominent topographic mountain range. The present Aleutian volcanic arc was superimposed on this area in early Tertiary and has persisted to the present.

South and east of the Alaska Peninsula, toward the edge of the continent, the geological history has been significantly different (Fig. 1). A great sequence of deep-water, eugeosynclinal sediments accumulated throughout the Mesozoic and the earliest Tertiary. Much of this sequence is exposed along the southern continental margin of Alaska for a distance of more than 1,500 km, from near the Canadian border to near the eastern edge of the Aleutian Islands. These eugeosynclinal sediments were intruded by early Tertiary granodiorite and uplifted by Eocene or Oligocene time, as indicated by local shallow-water sediments of that

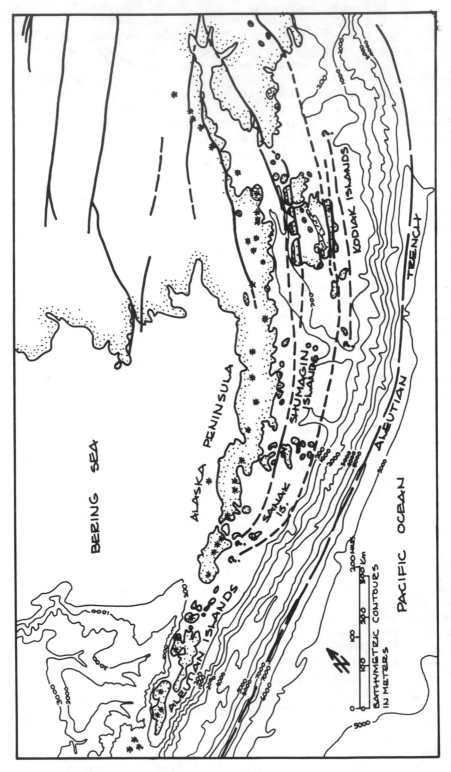

Figure 1. Regional geology and geography of southwestern Alsaka (modified *from* Burk, 1965; King, 1969; Plafker, 1969).

age. One of the most complete eugeosynclinal sequences in southern Alaska is exposed on the Kodiak Islands, approximately midway along this continental margin. These rocks have been studied most recently by Moore (1967, 1969), who reports the following sequence across the islands, aggregating a cumulative thickness reported to be at least 40 km (Fig. 2).

The northwest side of the Kodiak Islands consists of about 6 km of black shale, turbidites, schistose green tuff with pillow basalts, finely bedded cherts, limestone lenses, and ultramafic rocks. Sparse marine fossils suggest that this sequence is Triassic in age, and it may also include Jurassic sediments. These rocks are separated by a major fault from the graywacke turbidite and argillaceous beds which constitute the bulk of the island, and which are considered to be largely Cretaceous on the basis of very sparse *Inoceramus* specimens. This sequence is reported to be 30 km thick, and again it is separated everywhere by major faults from the next younger sequence to the southeast, of possible Paleocene to early Eocene age. These lower Tertiary rocks are described as 5 km of tuffaceous sandstone and graywackes, pillow basalts, turbidites, and wildflysch. Eugeosynclinal conditions ended by late Eocene or early Oligocene time, when shallow-water, coal-bearing sediments first appeared.

Nowhere is there any evidence that the early Tertiary eugeosynclinal sequence was deposited directly on the late Mesozoic turbidites, or that these were deposited

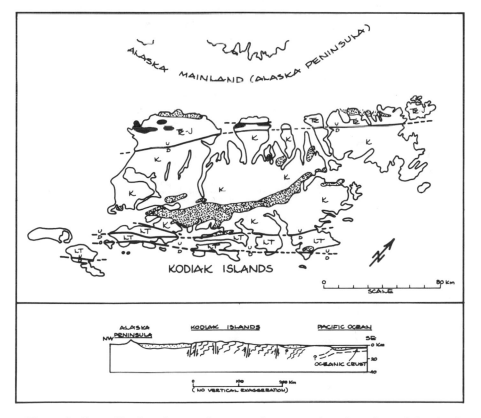

Figure 2. Generalized geology and structural cross section through Kodiak Islands (*after* unpub. maps by G. Moore, 1967).

directly on the earlier Mesozoic ophiolite-bearing sequence. Plafker and MacNeil (1966) described similar relations in this same sequence in the Prince William Sound area, 500 km to the northeast. The ophiolites and bedded cherts occur in the oldest part of this sequence, as noted in many other geosynclines (Hess, 1955), but submarine pillow lavas are found in rocks as young as early Tertiary.

DEFORMATION OF THE EUGEOSYNCLINAL SEQUENCE

The entire eugeosynclinal sequence of southern Alaska is deformed into many small folds, having a wavelength of a few meters to a few hundred meters, and locally to 1 km. Part of this is the result of penecontemporaneous soft-sediment deformation, and part is due to later compressional stresses (J. C. Moore, 1972). Inasmuch as there is considerable evidence of reworking of unconsolidated sediments (Burk, 1965, p. 68), much of the small-scale deformation in the thick flysch sequence may be due to mass movement resulting from continual oversteepening of the flanks of the subsiding trough. However, tectonic compression is obvious in the younger Tertiary sediments along the southeastern coast of the Kodiak Islands.

Considering even the complex minor folds in this sequence, the prevailing regional dip is toward the continent (northwest), striking northeast approximately parallel to the continental margin (G. Moore, 1969). This is also true of the Shumagin Islands, 500 km to the southwest (Burk, 1965; J. C. Moore, 1972), and of the Kenai–Prince William Sound areas, 500 km to the northeast (Martin and others, 1915, p. 101; Moffit, 1954).

The dominant major structural features of the area are the faults which tend to separate strata of different ages and different lithologies (Figs. 1 and 2). All of these large faults are steeply dipping, marked by nearly straight traces across the mountainous Kodiak terrain. G. Moore (1969) describes the northern fault as locally dipping steeply to the northwest, and the remainder as essentially vertical. In all cases the relative movement is up on the northwest and down on the southeast. The boundary between the Kodiak sequence and the shallow-water rocks of the Alaska Peninsula is unknown, but it probably represents a major fault zone, as on the Kenai Peninsula.

UPLIFT OF THE SOUTHERN ALASKA CONTINENTAL MARGIN

We can reasonably construct the following concise history of the southern continental margin of Alaska. Throughout the Mesozoic, great sequences of deep-water eugeosynclinal sediments accumulated off the southern margin of Alaska, all of which were uplifted by early Tertiary to form the present margin of the continent. These and other great uplifts in southwestern Alaska seemed related in time to major plutonic intrusions (Burk, 1965), but in detail the structural history appears to be much more complex.

The highly deformed eugeosynclinal sequence dips in general toward the northwest (toward the continent), but the age of the rocks becomes younger in general toward the southeast (toward the ocean). The sequence is interrupted by several long and steep faults, which are relatively downthrown on the southeast. The fault separations suggest a relative down-to-the-ocean type of faulting following a major broad uplift. However, simple reconstructions show that this is not possible, since the mountain range from which the youngest rocks were dropped would apparently have exceeded 40 km in elevation (Fig. 3a).

A more reasonable explanation would be to recognize the fault separation as merely relative, with both sides moving up, but with the northwest side having greater net displacement (Fig. 3b). This reasonably eliminates the need for a pre-existing and obviously extravagant mountain range, and is in accord with the net uplift required of the deep-ocean sediments which make up this continental margin.

Faults where both the hanging wall and footwall have moved up relative to a fixed datum within the earth might be referred to as *uplift faults* as opposed to *subsidence faults,* where both sides of a fault have moved downward relative to a fixed datum. In both cases the relative sense of movement might be normal or reverse.

Such actual displacements certainly occur in many other geological environments of the world. A recent example of uplift faulting in southern Alaska is Plafker's (1967) description of the Patton Bay and Hanning Bay faults on Montague Island, following the earthquake of 1964, in which both sides of the fault moved upward, but the northwest side moved farther upward (Fig. 4).

The net uplift of both sides of these *uplift faults* seems inescapable, but the details of their nature are yet to be resolved. It is conceivable that such faults were initially of reverse sense, dipping landward, and have since been rotated to their steeper attitude. Restoration of the generally northwest dip of the Kodiak sediments would indeed yield landward-dipping reverse faults, but if such successive rotation had occurred, the least-rotated (seaward) blocks should retain most of any large

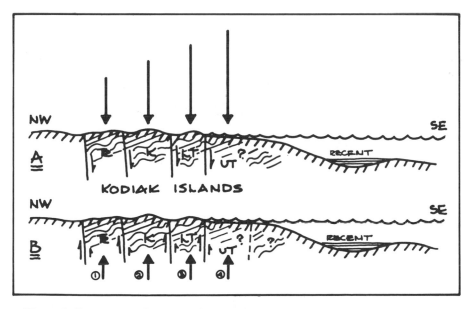

Figure 3. Reconstructed sections through the continental margin of southern Alaska, interpreted as (a) large initial uplift and subsequent down-to-the-ocean faulting, and as (b) continuous uplift along several successive *uplift faults.* (The dip of the fault planes is diagrammatic only.)

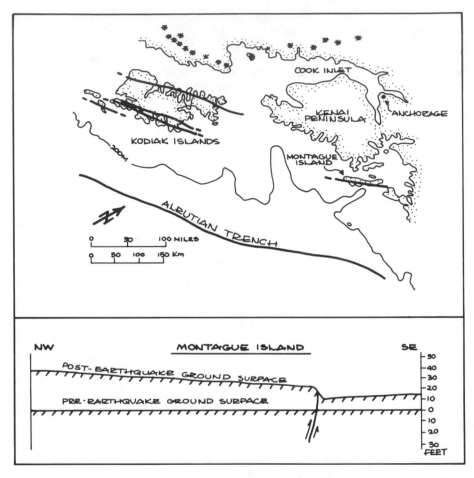

Figure 4. Tectonic uplift and the Patton Bay fault at Montague Island, an example of *uplift faulting* associated with the Alaska earthquake of March 1964 (*after* Plafker, 1967).

initial landward dip of the fault planes. This does not seem to be apparent in the field (G. Moore, 1967, 1969).

It is also apparent that any strike-slip or wrench displacement along old zones of tectonic weakness in the earth's crust, regardless of original dips and displacements, will tend to yield very steep fault zones (Moody, 1966). Any strike-slip displacement is much more difficult to establish in this area than is the apparent dip-slip separation.

TECTONIC HISTORY OF THE SOUTHERN ALASKA EUGEOSYNCLINE

The uplift and present broad structure of the eugeosynclinal sequence of southern Alaska seems best accounted for by *uplift faults,* where the landward side has moved upward to a greater extent than the seaward side. These faults are not necessarily all of the same age and could become younger seaward, just as the age of the stratigraphic sequence in this area does.

The very large apparent thickness of the eugeosynclinal sequence can perhaps thus be accounted for most reasonably by considering that this sequence was not deposited within a single depression, but in a series of depressions; each one formed, filled, and uplifted, yielding an incremental and younger addition to the southern continental margin of Alaska. The major *uplift faults* of the Kodiak Islands separate sequences of different ages.

It has already been suggested in various contexts that oceanic trenches at continental margins may migrate seaward with time (for example, Hess, 1965, p. 323). Certainly, the present Aleutian Trench did not exist prior to the Tertiary, and any Mesozoic trench (or trenches) in this area must have been in the position now occupied by the continental margin of southern Alaska (Burk, 1965).

The formation and dissolution of such an oceanic trench (or trenches) may be continuous—essentially a slowly rolling seaward wave in the earth's oceanic crust, with continuous uplift on its inner margin and continuous subsidence on its oceanic edge (Fig. 5b); but it would be difficult in this case to account for the discrete major *uplift faults* now exposed at Kodiak Islands. Also, at least 3 km of homoclinal sediments have been described on the Shumagin Islands (Burk, 1965), suggesting that the successive filling of the trenches can accumulate continuous sequences of at least this thickness. A more reasonable explanation is that the Aleutian Trench has migrated seaward in a series of discrete pulses, each involving subsidence, sedimentation, and uplift (Fig. 5a).

It seems reasonable, then, to expect that the unexposed rocks beneath the continental slope and inner wall of the Aleutian Trench consist of middle and upper Tertiary eugeosynclinal sequences presently being uplifted in the manner described above.

PACIFIC EUGEOSYNCLINAL SEQUENCES AND POSSIBLE RELATION TO GLOBAL TECTONICS

The regional relations described above seem most reasonably accounted for by successive uplifting of sediments accumulated in deep oceanic trenches, which cumulatively have provided the very thick Mesozoic and earliest Tertiary eugeosynclinal sequences now exposed on the southern Alaska continental margin. Similar

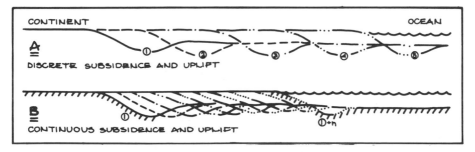

Figure 5. (a) Repeated sequence of tectonic depression of the oceanic trench, filling with sediments, and uplift; and (b) the possibly continuous sequence of oceanward migration of the trench with continuous uplift of the landward side of the trench.

histories may basically be applicable to other Pacific margins (always with necessary local modifications).

It is difficult, however, to reconcile these long steep faults of southern Alaska, spaced at irregular intervals and separating rocks of apparently discrete ages, with the postulated continuous filling of trenches as oceanic crust is continuously thrust underneath the continents. Marine magnetic surveys indicate that this Aleutian trench system has been the site of continued lithospheric plate consumption (for example, Grim and Erickson, 1969) and is not uniquely different from other Pacific margins. The high-pressure blueschist metamorphism typical of much of the peripheral Pacific has not yet been reported in the eugeosynclinal sequence of southwestern Alaska (the regional metamorphic grade does not appear to exceed zeolite and hornfels facies). It is also apparent that the oldest part of the eugeosynclinal sequence is adjacent to contemporaneous, shallow-water early Mesozoic sediments of the Alaska Peninsula.

Holmes and others (1970) have pointed out that recent reflection seismic surveys of the Aleutian Trench south of Amchitka Island have shown deformation of the trench sediments at the northern wall, yielding local folds of 500 m wavelength and 30 m amplitude, comparable to those observed in present exposures (J. C. Moore, 1971); and von Huene and Shor (1969) have shown that the oceanic crust on the seaward edge of the trench has been depressed, with its overlying sequence of pelagic deposits, to unconformably underlie underformed turbidite deposits of the present Aleutian Trench.

All of these factors are compatible with present concepts of sea-floor spreading and global plate tectonics, and with the observed data along the continental margin of the North Pacific Ocean. A major problem would be the apparent effect of seaward migration and uplift of the related trench system, as described above, with the geometry of subduction and Benioff Zones in the deep crust and upper mantle. It is obviously possible that the periodic uplift indicated in the successive trenches off southern Alaska is isostatic and represents the relaxation of tectonic stresses.

In any case, the seaward migration and successive marginal uplift of oceanic trenches and the presumed decrease in the dip of the Benioff zone must be considered as a major factor in any interpretation of global tectonics.

ACKNOWLEDGMENTS

I gratefully acknowledge useful discussions with J. Casey Moore and George Plafker, in addition to discussions with my associates at Princeton University and at Mobil Regional Geology.

REFERENCES CITED

Bailey, E. H., Erwin, W. P., and Jones, D. L., 1964, Franciscan and related rocks, and their significance in the geology of western California: California Div. Mines and Geology Bull. 183, 177 p.

Burk, C. A., 1965, Geology of the Alaska Peninsula, island arc and continental margin: Parts I and II: Geol. Soc. America Mem. 99, 250 p., 3 maps.

—— 1966, The Aleutian Arc and Alaska continental margin, *in* Continental margins and island arcs: Canada Geol. Survey Pub. 66–15, p. 206–215.

Drake, C. L., Ewing, M., and Sutton, G. H., 1959, Continental margins and eugeosynclines; the east coast of North America north of Cape Hatteras, *in* Ahrens, L. H., and others, eds., Physics and chemistry of the Earth, v. 3: New York, Pergamon Press, p. 110–198.

Grim, P. J., and Erickson, B. H., 1969, Fracture zones and magnetic anomalies south of the Aleutian Trench: Jour. Geophys. Research, v. 74, no. 6, p. 1488–1494.

Hess, H. H., 1955, Serpentines, orogeny, and epeirogeny, *in* Poldervaart, A., ed., Crust of the Earth: Geol. Soc. America Spec. Paper 62, p. 391–407.

—— 1962. History of the ocean basins, *in* Engel, A. E. J., and others, eds., Petrologic studies: A volume in honor of A. F. Buddington: Geol. Soc. America, p. 599–620.

—— 1965, Mid-oceanic ridges and tectonics of the sea-floor *in* Wittard, W. F., and Bardshaw, R., eds., Geophysics: London, Butterworths, p. 335–362.

Holmes, M. L., von Huene, R., and McManus, D. A., 1970, Underthrusting of the Aleutian Arc by the Pacific Ocean floor near Amchitka Island [abs.]: Am. Geophys. Union Trans., v. 51, no. 4, p. 330.

King, P. B., 1969, Tectonic map of North America: Washington, D. C., U. S. Geol. Survey.

Kuenen, Ph. H., 1964, Deep-sea sands and ancient turbidites, *in* Developments in sedimentology, v. 3: Amsterdam, Elsevier, p. 1–33.

Martin, G. C., 1915, Geology and mineral resources of Kenai Peninsula, Alaska: U. S. Geol. Survey Bull. 587, 243 p.

Moffit, F. H., 1954, Geology of the Prince William Sound region, Alaska: U. S. Geol. Survey Bull. 989–E, p. 225–310.

Moody, J. D., 1966, Crustal shear patterns and orogenesis: Tectonophysics, v. 6, no. 3, p. 479–522.

Moore, G. W., 1967, Geologic map of Kodiak Island and vicinity, Alaska: U. S. Geol. Survey open-file map.

—— 1969, New formations on Kodiak and adjacent islands, Alaska: U. S. Geol. Survey Bull. 1274–A, p. 27–35.

Moore, J. Casey, 1971, Geologic studies of the Cretaceous(?) flysch, southwestern Alaska [Ph.D. thesis]: Princeton, Princeton Univ.

—— 1972, Uplifted trench sediments: southwestern Alaska–Bering shelf edge: Science, v. 175, p. 1103–1105.

Murray, C. G., 1969, The petrology of the ultramafic rocks of the Rockhampton district, Queensland: Queensland Geol. Survey Pub. no. 343, 9 p.

Mutch, A. R., 1957, Facies and thickness of the upper Paleozoic and Triassic sediments of Southland: Royal Soc. New Zealand Trans., v. 84, no. 3, p. 499–511.

Plafker, G., 1967, Surface faults on Montague Island associated with the 1964 Alaska earthquake: U. S. Geol. Survey Prof. Paper 543–G, 42 p.

—— 1969, Tectonics of the March 27, 1964 Alaska earthquake: U. S. Geol. Survey Prof. Paper 543–I, 74 p.

Plafker, G., and MacNeil, F. S., 1966, Stratigraphic significance of Tertiary fossils from the Orca Group in the Prince William Sound region, Alaska: U. S. Geol. Survey Prof. Paper 550–B, p. 62–68.

Stoneley, R., 1969, Sedimentary thicknesses in orogenic belts, *in* Kent, P. E., and others, eds., Time and place in orogeny: London Geol. Survey Spec. Pub. 3, p. 215–238.

von Huene, R., and Shor, G. G., Jr., 1969, The structure and tectonic history of the eastern Aleutian Trench: Geol. Soc. America Bull., v. 80, no. 10, p. 1889–1902.

MANUSCRIPT RECEIVED BY THE SOCIETY MARCH 29, 1971

THE GEOLOGICAL SOCIETY OF AMERICA, INC.
MEMOIR 132
© 1972

Freeboard of Continents Through Time

DONALD U. WISE
University of Massachusetts
Amherst, Massachusetts 01002

ABSTRACT

Evidence for freeboard of continents (relative elevation with respect to sea level) as a function of time is evaluated. Eyged's interpretation of continental emergence with time, based on changing areas of flooding shown on global paleogeographic atlases, seems unfounded on grounds of inherent biases in the original maps, biases associated with changing time segments between successive maps, and by comparison with a freeboard versus time plot for North America compiled from Schuchert's more detailed atlas. Instead, Hess's simple assumption of constant average freeboard seems correct. The North American plot is used as a basis for a quantitative estimate of the distribution in time of deviations in freeboard. For over 80 percent of post-Precambrian time, freeboard has remained within ±60 m of a normal value 20 m above present sea level. A constant freeboard model of the earth is suggested with various feedback mechanisms continually maintaining this fine adjustment between volume of ocean basins and volume of ocean waters. From the model, a number of calculations and implications are drawn for continental and oceanic accretion, as well as for some rate relations in a global tectonic system.

INTRODUCTION

The relative elevation of a continent with respect to sea level is sometimes termed its freeboard (Kuenen, 1939). The most easily measured parameter related to past freeboard is the relative areal extent of marine flooding. This parameter is non-linear, being dependent on average slopes as a function of altitude.

Direct conflict exists betwen Hess's (1962) apparently simple assumption that continental freeboard has remained constant with time verus Egyed's (1956a and b, 1969) quantitative measurements apparently indicating emergence of continents with time. Each used his "observation" as a cornerstone for far-reaching tectonic theories. Hess used it for his theory of evolution of ocean basins, a theory which has since grown into the new plate tectonics. Egyed built his observation into a theory of global expansion in radius of 0.5 mm per yr, a theory which is still being considered quite seriously (Holmes, 1965; Jordan, 1969; Termier and Termier.

1969; Dearnley, 1969). Clearly, if such far-reaching theories rest on the problem of determination of freeboard, a more detailed examination of it is in order.

FREEBOARD FROM PALEOGEOGRAPHIC MAPS

Hess's assumption of constant freeboard is based on paleogeographic observation by a long line of geologists including Schuchert (1916), who presented a first plot of areas of North America emergent as a function of time; Stille (1924), with his ideas of periodic transgressions and regressions; Grabau (1933), with a theory of pulsating sea levels; Kuenen (1939), who placed some quantitative estimates on the magnitude of eustatic changes; and Umbgrove (1939, 1947), who stamped his book's title, *Pulse of the Earth,* on these eustatic changes. All interpreted the record as one of essentially constant freeboard.

Egyed's method of determination of past freeboard is by measurement of areas of flooding on successive maps of a paleogeographic atlas. Egyed's plots (reproduced as Fig. 1) for the two available global atlases (Termier and Termier, 1952; Strakhow, 1948) suggest a global emergence of the continents. Fitting the first few Paleozoic points to the Termier plot (Fig. 1) seems difficult, a problem which Holmes (1965, p. 969) resolves by assuming that required areas of Cambrian and Ordovician rocks were in the Antarctic and hence did not appear on the original atlases.

Unfortunately, measurement of secular change in freeboard by planimetry of paleogeographic maps is fraught with pitfalls and biases. (1) The record becomes increasingly blurred by erosion, burial, and metamorphism as a function of time. (2) Continental and upland deposits are more subject to erasure by erosion than

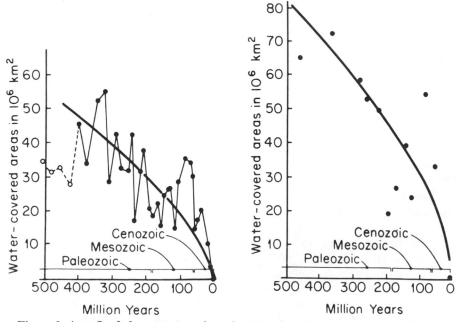

Figure 1. Area flooded versus time plots after Egyed (1956) for two global paleogeographic atlases. (A) Egyed plot based on Termier and Termier Atlas (1952). (B) Egyed plot based on Strakhow Atlas (1948).

deposits of marine basins. (3) Methods used do not take palinspastic considerations into account, an effect which is minor in cratonic areas but which may be significant in orogenic belts. (4) Use of global atlases insures inclusion of little-known regions and large segments of time so that shorelines become simpler and extrapolations grander as a function of age. As age increases the atlas maker's bias toward a marine or a continental world is less and less hindered by fact.

A more subtle pitfall is the length of time unit chosen for individual paleogeographic maps. In the extreme, one can imagine a plot of North America with all areas containing any marine fossil of Cambrian to Recent age encircled as part of a flooded region. This paleographic map marked "post-Precambrian" would indicate 100 percent flooding of North America. With this phenomenon in mind, Egyed's curves for flooding can be re-examined. The Strakhow Atlas, with post-Precambrian time broken into 11 maps, shows a systematic flooding 30 to 40 percent greater at any given time than the Termier Atlas, with its 34 maps for the same over-all time span. In addition, the slicing of time units on successive maps of both these global atlases becomes longer with increasing age (Table 1). It is obvious that correction for a factor of increased apparent flooding with longer time segments would reduce the apparent Paleozoic flooding and decrease the slopes of the curves.

A method of reducing some of these inherent biases in Egyed's global treatment is the use of a single continent for which good control and finely sliced paleogeographic time segments are available. Schuchert's monumental paleographic work on North America resulted in an atlas of the continent (Schuchert, 1955) with 84 successive paleogeographic maps. Some details on some of the maps are certainly open to revision, but on the whole it must be among the most complete coverages of this type of any extensive region of the world. The nearly uniform length of time segments on this atlas contrasts sharply with the changing length of time segments on the global atlases as indicated in Table 1.

Extent of flooding on each of the 84 maps of the Schuchert North American Atlas (1955) was measured to produce the curve of Figure 2 for North American freeboard, plotted as a function of time using the revised (Harland, and others, 1964) Holmes time scale. The area measurements included the Gulf of St. Lawrence, Hudson Bay, and the Arctic Archipelago as having reasonably reliable geologic control but excluded all continental shelf data on the maps as being too speculative. Using this definition, North America is presently 8.9 percent flooded; variations have ranged from 2 to 39 percent flooding since the Precambrian with an average value of 14.9 percent flooding. This curve has similarities to Schuchert's own curve (1916) of continental area with time. However, his later refinements of the maps make the regressions of the Silurian, Mississippian, Triassic, and Cenozoic appear much more prominent on this plot (Fig. 2) than on his older curve.

TABLE 1. AVERAGE NUMBER OF YEARS (IN MILLIONS) BETWEEN SUCCESSIVE
MAPS IN SEVERAL PALEOGEOGRAPHIC ATLASES

	Schuchert (1955) North America	Strakhow (1948) Global	Termier and Termier (1952) Global
Cenozoic	7.5	30.0	10.0
Mesozoic	8.9	40.0	14.6
Paleozoic	6.2	60.0	21.2

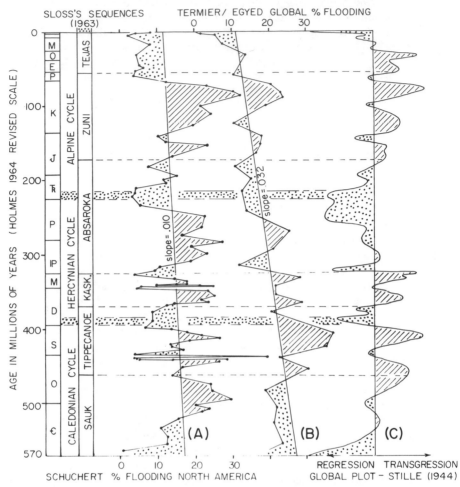

Figure 2. Three versions of plots of freeboard versus time. (A) Schuchert-Wise plot for North America: (B) Termier-Egyed global plot with Egyed figures changed from actual area flooded to percent flooding of surficial area of continents; (C) Stille (1944) qualitative plot of transgressions-regressions with no scale implied. The lines of least-squares fit of the data are regarded as average sea level. As discussed in the text, the apparent slopes are interpreted as biases inherent in the data.

RELEVANCE OF NORTH AMERICAN TO GLOBAL FREEBOARD

The extent to which the North American freeboard, or eustatic tidal gauge, reflects global sea level is explored in Figure 2. On it, Egyed's plot of the Termier Atlas is readjusted to the new time scale, and his areas are shown as percent of present continental area flooded. All points are plotted at the midpoints of their time span. Also included is Stille's (1944) qualitative and debatable curve estimating global flooding and regressions based on paleogeographic analyses. Various ages and cycles are included along the left margin. The curves, for the most part, are easily correlated one with another even to many of their details. The three great cycles of European geology have their orogenic separations marked by major continental emergence, a fact long recognized by Stille (1924), Bucher (1933), and Umbgrove (1947). Many of Sloss's (1963) North American cratonic sequences, marked by regional emergences and unconformities, are detectable in the global plots to an extent at least as great as the traditional time boundaries between standard geologic periods. There are also differences between the North American and the global plots. The Sauk sequence seems either mistimed or nonexistent on the global plots; the North American Cenozoic seems to have a greater average emergence than the global plots.

On the whole, the oscillations of the North American freeboard plot seem to be a slightly more detailed version of a global pattern. If details of this trend match the global one, it seems justifiable to consider the relation of its average slope to the global one. Least-squares linear fits are shown for both plots on Figure 2. Slope of the line based on Schuchert's (1955) maps is only one-third that of the Termier-Egyed line (0.010 versus 0.032). Application of a student's "t" test on the regression coefficients and the assumption of a true population zero slope yields probabilities of .10 and .001, respectively, that the apparent slopes or greater slopes are due to random sampling. In that a .05 or less probability is ordinarily required before a slope is considered to be real, one concludes that the Termier-Egyed line has significant slope but that the North American slope has not been established.

An Egyed-style interpretation of this difference between the global versus the North American plot would be that the earth has been expanding and average global sea level falling. Therefore, within this system North America has been systematically sinking with respect to other continents, decreasing its apparent rate of emergence by two-thirds. A simpler explanation is that detailed examination of one continent has eliminated two-thirds of the biases inherent in a "broad brush" global treatment. It is reasonable to expect that significant biases remain in a continent-wide treatment so that further detail would tend to reduce the remaining apparent minimal and unestablished slope of the line for North America.

In summary, the evidence does not favor long-term emergence or submergence of the continents. Instead, it seems that our planet has maintained a constant relation between elevation of continents and sea level at least since Cambrian times, and that Harry Hess's apparently bald assumption of constant freeboard was correct.

RELATION OF SEA LEVEL TO PERCENT FLOODING

Kuenen (1939) first related the percent flooding of paleogeographic maps to present hypsometric curves of the continents to conclude that the magnitude of marine transgressions-regressions requires at least ±40 m of eustatic change. A similar analysis can be performed with the Schuchert data from Figure 2 and the construction of a percent area versus elevation curve for present North America by means of a planimeter and a topographic map (Fig. 3). The flooding scale of Figure 3 was adjusted to have present sea level coincide with the "present flooding" of 8.9 percent, in order to fit the same definition of North America as used in measuring all the paleogeographic maps. Applying the Schuchert data (Fig. 2) to this hypsometric plot suggests that the minimum recorded flooding of 2 percent corresponds to −50 m sea level (a false value in that the paleogeographic definitions set an absolute limit at 100 percent emergence); maximum recorded flooding of 39 percent is a 180 m rise in sea level; intersection of least-squares fit with the present at 11.8 percent suggests normal freeboard of the continent is 20 m above present sea level. Plotting the deviations from the least-squares fit of the Schuchert-derived curve of Figure 2 for small time increments against the elevation distribution of Figure 3 results in the curve of Figure 4 showing the fraction of time that certain freeboard deviations have existed. The plot suggests that for 80 percent of post-Precambrian time, oscillations have remained within ±60 m of normal freeboard.

The data of Figure 4 apply only insofar as the present North American curve of elevation is similar to past elevations, and insofar as North American slopes and floodings are representative of past averages for the globe. The floodings are for periods averaging 7 m.y. in length so that larger short-term variations in sea level could exist, as, for example, with glacial eustatic or isostatic changes. Continued isostatic rise of the Hudson Bay region will decrease the defined flooding of North America by about 5 percent, whereas the melting of 25×10^6 km³ of global

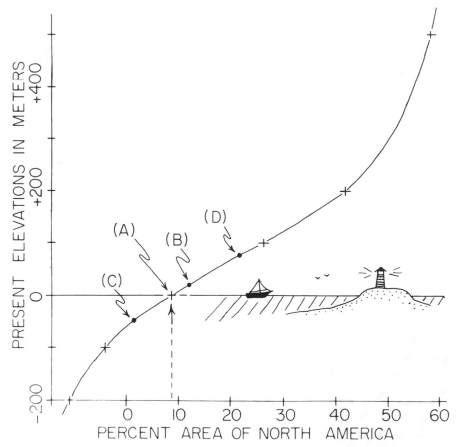

Figure 3. Relation of potential area flooded to sea level changes in present North America. The percentage areas correspond to the same total area as defined for the Schuchert paleogeographic maps, a definition under which North America is presently flooded by 8.9 percent. Accordingly, the horizontal scale is shifted so that 8.9 percent (A) corresponds to present sea level. Least-squares fit of the North America line at 11.8 percent flooding from Figure 2A is indicated by (B), with deviations of flooding of 10 percent from that line indicated by (C) and (D).

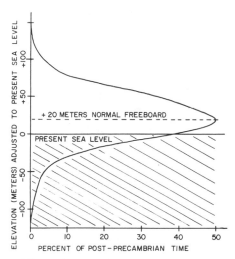

Figure 4. Deviations in North American freeboard as a function of time.

ice (Holmes, 1965) will raise sea level by 65 m and increase flooding by about 10 percent, for a net gain of about 5 percent flooding. That a nonglacial present North America would be very close to the normal long-term average freeboard runs counter to common opinion that the present time is one of great emergence. This disturbing result may be a reflection of systematic errors in the curves. On the other hand, we might recheck our assumptions, in that this same conclusion (that the maximum average emergence of the Cenozoic is now behind us) was reached by Umbgrove (1947, p. 94).

Converted to volumes of water, the ±60 m general range in freeboard corresponds to ±1.6 percent of the volume of oceans. The extreme Ordovician flooding with its 39 percent coverage and questionable 180 m rise above present sea level corresponds to a deviation of 4.2 percent of ocean volume. Thus, the constant freeboard data suggest that ocean waters have always just about filled ocean basins, the maladjustment being within 2 percent at most times and only rarely and briefly reaching 4 percent.

The North American freeboard curve (Fig. 2) also suggests that there may be a maximum time period of 60 to 70 m.y. before the system is again brought back to adjustment. The three longest floodings are 65, 73, and 52 m.y.; the three longest complete emergence cycles are 65, 20, and 35 m.y., whereas the partially completed (?) Cenozoic cycle is 65 m.y.; the incomplete lower Cambrian cycle is 57 m.y. The shape of most of these major deviations is relatively symmetric, suggesting that the system is not one of sudden perturbation and slow return to normal, but rather is one of slow perturbation and slow return.

CONSTANT FREEBOARD MODEL

If one has a static view of the earth, a constant freeboard with time seems anomalous on a globe evolving surface water and crust with time, eroding continents at a geologically catastrophic rate, and dumping the debris into the oceans to displace additional water. If continents are not to submerge with time, a dynamic model is demanded with massive self-adjusting or negative feedback systems continuously regulating the volume of ocean basins to coincide with the volume of ocean waters, an adjustment which the preceding section suggests has been maintained to within about 2 percent volume over most of post-Precambrian time.

A constant freeboard model capable of these self-adjustments is illustrated in Figure 5. Its debt to Harry Hess and a host of recent workers is obvious. It is designed fundamentally as an illustration of principles involved in constant freeboard feedback mechanisms, following more closely the model philosophy of Rube Goldberg than the model scaling methods of M. King Hubbert (1937). The earth's surface area is represented by the surface area of materials in a tank; a movable end permits "earth expansion." Mud "continents" are supported by a viscous paraffin "mantle" and separated by water-filled "ocean basins." A floating or isostatic link dependent on the relative density contrasts of mud and water with respect to "mantle" controls the ratio of thickness of "continents" to depth of "oceans." Submerged tank treads shift surface plates away from basins, causing continents to collect at collision points of these moving surface plates. (The treads are an easy way of representing surface motions and are *not* meant to imply a necessity for convection in the mantle. The author feels present geologic evidence favors gravitative sliding of lithospheric plates from off mantle bulges as the dominant driving force.) A solar powered "hydrologic cycle" is capable of eroding the continents to a limiting depth of sea level and spreading the debris into the ocean basins. Provision is made for injecting new water or crust through the mantle to the surface, whereas minor amounts of the surface water and crust may be ingested back into mantle at the descending collision points of the plates. Speeds of operation of the various components of the system are variable.

The most extreme of the changes possible in the model is "global expansion" by movement of the tank wall. This increases the area and hence the volume of ocean basins, leaving the continents standing higher. Increased continental elevation increases area and stream gradients. This increases regional rates of erosion, with the debris building continents laterally at sea level, displacing ocean waters, and raising average sea levels. Rising sea levels reduce areas of erosion until such time as the erosion rate catches up with the mantle processes which rebuild the continents. The ultimate result will be a return to equilibrium at constant freeboard

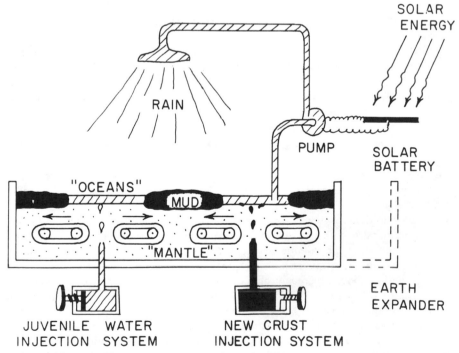

Figure 5. Constant freeboard global tectonic model. Use of conveyor belts in the model is a way of shifting crust rather than of implying that mantle convection is required.

with thinner broader continents and shallower wider oceans having the same ratio of continental depth to oceanic depth as before (in this overly simplified isostatic system), so that the volume of ocean basins again just equals the volume of ocean waters.

The reader is invited to play with the model mentally, making any of the possible changes to demonstrate for himself that any perturbation will result in short-term change in freeboard of the model but will trigger long-term feedbacks, thereby returning the system to normal. Some of the possible changes or limitations on the system are included in the work of Armstrong (1968), Birch (1968), Dietz and Holden (1966), Khain and Muratov (1969), Maxwell (1968), Menard (1964), and Weertman (1963).

IMPLICATIONS OF THE CONSTANT FREEBOARD MODEL

The constant freeboard model provides a convenient basis for visualizing the interdependence of some tectonically interesting variables: area of earth, A_e; area of continents, A_c; area of ocean basins, A_o; depth of continents, D_c; depth of oceans, D_o; volume of continents, V_c; volume of oceans, V_o; and the isostatic link, I_L, relating D_o to D_c. These 8 variables are governed by four basic equations:

(1) $A_e = A_c + A_o$, (2) $V_o = A_o \times D_o$, (3) $V_c = A_c \times D_c$, (4) $D_c = I_L \times D_o$.

One of the variables, A_e, is ordinarily assumed constant, leaving 7 variables, of which three are independent: V_o, V_c, and I_L.

The last variable, the isostatic link, I_L, is probably the most troublesome, requiring some geologic assumptions about the behavior of crust and mantle layerings and densities. A simple model for I_L includes constant-density, single-layered continents in isostatic balance with an oceanic crust of constant thickness and density at present values through geologic time. If we use Hess's (1962) crustal thicknesses and average densities with a constant .3 km of continent above sea level, then 9.1 km of continental crust is needed to counterbalance the oceanic crust and sediments with no water present. Beyond this thickness an increase in continental Moho depth must be balanced by increase in water depth controlled by the ratio of density contrasts of water and continent with respect to mantle. Accordingly, a simple isostatic link would be: $D_c = 9.1$ km $+ 4.98$ D_o. A more sophisticated link might include a two-layer continent for which additional assumptions of layer thickness, density, and method of variability would be required. Additional complexities of phase change limits on Moho, variable density mantle, and variable thickness or densities of multiple continental and oceanic layers are possible. Under any circumstances, some isostatic link of D_o to D_c must be devised until such time as past values of the other variables are known to a perfection that I_L can then be calculated.

If the simple I_L and constant A_e assumptions are made, the system is soluble for any past time at which two additional variables or certain ratios of variables are known. Graphic solutions of the four basic equations are plotted in Figure 6 using the above assumptions, Hess's average depth of continents and oceans, and the division in area between continents and oceans at the 2 km isobath. Lines at 100 percent V_o and V_c are drawn as limits on possible ranges of past A_c and D_c, such that the system does not involve oceans or continents decreasing in volume in more recent geologic time. The minimum D_c of 26 percent of present is defined by the requirement of isostatically balancing oceanic crust. Depending on the mode of formation of the earliest continents, this lower limit may or may not apply. There are clearly many possible paths on Figure 6 leading from the primordial earth to the present 100 percent values of all the variables. Insofar as the assumptions behind Figure 6 are valid, then values of areas, depths, and volumes are uniquely defined at any point along any path chosen. A major task of tectonics is the definition of the true path representing the simultaneous evolution of continents and oceans through geologic time.

An example of the way constant freeboard may be used to check various types of data paths and assumptions against one another is presented in Figure 7, using the system of assumptions and values inherent in Figure 6. Hurley's (1968) linear increase in V_c is plotted on Figure 7A, with time starting at 3.8 b.y., along with Condie and Potts's (1969) estimate of 10 to 25 km crust at 2.6 b.y. (based on chemistry of a Canadian Shield rock suite). The 10

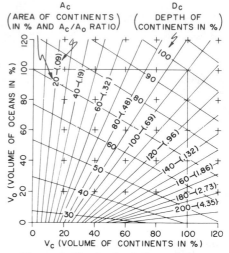

Figure 6. Graph of solutions of the constant freeboard equations. Assumptions are constant area of earth, constant oceanic crust, and simple isostatic link given in text. All values except A_c / A_o are percentages of present values here defined as $A_c = 2.07 \times 10^9$ km², $D_o = 5$ km, $D_c = 34$ km, $V_c = 7.04 \times 10^{10}$ km³, and $V_o = 1.52 \times 10^{10}$ km³. The use of normal oceanic depth as 5 km (for isostatic reasons) results in a V_o 10 percent larger than the presently accepted values.

km lower limit of D_c falls on a linear increase, whereas the highest curve drawn through the upper limit of D_c is constrained by an upper limit of 100 percent V_o (as drawn). The A_c curve for these upper limits is indicated. If one wished to lower the V_o curve to permit a continuing evolution of oceans, the D_c curve could be lowered and the A_c curve raised, as shown by the dotted lines. The resulting 50 percent A_c for 2.6 b.y. is in rough accord with that of North America as implied by Muehlburger and others (1967, Fig. 13 A). The "geopoetry" in this system seems to have internal rhyme, including a reasonable evolution of volume of oceans with time.

Two examples of nonrhyming "geopoetry" are plotted in Figures 7B and 7C. Hurley's linear V_c curve is included in each. Figure 7B shows Hurley and Rand's (1969) estimate of A_c based on areas of Precambrian age provinces, whereas Figure 7C is Engel's (1963) estimate of linear growth in the area of North America, here extrapolated to a global representation. Each of these plots seems reasonable in itself until one includes V_o and D_c, calculated with the same assumptions as above. Under this system the data would require an unlikely decrease in volume of oceans in later geologic time. In Figure 7B the Moho would have to rise with time, whereas 7C has the attractive feature of constant D_c, wherein shield would be created and maintained at essentially constant thickness. At least one of the assumptions in data or in model in each of these two nonrhyming plots must be incorrect. More sophisticated isostatic links could be chosen; the effect of obliteration of older crust by younger overprints could bias the area plots; or an expanding A_e could solve the V_o discrepancy.

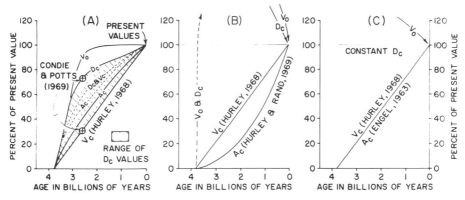

Figure 7. Some examples of rhyming (A) and nonrhyming (B and C) "geopoetry." Decrease in volume of oceans with time in B and C suggests inconsistency in data or in assumptions.

Similar plots can be prepared and checked for internal consistency for a wide range of assumptions or measurements of continental and oceanic accretion as a function of time. Either by use of Figure 6 or by intuition with the physical model of Figure 5, a number of basic tectonic rules can be established.

(1) The relative *area* of continents to oceans is a function of relative *volumes* of continents to oceans. Accordingly, continental accretion cannot be considered independently of oceanic evolution.

(2) Any net addition of juvenile water must deepen oceans, and, through isostasy, this forces continents to thicken. If a proportional amount of net continental material is not added, volume of continent will be maintained by *decrease* in area of continents with time. In other words, negative area of continental accretion is possible.

(3) Further areal accretion of continents is possible only if the ratio of new V_c to new V_o exceeds some ratio determined by isostasy and the existing volume ratios of these materials at the surface. If new material is added at precisely this ratio, the area of continents is constant. For the simple model of Figure 6, this ratio for the line of 100 percent A_c is 3 percent increase in V_c to 4 percent increase in V_o, or one part new V_c to 2.87 parts new V_o in the present earth.

(4) Continents and oceans are thickening with time if any net increase in new surface material takes place, be it either water or continental crust or both (provided A_e remains constant).

(5) The model does not preclude a variable A_e with time. It merely says that on an expanding earth areas of both continents and oceans will expand to maintain freeboard while depth ratios will be maintained according to isostasy. The change will be undetectable by ordinary paleogeographic flooding measurements. Of greater significance might be the fate of the average level of the sub-Cambrian unconformity, for if continents are to be progressively thinned, this will be systematically raised and eroded. Unfortunately, there are a host of other factors influencing average elevation of this surface so that the A_e factor cannot be separated with ease.

In addition to these evolutionary type implications of the constant freeboard model, another set of interdependences may be established based on rate phenomena.

(1) Average rates of regional erosion (Judson and Ritter, 1964; Menard, 1961)

over periods of a few tens of millions of years must equal the net rate at which continents are rebuilt, assuming negligible volume changes of crust and ocean on this time scale.

(2) Long-term average elevation of continents will be determined by the gradients necessary to move material off a continent at the same long-term rate at which it is being rebuilt. For two continents enjoying the same rate of rebuilding, the more arid continent with its less efficient streams would stand higher in elevation to provide higher gradient for the streams.

(3) Orogenic periods separating the great cycles of European geology seem to coincide with the periods of maximum lowering of sea level on Figure 2, a relation recognized by many early workers. Gilluly (1949) raised many pertinent objections to Stille's (1944) numerous global orogenies affecting all continental geology. He argued instead for periodicity of orogeny when viewed from any single spot, but a constant orogenic rate when viewed for the entire globe. Nevertheless, the correlations among the various freeboard plots suggest that many aspects of ocean basin tectonics are recorded as unconformities or transgressions in local geologic records. A few of the greatest periods of oceanic tectonics are recorded as global marine regressions and far-ranging contractions or deformations of continental plates. To this minor extent Gilluly's lucid arguments against periodic orogenic rates might well be modified. The variable rates suggested for the model (Fig. 5) may well be part of the real total system.

(4) Some interesting Cenozoic volumetric and areal figures can be derived, as shown on Table 2, if volumetric changes are negligible on that time scale. The correlation between water displaced by a well-developed Cenozoic ridge system and *lower* Cenozoic sea level seems anomalous at first glance. If we assume that only half of the present ridge system's volume came into being in the Cenozoic, item (A) on Table 2 becomes +150 m and item (C) is −100 m change in sea level. If this value of item (C), representing 2.6 percent of the total volume of oceans, was exclusively a change in area of deep ocean basins at the expense of area of continental blocks, as in the model, then continental blocks could have decreased in net area by 5.1 percent during Cenozoic orogenies. The correlation among wholesale orogenic disturbance, thrusting, and subduction of the continents with lowered sea level during an orogenic period seems expectable. Periods of world-wide molasse development seem a logical corollary (Van Houten, 1969).

CONCLUSIONS

Constant freeboard is the best evidence that the continuing struggle between oceanic waters and continental crust for the total surface area of the globe has

TABLE 2. SOME CENOZOIC SEA LEVEL CONSIDERATIONS

(A)	+300 meters:	sea level rise from present ridge system (Menard and Smith, 1966)
(B)	−100 meters:	by subsidence of Cretaceous Darwin Rise (Menard and Smith, 1966)
(C)	−250 meters:	required volume change of ocean basins by mechanisms other than the ridge system
(D)	−50 meters:	average drop in Cenozoic freeboard below normal (Figs. 2 and 3)

been a stalemate over much of geologic time. The stalemate results in the isostatic balance between typical oceanic and continental crusts with their relative areas a function of their respective volumes and volume ratios. This dynamic equilibrium is independent of shape or number of continents or ocean basins. Its rate of operation is controlled by the speed at which tectonic processes of the mantle drive the deeper half of the equilibrium toward wider ocean basins and thicker, smaller continents. Its surficial half pushes the system toward thin and widespread continents separated by shallow ocean basins. The surficial half adjusts its rates of operation by changing average continental elevations or by changing the extent of eroding area above sea level so as to counterbalance any effect of the deeper half. The maintenance of this volume-area-isostasy equilibrium to within a few percent by volume is the unifying theme behind an endless number of seemingly random, seemingly purposeless geologic changes. The tortured geology of continents is a partial record of adjustments to this equilibrium and shifts of equilibrium position as a function of time. If ocean waters carried in their fabric a similar memory of past configurations, an equally complex history of ocean basin adjustment would appear. There is little in the first-order features of our planet's crustal and surficial geology which is not linked in some manner to this underlying theme of constant freeboard in a stalemated struggle for area. It might well be called the master equilibrium of tectonics.

ACKNOWLEDGMENTS

As a student the author was introduced to this problem by Harry Hess and John Maxwell. Much of the thought expressed here was done as a visiting scientist with H. W. Menard in 1961. This paper has benefited by many discussions with these men and with other friends and colleagues through the years. It includes insights from a large number of old and recent authors, only some of whom could be cited directly in this short summary. Edward Butner, John Hubert, and George McGill acted as critical readers.

REFERENCES CITED

Armstrong, R. L., 1968, A model for the evolution of strontium and lead isotopes in a dynamic earth: Rev. Geophysics, v. 6, p. 175–200.

Birch, F., 1968, On the possibility of large changes in the earth's volume: Physics Earth and Planetary Interiors: v. 1, no. 3, p. 141–147.

Bucher, W. H., 1933, The deformation of the Earth's crust: Princeton Univ. Press, 518 p.

Condie, K. C., and Potts, M. J., 1969, Calc-alkaline volcanism and the thickness of the early Precambrian crust in North America: Canadian Jour. Earth Sci., v. 6, p. 1179–1184.

Dearnley, R., 1969, Crustal tectonic evidence for earth expansion, in Runcorn, K., ed., Application of modern physics to earth and planetary interiors: New York, Interscience Press, p. 103–110.

Dietz, R. S., and Holden, J. C., 1966, Miogeoclines (miogeosynclines) in space and time: Jour. Geology, v. 74, p. 566–583.

Egyed, L., 1956a, Determination of changes in the dimensions of the earth from paleogeographical data: Nature, v. 178, p. 534.

—— 1956b, Change of earth dimensions as determined from paleogeographical data: Geofisica Pura e Applicata, v. 33, p. 42–48.

—— 1969, The slow expansion hypothesis, in Runcorn, K., ed., Application of modern physics to earth and planetary interiors: New York, Interscience Press, p. 65–75.

Engel, A. E. J., 1963, Geologic evolution of North America: Science, v. 140, no. 3563, p. 143–152.

Gilluly, J., 1949, Distribution of mountain building in geologic time: Geol. Soc. America Bull., v. 60, no. 4, p. 561–590.

Grabau, A., 1933, Oscillation or pulsation [with dis.]: 16th Internat. Geol. Cong. Rept., v. 1, p. 539–533 [1936].

Harland, W. B., Smith, A. G., and Wilcock, I. B., eds., 1964, The Phanerozoic time scale, a symposium in honor of Arthur Holmes: Geol. Soc. London Quart. Jour., v. 1205, p. 260–262.

Hess, H. H., 1962, History of ocean basins, in Engle, A. E. J., James, H. L., and Leonard, B. F., eds., Petrologic Studies (Buddington Volume): Geol. Soc. America, p. 599–620.

Holmes, A., 1965, Principles of physical geology (2d ed.): New York, Ronald Press Co., 1288 p.

Hubbert, M. K., 1937, Theory of scale models as applied to study of geologic structures: Geol. Soc. America., Bull. v. 48, no. 10, p. 1459–1519.

Hurley, P. M., 1968, Absolute abundance and distribution of Rb, K and Sr in the earth: Geochim. et Cosmochim. Acta, v. 32, p. 273–283.

Hurley, P. M., and Rand, J. R., 1969, Pre-drift continental nuclei: Science, v. 164, no. 3885, p. 1229–1242.

Jordan, P., 1969, On the possibility of avoiding Ramsey's hypothesis in formulating a theory of earth expansion, in Runcorn, K., ed., Application of modern physics to earth and planetary interiors: New York, Interscience Press, p. 55–63.

Judson, S., and Ritter, D., 1964, Rates of regional denudation in the United States: Jour. Geophys. Research, v. 69, p. 3395–3401.

Khain, V. E., and Muratov, M. V., 1969, Crustal movements and tectonic structure of continents, in Hart, P. J., ed., Earth's crust and upper mantle: Am. Geophys. Union Geophys. Mon. 13, p. 523–538.

Kuenen, P. H., 1939, Quantitative estimations relating to eustatic movements: Geologie en Mijnbouw, v. 18, no. 8, p. 194–201.

Maxwell, J. C., 1968, Continental drift and a dynamic earth: Am. Scientist, v. 56, no. 1, p. 35–51.

Menard, H. W., 1961, Some rates of regional erosion: Jour. Geology, v. 69, p. 154–161.
—— 1964, Marine geology of the Pacific: New York, McGraw-Hill Book Co., 271 p.

Menard, H. W., and Smith, S., 1966, Hypsometry of ocean basin provinces: Jour. Geophys. Research, v. 71, no. 18, p. 4305–4325.

Muehlberger, W. R., Denison, R. E., and Lidiak, E. G., 1967, Basement rocks in continental interior of United States: Am. Assoc. Petroleum Geologists Bull., v. 51, no. 12, p. 2351–2380.

Schuchert, C., 1916, Correlation and chronology on the basis of paleogeography: Geol. Soc. America, Bull., v. 27, p. 491–513.
—— 1955, Atlas of paleogeographic maps of North America: New York, John Wiley & Sons, 177 p.

Sloss, L. L., 1963, Sequences in the cratonic interior of North America: Geol. Soc. America Bull., v. 74, p. 93–114.

Stille, H., 1924, Grundfragen der vergleichenden tektonik: Berlin.
—— 1944, Geotektonische gliederung der erdgeschichte: Abh. Preuss. Akad. Wiss. Math. Nat. Klasse, v. 3, 80 p.

Strakhow, N. M., 1948, Outlines of historical geology: Moscow, U.S.S.R. Govt. Publ.

Termier, H., and Termier, G., 1952, Histoire geologique de las biosphere: Paris, Masson.
—— 1969, Global paleogeography and earth expansions, in Runcorn, K., ed., application of modern physics to earth and planetary interiors: New York, Interscience Press, p. 87–101.

Umbgrove, J. H. F., 1939, On rhythms in the history of the earth: Geol. Mag., v. 76, p. 116–129.
—— 1947, The pulse of the earth: The Hague, M. Nijhoff, 357 p.

Van Houten, F. B., 1969, Molasse facies: Records of worldwide crustal stresses: Science, v. 166, p. 1506–1508.

Weertman, J., 1963, The thickness of continents: Jour. Geophys. Research, v. 68, no. 3, p. 929–932.

MANUSCRIPT RECEIVED BY THE SOCIETY MARCH 29, 1971   PRINTED IN U.S.A.

THE GEOLOGICAL SOCIETY OF AMERICA, INC.
MEMOIR 132
© 1972

Mid-Cenozoic Activity on Lithospheric Plates

FRANKLYN B. VAN HOUTEN
Princeton University, Princeton, New Jersey 08540

ABSTRACT

Widespread mid-Cenozoic (25 to 40 m.y. ago) changes in deformation, deposition, and volcanism on continents and island arcs reflect discontinuities in the behavior of lithospheric plates. Varied mid-Cenozoic activity near leading plate edges includes major orogeny and molasse accumulation along most of the western Tethyan belt and the southern Andes (with prevalent volcanism), and a distinct phase of deformation and igneous activity on the Alaska Peninsula, in Central America, and on cratonic Middle Europe. Tectonic stability following early Cenozoic orogeny was established in most of the West Indies and in the eastern ranges of the Rocky Mountains. At the end of mid-Cenozoic time, major orogeny, a new style of deformation or of sedimentation, or a major episode of volcanism occurred in the Red Sea–African rift zone, along most of the Pacific border of the United States and Mexico, in northern South America, in New Zealand and New Guinea, on the large island arcs of the Indian and western Pacific basins, and along the eastern Tethyan belt (with only minor volcanism). On trailing margins of continents most of the marine embayments underwent extensive regression during mid-Cenozoic time while interior continental Eurasia was flooded by a broad sea.

GENERAL STATEMENT

Dott (1969) has aptly called attention to profound mid-Cenozoic changes in land-sea relations, sedimentation, and structural disturbances (Fig. 1) that imply roughly synchronous changes in sea floor development. Current tectonic speculation postulates that volcanism, deformation, and deposition along edges of lithospheric plates (Fig. 2) are controlled largely by their interaction on a global scale, and especially by the direction, rate, and duration of their movements (Mitchell and Reading, 1969; Coney, 1970; Dewey and Bird, 1970; McBirney, 1970).

101

The present review sorts out the principal kinds of mid-Cenozoic events (Figs. 1 and 2) to provide information about the behavior of the plates about 25 to 40 m.y. ago. The data are from Gignoux (1955) and Kummel (1961) unless other references are cited. In this effort, assignment of some items to a particular category has been rather arbitrary because current information is scanty, or published interpretations of the available data differ. Consequently, a few items will have to be reassigned as new information is assembled. Even so, the focus is on 15 m.y. in the middle of the Cenozoic Era.

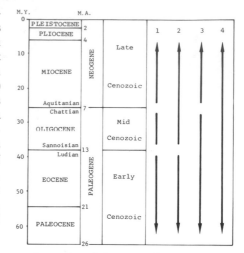

Figure 1. Subdivisions of Cenozoic Era and simplified patterns of mid-Cenozoic events. (1) Unlike or not continuous with early and late Cenozoic activity. (2) Unlike early Cenozoic activity; continuous into late Cenozoic time. (3) Continued preceding Eocene activity; unlike late Cenozoic events. (4) No distinctive activity. M.Y. = million years; M.A. = magnetic anomaly.

Unlike or Not Continuous with Early and Late Cenozoic Activity

a. Extensive regression during or near the end of mid-Cenozoic time, followed by late Cenozoic transgression, marked the trailing margins of eastern North America (Murray, 1961; Gibson, 1970; Owens, 1970), southeastern South America (Zambrano and Urien, 1970), western and southeastern Africa (Haughton, 1963), southern India (Wadia, 1953), and western and southern Australia (Brown and others, 1968), as well as notheastern cratonic Africa (Said, 1962).

b. Major marine transgression extending from northwestern to eastern Europe joined an early to mid-Cenozoic seaway along the eastern side of the Urals.

c. Deformation (commonly block-faulting with igneous activity) occurred along the Alaska Peninsula (Burk, 1965) and along the central Aleutian arc (Grow and Atwater, 1970), in Central America, followed in late Cenozoic time by the main episode of volcanism (Dengo and Bohnenberger, 1969; Williams and McBirney, 1969), and on cratonic Middle Europe with major sedimentation in the grabens (Sittler, 1969) and with increased volcanism in late Cenozoic time. During the mid-Cenozoic interval, major deformation and molasse accumulation were dominant in the Pyrenees and in the Andes of southern Chile—with active volcanism (Muñoz-Cristi, 1956; Harrington, 1962), and the Coral Sea basin apparently was opened by rotational spreading followed by late Cenozoic faulting and subsidence (Gardner, 1970).

d. Extrusion of extensive ignimbrites and an episode of plutonism were replaced by a late Cenozoic change in volcanic style (McKee and Silberman, 1970a, 1970b) and block-faulting in the Great Basin province of the United States (category 3 in terms of structural history).

e. A tectonic lull and marine transgression intervened between Eocene and Miocene deformation in the Tellian belt of northern Algeria and Tunisia (Wezel, 1970).

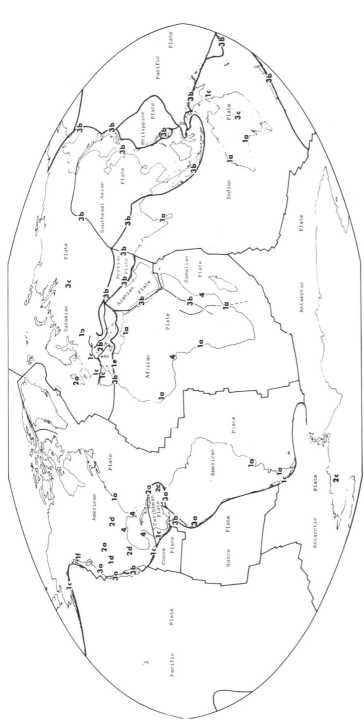

Figure 2. Aseismic lithospheric plates outlined by belts of shallow-focus earthquakes (mainly after Dewey and Bird, 1970, Fig. 1), with categories of mid-Cenozoic activity. (1) Unlike or not continuous with early and late Cenozoic activity: (a) extensive regression along continental margins; (b) major transgression of continental interior; (c) distinct episode of deformation, commonly with igneous activity; (d) distinctive style of volcanism with associated plutonism; (e) tectonic lull in orogenic belt; (f) volcanic lull. (2) Unlike early Cenozoic activity; continued into late Cenozoic time: (a) tectonic stability established or volcanism waned; (b) major deformation and molasse accumulation began; (c) oceanic volcanism began; (d) regression and waning sedimentation in embayments. (3) Continued preceding Eocene activity; unlike late Cenozoic events: (a) major deformation completed; new style or phase of deformation and sedimentation in late Cenozoic time; (b) major deformation, commonly with volcanism, began in late Cenozoic time; (c) major transgression of continental interior. (4) No distinctive activity.

f. A volcanic lull between early Cenozoic acidic volcanism and late Cenozoic basaltic extrusion prevailed throughout the central Canadian Cordillera (Souther, 1970).

Unlike Early Cenozoic Activity; Continued into Late Cenozoic Time

a. Tectonic stability was established throughout most of the West Indies (Senn, 1940; Donnelly, 1964; Eames and others, 1962, apparently assign the preceding orogeny to the entire Oligocene Epoch) and in the eastern ranges of the Rocky Mountains where volcanism continued into late Cenozoic time (Love, 1960). In the Thulean volcanic province of northwesternmost Europe, basaltic volcanism was greatly reduced following widespread early Cenozoic extrusion.

b. Major deformation and molasse accumulation predominated in the western Tethyan belt (with only local volcanism) and in Central Iran (Stöcklin, 1968).

c. Oceanic volcanism was initiated along the inner arc of the Lesser Antilles and in West Antarctica (Wade, 1970).

d. The sea began to retreat from the Rio Grande and Mississippi Embayments of the Gulf of Mexico as the major site of deposition shifted to the central Gulf Coast area by late Cenozoic time (Murray, 1961).

Continued Preceding Eocene Activity; Unlike Late Cenozoic Events

a. Major deformation was completed during or by the end of mid-Cenozoic time along the Pacific border of the United States (Dott, 1965; Yeats, 1968; Swe and Dickinson, 1970; Page, 1970), accompanied by mid-Cenozoic molasse in southern California and succeeded by a new style of late Cenozoic deformation, sedimentation, and prevalent volcanism. Middle or late Eocene to mid-Cenozoic deformation and associated change in sedimentation in southern Trinidad (Kugler, 1956) and along the ranges of northern Venezuela (Renz, 1957; Mencher, 1963; Stainforth, 1969; Bell, 1970) were followed by late Cenozoic deformation involving east-trending transcurrent faults. The Andes underwent a succession of Cenozoic orogenic episodes with accompanying volcanism and molasse accumulation. Assignment of some of this activity to a particular category has been arbitrary. From Colombia to Bolivia both volcanism and a new phase of deformation and sedimentation began in Miocene time (Ahlfeld, 1956; Jacobs and others, 1963; Van Houten, 1969; Lohmann, 1970), whereas these apparently occurred in mid-Cenozoic time in the southern Andes.

b. Major deformation, commonly with the main episode of volcanism, began near the end of Oligocene or in early Miocene time in the east African (Matsuzawa, 1966), Gulf of Aden (Azzaroli, 1968), and Red Sea (Girdler, 1969; Hassan and El-Dashlouty, 1970) rift zones; in the Rif and Betic belts of the westernmost Mediterranean basin, the central (Stöcklin, 1968) and eastern (Wadia, 1953; Gansser, 1964) Tethyan belt with accompanying molasse and subordinate volcanism, the Assam-Andaman-Indonesian arc (Karunkaran and others, 1968; Raju, 1968; Rodolpho, 1969), and in the Baikal Rift (Nalivkin, 1960; Florensov, 1969) of central Asia; as well as along most of the Circum-Pacific belt (Dott, 1969), including New Zealand (Brown and others, 1968) and northern New Guinea (Thompson, 1967), Fiji (Dickinson, 1967), and New Hebrides (Mitchell, 1970), the Philippines and Japan (Beloussov and Ruditch, 1961; Minato and others, 1965;

Wageman and others, 1970), Sakhalin and Kamchatka (Matsumoto, 1967; Push-charovsky, 1967) arcs; and in Baja and the Gulf of California (Allison, 1964; Rusnak and Fisher, 1964; McFall, 1968), and the Bolivar belt of northwestern Colombia and adjacent Ecuador (Bürgl, 1967; Stoneley, 1969).

c. An extensive early to mid-Cenozoic seaway along the eastern side of the Urals, interrupted by only a minor regression at the end of Eocene time, withdrew near the end of the Oligocene Epoch. A great lake within Australia waned after its maximum expansion in mid-Cenozoic time.

No Distinct Activity

Deposition was essentially continuous on trailing continental margins along the Gulf Coast (Gibson, 1970; but Eames and others, 1962, maintain that there was an Oligocene hiatus), on the Florida and Yucatan carbonate banks (Murray, 1961), and on the Niger Delta (Short and Stäuble, 1967), as well as in the Tanzania Embayment (Eames and others, 1962) within the Somalian plate.

SUMMARY

Mid-Cenozoic time was marked by discontinuities in tectonic, volcanic, and depositional activity. In fact, there are few places on the continents where these processes persisted essentially unchanged from the Eocene to the Miocene Epoch. It was also an interval of widespread shallower water sedimentation and cooler oceanic temperatures (Frerichs, 1970). This brief review provides both a calendar of the mid-Cenozoic events on continents and island arcs and a tally of the different kinds of activity along their margins. Even though the data form several patterns (Fig. 1, 1 through 4), discontinuities along the edges of particular lithospheric plates occurred at about the same time. In this span of about 15 m.y. most of the major deformation took place in late Oligocene to earliest Miocene time (Eames, 1970).

Patterns of activity in the broad global belt of the Caribbean plate and the Tethyan region differ from those along the Circum-Pacific belt. Orogeny began somewhat earlier in the east-trending girdle, generally progressing from late in early Cenozoic time on the Caribbean plate and at the west end of the Alpine chain to early in late Cenozoic time in the eastern Tethyan zone, and volcanism was minor or absent except in the West Indies, which are on oceanic crust. Around most of the Pacific basin, in contrast, major deformation and volcanism started near the beginning of late Cenozoic time.

Detailed differences in the time of disruption and kinds of sedimentation, especially in the western and central Tethyan zone, apparently resulted from convergence of irregular plate edges and the interaction of numerous microcontinents (Dewey and Bird, 1970, Fig. 14). The complex and varied patterns of activity along the length of the Andes probably reflect differing behavior of the several oceanic plates that impinged on South America.

Predominantly nonmarine molasse in troughs on the craton was a characteristic end product along those edges where deformation migrated from an ocean basin toward a continent—in the Andes, the ranges of northern Venezuela, and the Central Highlands of New Guinea—and where two continents converged or collided, as in the Tethyan belt. In each of these chains the molasse facies records the end of oceanic plate consumption.

Cenozoic events point to a correlation between activity on leading edges of continents and that along their trailing margins and within them. While leading edges of the larger continent-bearing plates were responding to varied conditions of convergence, the trailing margins were subjected to transgression and regression. Mid-Cenozoic regression along margins of North and South America, Africa, India, and Australia coincided in a general way with episodes of reduced tectonism along their leading edges. Transgressions of these trailing margins took place during early and late Cenozoic deformation of the leading edges. Similarly, major flooding within continents, as in large embayments along the Gulf Coast and in a broad seaway across the interior of northern Europe and adjacent Asia, occurred at the time of orogeny along corresponding leading edges. These transgressions and regressions were imposed on a progressive emergence of the continents during the Cenozoic Era, which may have resulted largely from convergence and assembly of continents along the Alpine-Himalayan belt (Valentine and Moores, 1970).

The results of this survey reaffirm the importance of continental records as sources of information about the behavior of still older lithospheric plates and the relations between their continents and ocean basins.

REFERENCES CITED

Ahlfeld, F., 1956, Bolivia, in Jenks, W. F., ed., Handbook of South American geology: Geol. Soc. America Mem. 65, p. 169–186.

Allison, E. C., 1964, Geology of areas bordering the Gulf of California: Am. Assoc. Petroleum Geologists Mem. 3, p. 3–29.

Azzaroli, A., 1968, On the evolution of the Gulf of Aden: 23rd Internat. Geol. Cong., Prague, 1968, Proc., v. 1, p. 125–134.

Bell, J. S., 1970, Tertiary global tectonics in the southern Caribbean area: Geol. Soc. America, Abs. with Programs (Ann. Mtg.), v. 2, no. 7, p. 492.

Beloussov, V. V., and Ruditch, E. M., 1961, Island arcs in the development of the earth's structure (especially in the region of Japan and the Sea of Okhotsk): Jour. Geology, v. 69, p. 647–658.

Brown, D. A., Campbell, K. S. W., and Crook, K. A. W., 1968, The geological evolution of Australia and New Zealand: New York, Pergamon Press, 409 p.

Bürgl, H., 1967, The orogenesis in the Andean system of Colombia: Tectonophysics, v. 4, p. 429–443.

Burk, C. A., 1965, Geology of the Alaska Peninsula–island arc and continental margin: Geol. Soc. America Mem. 99, pt. 1, 250 p.

Coney, P. J., 1970, The geotectonic cycle and the new global tectonics: Geol. Soc. America Bull, v. 81, p. 739–748.

Dengo, G., and Bohnenberger, O., 1969, Structural development in northern Central America: Am. Assoc. Petroleum Geologists Mem. 11, p. 203–220.

Dewey, J. F., and Bird, J. M., 1970, Mountain belts and the new global tectonics: Jour. Geophys. Research, v. 75, p. 2625–2647.

Dickinson, W. R., 1967, Tectonic development of Fiji: Tectonophysics, v. 4, p. 543–553.

Donnelly, T. W., 1964, Evolution of eastern Antillean island arc: Am. Assoc. Petroleum Geologists Bull., v. 48, p. 680–696.

Dott, R. H., 1965, Mesozoic-Cenozoic tectonic history of the southwestern Oregon coast in relation to Cordilleran orogenesis: Jour. Geophys. Research, v. 70, p. 4687–4707.

—— 1969, Circum-Pacific late Cenozoic structural rejuvenation: Implications for sea-floor spreading: Science, v. 166, p. 874–876.

Eames, F. E., 1970, Some thoughts on the Neogene/Palaeogene boundary: Paleogeography, Paleoclimatology, Paleoecology, v. 8, p. 37–48.

Eames, F. E., Banner, F. T., Blow, W. H., and Clark, W. J., 1962, Fundamentals of mid-Tertiary stratigraphical correlation: Cambridge, Cambridge Univ. Press, 163 p.

Florensov, N. A., 1969, Rifts of the Baikal Mountain region: Tectonophysics, v. 8, p. 443–456.

Frerichs, W. E., 1970, Paleobathymetry, paleotemperature and tectonism: Geol. Soc. America Bull., v. 81, p. 3445–3452.

Gansser, A., 1964, Geology of the Himalayas: New York, Interscience Press, 289 p.

Gardner, J. V., 1970, Submarine geology of the western Coral Sea: Geol. Soc. America Bull., v. 81, p. 2599–2614.

Gibson, T. G., 1970, Late Mesozoic-Cenozoic tectonic aspects of the Atlantic coastal margin: Geol. Soc. America Bull., v. 81, p. 1813–1822.

Gignoux, N., 1955, Stratigraphic geology: San Francisco, W. H. Freeman and Co., 682 p.

Girdler, R. W., 1969, The Red Sea—A geophysical background, in Hot brines and recent heavy metals in the Red Sea: New York, Springer-Verlag, p. 38–58.

Grow, J. A., and Atwater, T., 1970, Mid-Tertiary tectonic transition in the Aleutian arc: Geol. Soc. America Bull., v. 81, p. 3715–3722.

Jacobs, C., Bürgl, H., and Conley, D. L., 1963 Backbone of Colombia: Am. Assoc. Petroleum Geologists Bull., v. 46, p. 1773–1814.

Hassan, F., and El-Dashlouty, S., 1970, Miocene evaporites of Gulf of Suez region and their significance: Am. Assoc. Petroleum Geologists Bull., v. 54, p. 1686–1696.

Haughton, S. H., 1963, The stratigraphic history of Africa south of the Sahara: Edinburgh, Oliver and Boyd, 365 p.

Jacobs, C., Burgl, H., and Conley, D. L., 1963, Backbone of Colombia: Am. Assoc. Petroleum Geologists Mem. 2, p. 62–72.

Karunkaran, C., Roy, K. K., and Saha, S. S., 1968, Tertiary sedimentation in the Andaman-Nicobar geosyncline: Jour. Geol. Soc. India, v. 9, p. 32–39.

Kugler, H. G., 1956, Trinidad, in Jenks, W. F., ed., Handbook of South American geology: Geol. Soc. America Mem. 65, p. 353–365.

Kummel, B., 1961, History of the earth: San Francisco, W. H. Freeman and Co., 610 p.

Lohmann, H. H., 1970, Outline of tectonic history of Bolivian Andes: Am. Assoc. Petroleum Geologists Bull., v. 54, p. 735–757.

Love, J. D., 1960, Cenozoic sedimentation and crustal movement in Wyoming: Am. Jour. Sci., v. 258A, p. 204–214.

Matsumoto, T., 1967, Fundamental problems in the circum-Pacific orogenesis: Tectonophysics, v. 4, p. 595–613.

Matsuzawa, I., 1966, A study on the formation of the African rift valley: Nagoya Univ. Jour. Earth Sci., v. 14, p. 89–115.

McBirney, A. R., 1970, Cenozoic igneous events of the Circum-Pacific: Geol. Soc. America, Abs. with Programs (Ann. Mtg.), v. 2, no. 7, p. 749–750.

McFall, C. C., 1968, Reconnaissance geology of the Concepcion Bay area, Baja California, Mexico: Stanford Univ. Pubs. Geol. Sci., v. 10, p. 1–25.

McKee, E. H., and Silberman, M. L., 1970a, Geochronology of Tertiary igneous rocks in central Nevada: Geol. Soc. America Bull., v. 81, p. 2317–2328.

——— 1970b, Periods of plutonism in north-central Nevada: Geol. Soc. America, Abs. with Programs (Ann. Mtg.), v. 2, no. 7, p. 613–614.

Mencher, Ely, 1963, Tectonic history of Venezuela: Am. Assoc. Petroleum Geologists Mem. 2, p. 73–87.

Minato, M., Gorai, M., and Hunahashi, M., 1965, The geologic development of the Japanese Islands: Tokyo, Tsukiji Shokan Co., 442 p.

Mitchell, A. H. G., 1970, Facies of an early Miocene volcanic arc, Malekula Island, New Hebrides: Sedimentology, v. 14, p. 201–243.

Mitchell, A. H., and Reading, H. G., 1969, Continental margins, geosynclines, and ocean floor spreading: Jour. Geology, v. 77, p. 629–646.

Muñoz-Cristi, J. M., 1956, Chile, in Jenks, W. F., ed., Handbook of South American geology: Geol Soc. America Mem. 65, p. 189–214.

Murray, G. E., 1961, Geology of the Atlantic and Gulf Coastal province of North America: New York, Harper & Row, Publishers, 692 p.

Nalivkin, D. V., 1960, The geology of USSR: New York, Pergamon Press, 170 p.

Owens, J. P., 1970, Post-Triassic tectonic movements in the central and southern Appalachians as recorded by sediments of the Atlantic Coastal Plain, in Fisher, G. W., Pettijohn, F. J., Reed, J. C., Jr., and Weaver, K. N., eds., Studies of Appalachian geology: Central and Southern: New York, Interscience Press, p. 417–427.

Page, B. M., 1970, Time of completion of underthrusting of Franciscan beneath Great Valley rocks west of Salinian block, California: Geol. Soc. America Bull., v. 81, p. 2825–2834.

Pushcharovsky, Y. M., 1967, The Pacific tectonic belt of the earth's crust: Tectonophysics, v. 4, p. 571–580.

Raju, A. T. R., 1968, Geological evolution of Assam and Cambay Tertiary basins of India: Am. Assoc. Petroleum Geologists Bull., v. 52, p. 2422–2437.

Renz, H. H., 1957, Stratigraphy and geological history of eastern Venezuela: Geol. Rundshau, v. 45, p. 728–759.

Rodolfo, K. S., 1969, Bathymetry and marine geology of the Andaman Basin, and tectonic implications for southeast Asia: Geol. Soc. America Bull., v. 80, p. 1203–1230.

Rusnak, G. A., and Fisher, R. L., 1964, Structural history and evolution of Gulf of California: Am. Assoc. Petroleum Geologists Mem. 3, p. 144–156.

Said, R., 1962, The geology of Egypt: Amsterdam, Elsevier Publ. Co., 377 p.

Senn, A., 1940, Paleogene of Barbados and its bearing on history and structure of Antillean-Caribbean region: Am. Assoc. Petroleum Geologists Bull., v. 24, p. 1548–1610.

Short, K. C., and Stäuble, A. J., 1967, Outline of geology of Niger Delta: Am. Assoc. Petroleum Geologists Bull., v. 51, p. 761–779.

Sittler, C., 1969, The sedimentary trough of the Rhine Graben: Tectonophysics, v. 8, p. 543–560.

Souther, J. G., 1970, Volcanism and its relationship to Recent crustal movements in the Canadian Cordillera: Canadian Jour. Earth Sci., v. 7, p. 553–568.

Stainforth, R. M., 1969, The concept of seafloor spreading applied to Venezuela: Asoc. Venezolana Geología, Minería y Petróleo., v. 12, p. 257–274.

Stöcklin, J., 1968, Structural history and tectonics of Iran: A review: Am. Assoc. Petroleum Geologists Bull., v. 52, p. 1229–1258.

Stoneley, R., 1969, Sedimentary thicknesses in orogenic belts, in Kent, P. E., Satterthwaite, G. E., and Spencer, A. M., eds., Time and place in orogeny: Geol. Soc. London, p. 215–238.

Swe, W., and Dickinson, W. R., 1970, Sedimentation and thrusting of late Mesozoic rocks in the Coast Ranges near Clear Lake, California: Geol. Soc. America Bull., v. 81, p. 165–188.

Thompson, J. E., 1967, A geological history of eastern New Guinea: Australian Petroleum Explor. Assoc. Jour., p. 83–93.

Valentine, J. W., and Moores, E. M., 1970, Plate-tectonic regulation of faunal diversity and sea level; A model: Nature, v. 228, p. 657–659.

Van Houten, F. B., 1969, Molasse facies: Records of worldwide crustal stresses: Science, v. 166, p. 1506–1508.

Wade, F. A., 1970, The geology of West Antarctica between longitudes 70° W and 158° W and north of the Byrd subglacial basin: Geol. Soc. America, Abs. with Programs (Ann. Mtg.), v. 2, no. 7, p. 712.

Wadia, D. N., 1953, Geology of India: London, MacMillan Co., 531 p.

Wageman, J. M., Hilde, T.W.C., and Emery, K. O., 1970, Structural framework of East China Sea and Yellow Sea: Am. Assoc. Petroleum Geologists Bull., v. 54, p. 1611–1643.

Wezel, F. C., 1970, Numidian flysch: An Oligocene–early Miocene continental rise deposit off the African platform: Nature, v. 228, p. 275–276.

Williams, H., and McBirney, A. R., 1969, Volcanic history of Honduras: California Univ. Pubs. Geol. Sci., v. 85, p. 1–101.

Yeats, R. S., 1968, Southern California structure, sea-floor spreading and history of the Pacific Basin: Geol. Soc. America Bull., v. 79, p. 1693–1702.

Zambrano, J. J., and Urien, C. M., 1970, Geological outline of the basins in southern Argentina and their continuation off the Atlantic shore: Jour. Geophys. Research, v. 75, p. 1363–1396.

MANUSCRIPT RECEIVED BY THE SOCIETY MARCH 29, 1971   PRINTED IN U.S.A.

THE GEOLOGICAL SOCIETY OF AMERICA, INC.
MEMOIR 132
© 1972

Sedimentary Evolution of Rifted Continental Margins

ERIC D. SCHNEIDER

*Office of Research and Monitoring, Environmental Protection Agency,
Washington, D.C. 20460*

ABSTRACT

Geologic studies of present ocean basins and continental rift systems suggest a four-stage sedimentary model for evolution of rifted continental margins. Each stage produces a diagnostic sedimentary record, and examples of each can be seen today. (1) Uplift, volcanism, rifting, and nonmarine sedimentation patterns characterize the rift valley stage. (2) In the Red Sea stage and later, restricted size and circulation of the proto-ocean create a reducing environment; chemical precipitates (halides, gypsum, and metallic sulfides) and sapropelitic muds are deposited. (3) Gravity-induced processes dominate the sedimentary regime during the turbidite-fill stage, and flat-lying beds of coarse-grained, highly reflective sediments are formed. (4) Strong thermohaline contrasts in the mature ocean basin indicate geostrophic deep-water circulation. This deep-ocean current stage results in current-controlled sediment deposition and redistribution.

INTRODUCTION

Since introduction of the theory of ocean basins widening by growth of new crust at the mid-ocean ridges (Heezen, 1959; Hess, 1962; Dietz, 1961), exciting new findings have been presented describing: (1) rates of sea-floor spreading using sea-floor magnetic data (Vine and Mathews, 1963; Heirtzler and others, 1968); (2) ages of prominent seismic reflectors within the ocean floor (Ewing and others, 1966); and (3) deep-ocean sedimentary processes (Heezen and others, 1966; Schneider and others, 1967). With this background, one can attempt to reconstruct the history of sedimentation in a widening and deepening ocean environment.

109

Heezen (1960) suggested a tectonic history of such growing ocean basins. He envisioned an ocean initiating as a rift valley, widening with accretion of new crust along a ridge to a stage similar to the present-day Red Sea, then spreading to the width of a major ocean basin such as the Atlantic. Such tectonically controlled changes in the size and form of a growing ocean must create recognizably different sedimentary environments. I propose four sedimentary stages of development for a rifting ocean basin: (1) the rift valley stage, (2) the Red Sea stage, (3) the turbidite-full stage, and (4) the deep-ocean current stage (Fig. 1). The unique and diagnostic sedimentary rocks formed during the early stages in a rift valley or in a narrow proto-ocean may prove to be economically valu-

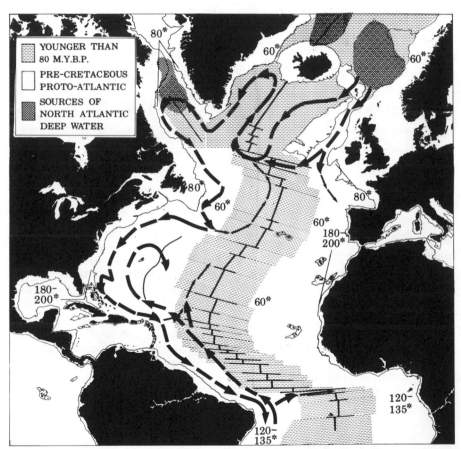

* = MILLION YEARS BEFORE PRESENT

Figure 1. Diagrammatic sedimentary evolution of a rifted continental margin: (A) *dormant rift valley stage* (for example, eastern North American Triassic basins) on flanks of *Red Sea stage*. Basin edges normal faulted, fringed with coral reefs. Basin floor covered by sapropelitic muds, slumped shallow-water biogenics, terrigenous turbidites, evaporites, metallic sulfides. (B) *Turbidite-fill stage*. Margins subside, fringing reefs survive, continental slope sediments prograde. Sapropels, pelagics, and gravity-induced turbidites mixed in flat, stratified layers filling basins. (C) *Deep-ocean current stage*. Deep geostrophic currents build continental rise, sediment ridges, and drifts above flat turbidite horizons. Currents redistribute pelagic and terrigenous deposits throughout basins. Ages in million years before present (M.Y.B.P.) refer to evolution of the North Atlantic Basin.

able. These deposits should be found buried along the margins of present-day major ocean basins, and perhaps in even older sequences where tectonic closing of ancient basins has caused their subsequent uplift.

Although the driving mechanisms for rifting and sea-floor spreading are still debated, evidence that oceans form through these processes is reflected in the sediments. I will describe present-day counterparts of each of the evolutionary stages, and present evidence that remnants of each stage can be found along the margins of a mature ocean basin.

RIFT VALLEY STAGE

Studies of East African rift valleys and of tensional grabens, such as the Rhine graben or the Triassic basins along the eastern coast of the United States, should give a clear insight into the early histories of sedimentation and tectonics in a proto-ocean.

Rifting initiates when upper mantle derivatives intrude into the crust, thermally expanding the surrounding country rock and forming a broad regional uplift. Tensional graben faulting results, which forms the initial rift valley. In the broad zone of thermal expansion and uplift, several near-parallel rift valley systems may form. This can be seen in East Africa today, where two rift systems are found, either of which might initiate the formation of a proto-ocean between East Africa and the rest of the African continent. The Atlantic apparently opened along one of at least two competing rift systems formed during the initial Triassic tension and rifting between the North American and Euro-African crustal blocks. The present-day Mid-Atlantic Ridge was one, and the system of North American Triassic fault basins found from Labrador to Alabama was another. Still other early rifts may lie buried beneath sediments of the continental margins.

As seen in East Africa, the rift valley is floored by a unique combination of basic and alkaline intrusive and extrusive rocks (Holmes, 1964). The rift valley walls become divides for local drainage; regional drainage is away from the broadly uplifted areas, and an independent internal drainage system forms within the rift valley system. Streams erode material of the original country rock from the rift valley walls, and fresh water and evaporating salt water lakes may form. In the North American Triassic basins, the volcanic rocks are found with granitic conglomerates whose presumed source is the local country rock which was rifted (Sanders, 1963); in the East African Rift Valleys, erosion of unusually sodium-rich volcanic rocks forms lakeshores rich in sodium carbonates.

The rift valley stage, then, forms a sedimentary mix of normal terrigenous material, basic and alkaline volcanic rocks, granitic conglomerates, river and lake deposits, and nonmarine evaporates.

RED SEA STAGE

After the initial rifting and graben faulting, basaltic mantle material extrudes to form a thin (eventually 4 to 5 km thick) "oceanic" crust, flooring the spreading rift basin. Heat dissipation is more rapid through this thin new crust than through the thick (20 to 30 km) continental crust on either side of the rift, so thermal expansion along the rift wanes, and the basin floor subsides more rapidly than its continental walls (Sleep, 1972). This differential subsidence preserves the original rift valley walls, which become the continental slopes seen along the margins of today's rifted ocean basins. The Red Sea stage exists when spreading through addition of new crust to the rift valley floor and subsidence due to heat dissipation make the rift wide and deep enough for the incursion of local oceans and seas.

The type of biogenic sedimentation in the proto-ocean basin will be determined by the gross climatic environment of the region. The present margins of the Red Sea support growth of shallow-water reefs. This reef growth might survive slow subsidence of the continental margins, and eventually create a sequence similar to material dredged from the Blake Escarpment. There Heezen and Sheridan (1966) found shallow-water corals, ranging in age from Lower Cretaceous through Miocene, at depths of 4,800 m. Analyses of well-hole data from sites along the entire eastern coast of the United States and in the Bahamas (Maher, 1965) have shown that the rifted continental margin displays a history of continuous, though decelerating, subsidence from the initial rift throughout the spreading of the basin (Vogt and Ostenso, 1967; Schneider and Vogt, 1968; Menard, 1969). Sleep (1972) has presented mathematical models for this subsidence. Since the continental margins sink at rates of 1 to 4 cm per 1,000 yrs (Vogt and Ostenso, 1967), such reef growth could easily survive the regional subsidence which accompanies sea-floor spreading.

Slumping also contributes shallow-water biogenic debris to bottom sediments in the Red Sea stage. In the present-day Red Sea, turbidites consisting of shallow-water calcareous material are found in cores taken from depths greater than 2,000 m (Schneider and Johnson, 1970). Cores raised from 5,100 m of water in the Horizon A outcrop north of San Salvador Island (Saito and others, 1966; Windisch and others, 1968) contain Lower Cretaceous through Miocene shallow-water material which slumped down from the surrounding Bahama Banks. This slumped material may have been deposited in the proto-Atlantic, during its narrow Red Sea stage.

The proto-ocean, during the Red Sea stage, is too restricted in size and depth to permit vigorous deep-ocean circulation. Such circulation, as is seen in a major ocean basin such as the present-day Atlantic, results from contrasts in surface-water salinities and temperatures which set up density gradients; cold saline waters sink and are replaced by warmer less saline waters. Without this sort of overturn, oxygenated surface waters do not replace stagnated bottom waters, and reducing conditions predominate at depth in the basin. Sediment cores raised from the present Red Sea contain sapropelitic, organic-rich muds indicative of restricted water circulation. Euxinic clays and sapropelitic muds should be deposited in any proto-ocean during its Red Sea stage. Piston cores collected by the Lamont-Doherty Geological Observatory, and cores drilled on Legs 1 and 11 of the JOIDES Deep-Sea Drilling project (JOIDES, 1968, 1970) in the western North Atlantic, contain middle and lowermost Cretaceous sediments rich in unoxidized organic clays. JOIDES Core 105 (from Leg 11, 1970) includes a sequence of Lower Cretaceous clays so rich in carbonaceous material that some samples will actually burn (JOIDES, 1970). The present-day Red Sea, then, appears to be not only a tectonic analog to the proto-Atlantic (Heezen, 1960), but a sedimentary analog as well.

Studies of the Red Sea "hot holes" suggest another sedimentary facies which might be indicative of deposition in a proto-ocean basin (Schneider, 1969b). Sediments cored from these brine basins include an unusual combination of dolomite, gypsum, halides, and metallic sulfides rich in iron, zinc, copper, lead, and gold (Degens and Ross, 1969). Theoretical calculations by Schmalz (1969) indicate the chemical feasibility of halite precipitation in a deep-water environment with restricted circulation. Features seen on seismic reflection records from the eastern North Atlantic appear to be diapiric in form. Several recent papers (Schneider, 1969b; Rona, 1969; Schneider and Johnson, 1970) suggest that these features are salt diapirs, and recent JOIDES drillings in the Gulf of Mexico and in the Mediterranean Sea support this hypothesis.

The Red Sea stage is characterized by deposition of diverse sedimentary materials: slumps and turbidites of shallow-water biogenic debris; turbidites of terrigenous material eroded from the subsiding basin margins; and sapropelitic muds associated with halide, dolomite, and metallic sulfide deposits.

TURBIDITE-FILL STAGE

As the growing ocean continues to widen, the thermally expanded lithosphere near the ridge axis preserves the height and topographic profile of the mid-ocean ridge. The oldest sea floor, near the continental margins, continues to subside, forming ever wider and deeper basins between the ridge crest and the two adjacent continents. Gravity-induced sedimentary processes fill the basins in this turbidite-fill stage. The ocean, at this stage, still lacks vigorous thermohaline circulation, and sapropelitic muds are intercalated with turbidite layers.

Early seismic profiling studies made by the Lamont-Doherty Geological Observatory identified a prominent series of reflectors, which was named Horizon A, occurring throughout the North Atlantic Basin (Ewing and Ewing, 1962). The high seismic reflectivity (generally indicative of coarse-grained sediments), and the highly stratified, flat-lying occurrence of these beds suggested to the Ewings that Horizon A is an ancient buried abyssal plain. JOIDES drilling and coring by the Lamont-Doherty Geological Observatory showed that in much of the western Atlantic, Horizon A contains turbidite material of Cretaceous to Eocene age (Saito and others, 1966; JOIDES, 1970). Sediments cored from Horizon A near San Salvador Island consist of shallow-water reef material slumped down from the adjoining Bahama Banks (Saito and others, 1966). Based on more recent analysis of JOIDES results, however, Ewing and others (1970) are now suggesting that the high reflectivity of Horizon A is actually caused by flat-lying chert layers, which may be lithified beds of siliceous ooze, rather than turbidites. Whether Horizon A itself is lithified pelagic ooze or turbidite fill forming a fossil abyssal plain, or both, the fact remains that sedimentation before and at the time of its formation created highly stratified, continuous, flat-lying deposits. Since seismic studies show that basement rock in the North Atlantic Basin is topographically irregular, pelagic draping alone could not produce flat sedimentary layers; gravity-induced sedimentation must have been predominant during this stage of ocean basin evolution.

Seismic reflection records from the westernmost North Atlantic also show that Horizon A extends as a flat sedimentary horizon beneath the continental rise and underneath a large sediment drift known as the Blake-Bahama Outer Ridge

(Schneider, 1970). This suggests that major reshaping of the continental margin has taken place since the turbidite-fill stage, when Horizon A turbidites were deposited. The present-day western Mediterranean, floored by a flat-lying abyssal plain, exemplifies a small ocean basin (like the proto-Atlantic) where gravity-induced sedimentation predominates and the continental margins are unmodified by major sedimentary drifting. In the Balearic Basin, continental slopes are steep and, except for the Rhone Delta and Fan, there is no noticeable formation of a continental rise.

The turbidite-fill stage of ocean development is recognized by accumulation of open-ocean pelagic material, sapropelitic muds, and flat-lying, highly stratified, terrigenous sediments filling the basins. The steep-walled continental margins continue to subside and build out by simple sedimentary progradation of the continental slopes.

DEEP-OCEAN CURRENT STAGE

In a mature ocean basin, strong thermohaline contrasts set up a geostrophic pattern of water circulation, and deep-ocean currents control sediment deposition and redistribution, especially along the continental margins.

Deep-water depositional processes in the present-day North Atlantic are influenced by the flow of two major water masses (Fig. 2): (1) the North Atlantic Deep Water, which forms in the Labrador, Norwegian, and Irminger Seas and follows a sinuous path down the flanks of the Mid-Atlantic Ridge, through fracture zones, and southward along the eastern continental margin of the United States; and (2) the Antarctic Bottom Water, which travels northward from near Antarctica, through the western South Atlantic, and into the western North Atlantic Basin (Heezen and others, 1966; Fox and others, 1968). Analysis of magnetic data (Avery and others, 1968; Vogt and others, 1970) has shown that the Atlantic Basin north of 53° did not open until 80 to 60 m.y. B.P. Thus, the northern seas which supply cold saline water for North Atlantic Deep Water circulation were nonexistant before that time. Seismic reflection records from the westernmost North Atlantic show that the continental rise and the Blake-Bahama Outer Ridge are both sedimentary features built over the flat-lying Cretaceous to Eocene reflectors in Horizon A (Schneider and others, 1967; Schneider, 1969a; Johnson and Schneider, 1969; Jones and others, 1970). The continental rise and this outer ridge were built by deep-ocean currents (Heezen and others, 1966) when rifting into Arctic latitudes allowed cold, saline waters to flow into the Atlantic Basin, thereby initiating vigorous, deep thermohaline circulation.

In this final stage of evolution, the ocean basin contains several different sedimentary environments. While deep-ocean contour currents control sedimentation along the basin margins, slumps and turbidity currents still carry terrigenous material to abyssal plains far from the continents (Schneider, 1970). On the flanks of the mid-ocean ridge, pelagic sediments are draped over the bottom topography. Much of this material, however, is scoured and redistributed by bottom currents, whose effects are controlled by water circulation patterns and by local and regional topography (Ruddiman and Schneider, 1971). Volcanic lavas are mixed with pelagic oozes in deposits along the ridge axis, far from the influx of terrigenous material from the distant rifted continental margins.

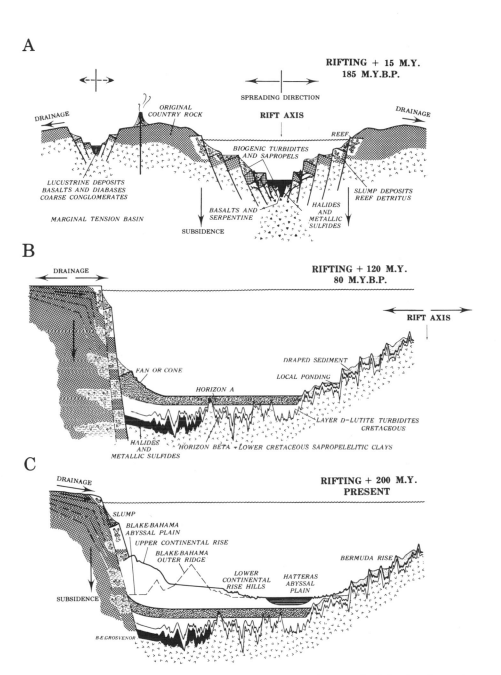

Fig. 2. Thermohaline circulation pattern in present-day North Atlantic. Antarctic Bottom Water flows from South Atlantic into western North Atlantic. North Atlantic Deep Water forms in Norwegian, Irminger, and Labrador Seas, and flows south along western margin of basin. Note sources of North Atlantic Deep Water were nonexistant before 80 m.y. ago.

DISCUSSION

A simple sedimentary model for evolution of a growing ocean basin is proposed above. Several additional variables, however, will undoubtedly complicate paleo-sedimentation patterns: (1) orientation of the basin, (2) climatic environment of the basin, (3) elevation of the mid-ocean ridge, isolating separate basins on either side, and (4) interaction of the growing basin with other oceans and seas.

The North Atlantic Basin spreads from a nearly north-south trending central rift. As the basin grew, the rift extended north and south across latitudinal climatic zones until thermohaline contrasts were great enough to initiate deep-ocean current circulation. If a basin were to spread from a mid-ocean ridge oriented in an east-west trend, the rift would extend longitudinally within similar climatic zones. Initiation of thermohaline circulation and its distinctive patterns of sedimentation would then be unlikely.

The proto-Atlantic appears to have evolved in an equatorial climatic belt (Irving, 1964; Schneider and Johnson, 1970), but changing polar positions (polar wandering) since Jurassic and Cretaceous times (de Boer, 1968) have resulted in the present latitudinal orientation of the basin. Such changing polar positions result in comparable changes in orientation of the global climatic belts, thus affecting thermohaline contrasts and locations of productivity belts in an ocean basin. The strong deep-sea circulation presently observed in the North Atlantic may also be unique to this present-day basin because of another climatic factor. The repeated glaciations of the late Pliocene and Pleistocene may represent a deterioration of usual climatic conditions, with the glacial cold periods producing unusually great thermohaline contrasts.

Rates of mid-ocean ridge subsidence depend primarily on rates of sea-floor spreading away from the ridge (P. R. Vogt, personal commun.). Elevation of the ridge would effect the relative isolation of basins on either side of it, which in turn would affect bottom current circulation and sedimentation of carbonates which are dependent on compensation depths and on water masses. Productivity belts, circulation patterns, and sedimentary distribution in a spreading basin could also be influenced by interactions with other seas. The North Atlantic Basin, for example, has undoubtedly been affected by past opening and closing of the Isthmus of Panama, and widening and closing of the Tethys Seas.

Influence of these four factors may have combined to create unique sedimentary facies in the present-day North Atlantic.

In order to present the over-all evolution of the Atlantic, I have deliberately avoided discussing all of the known details. Work now in progress will clarify some of the asymmetries and variations in this generalized model.

ACKNOWLEDGMENTS

I owe thanks to L. K. Glover for clarity of writing and editing, and for encouragement; to Dr. P. R. Vogt for critical comments which improved the content of this manuscript; and to B. E. Grosvenor for imaginative illustration.

REFERENCES

Avery, O. E., Burton, G. D., and Heirtzler, J. R., 1968, An aeromagnetic survey of the Norwegian Sea: Jour. Geophys. Research, v. 73, no. 14, p. 4583–4600.

de Boer, J., 1968, Paleomagnetic differentiation and correlation of the Late Triassic volcanic rocks in the central Appalachians (with special reference to the Connecticut Valley): Geol. Soc. America Bull., v. 79, p. 609.

Degens, E. T., and Ross, D. A., 1969, Hot brines and recent heavy metal deposits in the Red Sea: New York, Springer-Verlag, 600 p.

Dietz, R. S., 1961, Continent and ocean basin evolution by spreading of the sea floor: Nature, v. 190, p. 854–857.

Ewing, J., and Ewing M., 1962, Reflection profiling in and around the Puerto Rico Trench: Jour. Geophys. Research, v. 67, p. 4729–4739.

Ewing, J., Worzel, J. L., Ewing, M., and Windisch, C., 1966, Ages of Horizon A and bore-hole results: Jour. Geophys. Research, v. 75, no. 29, p. 5645–5653.

Ewing, J., Windisch, C., and Ewing, M., 1970, Correlation of Horizon A with JOIDES the oldest Atlantic sediments: Science, v. 154, p. 1125–1132.

Fox, P. J., Heezen, B. C., and Harian, A. M., 1968, Abyssal anti-dunes: Nature, v. 220, no. 5166, p. 470–472.

Heezen, B. C., 1959, Dynamic processes of abyssal sedimentation: Erosion transportation and redeposition on the deep-sea floor: Royal Astron. Soc. Geophys. Jour., v. 2, p. 142–163.

—— 1960, The rift in the ocean floor: Sci. American, v. 203, p. 98–110.

Heezen, B. C., and Sheridan, R. E., 1966, Lower Cretaceous rocks (Neocomian/Albian) dredged from the Blake Escarpment: Science, v. 154, p. 1614–1617.

Heezen, B. C., Hollister, C. D., and Ruddiman, W. F., 1966, Shaping of the continental rise by deep geostrophic contour currents: Science, v. 152, p. 502–508.

Heezen, B. C., Schneider, E. D., and Pilkey, O. H., 1966, Sediment transport by the Antarctic Bottom Current on the Bermuda Rise: Nature, v. 211, no. 5049, p. 611–612.

Heirtzler, J. R., Dickson, G. O., Herron, E. M., Pitman, W., and Le Pichon, X., 1968, Marine magnetic anomalies, geomagnetic field reversals, and motions of the ocean floor and continents: Jour. Geophys. Research, v. 73, p. 2119–2136.

Hess, H. H., 1962, History of the ocean basins, in Engel, A.E.J., and others, eds., Petrologic Studies: A volume in honor of A. F. Buddington: Geol. Soc. America, p. 599–620.

Holmes, A. D., 1964, Principles of physical geology: New York, The Ronald Press Co., 1288 p.

Irving, E. S., 1964, Paleomagnetism, New York, John Wiley & Sons, 399 p.

Johnson, G. L., and Schneider, E. D., 1969, Depositional ridges in the North Atlantic: Earth and Planetary Sci. Letters, v. 6, p. 416–422.

JOIDES (Joint Oceanographic Institutions for Deep Earth Sampling), 1968, Initial reports of Deep-Sea Drilling Project: Washington, D.C., U.S. Government Printing Office, v. I, 672 p.

—— 1970, Deep-Sea Drilling Project: Leg 11: Geotimes, p. 14–16.

Jones, E.J.W., Ewing, M., Ewing, J. I., and Eittreim, S. L., 1970, Influences of Norwegian Sea overflow water on sedimentation in the northern North Atlantic and Labrador Sea: Jour. Geophys. Research, v. 75, no. 9, p. 1655–1680.

Maher, J. C., 1965, Correlations of subsurface Mesozoic and Cenozoic rocks along the Atlantic coast: Am. Assoc. Petroleum Geologists Bull., 18 p.

Menard, H. W., 1969, Elevation and subsidence of oceanic crust: Earth and Planetary Sci. Letters, v. 6, p. 275–284.

Rona, P. A., 1969, Possible salt domes in the deep Atlantic off northwest Africa: Nature, v. 224, no. 5215, p. 141–143.

Ruddiman, W. F., and Schneider, E. D., 1971, Sediment cover on the lower Reykjanes Ridge [abs.]: Am. Geophys. Union Mtg.

Saito, T., Burkle, L. H., and Ewing, M., 1966, Lithography and paleontology of the reflective layer Horizon A: Science, v. 154, no. 3753, p. 1173–1176.

Sanders, J. E., 1963, Late Triassic tectonic history of the northeastern United States: Am. Jour. Sci., v. 261, p. 501–524.

Schmalz, R. F., 1969, Deep-water evaporite deposition: A genetic model: Am. Assoc. Petroleum Geologists Bull., v. 53, no. 4, p. 798–823.

Schneider, E. D., 1969a, Evolution of the continental margins [abs.]: Am. Assoc. Petroleum Geologists Meetings.

—— 1969b, The deep sea—a habitat for petroleum? Undersea Technology, Reprint R–254, 5 p.

—— 1970, Downslope and across-slope sedimentation as observed in the westernmost North Atlantic [Ph.D. thesis]: New York, Columbia Univ., 301 p.

Schneider, E. D., and Johnson, G. L., 1970, Deep-ocean diapiric occurrences: Am. Assoc. Petroleum Geologists Bull., v. 54, no. 2151, p. 2169.

Schneider, E. D., and Vogt, P. R., 1968, Discontinuities in the history of sea-floor spreading: Nature, v. 217, p. 1212–1222.

Schneider, E. D., Fox, P. J., Hollister, C. D., Needham, D., and Heezen, B. C., 1967, Further evidence for contour currents on the United States continental margin: Earth and Planetary Sci. Letters, v. 2, p. 351–359.

Sleep, N. H., 1972, Thermal effects of the formation of Atlantic continental margins by continental breakup: Am. Geophys. Union Trans. (in press).

Vine, F. J., and Matthews, D. H., 1963, Magnetic anomalies over oceanic ridges: Nature, v. 199, p. 947–949.

Vogt, P. R., and Ostenso, N. A., 1967, Steady-state crustal spreading: Nature, v. 215, p. 810–817.

Vogt, P. R., Anderson, C. N., Bracey, D. R., and Schneider, E. D., 1970, North Atlantic magnetic smooth zones and sea-floor spreading in low latitudes: Jour. Geophys. Research, v. 75, p. 3955–3968.

Windisch, C., Layden, R. L., Worzel, J. L., Saito, T., and Ewing, J., 1968, Investigation of Horizon Beta: Science, v. 162, p. 1473–1479.

Manuscript Received by the Society March 29, 1971

THE GEOLOGICAL SOCIETY OF AMERICA, INC.
MEMOIR 132
© 1972

Uncoupled Convection and Subcrustal Current Ripples in the Western Mediterranean

WALTER ALVAREZ

Accademia Britannica, Via Gramsci 61, 00197 Roma, Italy

ABSTRACT

The basic tectonic features of the western Mediterranean are explained by a model invoking the activity of a toroidal convection cell in the upper mantle, uncoupled from the base of the lithosphere (a situation different from the usual plate tectonic conditions). The key feature of this model is that the current is visualized as eroding the base of the lithosphere in a manner analogous to the erosion of sand by running water or wind, producing a pattern of inverted subcrustal ripples concentric about the rising mantle column. Isostatic subsidence of zones thinned in this manner would have given rise to the geosynclinal furrows which were present throughout most of the Mesozoic and the Tertiary. Continued removal of lithosphere beneath such a furrow would eventually bring it to the point where it would no longer be capable of resisting compressive stresses applied by the thick zone behind it, which would be pressing outward as a result of the drag of the mantle current. At this point the furrow would be crushed between the two adjacent thick zones, and its sedimentary fill would be bulldozed out and slide away as great gravity nappes. Subcrustal erosion and the episodic outward movements as furrows collapsed would lead to the removal of continental crust from above the rising mantle column. When convection ceased, new oceanic crust in the central area would subside to its normal isostatic level, while the surrounding regions would rise in compensation, producing the pattern seen today.

INTRODUCTION

A revolutionary decade in the earth sciences began when Harry Hess (1962) introduced the theory of sea-floor spreading. Together with its corollaries of plate tectonics (Morgan, 1968) and the ocean floor "magnetic tape recorder" (Vine,

1966), this theory has provided the insights through which many of the long-standing problems of tectonics have been at least partly solved. Among the more interesting features which are not clearly explained by the "new global tectonics" are small oceanic basins, which in some instances are partly surrounded by mountain systems. A case in point is the western part of the Mediterranean Sea.

To review briefly, the western Mediterranean is an area of oceanic crust about 700 km across from north to south and 1,500 km from east to west (Fig. 1). A spur of continental crust, which includes Corsica and Sardinia, divides the area into two basins. The Balearic Basin in the west reaches depths of over 2,500 m, while the smaller Tyrrhenian Basin in the east exceeds 3,000 m in depth. There is convincing paleomagnetic, morphological, and stratigraphic evidence that the Corsica-Sardinia block once fit against the coast of southern France, and has rotated counterclockwise to its present position, probably during the Oligo-Miocene (van der Voo and Zijderveld, 1969, p. 128–129, Fig. 9; Pannekoek, 1969, p. 55–57, Fig. 1; Alvarez, 1972). An arc of mountains forming part of the Alpine system nearly encircles the western Mediterranean, beginning at Genoa and passing through Italy, Sicily, Tunisia, Algeria, Morocco, southern Spain, and the Balearic Islands, where it is abruptly cut off at the continental margin. The presence in northeastern Corsica of a small fragment of this mountain system suggests that it once completely enclosed the western Mediterranean area.[1]

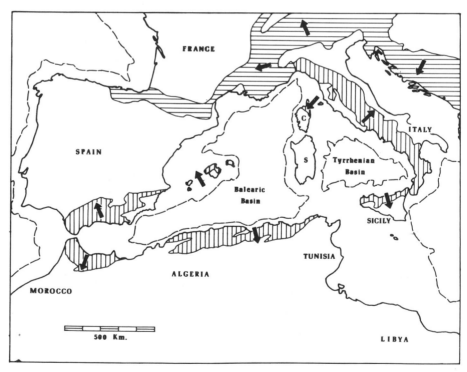

Figure 1. Tectonic sketch of the western Mediterranean and surrounding regions. Vertical ruling: exposed parts of the Alpine tectogenetic province encircling the western Mediterranean. Horizontal ruling: other Alpine mountain belts. Dashed line: 2,000-m depth contour. C: Corsica. S: Sardinia. Arrows show direction of polarity, from interior to exterior.

The history of this mountain chain goes back to the early Mesozoic, when a system of geosynclinal ridges and furrows gradually formed across a continental mass that had been consolidated during the late Paleozoic Hercynian orogeny. These zones of differential subsidence persisted throughout the remainder of the Mesozoic and into the Tertiary, although with occasional changes in the behavior of individual zones.

All around the western Mediterranean this geosynclinal system showed a distinct polarity. The terms "internal" and "external" are used to denote this polarity, the former referring to the side presently facing the Mediterranean, and the latter indicating the side nearest the adjoining continent. The internal zones have been more active tectonically and, during deformation of the geosyncline, have been thrust toward and over the external zones. Thus, in a less specific way, the terms carry the same connotation as "eugeosynclinal" and "miogeosynclinal" (Fig. 2). This geosynclinal system was affected by sporadic tectogenic movements from the Cretaceous until the late Miocene, with large gravity nappes moving from the internal toward the external part of the chain. Three lines of evidence indicate that until the late Miocene there was land—and, by inference, continental crust— where the deep oceanic basins of the western Mediterranean are now located: (1) there are Tertiary marine sedimentary units in southeastern France, Sicily, North Africa, and Spain, which were derived from source areas now forming part of the deep basins; (2) Miocene rivers in southeastern France drained northward, away from the present coast (reflecting the former position of Corsica against this coast); (3) all around the western Mediterranean, gravity-sliding olisto-stromes and nappes moved away from what is now a deep oceanic basin (Pannekoek, 1969, p. 54–60; de Booy, 1969). This basin is apparently the result

---

[1] For an introduction to the geology of the mountain system surrounding the western Mediterranean, the reader is referred to the following works: Italy, Manfredini, 1963; Sestini, 1970; Calabria, Ogniben, 1969; Sicily, Alvarez and Gohrbandt, 1970; Tunisia, Martin, 1967; Algeria, Durand Delga, 1967.

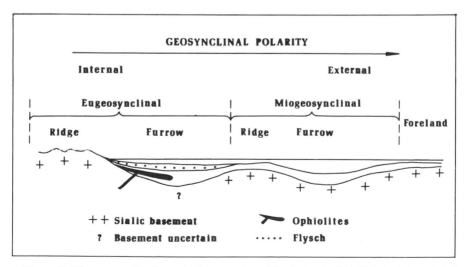

Figure 2. The geosynclinal terminology employed by Aubouin (1965, Fig. 16, modified slightly).

of collapse during latest Miocene, Pliocene, and Quaternary time, accompanied by uplift of former submerged tracts to form the land now rimming the basin. This striking reversal of land and sea has been called the "Mediterranean Revolution" (Bourcart, 1962; Pannekoek, 1969).

A classification of the earth's major tectonic features, based on insights provided by the new global tectonics, has been emerging in a series of recent papers (Dewey, 1969; Mitchell and Reading, 1969; Dewey and Bird, 1970; Dewey and Horsfield, 1970). Further refinements are to be expected, but at this point the classification can be summarized as follows:

    I. *Ocean basins* are in one of three phases of development (Dewey, 1969, p. 191):
        a. *Atlantic phase:* spreading—expanding;
        b. *Pacific phase:* spreading—contracting; or
        c. *Eastern Mediterranean phase:* nonspreading—contracting.
    II. *Continental margins* are of two types:
        a. *Atlantic type:* aseismic, nonvolcanic margins with no relative motion between the continent and the adjacent ocean floor; and
        b. *Pacific type:* consumption of oceanic lithosphere at or near the continental margin with a resulting orogenic belt of cordilleran or island arc type.
    III. *Orogenic belts* are of at least four kinds (Dewey and Horsfield, 1970):
        a. *Cordilleran type:* oceanic lithosphere descends near the foot of a continental rise;
        b. *Island arc type:* lithosphere descent occurs at some distance from the continental margin;
        c. *Himalayan type:* resulting from the collision of two continents; and
        d. *New Guinea type:* resulting from the collision of an island arc with a continent.

Geosynclinal systems are not subdivided in these papers. Thick, elongated prisms of sediment are presently accumulating on continental margins of Atlantic type; ancient features of this kind are invoked as the geosynclines ancestral to orogenic belts. Continental shelf sediments form the miogeosyncline or "miogeocline" (Dietz and Holden, 1966), while the eugeosyncline is composed of continental rise sediments. This picture, although satisfactory for many orogenic belts, omits the critical difference noted by Aubouin (1965, p. 17):

    IV. *Geosynclinal systems:*
        a. *American type:* "The geosyncline was marginal to the continent, and contained a very thick series of shallow-water sediments: i.e., it was not a trough." This type of geosynclinal system is well explained by the continental shelf and rise sediment prism described above;
        b. *European (Alpine-Mediterranean) type:* "The geosyncline was situated between two continental masses and contained a thick series of deep-water sediments: i.e., a trough."

The western Mediterranean is conspicuous as a region that does not fit into a tectonic classification based on the new global tectonics. It has a small ocean basin formed by late Tertiary subsidence, an encircling orogenic continental margin not apparently related either to lithosphere consumption or to collision, and its ancestral geosynclinal system was of the European type. Nevertheless, one can find attempts to explain the western Mediterranean through the usual effects of plate tectonics: fragmentation of an original continent (Smith, 1971), descent of the African lithosphere beneath that of Europe (Caputo and others, 1970), alternating compression and tension (Sander, 1970, p. 122), and rotation associated with the movement of lithosphere plates (Boccaletti and Guazzone, 1970). Various possibilities have been summarized by van Bemmelen (1969). The current lack of consensus on this topic suggests either that the western Mediterranean is due to a combination and interplay of processes resulting from plate tectonics (but so complex that its details have yet to be worked out), or else that it is due to a different process altogether. In this paper the second possibility will be explored.

Subsidence in the western Mediterranean took place precisely in the center of the surrounding geosynclinal system and followed immediately after the tectogenic paroxysm of the Miocene. This suggests that the development of the geosyncline, the tectogenesis, and the later subsidence are related phenomena. A mechanism to explain the subsidence and tectonism should therefore also explain the earlier geosynclinal phase, which, it should be noted, represents 80 to 90 percent of the time involved in the cycle. Conversely, the earlier behavior of the geosyncline offers a guide to the kind of process that should be sought.

The following sections describe a model for the origin of the western Mediterranean, based on erosion by a subcrustal convection current. Gilluly (1955, 1964) has invoked subcrustal erosion to explain the thinning of sialic crust beneath the sedimentary cover of continental shelves and slopes. Hsü (1965) has offered an alternative suggestion: thinning and disappearance of sialic crust, he suggests, could be due entirely to subaerial erosion and tectonic denudation, after uplift caused by changes in mantle density. Gidon (1963) suggested that subcrustal currents might remove sialic material from beneath the center of a continental mass and deposit it in marginal, accreting orogenic belts. Both Gilluly and Gidon applied the idea of subcrustal flow to marginal, circumcontinental orogenic belts; here it is used only in the special case of a small, circular intracontinental orogen. Van Bemmelen (1969, p. 16–19) has mentioned the possibility of subcrustal erosion in the western Mediterranean, but only as a means of removing sialic crust from the central area. The present model calls on subcrustal currents as the driving force for virtually all the tectonic activity of the region since the Triassic.

MODEL BASED ON UNCOUPLED CONVECTION AND SUBCRUSTAL CURRENT RIPPLES

The Hercynian orogenic climax in the late Paleozoic gave rise to a thick, relatively rigid and immobile continental crust which was reduced by erosion to low topographic relief. Sometime in the early Mesozoic a column or plume of mantle material began to rise under this thick crust, approximately in the middle of what is now the western Mediterranean. Whether this original rising was due to the heating and insulating effect of a thick sialic crust (Gidon, 1963; Schuiling, 1969), to the presence of a "hot spot" in the mantle, or to some other cause is not critical to the model. The rising mantle material, on reaching the base of the lithosphere, spread radially outward, eventually descending to fill in at the bottom of the rising column. The convection cell visualized is thus mushroom-shaped or toroidal, as opposed to the linear, bilateral spreading of Atlantic type. The size and worldwide movement pattern of lithosphere plates at this time must have been such that crustal separation could not take place above the ascending mantle column. The cell, however, continued to circulate, with the result that convecting mantle material moved along the base of the lithosphere for a long period of time.

Thus, the driving force for the tectonic development of the western Mediterranean is thought to have resulted from the activity of a convection current uncoupled from the base of the overlying lithosphere. This is not to deny the importance of coupled convection, for within any of the several global tectonic plates (Morgan, 1968), the lithosphere is coupled to—that is, passively riding on—the convecting mantle beneath. Coupling of the lithosphere to the upper mantle must take place through viscous frictional drag in the relatively fluid low-velocity layer. Therefore, the strength of the bond between the two will be proportional to their area of contact so that a small convection cell, such as that postulated for the western Mediterranean, can be uncoupled from the lithosphere above. The area of contact over this cell was relatively small (about 10^6 km²) compared with that of the Eurasia-Africa block within which it was located (about 10^8 km²). The Eurasia-Africa block (now fragmented into several smaller blocks) was part of a world-wide convective system in which the motions of the component plates were reasonably well adjusted to each other. The lithosphere of such a plate is firmly coupled to its underlying mantle by viscous drag over an enormous area of contact, and therefore will not be broken up by the activity of an internal parasitic cell not large enough to develop a strong viscous drag.

Formation of Current Ripples and Geosynclinal Features

At the surface of the earth, when a fluid such as air or water flows over a material that is capable of being eroded, such as sand, it does not simply carry the material away, thus progressively lowering a flat surface. Rather, it erodes more rapidly in some places than in others, forming a pattern of bed forms (ripples or dunes) trending roughly normal to the direction of flow. Such undulating surfaces are characteristic of erosion by a fluid moving over easily erodable material (Simons and others, 1965; Harms and Fahnestock, 1965, esp. Pl. I). Individual ripples generally extend a fair distance, but here and there a ripple may divide or terminate (Pettijohn and Potter, 1964, Pls. 80, 81B, 83A, 86B, 87B).

The fundamental feature of the tectonic model proposed here is a somewhat similar process of ripple formation. Very large ripples are thought to form at the base of the lithosphere, in an inverted sense, as the result of uncoupled convective motion of the asthenosphere. If the heat supplied by mantle convection is capable of mobilizing the base of the lithosphere, this is a case of erosion by a fluid moving *under* an easily erodable material. Despite the enormous difference in scale, it seems probable that such a process, like stream or wind erosion, would be accompanied by the development of an undulating surface. These ripples on the base of the lithosphere would have a trend normal to the radial flow lines in the mantle; that is, they would form concentric arcs around the center of the ascending mantle column. For a given set of mantle and lithosphere properties and a given rate of mantle movement, there would be a steady state equilibrium amplitude for the ripples, as is the case with running water and wind ripples. Part or all of the material eroded from the base of the lithosphere would be carried away by the mantle current, possibly as envisioned by van Bemmelen (1969, p. 17): "The sialic crust is corroded at its base by circuits of matter in the upper mantle, transported to greater depths as mantle-crust mixture, and then comes to rest at depth in the state of high density–high temperature mineral phases. . . ."

Two theoretical possibilities are apparent for the initiation and subsequent behavior of this system of ripples. The first possibility (Fig. 3A) is that subcrustal erosion would make certain zones in the lithosphere thinner, while part or all of the material removed would be redeposited in thickening zones between the eroded belts. The equilibrium amplitude for the ripples would in this case be unattainable, for isostatic uplift of thickening zones would expose their upper parts to surficial erosion. This would result in a continual transfer of sialic material at the surface from rising thick zones to subsiding thin ones, which would counteract the effects of subcrustal erosion. The manifestation of this process at the surface would be a tendency for certain linear zones to rise continuously while others continuously subsided. The rising zones would all the while be supplying terrigenous clastic detritus to the adjacent subsiding furrows.

The second possibility (Fig. 3B) is that only erosion would take place beneath the crust, without concurrent redeposition. (Deposition alone, of course, is impossible without massive influxes of new sial from below.) In the second case, certain

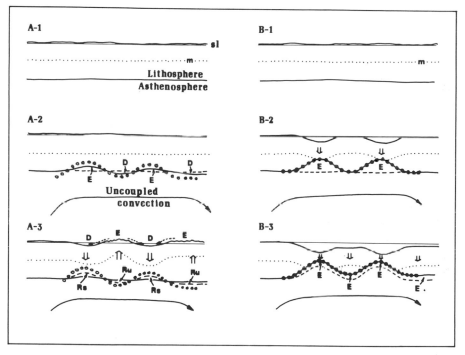

Figure 3. Two ways in which a geosynclinal system might evolve through the action of an uncoupled convection cell which forms subcrustal current ripples. A: Both erosion and redeposition take place at the base of the lithosphere (A-2). Isostatic uplift and surficial erosion of thickened zones prevent the subcrustal ripples from reaching their equilibrium amplitude, while clastic sediments are deposited in furrows formed at the surface (A-3). B: Only erosion takes place at the base of the lithosphere. This is concentrated in thinning, subsiding zones until the equilibrium ripple amplitude is reached (B-2), after which subcrustal erosion and subsidence take place throughout the geosynclinal system, thus maintaining the equilibrium ripple amplitude (B-3). In case B, little or no clastic sediment is deposited in the surface furrows. The stratigraphic history of the Mediterranean geosynclines indicates that they behaved in the manner shown in B.

Symbols: solid lines: top and base of lithosphere; dotted lines: Mohorovičić discontinuity; dashed lines: base of lithosphere from previous profile, for comparison; small circles: equilibrium amplitude for subcrustal ripples under the prevailing conditions; double arrows: isostatic uplift and subsidence; E: eroded; D: deposited; Ru: amplitude reduced through isostatic uplift; Rs: amplitude reduced through isostatic subsidence; sl: sea level.

linear zones would be attacked first by subcrustal erosion. As they were thinned from below, they would subside isostatically. This process would continue, without affecting the intervening zones, until the point was reached where the subcrustal ripples had achieved their equilibrium amplitude and the thick and thin belts of lithosphere were in isostatic balance. This state would be stable, for no zone would be above sea level, exposed to surficial erosion. From here on all zones would be equally attacked by subcrustal erosion, in order to maintain the equilibrium amplitude of the ripples, and thus all zones would subside isostatically. Since subcrustal currents with unchanged transportation capacity are now attacking an area roughly twice as large, the rate of erosion, and hence of isostatic subsidence, will be reduced to about half that which earlier affected the zones of thinning. Subcrustal effects, in this second case, would produce a different behavior at the surface. Initially certain linear zones would subside, while the intervening tracts would show no vertical movement at all. After these initial furrows had formed, the entire system, thick zones as well as thin, would begin to subside, although probably at a reduced rate. Throughout this process the thick zones would rarely, if ever, emerge above sea level. Thus, there would be little or no clastic detritus supplied to the adjoining furrows. This is in striking contract to the abundant detrital sedimentation which would characterize the first type of behavior.

Isopic zones—linear ridges and furrows characterized by the deposition of distinct facies—did in fact exist, maintaining their individual characteristics for long periods of time. This is perhaps the most basic conclusion reached by students of the stratigraphy of the Mediterranean orogenic belts. Structural complications imposed by later tectogenesis usually make it difficult to interpret the earlier paleageographic pattern, but the ridges and furrows seem generally to have had widths in the range of a few tens to a few hundreds of kilometers, while the total thickness of sediments deposited in the furrows was a few thousand meters. Individual zones may extend for thousands of kilometers.

The isopic zones in the geosynclinal system which surrounded what is now the western Mediterranean (Ogniben, 1970; Caire, 1970) show a history of sedimentation precisely corresponding to that predicted in the second of the two cases described above (erosion without redeposition at the base of the lithosphere). This behavior is in fact typical of Alpine Mediterranean geosynclines in general (Aubouin, 1965, p. 83–101). Deep furrows were rapidly generated, after which ridges, as well as furrows, began to subside. Sedimentation (almost entirely non-terrigenous) took place throughout the system, with facies of apparently deep-water origin in the furrows and shallow-water facies on the ridges. It was not until the beginning of tectogenic activity in the geosyncline that terrigenous or flysch sedimentation began. Here we must recall the difference between the European type and the American type of geosyncline.

Some students, such as Aubouin (1965), believe that the zones always appear in a certain order across the chain (from interior to exterior: eugeanticlinal ridge, eugeosynclinal furrow, miogeanticlinal ridge, miogeosynclinal furrow, foreland platform), each feature with unique and characteristic properties (Fig. 2). Others feel that this is an oversimplification and that zones may have varying characteristics, and may divide or terminate (*see* the critique of Aubouin's 1965 book by Debelmas and others, 1967). In an example from the western Mediterranean, Ogniben (1970) interprets the paleogeography of Sicily as corresponding to the pattern of Aubouin, while Caire and his students present a more unorthodox reconstruction, with dividing and terminating zones (Caire, 1970, Figs. 8, 10; Broquet, 1970; Duée, 1970; Mascle, 1970; Truillet, 1970). Such isopic zones were active features in the western Mediterranean and in other Alpine belts from some time in the early Mesozoic until as late as the Miocene, or for nearly 200 m.y. Again there is a contrast between the simpler view that a tendency once established would persist, and the less orthodox view that major changes in paleogeography could occur with the establishment of a new set of isopic zones (for the case of Sicily, see Broquet, 1970, p. 205–210).

The impartial observer of the two views would probably conclude that there is a strong tendency for zones to be long and continuous, but that they may, in some instances, divide or terminate. Similarly, he would say that zones tend to maintain their ridge or furrow character, but that alterations in the paleogeography may indeed occur. The first conclusion implies an areal pattern quite analogous to that of sand ripples. The second conclusion leads to a comparison with the alteration of ripple patterns that would accompany a change in the flow of running water.[2]

There is general agreement among all workers, however, that the internal zones are tectonically more active than the external. Aubouin (1965, p. 104) has summarized this greater activity: "The eugeosynclinal [internal] domain . . . is the first to materialize, is associated with ophiolitic emissions, includes early flysch, is affected by an early orogenesis, and is thrust towards the exterior. . . ."

The subcrustal erosion model accounts well for this orogenic polarity. However the process of subcrustal erosion be visualized, mobilization of the base of the lithosphere by the heat of the convecting mantle certainly plays an important part, perhaps in the manner suggested by Vogt and others (1969, p. 606): "A fruitful model to which theoretical approaches might be directed could assume that the phase interface occurs at the top of the low-velocity layer and involves melting and refreezing of the basaltic fraction in an ultrabasic crystal slush." Dissipation of heat and slowing of the rate of horizontal movement as the mantle material moves radially outward would thus result in a slower rate of subcrustal erosion, and hence of isostatic subsidence, in the external zones. Externally directed thrusting is a consequence of mantle drag in that direction, as described in the next section.

Mantle Drag and Tectonic Activity

As the mantle current continued to flow during the Mesozoic and early Tertiary, more and more material was removed from thin areas and presumably carried to depth. Thus, in the furrow zones the relatively strong crystalline portion of the crust became thinner and thinner, and an ever-increasing percentage of the remaining crust was made up of weaker sedimentary rocks. Eventually the point was reached where this reduced crust was no longer strong enough to resist the compression applied by the thick zone behind it, which was pressing outward as a result of drag due to the mantle current (Fig. 4).

The innermost (eugeosynclinal) furrow was the most actively eroded and was thus the first zone to be reduced to this critical thickness. When it was, the

---

[2] Throughout the remainder of this paper, Aubouin's terminology is used for convenience, with the recognition that it may, in some cases, represent an oversimplification.

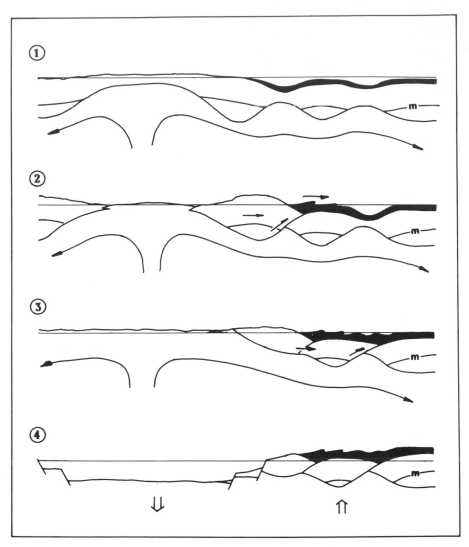

Figure 4. Subsequent development of the orogenic belt. Section 1 is the stage reached in Figure 3, section B-3, also showing the central area. Here, subcrustal erosion is very rapid, but the ever thinner crust is held up, out of isostatic equilibrium, by the ascending mantle current. In section 2 the crust beneath the eugeosynclinal furrow, having become too thin to resist the compressive stress generated by mantle drag, has sheared through, and the sedimentary fill of the furrow (black) has slid to the right. In section 3 the same thing happens to the miogeosynclinal furrow. Meanwhile, the lithosphere beneath the eugcanticlinal ridge has been further eroded. In section 4 the mantle current has ceased to flow. Lacking this support, the central area, with its new oceanic crust, is subsiding to oceanic depths, while the surrounding continental crust is rising in compensation. This is the situation at present in the western Mediterranean.

crust sheared through on surfaces dipping toward the interior. The eugeanticlinal ridge was thrust over the miogeanticlinal ridge, while the sedimentary filling of the eugeosynclinal furrow was bulldozed up and out, and then moved over the miogeanticlinal ridge as great gravity-sliding nappes. Aubouin (1965, p. 175) describes the process, reconstructed from field studies and generalized for the Mediterranean area, as follows: "These structures of the internal zones, peculiar to the geosynclinal chains, testify to the crushing of the eugeosynclinal furrow between the external miogeanticlinal ridge and the internal eugeanticlinal ridge—a process culminating in the ejection of the sedimentary material from the eugeosynclinal furrow to ride over the external zones in the form of spectacular superficial nappes characteristic of geosynclinal chains."

As the eugeanticlinal ridge rode up onto the miogeanticlinal ridge, the crust was thickened to the point where it could again withstand compressive stresses. After a further interval of convective erosion, the crust below the more external furrow was reduced to the point where it, too, collapsed, although generally with less intense deformation than took place in the internal furrow (Aubouin, 1965, p. 168–171).

This mechanism for deformation of the geosynclinal belt provides an explanation for a number of important observations. As noted above, it results in a squeezing-out of the contents of the troughs identical to that observed in field studies. Gravity sliding on an enormous scale is a generally accepted feature of Alpine deformation, but rock masses sliding off the sides of simple uplifts should move in both directions, whereas field studies show that almost invariably they have moved only toward the exterior. This is explained by the bulldozing action of the more internal ridge as it is pushed up over the more external one. The model also explains the presence, in addition to the superficial gravity-sliding nappes, of compressional thrust faults in the basement—features which would not be present if the gravity nappes were a response to simple uplift.

Collapse of the Central Area

The result of continuing erosion by the radially spreading mantle column and of the episodic outward movements of the lithosphere as the geosynclinal troughs collapsed was that the sialic crust above the ascending column, as well as the crust of the internal ridge and furrow in most places, was completely removed. This area remained elevated, however, since it was held up by the ascending mantle column. When the mantle current ceased to flow, a profound change took place. The central area, which had been held up out of isostatic equilibrium by the rising current, sank to the normal depth of the new oceanic crust which was now present. Subsidence of this central area was accompanied by uplift in the surrounding geosynclinal chain. In other words, the normal isostatic levels of oceanic and continental crust were restored, after having been held out of equilibrium by the action of the mantle current.

This restoration of isostatic levels after the mantle current ceased to flow is vividly documented in the western Mediterranean. As mentioned in the introduction to this paper, there is conclusive stratigraphic evidence that, beginning in the latest Miocene, the center of the western Mediterranean basin sank to oceanic depths, while the surrounding orogenic belts were elevated above sea level; this was the "Mediterranean Revolution." The western Mediterranean Sea is characterized by positive isostatic gravity anomalies (de Bruyn, 1955). Although a unique interpretation of such anomalies is not possible, they suggest that, if active forces are not holding the area out of equilibrium, the western Mediterranean Basin is still subsiding. The 50 mgal positive anomaly would be removed by about a 400 m subsidence (Collette, 1969, p. 140).

Important evidence bearing on the removal of crust from the central area comes from the nature of the present crust and upper mantle beneath the western Mediterranean Sea. Menard (1967) has shown profiles illustrating the wide range of crustal types under different small oceanic basins. Vogt and others (1969, p. 607–608, Fig. 26) have pointed out that the crust of the Tyrrhenian and Balearic Basins very closely resembles that found on the upper flanks of the Mid-Atlantic Ridge within 300 km of the crest. They suggest "that the western Mediterranean may be a complex rift of the mid-ocean type . . ." (Vogt and others, 1969, p. 608). The fact that the western Mediterranean Basin is a rounded triangle nearly encircled by a continuous orogenic belt makes it most unlikely that it was formed by sea-floor spreading of the Atlantic type, with the crust "bolted" or coupled to the mantle, as shown by Hess (1965, Fig. 123). On the other hand, the model proposed here—radial spreading with the mantle uncoupled from the crust, followed by subsidence as the convection ceases—would also account for the presence of crust of the type found on the flanks of a mid-oceanic ridge. In both cases new oceanic crust has subsided part of the way to oceanic depths because it is no longer held up by ascending mantle material. The situations differ only in the manner in which the pre-existing continental crust was removed.

The similarity to the mid-oceanic ridges continues downward into the upper mantle. Berry and Knopoff (1967) carried out a study of phase velocities in the Balearic Basin area that gave structural information down to a depth of 200 to 300 km. The most striking feature of the area is the presence of a low-velocity layer, the top of which is at about 50 km below the central part of the basin (shear velocity $4.10 \pm .05$ km/sec) and descends to about 100 km below the margins (shear velocity $4.43 \pm .10$ km/sec). Ritsema (1969, p. 105–110) and Vogt and others (1969, p. 609) have pointed out the detailed similarity between the upper mantle structure of the Balearic Basin and that of the Mid-Atlantic Ridge. It must be emphasized that the geology and geography of the continental areas ringing the western Mediterranean cannot be explained by Atlantic-type spreading; convection with coupled and with uncoupled crust could, however, result in similar crustal and mantle structure in the central areas. The top of Berry and Knopoff's (1967) low-velocity layer would represent the lithosphere-asthenosphere contact. The higher level of this contact under the center of the basin would be the remains of the uplift over the ascending mantle column, now in the process of subsiding to a normal depth.

CONCLUSIONS

The principal advantage of the model proposed here is that it postulates a simple type of mantle behavior which is capable of explaining most, if not all, of the geological and geophysical features, both past and present, of the western Mediterranean area. It will probably be possible to apply these ideas to a number of other problem areas as well, such as much of the rest of the Tethyan orogenic system and the Caribbean.

Finally, it should be noted that in this model the operation and termination of a single process account for all three phases of the widely recognized geotectonic cycle, a concept which has recently been criticized by Coney (1970).

REFERENCES CITED

Alvarez, W., 1972, Rotation of the Corsica-Sardinia microplate: Nature Phys. Sci., v. 235, p. 103–105.

Alvarez, W., and Gohrbandt, K.H.A., eds., 1970, Geology and history of Sicily: Tripoli, Petr. Exp. Soc. Libya, 291 p.

Aubouin, J., 1965, Geosynclines: Amsterdam, Elsevier Publ. Co., 335 p.

Berry, M. J., and Knopoff, L., 1967, Structure of the upper mantle under the western Mediterranean basin: Jour. Geophys. Research, v. 72, no. 14, p. 3613–3626.

Boccaletti, M., and Guazzone, G., 1970, La migrazione terziaria dei bacini toscani e la rotazione dell'Appennino Settentrionale in una "zona di torzione" per deriva continentale: Mem. Soc. Geol. Ital., v. 9, p. 177–195.

Bourcart, J., 1962, La Méditerranée et la révolution du Pliocène, in Livre P. Fallot, Vol. 1: Soc. Géol. France, v. 1, p. 103–116.

Broquet, P., 1970, The geology of the Madonie Mountains of Sicily, in Alvarez, W., and Gohrbandt, K.H.A., eds., Geology and history of Sicily: Tripoli, Petroleum Exploration Soc. Libya, p. 201–230.

Caire, A., 1970, Sicily in its Mediterranean setting, in Alvarez, W., and Gohrbandt, K.H.A., eds., Geology and history of Sicily: Tripoli, Petroleum Exploration Soc. Libya, p. 145–170.

Caputo, M., Panza, G. F., and Postpischl, D., 1970, Deep structure of the Mediterranean Basin: Jour. Geophys. Research, v. 75, no. 26, p. 4919–4923.

Collette, B. J., 1969, Mediterranean oceanization—a comment: Verhandelingen Kon. Ned. Geol. Mijnbouwk. Gen., v. 26, p. 139–142.

Coney, P. J., 1970, The geotectonic cycle and the New Global Tectonics: Geol. Soc. America Bull., v. 81, p. 739–748.

Debelmas, J., Lemoine, M., and Mattauer, M., 1967, Geosynclines, by J. Aubouin: A review: Am. Jour. Sci., v. 265, no. 4, p. 292–300.

de Booy, T., 1969, Repeated disappearance of continental crust during the geological development of the western Mediterranean area: Verhandelingen Kon. Ned. Geol. Mijnbouwk. Gen., v. 26, p. 79–103.

de Bruyn, J. W., 1955, Isogram maps of Europe and North Africa: Geophys. Prosp. [Netherlands], v. 3, p. 1–14.

Dewey, J. F., 1969, Continental margins: A model for conversion of Atlantic type to Andean type: Earth and Planetary Sci. Letters, v. 6, no. 3, p. 189–197.

Dewey, J. F., and Bird, J. M., 1970, Mountain belts and the New Global Tectonics: Jour. Geophys. Research, v. 75, no. 14, p. 2625–2647.

Dewey, J. F., and Horsfield, B., 1970, Plate tectonics, orogeny and continental growth: Nature, v. 255, p. 521–525.

Dietz, R. S., and Holden, J. C., 1966, Miogeoclines (miogeosynclines) in space and time: Jour. Geology, v. 75, no. 5, p. 566–583.

Duée, G., 1970, The geology of the Nebrodi Mountains of Sicily, in Alvarez, W., and Gohrbandt, K.H.A., eds., Geology and history of Sicily: Tripoli, Petroleum Explorations Soc. Libya, p. 187–200.

Durand Delga, M., 1967, Structure and geology of the northeast Atlas Mountains, in Martin, L., ed., Guidebook to the geology and history of Tunisia: Tripoli, Petroleum Explorations Soc. Libya.

Gidon, P., 1963, Courants magmatiques et évolution des continents (l'hypothèse d'une érosion sous-crustale): Paris, Masson, 155 p.

Gilluly, J., 1955, Geologic contrasts between continents and ocean basins, p. 7–18, in Poldervaart, A., ed., Crust of the earth: Geol. Soc. America Spec. Paper 62, 762 p.

—— 1964, Atlantic sediments, erosion rates, and the evolution of the continental shelf: Geol. Soc. America Bull., v. 75, p. 483–492.

Harms, J. C., and Fahnestock, R. K., 1965, Stratification, bedforms, and flow phenomena (with an example from the Rio Grande), in Middleton, G. V., ed., Primary sedimentary structures and their hydrodynamic interpretation: Soc. Econ. Paleontologists and Mineralogists Spec. Pub. no. 12, p. 84–115.

Hess, H. H., 1962, History of ocean basins, *in* Engel, A.E.J., James, H. L., and Leonard, B. F., eds., Petrologic studies: A volume in honor of A. F. Buddington: Geol. Soc. America, p. 599–620.

―――― 1965, Mid-ocean ridges and tectonics of the sea-floor, *in* Submarine geology and geophysics, Colston Papers: London, Butterworths, p. 317–332.

Hsü, K. J., 1965, Isostasy, crustal thinning, mantle changes, and the disappearance of ancient land masses: Am. Jour. Sci., v. 263, p. 97–109.

Manfredini, M., 1963, Schema dell'evoluzione tettonica della penisola Italiana: Servizio Geol. Italia Boll., v. 84, p. 101–130.

Martin, L., ed., 1967, Guidebook to the geology and history of Tunisia: Tripoli, Petroleum Exploration Soc. Libya, 293 p.

Mascle, G. H., 1970, Geological sketch of western Sicily, *in* Alvarez, W., and Gohrbandt, K.H.A., eds. Geology and history of Sicily: Tripoli, Petroleum Exploration Soc. Libya, p. 231–243.

Menard, H. W., 1967, Transitional types of crust under small ocean basins: Jour. Geophys. Research, v. 72, no. 15, p. 3061–3073.

Mitchell, A. H., and Reading, H. G., 1969, Continental margins, geosynclines, and ocean floor spreading: Jour. Geology, v. 77, p. 629–643.

Morgan, W. J., 1968, Rises, trenches, great faults and crustal blocks: Jour. Geophys. Research, v. 73, p. 1959–1982.

Ogniben, L., 1969, Schema introduttivo alla geologia del confine calabro-lucano: Mem. Soc. Geol. Ital., v. 8, p. 453–763.

―――― 1970, Paleotectonic history of Sicily, *in* Alvarez, W., and Gohrbandt, K.H.A., eds., Geology and history of Sicily: Tripoli, Petroleum Exploration Soc. Libya, p. 133–143.

Pannekoek, A. J., 1969, Uplift and subsidence in and around the western Mediterranean since the Oligocene: a review: Verhandelingen Kon. Ned. Geol. Mijnbouwk. Gen., v. 26, p. 53–77.

Pettijohn, F. J., and Potter, P. E., 1964, Atlas and glossary of primary sedimentary structures: Berlin, Springer-Verlag, 370 p.

Ritsema, A. R., 1969, Seismic data of the west Mediterranean and the problem of oceanization: Verhandelingen Kon. Ned. Geol. Mijnbouwk. Gen., v. 26, p. 105–120.

Sander, N. J., 1970, Structural evolution of the Mediterranean region during the Mesozoic Era, *in* Alvarez, W., and Gohrbandt, K.H.A., eds., Geology and history of Sicily: Tripoli, Petroleum Exploration Soc. Libya, p. 43–132.

Schuiling, R. D., 1969, A geothermal model of oceanization: Verhandelingen Kon. Ned. Geol. Mijnbouwk. Gen., v. 26, p. 143–148.

Sestini, G., ed., 1970, Development of the Northern Apennines geosyncline: Sed. Geol. ogy, v. 4, p. 203–647.

Simons, D. B., Richardson, E. V., and Nordin, C. F., Jr., 1965, Sedimentary structures generated by flow in alluvial channels, *in* Middleton, G. V., ed., Primary sedimentary structures and their hydrodynamic interpretation: Soc. Econ. Paleontologists and Mineralogists Spec. Pub. no. 12, p. 34–52.

Smith, A. G., 1971, Alpine deformation and the oceanic waves of the Tethys, Mediterranean and Atlantic: Geol. Soc. America Bull., v. 82, p. 2039–2070.

Truillet, R., 1970, The geology of the eastern Peloritani Mountains of Sicily, *in* Alvarez, W., and Gohrbandt, K.H.A., eds., Geology and history of Sicily: Tripoli, Petroleum Exploration Soc. Libya, p. 171–185.

van Bemmelen, R. W., 1969, Origin of the western Mediterranean Sea: Verhandelingen Kon. Ned. Geol. Mijnbouwk. Gen., v. 26, p. 13–52.

van der Voo, R., and Zijderveld, J.D.A., 1969, Paleomagnetism in the western Mediterranean area: Verhandelingen Kon. Ned. Geol. Mijnbouwk. Gen., v. 26, p. 121–138.

Vine, F. J., 1966, Spreading of the ocean floor: New evidence: Science, v. 154, p. 1405–1415.

Vogt, P. R., Schneider, E. D., and Johnson, G. L., 1969, The crust and upper mantle beneath the sea, *in* Hart, P. J., ed., The earth's crust and upper mantle: Am. Geophys. Union, Geophys. Mon. Ser., 13, p. 556–617.

MANUSCRIPT RECEIVED BY THE SOCIETY MARCH 30, 1971
PRESENT ADDRESS: LAMONT-DOHERTY GEOLOGICAL OBSERVATORY OF
COLUMBIA UNIVERSITY, PALISADES, NEW YORK, 10964

The Geological Society of America, Inc.
Memoir 132
© 1972

Rotational Pole Determination from Biological Diversity Gradients

R. A. Duncan

*Department of Geological and Geophysical Sciences,
Princeton University, Princeton, New Jersey 08540*

ABSTRACT

A new method for predicting rotational poles from biological diversity gradients consists of fitting a fourth degree polynomial in latitude to the data for each of a grid of trial "north poles" and measuring the "goodness of fit" for each tested pole. A confidence area for the predicted pole can then be contoured on the array of trial north poles. This method was tested against modern diversity gradients, and when applied to Stehli's Permian brachiopod data, it showed many advantages over the spherical harmonics method. A hypothetical Permian continental reconstruction and the present continental arrangement were tested, but the paleoclimatic data used were unable to discriminate adequately between the two.

INTRODUCTION

Paleomagnetic data are used widely to reconstruct continental positions in the past. These continental reconstructions all implicitly assume the validity of the "axial dipole" hypothesis, that the (averaged) magnetic field is, and has been, dipolar and axially symmetric about the rotational pole. The success of the recent developments in continental drift and sea-floor spreading hypotheses is compelling but does not lessen the importance of this assumption in reconstructing paleo-positions. A test of the axial dipole hypothesis can be made in two ways: either (1) directly, through the paleomagnetics, with well sampled magnetic data for the same period, or (2) indirectly, by comparing paleomagnetic with paleoclimatic data. By the first method, simultaneous worldwide observations should show the dipole pattern of the field. Accurate continental reconstructions are crucial for this, and such testing has been done only for large, supposedly unbroken continental units. For the second test we need a method of locating the position of the paleorotational pole, which is independent of the magnetics. Stehli (1968a,

133

1968b) has proposed a model based on diversities of organisms and a method of predicting the rotational pole position from these data. This report presents an alternate method of computing ancient rotational pole positions and shows, contrary to Stehli's conclusions, that the data are ambiguous and do not favor the present continental configuration over a Permian paleomagnetic fit.

DIVERSITY GRADIENT MODEL

The taxonomic diversity gradient model is based on the dependence of temperature on latitude. That is, maximum temperature (solar energy capture) occurs at the equator and decreases monotonically with latitude toward either pole. Temperature, in turn, is directly related to the rate of chemical reaction and hence to rate of living, including growth, reproduction, maturation, and duration of life. Moreover, the increase in bio-mass toward the tropics is due to an increasing number of different species rather than increasing individuals within a species (Fischer, 1960). Hence, diversity should be greatest at the equator and least at the poles. Such taxonomic diversity gradients have been observed for fossil as well as for living organisms.

Several assumptions must be invoked in order to use paleoclimatic data: (1) In the past, the earth has maintained equatorially centered diversity gradients similar to those of the present; (2) species of fossil organisms vary with temperature as do present organisms; (3) the particular organisms chosen have a wide latitudinal range; and (4) a standard of uniformity has been maintained in the collection of the fossil data (whether uniformly good or bad). Failure of any one of these four assumptions would invalidate the results of this kind of analysis.

Stehli (1968a, 1968b, 1970) has proposed that suitable diversity observations on fossil organisms can be used to predict ancient rotational poles independently of the paleomagnetics. If all the poles predicted by this diversity gradient model are not appreciably different from the present rotational pole, one may conclude that the magnetic pole has moved independently of the rotational pole. From this it follows that paleomagnetic data are not valid in testing the continental drift hypothesis. If, on the other hand, there is variation among the predicted rotational pole positions for various geologic ages, we may conclude that continental drift has occurred and that paleomagnetic evidence does provide information about the movements of the continents.

SPHERICAL HARMONICS METHOD

Given diversity observations at points well scattered about the globe, the most reliable prediction of the rotational pole position may be obtained from a mathematical model best fitting this raw data distribution. The model used in many analyses of spherically distributed observations is the set of orthogonal functions known as spherical harmonics. This approximates the three-dimensional diversity gradient which is a function of two independent variables, co-latitude and longitude, and which is confined to lie on the unit sphere. According to the theory of orthogonal functions, if $f(\theta, \phi)$ is a rational integral function of the coordinates of a point on the unit sphere ($\theta =$ co-latitude, $\phi =$ longitude), then it can be expressed as the sum of a finite series of the spherical harmonics.

The surface of biological diversity may now be expressed as a sum of the spherical harmonics,

$$y = \beta_0 P_0 + \beta_1 P_1 + \ldots + \beta_i P_i + \ldots ,$$

where P_i are the individual spherical harmonics (sometimes called Legendre functions), the β_i are coefficients which weight each P_i properly, and y is the predicted diversity. The spherical harmonics up to the second order are:

$P_0 = 1$ (constant) $P_3 = -\sin\phi\sin\theta$ $P_6 = -\sin\phi 3\sin\theta\cos\theta$
$P_1 = \cos\theta$ $P_4 = \frac{1}{2}(3\cos^2\theta - 1)$ $P_7 = \cos 2\phi 3\sin^2\theta$
$P_2 = -\cos\phi\sin\theta$ $P_5 = -\cos\phi 3\sin\theta\cos\theta$ $P_8 = \sin 2\phi 3\sin^2\theta.$

Then y, the diversity predicted by the model, can be expressed as an exact function of θ and ϕ. We need to know the β_i, though, in order to determine y. These are found analytically by solving integrals containing the known function $f(\theta, \phi)$. Since we do not know $f(\theta, \phi)$, but only a collection of scattered observed diversities, we must instead estimate the coefficients statistically. Thus,

$$y = \beta_0 P_0 + \beta_1 P_1 + \beta_2 P_2 + \ldots + \beta_8 P_8 + \text{error}.$$

The coefficients are estimated by multiple regression of y on $P_0, P_1, P_2, \ldots, P_8$. The choice of the second order is arbitrary (the higher the order, the greater the precision), but the total number of regressors (P_i's) cannot exceed the number of observations.

In matrix form this equation becomes:

$$Y = XB + e$$

$$\begin{bmatrix} y_1 \\ y_2 \\ \cdot \\ \cdot \\ \cdot \\ y_n \end{bmatrix} = \begin{bmatrix} 1 & P_{11} & P_{21} & \ldots & P_{81} \\ 1 & & & & \\ \cdot & & & & \\ \cdot & & & & \\ \cdot & & & & \\ 1 & & & & \end{bmatrix} \begin{bmatrix} \beta_0 \\ \cdot \\ \cdot \\ \cdot \\ \cdot \\ \beta_8 \end{bmatrix} + \begin{bmatrix} e_1 \\ e_2 \\ \cdot \\ \cdot \\ \cdot \\ e_n \end{bmatrix}.$$

The β_i can be estimated by b_i so that the predicted diversity is

$$\hat{y} = b_0 P_0 + b_1 P_1 + \ldots \ldots + b_8 P_8$$

Now the error is $e_i = (y_i - \hat{y}_i)$, and the parameters (b_i) will best fit the data when Σe_i^2 is a minimum. This is the case when

$$B = (X'X)^{-1}X'Y,$$

where B is the vector of parameters (b_i), X is the matrix of regressors, and Y is the vector of observed diversities. The result is a best-fitting diversity function of co-latitude and longitude from which the polar and equatorial regions can be extracted.

Given this model, do present biological diversity gradients predict the present rotational pole? To test this proposition, Stehli (1968a) chose diversity gradients for some 15 different living organisms and found best-fitting surfaces using the spherical harmonics method. The diversity observations used here are living species of mosquitos at 86 locations about the earth, as reported by Stehli (1968a). A FORTRAN program was written to compute the best-fitting spherical harmonics surface. In the program was a routine to print out the predicted diversity at every 6° longitude and every 2° co-latitude. This plot for mosquito diversities is contoured to emphasize the areas of maximum and minimum diversities. The predicted equator was located by connecting the points of maximum predicted diversity for each longitude. The predicted pole based on mosquito diversity, then, is 90° away from this equator, and is very close to the present north pole. Stehli has found that the poles for all 15 organisms he tested fall within 15° of the present north pole. Apparently the model works—at least for recent species.

An application of the spherical harmonics technique to Permian organisms (Fig. 2) uses Stehli's (1970) brachiopod data. Because of the scarcity of data (only 27 sample points), not as much confidence can be put in the predicted pole. The pole position predicted by the spherical harmonics for these Permian brachiopods is significantly different from the present north pole, being nearly 25° away. From this result we can tentatively conclude continental movement. The north pole for Permian times (Fig. 2) is at 65° N., 150° W., expressed relative to today's rotational pole.

This method of fitting with a surface of spherical harmonics is reliable for large quantities of data, randomly spaced about the sphere. Difficulties arise when only small numbers of sample points, nonrandomly distributed (in this case, confined to the continents and mainly from the Northern Hemisphere), are available. There is much uncertainty in fitting the surface, especially in the large unsampled areas of the sphere where there is no control. In the case of the Permian brachiopods, the 9 coefficients of the second order spherical harmonics were estimated with only 27 sample diversities, while 86 observations of mosquito diversity estimated the present diversity gradient surface. Yet the difference in resolution between the two is undiscernable. In other words, from the predicted pole alone, there is no indication of its reliability. Perhaps another method is better suited to this problem.

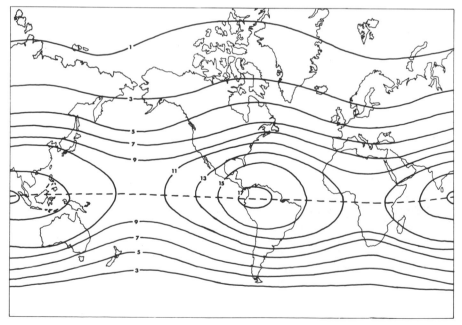

Figure 1. Second-order spherical harmonics surface best fitting the 86 observations of diversity for present-day species of mosquitos. The predicted equator is dotted.

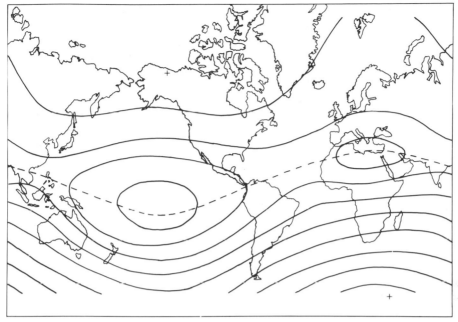

Figure 2. Second order spherical harmonics surface best fitting the 27 observations of diversity for species of Permian brachiopods. The predicted Permian equator and rotational poles are indicated by the dotted line and the crosses, respectively. The position of the rotattional pole is 25° (+) away from today's pole.

POLYNOMIAL TECHNIQUE

Consider instead the model

$$y = k_1 + k_2\theta^2 + k_3\theta^4 + \text{error},$$

where y, the diversity, is now a fourth degree polynomial in θ, the latitude. With this choice, the number of unknowns is reduced from $9\beta_i$'s to $3k_i$'s, thereby reducing the variance in the estimation of each coefficient. In effect, we have more sample observations estimating each of the parameters k_i. Notice, however, that the diversity model is a function of only one variable, the co-latitude, θ. To preserve symmetry about the equator, only even powers of θ are used. The procedure which allows us to take advantage of this model is as follows.

A particular point on the earth's surface is chosen as the north pole, $(\theta^*, \phi^*) = (0, 0)$. Then the coordinates of the observations (θ_i, ϕ_i) are expressed in terms of this new coordinate system by means of the following linear transformation,

$$\begin{bmatrix} \sin\theta'_i \cos\phi'_i \\ \sin\theta'_i \sin\phi'_i \\ \cos\theta'_i \end{bmatrix} = \begin{bmatrix} \cos\theta^* & 0 & -\sin\theta^* \\ 0 & 1 & 0 \\ \sin\theta^* & 0 & \cos\theta^* \end{bmatrix} \begin{bmatrix} \sin\theta_i\cos(\phi_i - \phi^*_i) \\ \sin\theta_i\sin(\phi_i - \phi^*_i) \\ \cos\theta_i \end{bmatrix},$$

from which $\theta' = \cos^{-1}(\cos\theta')$. At this point, longitude is unimportant and diversity can be estimated as a function of latitude only.

The model is fitted by least-squares linear regression, as in the spherical harmonics model. The success of the fit is measured by computing the sum of squares of residuals (observed diversity minus predicted diversity) for the 27 observations. Now another north pole is chosen and the process is repeated, finishing with a sum of squares of residuals as a measure of the goodness of fit. If this process is continued in a systematic way over a grid of sample north poles covering the present Northern Hemisphere, the resulting array of sums of squares of residuals can be studied and contoured to determine the pole which best fits the observations. An added advantage of this search and measure method is that we not only determine the absolute best-fitting pole (or as close to it as desired), but we also get a good idea of what happens around it. In other words, contouring the array reveals whether the pole is clearly marked by a significant rise in sums of squares of residuals around it, or whether there is a rather large area in which any one pole is as good as the next. The method enables us to outline a confidence area[1] which contains the pole, rather than having to pick one pole in particular.

One meaningful way of constructing this confidence area is to draw contours along which the sum of squares of residuals is 10 percent, 20 percent, 30 percent, and so forth, larger than the absolute minimum sum of squares. The absolute minimum for present mosquitos (Fig. 3) is at $\theta = 10°$, and the 20 percent contour includes the present north pole. Also, the contours define a rather steep-sided bowl which pinpoints the prediction of the pole position to a small area.

This polynomial method, then, predicts the present pole (as it lies within the chosen confidence area) which was expected, and the method may now be applied to a valid set of fossil diversities. The result of a test on 27 samples of Permian brachiopod diversities (Stehli, 1970) plotted on the present continental configuration is illustrated in Figure 4. Comparison with Figure 3 illuminates differences between the mosquito data and the brachiopod data. Three distinct minima are evident in Figure 4, and while today's north pole is a candidate for the Permian

---

[1] Not in the usual statistical sense.

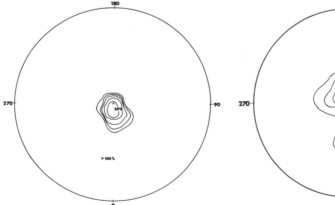

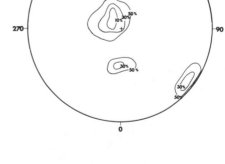

Figure 3. Polar projection of contoured sums of squares of residuals for an array of trial north poles tested with diversities of present-day mosquitos. The present rotational pole is indicated by the cross and is within 10° of the absolute best-fitting trial pole. (The contours are in intervals of 20 percent of the absolute minimum.)

Figure 4. Polar projection of sums of squares of residuals for an array of trial north poles tested with diversities of Permian brachiopods (present continental configurations). The present-day north pole is indicated by the cross.

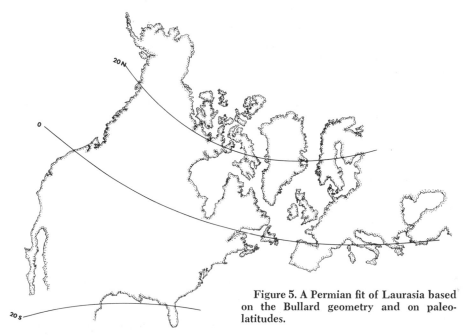

Figure 5. A Permian fit of Laurasia based on the Bullard geometry and on paleo-latitudes.

rotational pole, a large number of other trial north poles do equally as well. In other words, many very diverse north poles fit the data in the Permian brachiopod test, whereas a small cluster of trial poles suits the present mosquito data.

There is good indication, then, from both the spherical harmonics and the polynomial methods, that the Permian continental configuration differed from the present one. The next step is to test other configurations. The clear method of accomplishing this is to fit the continents into their assumed Permian positions and then to find the pole and confidence area predicted by the brachiopod diversity data. Paleomagnetics gives us the best idea of how this might be done (Irving, 1964). The reconstruction of the continents for Permian times (Fig. 5) was determined from the paleomagnetic latitudes of Permian continents given by Irving (1964), and is very nearly identical to the Bullard fit for the areas from which the data come.

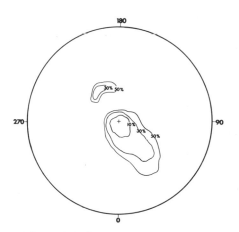

If these were the positions of the continents during Permian times, then the paleoclimatic data should produce results much like those for recent mosquitos. That is, the brachiopod diversities, when plotted on their (drift corrected) continental construction, should predict the north pole. Obviously the

Figure 6. Polar projection of contoured sums of squares of residuals for an array of trial north poles tested by diversities of Permian brachiopods relocated onto the paleo-latitudinal continental fit (Fig. 5). The cross indicates (O,O), the expected best-fitting trial pole if the continents are positioned correctly.

prediction will not be nearly as precise owing to much scarcer data and greater error in data collection. The resulting contoured projection (Fig. 6) of the Northern Hemisphere (Permian) reveals that a neat compact confidence area about one absolute minimum did not result. Instead there are two distinct areas of good fit, one centered very near the paleomagnetic north pole and the other centered 40° away. In fact, the result resembles the brachiopod data plotted on the present-day arrangement of continents. In this case the Permian magnetic pole may well be the rotational pole (being within 10 percent of the absolute best-fitting pole), but so may several other trial poles. Although the brachiopods seem to fit slightly better when plotted on the Permian paleomagnetic fit than on the present-day arrangement, the data do not clearly discriminate between the two alternatives. One conclusion from the comparison of the tests involving Permian diversity data with those involving present diversity data is that 86 sample diversities reliably predict a rotational pole, but that 27 or fewer lead to ambiguous and inconclusive results.

CONCLUDING REMARKS

Two tests of the Permian diversity data (Figs. 4 and 6), each using very different continental configurations, fall short of answering the axial dipole question. Permian brachiopod data compare well with Permian magnetic data, which implies that the Permian rotational and magnetic poles were coincident and that continental drift has occurred since Permian times. On the other hand, the brachiopod data behave nearly as well in the present continental configuration and thus lend no support to either possibility. The data themselves are not powerful enough to determine which Permian continental arrangement is correct. Diversity data offer a quantitative check on the position of the rotational pole, as demonstrated by the mosquito diversities, and solution of the axial dipole problem is mainly a matter of collecting adequate data.

ACKNOWLEDGMENTS

I thank W. J. Morgan, K. S. Deffeyes, F. J. Vine, R. B. Hargraves, and H. D. Holland for helpful discussion and critical reading of the manuscript. Grateful acknowledgment is made to E. Schneider and Project GOFAR of the U.S. Naval Oceanographic Office for computing facilities at Chesapeake Beach, Maryland. This research was partially supported by National Science Foundation Grant 195–6064.

REFERENCES CITED

Fischer, A. G., 1960, Latitudinal variation in organic diversity: Evolution, v. 14, p. 64–81.
Irving, E., 1964, Paleomagnetism and its application to geological and geophysical problems: New York, John Wiley & Sons, 399 p.
Stehli, F. G., 1968a, Taxonomic diversity gradients in pole location: The recent model, in Drake, E. T., ed., Evolution and environment: New Haven, Yale Univ. Press, p. 163–227.
—— 1968b, A paleoclimatic test of the hypothesis of an axial dipolar magnetic field, in Phinney, R. A., ed., The history of the earth's crust: Princeton, Princeton Univ. Press, p. 195–207.
—— 1970, A test of the earth's magnetic field during Permian time: Jour. Geophys. Research, v. 75, p. 3325–3342.

MANUSCRIPT RECEIVED BY THE SOCIETY MARCH 29, 1971
PRESENT ADDRESS: DEPARTMENT OF GEOPHYSICS, STANFORD UNIVERSITY,
STANFORD, CALIFORNIA 94305   PRINTED IN U.S.A.

THE GEOLOGICAL SOCIETY OF AMERICA, INC.
MEMOIR 132
© 1972

Paleomagnetism of Anorthosite in Southern Norway and Comparison with an Equivalent in Quebec

R. B. HARGRAVES

AND

J. R. FISH

Department of Geological and Geophysical Sciences,
Princeton University, Princeton, New Jersey 08540

ABSTRACT

Paleomagnetic measurements (including alternating field demagnetization) on 16 oriented samples from five sites in the Precambrian anorthosite massif at Egersund, southern Norway, revealed a stable, consistent reversed remanent magnetization (after 150 oersted AF demagnetization, D = 315°, I = −82°, k = 92, α_{95} = 8°). The calculated paleomagnetic (south) pole lies at 42° N., 20° E. and is more than 45° from the pole (39° N., 40° W.) previously obtained from anorthosite of similar age at Allard Lake, Quebec. If Eurasia is fitted to North America according to the model of Bullard and others (1965), the rotated Egersund pole is at 42° N., 18° W. Allowing for the uncertainties in the data, this is not significantly different at the 5-percent level from the Allard Lake pole, and suggests that about 1,000 m.y. ago North America and Europe were contiguous as in Laurasian reconstructions.

INTRODUCTION

Anorthosite massifs (as distinct from anorthosites in stratiform complexes) are an uncommon plutonic rock series (see compilation of known occurrences by Anderson, 1969) of generally similar petrology throughout the world. Herz (1969) gave particular significance to this similarity by showing that all the anorthosite bodies of the world have ages between 1,000 and 1,700 m.y. and that they tended to fall in two distinct belts when Laurasia and Gondwana were reassembled.

141

Hargraves and Burt (1967) had obtained a paleomagnetic pole for the anortho-site series rocks at Allard Lake, Quebec. In view of the petrological and age simi-larity of the Quebec anorthosites to those in southern Norway, Hargraves thought to test the possibility of their predrift contiguity by obtaining paleomagnetic data from the Norwegian rocks for comparison. A suite of oriented samples was col-lected from the Egersund anorthosite massif in July 1968. This paper reports the paleomagnetic results obtained, and their comparison with the data from Allard Lake, Quebec.

SAMPLING AND MEASUREMENT

Sixteen hand samples were collected from fresh roadcuts at five sites in the Egersund area (Fig. 1). The samples were oriented by means of an astrocompass (La Rochelle, 1961). The remanent magnetism of at least two 1 in. long × 1 in. diameter specimen cores drilled from each sample was measured on a Princeton Applied Research SM1 spinner magnetometer. Reconnaissance AF demagnetization of selected cores up to fields of 500 oersted revealed that the anorthosites possessed NRMs varying from 10^{-2} to 5×10^{-5} emu/cc, which was, however, in all cases relatively stable (see Fig. 2). In light of these data, all the remaining cores were demagnetized at 50, 150, and 300 oer-sted. The mean vectors, at sample and site and locality levels, were negligibly different at 150 and 300 oersted (*see* Table 1), but the statistics for the re-sults obtained at 150 oersted are slightly better, and the 150 oersted mean vector has been used for the pole determina-tion.

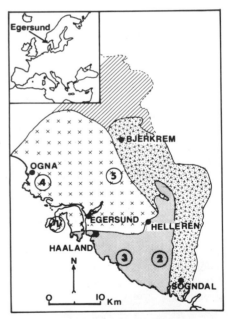

PETROLOGY AND MINERALOGY AND RELATION TO MAGNETIC PROPERTIES

The geology of the Egersund area (Fig. 1) has been described in several papers by P. Michot and J. Michot. Apart from other intrusive units, they distinguish two separate anorthosite massifs, the Egersund-Ogna (P. Michot, 1960, 1969; sample sites 1, 4, and 5), and the Haaland-Helleren (J. Michot, 1961; sample sites 2 and 3). In general the massifs consist of coarse, massive, leuco-anorthosite cores, with more gneis-sic, noritic border zones. The anortho-sitic rocks are cut by somewhat younger

Figure 1. Map showing location of Eger-sund in Norway, and generalized geology of the area (modified *from* P. Michot, 1960). The two principal anorthosite bod-ies—Egersund-Ogna (x-pattern) and Haa-land-Helleren (stippled)—are shown, and the paleomagnetic sample sites (circled numbers) within them; the v-dot pattern marks the Bjerkrem Sogndal mangerite massif; the diagonal lines signify charnock-ites.

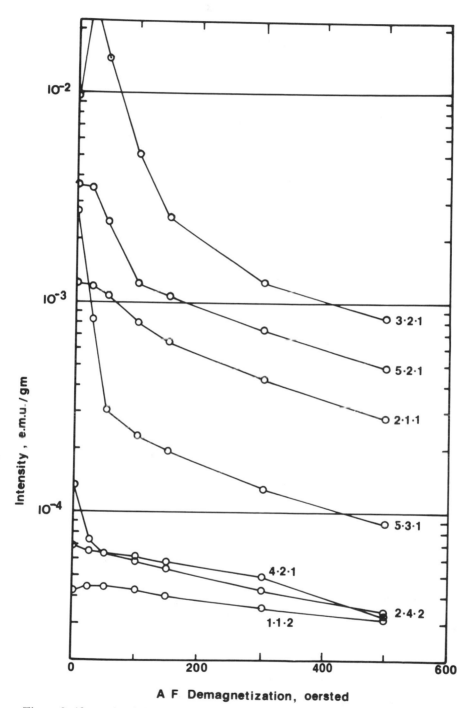

Figure 2. Alternating field demagnetization curves for representative specimens of the Egersund rocks sampled. Specimen 3-2-1 is from a pyroxene megacryst; 2-4-2 from a plagioclase megacryst; 5-3-1 is monzonorite; the others are anorthosite from sites 1, 2, 4, and 5. Note high intensity of the pyroxene and the relative instability of the monzonorite.

TABLE 1.   PALEOMAGNETIC RESULTS OBTAINED IN EGERSUND STUDY

Site	N	n	Peak Field oersted	D	I	K	α95
1	3	7	NRM	311.4	−75.3	44.7	18.7
			150	321.1	−78.8	173.7	9.4
			300	325.2	−77.1	192.9	8.9
2	4	9	NRM	303.1	−50.9	4.9	46.3
			150	295.1	−70.0	231.4	6.1
			300	298.6	−69.8	98.9	9.3
3	2	4	NRM	270.3	−34.7	4.1	90.0
			150	337.9	−83.2	12.7	77.2
			300	320.8	−85.9	13.0	76.1
4	3	6	NRM	58.2	−85.1	368.2	6.4
			150	64.7	−86.3	975.9	3.9
			300	71.7	−85.4	1074.9	3.8
5	4	11	NRM	0.2	−73.2	3.3	60.3
			150	311.1	−84.3	65.9	11.4
			300	298.3	−84.3	51.6	12.9
Mean of the			NRM	300.1	−68.3	9.8	25.8
5 sites:			150	315.4	−81.9	91.9	8.0
			300	314.4	−82.1	79.1	8.7

Sites: *see* Figure 1; N = number of samples;
n = number of specimens actually measured; Peak field oersted signifies intensity of AF demagnetization;
D = declination, I = inclination; k = estimate of Fisher's dispersion parameter K; α95 = half angle of cone of 95% confidence.

(although judged to be consanguineous) dikes and sheets of quartz norites or monzonorites (P. Michot, 1960, p. 28). The petrological similarity between this and the North American anorthosite massifs in the Adirondacks (New York State) and Quebec has been pointed out by several authors (for example, Hargraves, 1962, p. 182).

The oriented samples used in this study are, with three exceptions, coarse-grained (up to 1 cm) leucocratic anorthosite. The rocks showed little evidence of crushing or peripheral granulation, and the plagioclase in several samples displayed the chatoyance of laboradorite. Of the three exceptional samples, one is a plagioclase megacryst, one is a hypersthene megacryst, and the third is a sample of monzonorite from a dikelike body at site 5.

In thin section, the anorthosites consist of a coarse aggregate of anhedral plagioclase with subsidiary, strongly pleochroic hypersthenes, and minor opaque oxide. The plagioclase is charged with fine rod-shaped inclusions ($2\mu \times 10\mu$) which seem to be mostly ferromagnesian silicates, but some are opaque. In some crystals the inclusions appear to have aggregated to form larger crystals, the immediate surroundings being clear.

The hypersthene has relatively poor cleavage, which in some crystals is lined with fine reddish lamellae (exsolution?), imparting a schiller appearance. In all samples examined the hypersthene crystals are traversed by prominent serpentinized cracks, with associated fine opaque minerals.

In reflected light, the accessory opaque mineral is seen to be a coarse hematite-ilmenite intergrowth. The opaque rodlike inclusions in the feldspars, when large enough to identify, are also intergrowths of hematite and ilmenite in approximately equal proportions. These are indentical to the inclusions in the Brulé Lake (Quebec) anorthosite-plagioclase described by Anderson (1966).

The pinkish-brown schiller lamellae in the hypersthene likewise appear to be extremely thin plates of hematite-ilmenite lining the cleavage planes. The opaque phase associated with the serpentinized cracks in the pyroxene is hematite. No spinel phase was seen in reflected light.

The most likely origin for the primary oxide rodlets and plates in the plagioclase and pyroxene is thought to be exsolution from the silicate, although the crystallo-chemical relations are not clearly understood. The bulk composition of the lamellae intergrowths ($\cdot$5 hem: $\cdot$5 ilm) implies that they must have exsolved from the silicate at a temperature above the crest of the hematite-ilmenite solvus ($\sim$900° C; Carmichael, 1961).

The monzonrite sample contains feldspar crystals consisting of a plagioclase core, zoning outward to mesoperthite (40 percent), hypersthene (40 percent), with subsidiary augite (5 percent), and about 15 percent opaques. These consist of about equal portions of magnetite and ilmenite, either in discrete crystals, or associated in very coarse, sandwich intergrowth.

Measurement of saturation magnetization (8 kilogauss) versus temperature (Js/T) curves, in air and in vacuum, on an anorthosite chip and a fragment of a plagioclase single crystal revealed identical Curie point inflections between 570° and 590°. Taken alone this result can be considered to indicate the presence of pure magnetite ($T_c \simeq 580$°). Titaniferous hematite, however, is the only ferrimagnetic phase observed microscopically in reflected light. From the hematite-ilmenite solvus (Carmichael, 1961), unmixed hematite can contain at least 10 percent ilmenite in stable solid solution at room temperature, and the Curie point of such a titaniferous hematite would be about 590° C, very similar to that of pure magnetite. The Curie point recorded on the anorthosite and plagioclase samples is therefore tentatively concluded to be due to the titaniferous hematite observed, and not to magnetite. It should be noted, however, that the hematite phase in any natural unmixed hematite-ilmenite intergrowth will always contain *at least* 10 percent of ilmenite still in solid solution. The Curie point will be 600° C or below, and can be misinterpreted to indicate the presence of magnetite.

The AF demagnetization curve of the monzonorite sample (5–3–1, Fig. 2) indicates a substantial unstable component, with coercivity less than 100 oersted, and a stable component of intensity approximately 1×10^{-4} emu/cc, similar to the anorthosites. It is noteworthy that the monzonorite contains in excess of five percent by volume of magnetite, whereas the anorthosites contain less than one percent of titaniferous hematite as the only observable ferrimagnetic phase. The relative unimportance of homogeneous coarse-grained magnetite as a carrier of stable remanent magnetism is well illustrated by this example.

DISCUSSION OF PALEOMAGNETIC RESULTS

The stable component of NRM in all samples is reversed (N-seeking pole upwards), inclined at approximately 160° to the present field. As a result, elimination of the soft (normal) viscous component resulted, in some samples, in an initial increase in RM intensity.

The consistency of direction, both within and between all sites, is very good (Table 1). The remanent magnetism of the plagioclase and hypersthene megacryst samples was likewise indistinguishable from the associated anorthosite. One specimen of two from the feldspar crystal sample seemed to possess a more stable secondary component which required 300 oersted to erase, whereas the other gave stable directions after 100 oersted. The pyroxene megacryst specimens, on the other hand, possessed a strong but initially relatively unstable remanence; what remained after 300 and 500 oersted, however, was the strongest of Egersund samples measured. Although the megacrysts gave consistent magnetic data, we judge the ordinary anorthosite to be superior material for paleomagnetic study because of their higher microscopic coercivity.

There is no evident difference between the results for the Egersund-Ogna and the Haaland-Helleren massifs, and similarly, the monzonorite sample is indistinguishable from the associated anorthosite. This agreement is consistent with the hypothesis that all the rocks sampled are consanguineous and that the NRM was

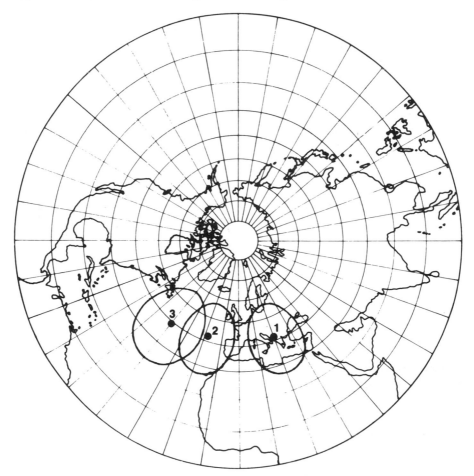

Figure 3. Northern hemisphere polar projection showing location of paleomagnetic poles. 1 = Egersund, this study; 2 = the Egersund pole rotated according to the continental reconstruction of Bullard and others (1965); 3 = Allard Lake (Hargraves and Burt, 1967).

acquired simultaneously on cooling, being effectively "blocked" at a relatively low temperature ($\leqq 300°$ C).

It is assumed that this low temperature is comparable to that at which the diffusive loss of Ar^{40} ceased, that is, the best estimate of the time of acquisition of the remanent magnetism is that given by the K-Ar radiometric age. This is reported to be $\simeq 1,000$ m.y. (Herz, 1969; Table 1). This pole is unlike that recorded from any European Paleozoic or younger rock, in keeping with the theory that it is a Precambrian paleomagnetism that has been preserved.

The Egersund anorthosite paleomagnetic pole is compared with that obtained from the Allard Lake, Quebec, anorthosite series (Hargraves and Burt, 1967) of similar 1,000 m.y. K-Ar age (Leech and others, 1963) as shown in Figure 3 and Table 2.

The poles are 45° apart when the Egersund pole is calculated according to the present-day coordinates of the sampling site. If the continents are juxtaposed according to the reconstruction of Laurasia given by Bullard and others (1965), however, the two are only 17° apart and are not significantly different (Watson and Irving, 1957; Table 2).

CONCLUSIONS

The results of the study are consistent with the hypothesis that the remanent magnetism of the anorthosite series at Allard Lake, Quebec, and Egersund, Norway, were acquired more or less contemporaneously approximately 1,000 m.y. ago, when the North American and European continents were contiguous, as in the Laurasian reconstruction. The original continuity of the "Grenville" belt from Quebec to Norway, as illustrated by Wynne-Edwards and Hasan (1970), is likewise favored by the data. The "proto-Atlantic Ocean," postulated by Wilson (1966) to have existed prior to the middle Paleozoic, must have, in turn, opened subsequent to the time of magnetization of the anorthosites.

TABLE 2. PALEOMAGNETIC POLES CALCULATED BY AVERAGING VIRTUAL POLES FOR EACH INDIVIDUAL SITE AND DATA FOR THE COMBINATION OF EGERSUND (ROTATED) PLUS ALLARD LAKE

Locality	Number of site poles averaged	Pole position	k	α95	R*
Egersund	5	42° N., 20° E.	28	15	4.86
Egersund (rotated)	5	42° N., 18° W.	28	15	4.86
Allard Lake (Hargraves & Burt, 1967)	4	39° N., 40° W.	27	18	3.89
Egersund & Allard Lake	9	41° N., 27.5° W.	24	10.8	8.66

*R = resultant vector; other symbols as in Table 1. A statistical test to determine whether the mean polar vectors for Allard Lake, and Egersund (rotated) are different (Watson and Irving, 1957) gave a result (see calculated parameters) indicating no significant difference at the 5-percent level:

$$\frac{N - 2(\Sigma R i - R)}{N - \Sigma R i} = 2.41$$

$$F_2, 2(N - 2) \ 5\% = 3.55.$$

ACKNOWLEDGMENTS

The samples were collected while Hargraves held a NATO post-doctoral fellowship at the Institut für Angewandte Geophysik, Munich; the hospitality of Professor G. Angenheister is warmly acknowledged.

We thank Mrs. Naoma Dorety for assistance with some of the remanence measurements and the Curie point determinations. This research was supported by National Science Foundation Grant GP 3451.

REFERENCES CITED

Anderson, A. T., 1966, Mineralogy of the Labrieville anorthosite, Quebec: Am. Mineralogist, v. 51, p. 1671–1711.
—— 1969, Massif-type anorthosite: A widespread Precambrian igneous rock, *in* Isachsen, Y.W., ed., Origin of anorthosite and related rocks: New York State Mus. and Sci. Service Mem. 18, p. 47–55.
Bullard, E., Everett, J. E., and Smith, A. G., 1965, The fit of the continents around the Atlantic: Royal Soc. London Philos. Trans., v. 258, p. 41–51.
Carmichael, C. M., 1961, The magnetic properties of ilmenite-hematite crystals: Royal Soc. [London] Proc., Ser. A., v. 263, no. 1315, p. 508–530.
Hargraves, R. B., 1962, Petrology of the Allard Lake anorthosite suite, Quebec, *in* Engel, A.E.J., and others, eds., Petrologic Studies (Buddington Volume): Geol. Soc. America, p. 163–189.
Hargraves, R. B., and Burt, D. M., 1967, Paleomagnetism of the Allard Lake anorthosite suite: Canadian Jour. Earth Sci., v. 4, p. 357–369.
Herz, N., 1969, Anorthosite belts, continental drift and the anorthosite event: Science, v. 164, p. 944–947.
La Rochelle, A., 1961, Adaptation of an astrocompass for collection of oriented rock specimens: Geol. Surv. Canada Topical Rept., no. 32, 4 p.
Leech, G. B., Lowden, J. A., Stockwell, C. H., and Wanless, R. K., 1963, Age determinations and geological studies: Canada Geol. Survey Paper, p. 63–117.
Michot, J., 1961, The anorthositic complex of Haaland-Helleren: Norsk. Geol. Tidsskr., v. 41, p. 157–172.
Michot, P., 1960, La Géologie de la Catazone: Le problème des anorthosites, la palingenèse basique et la tectonique catazonale dans le Rogaland méridional (Norvège Méridionale): 21st Internat. Géol. Congress, Copenhagen, 1960, Guide de l'excursion A9.
—— 1969, Geological environments of the anorthosites of South Rogaland, Norway, *in* Isachsen, Y. W., ed., Origin of anorthosites and related rocks: New York State Mus. and Sci. Service Mem. 18, p. 411–423.
Watson, G. S., and Irving, E., 1957, Statistical methods in rock magnetism: Royal Astron. Soc. Monthly Notices, Geophys. Suppl. 7, p. 289–300.
Wilson, J. T., 1966. Did the Atlantic close and then re-open?: Nature, v. 211, p. 676–681.
Wynne-Edwards, H. R., and Hasan, Z., 1970, Intersecting orogenic belts across the North Atlantic: Am. Jour. Sci., v. 268, p. 289–308.

MANUSCRIPT RECEIVED BY THE SOCIETY MARCH 29, 1971

THE GEOLOGICAL SOCIETY OF AMERICA, INC.
MEMOIR 132
© 1972

Kenoran Orogeny in the Hanson Lake Area of the Churchill Province of the Canadian Shield

L. C. COLEMAN
Department of Geological Sciences,
University of Saskatchewan, Saskatoon, Saskatchewan, Canada

ABSTRACT

The Hanson Lake area, Saskatchewan, lies within the Churchill structural province of the Canadian Shield, about 150 mi west of the Churchill-Superior boundary. On the Tectonic Map of Canada (1969) its rocks are designated as "mainly Archaean, folded during the Kenoran and refolded during the Hudsonian (orogenies)." Detailed geological studies indicate that these ideas need some revision.

The area is underlain by Kisseynew-type felsic gneisses that are overlain with no evident disconformity by Amisk-type metavolcanic and metasedimentary rocks, for which an isochron drawn on the basis of Rb/Sr isotope determination indicates an Archaean age of 2,521 ± 60 m.y. The metavolcanic rocks were intruded by four bodies of granite, for one of which a similar Rb/Sr isochron gives an Archaean age of 2,446 ± 16 m.y.

The Kisseynew- and Amisk-type rocks were complexly folded during one continuous cycle of deformation, toward the end of which the granites were emplaced. The age of the granites indicates that the deformation took place during the Kenoran orogeny.

The rocks were faulted and intruded by unmetamorphosed, discordant pegmatite dikes with an Rb/Sr age of 1,799 ± 2 m.y. However, there is no evidence of penetrative deformation later than the Kenoran and, therefore, no direct evidence of Hudsonian folding in the area.

INTRODUCTION

The Hanson Lake area of Saskatchewan lies within the Churchill structural province of the Canadian Shield, about 150 mi west of the Churchill-Superior boundary as defined by Stockwell (1964). Figure 1 shows an enlarged portion of the Tectonic Map of Canada (1969), within which this area is located. From this figure it can be seen that Precambrian rocks of the area are designated as being mainly Archaean, folded during the Kenoran orogeny and refolded during the Hudsonian orogeny. Detailed geological mapping of this area and geochronologic investigations indicate that this designation is incorrect and that this area, and possibly many others within the Church province, have undergone no structural deformation related to orogeny since Kenoran time.

Figure 2 is a generalized geological map of the area based on work by the author and J. W. Gaskarth which was part of a cooperative program of detailed geological

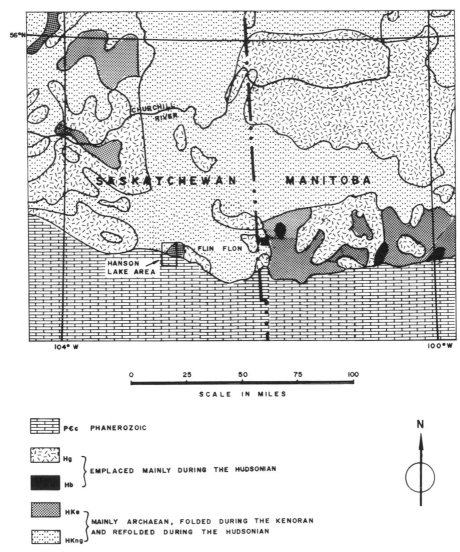

Figure 1. Enlargement of a portion of the Tectonic Map of Canada (1969) in the region surrounding Hanson Lake.

and geochemical studies sponsored by the Saskatchewan Research Council (Coleman and others, 1970). These investigations included the examination of all outcrops in the area and the collection of some 10,000 bedrock samples at 200-ft intervals for analysis for Fe, Cu, Zn, and Ni by x-ray fluorescence methods. While the results of these analyses were used primarily in the geochemical investigations carried out by J. R. Smith (Coleman and others, 1970, Pt. II), they were also used

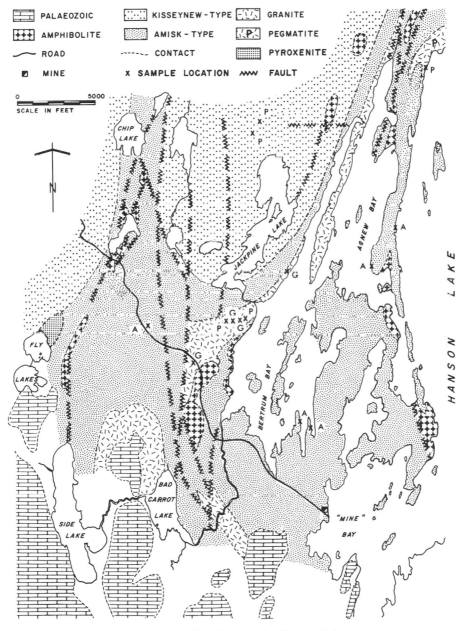

Figure 2. Generalized geology of the Hanson Lake area.

extensively by the author and Gaskarth in helping classify some of the rocks, particularly the metavolcanic rocks. Many of the collected specimens were also examined in thin section, and bulk chemical analyses were obtained for 61 of them.

GENERAL GEOLOGY

The Precambrian rocks of the area lie north of a Paleozoic escarpment that has a height of 20 to 60 ft. These rocks consist of four main groups.

1. A group of gneissic, porphyroblastic, migmatitic, agmatitic and, in a few places, massive phaneritic rocks occupies the cores of two major south-plunging anticlinal structures—one in the central part of the area and the other exposed in part along the area's western boundary. These rocks generally have a quartz-dioritic composition and are similar to rocks that elsewhere in the region are designated as being of Kisseynew type.

2. Overlying the Kisseynew-type rocks with apparent conformity is a group of metavolcanic and metasedimentary rocks. The metavolcanic rocks are dominantly fragmental and range from mafic rocks, whose Fe contents are usually greater than 7.5 percent, to felsic rocks, whose Fe contents are usually less than 5.5 percent, and soda rhyolite porphyries, whose Fe contents are usually less than 3.5 percent. The metasedimentary rocks are mainly graywackes that exhibit graded bedding in a few places and in which there are some conglomeratic horizons. Also included in this group is a number of bodies of calc-silicate rocks which are believed to represent metasomatized pyroclastics, lavas, and calcareous sediments, mixed together in varying proportions. Rocks of this group appear to be equivalent to ones which elsewhere in this region are referred to as being of Amisk type.

3. Several bodies of granite, granitic pegmatites of several types (Farquharson, 1971; Coleman, 1971), and two small bodies of metapyroxenite are intrusively related to other rocks of the area. One type of granitic pegmatite consists of small, irregular veins or lenses and discontinuous layers within Kisseynew-type rocks. The contacts of these bodies are characteristically gradational, and their development is believed to have accompanied the migmatization of the enclosing rocks. A second type, only one body of which is large enough to show in Figure 2, occurs as dikes and lenses in Kisseynew-type rocks, in Amisk-type rocks, and in granite, particularly in the granite body in the central part of the area. These bodies generally have sharp contacts and appear to be the products of pegmatitic differentiates intruded after, or late in, the development of the major folds in the area. A third type of granitic pegmatite is beryliferous and occurs in dikes which for the most part cross cut Kisseynew-type rocks in the north-central part of the area. These dikes range up to 20 ft in thickness, are exposed for distances up to 2,000 ft along strike, generally strike slightly east of north, dip nearly vertically, and have sharp contacts. The largest body of metapyroxenite occurs near the western margin of the area, and a second, much smaller, lenslike body occurs on an island in Hanson Lake near the eastern margin. While no contact relations can be observed in the former, in the latter occurrence the metapyroxenite exhibits sharp and discordant contacts with metagraywacke and with two small granitic pegmatite dikes which either terminate at the contacts or are truncated.

4. Rocks of several types which are all rich in amphibole occur in a number of bodies which are most common near major faults. Most bodies probably repre-

sent either basaltic flows that have been subjected to shearing, recrystallization, and metasomatism acompanying faulting, or basic intrusions which have also been extensively modified. The northern half of the body, immediately south of the granite in the central part of the area, consists of rocks rich in magnetite and appears to be a skarn.

All of the Precambrian rocks in the area, except for the last two types of pegmatite mentioned and the metapyroxenite, have been subjected to regional metamorphism. Petrographic and structural studies also indicate that, apart from the same exceptions, the Precambrian rocks have been subjected to extensive deformation which has produced in them several styles of folds, as well as numerous faults.

Amisk-type volcanic rocks are characteristically fine grained, commonly exhibit well preserved primary structures, contain relatively few pegmatitic segregations, and are generally not migmatized. In contrast, Kisseynew-type rocks are generally much coarser grained, exhibit no primary structures except for compositional layering, and contain numerous pegmatitic segregations and zones of migmatization. Despite these marked differences, mineral assemblages in both groups indicate no great difference in grade of metamorphism, which in both cases corresponds to the amphibolite facies of Turner (1968). The last two types of pegmatite mentioned exhibit no effects related to this regional metamorphism. While one metaphyroxenite body exhibits serpentinization and, in the other, large pyroxene crystals have been altered to hornblende, these effects cannot be definitely correlated with regional metamorphism. Evidence of retrograde metamorphism is common in all Amisk-type rocks and Kisseynew-type rocks and some granite, particularly near faults and where the rocks have been sheared.

Structural studies by Gaskarth (1967) suggest that deformation of Precambrian rocks in the area has involved the development of three successive fold styles. Diagrammatic representations of these fold styles and of some of their relations are shown in Figure 3. The first, which Gaskarth has designated as F_1 folds, are small and isoclinal with long limbs and are best seen in layered mafic volcanics adjoining the Kisseynew-type rocks of the two large anticlinal structures. F_2 folds are occasionally open but more commonly are nearly closed or isoclinal with shorter limbs than those of the F_1 folds. They are the most common type of minor folds in the area, and some are refolded by F_3 folds. In the layered mafic volcanics previously mentioned, they are commonly observed to have refolded F_1 folds. The contacts of one large unit of soda-rhyolite porphyry on the headland between Bertrum Bay and the main body of Hanson Lake appear to have been folded with the style of either F_1 or F_2 folds. F_3 folds range in size from microscopic to the magnitude of the large south-plunging anticlinal structure in the north-central part of the area. They are either gentle, open folds or chevron folds and are commonly observed refolding earlier F_1 and F_2 folds. Gaskarth suggests that the relation exhibited by the several fold styles can be best explained by a continuous process of deformation during which the successive fold styles resulted from the continuously changing structural positions and conditions of the rocks in which they were developed.

While all three fold styles can be observed in Kisseynew-type and Amisk-type rocks, the granite bodies of the area exhibit some features that can be most readily related to F_3 folding and others that indicate they were intruded into still older Amisk-type rocks. While in many places the boundaries of the granite bodies are broadly concordant with structures in the adjoining rocks, in some places they are

definitely discordant. In the granite body in the central part of the area, xenoliths are common, particularly in a zone extending northeasterly across the center of the body. Also, along the northwestern boundary of the southwestern body of granite is a zone of more mafic rock whose texture and composition suggest that it was formed by contamination of granite magma by material derived from adjoining mafic volcanic rocks to the west. A block of mafic volcanic rocks immediately west of the granite body in the center of the area is intimately intruded by a stockwork of granite dikes and stringers that is believed to have resulted from shattering of the country rocks and intrusion of magma during emplacement of this granite. The occurrence of skarn mentioned previously is also consistent with an intrusive origin for this body. The positions and outlines of all granite bodies suggest that they are phacolithic. Wherever their boundaries are concordant, they subparallel major F_3 folds in the surrounding rocks. The above evidence indicates that the granite bodies were intruded into older Amisk-type rocks and that their emplacement probably took place during the late F_3 folding. In all bodies, a well-defined penetrative mineral foliation can be seen which is generally either subparallel to the boundaries of the body or the axial planes of major F_3 structures. These relations would be consistent with a primary foliation developed during emplacement accompanying F_3 folding. Admittedly the foliation might also have developed at a much later time and, in such case, could be dated by the K/Ar method. Unfortunately, no K/Ar dates have been obtained for any of the rocks in this area.

Faults and zones of shearing or mylonitization are very common in Precambrian rocks of the area. One of these faults has been traced by Byers (1962) for some 170 mi to the north, and another system of faults may extend as much as 10 to 15 mi north-northeast of the area. Attitudes of schistosity associated with these faults indicate that they have a nearly vertical dip. Many of the faults are close to the borders of competent rock types, and their relations to other structures in the area suggest that they were either formed during late F_3 folding or after all folding had ceased.

GEOCHRONOLOGY

During the summer of 1966, a suite of fresh specimens was collected from rocks of the area for Rb/Sr isotope analyses, which were performed by Dr. H. Baadsgaard at the University of Alberta, Edmonton. Details of this work have been reported earlier by the author (Coleman, 1970). Unfortunately, Rb/Sr ratios in samples of Kisseynew-type rocks were found to be unsuitable for age determina- Amisk-type rocks are designated A; granite, G; and pegmatite, P. The isochron were obtained—one for the Amisk-type volcanic rocks, one for the body of granite in the central part of the area, and one for granitic pegmatites. Locations for the samples used in obtaining these isochrons are shown in Figure 2: samples of Amisk-type rocks are designated A; granite, G; pegmatite, P. The isochron

◄————————————————————————————————————

Figure 3. Sketches of various fold styles and patterns formed by refolding which are found in rocks of the Hanson Lake area (a) and (b), F_1 folds alone; (c), (d), and (e), F_2 folds alone; (f) and (g), F_3 folds alone; (h), (i), and (j), F_2 folds refolding F_1 folds; (k), F_3 folds refolding F_2 folds; (m), F_1 and F_2 folds wrapped around a large south-plunging F_3-type anticlinal structure. S1, S2, and S3 are axial planes of F_1, F_2, and F_3 folds, respectively.

TABLE 1. APPARENT AGES OF ROCK TYPES IN THE HANSON LAKE AREA

Rock Type	Apparent Age*
Amisk-type metavolcanics	2,521 ± 60 m.y.
Granite	2,446 ± 16 m.y.
Pegmatites	1,799 ± 2 m.y.

*$^{87}Rb = 1.39 \times 10^{-11}y$, $^{85}Rb/^{87}Rb = 2.600$(atomic),
$^{86}Sr/^{88}Sr = 0.1194$(atomic).

for the pegmatites was based both on samples from the larger discordant granitic pegmatites and on samples from the beryliferous granitic pegmatites. The isochrons obtained, listing plus and minus values after each age within the 95-percent confidence limits, are listed in Table 1.

Baadsgaard's work indicated that the data from the Amisk-type rocks and from the granite could be fitted to the same isochron with a somewhat greater scatter of points than that displayed when the data from each group of rocks were treated separately. The resulting apparent age for the two groups of rocks treated together is 2,425 ± 30 m.y.

There is a possibility that the bodies of granite in the area, which seem to have a phacolithic form, as has been previously mentioned, are of subvolcanic origin. In such a case, the single isochron would be appropriate. However, when the results of bulk chemical analyses for samples of Amisk-type rocks and for granites (Coleman and others, 1970) are plotted on variation diagrams, they exhibit different trends that suggest the two groups are not cogenetic and that the individual isochrons reported in Table 1 should be used.

Probably the most important result of the present work is that it provides unmistakable evidence for the presence of Archaean volcanic rocks and intrusive granites within the Churchill province some 150 mi west of its boundary. However, while Amisk-type rocks in the Hanson Lake area are shown to be Archaean, this idea should not be extended to the Amisk series in its type locality about 40 mi to the east. Preliminary Rb/Sr whole-rock isochrons indicate an age for the Amisk series of about 2,000 m.y. (M. R. Stauffer, 1970, personal commun.). Because of this great discrepancy in age, the observed relations between Amisk-type and Kisseynew-type rocks in the Hanson Lake area have no bearing on the actual relations that exist between rocks of the Amisk, Missi, and Kisseynew series in the type area.

SUMMARY

Geological studies in the Hanson Lake area combined with isotope analyses of some of the groups of rocks indicate that the Amisk-type rocks have an age of about 2,520 m.y. and are underlain by older Kisseynew-type rocks. Both groups were subjected to an extensive period of folding and metamorphism toward the end of which, about 2,450 m.y. ago, they were intruded by a number of bodies of granite. During this period, some of the Kisseynew-type material was melted locally to produce minor amounts of quartz-dioritic magma. When the results of chemical analyses of Kisseynew-type rocks and granite (Coleman and others, 1970) are plotted on variation diagrams, the resulting trends indicate that such magma and that producing the granite bodies are not related. After all deformation had ceased, and more than 600 m.y. after the intrusion of granite, pegmatite dikes were emplaced within structures that originally developed during late folding. This emplacement took place about 1,800 m.y. ago, after compressional forces producing these structures had been released.

It can be seen that Kisseynew-type rocks, Amisk-type rocks, and granite in the Hanson Lake area all are of Archaean age and that the only far-reaching effects of orogenesis that have acted upon them are of Kenoran age. While the time of pegmatite emplacement in the area was Hudsonian, this event does not appear to have been accompanied by any major folding.

Therefore, it appears that that part of the Churchill structural province within which the Hanson Lake area is located has remained as a stable crustal segment since Kenoran time—a situation analogous to that reported by Wynne-Edwards (1969) for some parts of the Grenville structural province.

REFERENCES CITED

Byers, A. R., 1962, Major faults in the western part of the Canadian Shield with special reference to Saskatchewan, *in* Stevenson, J. S., ed., The tectonics of the Canadian Shield: Royal Soc. Canada, Spec. Pub. No. 4, p. 40–59.

Coleman, L. C., 1970, Rb/Sr isochrons for some Precambrian rocks in the Hanson Lake area, Saskatchewan: Canadian Jour. Earth Sci., v. 7, p. 338–345.

—— 1971, Rb/Sr isochrons for some Precambrian rocks in the Hanson Lake area, Saskatchewan [Reply]: Canadian Jour. Earth Sci., v. 8, p. 182–183.

Coleman, L. C., Gaskarth, J. W., and Smith, J. R., 1970, Geology and geochemistry of the Hanson Lake area, Saskatchewan: Saskatchewan Research Council Geology Div. Rept. no. 10, 156 p.

Farquharson, R. B., 1971, Rb/Sr isochrons for some Precambrian rocks in the Hanson Lake area, Saskatchewan [Disc.]: Canadian Jour. Earth Sci., v. 8, p. 181–182.

Gaskarth, J. W., 1967, Petrogenesis of Precambrian rocks in the Hanson Lake area, east-central Saskatchewan [Ph.D. thesis]: Saskatoon, Univ. Saskatchewan.

Stockwell, C. H., 1964, Fourth report on structural provinces, orogenics, and time-classification of rocks of the Canadian Precambrian Shield: Geol. Survey Canada Paper 64–17, Pt. II, p. 1–21.

Tectonic Map of Canada, 1969: Geol. Survey Canada Map No. 1251A.

Turner, F. J., 1968, Metamorphic petrology: New York, McGraw-Hill Book Co., 403 p.

Wynne-Edwards, H. R., 1969, Tectonic overprinting in the Grenville province, southwestern Quebec, *in* Wynne-Edwards, H.R., ed., Age relations in high-grade metamorphic terrains: Geol. Assoc. Canada Spec. Paper No. 5, p. 163–182.

MANUSCRIPT RECEIVED BY THE SOCIETY MARCH 29, 1971

THE GEOLOGICAL SOCIETY OF AMERICA, INC.
MEMOIR 132
© 1972

Possible Fracture Zones and Rifts in Southern Africa

ARTHUR O. FULLER
Department of Geology,
Michigan Technological University, Houghton, Michigan 49931
and University of Cape Town, South Africa

ABSTRACT

Three major east-trending structures in Southern Africa are described. They are referred to as the 16°, 23°, and 29° S. lat belts, and they correspond in part to what are commonly referred to in Southern African literature as the Zambesi, Limpopo, and Orange River belts. It is shown that the crustal blocks defined by these belts have displayed a more or less independent mobility at intervals through much of geologic time, and that the belts themselves have acted as loci for periodic volcanism. It is proposed that the belts are analogous to present-day oceanic fracture zones and played a part in the fragmentation of Gondwanaland.

An extension of the African Rift System through South West Africa is postulated on the basis of some modern seismic activity and a line of Cretaceous alkaline volcanic centers. Opening of this rift in late Precambrian time provided a site for the accumulation of the Swakop Facies of the Damara System, and the extreme folding and thrusting of these rocks is related to closing in the Paleozoic. Rifting along the present west coast probably occurred simultaneously, leading to the formation of a proto-Atlantic Ocean. Subsequent closure of this proto-ocean is registered by intensely deformed and metamorphosed marginal facies of platform sediments underlying the Damara System.

Further, it is suggested that the hydrothermal activity responsible for ore mineralization at Tsumeb is related to these events.

INTRODUCTION

Developments during the past decade have prompted re-examination of certain broad structural and stratigraphic features of Southern Africa. In doing this, the author has adopted the models outlined below, considered appropriate to the tectonic circumstances prevailing on the subcontinent through much of geologic time.

AFRICAN UNITS, OR TESSERAE

It has long been recognized that Africa can be regarded as being made up of units, or tesserae, usually of polygonal outline, each unit showing a degree of independent mobility and separated from its neighbors by what have been referred to as hinge zones or lineament belts. This model has been adopted especially by Krenkel (1938) and Brock (1956).

The crustal units defined according to this model correspond to what are often referred to as cratons and orogens. The author takes the view that undue emphasis has been placed on the uniqueness and permanency of these features, preferring the modified view of Crockett and Mason (1968, p. 535), who write, "Taking an extremely long view of crustal history, what are now classified as cratons have merely remained unaffected by orogeny for a longer period than what are now classified as orogens. In the author's opinion, there is no inherent tendency for one part of the crust to become an orogen and another a craton." This point is well illustrated by the aspect of Zambian and adjacent Central African Precambrian geology presented by Vail and others (1968), where the Tanzanian, Bangweulu, Kasai, Rhodesian, and Transvaal cratons are defined by the crisscrossing Ubendian, Limpopo, Kibaride, Irumide, Mozambique, Lufilian, and Zambesi orogenic belts. Certain of these belts are given special significance in the analysis presented here, so that the model adopted represents a compromise between a unique and permanent crustal segment partition and one which is apparently more or less random.

EXTENSION OF OCEANIC FRACTURE ZONES INTO CONTINENTS

The first attempts to recognize extensions of oceanic fracture zones into the adjacent continental blocks were made for the northeastern Pacific, western North America area. Vacquier and others (1961, p. 1253) summarized their results when they wrote, "Evidence of the extension of the oceanic faults into the continent is weak and contradictory." Their analysis was made prior to the development of the plate tectonics model, yet it is interesting to note their tentative conclusions that "the displacements along oceanic faults propagate into the continents at the time they happen," and that "the present discussion has postulated without proof the extension of the oceanic displacements into the continent at the time of their formation." They were particularly concerned with apparent lateral displacement of magnetic anomalies mapped in the Pacific and were searching for evidence of similar displacement on land. Why they were unsuccessful has now been explained by the plate tectonics model, but their work did serve to show that the landward extension of the fractures influenced crustal development to an extent, however "weak and contradictory."

More recently Kutina (1969) has re-examined the same area and has concluded that preferential accumulation of large ore deposits occurs along landward prolongation of the main fracture zones of the northeastern Pacific.

An important aspect of great oceanic fracture zones such as the Mendocino and Murray is that they have been shown to divide segments of the crust which possess contrasting features, such as depth, sediment thickness, volcanicity, topography, and in places heat flow (Menard, 1964). It is a basic premise in the model adopted here that similar contrasts may be expected between crustal segments on continents where it can be demonstrated, or reasonably inferred, that the oceanic fractures do have landward extensions.

The presence of such continental fractures was proposed by Wilson (1965) in his now classic paper on transform faults. Figure 6A of Wilson's article showed a set of "lines of old weakness" which, after the onset of rifting, became the sites of active transform faults whose lengths were determined by the geometry of the rift fracture. The greater part of the "lines of weakness," inactive as far as transform movement is concerned, should presumably be preserved on the continental blocks bordering the spreading ocean.

Support for the extension of oceanic fractures into continental blocks is provided by the work of de Loczy (1970) in South America and Burke (1969) in Africa. De Loczy describes a number of structural zones which appear to line up with Atlantic fractures, possibly the best example being the direct alignment of the Amazonas trough with the Romanche fracture. Particularly significant is his observation that the faults were active during Precambrian-Eocambrian time, as shown by stratigraphic changes in the strata adjacent to them. He concludes that, "because the east-west striking geosutures and transcurrent faults seem to extend into the asthenosphere and even into the upper mantle, they possibly are primordial lineaments of the infra-Precambrian basement." Burke's study is important in demonstrating support for Wilson's contention that the offset of a continental margin by a fracture zone transform fault must be the same as the offset at the mid-ocean ridge. Some of the fractures Burke describes (for example, the Akwapin) cannot be traced farther than 300 km inland, but their importance is emphasized by the observation that they have controlled the distribution of Cretaceous and Tertiary sedimentary basins. It is observations of this type which the author regards as being significant in the analysis of aspects of Southern African stratigraphy and structure presented here.

PLATE TECTONICS MODEL

The basic premise of the plate tectonics model, in which essentially rigid lithosphere plates are created at spreading centers of oceanic ridges and consumed in trenches, is accepted for the analysis to follow. Further, following the argument of Dewey and Horsfield (1970), several riders to the model are considered pertinent to the structural evolution of Southern Africa. (1) The relation of continents, island arcs, and oceans has been determined by an ocean-based plate mechanism for at least 3×10^9 yr. (2) Oceans may expand and contract, expansion occurring during continental disruption (plate growth) and contraction during convergence and collision. (3) Continental margins will be the site of thick extensive sedimentation during continental separation and subsequently the site of orogeny during continental convergence. (4) Continental accretion involves a continuous sequence of separation and amalgamation of continents, sometimes along old lines of earlier orogenic belts, and sometimes across them to open new continent/ocean interfaces.

AFRICAN RIFT MODEL

The modern African Rift valleys are taken to be the products of the first stage of spreading (Wilson, 1969). However, it is accepted that "an extensive dislocation of Africa, originating the East African Rift System, took place in the Shamvaian Cycle" (2.5 to 2.8 b.y. ago) and "that movements have occurred during major orogenic cycles down to the present" (McConnell, 1967).

An important consideration in the analysis to follow concerns the southward extension of the rift system. This has recently been examined by Fairhead and Girdler (1969), who applied the technique of Joint Epicentral Determination to relocate earthquake epicenters relative to a master event which occurred in Tanzania on May 7, 1964. These epicenters are shown on Figure 1, together with the zones of possible incipient rifting suggested by Fairhead and Girdler. An alternative extension of the rift system, following the indicated epicenters, can be drawn (Fig. 1) in a southwesterly direction, striking the South West African coast in the vicinity of Walvis Bay. This is considered a more satisfactory extension because (1) it fits the fracture system of the African continent better (Furon, 1963), and (2) it agrees closely with the line of alkaline volcanic centers in South West Africa

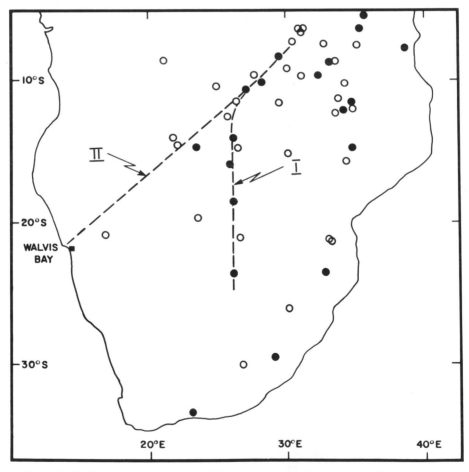

Figure 1. Epicenters for Southern Africa for the period January 1955 to June 1968. Solid circles are Joint Epicentral Determination redetermined epicenters. Possible "manmade" quakes have been omitted. Fairhead and Girdler's tentative southern extension of the rift is shown (I), as well as the extension proposed in this text (II). Based on Fairhead and Girdler (1969, Fig. 1, p. 1019).

(Fig. 2) and hence displays the well-known association between rift faulting and alkaline volcanism.

Significantly, this trend is in line with the major part of the Walvis Ridge, which may therefore very tentatively be interpreted as marking a line of volcanic centers on an extension of the rift system into the young crust of the southeast Atlantic; on the "hot-spot" hypothesis (Dietz and Holden, 1970), it would emphasize the junction of East African Rift extension with the site of the incipient Mid-Atlantic Rift.

EVIDENCE FOR ANCIENT EAST-TRENDING FRACTURE ZONES IN SOUTHERN AFRICA

Three well-known east-trending belts will be examined. They occur close to 16°, 23°, and 29° S. lat, and will be identified below by latitude rather than name. They are usually referred to as orogenic belts, mobile belts, lineaments, or zones, these names being preceded by the geographical location so that the 16° belt is commonly prefixed by Zambesi, the 23° by Messina or Limpopo, and the 29° by Orange River or Kheis.

The three belts will be examined here for evidence of continuity across the subcontinent and the contrasting geologic records preserved in crustal blocks defined by them will be noted. This analysis is somewhat handicapped by the extensive cover of Kalahari deposits (Fig. 2) which blanket so much of the west-central region of Southern Africa. This demands that two major regions be examined, the

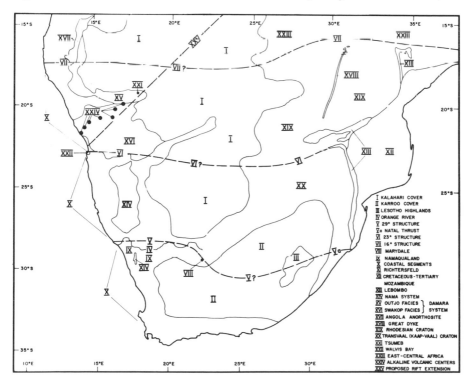

Figure 2. Index to location of features discussed in the text.

one lying between the west coast and 20° E. long, and including most of South
West Africa and Namaqualand, the other mostly east of 26° E. long and including
parts of Zambia, all of Rhodesia, and the Transvaal and territory to the east.

The most well-documented and most extensive of the three belts is the southernmost, and it will be dealt with first.

29° Belt

In defining the position of this belt and the two others to be described, it must
be borne in mind that if the oceanic fracture zones are to be used as a model for
their supposed counterparts on the continent, then representation of the land structure by a line rather than a zone perhaps some 60 km wide is somewhat misleading.
Nonetheless, lines are used on the maps here (Fig. 2) and should therefore be
viewed as indicating only the general location of the feature described.

Nicolaysen and Burger (1965) were the first to suggest that the 29° belt continued from the west to east coast of South Africa. Age data, coupled with the
mapping of Matthews (1959), showed clearly that crustal segments of vastly different ages were separated by east-trending thrusts in Natal (Fig. 2). The block to
the south of the 29° belt showed persistent 1 b.y. age, and this pattern was repeated
in the western sector of the country, where again the structure appears to represent a boundary between provinces of ages differing by 1 b.y. or more. In the
Marydale area (Fig. 2) two granites immediately north of the belt yielded ages
of 2.6 b.y., whereas to the south uniform ages of about 1 b.y. were recorded.

In the western sector, the 29° belt is marked by a distinctive lithology rather
than fractures, with thick lava sequences and intensely metamorphosed, fine-
grained quartzites (Kheis System) being characteristic.

Attempts to interpret this region as marking the infrastructure of an orogenic
belt in the classical sense have been unsatisfactory. Martin (1969, p. 20) summarizes the problem as follows: "There would be no difficulty in interpreting the Kheis
System as a normal geosynclinal suite which after deposition became involved in
an orogeny, if only an overlap of Kheis rocks onto an older basement or a transition into a platform facies existed. This, however, is not the case." Clearly the 29°
structure marks the boundary of a 1 b.y.-old province to the south with an older
block to the north. The view held here is that the belt marks a zone of ancient
fracture, locus of volcanic activity and repeated mobility.

Evidence supporting this interpretation is provided by Crockett and Mason
(1968). They point out that the belt is marked by a dense concentration of kimberlite pipes, the site of the Karroo lavas in the Lesotho highlands (Fig. 2) and site
of recent earthquakes. Furthermore, "the Namaqualand trend is of interest in
view of the possible recent recrudescence of movement along this line. Old drainage
channels running southwards towards the Orange River have been interrupted along
an approximately east-west axis of uplift (G.T. Lamont, personal communication)
which may be related to the Namaqualand trend." This youthful aspect of the
ancient structure is possibly also supported by certain features of the west coast
and continental shelf. The western margin of Southern Africa shows that the coastline can be divided into geometrically distinctive segments (Fig. 2), the divisions
occurring at approximately 23° and 29° S. lat, and Simpson and du Plessis (1968)
show a distinct and abrupt narrowing of the continental shelf, as defined by the
shelf break, north of 29° S. lat.

All the features mentioned above are regarded as evidence of independent mobility of the crustal blocks separated by the 29° belt, which in this analysis takes on the aspect of a deep fracture zone which has been active repeatedly throughout much of geologic time. Certain contrasting features of the geology of the blocks separated by the structure have already been dealt with; others must be mentioned. Firstly, Martin (1965) points out that the Kaaien quartzite, an important formation in the Kheis System in Namaqualand and northern Cape Province (that is, south of the 29° belt in the 1 b.y.-old province) is not found in the adjoining parts of South West Africa (that is, north of the 29° structure). Secondly, in the extreme western sector, Martin indicates several important discontinuities across the 29° belt (marked by the Orange River) in the region known as the Richtersveld (Fig. 2) and adjoining parts of South West Africa to the north. Thus the Black Hills Formation (700 m of carbonates and clastics) and the Holgat Formation (9,000 m, mostly clastics), both of the Gariep System, are not found north of the river, and there are still uncertainties as to whether the Hilda Formation (3,500 m of carbonates and clastics) does cross the river. The Stinkfontein Formation of the Richtersveld, which has an estimated thickness of 9,000 m, consisting mostly of clean clastics, does not occur north of the river. The Grootderm Formation, which includes an enormous thickness of lava (4,000 m), appears to be represented north and south of the river, and it is significant that here as in the Kheis System the volcanics show greater persistence across the 29° belt than the associated sediments. Presumably, eruptive fissures were located in the 29° belt in these late Precambrian times as they apparently were during Karroo times (Stormberg volcanics).

23° Belt

This belt was reviewed by Nicolaysen (1962). It is marked by an east-northeast-trending sequence of flow-folded marbles, calc-silicate rocks, quartzites, magnetite-quartzites, quartz-chlorite-sericite schists, and associated anorthosite, serpentinite, granites, and gneisses, the sequence being referred to as the Messina Formation. Age data from a radioactive concentrate (mostly sphene) from commonly occurring metamorphic schlieren indicated an age of about 1,940 m.y. for the major thermal event affecting the area. The belt is about 70 km wide and can be traced for some 500 km from east to west, being obscured in the west by the Kalahari cover and in the east by Cretaceous and younger formations of Mozambique (Fig. 2).

Cox and others (1964) have summarized several important structural features of the area and have emphasized the fact that vertical movements on fractures have repeatedly controlled the north-south distribution of sedimentary formations and have served as loci of basaltic volcanism. Several of their observations concerning faulting reveal periodic reactivation, and they conclude that "the Limpopo-trend faults are all probably genetically related to each other and are controlled by a common Basement structure." They show that during the deposition of the Dominion Reef (2.8 b.y.) Loskop and Waterberg Systems (1.4 to 1.8 b.y.) the area was the site of basaltic volcanism. Further, they comment that, just as in the Karroo period, the Messina block at least locally marked the northern limit of Karroo sedimentation, so it marked the northern limit of the pre-Karroo volcanic and sedimentary formations.

The comments above refer to observations made in the 23° belt itself, or close to it, and clearly reveal its importance in serving as a site for volcanic activity and limiting the extent of major sedimentary units. On a regional basis its influence on crustal mobility is even more striking. Firstly, the belt marks the point where the Lebombo monocline (Fig. 2) makes an abrupt change of strike, from north-south (to the south) to northeast-southwest (to the north). Secondly, it marks the northern limit of the crustal block lying between the 23° and 29° belts—including the whole of the so-called Transvaal or Kaap-Vaal craton—which contains the magnificent assemblage of Precambrian sequences including the Dominion Reef, Witwatersrand, Ventersdorp, Transvaal, and Waterberg Systems. Most of the Pre-cambrian history of South Africa is recorded on this block, a history which is not repeated or represented north of the 23° belt. Thus, on a regional scale, the observations of Cox and others (1964) are strikingly confirmed.

Apparently, as with the 29° belt, the 23° belt marks the site of a deep fracture system that has been active repeatedly through most of geologic time. The evidence of plastic deformation and metamorphism can be accounted for by assuming that what is seen on the surface today represents previous deep levels of this fracture system which have been elevated by localized vertical movement. This interpretation obviates involving an ancient geosynclinal belt, an involvement which is fraught with difficulties such as locating contemporary stable platform correlatives of the Messina Formation and post-orogenic derivatives.

The westward extension of the 23° belt is obscured in the central regions of Southern Africa by the blanket of Tertiary sediments of the Kalahari formations; evidence for its continuity across the subcontinent must be sought west of 18° W. long. in the territory of South West Africa. The first indication that the 23° belt may exist in this region is the break in the west coast profile in the vicinity of Walvis Bay (Fig. 2). A similar break was noted in the vicinity of the Orange River where the 29° belt is considered to reach the coast. The break at Walvis Bay is regarded as no more than suggestive that a major east-trending structure, which separates crustal blocks with more or less independent mobility, exists in this region. Stronger evidence is provided by the contrasting geologic record and structural style of the blocks lying north and south of 23° S. lat in South West Africa. To the south, the block lying between 23° and 29° S. lat (Fig. 2) is dominated by the almost undisturbed Eocambrian-Cambrian Nama System. This extensive succession of platform sediments, with carbonates in the lower half and fine-grained clastics above, totaling some 1,300 m in thickness, extends through the entire length of the block, and fragments of at least certain parts of it occur in the block to the south of 29° S. lat, indicating that this latter structure was not active during the probably eustatic sea-level rise which flooded the western sectors of South Africa near the end of the Precambrian. This rise may well have been associated with a decrease in ocean volume concomitant with the appearance of oceanic ridges during an Eocambrian drift episode. Underlying the Nama System is a north-striking sequence of acid and basic volcanics and clastics whose age and stratigraphy are as yet poorly understood. Work by van Niekerk and Burger (1969) suggests a late lower Pro-terozoic age of about 1,100 m.y. for at least part of these underlying sequences. To the north of 23° S. lat, the geology of South West Africa is dominated by rocks of the Swakop Facies of the Damara System (Fig. 2), to use the nomenclature of Martin (1965). The view presently held is that deposition of the Damara System overlapped in time the formation of Nama rocks to the south, but the lithology of the Damara and its subsequent folding about northeast-trending axes and metamorphism points to a major contrast in the tectonic style of the blocks north and south of 23° S. lat.

16° Belt

This feature marks the northern boundary of the Rhodesian "craton" and the southern boundary of the East and Central African block (Fig. 2). The latter is characterized by crisscrossing "orogenic" belts which in turn define a number of stable nuclei. This aspect has been referred to previously in connection with the work of Vail and others (1968). It is not repeated south of the 16° belt, where the Rhodesian block is characterized by numerous ancient greenstone belts in a sea of early to middle Precambrian granites which display the "gregarious batholith" style. Several groups of Precambrian sediments of uncertain age and interregional correlation (Piriwiri, Deweras, Lomagundi, and Hunyani) occur in the Rhodesian block, and their relation to units in South Africa and East-Central Africa is still the subject of active debate.

Further evidence that the 16° belt separates blocks of the crust which have displayed at least partially independent mobility through most of geologic time is seen in the termination of the Lebombo group of volcanics against it from the south and similar termination of the Great Dike (Fig. 2), which is entirely restricted to the Rhodesian block. The contrasts mentioned here are well known and have been the subject of comment by workers such as Krenkel (1938) and Brock (1956); however, the present interpretation relates them to what are regarded as fundamental east-trending fractures which possibly extended across Southern Africa and played a part in determining the site of oceanic fractures.

The most informative account of what is here referred to as the 16° belt is provided by de Swardt and others (1965). They describe several "dislocation zones" (Kapiri, Mposhi-Mkushi, and Mwembeshi) which show evidence of periodic movement originating in pre-Katanga (late Precambrian) time. Vertical movement has affected the north-south extent of both Katanga and pre-Katanga formations, and across the Mwembeshi zone juxtaposition of contrasting facies of Katanga rocks suggests a horizontal displacement of about 200 km. Analysis of fold structures suggests (de Swardt and others, 1965, p. 99) that "the Rhodesian block moved northeast and east for some 100 miles along arcuate dislocation zones that coincide roughly with the Zambesi trough."

In attempting to trace this belt across Southern Africa, the obscuring influence of the Kalahari cover again presents problems. However, scattered outcrops of pre-Damara formations with pronounced northeast grain in Botswana, parallel to the strike of the Damara belt, have suggested a continuation of the Zambesi trends in a southwesterly rather than westerly direction. The suggestion is made here that the 16° belt parallels the 23° and 29° belts, and that it emerges on the western margin of the subcontinent at the southern end of the huge Angola anorthosite massif (Fig. 2), which has been shown to be partly fault bounded (Martin, 1965, p. 14). Evidence for its extension into this region is not strong, but is supported by the location of the Outjo and Swakop Facies of the Damara System, which have their fullest development in the South West African block lying between 16° and 23° S. lat (Fig. 2). The Outjo Facies does extend north of 16° S. lat being represented by the Bembe System of the Chella Plateau, Angola, and this is interpreted as supporting the widespread Eocambrian eustatic transgression referred to previously.

EVIDENCE FOR A PROTO-ATLANTIC
AND PRE-MESOZOIC DRIFT

For the analysis presented here it is assumed that pre-Mesozoic drift occurred, and that opening and closing of a proto-Atlantic along a rift closely corresponding to the present west coast of Southern Africa took place in late Precambrian time and involved contemporaneous spreading of an extension of the African Rift System through the Damara belt of South West Africa.

Evidence for the existence of a proto-Atlantic has come from various sources in recent years. Drake and others (1968) have proposed the existence of a small Paleozoic North Atlantic Ocean to account for features of the continental margin of eastern North America. They saw no evidence for the existence of a proto-South Atlantic Ocean, in that there is "a lack of orogenic belts such as are formed in the Paleozoic in the North Atlantic." However, recent reconnaissance surveys along the coastal regions of northern South West Africa have revealed a marked coastward increase in deformation and metamorphism along north-south lines of Eocambrian formations, with pronounced thickening, migmatization, and granitization of sediments which farther to the east resemble a stable shelf facies (P. Guj, oral commun.). This phenomenon is interpreted to indicate an Eocambrian-Cambrian continental margin of convergence, and hence to mark the site of an orogenic belt.

Smith and Hallam (1970), in a thorough analysis of a computer fit of the southern continents, have suggested the possibility of pre-Mesozoic drift. They provide evidence "for the past existence of a small oceanic area next to southern South America, South Africa and north-west Antarctica." Their conclusions are especially significant to the present analysis. They write of Gondwanaland as follows: "Whether or not it existed as a plate in Lower Paleozoic and Precambrian time is debatable. Hurley and Rand have suggested that Gondwanaland marks the site of an ancient stable area (that is, part of a plate) that has grown peripherally since at least early Precambrian time. Matching orogenic belts of Precambrian age certainly support this idea, but Dewey's and Wilson's interpretation of the Caledonides and Appalachians implies that major orogenic belts mark the site of ancient oceans which have been eliminated by the approach of two continental masses. That is, orogenic belts are ancient compressive plate margins sited near the borders of vanished oceans. Applied to Gondwanaland, this interpretation would mean that it has existed as part of a single plate only since the time of the last orogeny to affect its interior—in the Lower Paleozoic." The basis for the comment above was derived from Wilson (1966), who proposed that a proto-Atlantic Ocean existed in lower Paleozoic time, and from Dewey (1969), who suggested that the evolution of the Appalachian/Caledonian orogen was related to a cycle of oceanic expansion and contraction, a cycle starting with a narrow North Atlantic in the Precambrian which disappeared in Silurian to Devonian time.

STRATIGRAPHIC IMPLICATIONS

The system of major east-trending belts at approximately 16°, 23°, and 29° S. lat, described here, are regarded as marking the sites of ancient abyssal fractures which divided the crust of Southern Africa into a number of blocks capable of movement more or less independent from their neighbors. As a result, each block has a distinctive geologic record. Although these fractures may date from the earliest geologic time, their influence on sedimentation patterns within the blocks they define probably did not commence until about 2.6 b.y. ago. Prior to that the record consists of ancient greenstone belts. In the model proposed by Anhaeusser and others (1969) these belts are considered to have formed by the outpouring of lavas and accumulation of sediments in elongate troughs which marked the sites of parallel fractures throughout the subcontinent. The trend of these fractures in Southern Africa is east-northeast and closely agrees with the strikes of the major 16°, 23°, and 29° belts. It is proposed, therefore, that the fundamental tectonic fabric of the subcontinent was established more than 3 b.y. ago, and that by the time the crust had evolved to a state similar to that which it displays today (that is, about 2.6 b.y. ago), the numerous deep fractures which marked the greenstone belts had been reduced to three.

Most of the thick Precambrian sequences in Southern Africa developed in cratonic basins with internal supply, and it is evident from the preserved and inferred distribution of these sequences that the role of the major east-west belts was to isolate blocks which displayed independent vertical mobility. Furthermore, if these continental belts are analogous to oceanic fractures, then they might be expected to contain sedimentary and volcanic deposits limited by the topographic restraints of the fracture zones themselves. The evidence of repeated movement along these fractures, periodic volcanicity, and presumably higher heat flow suggests that the preserved record of these localized sediments and volcanics will be difficult to decipher. It is precisely this situation which is thought to occur in the three belts described.

The second tectonic element that was significant in shaping the geology of Southern Africa is the African Rift System. Apparently an incipient southwesterly extension of this rift passes through South West Africa, cutting the coast in the vicinity of Walvis Bay and possibly extending into young Atlantic crust along the Walvis Ridge. The slight seismic activity associated with this trend justifies the term "incipient," but it is suggested that this trend is an ancient one which is tending to reopen. The model proposed is that in late Precambrian time this part of the rift became the site of active spreading, and that a stage was reached similar to that displayed by the modern Red Sea. Within the trough thus formed was deposited the Swakop Facies of the Damara System, a sequence of carbonates and clastics of unknown actual thickness, but estimated to be about 8,000 m thick. At present this facies of the Damara System is tightly folded about northeast-trending axes. The deformation is more severe toward the southeast margin where overturning, thrusts, and nappes occur along the borderland. Intense metamorphism predominates in the southwest sectors with extensive granitization and massive pegmatite emplacement.

According to the models adopted in this analysis, the Damara Rift opened and closed and little new oceanic crust developed during the spreading phase. As a result, the Damara System is underlain by sialic crust, as observed by Martin (1969).

The extension of the African Rift into South West Africa is supported by the line of alkaline volcanic centers. In view of the recent discoveries of metallization in the Red Sea area, it is interesting to note that the extraordinary lead-zinc-copper mineralization at Tsumeb (Fig. 2) lies close to the projected line of the South West African Rift, and farther to the southwest, in the vicinity of Walvis Bay, extensive uranium, copper, and lead deposits are known to occur.

The development of rifts in Southern Africa introduced a new phase of sedimentation involving continent-ocean interfaces, although the sediment sources
probably remained internal during the Damara event. This style of sedimentation
is again evident along the western boundary of the subcontinent, which probably
was the site of spreading in late Precambrian time. Whether this spreading occurred
simultaneously with the Damara rifting is not known. Radiometric ages for the
Pan-African event span from 450 to 680 m.y. ago. If the Damaran rifting can be
correlated with the Katangan interval episodes (580 to 680 m.y.), then opening
and closing of the Damara Rift may have occurred prior to the spreading which
split Gondwanaland on a line closely following the present position of the west
coast of the subcontinent. According to this view, the closing episode would be
related to a shift of the spreading center from the African Rift to the west coast
line. On the other hand, evidence from the coastal regions of northern South West
Africa contradicts this interpretation and suggests that the two rifts opened more
or less contemporaneously, with a junction in the vicinity of Walvis Bay.

On the western coastal regions of northern South West Africa (Kaokoveld),
widespread coarse- to fine-grained clastics of the Nosib Formation underlie the
Damara System. They increase in thickness westward and become progressively
more metamorphosed and deformed, grading into gneisses. These deposits point
to the former presence of a continental margin, and presumably indicate a proto-
Atlantic which opened in the late Precambrian and subsequently closed. The record
suggests that this rifting was on a larger scale than that which produced the trough
in which the Swakop Facies of the Damara was deposited, allowing thick accumulation of clastics on the continental margin, which subsequently went through a
cycle of deformation and metamorphism associated with a zone of convergence.
Nosib rocks are also present within the area of the proposed Damara Rift, so that
it seems likely that both spreading centers were active at the same time.

CONCLUDING REMARKS

The analysis of Southern Africa presented here has emphasized the role of two
apparently independent sets of fractures. The east-trending set divides the subcontinent south of 16° S. lat into three blocks on the basis of distinctive geologic
records. The second set is related to the African Rift System. No special significance
is attached to their present orientations.

Both sets are regarded as lines of weakness which have been in existence through
most of geologic time, and therefore permitted more or less independent vertical
or horizontal movement of adjacent blocks in response to forces causing crustal
deformation.

ACKNOWLEDGMENTS

The manuscript was critically read by Franklin B. van Houten and Robert B.
Hargraves of Princeton University, and grateful acknowledgment is made to them
for many helpful suggestions with regard to style and content. These have resulted
in an improvement in the text, but the author remains responsible for the ideas
presented.

REFERENCES CITED

Anhaeusser, C. R., Mason, R., Viljoen, M. J., and Viljoen, R. P., 1969, A reappraisal of some aspects of Precambrian shield geology: Geol. Soc. America Bull., v. 80, p. 2175–2200.

Brock, B. B., 1956, Structural mosaics and related concepts: Geol. Soc. South Africa Trans. and Proc., v. 59, p. 149–157.

Burke, K., 1969, Seismic areas of the Guinea coast where Atlantic fracture zones reach Africa: Nature, v. 222, p. 655–657.

Cox, K. G., Johnson, R. L., Monkman, L. J., Stillman, C. J., Vail, J. R., and Wood, D. N., 1964, The geology of the Nuanetsi igneous province: Royal Soc. London Philos. Trans., Ser. A, v. 257, p. 71–218.

Crockett, R. N., and Mason, R., 1968, Foci of mantle disturbance in Southern Africa and their economic significance: Econ. Geology, v. 63, p. 532–540.

de Loczy, L., 1970, Role of transcurrent faulting in South American tectonic framework: Am. Assoc. Petroleum Geologists Bull., v. 54, no. 11, p. 2111–2119.

de Swardt, A. M. J., Garrard, P., and Simpson, J. G., 1965, Major zones of transcurrent dislocation and superposition of orogenic belts in part of Central Africa: Geol. Soc. America Bull., v. 76, no. 1, p. 89–102.

Dietz, R. S., and Holden, J. C., 1970, The breakup of Pangaea: Sci. American, v. 223, p. 30–41.

Dewey, J. F., 1969, Evolution of the Appalachian/Caledonian orogen: Nature, v. 222, p. 124–129.

Dewey, J. F., and Horsfield, B., 1970, Plate tectonics, orogeny, and continental growth: Nature, v. 225, p. 521–525.

Drake, C. L., Ewing, J. I., and Stockard, H., 1968, The continental margin of the eastern United States: Canadian Jour. Earth Sci., v. 5, p. 1006–1009.

Fairhead, J. D., and Girdler, R. W., 1969, How far does the rift system extend through Africa?: Nature, v. 221, p. 1018–1020.

Furon, R., 1963, Geology of Africa: Edinburgh, Oliver and Boyd.

Krenkel, L., 1938, Geologie Afrikas Pts. 1, 2, and 3: Berlin, Gebrüder Borntraeger.

Kutina, J., 1969, Hydrothermal ore deposits in the western United States: A new concept of structural control of distribution: Science, v. 165, p. 1113–1119.

Martin, H., 1965, The Precambrian geology of South West Africa and Namaqualand: Univ. Cape Town, Dept. Geology Chamber Mines Precambrian Research Unit Ann. Rept.

—— 1969, Problems of age relations and structure in some metamorphic belts of Southern Africa: Geol. Assoc. Canada Spec. Paper No. 5, p. 17–26.

Matthews, P. E., 1959, The metamorphism and tectonics of the pre-Cape formations in the post-Ntingwe thrust-belt, S. W. Zululand, Natal: Geol. Soc. South Africa Trans. and Proc., v. 62, p. 257–322.

McConnell, R. B., 1967, The East African rift system: Nature, v. 215, p. 578–581.

Menard, H. W., 1964, Marine geology of the Pacific: New York, McGraw-Hill Book Co.

Nicolaysen, L. O., 1962, Stratigraphic interpretation of age measurements in Southern Africa, in Engel, A. E. J., and others, eds., Petrologic studies: A volume in honor of A. F. Buddington: Geol. Soc. America, p. 569–598.

Nicolaysen, L. O., and Burger, A. J., 1965, Note on an extensive zone of 1000 million-year old metamorphic and igneous rocks in Southern Africa: Sci. Terre, v. 10, no. 3–4, p. 497–516.

Simpson, E. S. W., and du Plessis, A., 1968, Bathymetric, magnetic, and gravity data from the continental margin of southwestern Africa: Canadian Jour. Earth Sci., v. 5, p. 1119–1123.

Smith, A. G., and Hallam, A., 1970, The fit of the southern continents: Nature, v. 225, p. 139–144.

Vacquier, V., Raff, A. D., and Warren, R. E., 1961, Horizontal displacements in the floor of the northeastern Pacific Ocean: Geol. Soc. America Bull., v. 72, p. 1251–1258.

Vail, J. R., Snelling, N. J., and Rex, D. C., 1968, Pre-Katangan geochronology of Zambia and adjacent parts of Central Africa: Canadian Jour. Earth Sci., v. 5, p. 621–632.

van Niekerk, C. B., and Burger, A. J., 1969, The uranium-lead isotopic dating of South African acid lavas: Bull. Volcanol., v. 32, no. 3, p. 481–498.
Wilson, J. T., 1965, A new class of faults and their bearing on continental drift: Nature, no. 4995, p. 343–347.
—— 1966, Did the Atlantic close and then re-open?: Nature, v. 211, no. 5050, p. 767–781.
—— 1969, Some aspects of the current revolution in earth sciences: Jour. Geol. Education, v. 17, p. 145–149.

MANUSCRIPT RECEIVED BY THE SOCIETY MARCH 29, 1971

THE GEOLOGICAL SOCIETY OF AMERICA, INC.
MEMOIR 132
© 1972

A Trench off Central California in Late Eocene–Early Oligocene Time

WILLIAM B. TRAVERS
Department of Geological and Geophysical Sciences,
Princeton University, Princeton, New Jersey 08540

ABSTRACT

An oceanic trench probably existed off central California in early and middle Cenozoic time, according to recently published interpretations of marine geophysical data. After allowance for late Cenozoic displacement along the San Andreas fault, a comparable part of onshore California was studied to determine if the geologic record there is compatible with these interpretations. Accumulation of mud at abyssal depths, chaotic deformation of sediments, the development of a submarine valley, thrust faulting, and basaltic and andesitic volcanism apparently all took place during late Eocene and early Oligocene time. Thus, onshore data support the concept of a late Eocene–early Oligocene trench. However, none of these phenomena occurred earlier in Cenozoic time in central California except the development of submarine valleys. Therefore, onshore data do not point to the presence of a trench off central California in Paleocene or early and middle Eocene time.

INTRODUCTION

The concept of plate tectonics has been applied recently to an analysis of the marine geophysical data from off the coast of central California (Atwater, 1970; McKenzie and Morgan, 1969). One of the conclusions of these papers is that a trench (subduction zone) existed off the coast of central California in early and middle Cenozoic time. The purpose of this paper is to show evidence for this trench in the onshore geologic record.

Raff and Mason (1961) first reported linear magnetic anomalies off the coast of central California. Subsequent identification and dating of the anomaly patterns (Bassinger and others, 1969) have shown that the youngest magnetic band is about

30 m.y. old. Moreover, the age of the anomalies increases with increasing distance from shore. This suggests that the continent has overridden oceanic crust some time within the last 30 m.y. (late Oligocene or later). The increasing age of the magnetic anomalies to the west also suggests that the subduction zone consumed considerable oceanic crust, including an oceanic spreading zone.

The area discussed is the western part of central California between 36° N. and 40° N. lat, or approximately from a latitude just south of Monterey Bay to that of the north end of the Great Valley of California. It is bounded on the east by the crest of the Sierra Nevada and on the west by the San Andreas fault (Fig. 1). The rocks west of the San Andreas fault are not considered because in Eocene and Oligocene time they probably were located a few hundred miles south of their present location (Dickinson and Grantz, 1968), and therefore did not lie within the region discussed in this paper. Late Eocene time is the Narizian stage of Mallory (1959) and the early Oligocene is the Refugian stage of Schenck and Kleinpell (1936).

Page (1970) concludes that in the Santa Lucia Range in southern California west of the San Andreas fault, Franciscan rocks were thrust beneath strata considered to be part of the Great Valley sequence in post–late Cretaceous (post-Campanian) and probably pre-latest Oligocene time. According to this interpretation, under-thrusting of oceanic material did occur along the continental margin of California in early Cenozoic time, although this area is well to the south of the area considered here.

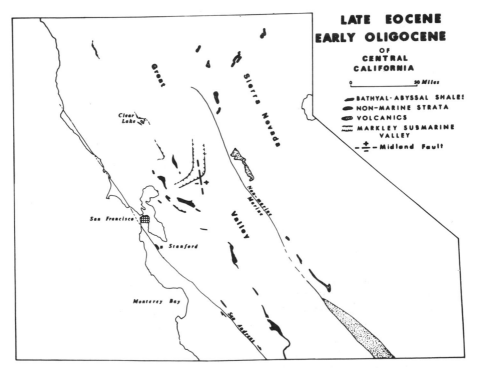

Figure 1. Map of part of central California showing the location of the geologic features related to the postulated late Eocene–early Oligocene trench.

Middle Cenozoic volcanism in the Great Basin and Rocky Mountain regions has been interpreted as related to trench activity off central California (Lipman and others, 1970). Recently, McKee and Silberman (1970) dated Cenozoic plutonism and volcanism in north-central Nevada and found that this activity began in latest Eocene and earliest Oligocene time (about 40 to 35 m.y. ago). The plutons are quartz diorite to quartz monzonite, and the intrusive emplacement was coincident with and followed by flows of dacite lava.

IDENTIFICATION OF ANCIENT TRENCHES

An oceanic trench is generally considered to be an unusually deep elongate trough with comparatively steep sides, characterized by shallow, intermediate, and deep-focus earthquakes, a very large negative gravity anomaly, and quite low terrestrial heat flow in the vicinity of the trench axis. Moreover, according to plate tectonic theory, trenches are developed where crustal rocks are destroyed or consumed by underthrusting of oceanic crust under continental crust into a subduction zone.

Modern trenches are characterized by certain other associated geologic features. Volcanism in adjacent continental areas produces rocks of the calc-alkaline suite. Hamilton (1969), among others, has suggested that the formation of granitic batholiths may be related to trench activity. Ernst (1965) has shown that the production of certain minerals (for example, lawsonite and jadeite) of the glaucophane schist metamorphic facies requires high pressures, but abnormally low temperatures. These conditions ought to be present down in the underthrust plate.

In attempting to recognize the effects of an ancient trench on geologic features of the adjacent continent, one does not expect to find the geophysical characteristics of localized low heat flow, deep seismic activity, and large negative gravity anomalies. But there may be evidence of deposition in unusually deep water, of crustal consumption, of the glaucophane schist facies, and of calc-alkaline igneous activity. Inasmuch as a trench has comparatively steep sides and the underthrust continental margin is the site of numerous earthquakes, "slope deformation structures" should be characteristic features. These structures include submarine landslides, olistostromes, and other types of slumped and crumpled strata that produce chaotic deposits and give evidence of deformation prior to or early in the process of dewatering and lithification.

Modern trenches are also characterized by numerous submarine valleys cut into the inner side of the trenches as well as into the adjacent continental slope (for example, see Fisher, 1961; von Huene and Shor, 1969; Grim, 1969). Even though submarine valleys are quite common along nontrench continental margins (Shepard and Dill, 1966), the Markley Submarine Valley is believed related to a trench margin for reasons discussed below.

No one of these features alone proves the former existence of a subduction zone, but because several of them developed in a region at the same time they are strong evidence of an ancient trench.

KREYENHAGEN SHALE

Mallory (1959) has noted that some of the Foraminifera of the Kreyenhagen Shale and correlative strata along the west side of the central part of the Great Valley of California (Fig. 1) probably lived in quite deep water. In fact, several

types suggest abyssal depths. Among them are the costate uvigerinids and bulimi-
nids, which are related to the living form *Bulimina rosata,* and abundant *Gyroidina
orbicularis* var. *planata, Asterigerina crassiformis,* and *Bathysiphon eocenicus.* The
presence of these forms suggests water depths of approximately 6,000 ft or greater.

The Kreyenhagen Shale is highly organic, very siliceous, and slightly silty mudstone with rare interbeds of graded sandstones with flute casts. The high silica content is due to abundant Radiolaria and some diatoms. The sandstones contain
abundant grains of glauconite which appear to be authigenic. It may be that these
"starved" organic mudstones were usually bypassed by turbidite sand on its way
down submarine canyons toward the trench floor (*see* Fig. 2 and the section on
Markley Submarine Valley). Correlative strata with similar lithology and microfauna include the Sidney Shale member of the Markley Formation, the Pereira
Shale member of the Alhambra Formation, and the unnamed late Eocene sandstones and mudstones described by Page and Tabor (1967) and discussed below.
The Kreyenhagen mudstones average about 900 ft thick and range up to about
2,000 ft thick. Rates of sedimentation were probably about 4 to 8 cm/1,000 yrs,
based on a 3 to 4 m.y. time span for Kreyenhagen deposition. The Kreyenhagen
Formation is underlain by sands and shales that were deposited in neritic to bathyal
depths, as indicated by the included microfauna. In an area east of San Francisco
(Pacheco Syncline) and in the southwest part of the Great Valley, the unit is
overlain by sandstones and shales whose microfauna suggest neritic and upper
bathyal water depths (Mallory, 1959). In the central part of the Great Valley,
particularly toward the eastern side of the valley, the deep water mudstones of the
Kreyenhagen Formation are overlain unconformably by nonmarine strata of
Miocene age.

The Eocene-Oligocene biostratigraphic boundary is contained within the Kreyenhagen Formation, as defined by changes in microfauna content. Where the strata

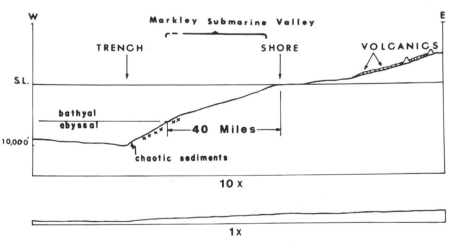

Figure 2. Cross section showing the location of geologic features in central California
related to the postulated trench. The x's indicate sites of Kreyenhagen shale deposition.
The volcanic outcrops are in the Sierra Nevada and the chaotic sediments are at Stanford
University. The site marked "Trench" indicates the easternmost possible location of the
trench axis. The axis could have been located farther to the west. The surface profile,
both the submarine and subareal portions, is taken from Fisher (1961, Fig. 2) and is of
the Middle America Trench and vicinity near Banderas Bay, Mexico (*see* Fig. 3).

containing this boundary have been preserved from subsequent erosion there is no evidence of any change in the depositional environment across the biostratigraphic boundary. Berggren (1969) dates the Narizian-Refugian boundary at about from 42 to 44 m.y.

The Kreyenhagen strata along the west side of the San Joaquin Valley generally dip eastward, forming a broad homocline. In the field, the author has found no evidence that suggests the formation was subjected to the chaotic folding and faulting usually associated with a subduction zone. Therefore, the trench axes must have been located to the west of these outcrops.

LATE EOCENE–EARLY OLIGOCENE VOLCANISM

On the western slope of the central Sierra Nevada, scattered outcrops of volcanic basalt and andesite of the Lovejoy Formation, Reeds Creek Andesite, and Ingalls Formation have been dated by stratigraphic methods, plant fossils, and mammal remains as either late Eocene or early Oligocene, or both (Durrell, 1944, 1959, 1966). These rocks are particularly significant because virtually no Cenozoic volcanic rocks older than these are known from this part of the Sierra Nevada. During the 25 m.y. without volcanism that preceded the outpouring of the latest Eocene basalts and andesites, apparently there was no trench activity adjacent to the central California coast.

The Lovejoy Formation contains olivine-bearing basalt flows that are interbedded with stream gravels and some lacustrine clays totaling about 600 ft thick. Durrell (1966) suggests that the olivine basalt flows may have originally been rather widespread in northern California. These basalts and those at Orland Buttes and the Putnam Peak basalts near Vacaville may represent remnants of once continuous sheets of basalt flows.

The Lovejoy Formation rests unconformably on the Ione Formation, on the "auriferous gravels," or on older rocks. The estimated age of the basalts is based on a fossil flora found by Pottsbury (1935) in the La Porte tuff which overlies lacustrine clays containing basalt fragments derived from the Lovejoy Formation. The fossil flora is considered to be latest Eocene or early Oligocene. Its field relations suggest that the Lovejoy basalts are only slightly older than the tuffs and therefore are of latest Eocene or earliest Oligocene age and younger than the late Eocene Ione Formation.

Evernden and James (1964), however, obtained radiometric ages indicative of middle Oligocene (32.4 m.y.) age from tuff beds underlying basalts that are considered to be part of the Lovejoy Formation. This difference in interpretation of the age of the early volcanic rocks raises some doubt about their temporal relation to the postulated late Eocene trench. Nevertheless, the volcanic age interpretations of Pottsbury (1935) and Durrell (1966) are supported by the age of volcanic debris in the fill of the Markley Submarine Valley (*see* below). Foraminifera found with the volcanic clasts in the lowest part of the valley fill are dated as no younger than late Refugian (A. A. Almgren, 1970, written commun.). Using Berggren's (1969) dating, the late Refugian is no younger than 36 m.y. and may be as old as 40 m.y. Because the Markley fill is clearly from a Sierran source, it is concluded that Evernden and James's radiometric age does not date the oldest Cenozoic volcanic material in the northern Sierra Nevada.

The Reeds Creek Andesite and the Ingalls Formation are predominantly andesite mudflow breccias. The Ingalls contains clasts of basalt derived from the Lovejoy Formation. No fossils are known from either of these formations. They are overlain by strata of early Miocene age and are therefore presumed to be Oligocene in age.

These relations are believed to indicate a latest Eocene–early Oligocene age for the early Cenozoic Sierran volcanism. If so, it is approximately contemporaneous with the central Nevada igneous activity (Lipman and others, 1970; McKee and Silberman, 1970). The former authors suggest that early Cenozoic Sierran volcanism was caused by a trench off the western United States. Similarly, recent volcanism adjacent to the margin of much of Central and South America is caused by the Middle America Trench and the Peru-Chile Trench. Apparently the volcanism of the western Sierra Nevada in California and of central Nevada are both related to a trench that existed about 35 to 45 m.y. ago.

CHAOTIC LATE EOCENE STRATA NEAR STANFORD UNIVERSITY

Excavation for the Two-Mile Linear Accelerator at Stanford University exposed a thick succession of late Eocene strata, some of which is chaotically deformed (Page and Tabor, 1967). The chaotic rocks contain soft-sediment deformation structures such as disordered mudstones and randomly rotated and irregular masses of sandstone. In some cases small, thin wisps of sand were apparently injected into soft masses of mudstone during deformation. These structures show that the deformation took place prior to the lithification of the sandstones and mudstones. Page and Tabor interpret the chaotic structures as due to submarine sliding. They believe that deformation and deposition were nearly contemporaneous. The slope down which the soft, only slightly lithified, late Eocene sediments slid may well have been a wall of the postulated trench.

There is, however, another interpretation of the chaotic nature of the late Eocene sediments at Stanford. It may be that these rocks owe their disturbed nature to tectonic activity in the trench. In this case the sediments may have been in the part of the trench where convergence of oceanic and continental crust in the subduction zone caused disruption of the sediments before lithification. If so, this outcrop of chaotic sediments was deformed near the axis of the late Eocene trench, which was probably located a few miles farther west, at or near the present trace of the San Andreas fault. No similar chaotic strata are known from the Paleocene or early and middle Eocene.

EVIDENCE OF UNDERTHRUSTING

Abundant subsurface data derived from exploratory wells drilled for oil and gas show that the Great Valley sequence of Cretaceous and Cenozoic strata, including the Kreyenhagen mudstones, are essentially unfaulted and continuous with equivalent strata that outcrop in the foothills of the Sierra Nevada. Thus, if crustal shortening has taken place it must have been by underthrusting.

There is little evidence of Eocene or Oligocene age underthrusting in central California. Nevertheless, Swe and Dickinson (1970) have recently reported that in the Clear Lake area (Fig. 1) Paleocene beds are in thrust contact with Lower Cretaceous rocks. The thrust zone contains discontinuous lenses of serpentine, metagabbro, and greenstone in the form of sheared and sheeted masses that are up to 100 ft thick. The Cretaceous rocks are shattered as much as 100 ft below the thrust contact. This thrusting moved younger rocks westward over older rocks after Paleocene time. Similar thrust relations apparently also occur in the northern Coast Ranges about 75 mi to the north of Clear Lake (Clark, 1940; Swe and Dickinson, 1970). There a large klippe (3 × 7 mi) of Late Cretaceous, Paleocene, and middle Eocene strata rests on the Jurassic (?) age Franciscan Formation. The klippe is separated by a distance of about 2 mi from the main part of the Great Valley sequence. Significantly, this thrust contact is overlain by undisturbed rocks that are considered by Clark to be of early Miocene age. Accordingly, the thrusting may have taken place in late Eocene or early Oligocene time. W. R. Dickinson (1970, oral commun.) suggests that the thrusting was due to gravity sliding. Therefore, the thrusting suggests that these Eocene sediments were east of the trench which lay near or west of the present San Andreas fault. The location of the chaos near Stanford is compatible with this interpretation.

THE MARKLEY SUBMARINE VALLEY

The Markley Submarine Valley (Fig. 1) was developed in late Eocene or early Oligocene time (Almgren and Schlax, 1957). The valley has a known length of about 50 mi; it averages 7 to 8 mi wide; and it was about 3,000 to 5,000 ft deep, although its deepest or westerly portion has not been penetrated by wells. Data from these wells, which were drilled for natural gas, show that the valley was eroded into strata as young as latest Eocene (Edmondson, 1967). Filling of the valley probably began in the Oligocene epoch (late Refugian; A. A. Almgren, 1970, written commun.), but it may have begun sooner. However, Almgren believes the older Foraminifera found in the valley fill probably were reworked. Significantly, the lowest deposits in the valley contain abundant volcanic debris. In contrast, no volcanic debris is known from the early and middle Eocene strata that compose the sides of the submarine valley.

It is suggested here that the cutting of the Markley Submarine Valley was caused by tectonic activity related to a nearby trench. Development of the trench probably increased the regional submarine slope, and this may have initiated the erosion of the valley. On the other hand, the Midland fault (Fig. 1) was most active in the late Eocene (Narizian) time, as indicated by large thickness differences of late Eocene sediments across the fault (Edmondson, 1967). The erosion that produced the Markley Submarine Valley may have developed across the fault trace. Significantly, the Midland fault has one of the largest displacements and is one of the longest of the few early Cenozoic faults in central California. Moreover, it is a normal fault downthrown about 2,000 ft on the west side. The fault may have been related to trench activity.

Apparently, shortly after the erosion ceased, the valley began to fill with sediments derived from sources to the east. Just before the cutting of the valley there had been no known volcanic activity in the Sierra Nevada and vicinity, but it appears that as the valley began to fill, volcanic activity also began.

The Banderas Bay Submarine Valley, east of the Middle American Trench

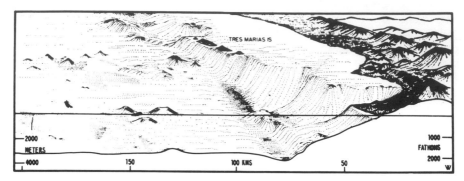

Figure 3. The north end of the Middle America Trench. Banderas Bay and the Banderas Bay submarine valley are shown in the right middle foreground. Taken from Fisher (1961, Fig. 2).

(Fig. 3), may be analogous to the Markley Submarine Valley, because both valleys began near shore, both are cut into the continental slope east of the inner trench wall, and both reach abyssal depths several miles east of the trench axis (Fig. 2).

It should be noted that the northernmost part of the Middle America Trench is not a typical trench. Intermediate and deep-focus earthquakes are not recorded inland from this part of the trench, although there is volcanic activity (Fisher, 1961, Fig. 1). Water depths are somewhat less than normal for trenches located adjacent to continents. This may be due to the proximity of the north end of the trench to its triple junction with the East Pacific Rise and the beginnings of the San Andreas fault. Perhaps a similar situation existed for the late Eocene and early Oligocene trench off central California, since water depths may not have been much more than 6,000 ft. It may be that strike-slip motion relative to dip-slip motion increased during middle and late Oligocene time, as the site of the trench lost its underthrusting character and became the San Andreas fault (McKenzie and Morgan, 1969). This may have occurred by the migration of a trench–transform fault junction as shown by Atwater (1970, Fig. 18).

CONCLUSIONS

Diverse and seemingly independent geologic features of onshore central California considered together suggest, but do not prove, that an oceanic trench existed off central California in late Eocene and early Oligocene time. Many of the geologic phenomena discussed could have developed without trench activity. However, taken together they suggest the presence of a trench. One of the geologic phenomena commonly associated with trench activity has not been found. No evidence for high-pressure, low-temperature metamorphic minerals (lawsonite and jadeite) of Eocene-Oligocene age is known.

Atwater (1970), McKenzie and Morgan (1969), and Hamilton (1969) have concluded that a trench existed off central California not only in latest Eocene and early Oligocene time, as pointed out in this paper, but also in Paleocene and mid-Eocene times. However, the pre–late Eocene trench activity is not supported by my interpretation of the onshore data from central California.

The onshore record suggests that either (1) a trench did not exist off central California in early Cenozoic before the late Eocene; or (2) that if one did exist it was located so far west of the continental margin or was of such a peculiar nature (possibly with mainly strike-slip motion) that it had little influence on sedimentation, bathymetry, and tectonics of the continental margin now east of the San Andreas fault in central California. Further, no related pre–late Eocene Cenozoic volcanism has been recognized in the central Sierra Nevada or in central Nevada.

ACKNOWLEDGMENTS

The author wishes to thank C. A. Burk, W. R. Dickinson, J. C. Moore, W. J. Morgan, and F. B. Van Houten for their reviews of the manuscript. Helpful suggestions and criticisms were offered by Tanya Atwater, J. C. Maxwell, B. M. Page, and F. J. Vine. A. A. Almgren and Lowell Redwine kindly supplied useful information. Figure 3 is used with the kind permission of R. L. Fisher.

REFERENCES CITED

Almgren, A. A., and Schlax, W. N. 1957, Post-Eocene age "Markley Gorge" fill, Sacramento Valley, California: Am. Assoc. Petroleum Geologists Bull., v. 41, p. 326–330.

Atwater, T., 1970, Implications of plate tectonics for the Cenozoic tectonics of western North America: Geol. Soc. America Bull., v. 81, no. 12, p. 3513–3536.

Bassinger, B. G., DeWald, D. E., and Peter, G., 1969, Interpretation of magnetic anomalies off central California: Jour. Geophys. Research, v. 74, p. 1484–1487.

Berggren, W. A., 1969, Cenozoic chronostratigraphy, planktonic Foraminiferal zonation, and the radiometric time scale: Nature, v. 224, p. 1072–1075.

Clark, S. G., 1940, Geology of the Covelo district, Mendocino County, California: Calif. Univ. Pubs. Geol. Sci., v. 25, no. 2, p. 119–142.

Dickinson, W. R., and Grantz, A., 1968, Tectonic setting of the San Andreas fault system: Stanford Univ. Pubs. Geol. Sci., v. 11, p. 117–119, 290–291.

Durrell, C., 1944, Andesite breccia dike near Blairsden, California: Geol. Soc. America Bull., v. 55, no. 3, p. 255–272.

—— 1959, The Lovejoy Formation of northern California: Calif. Univ. Pubs. Geol. Sci., v. 34, no. 4, p. 193–220.

—— 1966, Tertiary and Quaternary geology of the northern Sierra Nevada, California: Calif. Div. Mines and Geology Bull., v. 190, p. 185–197.

Edmondson, W. F. (chm.), 1967, Correlation Section 15, Sacramento Valley, Suisun Bay to Lodi, California: Pacific Sec., Am. Assoc. Petroleum Geologists.

Ernst, W. G., 1965, Mineral paragenesis in Franciscan metamorphic rocks, Panoche Pass, California: Geol. Soc. America Bull., v. 76, p. 879–914.

Evernden, J. F., and James, G. T., 1964, Potassium-argon dates and the Tertiary floras of North America: Am. Jour. Sci., v. 262, p. 945–974.

Fisher, R. L., 1961, Middle America Trench: Topography and structure: Geol. Soc. America Bull., v. 72, p. 703–720.

Grim, P. J., 1969, Seamap deep-sea channel: ESSA Tech. Rept., ERL 93-POL 2, 27 p.

Hamilton, W., 1969, Mesozoic California and the underflow of the Pacific mantle: Geol. Soc. America Bull., v. 80, p. 2409–2430.

Lipman, P. W., Prostka, H. J., and Christiansen, R. L., 1970, Cenozoic volcanism and tectonism in the western United States and adjacent parts of the spreading ocean floor, Pt. 1, Early and middle Tertiary: Geol. Soc. America, Abs. with Programs (Cordilleran Sec.), v. 3, no. 2, p. 112–113.

Mallory, V. S., 1959, Lower Tertiary biostratigraphy of the California Coast Ranges: Am. Assoc. Petroleum Geologists Mem., 416 p.

McKee, E. H., and Silberman, M. L., 1970, Geochronology of Tertiary igneous rocks in central Nevada: Geol. Soc. America Bull., v. 81, p. 2317–2328.

McKenzie, D. P., and Morgan, W. J., 1969, Evolution of triple junctions: Nature, v. 224, no. 5215, p. 125–133.

Page, B. M., 1970, Time of completion of underthrusting of Franciscan beneath Great Valley rocks west of Salinian block, California: Geol. Soc. America Bull., v. 81, p. 2825–2834.

Page, B. M., and Tabor, L. L., 1967, Chaotic structures and decollement in Cenozoic rocks near Stanford University, California: Geol. Soc. America Bull., v. 78, p. 1–12.

Pottsbury, S. S., 1935, The La Porte flora of Plumas County, California: Carnegie Inst. Washington Pub. 465, p. 29–81.

Raff, A. D., and Mason, R. G., 1961, Magnetic survey off the west coast of North America, 32° N. to 42° N. latitude: Geol. Soc. America Bull., v. 72, p. 1267–1270.

Schenck, H. G., and Kleinpell, R. M., 1936, Refugian stage of the Pacific Coast Tertiary: Am. Assoc. Petroleum Geologists Bull., v. 20, p. 215–225.

Shepard, F. P., and Dill, R. F., 1966, Submarine canyons and other sea valleys: Rand McNally, 381 p.

Swe, W., and Dickinson, W. R., 1970, Sedimentation and thrusting of late Mesozoic rocks in the Coast Ranges near Clear Lake, California: Geol. Soc. America Bull., v. 81, p. 165–188.

von Huene, R. E., and Shor, G. G., Jr., 1969, The structure and tectonic history of the Eastern Aleutian Trench: Geol. Soc. America Bull., v. 80, p. 1889–1902.

Manuscript Received by the Society March 29, 1971
Present Address: Department of Geological Sciences,
Cornell University, Ithaca, New York 14850

THE GEOLOGICAL SOCIETY OF AMERICA, INC.
MEMOIR 132
© 1972

Origin of Some Flat-Topped Volcanoes and Guyots

TOM SIMKIN

Smithsonian Institution, Washington, D.C. 20560

ABSTRACT

Many ancient ring complexes give evidence of having fed surface flows from rows of circumferential vents around a broad circle many kilometers in diameter. The volcanoes of the Galápagos Islands provide contemporary illustrations of this process. Whenever flows from these vents are viscous or small in volume, the circumferential feeder zone will grow vertically at the expense of the outer flanks. The area within this zone, however, being limited in extent and receiving flows from all sides, will soon fill and grow vertically along with the circumferential zone. The result should be a flat-topped volcano. The Galápagos volcanoes have gently sloping outer flanks which steepen to 35° approaching the circumferential vent zone, but the zone itself is nearly horizontal as the full top appears to have been before the development of central calderas. Filling of calderas in the waning stages of volcano growth may then return the flat-topped morphology.

The shapes of several Galápagos volcanoes are strikingly similar to those of many flat-topped seamounts on the ocean floor, and there seem to be no obvious reasons why the processes that built the Galápagos volcanoes could not operate below sea level as well as above. Wave base truncation and subsidence are clearly responsible for many flat-topped seamounts, or guyots, but others may simply exhibit primary shapes controlled by circumferential vents during growth of the volcano.

INTRODUCTION

Most familiar volcanoes have been fed by a central conduit system, and the resulting volcano shape is roughly conical with slope angles depending on the viscosity and volumes of the extruded lavas. Other volcanoes, however, were fed not from a central conduit but from rows of vents forming rings many kilometers

in diameter. Such circumferential feeders are well known in the eroded roots of
ancient volcanoes (for example, *see* Richey, 1961), and the Galápagos Islands
provide contemporary illustrations of such an eruptive process. The distinctive
volcano shape to be expected from ring feeders is flat topped rather than conical,
and this flat top is a primary growth feature. Growth by this process will be discussed with particular reference to Galápagos volcanoes and, since flat-topped
volcanoes are common on the sea floor, comparison will be drawn to possible
oceanic examples.

CIRCULAR FEEDERS

As outlined by Anderson (1951), steeply dipping circular fractures can form
above a near-surface magma chamber during its development, and these fractures may then be utilized by magma moving from the chamber to the surface.
Figure 1 illustrates the surface expression of these feeders—several concentric rows
of vents that have fed recent flows down the outer flank of Volcan Fernandina
(Narborough) in the Galápagos. If such circular feeders dominate the growth of
a volcano, then several stages of development should be expected, as outlined
schematically in Figure 2. Initially, as in Figure 2A, the ring feeders might be

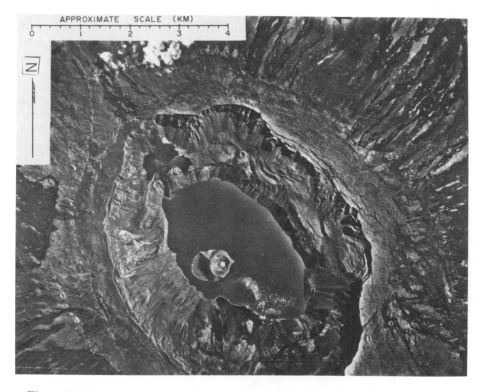

Figure 1. Mosaic of vertical air photographs taken in 1947 by the U.S. Air Force over
the summit of Volcan Fernandina, Galápagos Islands. The caldera has since been
changed by two eruptions and the caldera collapse of 1968, but the photograph is included to illustrate the several concentric rows of circumferential vents that have fed
recent flows outward down the flanks of the volcano.

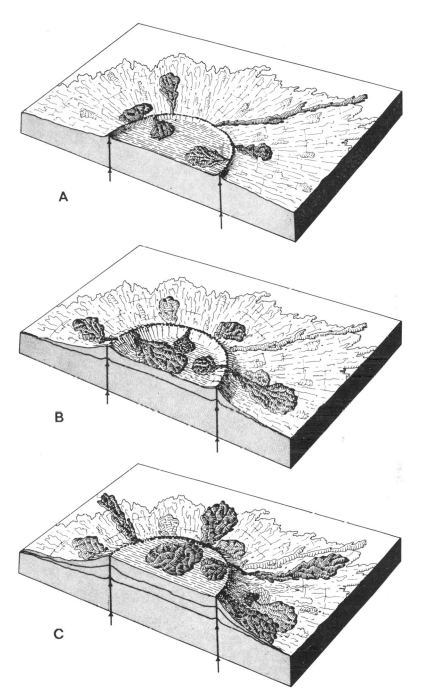

Figure 2. Schematic block diagram illustrating proposed development of flat-topped volcano from circumferential feeder vents. If ring (rather than central) feeder conduits are superimposed on a gentle shield (A), lava flows will soon pond within the ring (B). If inward-directed flows balance the volume of outward-directed flows, the much smaller area within the ring should fill to yield a roughly flat-topped volcano (C).

imposed on a gently domical shield morphology. Copious fluid flows should build both vents and outer flanks, but flows that are viscous or small (either small in total volume or small pulses of the eruption that have time to cool near the vent) will build the ring of vents vertically at the expense of the outer flanks. However, the area within the ring, being limited in extent and receiving flows from all sides, will soon fill and grow vertically with the concentric zone (Figs. 2B, C). Erosion following any explosive eruption from the ring of vents should heighten the topographic contrast between its two sides: redistribution of fragmental material should smooth and build the flattish summit surface while stripping at least some material from the steep side slopes.

Simplified calculations balancing the volumes inside and outside of the ring of feeders show that, given a ring diameter of 10 km and an average outer slope of 15°, the interior should be filled to the level of vents when the ring has grown to an elevation of 1.1 km above the base. Increasing or decreasing the average outer slope angle by 5° changes this elevation by approximately 0.3 km. The flat interior should then keep pace with continued vertical growth of the circumferential feeder zone and the majority of new flows would be forced outward down the flanks. Average slopes certainly vary and the schematic geometry is only an approximation of natural ring feeders, but these calculations suggest that a flat-topped morphology should be expected whenever circumferential feeders build a volcano to heights roughly one-tenth of the ring diameter. The internal structure of volcanoes built by this process should consist of a zone of high-angle, concentric feeder dikes separating steeply outward dipping flows from gently inward dipping flows and lava lakes on the inside of this broadly circular zone.

GALAPAGOS VOLCANOES

Two outstanding characteristics of active Galápagos volcanoes are: (1) their distinctive morphology, likened to "overturned soup-plates" by McBirney and Williams (1969) in contrast to the gently dipping "overturned saucers" of Hawaiian shields, and (2) their well developed circumferential feeder vent systems. The circumferential vents of Fernandina (Fig. 1) show well in a frequency count (Fig. 3A) that emphasizes the strong concentration of vents 3 to 5 km from this volcano's center. Beyond the concentric rows (beyond 5 km), the isolated and radial vents decline only slightly in number with distance from the center, but the number per unit area declines greatly. Figure 3B is a plot of beginning and end points for 104 surface flows that issue outward from the concentric zone of vents on Fernandina. If these flows now visible on the surface of the volcano are representative of those below, if the thickness of each flow is roughly uniform, and if the flows maintain constant angular width, then this plot can be read as a constructional cross-section of the volcano. For convenience, the flows have been ranked by distance of the end point from the volcano's center, but the order is unimportant: flows with these end points piled on top of each other in any sequence should yield the profile shown by this plot, and an equivalent volume poured inward would easily fill the volume inside the circumferential feeders. These inward flows could be thought of, in Figure 3B, as white lines issuing to the left from each of the vents that feeds a black flow outward to the right. Figure 3B demonstrates that there are enough short and intermediate length flows to build up

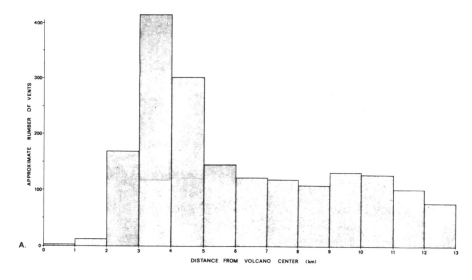

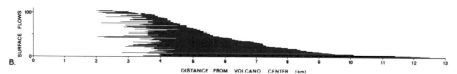

Figure 3. (A) Distribution of vents as a function of distance from center of Fernandina volcano. (B) Beginning and end points of all surface flows issuing from the circumferential vents shown in Figure 1, also plotted as a function of distance from center of Fernandina.

Fernandina's concentric feeder zone at the expense of the outer flanks, and the hypothetical profile obtained from these surface flows is closely similar to the natural profile of the volcano (Figs. 4A, B). Since magma that would otherwise run down outer slopes is trapped within the summit ring, this constructional process should be capable of building volcanoes with a higher height/diameter ratio than those built by equally fluid lavas from central feeders alone.

While we expect a symmetrical, roughly conical volcano from a central feeder vent system, the symmetry of the non-conical, ring-fed Galápagos volcanoes requires a different explanation. Recent work on Hawaii (Fiske and Jackson, 1972) may explain why Galápagos ring feeders maintain a roughly equal elevation as they build vertically. When a crude circular ridge has been constructed (Figs. 2A, B), and especially after central caldera development has emphasized this outer ridge, magma injected within the edifice should migrate laterally within the ridge to erupt at a topographic low point on the ridge axis. The eruption then builds up that low part of the ridge and future magma migration must seek the next low point for an eruption. Fiske and Jackson have shown, with the aid of gelatin models, that lateral magma migration within the ridge axis is a natural response to the gravitational stress on the ridge itself, and that such migration is responsible for the development and extension of Hawaiian ridges. In the Galápagos volcanoes, however, the ridge makes a crude circle rather than an arcuate

line, and this migration would be expected to act as a feedback mechanism en-
suring a roughly constant elevation for the circular ridge as it builds vertically.

The internal structure of Galápagos volcanoes appears consistent with the
growth model proposed here. At Cape Berkeley, on Isla Isabela, faulting has exposed a full cross-section of a volcano that was at least 6 km in diameter and
800 m above sea level. Although surface circumferential vents are not well developed here, there is a clear zone of near-vertical circumferential dikes and the

GALAPAGOS VOLCANOES

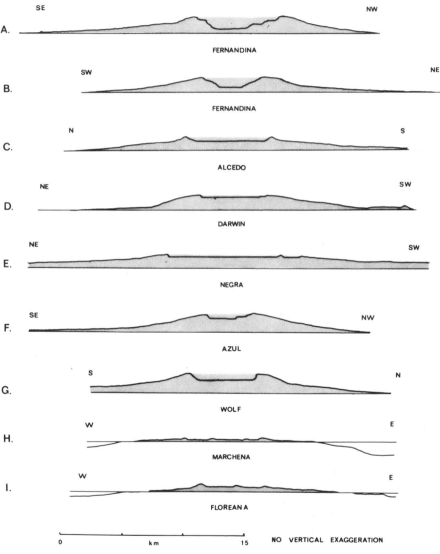

Figure 4. Profiles A to I are based on Hydrographic Office and American Geographical
Society charts of the Galápagos, with modifications added from field observations. Profiles
J to M are from Murray (1941), N from Budinger (1967), O from Carsola and Dietz

thin, ropy, flat-lying flows to the inside are in marked contrast to the more fragmental outer flows which reach dips of 35°. In this exposed section, neither dikes
nor sills provide sufficient dilation to account for the "overturned soup plate"
shape by distension of an "overturned saucer" as has been suggested by McBirney
and Williams (1969) to explain Galápagos shield morphology.

Other less well exposed Galápagos volcanoes also provide evidence indicating
growth from circumferential feeders. Summit flows dip gently inward from the

FLAT-TOPPED SEAMOUNTS

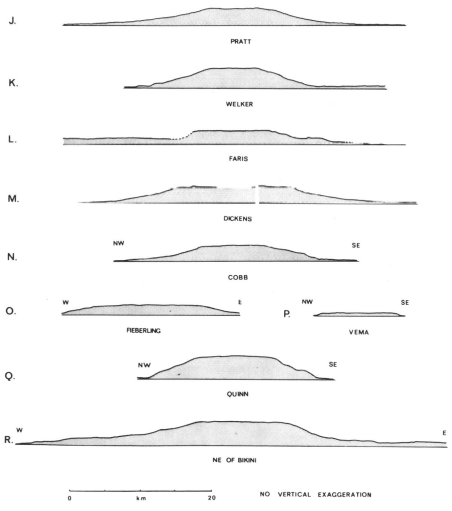

(1952), P from Simpson and Heydorn (1965), Q from McManus and Morrison (1969),
and R from Menard and Ladd (1963). Galápagos volcanoes are shaded above sea level
and seamounts are shaded above their local base level.

zone of concentric vents on the northeast side of Fernandina, north and west sides of Cerro Azul, and northeast side of Sierra Negra. Circumferential dikes are exposed on the caldera walls of these volcanoes and Keith Howard (written commun., 1970) reports that the deep caldera of Volcan Wolf exposes steeply inward-dipping fragmental flows overlain transgressively by flat-lying lava lake flows.

Prominent features of most Galápagos volcanoes are the central calderas, but field evidence suggests that these modify the basic morphology for only a geologically short time. In the well-exposed section at Cape Berkeley, tentative correlation of a picritic flow unit on both sides of the caldera boundary fault indicates that the block has dropped approximately 300 m, or the same distance that the caldera block is now below the rim. In other words, this volcano appears to have achieved its primary flat-topped morphology before subsidence of its central caldera. At the other end of caldera development is Isla Marchena (Fig. 4H), an old caldera now nearly completely filled by recent lava lakes. Apparently Galápagos calderas, like those in Hawaii, are filled in the waning stages of volcanic activity.

COMPARISON WITH GUYOTS

The foregoing has been an attempt to show that primary flat-topped volcanoes can develop from the circumferential feeder dike configurations known to have existed in ancient as well as modern volcanoes. A "sailor's-eye view" of these Galápagos volcanoes, however, shows profiles so similar to those of submarine flat-topped volcanoes, or guyots, that it is worth exploring the possibility that some guyots may result from the same process.

Figure 4 compares Galápagos volcanoes with flat-topped seamounts by profiles drawn with no vertical exaggeration and at similar scales. While the shapes of the two groups are very similar, the guyots show a much greater size range. However, guyots from a particular region may tend toward similar sizes and Menard (1964) notes that those from the northeast Pacific have flat tops with diameters in the 2 to 10 km range of the Galápagos volcanoes. Certainly larger and smaller feeder diameters are known from other regions which may have produced flat tops of a greater diameter range than that shown in the Galápagos.

Like Guyots, most Galápagos slopes average 15° to 20° (maximum around 35°) and their heights are at least one-tenth the diameter of their once-flat tops. Base level for both sets of profiles is the nearly horizontal outer slope of the volcano. This has been taken as sea level in the Galápagos despite the fact that most profiles show a "nick point" at this level, and submarine slopes range from gentle (on the shallow Galápagos platform) to steep (at the edge of the platform). The shape comparison seems valid, however, because the dominant and similar shape in both is the steep, flat-topped edifice above the nearly horizontal outer slopes.

The major shape distinction between Galápagos volcanoes and guyots is the presence of calderas at the summit of most Galápagos volcanoes. Calderas are apparently absent from submarine volcanoes and Menard (1964) suggests that the greater hydrostatic pressure may inhibit their development on the deep-sea floor. Even if calderas formed, however, they would probably be as geologically short-lived as their subaerial counterparts, being filled with late stage flows (and sediment) so as to be undetectable by standard bathymetric profiles throughout most of their lives.

To summarize the comparison, there is striking similarity in shape between submarine flat-topped seamounts and some subaerial volcanoes believed to have formed by growth from circumferential ring feeders. Development of ring fractures needs the presence of near-surface magma chambers, and these are known to exist in the oceanic crust. If submarine flows have sufficient fluidity to fill the interior of a circumferential ring of feeders, there seems to be no clear reason why primary flat-topped volcanoes could not grow on the deep-sea floor in much the same way that they have grown on the Galápagos platform. Even at shallower depths, where explosive volcanism would be expected, a ring of vents should be capable of producing a broad, roughly flat-topped mass of fragmental material.

Guyots were first recognized by Hess (1946) who suggested that they represented wave-base truncation and later drowning of former islands. This concept has been confirmed by the work of Hamilton (1956), Menard (1964, 1969) and others, and this evidence of vertical sea-floor movement was a valuable element in Hess's revolutionary concept of sea-floor spreading. It would be wrong to imply that *all* guyots are the primary products of ring feeders, but the above comparison suggests that *some* guyots may be primary features, and that bathymetry alone is unable to distinguish them from secondary, truncation-subsidence features. As increasing importance is being placed on indicators of former sea-floor movement, alternatives to the truncation-subsidence concept should be considered along with lines of evidence that might permit a choice between hypotheses for a particular guyot.

ALTERNATIVE GUYOT HYPOTHESES AND TESTS

Mathews (1947) and several workers in Iceland (*see* Kjartansson, 1967) have described flat-topped volcanoes that formed subaqueously of fragmental material and were then capped by lava flows at, and just above, water level. These flows contributed to, and helped preserve, the flat top, indicating that such forms on the deep-ocean floor would still require substantial subsidence. Nayudu (1962) argued that the hyaloclastites commonly dredged from guyot surfaces were inconsistent with truncation of a large subaerial volcano, and he suggested that some flat-topped seamounts might be primary features. Flat-topped volcanoes of subaqueous origin have since been reported by Christensen and Gilbert (1964), Tazieff (1969), and Bonatti and Tazieff (1970). These volcanoes consist largely of horizontally bedded fragmental deposits, and they are small (less than 2 km diameter). It is not clear whether the same process could build the guyots that are an order of magnitude larger, but these examples clearly show that flat tops can be obtained by primary growth below sea level. Menard (1964) cites several reasons why the relatively pointed tops of submarine volcanoes become much broader as they grow into islands. Later submergence of such islands would produce crudely flat-topped seamounts without invoking truncation. Subsequent sedimentation would smooth irregularities on this surface as found by the seismic work of Karig and others (1970) on guyots of the Mid-Pacific Mountains.

Evidence commonly cited for guyot subsidence is provided by material dredged from the flat summit surface. Near-surface organic remains, highly vesicular rocks, subaerially weathered flow surfaces, and large proportions of well-rounded cobbles all indicate that the guyot has been at or near the sea surface in the past, but they do not distinguish truncated volcanoes from those formed by processes involving subsidence without truncation. Presumably the above objects (suggesting former sea-level origin) could be dredged from a drowned Galápagos volcano.

Evidence for truncation is less easily found. Accordant summits of nearby guyots would suggest wave-base erosion, and deep truncation of volcanoes (leaving flat surfaces 20 to 30 km across) might be expected to expose plutonic rocks that would later appear in dredges from the submerged surface.

Evidence for a primary flat-top developed on the sea floor by the Galápagos model should include none of the above. If formed by ring feeders, the map outline would have to be broadly circular, and the height would have to be at least one-tenth of the ring diameter in order to build the interior to a flattish surface. Unless the flows are extremely fluid, there should be an inward dip on the final upper surface, but detailed seismic profiling might be required to distinguish this from the gently outward-dipping surface to be expected from truncation. Karig and others (1970) report sediment forming a domical cap over 100 m thick on some guyots, and this sedimentation pattern could change a primary, concave-up surface to a flat, or slightly domed, surface in areas of significant sedimentation rates. Doming might also occur through feeding from central vents. Ring and central feeders are not mutuallly exclusive, and some central activity could dome a volcano built mainly from ring feeders, just as circumferential venting could flatten a cone built by central feeders. The important point is that any circumferential venting will tend to flatten a volcano summit.

Ring feeders have played an important role on land and are believed able to build primary flat-topped volcanoes. If future evidence shows that the same process can operate on the sea floor, then separation of these guyots from others that certainly indicate subsidence will strengthen our data on vertical movements of the oceanic crust.

ACKNOWLEDGMENTS

K. A. Howard, B. O. Nolf, and B. Nordlie have contributed field observations as well as, with R. S. Fiske, W. G. Melson, and S. Francisco, helpful comments on the manuscript. I thank them and the Smithsonian Research Foundation. It is more difficult to express my debt to Harry Hess—for his warm and wise advice; but particularly for his rare blend of humor, questioning skepticism, and faith in the order of things. These touched all who knew and loved him.

REFERENCES CITED

Anderson, E. M., 1951, The dynamics of faulting and dyke formation: Edinburgh, Oliver and Boyd.
Bonatti, E., and Tazieff, M. H., 1970, Exposed guyot from the Afar Rift, Ethiopia: Science, v. 168, p. 1087–1089.
Budinger, T. F., 1967, Cobb Seamount: Deep-Sea Research, v. 14, p. 191–201.
Carsola, A. J., and Dietz, R. S., 1952, Submarine geology of two flat-topped northeast Pacific seamounts: Am. Jour. Sci., v. 250, p. 481–497.
Christensen, M. N., and Gilbert, C. M., 1964, Basaltic cone suggests constructional origin of some guyots: Science, v. 143, p. 240–242.

Fiske, R. S., and Jackson, E. D., 1972, Orientation and growth of Hawaiian volcanic rifts: Regional structure vs gravitational stresses: Royal Soc. London Philos. Trans., Ser. A.

Hamilton, E. L., 1956, Sunken islands of the Mid-Pacific Mountains: Geol. Soc. America Mem. 64, 97 p.

Hess, H. H., 1946, Drowned ancient islands of the Pacific Basin: Am. Jour. Sci., v. 244, p. 772–791.

Karig, D. E., Peterson, M. N. A., and Shor, G. G., 1970, Sediment-capped guyots in the Mid-Pacific Mountains: Deep-Sea Research, v. 17, p. 373–378.

Kjartansson, G., 1967, Volcanic forms at the sea bottom, in Bjornsson, S., Iceland and Mid-Ocean Ridges: Reykjavik, Societas Scientarum Islandica, XXXVIII.

Mathews, W. H., 1947, "Tuyas," flat-topped volcanoes in northern British Columbia: Am. Jour. Sci., v. 245, p. 560–570.

McBirney, A. R., and Williams, H., 1969, Geology and petrology of the Galápagos Islands: Geol. Soc. America Mem. 118, 197 p.

McManus, D. A., and Morrison, D. R., 1969, Quinn Guyot (GA-3) not tilted toward Aleutian Trench: Marine Geology, v. 7, p. 365–368.

Menard, H. W., 1964, Marine geology of the Pacific: New York, McGraw-Hill, 271 p.

—— 1969, Growth of drifting volcanoes: Jour. Geophys. Research, v. 74, p. 4827–4837.

Menard, H. W., and Ladd, H. S., 1963, Oceanic islands, seamounts, guyots and atolls, in Hill, M. N., ed., The sea, v. 3: New York, Interscience Publishers, p. 365–387.

Murray, H. W., 1941, Submarine mountains in the Gulf of Alaska: Geol. Soc. America Bull., v. 52, p. 333–362.

Nayudu, Y. R., 1962, A new hypothesis for origin of guyots and seamount terraces, in MacDonald, G. A., and Kuno, H., eds.: Am. Geophys. Union Geophys. Mon. No. 6.

Richey, J. E., 1961, Scotland: The tertiary volcanic districts: Edinburgh, British Regional Geology Handbook, H. M. Stationery Office, 3rd Ed.

Simpson, E. S. W., and Heydorn, A. E. F., 1965, Vema Seamount: Nature, v. 207, p. 249–251.

Tazieff, M. H., 1969, Volcanisme sous-marin de l'Afar (Ethiopie): Paris, C. R. Acad. Sci. Paris, v. 168, p. 2657–2660.

Manuscript Received by the Society March 29, 1971

Contribution Number 131 of the Charles Darwin Foundation

THE GEOLOGICAL SOCIETY OF AMERICA, INC.
MEMOIR 132
© 1972

A Model for the Gross Structure, Petrology, and Magnetic Properties of Oceanic Crust

F. J. VINE

*School of Environmental Sciences, University of East Anglia,
Norwich, NOR 88C. United Kingdom*

E. M. MOORES

Department of Geology, University of California, Davis, California 95616

ABSTRACT

The model is derived by equating the Troodos igneous massif of southern Cyprus with oceanic crust formed by sea-floor spreading. Comparison of the thicknesses and physical properties of the units of the Troodos massif with those deduced for the oceanic crust by seismic refraction experiments suggests the following correlation: layer 1—sediments; layer 2—pillow lavas and dikes, the lower part being predominantly dikes; layer 3—a layered plutonic complex of gabbros and minor diorites overlain by dikes; layer 4 (upper mantle)—pyroxenites, interpreted as accumulate phases of the gabbroic complex, grading downward into dunites and harzburgites thought to represent depleted mantle.

Despite the fact that the pyroxenites and dunites are as strongly magnetized as the unaltered pillow lavas, the record of reversals of the earth's magnetic field frozen into the oceanic crust at ridge crests is thought to be written largely within the uppermost 500 m or so of unaltered pillow lavas, the lower pillow lavas and the dikes having been subjected to a greenschist facies metamorphism. The total thickness of the dike and gabbroic complexes may well be inadequate to account for the whole of seismic layer 3, leaving the possibility that the lower crust consists of partially serpentinized peridotite and that the Moho is a transition from partially to unserpentinized peridotite as suggested by Hess.

CONSTRAINTS AND UNCERTAINTIES

Knowledge of the thickness, composition, and physical properties of the oceanic crust is derived almost entirely from seismic refraction experiments and dredging in the deep sea. Minor additional constraints are supplied by measurements of the earth's gravity and magnetic fields over the ocean basins and by models for the composition of the upper mantle beneath them. Refraction seismics have resolved three layers within the oceanic crust: layer 1—sediments, very variable in thickness and seismic velocities; layer 2—typically 1 to 2 km thick and having a compressional wave velocity (Vp) between 4.3 and 5.8 km/s; layer 3—4.5 to 5 km thick, Vp = 6.7 ±0.3 km/s. Beneath the Moho, Vp is commonly 8.1 ± 0.2 km/s. (Raitt, 1963).

Submarine volcanoes, volcanic islands, and the predominance of pillowed lava flows in bottom photographs and as fragments in dredge hauls taken over basement outcrops all indicate that the ocean basins, beneath any more recent sediments, are floored by basaltic extrusives. Similarly, at depth the upper mantle is generally considered to be peridotitic in composition (Ringwood, 1969). Historically, and to the present day, the essential problem regarding the composition of the oceanic crust is the definition of the depth at which the mafic carapace gives way to ultramafic material. Is this level within the crust, at the Moho, or even within the mantle? The latter would be possible if one considered the Mohorovicic discontinuity to be a phase change from basalt to eclogite (*after* Kennedy, 1959).

Following the measurement of the depth to the Moho and hence the thickness of the crust beneath oceanic areas (Ewing and Press, 1955), Hess, in common with most other authors, postulated that the oceanic crust is basaltic in composition and that the Moho is produced by the transition from basalt to peridotite. However, in 1959 and thereafter Hess maintained that the main crustal layer beneath the ocean basins is serpentinized peridotite (Hess, 1959, 1962; Vine and Hess, 1971). This he considered to result from the elevation of the geotherms and the availability of juvenile water associated with the upwelling mantle at a ridge crest, the thickness of the serpentinite layer being determined by the elevation of the 500° C isotherm at ridge crests, and being maintained away from ridge crests, despite the lower thermal gradient, because of the nonavailability of water. In order to satisfy the measured seismic P-wave velocity of 6.7 km/s for layer 3, Hess invoked 70 percent serpentinization of mantle material. To many this seemed to be implausibly fortuitous but Hess would counter this objection by maintaining that this was the degree of serpentinization in the Alpine-type ophiolites, which he considered to be fragments of oceanic crust caught up in the process of mountain building (Hess, 1965). In his classic 1962 paper on the history of the ocean basins Hess went so far as to equate seismic layers 1 and 2 with sediments, and layer 3 with serpentinite, thereby reducing the role of basaltic volcanism to the construction of submarine volcanoes and oceanic islands. Subsequently, on the basis of the magnetic anomalies, bottom photographs, and dredge hauls he acknowledged that layer 2 was almost certainly basaltic in composition (Hess, 1965; Vine and Hess, 1971).

MAGNETIC PROPERTIES

Early attempts to interpret the pronounced linear magnetic anomalies recorded over oceanic areas invoked magnetization contrasts within layer 2 which were correlated with basaltic extrusives (Mason, 1958; Vine and Matthews, 1963). This was in contrast to the ideas of Hamilton (1959) and Hess (1962), who at that time suggested that seismic layer 2 was consolidated sediments. More recently the results of the JOIDES drilling program have confirmed beyond reasonable doubt that the top of layer 2, or the basement reflector of continuous reflection profiling, is igneous, and basaltic in composition. However, the inherent ambiguity of magnetic interpretation does not permit one to deduce the vertical extent of these magnetization contrasts, that is, whether they extend down into layer 3 or even to the Moho or below. In their 1963 paper, for example, Vine and Matthews considered magnetization contrasts down to the depth of the maximum Curie point isotherm for the rock-forming ferromagnetic minerals, that is, within the upper mantle. Recently Peter (1970) has reiterated the potential importance of magnetization contrasts to depths as great as 20 km beneath the deep-ocean basins. On considering the intensities of magnetization fitted to seamounts and the intensities measured on dredged basalts, both Vine and Matthews (1963) and Vine and Wilson (1965) concluded that the near surface extrusives were almost certainly more strongly magnetized than the gabbroic or peridotitic intrusives at depth and therefore were the most effective, if not the sole contributors to the production of the linear magnetic anomalies. Moreover, it was demonstrated that such a distribution of magnetization gives better simulations of the observed anomalies (Vine and Wilson, 1965). Thus, in illustrating the general validity of the Vine and Matthews hypothesis, Pitman and Heirtzler (1966) and Vine (1966) assumed uniformly magnetized blocks of normally and reversely magnetized material confined to seismic layer 2. The authors would have been the first to agree that these assumptions are naive and geologically unreasonable. However, the remarkable success of these simple models in reproducing the observed anomalies should not be overlooked, and is presumably very significant. It may well mean that the zone of intrusion or emplacement of that portion of the crust in which the magnetic record is written is very narrow, as subsequently deduced by Matthews and Bath (1967), Harrison (1968), and Vine (1968); also that the vertical extent of these magnetization contrasts *is* very limited. Perhaps the most extreme suggestion along these lines is that of Talwani and others (1971) who deduced, as a result of a very detailed study of the Reykjanes Ridge, south of Iceland, that the highly magnetized layer is only a few hundred meters thick.

COMPARISON WITH ALPINE-TYPE OPHIOLITES

For many years fresh or weathered basalts and serpentinites were the only nonsedimentary rock types dredged from the ocean floor, but in the mid-1960s metabasalts and metagabbros were dredged (Melson and Van Andel, 1966; Cann and Funnell, 1967) and even more recently diorites, gabbros, anorthosites, and relatively unserpentinized peridotites have been obtained (Aumento, 1969; Engel and Fisher, 1969; Melson and Thompson, 1970). These findings provided further support for the suggestion of Hess (1965) that the Alpine mafic-ultramafic complexes, or ophiolites, are fragments of deep-sea floor exposed subaerially as a result of orogenesis and subsequent erosion. In the meantime the present authors decided to take Hess's suggestion more literally and re-study such a complex in the hope that this might yield more detailed information on the structure, petrology, and physical properties of the oceanic crust than could ever be obtained at sea.

We chose the Troodos massif of southern Cyprus in that it seemed to be relatively undisturbed tectonically, contained the basic elements of the ophiolite suite, and, presumably, deep-ocean floor (ultramafic units overlain in turn by gabbros, diabase, pillow lavas, and deep-sea sediments), and had previously been interpreted as an upthrust slice of oceanic crust and mantle by Gass and Masson-Smith (1963). It also included what we thought, in 1967, to be a unique feature—a "sheeted" dike complex; a horizon consisting of 100 percent dikes, implying 100 percent extension of the crust during emplacement, and hence making the whole massif an ideal candidate for oceanic crust formed by sea-floor spreading.

THE TROODOS IGNEOUS MASSIF, CYPRUS

The internal structure and petrology of the Troodos massif is summarized and equated with the layering of the oceanic crust (as defined by refraction seismics) in Figure 1. It will be seen that the sheeted dike complex grades into pillow lavas above and diorites and gabbros beneath. The lowermost unit containing pillow lavas (the Basal Group of the Troodos massif) is thought to be equivalent to the lower part of layer 2 and is predominantly dikes (90 to 100 percent). Within layer 2 this unit gives way somewhat abruptly to a pillow lava unit with a much smaller percentage of dikes (<50 percent). The contact between the two units is characterized by sill-like intrusions. The dikes and pillows of the Basal Group and the dikes of the sheeted complex beneath appear to have been subjected to a greenschist facies metamorphism, although the use of this regional metamorphic term is presumably inappropriate in the oceanic setting (Miyashiro and others, 1971). Thus, only the top 0.5 to 1 km of the complex consists of unmetamorphosed pillowed basalts.

At depth the greenstone dikes of the sheeted complex intrude and are intruded by the diorites and gabbros of the upper plutonic units, as shown schematically in Figure 1. The quartz diorites appear to be the most acidic differentiates of a roughly layered sequence which passes downwards through uralitized gabbro, hypersthene gabbro, anorthositic gabbro (locally), and olivine gabbro (commonly exhibiting phase layering) to pyroxenites. The gabbro unit may well be built up as a result of several phases of intrusion and therefore consists of a number of intrusive bodies as illustrated in Figure 1. The mafic-ultramafic contact is another horizon showing complex intrusive relations and is characterized by interfingering pyroxenite, olivine, or felspathic olivine pyroxenite, and layered olivine gabbro. Rodingitized pegmatitic gabbro dikes are also characteristic of this horizon. Below this, the pyroxene-rich rocks give way to dunite with complexly folded bands of chromite and harzburgite. The lowermost units (dunite and harzburgite) contain 80 to 100 percent olivine with interlayered enstatite, and exhibit a strong foliation; they have been interpreted as depleted (or residual) mantle material. On petrologic grounds one might wish to define the Moho as the base of the pyroxenite layer, in that this would appear to be the lowest unit or accumulate of a pseudo-stratiform complex forming the lower part of layer 3. However, on geophysical grounds (and one must remember, of course, that the Moho is only defined as yet by geophysical techniques) it seems more probable that the major discontinuity in physical properties will occur at the mafic-ultramafic contact unless the ultramafic units are partially serpentinized as discussed below.

This model for the gross structure and petrology of the oceanic crust is remarkably similar to that derived independently by Cann (1970) from considerations of the theory of plate tectonics and the results of geophysical experiments and dredg-

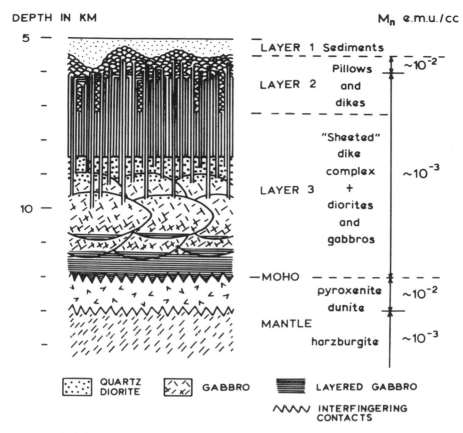

Figure 1. Schematic cross section summarizing the structure, petrology, and remanent magnetization of oceanic crust as deduced from a study of the Troodos igneous massif, Cyprus.

ing on mid-ocean ridges. It has been discussed at greater length elsewhere (Moores and Vine, 1971).

MAGNETIC PROPERTIES OF THE TROODOS ROCKS

Figure 2 summarizes measurements of the magnetic susceptibility and intensity of natural remanent magnetization (NRM) made on specimens cut from 170 oriented drill cores from the various units of the Troodos massif. Each value plotted is the average of measurements on up to four specimens from each sample. A logarithmic plot is presented following the suggestion of Irving and others (1966) that intensities and susceptibilities tend to fit a logarithmic normal rather than a Gaussian normal distribution. It will be seen that the NRM for the pillow lavas is typically on the order of 10^{-2} e.m.u./cc and that their Königsberger ratios ~ 10. Thus, in general, the natural remanent magnetization predominates, as is commonly assumed in interpreting oceanic magnetic anomalies, but the intensity of NRM is typically a factor of 2 greater than that assumed by authors who invoke magnetization contrasts throughout layer 2; for example, a thickness of 1.7 or 2 km, as assumed by Vine (1966) and Pitman and Heirtzler (1966), respectively. Thus a layer with an average NRM intensity of 10^{-2}, as measured on the Cyprus pillow lavas, need only be 0.5 to 1 km thick to produce a comparable effect. The green-stone dikes and the diorities and gabbros of the Troodos complex have NRM intensities on the order of 10^{-3} although they range over 3 orders of magnitude. The Königsberger ratios of these rock types are typically $\sim$1 except for many of the gabbros for which it is $\sim$10. As can be seen from the histograms this is not a result of higher NRM intensities but is because of systematically lower susceptibil-ities. The greenschist facies alteration of the pillow lavas and dikes, which produces a new mineral assemblage of albite, epidote, actinolite, chlorite, and sphene, lowers the NRM intensity by an order of magnitude, presumably because of the destruc-tion of the primary iron-titanium oxides which carry the high thermo-remanent magnetization. The iron is taken up in actinolite and chlorite, and the titanium in sphene.

The lower intensities of NRM for greenstones and gabbros are well known, and commonly assumed; a more surprising feature of these measurements is the fact that the NRM intensities and Königsberger ratios for the pyroxenites and dunites are so high, directly comparable with those obtained for the pillow lavas. This is in contrast to earlier measurements on partially serpentinized peridotites (Saad, 1969; Watkins and Paster, 1971), in which NRM intensities have rarely exceeded 10^{-3}, but it is compatible with the results of the aeromagnetic survey of the Troodos massif on which there is an intense magnetic anomaly associated with the outcrop area of pyroxenites and dunites in the vicinity of Mt. Olympus (Vine and Gass, in prep.). These high values are of great interest and require further investi-gation particularly in terms of the opaque mineralogy of these rock types. Presum-ably, in common with other serpentinized peridotites, the NRM is largely a chem-ical remanence acquired during serpentinization as a result of the formation of magnetite. In general there is a systematic increase in NRM intensity, and decrease in magnetic stability with increasing degree of serpentinization, although clearly other factors, such as the proportion of iron in the pyroxenes and olivine, will deter-mine the absolute value (Saad, 1969). The decrease in stability is attributed to an increase in grain and magnetic domain size. It may simply be the higher iron con-tent in the olivines and pyroxenes of these rocks, approximately 15 mol. percent (Moores and Vine, 1971), which accounts for their higher intensities of magnetiza-tion.

Because of their high NRM intensities the question arises as to whether these units will contribute significantly to the magnetic anomalies observed at sea level. In the first place, of course, their effectiveness is greatly reduced because of their greater depth, even if they did preserve "avenues" of essentially normally and re-versely magnetized material as is envisaged for the pillow lavas. However, there are several uncertainties which cast doubt on their ability to contribute to the de-velopment of the linear anomalies: (1) the thickness of these units on Cyprus is not well known but may well be only a few hundred meters; (2) the time of their emplacement is unknown; and (3) the time of their serpentinization is unknown. As mentioned above it seems probable, on the basis of the magnetic anomalies, that the emplacement of most of the pillow lavas and their associated feeder dikes occurs within a very narrow zone at a ridge crest. It is apparent from the intrusive rela-tions within the Troodos complex that many of the plutonic units may have been emplaced subsequently, that is, off axis. Deffeyes (1970) invoked this subsequent

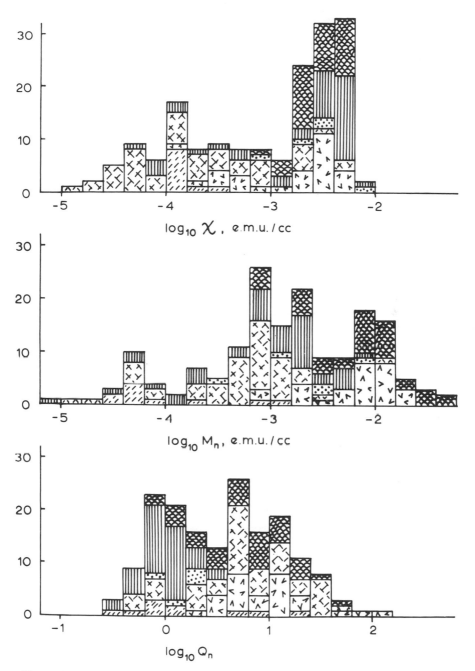

Figure 2. Histograms of susceptibility (χ), intensity of natural remanent magnetization (Mn), and Königsberger ratio (Qn) summarizing measurements on 170 samples from the Troodos igneous massif, Cyprus. The calculation of Qn ($= \mathrm{Mn}/\chi\mathrm{F}$) assumes that F, the intensity of the earth's magnetic field on Cyprus, is 0.45 oersted. The ordinate of each histogram is the number of samples, and the conventions used for the various rock types are as in Figure 1.

thickening of the lower crust to account for the development of a median valley on rifted ridge crests. If this is the case the normal/reverse blocks, if developed at all, will be out of phase with respect to those developed in the pillow lavas. The pyroxenites and dunites of the Troodos are approximately 50 percent serpentinized; their remanent magnetizations are very stable and directed subparallel to the present earth's field. This has been taken to indicate that the time of serpentinization was recent, probably associated with the uplift of the massif, and that prior to this these units were essentially unserpentinized (Moores and Vine, 1971). This is a moot point and may have important implications for the nature of the lower crust and the Moho as discussed below. It must be admitted that the directions of the remanent vectors deduced for these units may not be very meaningful because of the difficulty of making a structural correction. If serpentinization occurs at or near a ridge crest there are again problems of timing and emplacement. If, for example, the environment at this depth at a ridge crest is too hot for serpentinization to occur (greater than 500° C), and serpentinization takes place as the lithospheric plate cools by conduction on moving away from the ridge axis, then reversals of the Earth's magnetic field would presumably be recorded across subhorizontal boundaries. Because of these complexities it seems unlikely that the pyroxenite/dunite layer contributes significantly to the production of the linear magnetic anomalies.

The lowermost layer of the complex, the serpentinized harzburgite, is very weakly magnetized, having NRM intensities of 10^{-3} e.m.u./cc or less, and this, coupled with difficulties regarding the time of serpentinization and acquisition of remanent magnetization, renders it ineffective as a contributor to the linear magnetic anomalies. In principle, Peter (1970) is correct to point out that magnetization contrasts might exist to a depth of 20 km beneath the deep-ocean basins, as did Vine and Matthews in 1963. However, one should take into consideration the magnetic properties and process of formation of the crust, and the temperatures at the Moho beneath ridge crests (which may well be at or above those temperatures at which the rock is capable of exhibiting ferromagnetism or being serpentinized). This leads one to conclude that the record of reversals of the earth's magnetic field which is frozen into the oceanic crust at ridge crests and revealed in terms of linear and symmetrical magnetic anomalies is written most effectively within the unaltered pillow lavas and dikes forming the upper part of layer 2.

Two units of the Troodos igneous massif have been omitted from Figures 1 and 2, and both are thought to have their parallels in the oceanic crust. One is the Upper Pillow Lava unit which is more alkaline in character and discontinuously developed on the Lower Pillow Lavas, and the other is the diapiric serpentinites of the Troodos and Limassol forest areas. On a mid-ocean ridge both are thought to be emplaced away from the axis of spreading and to be associated with faults and fractures. Both have intensities of NRM $\sim 10^{-3}$ e.m.u./cc, and since they are emplaced off-axis will tend to confuse the pattern of linear anomalies but in different ways: the Upper Pillow Lavas, because they may be of opposite polarity to the block on which they are extruded, and the serpentinites, because their remanence is extremely unstable and randomly oriented. One might expect serpentinite diapirs, which appear to be common on the rifted ridges, to reduce the anomaly produced by the block they invade, whether this be positive or negative; one would expect them therefore to "punch holes" in the linear anomaly pattern.

RESERVATIONS

The correlation of the various units of the Troodos massif with the seismic refraction section for the oceanic crust in Figure 1 is tentative and equivocal both in terms of the thicknesses and seismic velocities assumed for the Troodos units. The seismic velocities are currently being investigated by means of both field and laboratory measurements and a larger scale seismic experiment is planned in an attempt to determine both velocities and thicknesses. At present there is considerable difficulty in assessing the thickness of the units; in the upper parts because they appear to vary greatly in thickness and in the lower parts because of the faulting associated with the central serpentinite diapir around which the gabbros and peridotites are best exposed. It must be admitted that most estimates of the thickness of the mafic section on Troodos, and of other ophiolites with which the oceanic crust is compared, fall short of the thickness of the oceanic crust as defined by seismic refraction studies (*see,* for example, Coleman, 1971, Fig. 1). This observation has led Reinhardt (1970, written commun.) to suggest that the lower part of layer 3 is composed of partially serpentinized peridotites. Reinhardt's suggestion is based on work on the Oman Mountains ophiolite which appears to be exactly analogous to the Troodos massif (Reinhardt, 1969). It implies that the ultramafics exposed at the base of these complexes today were partially serpentinized at or soon after their time of emplacement at a ridge crest, and represent lower crust rather than upper mantle as assumed here. Depending on the availability of water and the rate of cooling of the lithospheric plate, this process of serpentinization might account for the apparent increase in thickness of layer 3 away from ridge crests, as noted by Le Pichon (1969). Reinhardt points out that partially serpentinized peridotites could explain more readily the high seismic velocities obtained by Maynard (1970) for the lower oceanic crust (up to 7.6 km/s); however, olivine gabbro might also be capable of yielding such velocities. The suggestion also implies that the Moho is not exposed in Cyprus and Oman and this leaves unresolved the rather difficult problem of the nature of the transition from partially to unserpentinized peridotite at the Moho; this must be quite a sharp transition to satisfy the seismic data. However, it quite rightly leaves open the possibility that Harry Hess, who alone championed the concept of a serpentinite-peridotite transition as the explanation of the oceanic Moho, might yet be proved correct.

ACKNOWLEDGMENTS

Our work on Cyprus during the summers of 1968 and 1969 was supported by NSF grants GA-1257 and GA-1395 administered by Princeton University and the University of California at Davis, respectively. On Cyprus, we benefited greatly from the advice and assistance of L. M. Bear, formerly a member and Director of the Cyprus Geological Survey and the United Nations Development Program in Cyprus; Y. Haji Stavrinou, present Director of the Cyprus Geological Survey Department; and Dr. T. Pantazis, assistant Director.

The indebtedness of both authors to Harry Hess, both as friend and mentor, will be well known to many. We consider ourselves privileged to have worked with such a great geologist, and shall long continue to recall his wisdom and good humor. Surprisingly, at the outset, Harry was not entirely in favor of our Cyprus project, perhaps because the thick mafic section on Troodos seemed to detract from his idea of a main crustal layer of serpentinite beneath the oceans basins. However, rather than being deterred, we were encouraged by the fact that he once said of someone "I don't think he can be much of a petrologist, he agrees with everything I say."

REFERENCES CITED

Aumento, F., 1969, Diorites from the Mid-Atlantic Ridge at 45° N.: Science, v. 165, p. 1112–1113.

Cann, J. R., 1970, New model for the structure of the ocean crust: Nature, v. 226, p. 928–930.

Cann, J. R., and Funnell, B. M., 1967, Palmer Ridge: A section through the upper part of the oceanic crust?: Nature, v. 213, p. 661–664.

Coleman, R. G., 1971, Plate tectonic emplacement of upper mantle peridotites along continental edges: Jour. Geophys. Research, v. 76, p. 1212–1222.

Deffeyes, K. S., 1970, The axial valley: A steady-state feature of the terrain, in Johnson, H., ed., Megatectonics of continents and oceans: New Brunswick, N.J., Rutgers Univ. Press.

Engel, C. G., and Fisher, R. L., 1969, Lherzolite, anorthosite, gabbro, and basalt dredged from the Mid-Indian Ocean Ridge: Science, v. 166, p. 1136–1141.

Ewing, M., and Press, F., 1955, Geophysical contrasts between continents and ocean basins, in The crust of the Earth: Geol. Soc. America Spec. Paper, v. 62, p. 1–6.

Gass, I. G., and Masson-Smith, D., 1963, The geology and gravity anomalies of the Troodos massif, Cyprus: Royal Soc. London. Philos. Trans., Ser. A, v. 255, p. 417–467.

Hamilton, E. L., 1959, Thickness and consolidation of deep-sea sediments: Geol. Soc. America Bull., v. 70, p. 1399–1424.

Harrison, C. G. A., 1968, Formation of magnetic anomaly patterns by dyke injection: Jour. Geophys. Research, v. 73, p. 2137–2142.

Hess, H. H., 1959, The AMSOC hole to the Earth's mantle: Trans. Am. Geophys. Union, v. 40, p. 340–345.

—— 1962, History of ocean basins, in Engel, A. E. J., James, H. L., and Leonard, B. F., eds., Petrologic studies, (Buddington Vol.): Geol. Soc. America, p. 599–620.

—— 1965, Mid-ocean ridges and tectonics of the sea floor: Colston Papers, Proc. 17th Symp. Colston Research Soc., Univ. Bristol, Butterworths, Sci. Pub., p. 317–333.

Irving, E., Molyneux, L., and Runcorn, S. K., 1966, The analysis of remanent intensities and susceptibilities of rocks: Geophys. Jour., v. 10, p. 451–464.

Kennedy, G. C., 1959, The origin of continents, mountain ranges and ocean basins: Am. Scientist, v. 47, p. 491–504.

Le Pichon, X., 1969, Models and structure of the oceanic crust: Tectonophysics, v. 7, p. 385–401.

Mason, R. G., 1958, A magnetic survey off the west coast of the United States (Lat. 32°–36° N.: Long. 121°–128° W.): Geophys. Jour., v. 1, p. 320–329.

Matthews, D. H., and Bath, J., 1967, Formation of magnetic anomaly pattern of Mid-Atlantic Ridge: Geophys. Jour., v. 13, p. 349–357.

Maynard, G. L., 1970, Crustal layer of seismic velocity 6.9 to 7.6 kilometers per second under the deep oceans: Science, v. 168, p. 120–121.

Melson, W. G., and Thompson, G., 1970, Layered basic intrusion in oceanic crust: Romanche fracture, equatorial Atlantic: Science, v. 168, p. 817–820.

Melson, W. G., and Van Andel, Tj. H., 1966, Metamorphism in the Mid-Atlantic Ridge, 22° N. latitude: Marine Geology, v. 4, p. 165–186.

Miyashiro, A., Shido, F., and Ewing, M., 1971, Metamorphism in the Mid-Atlantic Ridge near 24° and 30° N.: Royal Soc. London Philos. Trans., v. A268, p. 589–603.

Moores, E. M., and Vine, F. J., 1971, The Troodos massif, Cyprus and other ophiolites as oceanic crust: Evaluation and implications: Royal Soc. London Philos. Trans., v. A268, p. 443–466.

Peter, G., 1970, Discussion of paper by B. P. Luyendyk, 'Origin of short-wavelength magnetic liberations observed near the ocean bottom': Jour. Geophys. Research, v. 75, p. 6717–6720.

Pitman, W. C., and Heirtzler, J. R., 1966, Magnetic anomalies over the Pacific-Antarctic Ridge: Science, v. 154, p. 1164–1171.

Raitt, R. W., 1963, The crustal rocks, in Hill, M. N., ed., The sea,: New York, Interscience Pub., v. 3, p. 85–102.

Reinhardt, B. M., 1969, On the genesis and emplacement of ophiolites in the Oman Mountains geosyncline: Schweizer. Mineralog. u. Petrog. Mitt., v. 49, p. 1–30.

Ringwood, A. E., 1969, Composition and evolution of the upper mantle, in The Earth's crust and upper mantle: Am. Geophys. Union Geophys. Mon., v. 13, p. 1–17.

Saad, A. H., 1969, Magnetic properties of ultramafic rocks from Red Mountain, California: Geophysics, v. 34, p. 974–987.

Talwani, M., Windisch, C. C., and Langseth, M. G., 1971, Reykjanes Ridge crest: A detailed geophysical study: Jour. Geophys. Research, v. 76, p. 473–517.

Vine, F. J., 1966, Spreading of the ocean floor: New evidence: Science, v. 154, p. 1405–1415.

—— 1968, Magnetic anomalies associated with mid-ocean ridges, in Phinney, R. A., ed., The history of the Earth's crust: Princeton, Princeton Univ. Press, p. 73–89.

Vine, F.J., and Hess, H. H., 1971, Sea-floor spreading, in Maxwell, A. E. ed., The sea: New York, Wiley-Interscience, v. 4, pt. 2, p. 587–622.

Vine, F. J., and Matthews, D. H., 1963, magnetic anomalies over oceanic ridges: Nature, v. 199, p. 947–949.

Vine, F. J., and Wilson, J. T., 1965, Magnetic anomalies over a young oceanic ridge off Vancouver Island: Science, v. 150, p. 485–489.

Watkins, N. D., and Paster, T. P., 1971, The magnetic properties of igneous rocks from the ocean floor: Royal Soc. London. Philos. Trans., v. A268, p. 507–550.

Manuscript Received by the Society March 29, 1971

Ultrabasic Rocks

THE GEOLOGICAL SOCIETY OF AMERICA, INC.
MEMOIR 132
© 1972

Types of Alpine Ultramafic Rocks and Their Implications for Fossil Plate Interactions

E. M. MOORES AND I. D. MACGREGOR

Department of Geology, University of California, Davis, California 95616

ABSTRACT

The occurrence of Alpine peridotites may reflect the interaction of two or more lithospheric plates, at least one of which bears continental crust. Bodies called Alpine ultramafic rocks include: (1) hot diapiric masses with thermal aureoles suggesting intrusion into or beneath crustal rocks at high temperatures; (2) mantle slabs, principally of oceanic crust and mantle, such as Tethyan ophiolitic complexes, formed by diapiric upwelling of mantle material at spreading centers, which were subsequently and independently emplaced into the continental margins; (3) disrupted mantle slabs incorporated into melanges; and (4) conformable bodies in regionally metamorphosed terranes representing recrystallization and deformation of occurrences of types 1, 2, or 3. Most mantle slabs possibly are emplaced by abortive subduction of a continental margin. This event is seen in mountain systems as the first deformation of a geosynclinal sequence. The P-T conditions of ultramafic assemblages, contact metamorphism, and regional metamorphism should provide data on the thickness of crust and geothermal gradients at times of mantle upwelling and/or crustal emplacement of ultramafic material.

INTRODUCTION

Since the pioneering work of Benson (1926), Bowen and Tuttle (1949), Hess (1939, 1955), and Steinmann (1905, 1926), Alpine peridotites have been a subject of controversy. Part of this controversy has stemmed from the argument over whether they were intruded in a magmatic or solid state (*see* Hess, 1965), part from confusing Alpine ultramafic rocks with those of classical stratiform igneous complexes, and part from generalizing features of one type of occurrence to all

209

others. Despite the confusion, some major salient characteristics have been recognized: (1) in contrast to most stratiform ultramafic intrusions, Alpine peridotites have higher $Mg/Mg + Fe$ ratios in olivines and pyroxenes (approximately 0.9; Thayer, 1960); (2) they display tectonic fabrics; and (3) as observed by Benson (1926) and Hess (1955), they are emplaced into the axial regions of mountain systems, apparently during the initial stages of their deformation.

In the light of the new global tectonics, one may gain new perspective on the Alpine ultramafic problem. An important corollary of plate theory is that mountain systems represent the interaction of plate boundaries (Dewey and Bird, 1970). Intracratonic or two-sided mountain systems may represent the results of collisions of two continental masses formerly separated by an ocean. Recently many investigators have interpreted Alpine peridotites as remnants of the lithosphere of the vanished ocean basin (Dewey and Bird, 1970; Moores and Vine, 1971; Vine and Hess, 1970), and doubtless many peridotites conform to this interpretation. Though many occurrences do not simply fall into this category, it seems clear that Alpine peridotites reflect the interaction of plate boundaries at the time of their emplacement, very commonly the interaction of plates, at least one of which contains continental crust.

In attempting to reinterpret Alpine ultramafic rocks in accordance with plate theory it has become necessary to separate carefully the time of diapiric upwelling or *intrusion* of the peridotite from below in the mantle, and its *emplacement* into its presently observed position. Ultimately all Alpine peridotites probably result from diapiric upwelling in the mantle at some time or other. The time of upwelling —or *intrusion*—clearly is not *per se* related to the emplacement of this peridotite into or on a continent. The time lapse between the two events is a function of the tectonic processes operating on the crust-mantle system subsequent to the diapiric activity which intruded the peridotite, and may be as great as millions of years. Seldom has this distinction received due emphasis.

Hence, in a given occurrence of peridotite three ages are present: the age of diapiric upwelling of the peridotite, the age of the rocks into which it is emplaced, and the age of its emplacement into those rocks.

TYPES OF ALPINE PERIDOTITE INTRUSIONS

In spite of their general similarities, many variations are present in Alpine peridotites which suggest differences in level of emplacement in the crust, thermal state at the time of emplacement, or subsequent history. Some of these variations are presented in Table 1 and Figure 1, and include the following features: temperature of the intrusion,

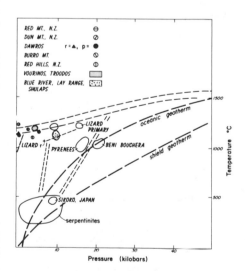

Figure 1. P-T diagram for ultramafic compositions, modified after O'Hara (1967), showing conditions of equilibration of selected ultramafic bodies. p = primary assemblage; r = recrystallized assemblage.

form and size of the intrusion, grade of regional and contact metamorphism, phase assemblages of the intrusion and associated mafic rocks, and chemical composition.

TABLE 1. SUMMARY OF CHARACTERISTIC FEATURES OF SOME
ULTRAMAFIC INTRUSIONS

| Area | Environment of Emplacement | | Phase Assemblages | |
	Temp.	Depth D/P	Ultramafic Composition	Facies of Mafic Associations
Ronda	>500°		Ol+opx+cpx+CrSp	Eclogite, Pyroxenite + Gabbro
Beni Bouchera	>500°	20 Kb	Ol+opx+G±opx	Eclogite
Mt. Albert	700–800	15	Ol+opx+cpx+CrSp	Pyroxenite
Lizard	≈700	10	Prim: Ol+opx+cpx+Sp Rexl: Ol+opx+CₗSp+Pg	Pyroxenite Gabbro
Tinaquillo	700	15	Ol+opx+CrSp	Gabbro
Lay Range	500	7–10	Ol+CrSp+cpx+opx+biot	Hornblende Gabbro
Blue River	400–500	5–10	Ol+opx+cpx+CrSp	Gabbro
Red Hills	500	0–1	Ol+opx+CrSp+cpx	Gabbro
Shulaps Range	500	5	Ol+opx+CrSp	Gabbro
Bay of Islands	0–600	0–10	Ol+opx+CrSp	Gabbro
Canyon Mountain	<200	<5	Ol+opx+CrSp	Gabbro
Vourinos	<200	<5	Ol+opx+CrSp±cpx	Gabbro
Troodos	<200	5	Ol+opx+CrSp	Gabbro
Norway-Caledonides			Ol+opx+cpx+G	Eclogite Pyroxenite
Penninic			Ol+opx+G	Eclogite Pyroxenite
Totalp	<200?	10 Km	Ol+opx+Cpx+Sp	Eclogite Pyroxenite
Calif. Coast Ranges	<200?	10–30 Km	Ol+opx+Cpx+Sp	Gabbro
Klamaths-Sierra	200?	5–10 Km?	Ol+opx+hb	Gabbro
Quetico-Shebandowan			Ol+cpx+hb	Gabbro
Burro Mtn., Calif.			Ol+opx+cpx+CrSp	Gabbro

Figure 1 shows a P-T diagram for ultramafic compositions with the positions of selected intrusions plotted on it. This diagram, modified from O'Hara (1967), shows that the temperature and pressure of equilibration of most ultramafic assemblages, as determined by O'Hara's petrogenetic grid, fall generally near the mantle solidus boundary. Some significant exceptions are present, particularly occurrences in Norway, Switzerland, southern Appalachians, and serpentinites; these exceptions will be discussed in more detail below. The diagram suggests, however, that the phase assemblages of most ultramafic compositions equilibrated during the diapiric upwelling which gave rise to these rocks, and that few of them subsequently recrystallized at lower temperatures and pressures.

The data on Alpine peridotite bodies indicate that they may be viewed as consisting of two broad categories: (1) bodies which exhibit thermal contact metamorphic aureoles, and (2) bodies which do not display thermal contact effects (much more common).

TABLE 1. SUMMARY OF CHARACTERISTIC FEATURES OF SOME
ULTRAMAFIC INTRUSIONS (*continued*)

Area	Chemistry		
	Al_2O_3 Perid.	CaO Perid.	Al_2O_3 opx/cpx
Ronda	2.5–3.6	2.42–3.29	.5–3.5
Beni Bouchera	1.9, 4.80	—	—
Mt. Albert	2–6	.29–1.9	2.5–5
Lizard	3.1–4.5	1.5–2.75	4.16
Tinaquillo	2.65	—	3.19
Lay Range	—	—	1.2
Blue River	—	—	—
Red Hills	0.9–1.2	0–.76	1.4–2.3/2.2
Shulaps Range	—	—	1.00
Bay of Islands	1.8	.05–1.25	—
Canyon Mountain	1.7	1.36	—
Vourinos	.6–1.5	.5–.7	.7–1.4
Troodos	—	—	—
Norway-Caledonides	6.7	7.9	1.1–1.6/1.7–2.7
Penninic	2.97/.11	3.30/.93	.38
Totalp	4.65	3.2	2.5–5
Calif. Coast Ranges	—	—	1.3–3
Klamaths-Sierra	—	—	—
Quetico-Shebandowan	2–3	9–14	—
Burro Mtn., Calif.	—	—	1.5–3

Bodies with Thermal Contact Aureoles

The first category includes the first ten bodies in Table 1 and generally represents diapiric masses which were intruded at high temperatures into pre-existing country rocks. Hence they represent bodies in which the processes of intrusion and emplacement were acting nearly simultaneously. These intrusions fall loosely into five categories.

Beni Bouchera. The Beni Bouchera, Morocco (Kornprobst, 1966, 1969; Milliard, 1959), and the Ronda, Spain bodies both contain spinel peridotite assem-

TABLE 1. SUMMARY OF CHARACTERISTIC FEATURES OF SOME
ULTRAMAFIC INTRUSIONS (*continued*)

| Area | Geology | | | Str. |
	Form	Size (Km²)	Age	
Ronda	Folded lens	300	post-Tr.	
Beni Bouchera	Elongate Dome	53	pre-Carbon.	Nappes
Mt. Albert	Stock	50	Ordov. (495 m.y.)	Isocl. Folds
Lizard	Stock	75	Ordov. (497 m.y.)	Complex Folds
Tinaquillo	Stock	75	pre-M.K.	Thrusts
Lay Range	Stock	64–128	m.P. p.-u.Tr.	Folds & Thrusts
Blue River	Lens	25	245 m.y.	Folds & Thrusts
Red Hills	Plug or Lopolith	100	Permian	Folds
Shulaps Range	Lensoid-pseudostratiform	150	1. Eocene	Folds & Thrusts
Bay of Islands	Pseudostratiform	1500	Ordov.	Thrust
Canyon Mountain	Pseudostratiform	150	u.P.-u.Tr.	Melange-Thrust
Vourinos	Pseudostratiform	270	1.K.	Melange-Thrust
Troodos	Pseudostratiform	>1000	pre-u.K.	Melange-Thrust
Norway-Caledonides	Lens	very small	pre-Silurian	Nappes
Penninic	Lenses	very small	Mesozoic	Nappes
Totalp	Exotic Block	13	Paleozoic	Melange
Calif. Coast Ranges	Mantle Slabs + Exotic Blocks	variable	Mesozoic	Melange
Klamaths-Sierra	Sheets + Folded Lenses	variable	Paleozoic-Mesozoic	Melange
Quetico-Shebandowan	Lenses	variable-small	early p€	Complex Folds
Burro Mtn., Calif.	Exotic Mass	25	Mesozoic	Melange

blages, grading in the case of Ronda into plagioclase-bearing types (Dickey, 1970). Both appear to be diapiric intrusions into deep-level continental crustal rocks. Dickey has interpreted the Ronda body as a lens folded within metamorphic rocks. The Beni Bouchera mass apparently is a domelike body. Both masses contain a variety of mafic layers, ranging from garnet to plagioclase-bearing varieties, which Dickey (1970) interprets as partial fusion products.

Lizard. The Lizard, Cornwall (Green, 1964), Tinaquillo, Venezuela (MacKenzie, 1960), and Mt. Albert, Quebec intrusions are stocklike masses of spinel lherzolite, generally 50 to 100 km² in size, and have pronounced thermal aureoles. In the Mt. Albert and Tinaquillo intrusions, rocks of the garnet granulite facies occur at the contact, whereas pyroxene hornfels rocks occur at the contact of the Lizard

TABLE 1. SUMMARY OF CHARACTERISTIC FEATURES OF SOME
ULTRAMAFIC INTRUSIONS (*continued*)

Area	Country Rocks Reg. Met.	Age	Contact Metam. Max. Facies	Max. Temp.
Ronda		Paleoz. + Mesoz.	Granulite	700–800
Beni Bouchera	Staur.	preCarbonif.	Granulite	700–800
Mt. Albert	Greensch.	Cambrian	Granulite	700–800
Lizard	Amphibolite	Cambrian	Granulite	700–800
Tinaquillo	Amphibolite	preCretaceous	Granulite	700–800
Lay Range	Unmet.	Paleozoic	Almandine Amphib.	500
Blue River	Unmet.	Paleozoic	Low Alman. Amphib.	400–500
Red Hills	Zeolite	Paleozoic	Px Hornfels	500–700
Shulaps Range	Unmet.	u.Paleoz.	Alman. Amphib.	500
Bay of Islands	Unmet.	p€-Ord.	Granulite Max.	0–600
Canyon Mountain	None	Paleoz.-Mesoz.	None	<200
Vourinos	None	Mesozoic	None	<200
Troodos	None	Mesozoic	None	<200
Norway-Caledonides	Granulite	€-Sil.	None	—
Penninic	Granulite			
Totalp	None	Paleoz.-Mesoz.	None	<200
Calif. Coast Ranges	Blueschist	Mesozoic	None	<200
Klamaths-Sierra	Greensch.	Paleoz.-Mesoz.	None	
Quetico-Shebandowan	Amphibolite	early p€		
Burro Mtn., Calif.	Blueschist	Mesozoic	None	

mass. Phase assemblages in the intrusions suggest equilibration conditions of the primary assemblages of 15 to 20 kbars and 1100 to 1300° C at or near the solidus. The metamorphic aureoles indicate maximum contact temperatures from 700° to 800° C, with the Mt. Albert and Tinaquillo aureoles indicating maximum pressures between 10 and 15 kbars, and the Lizard pressures up to 10 kbars.

Blue River. The Blue River (Wolfe, 1967), Lay Range, and Shulaps Range (Leech, 1953) intrusions, British Columbia generally are of lensoid form and are intruded into relatively unmetamorphosed rocks. They exhibit almandine amphibolite contact metamorphic aureoles, suggesting pressures from 2 to 4 kbars and 400° to 500° C contact temperature.

Red Hills. The Red Hills intrusion, New Zealand (Challis, 1965), is intrusive into zeolite-grade metamorphic rocks, is a plug or lopolith in form, and exhibits a low pressure, pyroxene hornfels aureole.

Bay of Islands. The Bay of Islands, Newfoundland (Smith, 1958), which is pseudostratiform in nature, but which has a garnet granulite aureole at its base, displays features similar to high-temperature intrusions and to mantle slabs (see below). The Feather River Body, Sierra Nevada, and the Trinity pluton, Klamath Mountains, California, also are large masses of ultramafic rocks spatially associated with amphibolite, the latter which may represent thermal aureoles or regionally metamorphosed rocks (Davis and others, 1965).

In Figure 2 are plotted the estimated P-T conditions of equilibration for several peridotites with thermal contact aureoles, together with the estimates of P-T range of their contact metamorphic aureoles. There is a general tendency for intrusions to lie near the mantle solidus, though Blue River and Beni Bouchera are exceptions to this rule. Also noted is that there is a tendency for the aureoles of the intrusion in question to lie near the field of geothermal gradients with the intrusion at both

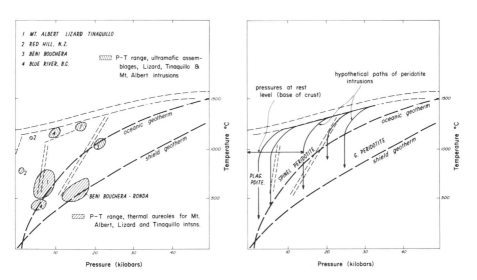

Figure 2. P-T diagram for ultramafic compositions, modified after O'Hara (1967), showing equilibration fields of ultramafic assemblages and associated thermal aureoles.

Figure 3. P-T diagram for ultramafic compositions, showing hypothetical equilibration trajectories for ultramafic bodies intruded at different levels.

higher temperatures and higher pressures than their contact aureoles. This relation suggests that the chemical equilibration of the intrusion lags somewhat behind its true physical conditions, or, in other words, that the reaction rates of equilibration are too slow to keep pace with the rise of the mantle diapir during its intrusion and emplacement.

The P-T conditions of a hot peridotite also afford a means of determining the thickness of crust into which it is intruded. One might expect that a mantle diapir will rise until an equilibrium level is reached, which should be at the base of lighter rocks, or in most cases the base of the crust, or Moho discontinuity. Its rise will be at first adiabatic until it intersects the mantle solidus (Figure 3), which it will then follow until it comes to rest. When it comes to rest, it should adjust to the ambient P-T conditions of its surroundings. Because equilibrium is not obtained throughout an intrusive body, a variety of P-T conditions should be displayed by a given intrusive mass. Analyses of various parts of the intrusion should give a curve which will follow the mantle solidus until the pressure corresponding to its level of rest has been reached, at which point the curve will turn off and trend toward the ambient P-T conditions. Figure 3 gives a representation of the paths of hypothetical intrusions. Figure 4 shows possible locations of these bodies at a mid-ocean ridge island arc or continental margin spreading center.

Bodies Which Lack Thermal Aureoles

The vast majority of peridotite occurrences lack thermal aureoles, and consequently they have probably been intruded as relatively cold, solid masses. These masses fall into several groups.

Mantle Slabs. These bodies are the ultramafic bodies of large areal extent which

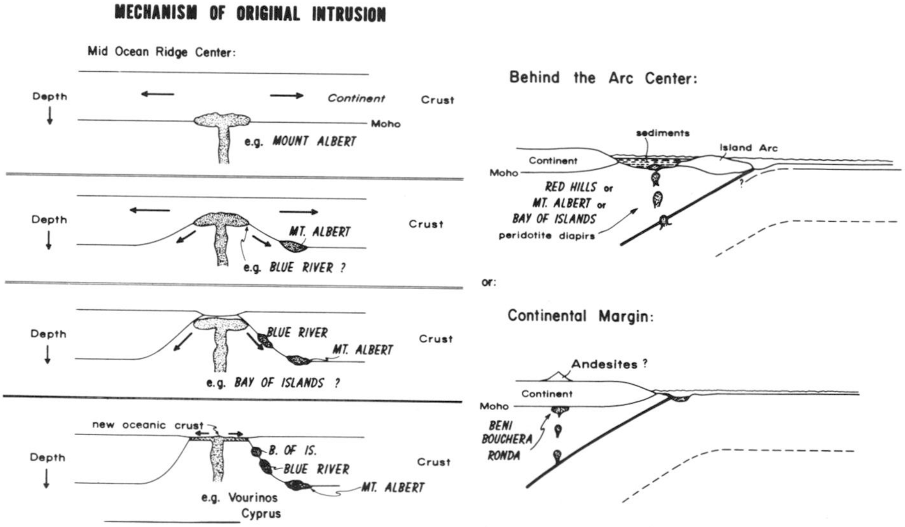

Figure 4. Hypothetical methods of intrusion of high-temperature bodies at a mid-ocean ridge spreading center or consuming plate margin (behind arc or continental margin). In mid-ocean ridge center, peridotite is intruded at higher and higher levels as crust thins. In consuming margin, depth of emplacement is a function of thickness of crust on overriding plate.

are present together with crustal rocks. Most bodies represent oceanic crust and mantle; sub-continental mantle appears to be rare.

1. Slabs of oceanic crust and mantle. These bodies contain a partial to complete section through the oceanic crust and into mantle rocks. Examples of such occurrences include Tethyan ophiolite complexes such as the Troodos complex, Cyprus (Moores and Vine, 1970), and the Vourinos complex, Greece (Moores, 1969); great occurrences in island arc regions such as New Caledonia (Avias, 1967), New Guinea (Davis, 1968), and Cuba (Thayer and Guild, 1947); and many masses of the western United States cordillera, such as the Canyon Mountain complex, Oregon (Thayer, 1963) and the Great Valley ultramafic-mafic sheet at the base of the Great Valley sequence (Bezore, 1969; Bailey and others, 1970). Many of these bodies show no thermal effects and many are associated with blueschist terranes. Many workers have now interpreted these masses as slices of oceanic crust and mantle. These masses are presumably formed at spreading centers, either at a mid-ocean ridge, or behind an island arc (Karig, 1970). The structures and equilibrium assemblages and types of associated rocks reflect the history of those centers. Subsequent tectonic activity has emplaced these slabs at continental margins.

2. Overthrust or otherwise exposed sheets of subcontinental mantle. These masses appear to be much less common than preserved oceanic crust-mantle slices. Only one example of this type is known—in the Ivrea and Lanzo zones of the Alps (Berckhermer, 1969; Nicoias, 1968).

Disrupted Mantle Overthrust or Underthrust Slabs Presently Incorporated into Melange Terranes. Examples of this may include smaller masses in the California-Oregon Coast Ranges (Hsü, 1969; Himmelberg and Loney, 1969; Medaris and Dott, 1970), in Italy (Maxwell, 1969, 1970), in the Arosa Schuppen zone (Peters, 1963), and in the Tethyan Belt (Temple and Zimmerman, 1969; Gansser, 1959). These bodies generally are discordant to surrounding rocks; most are serpentinized and possibly have undergone remobilization as serpentinite diapirs. Some show relations with mafic rocks, but ophiolite sequences are commonly disrupted. Mineralogic studies indicate widely varying P-T conditions of equilibration.

Conformable Bodies in Regionally Metamorphosed Terranes. Examples of this type include occurrences in the Sierra Nevada, California (Hietanen, 1951), Appalachians (Miller, 1951; Lapham, 1967), Penninic zone of the Alps (Dal Vesco, 1953; O'Hara and Mercy, 1963), and the Caledonian core region (Bruckner, 1969). These masses may represent metamorphosed melanges or overthrust sheets. As shown in Figure 1, many bodies of this category show equilibrium assemblages approaching that of the surrounding regional metamorphic rocks, and fall near the field of geothermal gradients. Most of these bodies show evidence of recrystallization, and their assemblages should reflect the ambient metamorphic conditions more than their original conditions of intrusion.

With such a separation, it becomes evident that both the original conditions of emplacement and the subsequent history will affect the observed features of a given body. Hence, in estimating the apparent conditions of emplacement, one may be either reading the conditions of original intrusion of emplacement or of subsequent burial and metamorphism of a body. With care, these different effects possibly may be separated, and it may be possible to estimate the P-T conditions extant at the time of intrusion and subsequent history, thereby ascertaining the paleogeothermal gradients at the time of their intrusion.

EMPLACEMENT MECHANISMS

Two features of Alpine intrusive rocks of critical interest which Hess (1955) noted are: (1) they were emplaced during the initial stages of deformation of a mountain belt, and (2) the time of intrusion, hence onset of orogeny, varied along strike of a mountain system. Figure 5 is a composite representation of various emplacement mechanisms proposed for the various types of Alpine peridotites outlined above.

Mantle and Oceanic Crustal Slabs

Two mechanisms of emplacement of these mantle-crust slabs are worthy of consideration.

A. Collision of a continent or microcontinent with a subduction zone which dips away from it, as proposed in several recent works (Temple and Zimmerman, 1969; Coleman, 1971; MacKenzie, 1969; Dewey and Bird, 1971; Moores, 1970) may achieve the observed field relations, and, in addition, provide for the emplacement of heavy mantle rock over lighter crustal rock. This process, sum-

MECHANISM OF EMPLACEMENT

Just prior to collision:

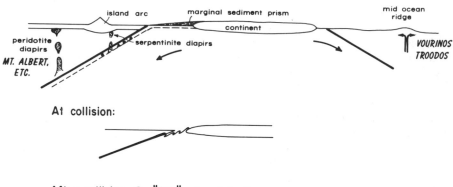

At collision:

After collision & "flip" of subduction:

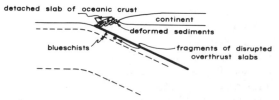

Figure 5. Mechanism of emplacement of peridotite, showing emplacement of slabs of oceanic crust and mantle emplaced by abortive subduction of a continental margin, followed by "flip" of subduction direction. Also depicted are disrupted fragments of overthrust or underthrust mantle rocks incorporated into melanges, formation of peridotite and serpentinite diapirs, and of Troodos-like structures at mid-ocean ridge. Triangles represent blueschists, squares represent disrupted fragments of slabs.

marized diagrammatically in Figure 5, involves the abortive incipient subduction of a continent. If a plate being consumed includes a continent, consumption will proceed until the continental margin collides with the consuming plate margin. It may go a small way into the trench, thus overcasting the overriding plate up onto the continental margin until the buoyant force of the lighter continental material takes over and arrests the subduction. Subsequently, the subduction may flip and continue operation under the continent, or the convergent plate motion will be taken up elsewhere. In the first case, the classic orogenic cycle of sedimentation and ultramafic intrusion, then granitic intrusion and regional metamorphism, will be observed. In the second case, possibly *zwischengebirgen* will form, which finally will result in crystalline nappes as continents converge and collide (Moores, 1970).

Examples of such situations in the present globe are, understandably, rare, but Fitch and Molnar (1970) document examples of two general types of collisions in the Indonesia-Philippines region—one where the Australian Plate is being consumed along the Java–New Hebrides trench system, and one in the Celebes-Halmahera region where two oppositely dipping zones are nearly intersecting.

The important implications of this mechanism for interpreting mountain ranges are as follows:

1. One can interpret the type of continental margin prior to collision—whether Atlantic style, Andes or Japan style—from the geology prior to ultramafic emplacement, by study of the pre-ultramafic "geosynclinal" sequences.

2. In those areas where the relation of the peridotite slabs to the foreland can be established, it follows that the direction of dip of the fossil subduction zone can be inferred. In each case, the zone must have dipped away from the continental margin toward the ocean basin.

3. The variation in time of emplacement and the rate of progression of emplacement along strike can be interpreted by fossil or radiometric dating. This variation will depend upon the rate and angle of convergence.

B. Collision of two subduction zones and accompanying island arcs dipping in the same direction. In this case presumably the plate bearing one arc is consumed at the trench of the other one, as may have happened in the Philippines or Japan (Miyashiro, 1961). Presumably some of the mantle would be caught up on the collision and subsequently exposed. This mechanism is possibly similar in effect to 1 above, though its implications are unexplored.

We feel that mechanism A. above is the one most likely to produce the observed relations, and we tentatively invoke it for emplacement of these oceanic crust-mantle slabs.

Melange and Metamorphosed Masses

Disrupted masses in melanges could be obtained in two possible ways: (1) by flip of subduction subsequent to collision postulated in mechanism A. and continuation of it under the continent; in this case the marginal sediments and overthrust wedge of mantle would be incorporated into the melange to some extent; (2) by diapiric rise of serpentinized ultramafic rock derived from the downgoing slab. Some serpentinized masses within the Franciscan terrane may have originated in this manner. Figure 5 gives a diagrammatic representation of these two mechanisms. If subsequent subduction takes place oceanward of these areas, thereby giving rise to regional metamorphism, intrusion, and deformation behind the zone of consumption, metamorphosed melange deposits will be formed. Possibly the bodies in the northern Sierra Nevada, the Caledonian core zone and the Penninic zone provide examples of this type of occurrence.

Hot Diapiric Masses

Ultimately all Alpine peridotite bodies probably are derived from diapiric mantle upwellings. These upwellings are presently primarily at spreading centers—mid-ocean ridges and behind island arcs or subduction zones (Karig, 1970; Fig. 4). The differences in thermal effect between hot and cold intrusions may stem directly from the time lapse between diapiric upwelling and emplacement. Hot diapiric intrusions may be formed at incipient spreading centers where the upwelling mantle material is emplaced directly into pre-existing crustal rocks. Of the two spreading centers where such incipient activity might take place, that of an incipient behind-the-arc center may be the most likely to be preserved. Thus, most observed diapiric masses may have originated on consuming plate margins, over down-going slabs. This conclusion implies that such bodies should be intruding behind present consuming plate margins, for example, Japan and the Andes.

If the spreading center activity goes beyond the incipient stage, then the diapiric material will intrude into pre-existing mantle and ocean crust. It will merely become part of a small ocean basin or sea floor. Its emplacement on the continental margin is then entirely subject to the vicissitudes of tectonic processes. Masses such as the Bay of Islands complex, the Trinity Pluton, and the Feather River peridotite body, which appear to be mantle and crust slabs with basal thermal aureoles, may represent an intermediate stage in this process, where spreading has gone beyond the incipient stage but has not proceeded so far that the overthrust mantle slab has completely cooled.

One might expect that these diapirs should rise until they find their own equilibrium level, which would probably be the base of the crust. When the crust is thin, or oceanic, as with the Red Hills body or possibly in St. Paul's Rocks (Melson and others, 1967), low pressure or shallow features and assemblages will be observed. When the crust is thick, or continental, the intrusion will show deep-level or high-pressure features (Beni Bouchera and Ronda). It should be possible to interpret from these diapiric intrusions not only the paleogothermal gradients, but also the thickness of crust at the time of their emplacement.

The above model is clearly hypothetical, but seems to explain the observed tectonic and petrologic features of most Alpine ultramafic bodies fairly well. It is amenable to testing by field and laboratory work.

SUMMARY AND CONCLUSIONS

Alpine ultramafic intrusions originate by diapiric upwelling of mantle material at spreading centers either at mid-ocean ridges or behind island arcs. The various features associated with these rocks are a function of depth of intrusion, of ambient geothermal gradients, and of post-intrusion emplacement and metamorphic histories. Oceanic crust-mantle slabs probably are emplaced onto or over continental rocks by abortive subduction of a continent. These continent-reverse subduction zone collisions are present but rare in the world today, but are recorded in Alpine mountain systems by the "first deformations" of "geosynclinal" sequences.

Classical "orogenic events" involve emplacement of ultramafic intrusions and tectonic transport of continental marginal deposits toward the craton. These effects cannot be produced by simple underthrusting tectonics presently observed around the Pacific, which give rise only to "eugeosynclinal" sequences. Hence "orogenic events" may reflect continent-island arc or continent-continent collisions, or abortive subduction of a continent.

ACKNOWLEDGMENTS

We have benefited from discussion with R. Coleman, D. Jackson, J. C. Maxwell, and F. J. Vine. This work was partly supported by National Science Foundation Grants GA-1395 and GP-1554.

REFERENCES CITED

Avias, J., 1967, Overthrust structure of the main ultrabasic New Caledonian massives: Tectonophysics, v 4, p. 531–541.

Bailey, E. H., Jones, D. L., and Blake, M. C., 1970, On-land Mesozoic oceanic crust in California Coast Ranges: U.S. Geol. Survey Prof. Paper 700C, p. C70–C81.

Benson, W. N., 1926, The tectonic conditions accompanying the intrusion of basic and ultrabasic igneous rocks: Natl. Acad. Sci., Memoirs, v. 19, 1st mem., p. 1–90.

Berckhermer, H., 1969, Direct evidence for the composition of the lower crust and the Moho. Tectonophysics, v. 8, p 97–105.

Bezore, S. P., 1969, The Mt. St. Helena ultramafic-mafic complex of the northern California Coast Ranges: Geol. Soc. America, Abs, with Programs for 1969 (Cordilleran Sec.), Pt. 3, p. 5–6.

Bowen, N. L., and Tuttle, O. F., 1949, The system MgO-SiO$_2$-H$_2$O: Geol. Soc. America Bull., v. 60, p. 439–460.

Bruckner, H., 1969, Timing of ultramafic intrusion in the core zone of the Caledonides of southern Norway: Am. Jour. Sci., v. 267, p. 1195–1212.

Challis, G. A., 1965, The origin of New Zealand ultramafic intrusions: Jour. Petrology, v. 6, p. 322–364.

Coleman, R. G., 1971, Plate tectonic emplacement of upper mantle peridotites along continental edges: Jour. Geophys. Research, v. 76, p. 1212–1222.

Dal Vesco, E., 1953, Genesi e metamorfosi delle rocce basiche e ultrabasiche nell'ambiente mesozonale dell'orogene Pennidico (Cantone Ticino): Schweizer Mineralog. u. Petrog. Mitt., v. 33, p. 173–480.

Davies, H. L., 1968, Papuan ultramafic belt: Internat. Geol. Cong., 24th Prague, Comptes Rendus, Sec. 1, p. 209–220.

Davis, G. A., Holdaway, M. J., Lipman, D. W., and Romey, W. D., 1965, Structure, metamorphism and plutonism in the south-central Klamath Mountains, California: Geol. Soc. America Bull., v. 76, p. 933–966.

Dewey, J. F., and Bird, J. M., 1970, Mountain belts and the new global tectonics: Jour. Geophys. Research, v. 75, p. 2625–2647.

—— 1971, Origin and emplacement of the ophiolite suite: Appalachian ophiolites in Newfoundland: Jour. Geophys. Research, v. 76, p. 3179–3206.

Dickey, J. S., Jr., 1970, Partial fusion products in Alpine-type peridotites: Serrania de la Ronda and other examples: Mineralog. Soc. America Spec. Paper 3, p. 33–49.

Fitch, T. J., and Molnar, P., 1970, Focal mechanisms along inclined earthquake zones in the Indonesian-Philippine region: Jour. Geophys. Research, v. 75, p. 1431–1444.

Gansser, A., 1959, Ausseralpine ophiolithprobleme: Eclogae Geol. Helvetiae, v. 52, p. 659–680.

Green, D. H., 1964, The petrogenesis of the high-temperature peridotite intrusion in the Lizard area, Cornwall: Jour. Petrology, v. 5, p. 134–188.

Hess, H. H., 1939, Island arcs, gravity anomalies, and serpentinite intrusions, a contribution to the ophiolite problem: XVII Internat. Geol. Cong., 17th Comptes Rendus Moscow, 17, v. 2, p. 263–283.

—— 1955, Serpentines, orogeny and epeirogeny: Geol. Soc. America Spec. Paper 62, p. 391–408.

—— 1965, Mid-oceanic ridges and tectonics of the sea floor, in Submarine geology and geophysics: Proc. 17th Symposium Colston Rearch Soc., p. 317–334.

Hietanen, A., 1951, Metamorphic and igneous rocks of the Merrimac area, Plumas National Forest, California: Geol. Soc. America Bull., v. 62, p. 565–607.

Himmelberg, G. R., and Loney, R. L., 1969, Mineralogy and petrology of the Alpine-type peridotite at Burro Mountain, California: Geol. Soc. America, Abs. with Programs for 1969 (Ann. Mtg.), Pt. 7, p. 101.

Hsu, K. J., 1969, Preliminary report and geologic guide to Franciscan melanges of the Morro Bay–San Simeon area, California: California Div. Mines and Geology Spec. Rept. 35, 46 p.

Karig, D. E., 1970, Ridges and basins of the Tonga-Kermadec Island arc system: Jour. Geophys. Research, v. 75, p. 239–254.

Kornprobst, J., 1966, A propos les peridotites du massif des Beni-Bouchera (Rif seplentrional, Maroc): Soc. Française Minéralogie et Christallographie Bull., v. 89 p. 399–404.

—— 1969, Le massif ultrabasique des Beni Bouchera (Rif Interne Maroc): Etude des peridotites de hautes temperature et de haute pression et des pyroxenolites, a grenat ou sans grenat, qui leur sont associees: Contr. Mineralogy and Petrology, v. 23, p. 283–322.

Lapham, D. M., 1967, The tectonic history of multiply deformed serpentinite in the piedmont of Pennsylvania, in Wyllie, P. J., ed., Ultramafic and related rocks: New York, John Wiley & Sons, Inc., p. 183–190.

Leech, G. B., 1953, Geology and mineral deposits of the Shulaps Range, southwestern British Columbia: British Columbia Dept. Mines and Petroleum Resources Bull., v. 32, 54 p.

MacKenzie, D. B., 1960, High-temperature alpine-type peridotite from Venezuela: Geol. Soc. America Bull., v. 71, p. 303–318.

MacKenzie, D. P., 1969, Speculations on the consequences and causes of plate motions: Royal Astronomical Society, Geophys. Jour., v. 18, p. 1–32.

Maxwell, J. C., 1969, "Alpine" mafic and ultramafic rocks—the ophiolite suite; a contribution to the discussion of the paper "The origin of ultramafic and ultrabasic rocks" by P. J. Wyllie: Tectonophysics, v. 7, p. 489–494.

—— 1970, The Mediterranean ophiolites and continental drift; in What's new on Earth; New Brunswick, Rutgers Univ. Press.

Medaris, L. G., and Dott, R. H., 1970, Mantle-derived peridotites in southwestern Oregon in relation to plate tectonics: Science, v. 169, p. 971–974.

Melson, W. G., Jarosewich, E., Bowen, V. T., and Thompson, G., 1967, St. Peter and St. Paul Rocks, a high-temperature, mantle-derived intrusion: Science, v. 155, p. 1532–1535.

Miller, R. W., III, 1951, The Webster-Addie ultramafic plug, Jackson County, North Carolina and secondary alteration of its chromite: Am. Mineralogist, v. 38, p. 1134–1147.

Milliard, Y., 1959, Les massifs metamorphiques et ultrabasiques de la zone paleozoique interne du Rif: Maroc Service Géol. Notes et Mém., v. 147, p. 125–160.

Miyashiro, A., 1961, Evolution of metamorphic belts: Jour. Petrology, v. 2, p. 277.

Moores, E. M., 1969, Petrology and structure of the Vourinos ophiolitic complex, northern Greece: Geol. Soc. America Spec. Paper 118, 74 p.

Moores, E. M., 1970, Ultramafics as keys to orogeny, with models for the U.S. Cordillera and the Tethys: Nature, v. 225, p. 837–842.

Moores, E. M., and Vine, F. J., 1970, The Troodos, Cyprus, and other ophiolites as oceanic crust; evaluation and implications: Royal Soc. London Philos. Trans., ser. A, 268, p. 443–466.

Nicolas, A., 1968, Relations structurales entre le massif ultrabasique de Lanzo, ses satellites et la zone de Sesia Lanzo: Bull. Suisse Minér. Pétrogr., v. 48, p. 145–156.

O'Hara, M. J., 1967, Mineral parageneses in ultrabasic rocks, in Wyllie, P. J., ed., Ultramafic and related rocks: New York, John Wiley & Sons, p. 393–403.

O'Hara, M. J., and Mercy, E. L. P., 1963, Petrology and petrogenesis of some garnetiferous peridotites: Royal Soc. Edinburgh Proc., v. 65, p. 251–314.

Peters, T., 1963, Mineralogie und Petrographie des Totalserpentins bei Davos: Schweizer Mineralog. u. Petrog. Mitt., v. 43, p. 529–685.

Smith, C. H., 1958, The Bay of Islands complex, western Newfoundland: Canada Geol. Survey Mem. 290, 132 p.

Steinmann, G., 1905, Geologische Beobachtungen in den Alpen II. Die Schardt'sche Überfaltungstheorie und die geologische Bedeutung der Tiefseeabsätze und der ophiolithischen Massengesteine: Freiburg Naturforschung Gesellschaft, Bd. 16, p. 44–65.

——— 1926, Die ophiolithischen Zonen in den mediterranean Kettengebirgen: 14th Internat. Geol. Cong., 14th, Madrid, Comptes Rendus 2, p. 638–667.

Temple, P. G., and Zimmerman, J., 1969, Tectonic significance of Alpine ophiolites in Greece and Turkey: Geol. Soc. America, Abs. with Programs for 1969 (Ann. Mtg.), Pt. 7, p. 221.

Thayer, T. P., 1960, Some critical differences between Alpine-type and stratiform complexes: 21st Internat. Geol. Cong., 21st, Copenhagen, Comptes Randus 13, p. 175–187.

Thayer, T. P., 1963, Canyon Mt. Complex, Oregon and the Alpine mafic magma stem: U.S. Geol. Survey Prof. Paper C 82, p. 475.

Thayer, T. P. and Guild, P. W., 1947, Thrust faults and related structures in eastern Cuba: Am. Geophys. Union Trans., v. 28, p. 919–930.

Vine, F. J., and Hess, H. H., 1971, Sea floor spreading, in Maxwell, A. E., ed., The sea, v. 4: New York, John Wiley & Sons, Inc.

Wolfe, W. J., 1967, Petrology, mineralogy and genesis of the Blue River ultramafic intrusion, Cassiar District, British Columbia [Ph.D. thesis]: New Haven, Yale Univ. [available from Univ. Microfilms, Ann Arbor, Mich.]

MANUSCRIPT RECEIVED BY THE SOCIETY MARCH 29, 1971

THE GEOLOGICAL SOCIETY OF AMERICA, INC.
MEMOIR 132
© 1972

Emplacement of the Vourinos Ophiolitic Complex, Northern Greece

JAY ZIMMERMAN, JR.
Esso Production Research Company
Houston, Texas 77001

ABSTRACT

The Vourinos Complex, an allochthonous alpine ophiolite, overlies a melange and a sequence of deformed, recrystallized Triassic-Jurassic neritic to intertidal carbonate rocks along a tectonic contact. Rocks subjacent to the complex, including Paleozoic metasedimentary and metaigneous rocks which underlie the Mesozoic sequence along a bedding-parallel fault, have been subjected to regional greenschist-grade metamorphism.

The Zygosti Ultramafic-Mafic Belt, separated from the Vourinos massif by Triassic-Jurassic marble, is probably equivalent to the upper portion of the Vourinos Ultramafic Zone.

The Vourinos Complex is lithologically similar to current models of oceanic crust and upper mantle. The four-stage mechanism proposed by Temple and Zimmerman (1969), in which the edge of a continental sialic plate is partially subduced beneath simatic oceanic crust, is used to explain emplacement of the Vourinos Complex. Fragments or thrust sheets of oceanic material are emplaced upon shelf sediments and the crystalline basement of the continental margin during subduction. Lower density of the sialic plate will inhibit and eventually halt its downward movement, and it will be raised by subsequent isostatic readjustment. Because continued subduction of the continental plate is no longer possible, the orientation of the subduction zone will flip to the opposite sense, or the zone will shift to a new location. During the subduction phase, pelagic sedimentation may supplant shallow shelf conditions along the continental margin, and the sediment and underlying sialic basement may be subjected to low-grade regional metamorphism.

Stratigraphic evidence, the orientation of fabric elements and large-scale structures in units directly beneath the complex, the direction of dominant regional thrusting, and the present distribution of ultramafic rocks in Greece suggest that the Vourinos Complex was thrust southwestward from the Vardar Zone during a period of subduction and related orogeny in latest Jurassic or Early Cretaceous time. Additional transport to the southwest, probably involving portions of the Pelagonian Zone and overlying rocks including the Vourinos Complex and its associated units, occurred during an Alpine orogenic phase.

INTRODUCTION

The Vourinos Ophiolitic Complex, located in north-central Greece, is composed primarily of serpentinized peridotite, dunite, pyroxenite, and associated mafic rocks. The complex overlies metamorphosed melange, Mesozoic shelf carbonates, and Paleozoic rocks along a thrust fault marked by a zone of highly sheared serpentinite. Subjacent units dip beneath the ultramafic rocks along an arcuate contact convex to the northeast (Figs. 1 and 2). There is no chilled margin at the base of the complex.

The area of the Vourinos massif was part of Brunn's (1956) larger study of the Pelagonian, Sub-pelagonian, and Pindos Zones (Fig. 3). The complex and subjacent rocks have recently received detailed petrological and structural consideration (Moores, 1969; Zimmerman, 1968) because of the possibility that it represents an autochthonous intrusive alpine ophiolite. Disagreement as to the mode of emplacement of the complex still exists. This paper suggests that the Vourinos Complex is an allochthonous fragment of oceanic crust and upper mantle thrust over neritic carbonates concurrent with partial subduction of continental lithosphere (Zimmerman, 1969; Temple and Zimmerman, 1969).

The reader is referred to Moores (1969) for a general summary of the alpine ophiolite problem.

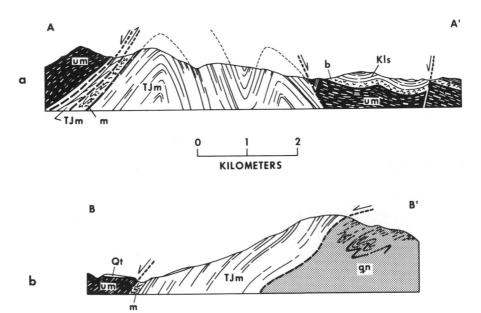

Figure 2. Cross sections keyed to Figure 1; horizontal scale equals vertical scale. Drawings by D. H. Roeder.

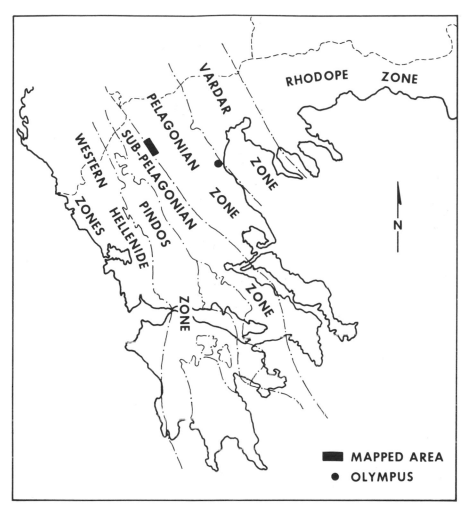

Figure 3. Generalized map of Greece showing major isopic zones, Olympus, and the location of the mapped area (*after* Aubouin, 1965).

SUBJACENT UNITS

Triassic-Jurassic Marble

The Vourinos Complex is underlain along a fault contact (Fig. 4) by approximately 2 km of Middle Triassic to Late Jurassic (Brunn, 1956) dolomitic calcite marble (Figs. 1 and 2). Although no microfauna were found in the severely recrystallized marble, dasycladacean algae, algal oncolites, and loferites (Fischer, 1964) distributed throughout the entire thickness of the unit suggest a neritic to littoral environment of deposition (Johnson, 1961; Fischer, 1968, oral commun.).

Bedding lamination from fractions of a millimeter to about 2.5 mm thick, defined by thin layers of detrital minerals, is characteristic of the unit and is preserved in the marble immediately below the fault contact with overlying ultramafic rocks.

Figure 4. Fault contact between basal serpentinite of the Vourinos Complex (left) and Triassic-Jurassic marble (right) about 5 km southeast of Paleokastro. View looking northwest.

Insoluble residues of the marble contain quartz, muscovite, chlorite, and albite. The $2M_1$ muscovite polymorph occurs individually or with minor amounts of 1M. Yoder and Eugster (1955) found the inversion from 1M to $2M_1$ to occur between $200°$ C and $350°$ C at $P_{H_2O} = 1$ kb (approximate), and Velde (1965) at $125°$ C, $P_{H_2O} = 4.5$ kb.

Tectonite fabrics and recrystallization textures similar to those produced during experimental deformation and annealing recrystallization of Yule marble by Griggs and others (1960b) and Griggs and others (1960a) are typical of the Triassic-Jurassic marble.

In the northern portion of the area, the marble is broadly folded along arcuate, southeast-trending axes. These folds tighten to the southeast, and the southernmost anticline is overridden by ultramafic rocks at the fault contact near the summit of the Vourinos massif. North of Chromion, a single anticline is preserved. Folds of this magnitude are not present in the area southwest of the Zavordhas monastery (Fig. 1).

Concentric flexural-slip folds about 5 m in wavelength are present adjacent to the contact with overlying ultramafic rocks east of Paleokastro. Fold axes plunge southeast and are overridden along the Vourinos fault (Fig. 1). These folds occur within about 50 m of the fault but die out rapidly to the north.

East of Paleokastro and northwest of Chromion, marble has been thrust upon overlying mélange. Rarely, smaller blocks of marble have been included in the melange. North of the Vourinos summit, small blocks of Triassic-Jurassic marble have been tectonically incorporated into the basal serpentinite of the complex and presently crop out up to about 100 m from the contact.

The Melange

A melange composed primarily of chlorite-sericite phyllite dips beneath the Vourinos Complex along its northern, eastern, and southern boundaries (Fig. 1). A matrix of predominant metasiltstone and some metatuff contains exotic blocks of mafic volcanic rock, marble, radiolarite, and minor amounts of serpentinite, ranging from less than 1 m to more than 100 m in long dimension. Contact relations between essentially homogeneous matrix and exotic blocks indicate lithification of the latter prior to inclusion into relatively plastic matrix material.

Quartz-albite-muscovite-epidote/clinozoisite(-sphene-calcite-biotite) and quartz-albite-muscovite-chlorite(-calcite) assemblages are typical of the metasilt matrix, while the quartz-albite-muscovite-chlorite-epidote/clinozoisite(-sphene-actinolite-calcite) assemblage dominates the metavolcanic matrix fraction.

Exotic blocks of dolomitic calcite marble contain deformed, thin-bedded, recrystallized white chert (Fig. 5), suggesting a pelagic environment of deposition, and are further distinguished from Triassic-Jurassic marble by the absence of bedding lamination. Insoluble residue of the carbonate includes quartz, muscovite, chlorite, and trace amounts of talc. The $2M_1$ muscovite polymorph occurs individually or with minor amounts of $1M$.

Exotic mafic rocks typically contain quartz, plagioclase (An_{47}-An_{49}), white mica, chlorite, sphene, relict augite, secondary calcite and pumpellyite, epidote minerals, and opaque oxides. Biotite is rare, occurring only in the Zavordhas area.

Red radiolarian chert is abundant only in the vicinities of Siatista and Paleokastro, where it is spatially associated with mafic exotic blocks. Exotic serpentinite is limited to single blocks at Siatista and about 6 km southeast of Chromion.

Stilpnomelane is present in several melange matrix samples, and biotite, although rare, is evenly distributed in the matrix over the entire outcrop area.

Bedding has been obliterated throughout the matrix except locally in the Zavordhas area where strongly folded and attenuated laminations have been observed at one location. Penetrative foliation (s_1), commonly folded into crenulations averaging 5 cm to 10 cm in wavelength, is developed over the entire outcrop area of the melange matrix but is not present in exotic blocks. Crenulation axes plunge steeply to the southeast near Paleokastro (Fig. 6 a,b). Tension surfaces (s_2), often filled with quartz, lie in the *bc* plane of a triaxial coordinate system (the *achsenkreuz* of Sander, 1948) oriented by arbitrarily equating *b* to crenulation B and placing *a* in the axial surface of the folds (Fig. 6c,d).

Mineral assemblages and deformation fabrics observed in the melange are compatible with formation by regional dynamothermal metamorphism of greenschist facies, possibly transitional between the quartz-albite-muscovite-chlorite and quartz-albite-epidote-biotite subfacies (Winkler, 1967).

Amphibolite

A thin, discontinuous amphibolite containing colorless to pale brown hornblende, plagioclase (An_{46}-An_{54}), sphene, relict clinopyroxene altering to actinolite, and rarely occurring garnet crops out locally at the ultramafite-melange contact about 6 km northwest of Chromion and about 0.5 km northwest of Zavordhas. A second-

Figure 5. Marble exotic block with bedded chert in melange east of Paleokastro.

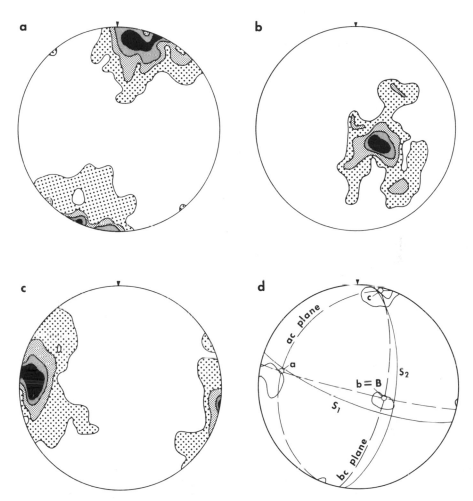

Figure 6. (a) Equal-area projection of 134 poles to melange matrix s_1 east of Paleo-kastro. Contours at 10 percent, 6 percent, 4 percent, and 1 percent per 1 percent area. (b) Equal-area projection of 60 crenulation B-axes in melange matrix east of Paleokastro. Contours at 10.8 percent, 8.3 percent, 5 percent, and 0.6 percent per 1 percent area. (c) Equal-area projection of 125 poles to melange matrix s_2 east of Paleokastro. Contours at 12 percent, 8 percent, 4 percent, and 1 percent per 1 percent area. (d) Spatial relation between crenulation B-axes, s_1, s_2, and a triaxial coordinate system. Explanation in text.

ary assemblage, possibly representing diaphthoresis of the rock, includes chlorite, sericite, albite, biotite, and epidote. Thickness of the amphibolite averages about 2 m. The contact between amphibolite and melange is sharp and shows no evidence of mineralogical gradation, suggesting that the formation of the amphibolite has no direct thermal relation to metamorphism of the melange material.

The origin of the amphibolite is unknown. Hess (1968, oral commun.) suggested essentially *in situ* recrystallization of melange constituents by continued serpentinization of overlying ultramafic units, an exothermal process (Schuiling, 1964). Alternatively, it may represent fragments of sedimentary or volcanic material, overridden by the base of a high-temperature Vourinos sheet during the initial stage of thrusting, which have survived long-distance tectonic transport. Amphibolite, where present, occurs consistently at the base of the complex between serpentinite and melange, parallel to the fault contact. It is not present as exotic inclusions in the melange.

Paleozoic Crystalline Rocks

Paleozoic crystalline rocks (Brunn, 1956) of the Pelagonian Zone including granitic augen gneiss, sericite phyllite, and mafic schist containing brown hornblende partially altered to actinolite underlie Triassic-Jurassic marble along a bedding-parallel fault marked by about 20 cm of gouge (Figs. 1 and 2b). Compositional layering in the gneiss up to about 10 m below the fault is consistently subparallel to bedding lamination in overlying marble from the Vounassa massif northward to the Aliakmom River, although farther down-section, to the south and east, the gneiss is characterized by tight similar folds which bear no geometric relation to the gently warped Mesozoic carbonate rocks.

Composition of the Paleozoic units indicates metamorphism to amphibolite facies probably coincident with deformation during a Hercynian orogenic phase (Brunn, 1956), followed by diaphthoresis corresponding to formation of the greenschist facies mineral assemblages presently dominant in the mafic and phyllitic units.

A serpentinite body located directly beneath but not cutting the marble-gneiss fault contact northwest of Tranovolto is principally composed of antigorite, suggesting that it has been subjected to high shear stress and dynamothermal metamorphism to at least albite-epidote amphibolite grade (Hess and others, 1952). The emplacement age of the serpentinite body is probably late Paleozoic.

THE ZYGOSTI ULTRAMAFIC-MAFIC BELT

A northwest-trending belt of ultramafic and mafic rocks, lying generally southeast of Metamorphosis and south and west of Rodhiani, is separated from folded Triassic-Jurassic marble along its southwestern flank by an oblique-slip eastdipping normal fault. The belt is overlain by unmetamorphosed Cretaceous, Tertiary, and Quaternary sedimentary rocks to the northeast (Figs. 1 and 2a). Interlayered serpentinized dunite and peridotite; irregular dunite bodies; massive serpentinite; and pyroxenite, gabbro pegmatite, rodingite, and serpentinite dikes are the major constituents, unconformably overlain by amygdalar basalt.

Moores (1969) divided the Vourinos Complex into a lower ultramafic zone, an intermediate transition zone, and an upper mafic zone on the basis of characteristic lithologies. The sheared serpentinite zone and melange characteristic of the base of the Vourinos Complex do not crop out at surface in the Zygosti Belt. In lithology, relative abundance, and areal distribution, the major units resemble the ultramafic zone of the Vourinos. The presence of gabbro pegmatite dikes characteristic of Moores's transition zone suggest that the Zygosti rocks correspond roughly to the upper ultramafic zone. Thick units of norite, diorite, and mafic extrusive rocks which mark the stratigraphic top of the Vourinos Complex do not crop out in the Zygosti area, although about 10 m of leucocratic, phaneritic igneous rock containing ortho- and clinopyroxene and hydrogrossular exposed south of Rodhiani may represent altered norite. Unlike the Vourinos volcanic extrusive rocks, basalts of the Zygosti Belt overlie ultramafic units along a sharp, non-gradational contact.

Major pyroxenite dikes, compositional layering, and fold axes in dunite-peridotite strike northwest. No preferred orientation of smaller pyroxenite, gabbro pegmatite, or rodingite dikes is apparent.

Units missing from the typical Vourinos succession may have been removed by displacement on thrusts subsidiary to the main Vourinos sole fault.

CENOMANIAN LIMESTONE

Erosional remnants of Cenomanian (Brunn, 1956) orbitoline and rudistid limestone nonconformably overlie ultramafic rocks southeast of Xirolimni and mafic and ultramafic units along the eastern boundary of the Zygosti Belt (Fig. 1). While deformed, the limestone is much less severely folded than the Triassic-Jurassic marble and is the oldest sedimentary unit in the area which shows no effects of metamorphism.

DISCUSSION OF METAMORPHIC AND STRUCTURAL EVIDENCE

A typical contact metamorphic aureole is not present beneath the Vourinos Complex. Consistent mineral assemblages in the melange, Triassic-Jurassic marble residues, and Paleozoic pelitic and mafic units have been mapped continuously up to 15 km from the base of the complex. Reconnaissance strongly suggests that they continue for at least an additional 10 km in the same direction. The assemblages, their spatial continuity, and tectonite fabrics in the melange and Triassic-Jurassic marble indicate formation by regional dynamothermal metamorphism.

An alternative to the suggested emplacement of the complex by overthrust of a portion of oceanic lithosphere is its intrusion as a solid-liquid mass and subsequent emplacement on the ocean floor. In order to evaluate this alternative, a model describing heat dissipation by an intrusive body with the approximate physical characteristics of the Vourinos Complex has been calculated using the technique of Jaeger (1957). The model (Fig. 7a, b) assumes a vertical slab of doleritic magma 11.5 km thick (W), 30 km long (L), and of infinite depth, intruding rocks of equivalent thermal characteristics. The width and length parameters are based on Moores's (1969) reconstruction and the length of the present outcrop, respectively. Thermal diffusivity (K) was calculated using specific heat and latent heat values from Jaeger (1957) and thermal conductivity from Clark (1966). Range of solidification of magma was taken as 1,100° C to 1,350° C, and density was again based on Moores's reconstructed complex. Temperature of the country rock was assumed to be 0° C. The model was calculated for initial intrusive temperatures (TI) of 1,235° C and 1,100° C to conform to Moores's suggested crystallization range compatible with a final liquid fraction of mafic composition.

Figure 7 illustrates heating of adjacent country rock as a function of time. Years subsequent to intrusion are shown as curves of decreasing amplitude. Distance from the center of the intrusion is recorded on the horizontal axis, and the vertical line in the interior of the diagrams represents both time = 0 and the geographical edge of the intrusive mass.

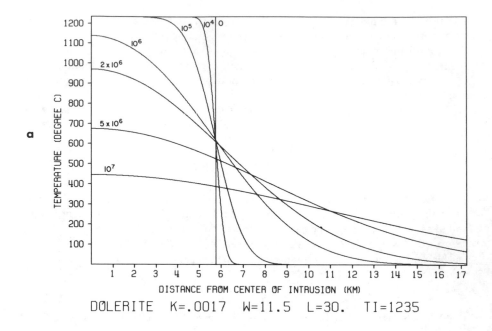

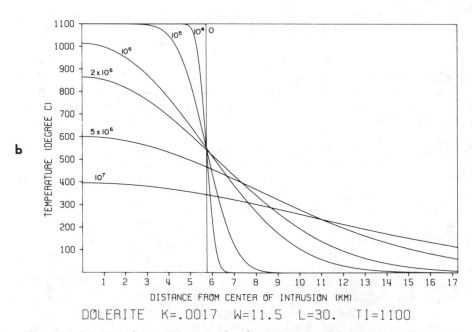

Figure 7. (a) Heat dissipation curves for the intrusive model at TI = 1,235° C. (b) Heat dissipation curves for the intrusive model at TI = 1,100° C. Explanation in text.

Figure 7 a and b indicates that during the cooling history of the model intrusion, temperatures of the order of 300° C would have extended from about 3.75 km (TI = 1,100° C) to 4.75 km (TI = 1,235° C) into the country rock. The model is admittedly crude but suggests that with an intrusive complex, relatively high-temperature calcium and magnesium silicates such as talc, tremolite, and diopside should be widely distributed in underlying siliceous calcite and dolomite marble. A detailed survey of these rocks does not reveal the presence of these minerals, nor does the model explain the consistent occurrence of greenschist assemblages and tectonite fabrics over their present area of distribution.

Preservation of bedding lamination adjacent to the principal fault contact, inclusion of discrete blocks of Triassic-Jurassic marble in the basal serpentinite and melange, the orientation of fold axes in marble, and orientation of crenulation axes and s_2 in the melange matrix suggest that the Vourinos Complex was originally thrust in a northeast direction over lithified Mesozoic shelf carbonates. The sense of transport is not specified by these features. Approximately 5 km southeast of Chromion, probable Neogene (Faugeres, 1967, oral commun.) valley fill has been overthrust by basal serpentinite of the complex, indicating that the latest movement has been to the northeast. Truncation and overriding of melange and large- and small-scale folds in Triassic-Jurassic marble may have occurred at this time. The Late Tertiary thrusting represents minor displacement of the complex and does not necessarily reflect the sense of earlier transport.

SIMILARITY OF THE VOURINOS COMPLEX TO OCEANIC CRUST AND UPPER MANTLE

The succession of units comprising the Vourinos Complex bears a striking resemblance to oceanic crustal layers 1, 2, and 3 (Temple and Zimmerman, 1969; Moores, 1969). Figure 8 compares the model developed by Hess (1955, 1962, 1965; Vine and Hess, 1970) to the reconstructed Vourinos. Seismic velocities in km/sec on the Vourinos column represent the ranges expected in these rocks at pressures of 500 to 1,000 atmospheres (Clark, 1966) and are comparable to the velocity range considered by Hess to be representative of partially serpentinized peridotite. Diorite and gabbro layers typical of the Vourinos Complex and other alpine ophiolites (Thayer, 1967) may be difficult to distinguish from serpentinized peridotite by seismic refraction methods, and these rocks may be more widespread in oceanic crust than generally supposed. Aumento (1969) dredged probable *in situ* diorites associated with basalt and serpentinized gabbro and peridotite from scarps on the Mid-Atlantic Ridge.

THE TECTONIC MODEL

Overthrust masses of oceanic crust have been discussed by de Roever (1956), Hess (1965), and other writers. Temple and Zimmerman (1969) have proposed a 4-stage model for the emplacement of ophiolites upon shallow-water sedimentary rock (Fig. 9).

Stage 1: A lithospheric plate containing continental crust preceded by oceanic crust approaches a subduction zone.

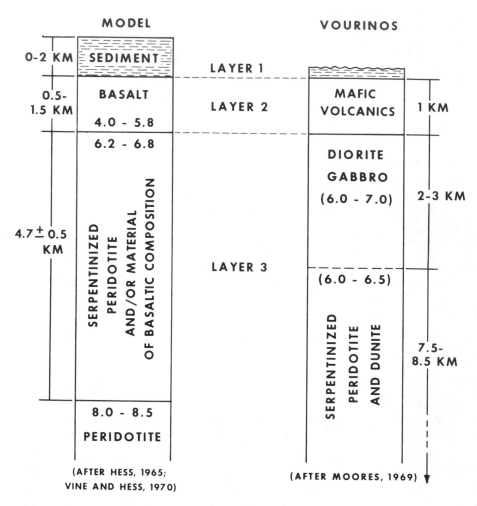

Figure 8. Comparison between Hess model for oceanic layers 1, 2, and 3 and Moores's reconstruction of the Vourinos Complex. Seismic velocities in km/sec.

Stage 2: The leading edge of the continent is partially subduced, and oceanic crust and upper mantle override the sedimentary apron. Because of the lower density of the sialic plate, incipient isostatic readjustment will add a vertical component of motion (broken arrow in Fig. 9).

Stage 3: With further subduction, isostatic readjustment becomes dominant. Discrete fragments of the overriding oceanic plate are tectonically incorporated upon and within the sedimentary apron. Alternatively, slices of simatic crust and upper mantle may override the continent for some distance (D. H. Roeder, 1971, written commun.) creating ophiolite "nappes."

Stage 4: Due to the inability to totally subduce continental under oceanic lithosphere, the subduction zone flips to an opposite orientation (discussed quantitatively by McKenzie, 1969) or, alternatively, maintains its original orientation and shifts to another location.

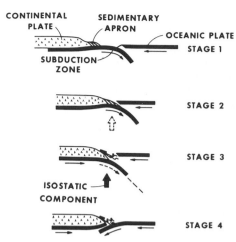

CONTINENTAL   SEDIMENTARY
PLATE   APRON
                              OCEANIC PLATE
                                 STAGE 1
SUBDUCTION
ZONE

                                 STAGE 2

                                 STAGE 3

ISOSTATIC
COMPONENT

                                 STAGE 4

Figure 9. Four-stage model of oceanic lithosphere thrust over a partially subduced continental edge (*after* Temple and Zimmerman, 1969). Explanation in text.

With partial subduction of the continent in stage 2, a marine transgression over crystalline basement or a comparatively rapid change from neritic to pelagic sedimentation in the sedimentary apron can be expected.

The latter may be recorded in the Vourinos area where neritic to littoral carbonates underlie the melange which contains exotic material of probable deeper water origin.

Widespread, low-grade regional metamorphism, such as affects units subjacent to the Vourinos Complex, appears to be related to the process of emplacement of the ultramafic rocks and may result from burial under more than 11 km of oceanic crust and upper mantle.

EMPLACEMENT OF THE VOURINOS COMPLEX

The Mafic Zone of the Vourinos Complex is overlain by radiolarite and by unmetamorphosed calpionellid limestone of Tithonian(?) or Neocomian(?) age (Moores, 1969). These rocks are, in turn, unconformably overlain by unmetamorphosed Cenomanian rudistid limestone, suggesting that in latest Jurassic or earliest Cretaceous time the igneous units were at depth compatible with pelagic sedimentation. Provided the upper beds of the Triassic-Jurassic marble represent Late Jurassic deposition and that regional metamorphism is associated with thrusting of the ultramafic rocks, the Vourinos Complex was most probably emplaced between early Neocomian and late Albian time.

In central Greece, peridotite and sepentinite are distributed over a northwest-trending belt approximately 200 km wide (Renz and others, 1954) which encompasses the Vardar, Pelagonian, and Sub-pelagonian Zones, and part of the Pindos Zone. The Vardar Zone is cited as the location of major diastrophism in Late Jurassic–Early Cretaceous time by Mercier (1966). This area is a probable location for the subduction which resulted in the emplacement of alpine ultramafic rocks during the late Mesozoic and may have been the site of a similar process in early Tertiary time. The sense of thrust transport of post-Paleozoic rocks in central and western Greece, where decipherable, has been from northeast to southwest, and the Vardar Zone presently occupies the possible site of a pre-Alpine Tethyan seaway (Ciric, 1963).

Approximately 175 km of lateral transport along thrust faults, such as that directly underlying the Vourinos Complex, is required to move ultramafic material from the western edge of the Vardar Zone to the eastern perimeter of its present distribution range. Long-distance displacement may also have occurred along the fault which separates Triassic-Jurassic marble from Paleozoic crystalline rocks (Fig. 2). A significant portion of the total transport distance may be younger than the initial Late Jurassic–Early Cretaceous phase. Godfriaux (1962) reports a tectonic window in the Olympus massif (Fig. 3) which exposes Eocene lime-stone and flysch, typical of the external western Hellenides, beneath Paleozoic crystalline rocks of the Pelagonian Zone. The Olympus window suggests an 80 km southwestward displacement of Paleozoic and overlying rocks, including the Vourinos Complex, Triassic-Jurassic marble, and melange during an Alpine oro-genic phase.

At present, detailed geological knowledge of central Greece is fragmentary. Any future regional synthesis must seriously consider the probability of periodic major southwest-directed thrusting from earliest Cretaceous to middle Tertiary time.

ACKNOWLEDGMENTS

The writer is indebted to J. C. Maxwell, H. H. Hess, L. Faugeres, E. M. Moores, R. Mullins, D. H. Roeder, J. Rodgers, the Institute for Geology and Subsurface Research in Athens, and Esso Production Research Company. The paper was critically reviewed by H. G. Ave'Lallemant, D. R. Baker, and J. J. W. Rogers, whose comments and suggestions are sincerely appreciated. The field portion of this study was supported by NSF Grant GP-1554.

REFERENCES CITED

Aubouin, J., 1965, Geosynclines: New York, Elsevier, 335 p.

Aumento, F., 1969, Diorites from the Mid-Atlantic Ridge at 45° N.: Science, v. 165, p. 1112–1113.

Brunn, J. H., 1956, Contribution a l'étude géologique du Pinde Septentrional et d'une partie de la Macédoine Occidental: Annales Géol. Pays Helléniques, 1e serie, t. VII, 358 p.

Ciric, B., 1963, Le développement des Dinarides Yougoslaves pendent le cycle Alpin, in Delga, M. D., ed., Livre à la memoire du Professor Paul Fallot, Tome II: Géol. Soc. France, p. 565–582.

Clark, S. P., Jr., 1966, ed., Handbook of physical constants: Geol. Soc. America Mem. 97, 587 p.

de Roever, W. P., 1956, Sind die Alpinotypen Peridotit-Massen vielleicht tektonisch verfrachtete Bruchstücke der Peridotitschale?: Geol. Rundschau, v. 46, p. 137–146.

Fischer, A. G., 1964, The Lofer cyclothems of the Alpine Triassic: Kansas Geol. Survey Bull., v. 169, p. 107–149.

Godfriaux, I., 1962, L'Olympe: une fenêtre tectonique dans les Hellénides internes: Acad. Sci. Comptes Rendus, v. 255, no. 14, p. 1761–1763.

Griggs, D. T., Paterson, M. S., Heard, H. C., and Turner, F. J., 1960a, Annealing re-crystallization in calcite crystals and aggregates, in Griggs, D., and Handin J., eds., Rock deformation; Geol. Soc. America Mem. 79, p. 21–37.

Griggs, D. T., Turner, F. J., and Heard, H. C., 1960b, Deformation of rocks at 500° to 800°C., in Griggs, D., and Handin, J., eds., Rock deformation: Geol. Soc. America Mem. 79, p. 39–104.

Hess, H. H., 1955, The oceanic crust: Jour. Marine Research, v. 14, p. 423–439.

——1962, History of ocean basins, in Engel, A. E. J., James, H. L., and Leonard, B. F., eds., Petrologic studies: A volume in honor of A. F. Buddington: Geol. Soc. America, p. 599-620.

—— 1965, Mid-oceanic ridges and tectonics of the sea-floor, in Submarine geology and geophysics: Proc. 17th Symposium Colston Research Soc., p. 317–333.

Hess, H. H., Smith, R. J., and Dengo, G., 1952, Antigorite from the vicinity of Caracas, Venezuela: Am. Mineralogist, v. 37, p. 68–75.

Jaeger, J. C., 1957, The temperature in the neighborhood of a cooling intrusive sheet: Am. Jour. Sci., v. 255, p. 306–318.

Johnson, J. H., 1961, Limestone-building algae and algal limestones: Golden, Colorado School of Mines, 297 p.

McKenzie, D. P., 1969, Speculations on the consequences and causes of plate motions: Royal Astron. Soc. Geophys. Jour., v. 18, p. 1–32.

Mercier, J., 1966, Mouvements orogéniques et magmatism d'age Jurassique Supérieur— Éocrétacé dans les zones internes des Hellénides (Macédoine, Grèce): Rev. Géographie Phys. et Géologie Dynam., v. VIII, no. 4, p. 265–278.

Moores, E. M., 1969, Petrology and structure of the Vourinos Ophiolitic Complex of northern Greece: Geol. Soc. America Spec. Paper 118, 74 p.

Renz, C., Liatsikas, N., and Paraskevaidis, I., 1954, Geologic map of Greece: Athens, Institute for Geology and Subsurface Research, scale 1:500,000.

Sander, B., 1948, Einführung in die Gefügekunde der Geologischen Körper, Teil I: Wien and Innsbruck, Springer-Verlag, 215 p.

Schuiling, R. D., 1964, Serpentinization as a possible cause of high heat-flow values in and near the oceanic ridges: Nature, v. 201, p. 807–808.

Temple, P. G., and Zimmerman, J., 1969, Tectonic significance of alpine ophiolites in Greece and Turkey: Geol. Soc. America, Abs. with Programs for 1969, Pt. 7 (Ann. Mtg.), p. 221–222.

Thayer, T. P., 1967, Chemical and structural relations of ultramafic and feldspathic rocks in alpine intrusive complexes, in Wyllie, P. J., ed., Ultramafic and related rocks: New York, John Wiley & Sons, p. 222–239.

Velde, B., 1965, Experimental determination of muscovite polymorph stabilities: Am. Mineralogist, v. 50, p. 436–449.

Vine, F. J., and Hess, H. H., 1970, Sea-floor spreading, in Maxwell, A. E., ed., The sea, Vol. IV: New York, Wiley-Interscience.

Winkler, H. G. F., 1967, Petrogenesis of metamorphic rocks: New York, Springer-Verlag, Revised 2nd ed., 237 p.

Yoder, H. S., and Eugster, H. P., 1955, Synthetic and natural muscovites: Geochim. et Cosmochim. Acta, v. 8, p. 225–280.

Zimmerman, J., Jr., 1968, Structure and petrology of rocks underlying the Vourinos Ophiolitic Complex, northern Greece [Ph.D. thesis]: Princeton, New Jersey, Princeton University, 99 p.

——1969, The Vourinos Complex—an allochthonous alpine ophiolite in northern Greece: Geol. Soc. America, Abs. with Programs for 1969, Pt. 7 (Ann. Mtg.), p. 245.

Present Address: Department of Earth and Space Sciences, State University of New York at Stony Brook, Stony Brook, New York 11790.

Manuscript Received by the Society March 29, 1971

THE GEOLOGICAL SOCIETY OF AMERICA, INC.
MEMOIR 132
© 1972

St. Paul's Rocks, Equatorial Atlantic: Petrogenesis, Radiometric Ages, and Implications on Sea-Floor Spreading

W. G. MELSON
Smithsonian Institution, Washington, D. C. 20560
S. R. HART
*Carnegie Institution of Washington, Department of Terrestrial
Magnetism, Washington, D. C. 20015*
G. THOMPSON
*Woods Hole Oceanographic Institution,
Woods Hole, Massachusetts 02543*

ABSTRACT

St. Paul's Rocks are an upper mantle-derived ultrabasic intrusion emplaced largely in the solid state. They consist of three major groups of mylonitized rocks which originally were peridotites, brown hornblende-rich alkaline plutonic rocks, and gabbros. Major, minor, and trace element compositions appear to rule out any simple single stage genetic relation between these three groups. Hornblendes are an important constituent of these rocks, and range from pargasitic in the peridotites to kaersutitic in the brown hornblende mylonites. The subgroups in the peridotites include hornblende peridotite, plagioclase peridotite, and aluminous pyroxene peridotites. Coexistence of these assemblages is attributed to (1) inhomogeneous distribution of water pressure in the intrusive mass, (2) mixing of assemblages which possibly equilibrated at different mantle depths and were subsequently mixed during emplacement, and (3) rapid ascent and emplacement, preventing re-equilibration. Maximum temperatures and pressures for the primary assemblages appear to be around 1,100° C, and 15 kbars. The rocks were mylonitized during emplacement at temperatures higher than 500° C.

Rb-Sr measurements indicate that the peridotites may have been emplaced less than 100 m.y. ago, an age consistent with the sea-floor spreading model ages for St. Paul's Rocks. Remarkably, a hornblende separate from Southeast Islet gives an apparent age of 835 m.y. by the K-Ar method. This age does not appear to be due to excess radiogenic argon. The intrusion appears to have incorporated older material, but it is not clear how such older material could have been left behind during sea-floor spreading.

St. Paul's Rocks cannot be representative of undifferentiated (chondritic) oceanic upper mantle, but they may be representative of the kind of mantle rock which yields alkali olivine basalts by partial fusion.

INTRODUCTION

St. Paul's Rocks (equatorial Atlantic, 0°56′ N., 29°22′ W.) have attracted much scientific attention because of a number of features which are anomalous compared to all other islands of the mid-oceanic ridges. Their location relative to ridge segments and fracture zones in the equatorial Atlantic is shown in Figure 1. These small (surface area of about 0.01 km²), barren islets have had prominent places in a number of important theories, ranging from the former existence of "Atlantis" (Renard, 1882) to the hypothesis that they may be fragments of the suboceanic mantle (Hess, 1954).

We herein review and present new data on these and other problems presented by these remarkable islets, islets to which H. H. Hess attached so much importance. Our new data include bulk major and minor element analyses by E. Jarosewich using classical methods; trace element analyses by G. Thompson using direct reading optical emission spectrometry; petrographic and electron probe analyses by W. Melson; and K, Rb, total Sr, and Sr87/Sr86 anlyses by S. Hart using isotopic dilution techniques. Two of us (Melson and Thompson) participated in the geologic reconnaissance and sampling of St. Paul's Rocks during cruise 20 of the R. V. *Atlantis II* in 1966.

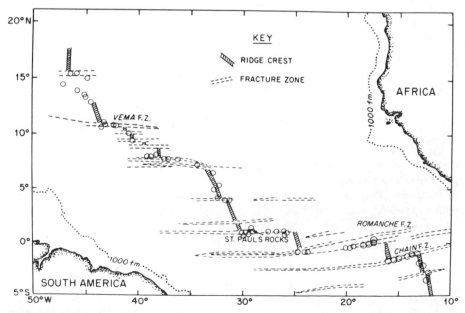

Figure 1. Location of St. Paul's Rocks, ridge segments, and fracture zones in the Atlantic. Open circles are earthquake epicenters for the period 1965–1969. Ridge crests and fracture zones from Heezen and others (1964a, b); epicenters from Sykes (1967).

PREVIOUS WORK

The Unique Islets

The first scientific accounts of St. Paul's Rocks were by Charles Darwin (1900), who visited the islets in 1843 during the voyage of the *Beagle*. He found that he could not identify nor classify the rocks because of their fine-grain size, but noted the presence of serpentine. Nonetheless, he made one of the most important observations about the islets: they are not volcanic, and thus differ from all other mid-oceanic islets.

A. Renard (1882) first described and analyzed the rocks, noting the abundance of fresh olivine samples collected during the *Challenger* expedition. He was also the first to recognize (1) the other minerals in the peridotite mylonites (clinopyroxene, orthopyroxene, amphibole, and chromite), and (2) the cataclastic nature of the groundmass.

Volcanic or Metamorphosed Plutonic Rocks

Renard (1882) initiated the idea that St. Paul's Rocks may be effusive, although he later favored a metamorphic origin. H. S. Washington (1930a) dismissed a volcanic origin, and believed the rocks were sheared plutonics. Around this time, N. L. Bowen had discovered the unreasonably high liquidus temperatures of peridotites, and put forth his views that peridotite magmas did not exist in nature. Washington (1930a) seized upon this evidence to dispute the volcanic origin for the peridotites of St. Paul's Rocks. Washington also presented new data, including a chemical analysis which he viewed as more reliable than the earlier ones by Renard (1882). He further identified what he believed to be jadeite augen in a sample from the Main Islet collected during the *Quest* expedition. He correlated the supposed jadeite augen with the unusually high Na_2O content (1.52 percent) of the bulk sample.

Character of the Peridotites and the Oceanic Upper Mantle

Tilley (1947), intrigued by the high Na_2O content in the peridotite analyzed by Washington (1930a) and by the reported jadeite augen, re-examined the analyzed sample. His analysis showed that the high Na_2O content was erroneous and that there was no jadeite in the sample. Tilley (1947) discovered that the fresh peridotites could be classified into two groups: those with abundant accessory pyroxene, and those with abundant accessory amphibole. He further noted that these two types had essentially identical chemical compositions. Hess (1954) appears to be the first petrologist to ascribe a mantle-origin to St. Paul's Rocks, viewing them as upper mantle fragments brought up along a line of fracture during basalt genesis beneath mid-ocean ridges. Later in 1966, Tilley drew attention to the very low alkali contents of the peridotites, and suggested that they were the barren mantle residue from which basalt had been extracted by partial fusion.

J.D.H. Wiseman (1966) sampled some of the islets in 1960 on an expedition of the H. M. S. *Owen*. His samples included varieties not previously reported, including some containing brown hornblende and some with up to 0.28 percent potassium. He coined the term "challengerites" for such samples, and mentioned other new rock names, although he gave no modal or chemical analyses to define the new rock types. Tilley (1966) dismissed these new names as unnecessary, and we concur. Wiseman, quoting unpublished work of S. Hart, reported low Sr^{87}/Sr^{86} values for all these samples, ranging between 0.7030 and 0.7054, with an average of 0.7044. He also reported U (0.14 ppm) and Th (0.395 ppm) contents of the amphibole-rich samples (the "challengerites"), values obtained by G. R. Tilton and in accord with the speculated abundance of these elements in the upper mantle. Wiseman concluded from his data that St. Paul's Rocks may well represent an exposure of the upper mantle, and recommended drilling on the islets.

H. H. Hess (1964) presented a new analysis of a peridotite mylonite from the Main Islet, and included this material with olivine nodules and other rock types which he thought were likely samples of the upper mantle.

An Ancient Age for the Peridotites

Hart (1964a) investigated the Rb-Sr relation in a number of samples. He interpreted these results as indicating a very ancient age, a result difficult to reconcile with current models of sea-floor spreading. Since that time, numerous other samples, some better suited for radiometric dating, have been examined. The new results are reported in this paper.

A Heterogeneous High-Temperature Mantle-Derived Intrusion

W. Melson and others (1967a) briefly described the largest variety of rock types yet collected from the islets. This collection, made during cruise 35 of the R. V. *Chain* in 1963 and cruise 20 of the R. V. *Atlantis II* in 1966, include samples obtained from the Southeast Islet (following islet names given by Tressler and others, 1956), an island found to differ from all other islets in the wide variety of rocks found there and in that the orientation of bands in the mylonites is readily measurable. The banding is nearly vertical, and strikes roughly north, indicating that shallow drilling may not give important new information. Samples from the Southeast Islets include the highly differentiated brown hornblende mylonites, which are highly sheared ultrabasic alkaline plutonic rocks, and plagioclase-clinopyroxene mylonites, which appear to be mylonitized olivine gabbros. In a brief description of these and other samples, which included a number of new bulk analyses and microprobe analyses of critical phases, St. Paul's Rocks were postulated to be an exposure of a high-temperature mantle-derived intrusion. The extreme heterogeneity was viewed as a result of mixing of various genetically unrelated mantle rock types during the ascent and emplacement of a plastic mass into the crust.

Electron probe analyses of the amphiboles in the peridotite mylonites revealed that they were pargasites, and not actinolites, as usually was inferred. Analyses of other phases, zoning, and application of experimental data revealed a sequence of incomplete equilibrations in the mylonites, probably reflecting successively lower pressure conditions in a rapidly ascending intrusion. Electron probe analyses further revealed that the dominant amphibole in the brown hornblende mylonites was kaersutitic hornblende, a phase recently postulated to occur in the upper mantle (Mason, 1966).

A Genetic Relation Among the Various Rock Types

F. Frey (1970) reported the rare earth element contents of fifteen samples from St. Paul's Rocks, including all the samples previously described by Melson and others (1967a, b). He discovered that the samples were markedly enriched in the light rare earth elements and concluded that the peridotites could not be undifferentiated mantle material, nor was it likely that the peridotites were the residua of partial fusion. He suggested that the peridotites were cumulates in a large volume of alkali basaltic magma already enriched in the light rare earths. The brown hornblende mylonites were shown to have REE contents similar to alkaline basic and ultrabasic igneous rocks, a view in accord with their major and minor element composition (Melson and others, 1967a).

THE THREE MAJOR GROUPS

We place the various mylonites into three major chemical-mineralogic classes (Melson and others, 1967a): peridotite mylonites, brown hornblende mylonites, and clinopyroxene-plagioclase mylonites. A number of new chemical analyses better define each of these groups. The first and second classes are based on a number of analyzed samples, but the third class is defined by only one sample. Table 1 gives analyses for major, minor, and trace element analyses of peridotite mylonites, and Table 2 gives the same information for the other two types. Rare earth analyses for all but one of these samples (SE-6) have been given by Frey (1970). Other recent analyses for major and minor elements in the peridotite mylonites are in agreement with those given here, and are in Tilley (1947, 1955), Hess (1955), and Otalora and Hess (1969).

The peridotite mylonites are by far the most abundant type; the other two types have been found in situ only on the Southeast Islet. There, brown hornblende mylonites compose about 30 percent of the surface area, but the clinopyroxene-plagioclase mylonites were noted in only one place. Figure 2 is a geologic sketch map of Southeast Islet, showing sample localities and orientation of brown hornblende schlieren in peridotite mylonite.

PETROGRAPHY: THE THREE STAGES OF PARTIAL EQUILIBRATION

The mylonites have a complex history, which can be partially unraveled petrographically. Minerals can be assigned, although in some cases with much difficulty, into three sequentially formed assemblages (Table 3). These assemblages appear to have moved toward equilibrium at successively lower temperatures and pressures.

Primary Assemblage

Minerals are assigned to this group if they occur as deformed augen. These minerals apparently crystallized at depth as large crystals, and might be expected to record the highest temperature and pressure conditions. Because preservation of a mineral of the primary assemblage as augen is related to its mechanical properties, some minerals which are no longer preserved as augen may also belong in

TABLE 1. MAJOR AND TRACE ELEMENT ANALYSES OF PERIDOTITE MYLONITES FROM ST. PAUL'S ROCKS AND NEARBY DREDGE HAULS

	NE-4	7–479	18–900	7–327	SE-22	Avg.*
SiO_2	44.38	43.80	42.22	44.35	43.81	43.55
Al_2O_3	2.36	2.40	4.42	3.41	3.25	3.69
Fe_2O_3	0.98	1.41	2.86	1.19	1.39	1.45
FeO	7.18	6.22	4.45	7.07	7.61	6.64
MnO	0.22	0.14	0.13	0.15	0.16	0.14
MgO	41.66	42.13	34.61	38.88	39.25	38.45
CaO	1.01	1.13	3.92	2.77	2.14	2.61
Na_2O	0.14	0.14	0.43	0.17	0.16	0.33
K_2O	0.04	0.07	0.11	0.05	0.05	0.10
H_2O^+	1.08	1.54	5.73	1.16	0.97	1.59
H_2O	0.09	0.10	0.19	0.13	0.10	0.12
TiO_2	0.08	0.07	0.30	0.08	0.10	0.28
P_2O_5	<0.05	< 0.05	0.05	< 0.05	< 0.05	0.14
Cl	0.11	0.05	0.20	0.09	0.14	0.14
CO_2	0.08	–	–	–	0.06	< .01
NiO	0.28	0.32	0.27	0.25	0.26	0.27
Cr_2O_3	0.35	0.54	0.50	0.53	0.33	0.51
Total	100.04	100.06	100.39	100.28	99.78	100.00
Cl=0	0.02	0.01	0.05	0.02	0.03	–
Total	100.02	100.05	100.34	100.26	99.75	100.00

Trace elements (ppm)

	NE-4	7–479	18–900	7–327	SE-22
B	6	9	10	9	5
Ba	12	15	18	10	9
Co	115	120	105	110	125
Cr	3490	3550	3050	3350	3300
Cu	15	5	19	10	28
Ga	15	17	13	14	13
Li	5	5	5	3	4
Ni	2110	2270	2130	1880	2030
Pb	1	2	1	1	1
Sr	26	39	23	40	35
V	35	11	105	40	65
Zn	60	8	35	60	75
Zr	8	12	34	< 5	< 5

The following elements were also determined but are present in amounts below their detection limit (ppm): Ag; Bi; Cd (<1); Mo (<2); Sn (<5); Rb (<10).

Analyses NE-4 and SE-22 are new. The remaining analyses were reported by Melson and others (1967).

*Weighted average of peridotite and brown hornblende mylonites (90 percent of 7–327, 5 percent of 18–900 and 5 percent brown hornblende mylonite SE-13; Melson and others, 1967).

NE-4. Extremely fine-grained with very few augen. Augen include "pargasite," olivine, and "aluminian chromite." Veinlets of "pargasite" and olivine. Enstatite, diopside, and pargasite probably present in granulated matrix material, but presence not confirmed. Also, possible that plagioclase is present. Samples from Northeast Islet, collected during cruise 20 of the R. V. Atlantis II.

7–479. Contains primary assemblage of olivine, enstatite, pargasite, chromian spinel, blue spinel ($Sp_{80}He_{20}$), and phlogopite. Recrystallized assemblage: olivine, pargasite, phlogopite, calcite, and "pyrite." Dredge 7, R. V. Chain, cruise 35.

18–900. Pargasite-rich mylonite. Pargasite average analysis in Table 3; also contains augen of olivine, enstatite, and chromian spinel. Unanalyzed portion of sample contains veinlet of "hydrogrossular," chlorite, and amphibole, possibly derived by hydrothermal alteration of a plagioclase-rich layer. Contains considerable serpentine. Dredge 18, R. V. Chain, cruise 35.

7–327. Primary assemblage: olivine, enstatite, endiopside, plagioclase (An_{08}), and chromian spinel. Recrystallized assemblage includes olivine and pargasite. Dredge 7, R. V. Chain, cruise 35.

SE-22. Primary assemblage: endiopside, enstatite, chromian spinel, olivine. At least one thin schlieren of brown kaersutitic hornblende. No well-developed recrystallized assemblage. Clinopyroxene analysis in Table 5. From Southeast Islet (Fig. 2). From cruise 20, R. V. Atlantis II.

TABLE 2. ANALYSES OF CLINOPYROXENE-PLAGIOCLASE MYLONITE (SE-6), BROWN HORNBLENDE MYLONITES (SE-30, 31, 13, AND 7–320), AND OF AN ALKALI OLIVINE BASALT ON THE NORTH SLOPE OF ST. PAUL'S ROCKS (43–49, MELSON AND OTHERS, 1967b)

	SE-6	SE-30	SE-31	SE-13	7–320	43–49
SiO_2	49.24	38.10	37.05	36.64	39.04	43.15
Al_2O_3	10.44	16.07	16.33	17.20	19.24	13.46
Fe_2O_3	1.20	1.56	1.64	2.78	4.43	4.52
FeO	4.46	6.83	8.26	8.88	4.81	8.22
MnO	0.10	0.10	0.09	0.13	0.11	0.11
MgO	13.21	12.24	11.69	6.48	5.82	10.80
CaO	14.37	12.20	11.87	13.30	10.83	9.80
Na_2O	1.99	2.82	2.95	3.85	4.14	3.47
K_2O	0.31	1.20	1.19	0.80	1.51	1.63
H_2O^+	0.92	1.73	2.02	1.88	2.84	1.21
H_2O^-	0.13	0.09	0.14	0.12	0.47	0.15
TiO_2	2.33	5.58	5.04	3.99	2.86	2.70
P_2O_5	0.23	0.27	0.77	2.64	1.35	0.75
Cl	0.67	1.22	1.25	1.47	1.86	n.d.
CO_2	0.10	0.48	0.23	0.08	0.24	n.d.
NiO	0.08	0.07	0.04	–	0.02	n.d.
Cr_2O_3	0.16	0.02	0.01	0.02	0.02	n.d.
Total	99.94	100.58	100.56	100.24	99.55	99.97
Cl=O	0.15	0.28	0.29	0.33	0.42	n.d.
Total	99.79	100.30	100.27	99.91	99.13	99.97

Trace elements (ppm)

	SE-6	SE-30	SE-31	SE-13	7–320	43–49
Ba	950	1700	2450	1500	3100	300
Bi	<1	4	4	2	<1	5
Co	105	60	60	105	50	75
Cr	1200	3	7	30	13	250
Cu	1900	120	450	125	550	55
Ga	20	22	24	25	15	18
Li	6	6	9	5	4	8
Mo	2	20	20	10	2	10
Ni	670	645	385	435	340	270
Pb	<1	<1	<1	<1	<1	3
Rb	n.d.	n.d.	n.d.	n.d.	n.d.	45
Sn	<5	11	12	8	<5	<5
Sr	760	1100	1000	1250	>2000	500
V	130	675	640	335	130	240
Zn	23	6	7	12	12	90
Zr	50	80	85	90	80	200

The following elements were also determined but are present below their detection limit (ppm): Ag; Cd (<1); B (<5).

Analyses SE-6, SE-30, SE-31, and 7-320 are new; SE-13 was previously reported.

SE-6. Primary assemblage: plagioclase and salite augen in deformed matrix largely of brown hornblende. Secondary assemblage: scapolite and other as yet unidentified phases. Salite analysis in Table 5, hornblende analysis in Table 4. From Southeast Islet (Fig. 2), R. V. *Atlantis II*, cruise 20.

SE-30. Primary assemblages: kaersutitic hornblende, Fe-Ti oxide, and possible traces of plagioclase. Secondary assemblage: zeolite and other as yet unidentified phases. Hornblende analysis in Table 4. From Southeast Islet, R. V. *Atlantis II*, cruise 20.

SE-31. Same data as for SE-30.

SE-13. Primary assemblage: brown hornblende, plagioclase, Fe-Ti oxide, scapolite, apatite, allanite. Recrystallized assemblage: scapolite, chloro-hornblende (dashkesanite), apatite, titanbiotite, analcite, sphene, natrolite. Hornblende analysis in Table 4. North side of Southeast Islet, R. V. *Atlantis II*, cruise 20.

7–320. Similar to SE-13. From dredge 7, R. V. *Chain*, cruise 35.

43–49. Alkali olivine basalt. Microlites of olivine, plagioclase, pyroxene, opaques, and brown hornblende, with minor interstitial glass. Contains olivine nodules, and peridotite mylonite xenoliths. From dredge 43, R. V. *Atlantis II*, cruise 20.

the primary assemblage. For example, in SE-6, a plagioclase-rich rock, plagioclase is preserved only rarely in large crystals, but is abundant in the granulated matrix. Plagioclase forms rare small augen (around An_{80}) in at least one peridotite mylonite (sample 7-327), a sample which has been shown to have a positive europium anomaly (Frey, 1970). This europium anomaly indicates that plagioclase has been added as a solid primary phase in this sample, most likely as a cumulate phase, and is not a product of solid-state recrystallization.

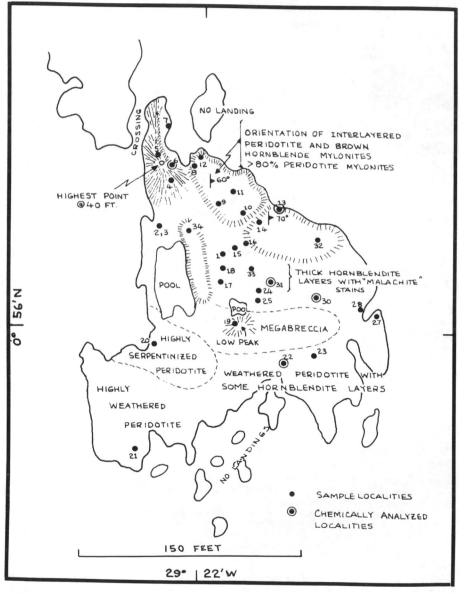

Figure 2. Geologic sketch map of Southeast Islet, St. Paul's Rocks. Data collected in 1966 during cruise 20, R. V. *Atlantis II*; islet sampled and mapped by T. Moore and W. G. Melson.

High-Temperature Neomineral Assemblage

These include those minerals which occur as undeformed grains, reflecting either continuous recrystallization during mylonitization or formation after mylonitization. To this assemblage belong a number of high-temperature minerals not found in the primary assemblage, as well as some which, except for recrystallization, are identical to those that form augen. Some of the neominerals contain elements like chlorine (in dashkesanite, chloroscapolite, and chloroapatite), which were probably derived by reaction between the primary minerals and a chlorine-rich pore fluid, or crystallized directly from the pore fluid.

TABLE 3. SUMMARY OF THE MINERALOGY OF THE MYLONITES

	Peridotites	Clinopyroxene-Plagioclase	Brown Hornblende
Olivine	P & R (A)	X	X
Pyroxenes			
orthopyroxene	P & R (AC)	X	X
clinopyroxene	P & R (AC)	P (A)	X
Oxides			
chromian spinel	P (AC)	X	X
aluminian chromite	P (AC)	X	X
spinel	P & R (AC)	X	X
Fe-Ti oxides	X	P (T)	P (AC)
Amphiboles			
pargasitic hornblende	P, R, & N (AC, T)	X	X
kaersutitic hornblende	X	R & N (?), (AC)	P & R (A)
dashkesanite	X	X	N (T)
Micas			
phlogopite	P & R (T)	X	X
titanbiotite	X	X	R (T)
Feldspathic Minerals			
plagioclase	P? & R? (T)	P & R (A)	P (T)
scapolite	X	?N? (T)	N (AC)
Zeolites			
analcite	X	X	N (T to AC)
natrolite	X	X	S?(T to AC)
Miscellaneous			
allanite	X	X	P (T)
zircon	X	X	P (T)
carbonate	R? (T)	R? (T)	R? (T)
sulfides	R? (T)	X	R? (T)
hydrogrossular	?N? (T)	X	X

Provisional assignment is to the primary assemblage (P, commonly occurred as deformed augen); recrystallized assemblage (R, not deformed, product of high-temperature recrystallization); new mineral (N, high-temperature neomineral formed after mylonitization, $> 500°C$); and secondary mineral (S, low-temperature, $< 500°C$, hydrothermal alteration). Abundance indicated by letter in parentheses: (A)—abundant, (AC)—accessory, —10 percent or less, (T)—trace, less than 1 percent, (X)—not found. Question mark preceding assemblage assignment means phase is probably present but presence not categorically established. Question mark after assemblage assignment means assignment to that assemblage is very uncertain.

Secondary Low Temperature Assemblage and Sample Contamination

This assemblage includes phases such as serpentine and zeolites, which most likely formed at temperatures of less than 500° C, and after mylonitization.

Surficial weathering in the presence of salt water and phosphate-rich solutions from guano have produced still another assemblage, including a number of calcium phosphates (Washington, 1930a) and other minerals. This assemblage is of no importance here, except that we have tried to avoid it in selecting samples for chemical analyses. Chlorine and phosphorous do occur in some of the neominerals in the brown hornblende mylonites, but these are in high-temperature assemblages, and high chlorine and high phosphorous are coherent with the major, minor, and trace element contents of this group.

The high deuterium-hydrogen ratio for some of the brown hornblende mylonites (−33 per mil for SE-13 whole rock, −36 for a hornblende concentrate from SE-13, and −40 for SE-30 whole rock) suggests hydrogen ion exchange with sea water (Sheppard and Epstein, 1970). However, the $^{18}O/^{16}O$ ratios (5.4 to 6.6, Sheppard and Epstein, 1970) do not reflect equilibration with sea water, and are more in accord with the postulated range for unaltered igneous rocks.

PERIDOTITE MYLONITES

The peridotite mylonites have similar major element compositions (Table 1). Modal analyses are difficult to determine except by x-ray diffraction techniques (Otalora and Hess, 1969), but Figure 3 shows that the peridotites cluster closely in their normative mineralogy.

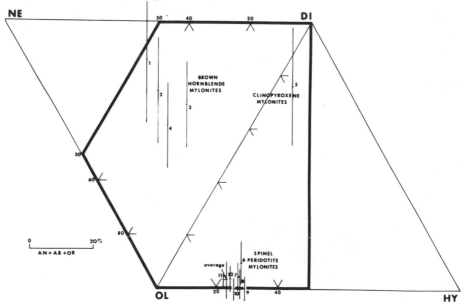

Figure 3. The mylonites fall into three major groups based on their normative mineralogy. Projections in the ternary system diopside (DI), olivine (OL), and hypersthene (HY) or nepheline (NE). Vertical line through each point is proportional to the total amount of normative feldspar.

The peridotite mylonites may be subdivided into three mineralogical subgroups: the "amphibole" and "pyroxene" types of Tilley (1947), and pargasite-rich mylonites (Melson and others, 1967). The latter type belong with the amphibole type of Tilley (1947), but are so enriched in pargasite compared to others of this type that they have significantly different chemical compositions (Table 1), and, of the various rock types, they most closely approach postulated compositions for undifferentiated mantle. Careful search with the microprobe reveals that small amounts of pargasite are present in the pyroxene type, and that traces of clinopyroxene occur in some of the amphibole type of Tilley. Enstatite, various spinel group minerals (Table 3), and abundant olivine occur in both types. Some amphibole and pyroxene types have essentially identical bulk compositions (Tilley, 1966). One pyroxene type (sample 7-327) contains rare augen and matrix granules of bytownitic plagioclase, which were overlooked during earlier examination (Melson and others, 1967a).

The amphibole in the amphibole type is characteristically pargasite (Table 4), and is the main host mineral for the alkalies. The pyroxenes in the pyroxene type include aluminous enstatite (Figure 4) and aluminous and sodic diopside (Table 5). Typically, these pyroxene augen have low alumina rims. Some of the enstatites contain blebs and sometimes short lamellae of presumably exsolved spinel. Phlogopite occurs rarely in both types of peridotites.

TABLE 4. AVERAGE COMPOSITION OF A PARGASITIC AND KAERSUTITIC HORNBLENDE IN THE MYLONITES FROM ST. PAUL'S ROCKS

		1	2	3	4	5
SiO_2		46.1	40.2	39.6	39.6	41.4
Al_2O_3		11.8	14.7	15.0	14.6	12.0
TiO_2		0.2	3.0	2.8	2.8	2.8
FeO		4.0	11.6	8.6	9.2	9.4
MgO		20.4	13.8	15.9	15.4	14.3
CaO		11.6	11.7	12.7	12.7	12.9
Na_2O		2.5	2.6	2.3	2.7	1.7
K_2O		0.2	1.5	1.7	1.5	1.1
Cations per 23 oxygens						
Si		5.5 ⎱8.0	5.87 ⎱8.00	5.78 ⎱8.00	5.80 ⎱8.00	6.20 ⎱8.00
Al		2.5 ⎰	2.13 ⎰	2.22 ⎰	2.20 ⎰	1.80 ⎰
Al		0.3 ⎫	0.42 ⎫	0.36 ⎫	0.32 ⎫	0.32 ⎫
Ti		0.2 ⎪4.4	0.33 ⎪5.19	0.30 ⎪5.17	0.30 ⎪5.11	0.31 ⎪5.00
Fe		1.3 ⎬	1.42 ⎬	1.05 ⎬	1.13 ⎬	1.17 ⎬
Mg		2.6 ⎭	3.02 ⎭	3.46 ⎭	3.36 ⎭	3.20 ⎭
Ca		2.2 ⎫	1.83 ⎫	1.99 ⎫	2.00 ⎫	2.07 ⎫
Na		0.7 ⎬3.2	0.74 ⎬2.85	0.64 ⎬2.94	0.75 ⎬3.02	0.49 ⎬2.76
K		0.3 ⎭	0.28 ⎭	0.31 ⎭	0.27 ⎭	0.20 ⎭

Analyses by electron probe. All Fe assumed to be as FeO.

1. Average pargasite augen in sample 18–900 (dredge 18, R. V. *Chain,* cruise 35).
2. Augen in brown hornblende mylonite (sample SE-13, Southeast Islet). Average composition.
3. Augen in brown hornblende mylonite (sample SE-30, Southeast Islet). Average composition.
4. Augen in brown hornblende mylonite (sample SE-31, Southeast Islet).
5. Interstitial fine-grained hornblende in clinopyroxene-plagioclase mylonite (sample SE-6, Southeast Islet).

Petrogenesis

Rocks of peridotitic compositions can normally be placed into one of four anhydrous or one hydrous facies. The primary assemblages in the peridotites from St. Paul's Rocks appear to fall in at least three facies: the aluminous two-pyroxene facies, plagioclase facies, and the amphibole facies (Green and Ringwood, 1963). These three facies border closely on one another, and one might imagine that all the assemblages equilibrated at close to the same P-T, but that P_{H_2O} varied in the rock mass. Alternatively, the three types may reflect equilibration or partial equilibration to a wide range of temperatures, total pressures, and water pressures. This later hypothesis was favored in earlier work on these peridotites (Melson and others, 1967a).

There are a number of specific reactions which help to delineate the stability fields of the various primary assemblages. The "pyroxene" and "amphibole" assemblages appear to be related to one another by a reaction of the type:

$NaCa_2Mg_2Al(SiO_3)_6 + 2Mg_2SiO_4$
Jadeitic diopside   Olivine
          (Pyroxene type)

$+ H_2O$

$= NaCa_2Mg_5AlSi_7O_{22}(OH)_2 + MgSiO_3$
          Edenite   Enstatite
          (Amphibole type)

This reaction occurred in the presence of excess olivine and enstatite, that is, in an environment saturated with regard to these two minerals. The two assemblages clearly coexist in some samples, suggesting formation under the same temperatures and total pressures but with only limited amounts of water available for reaction.

The approximate $P_{H_2O}-T$ boundary for the two assemblages probably corresponds very closely to the experimentally determined reaction for the breakdown of pargasite (Boyd, 1959) at low water pressures (Figure 5), but an important difference should be noted between the real and synthetic systems. The breakdown of pure pargasite at less than 750 bars produced nepheline and other phases. If pargasite is in an ultramafic rock containing enstatite, as at St.

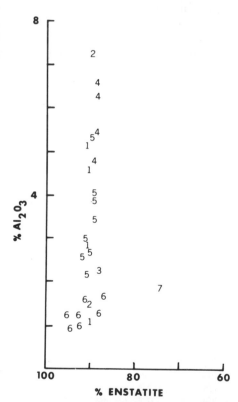

Figure 4. Alumina content of orthopyroxenes from the peridotites of St. Paul's Rocks (1), compared with the following: (2) Vema Fracture Zone (Melson and Thompson, 1971), (3) the Romanche Fracture Zone (Melson and Thompson, 1970), (4) the Lizard Complex, England (Green, 1964), (5) olivine nodules (various sources listed by Green, 1964), (6) various continental peridotite intrusions (*in* Green, 1964), and (7) in laminated gabbro, Romanche Fracture (Melson and Thompson, 1970).

Paul's Rocks, it is likely that the albite molecule rather than nepheline will form, according to a reaction of the type:

$$NaAlSi_2O_6 + 2MgSiO_3 \longrightarrow NaAlSi_3O_8 + Mg_2SiO_4$$

Nepheline	Enstatite		Albite	Olivine

(in plagioclase)

Pargasite begins to melt incongruently at a minimum temperature of about 1,040° C for $P_{H_2O} = 750$ bars. Its melting temperature rises but slightly with increasing water pressure, reaching 1,060° C for $P_{H_2O} = 1,500$ bars. A rough extrapolation to even higher water pressures shows that pargasite is probably unstable at any water pressure for temperatures in excess of 1,100° C. This places a rough upper limit on the stability fields of the "amphibole" peridotites. This temperature is reached at around 70 km depth (P = 22 kbars) based on postulated oceanic geotherms. The presence of phlogopite also limits the upper temperature limit for the peridotites: about 1,250° C, assuming no fluorine is present (Kushiro and others, 1967).

Pargasite peridotites like those from St. Paul's Rocks appear to yield critically undersaturated magmas as first liquids on melting under high water pressures. This view arises from the experiments of Boyd (1959), where pargasite melts to critically undersaturated liquids plus an assemblage of olivine and aluminous diop-

TABLE 5. COMPOSITION OF CLINOPYROXENES IN MYLONITES FROM ST. PAUL'S ROCKS

	1	2	3	4	5
SiO_2	53.4	53.4	52.7	54.5	52.2
Al_2O_3	5.0	3.5	4.5	2.0	7.0
FeO	3.0	2.8	6.7	6.0	2.4
MgO	14.3	15.8	13.0	14.1	16.1
CaO	22.8	22.5	22.7	22.9	22.6
Na_2O	1.1	1.0	1.0	0.6	1.6
K_2O	0.1	0.01	0.02	0.01	0.03
TiO_2	0.4	0.2	0.58	0.18	0.20
Cr_2O_3	n.d.	n.d.	n.d.	n.d.	0.60
MnO	n.d.	n.d.	n.d.	n.d.	0.10

Number of ions on the basis of 6 oxygens

	1		2		3		4		5	
Si	1.93	2.00	1.95	2.00	1.92	2.00	1.99	2.00	1.79	2.00
Al	0.07		0.05		0.08		0.01		0.21	
Al	0.15		0.10		0.19		0.08		–	
Fe	0.09		0.09		0.20		0.18		0.07	
Mg	0.77		0.86		0.71		0.77		0.82	
Ca	0.89		0.88		0.89		0.90		1.01	
Na	0.07	1.99	0.07	2.01	0.07	2.08	0.04	1.99	0.11	2.05
K	0.005		0.01		0.01		0.01		0.01	
Ti	0.01		0.01		0.02		0.01		0.01	
Cr	–		–		–		–		0.01	
Mn	–		–		–		–		0.02	
Ca	50.6		48.3		49.3		48.5		48.2	
Mg	44.1		47.0		39.3		41.6		47.7	
Fe	5.20		4.7		11.4		9.9		4.1	

Analyses by electron probe.

1 and 2. Aluminous diopside in peridotite mylonite SE-22 (Southeast Islet).
3 and 4. Aluminous diopside in mylonite SE-6 (Southeast Islet).
5. Aluminous diopside in 7–327 (dredge 7, R. V. *Chain,* cruise 35).

side which, except for the lack of enstatite, is similar to the assemblages in olivine nodules of alkali basalts. At high pressures, excess enstatite would not lead to production of liquids without normative nepheline. This is because enstatite is a liquidus phase at moderate to high pressures in critically undersaturated basaltic liquids (Green and Ringwood, 1967). Also, excess enstatite would probably not substantially change the solidus for pargasite peridotites, which should be close to curve b (Figure 5).

The compositions of pyroxenes in the pyroxene type peridotites are indicators of the temperatures and pressures of equilibration. The provisional P-T grid for clinopyroxenes proposed by O'Hara (1967) has been applied to the analyzed pyroxenes in Table 5, assuming an Fe_2O_3/FeO of 0.3. Although this grid has been subject to some criticism (O'Hara, 1971, oral commun.), it suggests that the aluminous interiors of the clinopyroxene augen equilibrated at around 1,100° C at about 12 kbars (or about 40 km depth). The low alumina margins appear to have recrystallized at around this temperature, but at pressures as low as 2 kbars. This wide range of indicated pressures for nearly constant temperatures has recently also been reported for some Oregon ultramafic intrusions (Medaris and Dott, 1970), and was noted in a peridotite from the Vema Fracture Zone (Melson and Thompson, 1971).

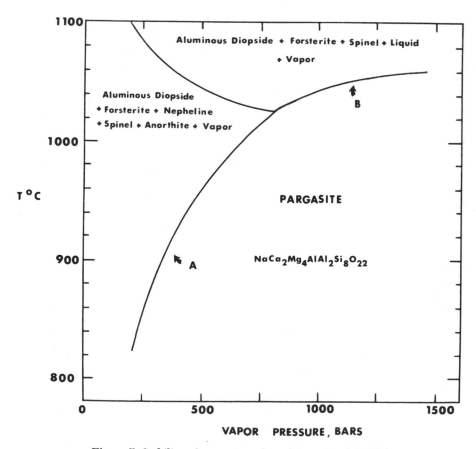

Figure 5. Stability of pargasite adapted from Boyd (1959).

Some large enstatite augen have alumina contents as high as 5.7 percent Al_2O_3 (Fig. 4). These augen typically have a low alumina rim. This zoning with regard to alumina is commonly attributed to a high-pressure equilibration stage, followed by partial re-equilibration to lower pressures (Green, 1963). Figure 4 compares the alumina contents of the St. Paul's enstatites with enstatites from both oceanic and continental peridotite bodies.

The pyroxene data appear to support the idea that St. Paul's Rocks are an exposure of a diapiric, mantle-derived intrusion. The primary assemblages crystallized at depths no greater than 70 km, and most likely on the order of 30 to 40 km at temperatures of 1,100° C or less. The assemblages only partially recrystallized during intrusion, suggesting that emplacement may have been rapid.

BROWN HORNBLENDE MYLONITES

The least abundant rocks, the brown hornblendes and clinopyroxene-plagioclase mylonites, contain substantial normative feldspar (Fig. 3) and may be loosely considered as mylonitized gabbroic rocks. The normative feldspar in the brown hornblende mylonites is largely contained in the brown hornblende in SE-30 and SE-31, and in scapolite as well in SE-13 and 7-320. These brown hornblende mylonites cannot be strictly classed with gabbros because of their low silica contents (less than 40 percent). Rather, we identify them chemically with various types of rare alkaline ultrabasic plutonic rocks. For example, SE-13 can be matched compositionally rather closely by three such rocks listed by Tröger (1935): theralite gabbro (berondite), melilite ijolite (melilite fasinite), and huayne gabbro (marevgite). Such rocks are similar to the brown hornblende mylonites even in regard to presence of certain trace elements, such as high barium, strontium, and chlorine. Tröger (1935) places such rocks in the strongly alkaline Mediterranean and Atlantic igneous rock suites. In addition to their critical undersaturation, such rocks are characterized by high TiO_2 contents, a feature of interest in comparison with the high TiO_2 content of Apollo 11 lunar rocks.

The brown hornblende mylonites appear to have been derived largely from coarse-grained kaersutitic hornblende-rich rocks. One (SE-13) appears to have contained plagioclase as well. Table 3 lists the various assemblages in these rocks. The kaersutitic hornblendes deserve special attention because they account for many of the compositional peculiarities of this group. These hornblendes (Table 4) have uniform TiO_2 (2.8 to 3.0), SiO_2 (39.6–41.4), Al_2O_3 (12.0–15.0), CaO (11.7–12.7), Fe (as FeO, 8.6 to 11.6), MgO (13.8–15.9), but range more widely in their Na_2O contents (1.7 to 2.7). The brown hornblende mylonites are uniformly higher in TiO_2 and Fe/Mg ratio than their kaersutitic hornblendes because Fe-Ti oxides are present.

Petrogenesis

The composition and primary minerals of the brown hornblende mylonites are similar to known igneous rocks. It is reasonable then to postulate that at some point early in their history they were magmas. Recently, Millhollen and Wyllie (1970) determined the liquidus and solidus on one of the brown hornblende mylonites (SE-13, Table 2). The solidus for this sample reaches a minimum of around 675° C at about 12 kbars P_{H_2O}. The liquidus stays around 1,100° C for a wide range of water pressures. Hornblende, as one might suspect from its occur-rence in relict large augen in the mylonites, is the first mineral to crystallize from melts compositionally like the brown hornblende mylonites. At temperatures around 1,000° C, liquids like SE-13 will coexist with solid spinel peridotites based on their experiments. Such temperatures and even slightly higher temperatures are suggested by the diopside augen in the peridotite, supporting the notion that at some time the peridotites could have coexisted with the liquid equivalents of the brown hornblende mylonites. Millhollen and Wyllie (1970) suggest that such liquids were interstitial to the peridotites, although the trace element data do not appear to be in accord with this view.

Hornblende has a pressure-induced breakdown which restricts its mantle stability even for very low temperatures. Under likely oceanic geotherms, common horn-blende is probably not stable below about 60 km, corresponding to a temperature of around 1,000° C (Lambert and Wyllie, 1968). The experimentally used horn-blendes were not kaersutitic, and the occurrence of kaersutite in nodules suggests that the stability field of this phase may extend deeper into the mantle than ordinary hornblendes (Mason, 1968).

Frisch and Schmincke (1969) report cumulate textures in inclusions of kaersu-tite hornblendites from the alkali basalts of Gran Canaria, Canary Islands. Brown hornblende mylonites like SE-30 and SE-31, which contain greater than 80 volume percent hornblende, could have such an origin, but any textural evidence has been obliterated by mylonitization.

Kaersutites also occur as primary phases in ultrabasic plutonic rocks, such as some of those which form the plutonic core of Tahiti (McBirney and Aoki, 1968). The hornblendes in these Tahitian rocks differ from those at St. Paul's Rocks though, in higher TiO_2 (5.06 percent to 6.88 percent), higher Fe/Mg ratios, lower alumina (10.38 to 12.71), and lower total alkalies (2.97 to 3.98 percent). These plutonic hornblendes probably crystallized at high temperatures (perhaps as high as around 1,150° C and as low as 900° C) within the core of the ancient volcanic complex of Tahiti.

Miyashiro and others (1970) report very high TiO_2 contents in gabbros dredged from various places on the Mid-Atlantic Ridge. Some of these contain brown horn-blende and these "gabbros," some of which have less than 40 percent SiO_2, are similar to the brown hornblende mylonites in major element composition. Analyses of the brown hornblendes and trace element analyses have not yet been reported for these dredge samples.

CLINOPYROXENE-PLAGIOCLASE MYLONITES

This rock type, of which we have found but one representative, (SE-6, Table 2), corresponds chemically to certain olivine gabbros. The silica content (49 percent) is not unusual, but the alkali content and Fe/Mg ratio appear low, and the TiO_2 content (2.3 percent) is high compared to most gabbros. The high TiO_2 content reflects the presence of about 20 volume percent accessory kaersutitic hornblende. The rock further has anomalously high Cu, Sr, Ni, and Cr compared to most gabbros.

The clinopyroxene in SE-6 is salite (Figure 6) which, except for a somewhat

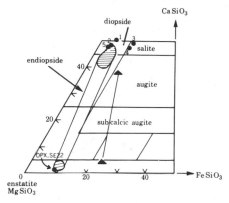

Figure 6. Pyroxene compositions (Table 5) projected in the system wollastonite, enstatite, ferrosilite. Ruled areas are compositions of pyroxenes from a partially serpentinized peridotite dredged from the Vema Fracture Zone (Melson and Thompson, 1971); solid triangles for pyroxenes from a sample of a probable layered complex in the Romanche Fracture Zone (Melson and Thompson, 1970). Some aluminous clinopyroxenes from peridotite mylonites (e.g. analysis 1, Table 5) project above 50 percent wollastonite.

higher Fe/Mg ratio, is quite similar to the diopsides in the peridotite mylonites (Table 5). The plagioclase, which occurs in relict deformed grains, is unusually sodic (An_{45}-An_{50}) for a rock with such a low Fe/Mg ratio.

Petrogenesis

Under high pressures, this mylonite would convert to eclogite. Other than this limiting condition, the assemblage salite-plagioclase provides little P-T information. The dry liquidus and solidus temperatures might be expected to be somewhat elevated compared to those for the brown hornblende mylonites, which have lower silica and higher alkali contents.

This mylonite may well have been a liquid at some time in its history, or it may be a cumulate from a liquid. Salite is common in plutonic and hypabyssal rocks derived from alkali basalt magmas, thus, the rock may be a cumulate from such magma.

Coexisting kaersutite and salite in nodules from Gran Canaria (Frisch and Schmincke, 1969) show partitioning of TiO_2 in the ratio of 2.5 (hornblende/pyroxene). A similar assemblage, but including pyrope and ilmenite, from Kakanui, New Zealand (Mason, 1968) gives a ratio of 3.3. This ratio is about 4.7 for these phases in the clinopyroxene-plagioclase mylonite. If these other ratios are a reasonable range of equilibrium partitioning of TiO_2 between hornblende and clinopyroxene, we can conclude that the hornblende and pyroxene in SE-6 are not an equilibrium assemblage. Their textural relations also indicate that they are not in equilibrium. The brown hornblende is interstitial to and partly replaces the pyroxene. Thus, SE-6 may be a hybrid rock, involving introduction of a late alkalic titania-rich magma, represented mainly by the kaersutitic matrix hornblende, into an already solid assemblage of plagioclase-salite.

RELATIONS BETWEEN THE GROUPS

Major Element Chemistry

The groups appear both related or unrelated genetically, depending on what chemical parameters are considered. On an AFM diagram (Figure 7), there appears to be a single differentiation trend, with the peridotites near the magnesian end, and with the brown hornblende mylonites at the most alkali and iron-enriched end. The clinopyroxene-plagioclase mylonite neatly falls between these extremes.

This single differentiation trend van-
ishes on silica versus alkalies or versus
Fe/Mg diagrams (Figure 8). When silica is thus included, the brown hornblende mylonites show increasing alkalies and increasing Fe/Mg ratios with
decreasing silica relative to the peridotites, whereas the clinopyroxene-plagioclase mylonites show increasing alkalies
and Fe/Mg with *increasing* silica relative to the peridotites, a more normal
differentiation scheme. These different
trends complicate the search for a single
process which will genetically relate the
three types of rocks.

One of the most reasonable genetic
relations between the peridotites and
brown hornblende mylonites is summarized by the following equation: [brown
hornblende liquid] = a [parental magma]
− b [aluminous enstatite] − c [alumi-
nous diopside] − d [forsteritic olivine].

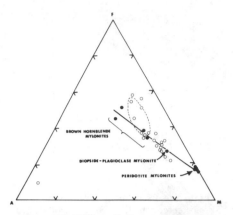

Figure 7. Differentiation trend line for
the mylonites from St. Paul's Rocks in an
AFM diagram (A = $Na_2O + K_2O$, M =
MgO, F = $FeO + Fe_2O_3$). Comparisons include gabbros and an aplite from a fracture zone near 23°, Mid-Atlantic Ridge
(Miyashiro and others, 1970). Most abyssal ocean ridge basalts plot within the
dash-bounded area.

This is reasonable because the crystalline phases (1) are likely liquidus phases of
an alkali basalt under upper mantle pressures, (2) would separate from the residual liquid by settling because of their higher densities, and (3) correspond to the
assemblages of the peridotite mylonites.

Such a hypothesis can be tested rigorously if the compositions of all phases
are known. Recently developed computer programs using least-squares approximations test the fit of various phases to the hypothesis, and provide the proportions (the coefficients a, b, c, and d above) in which the components must be
mixed to give the best fit (Bryan and others, 1969; Wright and Doherty, 1970).
Frey (1970) demonstrated that the above model accounted for the rare earth
elements in a satisfactory way if we assume the alkali basalt dredged north of
St. Paul's Rocks (analysis 43–49, Table 2) was chosen as a parent. Using this
composition, and likely mineral compositions, we obtain the fits listed in Table 6,
where three brown hornblende mylonites were tested. The results show that there
are serious problems with this model. Specifically, the alkali olivine basalt parent
is too high in total iron and alkalies, and too low in MgO and CaO to make a
suitable parent. The model is nonetheless suggestive of a few significant features.
Best fits are obtained with subtraction of but 20 to 25 percent crystals. Further,
olivine, and even diopside in one case, need not be subtracted to give the best fit.
Aluminous enstatite, on the other hand, is always subtracted from the parent, in
amounts ranging from 7 to 21 percent. Furthermore, cumulate materials would
correspond to pyroxenites rather than peridotites.

K, Rb, and Sr Relations

Isotope dilution analyses for K, Rb, and Sr in a number of the mylonites are
reported in Table 7 and shown graphically in Figure 9. The hornblende mylonites
are easily distinguished from the peridotites by K, Rb, and Sr concentrations

which are fairly tightly grouped at a level 10 to 100 times the values in the spinel peridotites. The peridotites are not so closely grouped and show a rather large variation in K, Rb, and Sr; this variation appears almost continuous in the K-Rb plot, whereas the K-Sr plot suggests that further subdivision of the peridotites may be possible. The one clinopyroxene-plagioclase mylonite on Figure 9 (No. 16, SE-6) is not especially distinctive in terms of these parameters and falls intermediate between the peridotite and hornblende mylonite groups. In terms of K/Rb ratio, the dredge samples are generally distinct from the trend of the other groups, being relatively enriched in Rb. While this may be a real characteristic, we feel it may also be due to exchange or contamination with sea water, as Rb is known to be strongly enriched relative to K and Sr in dredge basalts which have undergone sea-water alteration (Hart, 1969). However, peridotites 7–479 and 18–900

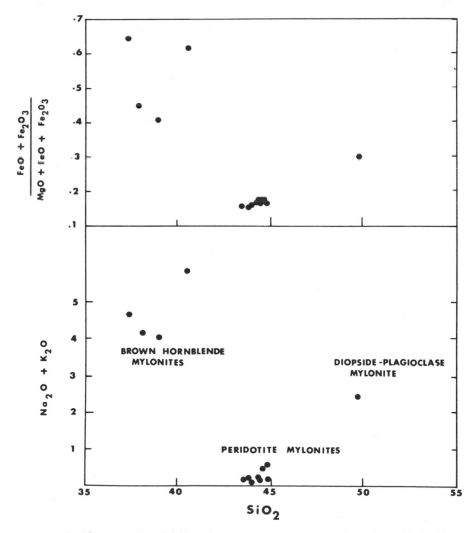

Figure 8. Silica variation diagrams, showing two different trends, one for the peridotites-brown hornblende mylonite, another for the peridotite-clinopyroxene-plagioclase mylonites.

contain accessory phlogopite, which may also account for their low K/Rb ratio. With the exclusion of the dredge samples, the K/Rb data form a rather cohesive trend suggesting that, while the mineralogical and major and minor element groups are distinct, there is nevertheless a degree of relatedness among all the rock types in regard to K/Rb.

There is generally a good correlation between the contents of K, Rb, and Sr shown in Figure 9 and the modal content of amphibole in the rocks. The peridotites with K contents of about 100 ppm have very little amphibole (pargasite) whereas the samples with 1 percent K are almost pure brown hornblende. It is thus natural to attempt to explain the K-Rb-Sr trends of Figure 9 entirely by a simple mixing of the two extreme end members. Mixing lines joining two such typical end members are shown in Figure 9 as solid lines; while most of the K-Rb data are very adequately represented by such a mixing hypothesis, it is clear that the Sr contents cannot be explained in this way. Even introducing the clinopyroxene-plagioclase mylonites as a third mixing component would not help, as this type (at least sample 16, Fig. 9) is enriched in Sr relative to the two-component mixing line, rather than depleted as would be necessary to account for the low Sr contents of those spinel peridotites which group around 1,000 ppm K. Frey (1970) presents similar arguments against a mixing hypothesis, based on the rare earth contents of some of these same rocks.

We have investigated various partial melting and crystal differentiation schemes in an attempt to model the trends of Figure 9. Models with amphibole as a primary

Table 6. Least-squares Fits Given by the Fractionation Model Outlined in the Text (Parent Liquid is 43–49)

	Postulated Parent	Calculated Parent for Each Residual Liquid with Least-Squares Best Fit to Postulated Parent		
	43–49	7–320	SE-13	SE-31
SiO_2	43.76	43.93	41.44	41.42
Al_2O_3	13.65	16.08	14.67	14.30
Fe_2O_3	4.59	3.64	2.40	1.46
FeO	8.34	4.95	8.18	7.43
MgO	10.96	12.42	13.05	14.01
CaO	9.94	10.90	10.94	12.84
Na_2O	3.52	3.40	3.15	2.58
K_2O	1.66	1.22	0.66	1.00
TiO_2	2.74	2.32	3.27	4.20
P_2O_5	0.77	1.08	2.16	0.64
MnO	0.12	0.12	0.14	0.09

Amount of crystalline phases subtracted (weight %) from parent to give best fit.				
Aluminous enstatite		9.93	21.08	7.00
Aluminous diopside		9.89	0	13.38
Forsteritic olivine		5.42	0	0
Amount of residual liquid		74.78	78.93	79.63

Cumulate phases were assigned the following compositions, given in order of (1) aluminian enstatite, (2) aluminian diopside, and (3) forsteritic olivine: $SiO_2 = 54.68$, 53.38, 40.92; $Al_2O_3 = 2.96$, 5.00, 0.20; $Fe_2O_3 = 0.60$, 0.50, 0.14; $FeO = 4.41$, 2.45, 8.20; $MgO = 36.78$, 14.30, 50.37; $CaO = 0.40$, 22.79, 0.00; $Na_2O = 0.00$, 1.10, 0.00; $K_2O = 0.03$, 0.10, 0.00; $TiO_2 = 0.03$, 0.40, 0.05; $P_2O_5 = 0.00$, 0.00, 0.00; $MnO = 0.15$, 0.00, 0.18.

igneous phase which is then allowed to separate from its parent liquid were not found to provide a satisfactory fit to the data. In these models the amphibole dominates the alkali partitioning, and the liquids invariably have lower K/Rb ratios and higher K content than the coexisting crystalline assemblage. This is quantitatively opposite to the observed trends in Figure 9. The brown hornblende mylonites thus cannot be viewed as a basaltic-type liquid which was either derived from the peridotites by partial melting or which was a residual liquid after crystallization of the spinel peridotites.

Frey (1970) noted that the rare earth elements in the brown hornblende mylonites are very similar to those in an alkali basalt which occurs on the north flank of St. Paul's Rocks (Melson and others, 1967b) and suggested that fractional crystallization of liquids of either of these types could generate the rare earth patterns observed in the peridotites. Our trace element studies thus also rule out Frey's model insofar as the brown hornblende mylonites are concerned. We note, however, that alkali basalts are quite different from the brown hornblende mylonites in having a higher K/Sr and lower K/Rb ratio, and that in terms of these

TABLE 7. K, RB, SR CONCENTRATION AND Sr^{87}/Sr^{86} RATIOS IN MYLONITES
FROM ST. PAUL'S ROCKS

Text No.	Sample No.	Rock Type*	K (ppm)	Rb (ppm)	Sr (ppm)	K/Rb	Sr^{87}/Sr^{86}
1	Hess 1 (WD55)	1	502.0	0.90	34.4	558	0.7051
2	Hess 2	1	99.3	0.250	9.06	397	0.7043
3	WC-2	1	135.5	0.321	8.04	422	0.7083
4	W-5	1	150.1	0.346	10.08	434	—
5	WPHi	2	1284.0	1.70	40.7	755	0.7028
6	WPIi	2	1447.0	1.46	90.4	991	0.7044
7	WB-7	1	63.0	0.079	14.62	798	0.7046
8	W65139	1	64.0	0.195	24.2	328	0.7044
9	WC 2–2	1	140.6	0.376	10.06	374	0.7052
10	18–900	1	971.0	2.65	16.7	366	0.7042
11	7–320	3	13,000.0	34.6	2960.0	376	0.7045
12	7–479	1	760.0	3.36	40.5	226	0.7047
13	SE-13 rock	3	6720.0	4.69	1293.0	1433	0.7047
14	SE-13 HBDE	–	8350.0	5.95	—	1402	—
15	SE-13 scapolite	–	6120.0	5.87	—	1075	—
16	SE-6	4	2790.0	5.60	—	498	—
17	SE-22	1	409.3	0.413	—	991	—
18	SE-30	3	9730.0	10.82	1030.0	899	—
19	NE-4	1	310.0	0.547	—	567	—
20	31–219	3	8620.0	11.2	1500.0	769	—

* Rock type: 1 = peridotite mylonite, 2 = intermediate between brown hornblende and peridotite mylonites, 3 = brown hornblende mylonite, 4 = clinopyroxene-plagioclase mylonite.

Samples 1–9 cleaned in hot 6N HCl with water rinse before powdering: samples 10–12 cleaned in warm 3N HNO_3; other samples powdered without acid cleaning. Weight loss during acid cleaning ranged from less than 1 percent (1, 2, 3, 4, 6, 10) to over 6 percent (9, 11). Sr concentrations in samples 11, 18, and 20 were obtained by XRF (± 10 percent); all other concentrations are by stable isotope dilution, with precision and accuracy of ± 2–4 percent. Sr isotope measurements spanned the period from 1962–1967, and have a precision of ±0.001 (2σ) relative to a value of 0.7080 for the Eimer and Amend standard. Samples 3 to 9 kindly supplied by J. D. H. Wiseman. Sr values for samples 10, 12, 13, and 18 were also obtained by emission spectrometry and reported in Tables 1 and 2, which agree well with those reported here.

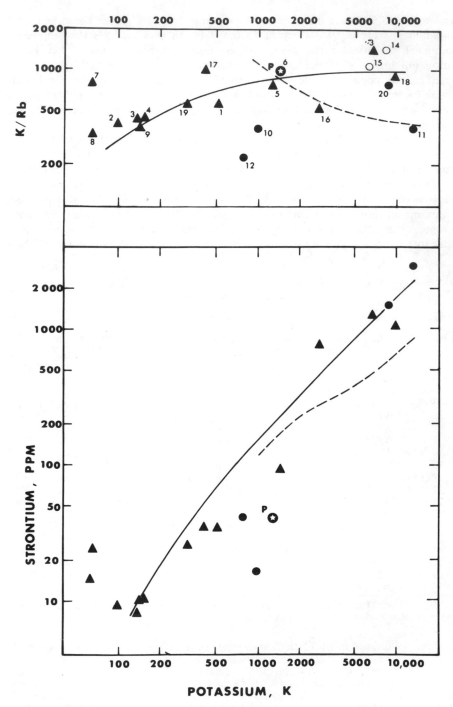

Figure 9. K/Rb ratio and Sr concentration versus K concentration of mylonites from St. Paul's Rocks. Triangles, whole-rock surface samples; solid circles, whole-rock dredge samples; open circles, mineral samples. Solid line is a two-component mixing line between the typical high and low potassium groups. Dashed line is trend of average oceanic basalt types (Hart and others, 1970). Numbers refer to the text numbers of Table 7. Three Sr values are taken from Tables 1 and 2.

elements an alkali basalt would be an adequate parent liquid for production of the peridotites by fractional crystallization. For example, Frey's model with 10 percent crystallization of olivine, enstatite, and amphibole from a typical alkali basalt would produce a peridotite cumulate with K, Rb, and Sr contents as shown by point P in Figure 9. The observed variations within the peridotite group might then be produced by different degrees of mineralogical segregation of the olivine and pyroxene from the amphibole. We emphasize that considerations of this sort still do not lead to an adequate explanation for the brown hornblende mylonites.

The spinel peridotites would provide a suitable mantle parent for production of alkali basalt by partial melting, at least in terms of K, Rb, and Sr (and the rare earths). Our calculations suggest that the liquid produced by a few percent partial melting of spinel peridotites (such as sample 5 or 6) is remarkably similar to alkali basalts in terms of these trace elements. Frey (1970) concluded that St. Paul's Rocks are not an example of undifferentiated mantle. We would like to emphasize that, whether or not it is differentiated relative to chondritic rare earth patterns, we believe that a composite sample (for example, "Avg.," Table 1) is still a reasonable example of the type of mantle which is required to produce typical oceanic alkali olivine basalts.

Other Trace Element Implications

The peridotites have a mean K/Ba of 41, K/Sr of 18, and Ba/Sr of 0.4, values which are quite similar to those predicted for a likely mantle composed of a peridotite with two percent hornblende (Griffin and Murthy, 1969). This tends to support the idea that St. Paul's Rocks may be representative of at least one kind of mantle composition.

Consideration of these ratios also argues against the idea that the brown hornblende mylonites are partial melts of the peridotite mylonites. The brown hornblende mylonites have mean K/Ba of 4.2, K/Sr of 6.5, and a Ba/Sr of 1.5. Griffin and Murthy (1969), based on distribution coefficients for common minerals of ultramafic rocks, suggest that 5 to 30 percent partial melting of a peridotite with 2 percent hornblende would yield a liquid with ratio K/Ba between 78 and 61, K/Sr between 17 and 13, Ba/Sr between 0.2 and 0.25, and a K/Rb between 1050 and 1117. These ratios, except for K/Rb, differ markedly from the values for brown hornblende mylonite. Even inclusion of an additional phase in the peridotites, such as phlogopite, which contains high concentrations of Ba, does not markedly change these values, since phlogopite is also rich in K.

PROBLEM OF ABSOLUTE AGE

Early work on the Rb-Sr relations of these rocks was interpreted in favor of a very ancient age (Hart, 1964a). This was due largely to limited sampling and to the high Sr^{87}/Sr^{86} ratio of sample 3, Table 6. With the many analyses now available (Table 6), it becomes apparent that the Sr^{87}/Sr^{86} ratio is essentially uniform within experimental errors and that the analysis of sample 3 is, in fact, distinctly anomalous. There is no evidence from the K/Rb data to suggest this sample has undergone contamination or exchange with sea water; however, we now conclude that some form of contamination, either natural or laboratory-derived, is responsible for this high ratio.

Taken at face value the constant Sr^{87}/Sr^{86} ratio, coupled with the variation in Rb/Sr ratio from 0.004 to 0.16, would lead to an isochron age for the time of last isotopic equilibration of about 0 ± 100 m.y. (Fig. 10). However, the two samples with highest Rb/Sr ratios are in dredge samples whose K/Rb relations (Fig. 9) suggest possible Rb addition during interaction with sea water, although the presence of phlogopite indicates that this K/Rb ratio may be a primary feature.

K-Ar ages were determined on a hornblende concentrate from one of the brown hornblende mylonites (SE-13, Table 2) and on three whole-rock samples of peridotite (Table 7). The hornblende age of 835 m.y. is truly surprising in view of all the tectonic evidence for a recent "emplacement" age for the St. Paul's mylonites. The easy way out of this problem is to ascribe the old hornblende age to inclusion of excess radiogenic argon; the fact that fluid inclusions are especially abundant in the amphiboles in these rocks would support this interpretation. However, the argon data on the other rocks, and especially the helium data, are difficult to reconcile with this view. Utramafic rocks and nodules commonly show very old K-Ar ages (Hart, 1966; Funkhauser and Naughton, 1968) and this is usually ascribed to the presence of excess argon. Such excess argon is strongly suggested for the St. Paul's peridotites by the relatively constant argon content (factor of two) in the three rocks coupled with the large variation (factor of 10) in K content. The He^4 is undoubtedly "excess" also, as one of the He ages is impossibly large.

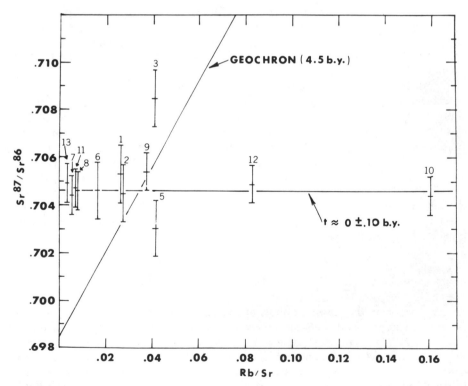

Figure 10. Isochron diagram for various mylonites from St. Paul's Rocks. Data in Table 6.

The important fact is that the He^4 content in the hornblende is similar to that in the rocks, and that the associated excess argon in the hornblende should be similar also to that in the rocks. Since the total argon content of the hornblende is 10 times that of the rocks, it is difficult to evade the conclusion that 90 percent of the argon is in situ in the hornblende, and that the K-Ar hornblende age is in some sense real. Recently, similar results were obtained on SE-13 by Umberto Cordani (1971, written commun.), confirming this apparent old age.

COMPARISONS WITH OTHER ULTRAMAFIC INTRUSIONS

The evidence of solid flow at high but sub-liquidus temperatures suggests that St. Paul's Rocks are an exposure of an intrusion. Its submarine surface area cannot be large, since dredging within a few miles of the islets revealed exposures of other rock types. These features, along with the petrology, suggest analogies to a number of continental ultramafic intrusions.

Two of the best analogies are to the Lizard and Tinaquillo intrusions. Both of these contain schlieren of basic gabbroic rocks and evidence of re-equilibration during emplacement. St. Paul's Rocks differ from these, though, in that they are predominantly unserpentinized; densities of the peridotite mylonites are on the order of 3.3 gm/cc. Such fresh peridotites (and dunites) do occur, such as at Dun Mountain, New Zealand, Webster area, North Carolina, and in the Twin Sisters intrusion, Oregon, but are rare compared to serpentinized peridotites. The freshness of St. Paul's Rocks is even more surprising in view of the presumably abundant saline solutions in the environment into which the intrusion moved and is now exposed.

Fresh or metamorphosed gabbroic rocks are commonly associated with ultramafic intrusions. Such is the case at St. Paul's Rocks, but the brown hornblende mylonites are more akin to the critically undersaturated rocks associated with carbonatites and with kimberlites than to gabbros. Such rocks, however, are associated with ultramafic rocks in the Miyamori intrusion (the pyroxene hornblendites, Onuki, 1965) and the Duke Island body (the hornblende clinopyroxenites, Irvine, 1967). Both these rock types have unusually high titania and alkalies, but not as high as in the brown hornblende mylonites.

Wyllie (1970) classifies St. Paul's Rocks in his oceanic peridotites group. This is simply a geographic grouping, and does not recognize the wide variety of oceanic peridotites now described (Melson and Thompson, 1970, 1971). From our studies, it appears that most of the varieties of peridotites described from continental exposures are also present in oceanic crust.

Thayer (1969) classifies St. Paul's Rocks with alpine peridotite-gabbro complexes of eugeosynclinal belts. He views the various rock types as co-magmatic. We find this hard to believe, if by co-magmatic one means that they are genetically related. If co-magmatic means only that they were emplaced together at high temperatures, we can accept a co-magmatic origin, although emplacement temperatures were much below the liquidus of even the gabbroic rocks. Thayer (1970) presents a classification which recognizes allogenic ultramafic intrusions. This applies to intrusions like that of St. Paul's Rocks, where the primary minerals crystallized elsewhere at depth, and were then transported in a largely solid (rheid) intrusion to their present position.

ST. PAUL'S ROCKS AND SEA-FLOOR SPREADING

Washington (1930b) postulated that the obvious evidence of "metamorphism by pressure" at St. Paul's Rocks proved that the equatorial Mid-Atlantic Ridge was a result of compression, and did not mark the zone along which Africa and South America were rifting apart, as had been proposed by Wegener (1924). The developing picture of plate tectonics allows quite another interpretation of St. Paul's Rocks. Its mylonitic texture can be viewed as a result of plastic flow during emplacement, and it can, although with some difficulties, be reconciled with current models of sea-floor spreading in the Atlantic.

We can calculate a sea-floor spreading model age for oceanic crust at the position of St. Paul's Rocks if we can identify the ridge crest from which the area has spread and if we know the average spreading rate. The complicated bathymetry and uncertainties about the development of the equatorial Atlantic make this a difficult task. Figure 1 shows that St. Paul's Rocks may have spread westward from the ridge crest about 260 nautical mi to the east. From this distance, and assuming an average drift rate of 1.4 cm/yr, a value often given for the northern Atlantic, we calculate a sea-floor spreading age of around 35 m.y. for the oceanic crust near St. Paul's Rocks. A much younger age, 9.3 m.y., is obtained if we assume St. Paul's Rocks spread eastward from the nearer ridge crest to the west.

The apparent 835 m.y. age for the brown hornblende is inconsistent with these model ages. We have tried to explain this old age by assuming fragments of the continental margins or uppermost mantle were somehow left behind, and portions of this old material were then incorporated in the intrusive mass during its ascent. These models are not very satisfactory as long as we regard the zone at which new crust is forming as very narrow, a notion well-founded on magnetic evidence over presently active ridge crests. We thus can offer no good explanation for this contradiction. As previously discussed, excess radiogenic argon is not a likely explanation for this particular sample. The sample from which the old hornblende was separated has a well-known history. It was collected in situ along the northeast side of Southeast Islet, and the locality is accurately located on Figure 2.

The possible young "ages" based on Rb/Sr ratios for two of the amphibole peridotites—less than 100 m.y.—is consistent with the predicted age from the sea-floor spreading model. Thus, it seems most likely that the peridotite did incorporate older material during its ascent. The old material could not have been stored at great depth, because high temperature would reset the K-Ar "clock" in the hornblende.

TABLE 8. APPARENT K-AR AND U, TH-HE AGES, ST. PAUL'S ROCKS

Sample	K (ppm)	AR^{40}*(ccstp/g)	He^4(ccstp/g)	K-Ar Age (m.y.)	He Age (m.y.)
H2	99.3	1.37×10^{-6}	1.69×10^{-5}	2000	6650
WPIi	1447	2.96×10^{-6}	2.06×10^{-5}	450	700
WPHi	1284	1.95×10^{-6}	——	350	——
SE-13 Hornblende	8350	3.50×10^{-5}	1.48×10^{-5}	835	——

Argon concentrations in H2 and WPHi measured by isotope dilution ($\pm$ 1 percent); other argon values and all He^4 values by calibration of ion beam intensities ($\pm$ 5–10 percent). U and Th value of WPIi (0.14 ppm and 0.40 ppm) from Tilton (1965); value for H2 estimated at 0.003 ppm U and 0.010 ppm Th by comparison with values reported for WC-2.

Thus a crustal or at best very uppermost mantle origin site of incorporation in the intrusion seems most likely. Although hornblendes have relatively high Ar-40 retentivity (Hart, 1964b), why did the hornblende not lose argon during intrusion at temperatures in excess of 500° C? Perhaps the explanation is that the hornblende was subjected to high temperatures for but a short time.

St. Paul's Rocks are a unique intrusion in regard to sea-floor spreading, for they may be the youngest sub-aerially exposed peridotite emplaced during formation of new oceanic crust. This may account in part for their freshness. Most orogenic belt peridotites are viewed by many as having formed on ocean ridges. They are then carried passively along, until some, where they meet trenches marginal to continents, become welded onto the continental margin, or descend back into the mantle. During this long history, opportunities for post-emplacement serpentinization are of course great, and few intrusions would be expected to survive unaltered.

The young basalt (no older than Quaternary, Melson and others, 1967b; Cifelli, 1970) which occurs on the submarine north slope of St. Paul's Rocks shows that volcanism has occurred in this region long after it was near an active segment of the Mid-Atlantic Ridge. Some intrusion thus must be occurring in these fracture zones away from ridge segments, and hence, the St. Paul's intrusion may be even younger than its sea-floor spreading model ages. The oldest sediments dredged from the slopes of St. Paul's Rocks are mid-Miocene (Cifelli, 1970) and too old to be consistent with spreading of St. Paul's Rocks from the ridge segment to the west.

Some extension of oceanic crust must occur normal to equatorial ridge segments to account for the present position of the continents which border the Atlantic. Extension of some sort has been postulated to be important in giving exposures of the plutonic rocks which are so often dredged from equatorial Atlantic fracture zones (Melson and Thompson, 1971). Perhaps the alkali basalt which occurs north of St. Paul's Rocks indicates that volcanism and plutonism associated with the opening of equatorial fracture zones is fundamentally different from that which occurs along actively spreading ridge crests. These latter rocks characteristically lack normative nepheline, whereas the former contain normative nepheline. Certainly the brown hornblende mylonites of St. Paul's Rocks and the young basalt on its north slope differ from ridge rocks in their critical undersaturation, and in their characteristic alkaline basalt trace element characteristics.

ST. PAUL'S ROCKS, LAYER 3, AND THE SUB-OCEANIC MANTLE

Much evidence has suggested that the lower zone of the oceanic crust—layer 3—is largely basic and metabasic rocks (Oxburgh and Turcotte, 1968). However, St. Paul's Rocks and the abundance of ultramafic rocks dredged from the mid-ocean ridge system, particularly from fracture zones, are not in accord with this view. Dredge samples suggest a complexity to the oceanic crust, involving both basic and ultrabasic rocks. It appears to us that without deep drilling in appropriate sites we will continue to debate the makeup of layer 3.

St. Paul's Rocks are perhaps an exposure of the sub-oceanic mantle. By this we mean that they may continue downward without interruption in density or composition into the mantle. If the intrusion is cylindrical, has a diameter of about 2 km, and is intruded into oceanic crust, as shown in Figure 11, a relatively small residual anomaly of 15 mgals would result (L. Cordell, 1970, written commun.).

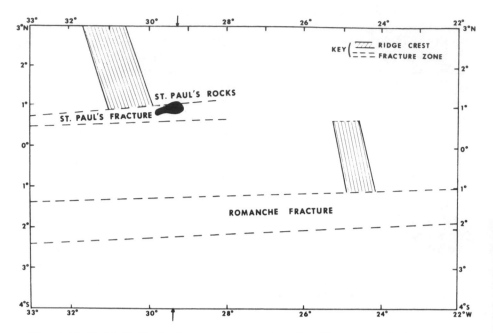

Figure 11. St. Paul's Rocks, fracture zones, and nearby ridge crests. Black area delineates the approximately east-trending ridge which is capped by St. Paul's Rocks.

This anomaly would be detectable by most marine gravity surveys. However, the anomaly is largely due to only the upper 2 km of the intrusion; the deeper extension to the mantle gives rise to but a small 3 mgal increment, an increment not resolvable by most marine gravity surveys. If the surface area of such an intrusion were much larger, or if 2 km wide but dike-shaped, a much larger anomaly would result. However, if the surface area were much larger, a view not in accord with the limited distribution of the peridotites in dredge samples from around the islet, isostatic compensation would give rise to a trench, rather than the pronounced approximately east-west ridge of which St. Paul's Rocks are a part.

SUMMARY AND CONCLUSIONS

1. St. Paul's Rocks are a solid-state intrusion (rheid) which contains at least three distinct rock types, but consists primarily of peridotite with average normative mineralogy 64 percent olivine, 22 percent enstatite, 2 percent diopside, and 8 percent plagioclase.

2. The intrusion was initially mobilized in the upper mantle at a depth of at least 45 km and at around 1,100° C, and moved upward rapidly, only partially equilibrating to the falling pressure and less rapidly falling temperature.

3. Intense mylonitization occurred during ascent and emplacement at temperatures in excess of 500° C. Very shallow depth of emplacement, giving rapid cooling and less time for recrystallization to a coarse aggregate, is in accord with preservation of the mylonitic texture.

4. The intrusion has structural and petrologic analogs among continental alpine peridotite intrusions, but differs from most of these in its extreme mylonitization,

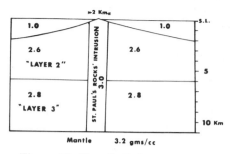

Figure 12. Hypothetical east-west cross section of the oceanic crust beneath St. Paul's Rocks. Densities and dimensions are those used in calculating hypothetical gravity anomaly by L. Cordell.

the occurrence of schlieren of brown hornblende-rich ultrabasic alkaline rocks, and certain trace element characteristics.

5. Major, minor, and trace element data show that the three main rock types from the islets are difficult to relate to one another by simple petrogenetic models. The brown hornblende mylonites do not appear to be derived from the coexisting peridotites by partial fusion, nor are they the last liquids from fractional crystallization of the peridotite assemblage from an alkali olivine basalt magma. Extensive modification of the existing models, or their extension to include other parent liquids and other phases, especially garnet, have not yet been attempted.

6. The trace element characteristics of the peridotites suggest that a composite sample or certain specific individual samples may be representatives of the kinds of compositions which will yield alkali olivine basalts with a few percent of partial melting. High water pressure is one condition that appears to guarantee that these liquids will have appropriate major element compositions, that is, remain critically undersaturated.

7. The samples most favorable for K-Ar dating, the brown hornblende mylonites, give very old apparent ages, up to 835 m.y., on a hornblende separate. Although excess radiogenic argon is present in some peridotites, such argon cannot account for the old hornblende age.

8. So far, Rb-Sr ages are not meaningful because of possible Rb enrichment from contact with sea water, and from low Rb/Sr ratios. The Sr^{87}/Sr^{86} ratio is low compared to many continental and other oceanic peridotites, ranging from about 0.703 to 0.708, and averaging around 0.705.

9. The St. Paul's Rocks intrusion may extend without discontinuity into the suboceanic mantle, but, because of its probable limited surface extent (less than several square kilometers?), geophysical determination of this possibility will probably prove difficult.

10. The section of oceanic crust of which St. Paul's Rocks is a part is a wide fracture zone which may or may not have formed at an ocean-ridge segment. If it did form at a ridge segment, it may have spread to either the east or west, giving sea-floor spreading model ages of 9.3 and 35 m.y. years, respectively. Alternatively, the St. Paul's Rocks intrusion may have been emplaced during opening and formation of the crust of the fracture zone. Some of its petrologic anomalies, especially its alkaline characters, may be intrinsic properties of fracture zone oceanic crust.

ACKNOWLEDGMENTS

Special assistance was given by the crew, officers, and scientific staff of the R. V. *Chain* and the R. V. *Atlantis II* during the 1963 and 1966 surveys of St. Paul's Rocks. V. T. Bowen served as chief scientist during these cruises and G. D. Nicholls supervised the geological studies during *Chain* cruise 35. H. H. Hess provided advice at various times during the investigation. T. P. Thayer, R. F. Fudali, and F. Frey reviewed and improved the manuscript. T. Wright performed least-squares approximation calculations by computer, and advised on the use of this method. S. R. Hart is grateful for a year's visit at the University of California, San Diego, where part of the K, Rb, and He work was carried out under support of NSF Grant GA-694. The research at the Smithsonian was performed with support of the Smithsonian Research Foundation. Mr. Banks performed mineral separations, and Larry Throup prepared illustrations at the Smithsonian. The research at Woods Hole was supported by the AEC, ONR, and NSF. We gratefully acknowledge all this support.

REFERENCES CITED

Boyd, F. R., 1959, Hydrothermal investigations of amphiboles, in Researches in geochemistry, Abelson, P. H., ed.: New York, John Wiley & Sons, p. 377–396.

Bryan, W. B., Finger, L. W., and Chayes, F., 1969, Estimating proportions in petrographic mixing equations by least-squares approximation: Science, v. 163, p. 926–927.

Cifelli, R., 1970, Age relationships of mid-Atlantic Ridge sediments: Geol. Soc. America Spec. Paper 124, p. 47–69.

Darwin, C., 1900, Geological observations on the Volcanic Islands and parts of South America during the voyage of H. M. S. Beagle: New York, Appleton, 3d ed.

Frey, F. A., 1970, Rare earth and potassium abundances in St. Paul's Rocks: Earth and Planetary Sci. Letters, v. 7, p. 351–360.

Frisch, T., and Schmincke, H. U., 1969, Petrology of clinopyroxene and amphibole inclusions from the Roque Nublo Volcanics, Gran Canaria, Canary Islands: Bull. Volcanol., v. 33–34, p. 1073–1088.

Funkhauser, J. G., and Naughton, J. J., 1968, Radiogenic helium and argon in ultramafic inclusions from Hawaii: Jour. Geophys. Research, v. 73, p. 4601–4607.

Green, D. H., 1963, Alumina content of enstatite in a Venezuelan high-temperature peridotite: Geol. Soc. America Bull., v. 74, p. 1397–1402.

—— 1964, Petrogenesis of the high-temperature peridotite intrusion in the Lizard Area, Cornwall: Jour. Petrology, v. 5, no. 1, p. 134–188.

Green, D. H., and Ringwood, A. E., 1963, Mineral assemblages in a model mantle composition: Jour. Geophys. Research, v. 68, no. 3, p. 937–945.

—— 1967, Genesis of basaltic magmas: Contr. Mineralogy and Petrology, v. 15, p. 103–190.

Griffin, W. L., and Murthy, R. V., 1969, Distribution of K, Rb, Sr and Ba in some minerals related to basalt genesis: Geochim. et Cosmochim. Acta, v. 33, p. 1389–1414.

Hart, S. R., 1964a, Potassium, rubidium, and strontium in the ultramafic rocks of St. Paul's Islands: Geol. Soc. America Abs. for 1964, p. 86–87.

—— 1964b, The petrology and isotopic-mineral age relations of a contact zone in the Front Range, Colorado: Jour. Geology, v. 72, no. 5, p. 493–525.

—— 1966, Radiogenic argon in oceanic ultramafic rocks: Trans. Am. Geophys. Union, v. 47, p. 147.

—— 1969, K, Rb, Cs contents and K/Rb, K/Cs ratios of fresh and altered submarine basalts: Earth and Planetary Sci. Letters, v. 6, p. 295–303.

Hart, S. R., Brooks, G., Krogh, T. E., Davis, G. L., and Nava, D., 1970, Ancient and modern volcanic rocks: a trace element model: Earth and Planetary Sci. Letters, v. 10, 17–28.

Heezen, B. C., Bunce, E. T., Hersey, J. B., and Tharp, M., 1964a, Chain and Romanche Fracture Zones: Deep-Sea Research, v. 11, p. 11–33.

Heezen, B. C., Gerard, R. D., and Tharp, M., 1964b, The Vema Fracture Zone in the equatorial Atlantic: Jour. Geophys. Research, v. 69, p. 733–739.

Hess, H. H., 1954, Geological hypotheses and the earth's crust under the oceans: Royal Soc. [London] Proc., Ser. A, v. 222, p. 341–348.

——1955, Serpentines, orogeny and epeirogeny, *in* Poldervaart, Afie, ed., Crust of the Earth: Geol. Soc. America Spec. Paper 62, p. 391–408.

—— 1964, The oceanic crust, the upper mantle and the Mayaguez serpentinized peridotite, *in* A study of serpentinite: NAS-NRC pub. 1188, p. 153–169.

Irvine, T. N., 1967, Duke Island ultramafic complex, southeastern Alaska, *in* Wyllie, P. J., ed., Ultramafic and related rocks: New York, John Wiley & Sons, p. 84–96.

Kushiro, I., Syono, Y., and Akimoto, S., 1967, Stability of phlogopite at high pressures and possible presence of phlogopite in the earth's upper mantle: Earth and Planetary Sci. Letters, v. 3, p. 197–203.

Lambert, I. B., and Wyllie, P. J., 1968, Stability of hornblende and a model for the low velocity zone: Nature, v. 19, no. 5160, p. 1240–1241.

Mason, B., 1966, Pyrope, augite, and hornblende, from Kakanui, New Zealand: New Zealand Jour. Geology and Geophysics, v. 9, no. 4, p. 474–480.

—— 1968, Eclogitic xenoliths from volcanic breccia at Kakanui, New Zealand: Contr. Mineralogy and Petrology, v. 19, p. 316–327.

McBirney, A. R., and Aoki, K., 1968, Petrology of the island of Tahiti, *in* Studies in volcanology: Geol. Soc. America Mem. 116, p. 523–556.

Medaris, L. G., and Dott, R. H., 1970, Mantle-derived peridotites in southwestern Oregon: Relation to plate tectonics: Science, v. 169, p. 971–974.

Melson, W. G., and Thompson, G., 1970, Layered basic complex in oceanic crust: Science, v. 168, p. 817–820.

—— 1971, Petrology of a transform fault zone and adjacent ridge segments: Royal Soc. London Philos. Trans., v. 268, p. 423–441.

Melson, W. G., Jarosewich, E., Bowen, V. T., and Thompson, G., 1967a, St. Peter and St. Paul's Rocks: a high-temperature mantle-derived intrusion: Science, v. 155, p. 1532–1535.

Melson, W. G., Jarosewich, E., Cifelli, R., and Thompson, G., 1967b, Alkali olivine basalt dredged near St. Paul's Rocks, Mid-Atlantic Ridge: Nature, v. 215, no. 5099, p. 381–382.

Millhollen, G. L., and Wyllie, P. J., 1970, Relationship of brown hornblende mylonite to spinel peridotite mylonite at St. Paul's Rocks: experimental melting study at mantle pressures: Geol. Soc. America, Abs. with Programs (Ann. Mtg.), v. 2, no. 7, p. 625.

Miyashiro, A., Shido, F., and Ewing, M., 1970, Crystallization and differentiation in abyssal tholeiites and gabbros from mid-oceanic ridges: Earth and Planetary Sci. Letters, v. 7, p. 361–365.

O'Hara, M. J., 1967, Mineral paragenesis in ultrabasic rocks, *in* Wyllie, P. J., ed., Ultramafic and related rocks: New York, John Wiley & Sons, p. 393–403.

Onuki, H., 1965, Petrochemical research on the Horoman and Miyamori ultramafic intrusives, northern Japan: Tohoku Univ. Sci. Repts., Ser. 3, no. 9, p. 217–276.

Otalora, G., and Hess, H. H., 1969, Modal analysis of igneous rocks by x-ray diffraction methods with examples from St. Paul's Rocks and an olivine nodule: Am. Jour. Sci., v. 267, p. 822–840.

Oxburgh, E. R., and Turcotte, D. L., 1968, Mid-ocean ridges and geotherm distribution during mantle convection: Jour. Geophys. Research, v. 73, p. 2643–2661.

Renard, A., 1882, On the petrology of St. Paul's Rocks: Appendix B, Narrative of the *Challenger* Report, v. 2.

Sheppard, S. M. F., and Epstein, S., 1970, D/H and $^{18}O/^{16}O$ ratios in minerals of possible mantle or lower crustal origin, Earth and Planetary Sci. Letters, v. 9, p. 232–239.

Sykes, L. R., 1967, Mechanism of earthquakes and nature of faulting on the mid-oceanic ridge: Jour. Geophys. Research, v. 72, p. 2131–2153.

Thayer, T. P., 1969, Peridotite-gabbro complexes as keys to petrology of mid-oceanic ridges: Geol. Soc. America Bull., v. 80, p. 1515–1522.

—— 1970, Authigenic versus allogenic ultramafic and gabbroic rocks as hosts for magmatic ore deposits [Abs.]: Geol. Soc. Australia, General Mtg., May, 1970.

Tilley, C. E., 1947, Dunite mylonites of St. Paul's Rocks (Atlantic): Am. Jour. Sci., v. 245, p. 483–491.

—— 1955, Contribution to the discussion on the origin of the earth: Geol. Soc. London Proc., v. 1521, p. 48–49.

—— 1966, A note on the dunite (peridotite) mylonites of St. Paul's Rocks (Atlantic): Geol. Mag., v. 103, no. 2, p. 120–123.

Tilton, G. R., 1965, Radioactive heat production in a periodotite: Carnegie Inst. Washington Year Book 64, p. 176–177.

Tressler, W. L., Bershad, S., and Berninghausen, W. H., 1956, Rochedos Sao Pedro e Sao Paulo (St. Peter and St. Paul Rocks): U. S. Navy Hydrographic Office Tech. Rept. Ho 31, p. 1–63.

Tröger, W. E., 1935, Spezielle Petrographie der eruptivgesteine: Berlin, Deutschen Mineralogischen Gesellschaft.

Washington, H. S., 1930a, The petrology of St. Paul's Rocks (Atlantic), in Report on the collections of the H. M. S. Quest expedition: London, British Museum.

—— 1930b, The origin of the mid-Atlantic Ridge: Jour. Maryland Acad. Sci., v. 1, no. 1, p. 1-29.

Wegener, A., 1924, The origin of continents and oceans: London, Methuen.

Wiseman, J.D.H., 1966, St. Paul's Rocks and the problem of the upper mantle: Royal Astron. Soc. Geophys. Jour., v. 11, p. 519–525.

Wright, T. L., and Doherty, P. C., 1970, A linear programming and least-squares computer method for solving petrologic mixing problems: Geol. Soc. America Bull., v. 81, p. 1995–2008.

Wyllie, P. J., 1970, Ultramafic rocks and the upper mantle: Mineralog. Soc. America Spec. Paper 3, p. 3–32.

Contribution No. 2617, Woods Hole Oceanographic Institution

Manuscript Received by the Society March 29, 1971

THE GEOLOGICAL SOCIETY OF AMERICA, INC.
MEMOIR 132
© 1972

Possible Mechanisms for the Emplacement of Alpine-Type Serpentinite

JOHN P. LOCKWOOD

U.S. Geological Survey, 345 Middlefield Rd., Menlo Park, California 94025

ABSTRACT

A model for the emplacement of alpine-type serpentinite has evolved from studies of an anomalous serpentinite outcrop discovered in northern Colombia during mapping under Harry Hess's Caribbean Project. According to the model, much alpine-type serpentinite is derived from oceanic crust or uppermost mantle that has been subducted beneath continental crust along continental margins. As the temperature of the subducted slab rises, serpentinite becomes mechanically unstable at about 300° to 350° C, owing to thermally induced dehydration weakening. Because of its low density relative to enclosing rocks and the lubricating effect of interstitial water, the weakened serpentinite migrates upward as protrusions along major faults, and owing to internal shearing and the expansion of water released during dehydration, its velocity may increase as it rises. Eclogite fragments and other high-grade metamorphic rocks torn from deep crustal levels are carried upward with the serpentinite. At shallow crustal levels, steam may be generated within the protrusions, possibly propelling serpentinite to the sea floor explosively. Once at the surface, sheared serpentinite and any associated tectonic inclusions flow downslope as turbidity currents and submarine landslides and are deposited in oceanic trench areas and interrise basins as exotic blocks, olistostromes, and detrital beds. With continuing subduction and sedimentation in the trench area, the sedimentary serpentinite is buried in a mélange of graywacke, deep-sea chert, and spilite. During deformation and metamorphism of this mélange, shearing of the serpentinite may mask original sedimentary features and make a later determination of origin extremely difficult.

Alpine-type serpentinites comprise four principal, overlapping groups: (a) serpentinite protrusions—mostly large, elongate bodies bounded by major faults, (b) sedimentary serpentinites—mostly small masses scattered helter-skelter throughout eugeosynclinal sequences, (c) serpentinites associated with large slabs of oceanic crust or upper mantle (ophiolite sequences), and (d) serpentinites associated with high-temperature peridotite protrusions. Serpentinites are highly mobile rocks, and individual masses could have belonged to several of these categories at different stages of their tectonic evolution.

INTRODUCTION

The debate over the origin of serpentinite has occupied geologists for well over a century. Most early geologists considered serpentinites to be aqueous deposits of some sort (for example, Hunt, 1883), owing to the lack of contact metamorphism and the fact that most serpentinite bodies are concordant with enclosing sedimentary rocks. Later, accumulating evidence indicated that the mineral serpentine formed from pre-existing igneous minerals, and W. N. Benson's (1918) widely read paper on serpentinite origin convinced most geologists that these rocks formed as intrusive igneous bodies. Hess adopted this view in his early papers on serpentinite emplacement (1939, 1955) and became the leading proponent of magmatic emplacement of all ultramafic rocks in the classic literature debate with N. L. Bowen and other experimentalists. However, in the face of new experimental and mineralogical evidence he eventually abandoned this position and accepted the view that alpine serpentinites and peridotites must reach upper levels of the earth's crust as solids (Hess, 1966).

As it became generally accepted that magmatic emplacement of serpentinites is unlikely, geologists began to view protrusion[1] as the likely mechanism of emplacement for most alpine serpentinite masses. Persuasive support for this view was found in the sheared contacts so typical of serpentinite bodies. The emplacement of serpentinites of all shapes and sizes, from narrow "sills" and small lens-shaped bodies to gigantic mountain-size blocks, has been attributed to tectonic processes. According to a popular view, serpentinites are tectonically incorporated in eugeosynclinal sedimentary rocks by unspecified mechanisms that tear serpentinite from oceanic crust or upper mantle and mix it with sedimentary rock during tectonism.

Many serpentinites in alpine orogenic areas have been emplaced by processes other than igneous intrusion or tectonic protrusion; these are the sedimentary serpentinites, for which the Neptunist views of early geologists are partially valid. Such serpentinites are coeval with enclosing rocks, and have been emplaced by sedimentary mechanisms directly on the earth's surface as detrital accumulations or gravity-slide masses.

The beginnings of this paper are traceable back to an outcrop of stratified serpentinite breccia and sandstone in northern Colombia, discovered during thesis mapping for Harry Hess's Caribbean Project (Lockwood, 1971a). Hess visited the area in August 1963 and, when told of the strange breccia outcrop, immediately opined that it was doubtless a nonsedimentary tectonic breccia. Once on the outcrop, however (Fig. 1), he became quickly convinced (by the lack of interclast

---

[1] *Protrusion* is a word coined by Lyell to describe masses of crystalline rock injected into sedimentary beds by tectonic processes (*see* Knipper and Kostanyan, 1964, p. 77; Lockwood, 1971b, p. 920). This word replaces "cold intrusion," "solid intrusion," "tectonic intrusion," and so forth.

Figure 1. Harry Hess examining outcrop of serpentinite breccia near Jureheruhu, Guajira Peninsula, Colombia. From left to right: K. W. Barr, Hess, W. S. Alvarez.

shearing) that the serpentinites were indeed formed by sedimentary processes. Harry was never slow to change his mind in the face of new evidence, especially field evidence.

SEDIMENTARY SERPENTINITES AND THEIR EMPLACEMENT

Sedimentary accumulations of serpentinite, ranging in age from early Paleozoic to Tertiary, are widespread along modern and fossil continent margins, where they are mostly interbedded with eugeosynclinal facies rocks (Zimmerle, 1968; Lockwood, 1971b). Most of these deposits consist entirely of serpentinite, although a few contain large amounts of partly serpentinized peridotite. Rock types foreign to the alpine-type ultramafic assemblage are usually rare or absent. Detrital deposits range in thickness from a few meters to over 1,000 meters, and typically consist of coarse, poorly sorted breccia together with lesser amounts of bedded serpentinite ranging in particle size from coarse conglomerate to shale. Graded bedding is common. The monolithologic nature of the deposits, the enormous size of some blocks (Abbate and others, 1970), the poor sorting, and the graded bedding each suggest very rapid deposition as submarine landslides, mudflows, or turbidity currents.

Deformation and shearing may make recognition of sedimentary serpentinites very difficult. Serpentinite gravity-slide blocks (olistoliths) will be marginally sheared during any postemplacement deformation, and may easily be mistaken for protrusions. As an added complexity, masses of sedimentary serpentinite may be plastically mobilized and protrude enclosing rocks (Moiseyev, 1970).

POSSIBLE EXAMPLES OF UNRECOGNIZED
SEDIMENTARY SERPENTINITE

Extensive eugeosynclinal terranes wherein small masses of serpentinite are scattered in rather helter-skelter fashion are found throughout alpine orogenic areas of the world. These masses are normally interpreted as protrusions, yet they parallel bedding of enclosing rocks, are not normally associated with identifiable faults, and are generally limited to one facies of enclosing sedimentary or metasedimentary rock; moreover, those that have been dated are of limited stratigraphic age. Examples of this type are found in the ultramafic belt of South Island, New Zealand, where serpentinite is limited to a narrow stratigraphic horizon in the middle Permian (Grindley, 1958; Waterhouse, 1964; Coleman, 1966); in the northern Guajira Peninsula, Colombia, where thousands of serpentinite bodies, ranging in maximum dimension from less than a meter to several kilometers, are scattered throughout a eugeosynclinal formation (Lockwood, 1971a); in the northern Appalachian ultramafic belt, especially Vermont, where serpentinites are limited to Ordovician eugeosynclinal strata (Fig. 2A; Stuckless, 1972); and in California, where smaller serpentinite bodies are widespread within the Franciscan Formation (Fig. 2B, 2C). Quite possibly serpentinite masses such as these are not younger protrusions or intrusions, but instead are coeval aggregates of serpentinite detritus and exotic slide blocks that were emplaced by sedimentary processes contemporaneously with sedimentation.

Consideration of sedimentary emplacement mechanisms may have important bearing on tectonic analysis. For example, Brueckner (1969) has argued that the restriction of ultramafic bodies to Caledonian metamorphic rocks in southwestern Norway indicates that the Caledonides were overthrust great distances into their present position, since the lack of ultramafic intrusions in underlying Precambrian rocks militates against ultramafic intrusion *in situ*. An alternative interpretation is that the widespread ultramafic bodies of this area are restricted to the Caledonian rocks because of initial sedimentary emplacement at this stratigraphic level as exotic blocks—possibly blocks of serpentinite which were later dehydrated during regional metamorphism. Support for this interpretation is offered by the well-documented deposits of sedimentary serpentinites that have been reported in Caledonian rocks at numerous localities in central Norway and Sweden (Hedström, 1930; Zachrisson, 1969).

SOURCE OF SEDIMENTARY SERPENTINE DETRITUS

Typical sedimentary serpentinites have formed by the very rapid influx and deposition of large amounts of serpentinite detritus from relatively distant sources. Very few deposits were derived from local erosion sources. Those deposits which were eroded from nearby serpentinite bodies are small, commonly contain abundant rounded clasts of rock types other than serpentinite, and formed in near-shore or subaerial environments (Okada, 1964).

Although distant, the sources of serpentinite debris for typical deposits were probably directly upslope from the area of deposition, since the gigantic size of some blocks indicates that downslope gravitational transport processes were more important than longshore currents. The monolithologic nature of typical sedimen-

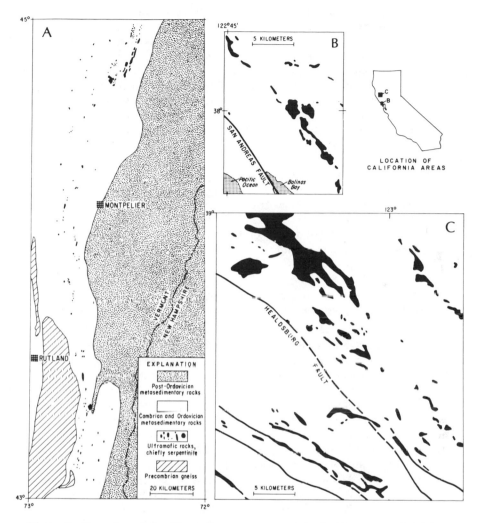

Figure 2. Examples of possible sedimentary serpentinite in Vermont and California. (A) Distribution of ultramafic rocks, chiefly serpentinite, in eastern Vermont (*from* Doll and others, 1961; Chidester, 1962, 1968). The cross-cutting contacts observed in a few of the localities (Chidester, 1968) can be explained by minor protrusive migration during metamorphism. (B) Distribution of serpentinite in Franciscan Formation graywackes 35 km northwest of San Francisco (*from* Jennings and Burnett, 1961; Koenig, 1963). (C) Distribution of serpentinite in Franciscan Formation rocks 110 km northwest of San Francisco (*from* Koenig, 1963).

tary serpentinites precludes an origin by normal subaerial erosion of the source area, since this would lead to contamination by sediments typical of the overall basin (graywackes, feldspathic siltstones, and so forth). The source areas of typical deposits had the capacity of suddenly providing a large volume of brecciated serpentinite very rapidly to adjoining basins.

Fortunately, actual sources of serpentinite detritus are known from California, and these serve as examples of the type of source and the kind of process that must have operated to provide serpentinite detritus elsewhere. The New Idria serpentinite protrusion of the central Coast Ranges has supplied serpentinite debris to adjoining basins in several episodes since Mesozoic time (Eckel and Myers, 1946). This protrusion is an intensely sheared, fault-bounded elliptical mass of asbestos–rich serpentinite 23 km long. Submarine mudflows and landslides from this protrusion formed a large deposit of sedimentary serpentinite in the Miocene (the Big Blue Formation as redefined by Adegoke, 1969), and subaerial mass movements have continued into historic time (Cowan and Mansfield, 1970). Another known Coast Range serpentinite source is at Table Mountain (Dickinson, 1966), where the Quaternary extrusion of serpentinite from a narrow fissure has formed a surficial blanket of serpentinite breccia with an original areal extent of over 75 km[2].

Although the shapes of the New Idria and Table Mountain protrusions differ, both consist of very highly sheared serpentinite and show evidence of prolonged upward migration. These protrusions supplied large amounts of weak, internally sheared serpentinite breccia to the earth's surface where, owing to its internal weakness, it moved downhill under gravitational stress. Protrusions such as these are believed to be the principal source of typical sedimentary serpentinite deposits. Other mechanisms for generating serpentinite gravity-slide material are discussed by Klemme (1958, p. 497–504).

PROTRUSIVE SERPENTINITES AND THEIR EMPLACEMENT

From numerous published descriptions (Milovanoviĉ and Karamata, 1960; Knipper, 1965; Dickinson, 1966; Oakeshott, 1968; Moiseyev, 1970) it is apparent that serpentinite has a great natural propensity for upward movement in orogenic zones, where it may protrude along thrust or normal faults or as fault-bounded diapirs. The principal reason for this upward migration is the relatively low density of serpentinite relative to enclosing rocks. The bulk density of unsheared serpentinite normally ranges from 2.40 to 2.44 (Huggins and Shell, 1965) to as high as 2.62 (unpub. data), whereas, for example, median values of 2.55 to 2.65 are reported for unmetamorphosed graywacke from the California Coast Ranges (Bailey and others, 1964). Metamorphosed rocks are much denser; mean values of bulk density range from 2.76 to 3.10 for metamorphosed eugeosynclinal rocks of California and Japan (Ernst and others, 1970). The overall density of the enclosing sedimentary section is further increased by any volcanic rocks present[2] and by load compaction processes at depth. Furthermore, shearing decreases serpentinite density, and values as low as 2.2 have been measured for highly sheared serpentinite of the New Idria mass (R. G. Coleman, 1970, written commun.). Thus, the velocity of protrusion may be self-accelerating owing to a lowering of serpentinite density as a result of internal shearing during movement.

Typical serpentinite protrusions consist of aggregates of highly polished, slickenside-bounded serpentinite blocks set in matrices of finely divided crushed serpentine. Movement of these masses occurs largely by interblock and intergranular displacements that result in plastic behavior for the body as a whole. Such sheared bodies have low shear strength and high internal mobility (Bailey and others, 1964, p. 87; Cowan and Mansfield, 1970, p. 2624–2625) and can readily respond to

---

[2] The bulk density of deep-sea pillow basalts and flows ranges from 2.8 to 3.0 (J. G. Moore, 1970, oral commun.).

slight density differences relative to the enclosing rocks or can migrate to areas of lower confining pressure. These factors are most important for large serpentinite masses, but are less effective for smaller bodies. Serpentinite masses less than a certain size (perhaps a few tens of meters) are not likely to undergo protrusion, although in regionally deformed terranes they may have sheared external contacts because of differences in rigidity relative to enclosing rocks.

Another probable cause of serpentinite protrusion at lower crustal levels, or in areas where frictional heat may be generated (fault zones) is serpentinite dehydration. Raleigh and Paterson (1965) have shown that at temperatures as low as 300° C (at 3.5 kb confining pressure), dehydration reactions begin to drastically lower rock strength of lizardite-chrysotile-brucite serpentinite (Fig. 3). They attribute the dehydration weakening to the following reaction:

$$\text{serpentine} + \text{brucite} \rightarrow 2 \text{ forsterite} + 3 \text{ H}_2\text{O}.$$

Early experimental work (Bowen and Tuttle, 1949) indicated that this reaction should occur at about 450° C at 3.5 kb, but later reexamination (Johannes, 1968) lowered the reaction temperature by about 60° C, and it is likely this temperature would be further lowered by the effects of differential stress and impurities such as Fe and CO_2 in natural systems. Specific volume relations for this reaction and for water (Kennedy and Holser, 1966; Burnham and others, 1969) indicate that in most parts of the earth's crust, serpentinite dehydration will result in a slight

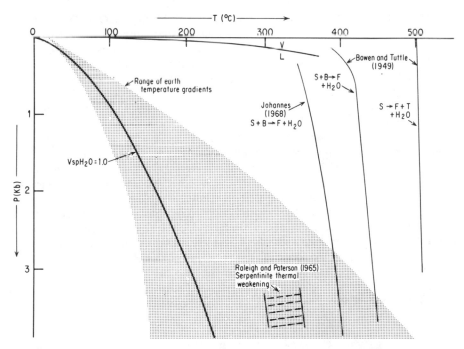

Figure 3. Pressure-temperature conditions of serpentinite dehydration, showing experimentally derived dehydration reactions, position of initial thermal weakening observed in serpentinite deformation studies, boiling curve of water, and loci of points where specific volume of water equals unity (after Kennedy and Holser, 1966; Burnham and others, 1969). For the dehydration reactions S = serpentine, B = brucite, F = forsterite, and T = talc.

volume increase,[3] although in areas of very low geothermal gradients (where the specific volume of water can be <1.0), volumes would remain about equal. At shallow levels of the earth's crust, however, this reaction would produce very large volume increases, as water would be released as steam. Since alpine serpentinites rarely, if ever, show evidence of secondary olivine formation (except where subjected to high temperature metamorphism), it is apparent that if these dehydration reactions have occurred, only a small amount of serpentine is normally dehydrated. The amount of serpentinite which could be dehydrated at low temperature would be limited by the amount of brucite originally present, and if only a few percent dehydrated, the minor amounts of olivine formed would likely be reserpentinized during cooling or under near-surface weathering conditions (Barnes and O'Neill, 1969).

The dehydration of even minor amounts of serpentinite would cause H_2O pore pressures within serpentinite that would promote fracturing, greatly reduce rock strength, and facilitate internal shearing, as demonstrated by Raleigh and Paterson (1965). Thus serpentinite protrusion is facilitated wherever massive serpentinite[4] is heated above $300°$ C.

RELATION OF SERPENTINITE EMPLACEMENT
TO ALPINE OROGENESIS

The intimate genetic relation of alpine-type (high Cr_2O_3, MgO/FeO; low CaO) serpentinites to the processes of alpine orogenesis is established by the geographic restriction of these unique rocks (in continental areas) to alpine orogenic belts. Furthermore, there is near-unanimity among geologists that this serpentinite is ultimately derived from peridotite of the earth's mantle, and that serpentinite or its parent peridotite is emplaced in the crust early in the geosynclinal cycle. This unanimity ceases, however, when geologic discussion turns to the means of ultramafic rock emplacement.

The new concepts of sea-floor spreading provide us with several emplacement models, especially if one accepts the view that serpentinite is an important constituent of the oceanic crust, as suggested by Hess (1955, 1962). Recent studies of Tethyan ophiolite–oceanic crust sequences however, and new oceanographic evidence (Cann, 1968) indicate that serpentinite is probably not as abundant in the lower oceanic crust as Hess postulated. Nonetheless, abundant serpentinite continues to be dredged from oceanic ridges, fault scarps, and trenches (Bowin and others, 1966; Miyashiro and others, 1969), and thus the oceanic lower crust or upper mantle must contain significant amounts of this rock.

Dietz (1963) suggested that alpine serpentinites are fragments of the oceanic crust which have been "tectonically incorporated" during underthrusting of oceanic crust below the continental rise (Fig. 4). However, the process of tectonic in-

---

[3] Exact computations of volume changes cannot be made, owing to the lack of compressibility data for serpentinite.

[4] It appears that of the two principal serpentinite varieties only "bastite serpentinite" (lizardite-chrysotile-brucite) is normally susceptible to protrusion. Antigorite serpentinite has a higher density (as high as 2.66), and does not undergo significant thermal weakening until much higher temperatures are reached ($500°$ C; Raleigh and Paterson, 1965).

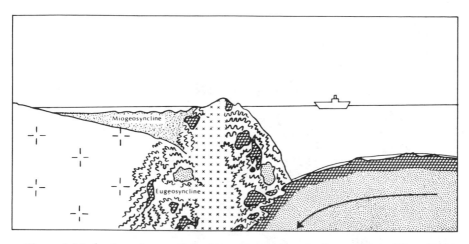

Figure 4. Mechanism of serpentinite "tectonic incorporation" according to Dietz (1963, Fig. 2, modified).

corporation, whereby small bodies of serpentinite are picked up from an underlying substrate and disseminated throughout older sediment during folding, as described by Dietz (1963) and Oakeshott (1968), is difficult to envision and, as pointed out by Maxwell (1969), is rather improbable mechanically, especially for smaller bodies. Hsu (1967, Fig. 8) has proposed another mechanism of tectonic incorporation whereby ultramafic rocks are protruded into a eugeosynclinal terrane and are later transported laterally (through thrusting or gravity-sliding) as fragments in complex mélange units. Hsu is careful to point out that such ultramafic bodies are "tectonic inclusions" (Hsu and Ohrbom, 1969), and that mélanges should not be confused with sedimentary breccias or olistostromes (Hsu, 1968). In contrast to these various mechanisms of tectonic incorporation, I prefer to postulate an alternate process of sedimentary emplacement contemporaneous with eugeosynclinal sedimentation (Fig. 5).

According to this model, oceanic lithosphere, consisting at least in part of serpentinite, is subducted beneath the continental rise during early phases of geosynclinal development and slowly rises in temperature as it descends. At temperatures as low as 300° C, partial dehydration of serpentinite may begin. As described in the preceding section, dehydration of even minor amounts of serpentinite will produce local water pore pressures that cause rock fracturing and abrupt lowering of rock strength. At this point an unstable situation exists wherein relatively low-density, weak serpentinite is overlain by higher density rocks. This situation is akin to sedimentary sections containing salt, and like salt, serpentinite is susceptible to upward protrusion. The mechanisms which trigger protrusion are unknown, although in an area of subduction faults in the overriding plates are to be expected and may provide protrusion channels. As shown by Oxburgh and Turcotte (1970) and others, temperatures in the overthrust plate will be higher than in the underthrust plate, so that serpentinite beginning to rise will undergo further thermal weakening as it rises. Volatiles from sedimentary rocks undergoing metamorphic dehydration in this zone are likely to be concentrated along faults and may also aid protrusion. As serpentinite rises higher in the Earth's crust, two principal factors can cause increased upward velocity: (a) shearing may further

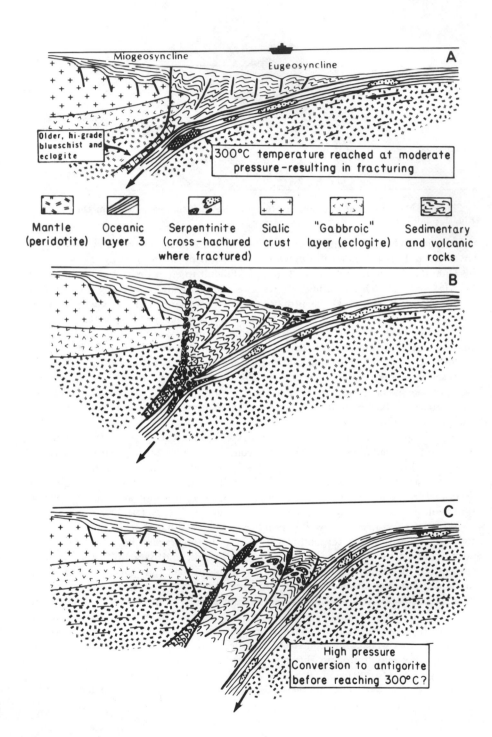

Mantle (peridotite) | Oceanic layer 3 | Serpentinite (cross-hachured where fractured) | Sialic crust | "Gabbroic" layer (eclogite) | Sedimentary and volcanic rocks

Miogeosyncline

Eugeosyncline

A

Older, hi-grade blueschist and eclogite

300°C temperature reached at moderate pressure-resulting in fracturing

B

C

High pressure Conversion to antigorite before reaching 300°C?

reduce serpentinite density, and (b) as lithostatic pressure is lowered, the specific volume of water generated within the serpentinite increases, causing a slight volume expansion most easily translated into upward movement. If high temperatures are maintained in the serpentinite, as is perhaps likely due to frictional heating, any water released by serpentine dehydration or admixed during uplift will flash into steam or supercritical vapor at somewhat less than 1 km depth with very large attendant increases in volume. The resultant gas-solid mixture could be propelled toward the surface at high velocities, and might resemble a diatreme in its mode of surficial emplacement. Once extruded on the earth's surface, serpentinite will flow downslope, especially if it is internally lubricated by large amounts of water derived from depth. If extruded on the continental slope or upper rise, the serpentinite will move downslope to form chaotic deposits (olistostromes) and exotic blocks (olistoliths) in a trench, in interrise basins, or on the deep-sea floor. These deposits will be buried by continental margin sediments, and may become mixed with any deep-sea sediments and volcanic rocks that are being scraped off the upper oceanic crust as subduction continues. It is in this mélange environment at the locus of subductive downthrusting that stratigraphically jumbled versions of Steinmann's Trinity (serpentinite, spilite, and radiolarian chert) may in some places come together. Exotic blocks of eclogite and other high-grade metamorphic rocks such as those associated with serpentinites of the Franciscan Formation (Bailey and others, 1964, Fig. 18; Ernst and others, 1970; Coleman and Lanphere, 1971) may be torn from the lower crust and carried to the surface with the serpentinite protrusions, where they are also emplaced by sedimentary processes in the mélange. Such an interpretation differs completely from the common view that such high-grade tectonic blocks are injected into mélanges from below (Dewey and Bird, 1970, p. 2637).

This subduction-protrusion model is especially applicable to terranes wherein high-pressure metamorphic rocks are associated with serpentinite (California, Japan). It cannot explain allochthonous gravity-slide ophiolite masses which include relatively undeformed spilite and chert (as in the Apennines); such ophiolites must have slid into their present locations directly from high-standing blocks of oceanic crust that were tectonically elevated by some means in adjacent areas.

Figure 5. Model for the emplacement of certain alpine serpentinites. (A) During an early stage of the alpine orogenic cycle, oceanic crust is subducted beneath the continental margin at a low angle, in a manner similar to the modern subduction of the Gorda Plate beneath North America (Silver, 1971, Fig. 19). Upon reaching 300°, downgoing slabs of serpentinite in layer 3 (schematically shown—actual shapes are conjectural) are fractured and weakened by partial dehydration. (B) Weakened serpentinite rises upward along a fault because of density contrast, carrying with it pieces of high-grade metamorphic rock from previously subducted and metamorphosed oceanic crust or from the lower continental crust. Upon reaching surface, serpentinite slides downslope, accumulating in the trench area as slide blocks, olistostromes, and detrital beds. (C) As subduction continues, angle of plate junction steepens and rate of subduction may quicken, so that down-going slabs of serpentinite reach higher pressures than before for any given temperature. Owing either to metamorphic antigoritization, or to lessening dehydration effect in this high-P/low-T environment, serpentinite no longer provides new material to continental crust. The original zone of serpentinite protrusion has now rotated and is the locus of reverse faulting. Serpentinite masses in the continental crust are now of two types: coeval bodies in a eugeosynclinal mélange, and protrusions along major faults (including remobilized sedimentary bodies).

CONCLUSIONS

Although all alpine ultramafic rocks must ultimately have been derived from the Earth's mantle, the paths that individual masses have followed to their present-day outcrops are exceedingly complex and varied. In this paper, attention has been focused on serpentinite and two principal modes of its emplacement: protrusion and sedimentation. Other emplacement mechanisms are discussed by Dewey and Bird (1971), and by Moores and MacGregor elsewhere in this memoir.

Protrusive serpentinites are typically large, elongate bodies located in narrow, linear belts along major faults. Such masses commonly truncate structures in enclosing rocks, separate rocks of differing facies or age, and may show evidence of protrusion over an extended period of geologic time. Examples of this type include the large ultramafic massifs of the Ural Mountains, serpentinites of the western Sierra Nevada, and possibly some of the serpentinite sheets separating the Great Valley and Franciscan sequences in California. An important source of such protrusions is postulated to be thermally weakened, low-density serpentinite in oceanic layer 3 that has been subducted beneath continental margins during early stages of alpine orogenesis. Protrusions reaching the Earth's surface provide a source of serpentinite for reemplacement by sedimentary processes.

Sedimentary serpentinites are relatively small bodies, may be widespread over broad but stratigraphically limited areas of a geosynclinal basin, and have contacts which always parallel the bedding foliation of enclosing rocks (unless they have been remobilized). The features most suitable for positive identification of such serpentinites (bedding, fossils, gradational contacts, and so forth) are easily destroyed by shearing, however, and for this reason most of these rocks have been described from somewhat atypical areas that have escaped major deformation. In deformed orogenic regions serpentinites originally emplaced by sedimentary processes may easily be misidentified as protrusions.

Alpine-type serpentinites may be loosely divided into four principal, overlapping groups: (a) serpentinites emplaced as low-temperature protrusions, (b) serpentinites emplaced by sedimentary means, (c) serpentinites associated with large slabs of oceanic crust or upper mantle (ophiolite sequences), and (d) serpentinites associated with high-temperature peridotite protrusions. Relations among these major categories are complex, and a single serpentinite body could have belonged to each of these categories during different stages of its tectonic evolution. For example, following high-temperature emplacement, peridotite could be completely serpentinized, undergo low-temperature protrusion to the surface, and be re-emplaced by sedimentary processes. Following burial, it could be protruded again. The potential complexity of emplacement history possible for any given mass of alpine serpentinite can thus hardly be over-emphasized.

ACKNOWLEDGMENTS

The speculations presented here have been nurtured and modified by profitable discussion with numerous geologists, especially E. H. Bailey, V. Bortolotti, R. G. Coleman, W. R. Dickinson, M. Galli, A. L. Knipper, E. M. Moores, P. Passerini, and J. Rodgers. Manuscript review comments by E. M. Moores, P. C. Bateman, J. C. Maxwell, and C. B. Raleigh resulted in numerous beneficial changes to this paper.

H. H. Hess introduced me to alpine serpentinites and provided the seeds for several ideas developed in this paper; my debt to him, however, goes far beyond the bounds of geology. Harry was much more than an esteemed professor—he was a good friend whose warm humanity served as an example of human goodness that continues to profoundly influence all who worked with him.

REFERENCES CITED

Abbate, Ernesto, Bortolotti, Valerio, and Passerini, Pietro, 1970, Olistostromes and olistoliths: Sed. Geology, v. 4, nos. 3–4, p. 521–557.

Adegoke, O. S., 1969, Stratigraphy and paleontology of the marine Neogene formations of the Coalinga region, California: California Univ. Pubs. Geol. Sci., v. 80, 269 p.

Bailey, E. H., Irwin, W. P., and Jones, D. L., 1964, Franciscan and related rocks and their significance in the geology of western California: California Div. Mines and Geology Bull. 183, 177 p.

Barnes, Ivan, and O'Neil, J. R., 1969, The relationship between fluids in some fresh alpine-type ultramafics and possible modern serpentinization, western United States: Geol. Soc. America Bull., v. 80, no. 10, p. 1947–1960.

Benson, W. N., 1918, the origin of serpentinite, a historical and comparative study: Am. Jour. Sci., 4th ser., v. 46, p. 693–731.

Bowen, N. L., and Tuttle, O. F., 1949, The system MgO-SiO$_2$-H$_2$O: Geol. Soc. America Bull., v. 60, no. 3, p. 439–460.

Bowin, C. O., Nalwalk, A. J., and Hersey, J. B., 1966, Serpentinized peridotite from the north wall of the Puerto Rico Trench: Geol. Soc. America Bull., v. 77, no. 3, p. 257–270.

Brueckner, H. K., 1969, Timing of ultramafic intrusion in the core zone of the Caledonides of southern Norway: Am. Jour. Sci., v. 267, no. 10, p. 1195–1212.

Burnham, C. W., Holloway, J. R., and Davis, N. F., 1969, The specific volume of water in the range 1000 to 8900 bars, 20° to 900° C: Am. Jour. Sci., v. 267–A, p. 70–95.

Cann, J. R., 1968, Geological processes at mid-ocean ridge crests: Royal Astron. Soc. Geophys. Jour., v. 15, no. 3, p. 331–341.

Chidester, A. H., 1962, Petrology and geochemistry of selected talc-bearing ultramafic rocks and adjacent country rocks in north-central Vermont: U.S. Geol. Survey Prof. Paper 345, 207 p.

—— 1968, Evolution of the ultramafic complexes of northwestern New England, in Zen, E-an, White, W. S., and Hadley, J. B., eds., Studies of Appalachian geology—northern and maritime: New York, John Wiley and Sons, p. 343–354.

Coleman, R. G., 1966, New Zealand serpentinites and associated metasomatic rocks: New Zealand Geol. Survey Bull. n.s. 76, 102 p.

Coleman, R. G., and Lanphere, M. A., 1971, Distribution and age of high-grade blueschists, associated eclogites, and amphibolites from Oregon and California: Geol. Soc. America Bull., v. 82, no. 9, p. 2397–2412.

Cowan, D. S., and Mansfield, C. F., 1970, Serpentinite flows on Joaquin Ridge, southern Coast Ranges, California: Geol. Soc. America Bull., v. 81, no. 9, p. 2615–2628.

Dewey, J. F., and Bird, J. M., 1970, Mountain belts and the new global tectonics: Jour. Geophys. Research, v. 75, no. 14, p. 2625–2647.

Dewey, J. F., and Bird, J. M., 1971, Origin and emplacement of the ophiolite suite—Appalachian ophiolites in Newfoundland: Jour. Geophys. Research, v. 76, no. 14, p. 3179–3206.

Dickinson, W. R., 1966, Table Mountain serpentinite extrusion in California Coast Ranges: Geol. Soc. America Bull., v. 77, no. 5, p. 451–472.

Dietz, R. S., 1963, Alpine serpentinites as oceanic rind fragments: Geol. Soc. America Bull., v. 74, p. 947–952.

Doll, C. G., Cady, W. M., Thompson, J. B., Jr., and Billings, M. P., 1961, Centennial geologic map of Vermont: Vermont Geol. Survey, scale 1:250,000.

Eckel, E. B., and Myers, W. B., 1946, Quicksilver deposits of the New Idria district, San Benito and Fresno Counties, California: California Jour. Mines and Geology, v. 42, no. 2, p. 81–124.

286   J. P. LOCKWOOD

Ernst, W. G., Seki, Y., Onuki, H., and Gilbert, M. C., 1970, Comparative study of low-grade metamorphism in the California Coast Ranges and the Outer Metamorphic Belt of Japan: Geol. Soc. America Mem. 124, 276 p.
Grindley, G. W., 1958, The geology of the Eglinton Valley, Southland: New Zealand Geol. Survey Bull. n.s. 58.
Hedström, H., 1930, Om ordoviciska fossil från Ottadalen i Det centrala Norge: Det Norske Vidensk. Acad. i Oslo, Avh. 1, Mat. Naturv. Klasse 1930, no. 10, p, 2–10.
Hess, H. H., 1939, Island arcs, gravity anomalies and serpentinite intrusions: Internat. Geol. Cong. Moscow, 1937, Comptes Rendus 17, v. 2, p. 263–283.
—— 1955, Serpentines, orogeny, and epeirogeny: Geol. Soc. America Spec. Paper 62, p. 391–408.
—— 1962, History of ocean basins, in Petrologic studies—A volume in honor of A. F. Buddington: Geol. Soc. America, p. 599–620.
—— 1966, Caribbean Research Project, 1965, and bathymetric chart: Geol. Soc. America Mem. 98, p. 1–10.
Hsu, K. J., 1967, Mesozoic geology of the California Coast Ranges—a new working hypothesis, in Étages Tectoniques, Colloque de Neuchâtel, Neuchâtel Univ. Inst. Geol., 1966: Neuchâtel, Switzerland, La Baconnière, p. 279–296.
—— 1968, Principles of mélanges and their bearing on the Franciscan-Knoxville Paradox: Geol. Soc. America Bull., v. 79, no. 8, p. 1063–1074.
Hsu, K. J., and Ohrbom, Richard, 1969, Mélanges of San Francisco Peninsula—geologic reinterpretation of type Franciscan: Am. Assoc. Petroleum Geologists Bull., v. 53, no. 7, p. 1348–1367.
Huggins, C. W., and Shell, H. R., 1965, Density of bulk chrysotile and massive serpentine: Am. Mineralogist, v. 50, p. 1058–1067.
Hunt, T. S., 1883, The geological history of serpentines, including notes on pre-Cambrian rocks: Royal Soc. Canada, Proc. Trans. 1, sec. 4, p. 165–215.
Jennings, C. W., and Burnett, J. L., 1961, Geologic map of California, Olaf P. Jenkins edition—San Francisco sheet: California Div. Mines and Geology, scale 1:250,000.
Johannes, W., 1968, Experimental investigation of the reaction forsterite $+ H_2O \leftrightarrows$ serpentine + brucite: Contr. Mineralogy and Petrology, v. 19, no. 6, p. 309–315.
Kennedy, G. C., and Holser, W. T., 1966, Pressure-volume-temperature and phase relations of water and carbon dioxide: Geol. Soc. America Mem. 97, p. 371–383.
Klemme, H. D., 1958, Regional geology of the circum-Mediterranean region: Am. Assoc. Petroleum Geologists Bull., v. 42, no. 2, p. 447–512.
Knipper, A. L., 1965, Osobennosti obrazovaniya antiklinaley s serpentinitovymi yadrami (Sevano-Akerinskaya zona Malogo Kavkaza) [The development of anticlines with serpentinite cores (Sevan-Akerin zone of the Lesser Caucasus)]: Moskov. Obshch. Ispytateley Prirody Byull. Otdel. Geol., v. 15, no. 2, p. 46–58.
Knipper, A. L., and Kostanyan, Yu. L., 1964, Vozrast giperbazitov severovostochnogo poberezh'ya ozera Sevan [The age of ultramafics of the northeastern coast of Lake Sevan]: Akad. Nauk SSSR, Izv. Ser. Geol., no. 10, p. 67–79.
Koenig, J. B., 1963, Geologic map of California, Olaf P. Jenkins edition—Santa Rosa sheet: California Div. Mines and Geology, scale 1:250,000.
Lockwood, J. P., 1971a, Detrital serpentinite from the Guajira Peninsula, Colombia, in Donnelly, T. W., ed., Caribbean geophysical, tectonic, and petrologic studies: Geol. Soc. America Mem. 130, p. 55–75.
—— 1971b, Sedimentary and gravity-slide emplacement of serpentinite: Geol. Soc. America Bull., v. 82, no. 4, p. 919–936.
Maxwell, J. C., 1969, "Alpine" mafic and ultramafic rocks—the ophiolite suite—a contribution to the discussion of the paper "The origin of ultramafic and ultrabasic rocks" by P. J. Wyllie: Tectonophysics, v. 7, nos. 5–6, p. 489–494.
Milovanoviê, Branislav, and Karamata, Stevan, 1960, Über den diapirismus serpentinischer Massen: Internat. Geol. Cong., 21st, Copenhagen 1960, Comptes Rendus, pt. 18, p. 409–417.
Miyashiro, Akiho, Shido, Fumiko, and Ewing, Maurice, 1969, Composition and origin of serpentinites from the Mid-Atlantic Ridge near 24° and 30° north latitude: Contr. Mineralogy and Petrology, v. 23, no. 2, 117–127.
Moiseyev, A. N., 1970, Late serpentinite movements in the California Coast Ranges—new evidence and its implications: Geol. Soc. America Bull., v. 81, no. 6, p. 1721–1732.

Oakeshott, G. B., 1968, Diapiric structures in Diablo Range, California: Am. Assoc. Petroleum Geologists Mem. 8, p. 228–243.

Okada, Hakuyu, 1964, Serpentine sandstone from Hokkaido: Kyushu Univ., Fac. Sci. Mem., ser. D., v. 15, no. 1, p. 23–38.

Oxburgh, E. R., and Turcotte, D. L., 1970, Thermal structure of island arcs: Geol. Soc. America Bull., v. 81, no. 6, p. 1665–1688.

Raleigh, C. B., and Paterson, M. S., 1965, Experimental deformation of serpentinite and its tectonic implications: Jour. Geophys. Research, v. 70, no. 16, p. 3965–3985.

Silver, E. A., 1971, Transitional tectonics and late Cenozoic structure of the continental margin off northernmost California: Geol. Soc. America, v. 82, no. 1, p. 1–22.

Stuckless, J. S., 1972, New evidence concerning the origin of ultramafic masses in southern Vermont: Geol. Soc. America, Abs. with Programs (North-Central Sec.), v. 4, no. 5, p. 350–351.

Waterhouse, J. B., 1964, Permian stratigraphy and faunas of New Zealand: New Zealand Geol. Survey Bull. n.s. 72.

Zachrisson, Ebbe, 1969, Caledonian geology of northern Jämtland—southern Väster-botten: Sveriges Geol. Undersokning Årsb., v. 64, no. 1, p. 1–33.

Zimmerle, Winfried, 1968, Serpentine greywackes from the North Coast basin, Colombia, and their geotectonic significance: Neues Jahrb. Mineralogie Abh., v. 109, nos. 1–2, p. 156–182.

MANUSCRIPT RECEIVED BY THE SOCIETY MARCH 29, 1971

THE GEOLOGICAL SOCIETY OF AMERICA, INC.
MEMOIR 132
© 1972

A Primary Peridotite Magma—Revisited: Olivine Quench Crystals in a Peridotite Lava

JOHN S. DICKEY, JR.
Smithsonian Astrophysical Observatory
Cambridge, Massachusetts 02138

ABSTRACT

Serpentinized peridotite layers in the Early Precambrian (3.4 b.y.) Onverwacht Group of the Swaziland Sequence, South Africa, appear to have been lava flows. These flows, which are interlayered with metamorphosed pillow basalts, display pillow structures and chilled, ultramafic margins. A specimen of the chilled peridotite has a quench texture consisting of serpentine pseudomorphs of bladed olivine crystals with interstitial plumose tremolite, probably pseudomorphous after dendritic clinopyroxene. Analyses of relict olivine (Fo_{88}) in the chilled and normal peridotites show unusually high Ca content (0.2 to 0.3 wt percent CaO).

INTRODUCTION

In 1938 Harry Hess published a paper titled "A Primary Peridotite Magma." Experimental petrologists, however, led by N. L. Bowen and his co-workers (Bowen and Anderson, 1914; Bowen and Schairer, 1935), had long denied the magmatic hypothesis because of the high temperatures required to melt peridotite (dry) and the apparent absence of peridotite lavas. Bowen (1927) believed, instead, that such olivine-rich rocks are formed by accumulation of olivine crystals from basaltic liquids. Hess acknowledged that some ultramafic rocks (such as dikes on the Isle of Skye and layers at the base of the Bushveld complex) formed in that way; however, field evidence persuaded him that other ultramafic rocks, found in narrow, strongly deformed zones of the crust, crystallized directly from peridotite magmas. These he called rocks of the ultramafic magma series. Today they are called alpine-type peridotites.

The field evidence which Hess cited for a primary peridotite magma was: (a) the presence of large peridotite intrusions in regions lacking significant volumes of less mafic, contemporaneous igneous rocks, (b) the ultramafic character of the border facies of the intrusions, (c) the existence of dike- and sill-like apophyses of serpentinized peridotite extending from the main intrusions into the country rocks, and (d) the undeformed interiors of some peridotite intrusions.

The strongest field evidence against the primary magma hypothesis was the absence of high temperature metamorphism around most peridotite intrusions. To explain the low intrusion temperatures and the apparent absence of peridotite lavas, Hess suggested that the magma contained 5 to 15 percent H_2O, which would lower the melting temperature and cause explosive brecciation of any surfacing magma. The possibility of a low temperature hydrous peridotite or serpentine magma provoked Bowen and Tuttle (1949) to investigate the system MgO-SiO_2-H_2O. They demonstrated that no serpentine melt exists below 1,000° C.

Following Bowen and Tuttle's report, the idea of peridotite magmas languished. Most petrologists favored the view that peridotite intrusions were not emplaced as melts but as crystalline masses, lubricated by interstitial fluids. Hess (1955) accepted this explanation for some peridotite intrusions but not for all. In particular, he believed that the high temperature peridotites, which have substantial contact metamorphic aureoles, were magmatic. He held this view until D. H. Green (1963, 1964) showed, by the high Al content of the pyroxenes and the presence of pyroxene + Al-spinel rather than olivine + plagioclase, that the high temperature peridotites of Tinaquillo (Venezuela) and the Lizard (Cornwall) area crystallized under high pressure and must have been intruded as largely solid masses. At that point Hess (1966) abandoned his advocacy of peridotite magmas. He did so with reluctance and, perhaps, prematurely; for other experimenters (notably Kushiro, 1969), by working at high pressures (20 kb) and expanding their systems to include CaO, soon found evidence of hydrous peridotite liquids below 1,400° C. These results indicated that, although the low temperature serpentine magma was unrealistic, hot hydrous peridotite melts might be generated in the deep crust and upper mantle.

In the late 1960s rumors began circulating that peridotitic pillow lavas had been found in the Precambrian Swaziland Sequence of South Africa. Hess was skeptical of these reports; nevertheless, he was determined to see the rocks in the field and did so in July of 1969. The following is a brief account of those lavas and a description of skeletal olivine crystals present in samples which were collected by Hess from a quenched flow margin.

PERIDOTITE LAVAS OF THE BARBERTON MOUNTAIN LAND[1]

The Barberton Mountain Land of the Eastern Transvaal, South Africa (31° E. long, 26° S. lat) is composed of early Precambrian sediments and volcanic rocks of the Swaziland Sequence, surrounded and intruded by younger Precambrian granites of the Kaapvaal Craton (Pretorius, 1964). From top to bottom the se-

---

[1] The geology of the Barberton Mountain Land has been studied and described as part of the Republic of South Africa's contribution to the Upper Mantle Project. Except where otherwise noted, this section is based *entirely* upon the published results of this study by M. J. and R. P. Viljoen (1970a, b, c, d).

quence comprises the Moodies, Fig Tree, and Onverwacht Groups. The Moodies and Fig Tree Groups are mainly sedimentary, the Onverwacht Group mainly volcanic. Radiometric dating indicates that the Onverwacht Group is at least 3.36 $\times 10^9$ years old (Van Niekerk and Burger, 1969), and the Moodies Group is intruded by 3.0×10^9-year-old granites (Allsopp and others, 1962). Although deformed and somewhat metamorphosed by the granite intrusions, the Swaziland Sequence is remarkably well preserved. From the Fig Tree sediments, for example, have come remains of the oldest known microorganisms in the geological record (3.1×10^9 years old; Barghoorn and Schopf, 1966). The unusual preservation of the rocks has also facilitated the recognition of peridotite lavas.

Layers of serpentinized peridotite alternate with metamorphosed basalt layers in the Onverwacht Group, particularly in its lower members: the Sandspruit, Theespruit, and Komati Formations. The Komati Formation contains the most abundant exposures of these layers.

The type section of the Komati Formation, exposed on the slopes of the Komati River Valley, is a steeply dipping succession (11,500 ft thick) of serpentinized peridotite and metabasalt layers. Although the ground is grass covered, one can trace the layers for miles along the slopes as a series of low crests (peridotites) and troughs (basalts). Good exposures are found in tributary streams which traverse the section.

The metabasalts, some of which form beautifully pillowed flows, have distinctive compositions (Fig. 1a) characterized by high Mg and Ca, low alkalies, and low Al concentrations. They are tholeiitic rocks, rich in normative hypersthene, diopside, and olivine. An average analysis appears in Table 1 (analysis a); as the H_2O+ content suggests, the basalts have been substantially altered to amphibole and other secondary minerals.

The peridotites are typically altered to serpentine, tremolite, chlorite, and talc, yet original textures and relicts of primary phases remain. They are rich in Si, Fe, and Ca and poor in Mg (Table 1, analysis b), and are chemically distinct from other peridotites, including (Fig. 1b) high-temperature peridotite intrusions, volcanic xenoliths, and Alaskan-type ultramafic complexes.

Because of their unusual compositions, the metabasalts and peridotites from the Komati Formation have been described as new classes of rocks, basaltic and peridotitic komatiites (Viljoen and Viljoen, 1970c).

Peridotite layers are most abuandant in the lower half of the Komati Formation. There are ultramafic sections up to 1,700 ft thick, consisting of several individual layers, varying from 15 to 500 ft in thickness. These layers are believed to be lava flows or, in some instances, sills. This magmatic interpretation rests upon the following observations (Viljoen and Viljoen, 1970d):

1. The peridotite layers are equally as continuous laterally as the basalt layers with which they occur.

2. Regionally, the peridotite layers thin in sympathy with the basalt flows, but there is no consistent local relation between the thicknesses of basalt and peridotite to suggest crystal accumulation.

3. Many of these peridotite layers are internally differentiated, with basal accumulations of idiomorphic olivine crystals grading up to more pyroxenous peridotite. Individual layers, however, do not grade upward into basalt, and contacts between basalt and peridotite layers are sharp, even when the latter are differentiated.

4. Volcanic pillow structures, not orbicular structures related to serpentinization,

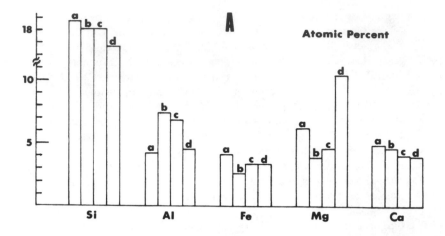

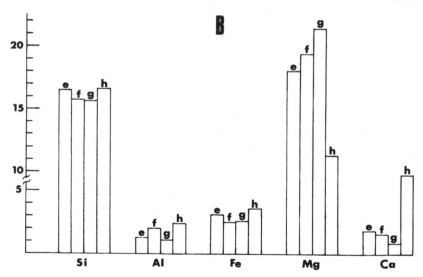

Figure 1. Major element concentrations (expressed as atomic percent) in metabasalts and serpentinized peridotites from the Komati Formation (basaltic and peridotitic komatiites) compared with some other varieties of basalt and peridotite. (a) Average basaltic komatiite (Table 1, analysis a); (b) average deep oceanic tholeiite (Engel and others, 1965); (c) average olivine tholeiite basalt and dolerite (Manson, 1967); (d) average Group II (noncumulitic) olivine tholeiite basalt, Svartenhuk Peninsula, west Greenland (Clarke, 1970); (e) average peridotitic komatiite (Table 1, analysis b); (f) estimated bulk composition of Serranía de la Ronda high-temperature peridotite intrusion (Dickey, 1970); (g) average peridotite nodule in basalts (Ito and Kennedy, 1967); and (h) average estimated bulk composition of Duke Island ultramafic complex, Alaska (Irvine, 1963).

Table 1. Average Rock Compositions

	Weight Percent			CIPW Norm	
	a	b		a	b
SiO₂	48.39	41.61	OR	0.53	0.18
TiO₂	0.94	0.31	AB	15.49	1.44
Al₂O₃	9.41	2.70	AN	17.20	7.39
Fe₂O₃	2.37	5.63	DI	32.99	13.02
FeO	10.71	4.35	HY	18.16	26.82
MnO	0.24	0.18	OL	6.95	48.86
MgO	10.91	30.58	MT	3.44	1.62
CaO	11.63	4.29	IL	1.79	0.66
Na₂O	1.83	0.15	Rest	2.52	0.00*
K₂O	0.09	0.03			
H₂O⁺	2.52	8.59			
Total	99.04	98.42			

* Composition recalculated H₂O-free; all but 1 percent Fe₂O₃ reduced to FeO.
a. Average of 3 metabasalts (basaltic komatiites).
b. Average of 8 serpentinized peridotites (peridotitic komatiites).
Analyses from Viljoen and Viljoen (1970c).

are present in the upper parts of some peridotite layers. These pillows are not as well formed or exposed as those in the basaltic flows, but diagnostic features such as pillow form, chilled margins, cuspate interstitial voids, and slump structures are found.

5. Peridotites near the margins of some layers have textures similar to those produced by experimental quenching of an ultrabasic silicate melt.

Each of these observations supports the magmatic interpretation; the last is commanding evidence of a peridotite melt.

Table 2. Olivines in Peridotite Lavas

	a	b
SiO₂	40.3	40.3
FeO	10.8	11.2
MnO	0.2	0.2
MgO	48.0	47.4
CaO	0.2	0.3
NiO	0.4	0.4
Sum	99.9*	99.8*
Cations Per 4 Oxygens		
Si	0.996	0.999
Fe	0.223	0.232
Mn	0.004	0.004
Mg	1.768	1.751
Ca	0.006	0.007
Ni	0.008	0.007
Sum	3.005	3.000

* Analyses by electron microprobe. Oxygen determined stoichiometrically, assuming all Fe to be divalent.
a. Skeletal olivine in quenched peridotite flow margin from the Komati Formation, Barberton Mountain Land, South Africa.
b. Normal olivine crystal from basal zone of peridotite flow in the Komati Formation, Barberton Mountain Land. Specimen VU 32A by courtesy of Viljoen and Viljoen.

Figure 2. Quenched peridotite from the Komati Formation. The large, intersecting blades are serpentinized olivine crystals. Sheaths of tremolite (pseudomorphous after dendritic clinopyroxene?) dominate the interstices. Accessory minerals are plagioclase, magnetite, Cr-spinel, and talc. Width of field: 14 mm. Transmitted light.

DESCRIPTION OF QUENCHED PERIDOTITE

The quenched peridotites, which occur near margins of ultramafic layers in the Komati Formation, consist of intersecting blades of serpentine (Fig. 2), which are pseudomorphous after olivine, with feathery interstitial sheaths of dendritic crystals (tremolite probably pseudomorphous after clinopyroxene), rare laths of plagioclase, and opaque matter. The primary oxide was a Cr-rich spinel which is present as rare octahedra, but most of the visible opaque matter is secondary magnetite. Relict olivine (Fo_{88}) is present in the serpentinized blades but is exceedingly rare. Before serpentinization olivine constituted more than 50 percent of the rock.

The elongated bladed habit is typical of olivine crystals grown under quench conditions. Figure 3 shows fresh olivine crystallized in this habit by melting and quenching a piece of the Murray meteorite (a Type II carbonaceous chondrite) in the laboratory. These artificial crystals are tabular on (010), and elongated parallel to the c crystallographic axis. They are smaller than the natural olivine pseudomorphs, but their habit appears to be similar.

The Ca content of the natural olivine also implies rapid crystallization. According to Simkin and Smith (1970) slower cooling plutons crystallize less calcic olivines

Figure 3. Skeletal olivine crystals in glass, grown by melting and quenching a portion of the meteorite, Murray, in the laboratory. Murray (Mason, 1962) is a Type II carbonaceous chondrite, consisting of olivine chondrules and aggregates in a groundmass of serpentine and carbonaceous matter. Width of field: 1 mm. Transmitted light.

(< 0.1 wt percent Ca) than faster cooling lavas and shallow intrusions (> 0.1 wt percent Ca). Analyzed olivines from these peridotite lavas (Table 2) fall in the high Ca group and contain nearly twice as much Ca as typical olivines from other kinds of peridotites (alpine-type, stratiform-type, xenoliths in basalts).

DISCUSSION

Bowen's interpretation (1928) of all olivine-rich igneous rocks, from picrites to peridotites, as the products of fractional crystallization of basaltic liquids is no longer without exception. Partial fusion of peridotite in high-pressure (40 kb) experiments yields picritic liquids (Davis and Schairer, 1965; Ito and Kennedy, 1967), and there is evidence that similar natural liquids exist, not only at depth (Dickey, 1970), but even as volcanic effusives (Clarke, 1970). Similarly, the case against the existence of peridotite magmas has been weakened by high-pressure experiments and by the discovery of peridotite lavas in the Barberton Mountain Land. From the evidence of field relations, petrographic textures, and olivine compositions it appears that many of these rocks crystallized directly from peridotite liquids.

To be sure, peridotite lavas are rare. Reports of such lavas in younger rocks have usually been discredited, although extrusive peridotitic crystal mushes, such as the serpentinite flows of Turkey (Bailey and McCallien, 1953) have been found. In view of the great age of the Barberton rocks it may be that peridotite lavas appeared only as submarine eruptions in the early Precambrian, when the upper mantle was probably hotter and more hydrous and the crust thinner and discontinuous.

An interesting question is whether such liquids have existed in the deep crust and upper mantle during the Phanerozoic and what roles they may have played in the formation of peridotites which later reached the outer crust as solid intrusions. Many perplexing mysteries, such as the upward displacement of large volumes of dense mantle material into the sial, the ultrabasic bulk compositions of many ophiolite complexes, and the sedimentary structures of podiform chromite deposits, could be clarified by magmatic hypotheses.

ACKNOWLEDGMENTS

The author was privileged to accompany Harry Hess on his 1969 trip to South Africa. Expenses were defrayed by the National Science Foundation (NSF GA-408, H. H. Hess and J. C. Maxwell, Principal Investigators), the Smithsonian Astrophysical Observatory, and the Bernard Price Institute for Geophysical Research, Witwatersrand University.

To C. R. Anhaeusser, M. J. Viljoen, and R. P. Viljoen the author acknowledges a substantial debt for published information, expert guidance in the field, and generous hospitality. The help of the following individuals is also gratefully acknowledged: R. B. Hargraves, C. Lundquist, Ursula B. Marvin, L. O. Nicolaysen, D. A. Pretorius, and J. A. Wood.

REFERENCES CITED

Allsopp, H. L., Roberts, H. K., Schreiner, G.D.L., and Hunter, D. R., 1962, Rb-Sr age measurements on various Swaziland granites: Jour. Geophys. Research, v. 67, p. 5307–5313.

Bailey, E. B., and McCallien, W. J., 1953, Serpentine lavas, the Ankara mélange and the Anatolian thrust: Royal Soc. Edinburgh Trans., v. 62, p. 403–442.

Barghoorn, E. S., and Schopf, J. W., 1966, Microorganisms three billion years old from the Precambrian of South Africa: Science, v. 152, p. 758–763.

Bowen, N. L., 1927, The origin of ultra-basic and related rocks: Am. Jour. Sci., v. 14, p. 89–108.

—— 1928, The evolution of the igneous rocks: Princeton, N. J., Princeton Univ. Press, 334 p.

Bowen, N. L., and Anderson, O., 1914, The binary system MgO-SiO_2: Am. Jour. Sci., v. 37, p. 487–500.

Bowen, N. L., and Schairer, J. F., 1935, The system MgO-FeO-SiO_2: Am. Jour. Sci., v. 29, p. 151–217.

Bowen, N. L., and Tuttle, O. F., 1949, The system MgO-SiO_2-H_2O: Geol. Soc. America Bull., v. 60, p. 439–460.

Clarke, D. B., 1970, Tertiary basalts of Baffin Bay: possible primary magma from the mantle: Contr. Mineralogy and Petrology v. 25, p. 203–224.

Davis, B.T.C., and Schairer, J. F., 1965, Melting relations in the join diopside-forsterite-pyrope at 40 kilobars and at one atmosphere: Carnegie Inst. Washington Year Book 64, p. 123–134.

Dickey, J. S. Jr., 1970, Partial fusion products in alpine-type peridotites: Serranía de la Ronda and other examples, *in* Morgan, B. A., ed., Fiftieth anniversary symposia: mineralogy and petrology of the upper mantle, sulfides, mineralogy and geochemistry of non-marine evaporites: Mineralog. Soc. America Spec. Paper 3, p. 33–49.

Engel, A.E.J., Engel, C. G., and Havens, R. G., 1965, Chemical characteristics of oceanic basalts and the upper mantle: Geol. Soc. America Bull., v. 76, p. 719–734.

Green, D. H., 1963, Alumina content of enstatite in a Venezuelan high-temperature peridotite: Geol. Soc. America Bull., v. 74, p. 1397–1402.

—— 1964, The petrogenesis of the high-temperature peridotite intrusion in the Lizard area, Cornwall: Jour. Petrology, v. 5, p. 134–188.

Hess, H. H., 1938, A primary peridotite magma: Am. Jour. Sci., v. 35, p. 321–344.

—— 1955, Serpentines, orogeny and epeirogeny, *in* Poldervaart, Arie, ed., Crust of the earth—a symposium: Geol. Soc. America Spec. Paper 62, p. 391–408.

—— 1966, Caribbean research project, 1965, and bathymetric chart, *in* Hess, H. H., ed., Caribbean geological investigations: Geol. Soc. America Mem. 98, p. 1–10.

Ito, K., and Kennedy, G. C., 1967, Melting and phase relations in a natural peridotite to 40 kilobars: Am. Jour. Sci., v. 265, p. 519–538.

Irvine, T. N., 1963, Origin of the ultramafic complex at Duke Island, southeastern Alaska, *in* Fisher, D. J., Frueh, A. J., Jr., Hurlbut, C. S., Jr., and Tilley, C. E., eds., Int. Min. Assoc. Papers and Proceedings, Third General Meeting: Mineralog. Soc. America Spec. Paper 1, p. 36–45.

Kushiro, I., 1969, The system forsterite-diopside-silica with and without water at high pressures: Am. Jour. Sci., v. 267–A, p. 269–294.

Manson, V., 1967, Geochemistry of basaltic rocks; major elements, *in* Hess, H. H., and Poldervaart, Arie, eds., Basalts: the Poldervaart treatise on rocks of basaltic composition, Vol. 1; New York, John Wiley & Sons, p. 215–269.

Mason, B., 1962, Meteorites: New York, John Wiley & Sons, 274 p.

Pretorius, D. A., 1964, The geology of the central Rand goldfield, *in* Haughton, S. H., ed., The geology of some ore deposits of southern Africa, Vol. 1: Johannesburg, Geol. Soc. South Africa, p. 275–311.

Simkin, T., and Smith, J. V., 1970, Minor-element distribution in olivine: Jour. Geology, v. 78, p. 304–325.

Van Niekerk, C. B., and Burger, A. J., 1969, A note on the minimum age of the acid lava of the Onverwacht Series of the Swaziland System: Geol. Soc. South Africa Trans., v. 72, p. 9–21.

Viljoen, M. J., and Viljoen, R. P., 1970a, Archaean vulcanicity and continental evolution in the Barberton region—Transvaal, *in* Clifford, T. N., and Gass, I., eds., African magmatism and tectonics: Kennedy Volume: Edinburgh, Oliver & Boyd, p. 27–49.

—— 1970b, An introduction to the geology of the Barberton granite—greenstone terrain: Geol. Soc. South Africa Spec. Pub. 2, p. 9–28.

—— 1970c, The geology and geochemistry of the lower ultramafic unit of the Onverwacht Group and a proposed new class of igneous rocks: Geol. Soc. South Africa Spec. Pub. 2, p. 55–85.

—— 1970d, Evidence for the existence of a mobile extrusive peridotitic magma from the Komati Formation of the Onverwacht Group: Geol. Soc. South Africa Spec. Pub. 2, p. 87–112.

MANUSCRIPT RECEIVED BY THE SOCIETY MARCH 29, 1971

PRESENT ADDRESS: DEPARTMENT OF EARTH AND PLANETARY SCIENCES, MASSACHUSETTS INSTITUTE OF TECHNOLOGY, CAMBRIDGE, MASSACHUSETTS 02139

The Geological Society of America, Inc.
MEMOIR 132
© 1972

Peculiar Inclusions in Serpentine from Hokkaido, Japan

EDWARD SAMPSON
Princeton University, Princeton, New Jersey 08540

ABSTRACT

At the Nozawa asbestos mine of Hokkaido, Japan, rounded inclusions of perfectly fresh harzburgite occur in a nearly completely serpentinized matrix. Their origin is uncertain but the writer favors inclusion of fresh harzburgite brought into place within a largely serpentinized harzburgite.

INCLUSIONS

The Nozawa asbestos mine, the largest in Japan at the time of my visit (1945), lies in the remarkable belt of ultramafic intrusions that runs north-south the entire length of Hokkaido. The mine is about 3½ km northeast of the village of Yamake in Yamake-mura, Sorachi-gun, and covers an area of 250 by 300 m.

A peculiar feature of the rock is the occurrence of numerous rounded harzburgite inclusions (Figs. 1, 2). The outer surfaces of the inclusions are smooth, rounded, and very sharply bounded. They consist of a relatively coarse-grained enstatite in a matrix of much finer-grained olivine, and they contain scattered grains of chromite. The small size of the olivine may be due to crushing. The enstatite in the specimen was determined by H. H. Hess to have a $2V = 79°$, indicating a composition of $En_{92\ 1/2}$. The material is completely unserpentinized in spite of the almost complete serpentinization of the enclosing rock.

A similar occurrence of fresh inclusions in a highly serpentinized peridotite has also been observed by H. Kuno (oral commun.) at a locality just south of the town of Kamogawa in Chiba Prefecture. In a mass of strongly serpentinized peridotite, Kuno (1941) found a single inclusion, about 10 cm in diameter. This was bounded against the serpentine by a smooth surface. The inclusion consists of idiomorphic

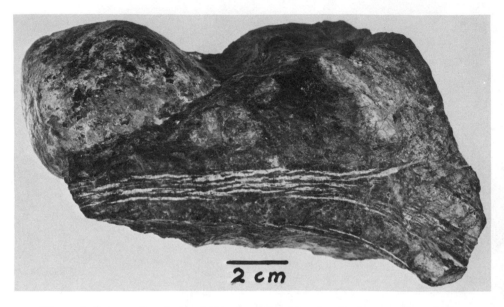

Figure 1. Outer surface (upper left) of nodule of harzburgite in serpentine which contains some veinlets of chrysotile.

Figure 2. Sawed section of same specimen as Figure 1.

enstatite (1 to 1.5 mm size) and a little interstitial olivine. The enstatite[1] has $2V = 68°$, indicating En_{95}, and the olivine has $\beta = 1.665$ corresponding to Fo_{94}. Significantly, this enstatite is of very nearly the same composition as that in the Nozawa inclusion. The remarkable manner of occurrence of the inclusions at the two localities, together with their very similar composition, implies a common origin.

The Nozawa inclusion is set in a matrix of antigorite with a few residual grains of olivine and enstatite. The antigorite contains veins of chrysotile asbestos and has a very minor amount of late brucite. It also contains accessory chromite of the same size and distribution as that of the harzburgite.

There is no conclusive evidence for the mode of emplacement of the inclusions. An origin by serpentinization in place is the simplest explanation, although the completely unaltered nature of the specimens seems somewhat improbable. An alternative manner of origin is that the inclusions were carried into place by a fully hydrated serpentine. In favor of this alternative is the occurrence described by Kuno in which only a single specimen was found. That only one inclusion, and that unaltered, should remain from the whole mass of rock seems improbable. I am inclined to favor an origin by which the inclusions were carried into place by a hydrated serpentine.

I know of no other such occurrence.

I am indebted to Professor R. B. Hargraves for assistance in mineral determinations.

MANUSCRIPT RECEIVED BY THE SOCIETY MARCH 29, 1971

---

[1] A chemical analysis of the enstatite is given by Kuno in (1941, Dispersion of the optic axes in the orthorhombic pyroxene series: Proc. Imperial Acad., Tokyo, v. 17, p. 204–209).

PRINTED IN U.S.A.

THE GEOLOGICAL SOCIETY OF AMERICA, INC.
MEMOIR 132
© 1972

Review of Caribbean Serpentinites and Their Tectonic Implications

GABRIEL DENGO

*Instituto Centroamericano de Investigación
y Tecnología Industrial (ICAITI),
Guatemala City, Guatemala, C.A.*

ABSTRACT

Serpentinites occur on three sides of the Caribbean tectonic plate, essentially along the boundary zones. The different geological environments in which they occur result from the crustal composition of the bounding plates and their interfaces. The masses that occur between plates of continental crust (northern Central America) are related to major rift zones. Those that occur in areas of oceanic crust (Greater Antilles and southern Central America) or at the zone of oceanic and continental boundaries (northwestern South America) are associated with volcanic and sedimentary rocks of the ophiolite suite. Others (northern Colombia and Venezuela) occur in association with metasedimentary rocks along the thin edge of a continental plate.

The serpentinites were emplaced mainly during Cretaceous time around the Caribbean plate. Their distribution and modes and ages of emplacement are compatible with modern concepts of plate tectonics.

INTRODUCTION

Much of our present knowledge and concepts on the nature and emplacement of serpentinite and its relation to Caribbean tectonics evolved as a result of Hess's (1938) pioneering work and interest in the Caribbean region. When the Princeton Caribbean Research Project was conceived and initiated by Hess (1966), one of the main aims was to understand the formation of island arcs and to study the highly deformed areas around the Caribbean. Consequently, the study of the tectonic relations and mechanisms of emplacement of the serpentinites was a fundamental aspect of the project.

This review demonstrates that the circum-Caribbean serpentinites occur essentially along the boundary zones of the Caribbean crustal plate. The different geological environments in which they occur are the result of the crustal composition of the bounding plates and their interfaces. This concept varies from, but also complements, the original concept presented by Hess (1938), that the serpentinites occur only in and immediately adjacent to the geotectoclinal zones and are commonly distributed in two parallel belts along the margins of the zone. The present paper, and perhaps even more its bibliography, stands as a testimonial to the vision and imagination of Hess, which have served so much to stimulate subsequent work in the Caribbean region. Other contributions by Hess on this subject are listed in the references.

I have discussed recently (Dengo, 1969b) some of the major tectonic problems dealing with the relations of Central America and the Caribbean, and pointed out that the serpentinites are distributed around the Central Caribbean tectonic plate, which has been a stable crustal block at least since middle Mesozoic time. The purpose of this article is to review the geographic distribution, tectonic relations, and age of emplacement of the serpentinites in the entire region in order to understand better their significance in Caribbean tectonics.

It is difficult to trace ideas on Caribbean tectonics to the original authors without making a thorough review of an extensive literature in several languages, published in many scientific journals. To avoid burdensome reading, only recent and more generalized references are cited. For more details on specific areas, I have included an extensive bibliography, which is nearly complete. The following recent maps will also be of use in localizing serpentinite areas: Tectonic Map of North America (King, 1969), Mapa Geológico-Tectónico del Norte de Venezuela (Smith, 1962), Mapa tectónico de Cuba (Zhiv, 1966), and Metallogenic Map of Central America (Dengo and others, 1969).

REVIEW OF SERPENTINITE OCCURRENCES IN THE CARIBBEAN

Northern Central America

Serpentinites in northern Central America occur in a wide zone, in several subparallel belts, from Chiapas, Mexico, to the Caribbean coast, a distance of more than 600 km (Fig. 1).

Several authors (McBirney, 1963; Bonis, 1967) have pointed out that the serpentinites occur along or near two major fault zones, namely, the Motagua and the Polochic. These faults seem to be the western extension of the Cayman Trough in the Caribbean. McBirney has suggested that the serpentinites represent materials derived from the upper mantle that have been mobilized along major faults. Bonis (1967) has found evidence that these rocks did not crop out until Maastrichtian time. Taking into consideration the complex tectonic history of the central ranges of Guatemala where the serpentinites occur, Dengo and Bohnenberger (1969) concluded that the history and mechanism of emplacement have been complex, and that emplacement started in Paleozoic and culminated in Late Cretaceous times. These authors also consider that the tectonic history of the fault zones and the emplacement of the serpentinites are closely related.

It is of interest to note that the serpentinites of northern Central America are not typically associated with the ophiolite suite of rocks. The only rocks of this type that have been found in the area are in the El Tambor Formation (McBirney and Bass, 1969). Others, which recently were found by Bonis (1969 oral commun.) in the Sierra de Santa Cruz, have not yet been studied.

The serpentinites in northern Central America occur in an area of continental crust. It has been indicated recently by the author (Dengo, 1969b), that the

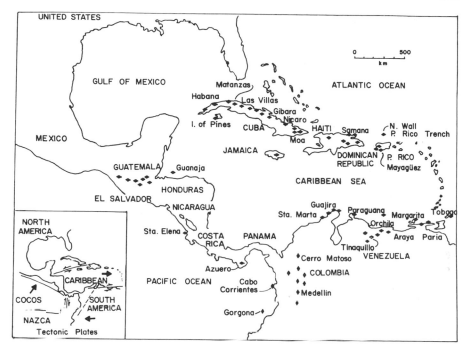

Figure 1. Serpentinite occurrence in Central America and on the Caribbean coast.

Motagua fault essentially separates two areas of different basement. These two blocks may actually represent parts of different crustal plates, in which case the zone of serpentinites is essentially located along the interface between these plates.

Greater Antilles

The basement of the Greater Antilles is considered to be largely, if not entirely, oceanic crust materials (Donnelly, 1964; Mattson, 1968; Arden, 1969). The serpentinite bodies that have not been subjected to post-emplacement deformations are associated with this basement, as part of it, or probably as domical uplifts of the upper mantle. Therefore, the geological environment is different from that of northern Central America.

Serpentinites are found throughout the Greater Antilles. In Cuba they have been studied extensively. Selected references to these works and new information are found in recent publications by Wassall (1956), Ducloz and Vaugnat (1962), Khudoley (1967), Hatten (1967), Kozary (1968), and Khudoley and Meyerhoff (1971). Kozary (1968, p. 2299) states that the serpentinites "extend 900 km along the entire length of the island. They are found in bands in pre-late-Eocene inliers and are restricted to the north central part of the island. Within any specified zone the largest bodies are along the southern edge of the ultramafic zone and the smallest and narrowest units are along the northern margins." Cuba presents the largest area of serpentinite outcrops and shallow-depth subcrops in the Caribbean region, estimated at 15,000 km^2 by Khudoley and Meyerhoff (1971). The tectonic significance of this fact thus far is not understood.

Apparently there are several types of tectonic relations between serpentinites and the enclosing rocks, which indicate different mechanisms of emplacement. There are two major types of relations that also correspond to the geographic distribution. According to Guild (1947), the northern belt has been extensively affected by thrusting, whereas the southern one is characterized by domical uplifts. Recent interpretations of the tectonics of Cuba, such as those by Pardo (1965), Kozary (1968), and Knipper and Puig (1967), consider that the intense thrusting in northern Cuba is due mostly to gravity sliding of allochthonous blocks.

The age of emplacement of the Cuban serpentinites, as in other serpentinite areas, has been the subject of much discussion. However, the present trend is to consider them either as part of the Mesozoic basement of the island (Mattson, 1968; Khudoley and Meyerhoff, 1971), in which case they could represent domical uplifts of the mantle, or as having been emplaced originally as igneous intrusive bodies, probably during Late Jurassic to Neocomian, and affected by intense deformation during the Eocene. Those of the Island of Pines, however, may be older, even Paleozoic (Khudoley and Meyerhoff, 1971).

In the Central part of the Dominican Republic serpentinized peridotite occurs along a northwest-trending zone that separates two formations of metavolcanic rocks. Its emplacement is considered by Bowin (1966) to be pre-middle Albian, and was controlled by a pre-existing fault zone. Post-Oligocene tectonic movements remobilized part of the mass along another fault zone. Weyl (1966, Fig. 37) has indicated small serpentinite outcrops in the Sierra del Seibo, also associated with metavolcanic rocks.

In the north coast of the Dominican Republic three serpentinite bodies occur along a belt 170 km long and trending northwest (written commun. from Frederick Nagle to W. D. MacDonald). They are associated with metavolcanic and metasedimentary rocks of unknown age.

Little is known of the occurrence of serpentinites in Haiti. Pebbles and outcrops of these rocks have been mentioned by Woodring and others (1924) and by Butterlin (1960). These localities are in the Massif du Nord.

In Puerto Rico, Mattson (1960, 1964) has studied three serpentinized peridotite bodies in the southwestern part of the island. They occur in cores of anticlines and are associated with chert and hornblende-feldspar gneiss of the Bermeja complex of pre-Campanian age. These occurrences were interpreted by Hess (1964) as part of the oceanic crust or perhaps the mantle. They have been studied petrographically in detail as part of the AMSOC core hole project (Burk, 1964).

North of Puerto Rico, serpentinites have been dredged along the north wall of the Puerto Rico Trench, together with sedimentary rocks of Cenomanian and Eocene ages (Bowin and others, 1966). These rocks are exposed in a submarine fault scarp and may represent the upper part of the oceanic crust.

In Jamaica, two serpentinite bodies associated with low-grade metamorphic rocks have been described by Matley (1951), Matley and Raw (1951), and Zans (1961). Mitchell (1955) concluded that they were emplaced between Senonian and early Tertiary times. The metamorphic rocks are considered by Zans and others (1963) to be of Cretaceous and Tertiary ages and to have been deformed during Late Cretaceous to Eocene times.

Northern South America

In northern South America serpentinites occur in a nearly east-trending belt for 1,300 km, along the southern margin of the Caribbean (Fig. 1). This belt consists of two parts which are not continuous: one that extends from the island of Tobago (Maxwell, 1948) to near the city of Barquisimeto in Venezuela (Bellizzia, 1967), and another from the Paraguaná Península in Venezuela to near the city of Santa Marta in Colombia.

In Margarita Island, Hess and Maxwell (1949) described pre-mid-Eocene, probably Cretaceous, serpentinized peridotites that intrude metasedimentary rocks. In the Paria-Araya penínsulas, serpentinite bodies have been studied by Christensen (1961), González de Juana and Muñoz (1968), Schubert (1971), and Seijas (1972). Carlos Schubert (written commun., 1970) concludes that most of the masses seem to be tectonic intrusions emplaced along faults or foliation planes of low-grade metasedimentary rocks. González de Juana and Muñoz (1968) state that they were emplaced during post-Albian to pre-Paleocene times.

An excellent summary of the serpentinite bodies in the western Caribbean Mountains of Venezuela has been presented by Bellizzia (1967). In this area the serpentinites are found as sills within metasedimentary rocks (Dengo, 1953), as lenticular tectonic intrusions along faults and foliation planes (Smith, 1953; Shagam, 1960; Bellizzia, 1967), as high-temperature intrusives such as those at Tinaquillo (MacKenzie, 1960), and as exotic masses associated with large allochthonous blocks of metavolcanic rocks (Menéndez, 1966; Bell, 1972).

The Paraguaná–Santa Marta group occurs in the Ruma metamorphic zone of Mesozoic, probably Cretaceous, metasedimentary rocks (MacDonald, 1964; Lockwood, 1965; Alvarez, 1967).

In the Colombian Andes there are serpentinites in the Central and Western Cordilleras and in the Pacific Coast Ranges. A review of the literature (Olsson, 1956; Wokittel, 1961; Botero Arango, 1963; Radelli, 1962, 1967; Nelson, 1962), shows that there are at least two major belts, and possibly three, although only one is well defined. These belts follow the structural trend of the Andes. According to Earl Irving (1970, written commun.) they occur in metasedimentary rocks of possible Paleozoic age in the Central Cordillera, and in folded Cretaceous sediments and volcanics in the Western Cordillera, and in the Coast Ranges. It should be pointed out, however, that some authors (Radelli, 1967; Butterlin, 1972), have indicated the possibility that the metasediments of the Central Cordillera may be of Mesozoic age. Irving believes that most of these serpentinites are of late Mesozoic age. Some cut dated Cretaceous rocks, but many cut metamorphic rocks and can be dated only as post-metamorphic. Most seem to be low-temperature tectonic intrusions, but none has been found associated with a major rift structure. In Irving's opinion, the serpentinites are part of a late Mesozoic eugeosyncline that extended along western Colombia and which probably curved northeastward to pass through the northwest corner of the Sierra Nevada de Santa Marta and thence continued through the Guajira Península. These rocks were folded during the Maastrichtian epoch.

Southern Central America

When the Santa Elena Península peridotite in Costa Rica was first described by Harrison (1953) it appeared to be an isolated occurrence not related to a major serpentinite belt. Consequently, it was interpreted as lying along the extension of the Clipperton fracture zone of the Pacific Ocean; it is aligned with it and it is bound on the north by a large east-trending fault (Dengo, 1962a, 1962b).

A later discovery of ultramafic rocks in the Azuero Península of Panama (Ferencic and others, 1968; del Giudice and Recchi, 1969), suggests that both occurrences may be part of a belt on the Pacific side of southern Central America. I have proposed (Dengo, 1967) that the basic igneous and associated sedimentary rocks of the Nicoya complex, on the Pacific side of southern Central America, are an extension of the Mesozoic basic volcanic rocks of western Colombia. If so, the Azuero and Santa Elena ultramafic masses, which are associated with the Nicoya complex, are part of the serpentinite belt that crops out along the Pacific coast of northwestern South America. The Santa Elena mass has been interpreted as a mantle outcrop brought up by faulting (Dengo, 1962a).

Both of these occurrences probably are of Cretaceous age, and certainly are older than Campanian, as indicated by their relations to dated sedimentary rocks (Dengo, 1962a; del Giudice and Recchi, 1969).

TECTONIC IMPLICATIONS

In a discussion of the Central American serpentinites, I have indicated elsewhere (Dengo, 1969a) that the serpentinites were probably emplaced around the Caribbean plate as one single circum-Caribbean zone which tectonic trends have separated into three parts: (1) northern Central America–Greater Antilles, (2) east-trending belt of Venezuela and Colombia, and (3) southern Central America. The term "circum-Caribbean serpentinite zone" already had been implied and used by Hess (1938) and by Hess and Maxwell (1949), but I extended it to include the ultramafic rocks of southern Central America, not known to Hess and Maxwell at that time. Hess considered that along the Lesser Antilles island arc the zone of intense deformation was younger and indicated by a gravity negative anomaly strip.

The relation of the serpentinites to major fault zones has been discussed elsewhere (Dengo and Bohnenberger, 1969), indicating that in some places, such as northern Central America, they are related in their history, even if the serpentinites do not everywhere occur along the faults.

This review permits further elaboration and the conclusion that the circum-Caribbean serpentinites occur essentially along the boundary zones of the Caribbean crustal plate. In some areas, particularly Cuba and Venezuela, later deformations (Maastrichtian to Eocene) have incorporated some of the serpentinites in allochthonous masses that moved away from the Caribbean plate, either by gravity sliding or by thrusting (Pardo, 1965; Bellizzia, 1967; Menéndez, 1966; Kozary, 1968).

Similarly, the serpentinite belts in the Colombian Andes probably indicate several stages of deformation along the boundary zone between the Nazca plate and the South American plate (Fig. 1, insert). If this is the case, the belt along the Cordillera Central must be older, perhaps even late Paleozoic, and the others are younger, the youngest ones of which are those of the Pacific Coast Ranges. Case and others (1968) have interpreted the Western Cordillera and Coast Ranges of Colombia as resulting from overriding toward the west of the continental mass of the Cordillera Central.

The review presented here indicates that most of the serpentinites in the Caribbean region are Cretaceous, although they are not exactly coeval. Some are without question related to the Laramide orogeny and were emplaced, or exposed to the surface, during Maastrichtian to Eocene times. Others were emplaced in earlier deformations during the Cretaceous Period. Only a few may be older than Cretaceous.

Only those masses between plates of continental crust are related to major rift zones, such as in northern Central America. Those associated with typical ophiolites are found either in zones of oceanic crust as in the islands of Hispañola, Puerto Rico, Jamaica, and southern Central America, or at the interfaces between oceanic and continental plates, such as those of the Pacific belt of northwestern South America and probably those of Cuba. The belt of northern South America is associated with metamorphosed sedimentary rocks of the Ruma metamorphic belt and the Caribbean Mountain system, along the thin edge of the continental crust. These different tectonic associations may account to a large extent for the various mechanisms and temperatures of emplacement of the serpentinites. The domal uplifts seem to be related to areas of oceanic crust where the mantle is closer to the surface.

There are no serpentinites in the Lesser Antilles island arc, which is the eastern edge of the Caribbean plate. According to Christman (1953), the tectonic activity in this arc, characterized by volcanism, may be as old as Eocene. The possibility should be considered, however, that the older history of the arc is not yet known for lack of more evidence. According to T. W. Donnelly (1970, oral commun.), the volcanic activity of the Lesser Antilles may be as old as Jurassic, since an age of 142 m.y. was obtained for the plutonic rocks of Desirade by Fink. Donnelly also thinks that the volcanic rocks from the Lesser Antilles are already quite evolved chemically—about where they should be if they were preceded by a Cretaceous igneous history like that of Puerto Rico. It is not possible, therefore, to know to what extent this part of the Caribbean plate was affected by the strong tectonic deformations of Late Cretaceous early Eocene times and if any serpentinites were emplaced there and are now buried by younger rocks.

Other authors have also proposed that a stable Caribbean crustal plate has localized the strongly deformed mobile belts around it (Ewing and others, 1967). MacGillavry (1970) and F. Nagle (1970, written commun. to W. D. MacDonald) also consider that the serpentinites were emplaced around the external part to the Caribbean plate during the Cretaceous Period.

The interpretation presented in this article is in agreement with modern ideas of plate tectonics (Molnar and Sykes, 1969), as well as with some of the earlier concepts of Meyerhoff (1954), Bucher (1952), and Hess (Hess and Maxwell, 1953; Hess, 1955a, 1955b). According to these ideas, the Caribbean plate is moving eastward in relation to North and South America, while the oceanic crust of the Pacific is underriding Central and South America. From the evidence presented here, the underriding of the Pacific crust beneath the western part of the Caribbean plate (southern Central America) probably dates at least from the Cretaceous and was related also to the deformations along the north and south boundaries of the plate. The eastward shift of the plate, as evidenced by the Lesser Antilles, is probably younger.

The areal distribution of the Caribbean serpentinites, together with their modes and ages of emplacement, tend to outline the structural history of the region and reinforce the concepts of Caribbean plate tectonics.

ACKNOWLEDGMENTS

Earl M. Irving, of the U.S. Geological Survey group in Colombia, Carlos Schubert of the Instituto Venezolano de Investigaciones Científicas, William D. MacDonald and Thomas W. Donnelly of the State University of New York at Binghamton, and Arthur A. Meyerhoff of the American Association of Petroleum Geologists provided new data and references. I wish to acknowledge their collaboration and to specially thank Samuel S. Bonis, Instituto Geográfico de Guatemala, for his critical review and help in editing this paper.

REFERENCES CITED

Alvarez, W., 1967, Geology of the Simarria and Carpindero areas, Guajira Peninsula [Ph.D. thesis]: Princeton, Princeton University, 168 p.

Arden, Daniel D., 1969, Geologic history of the Nicaraguan rise: Gulf Coast Assoc. Geol. Socs. Trans., v. XIX, p. 294–309.

Bell, J. S., 1972, El significado de la faja piemontina de la Cordillera de la Costa de Venezuela: Fourth Venezuelan Geol. Cong. Proc., Caracas, Ministerio de Minas e Hidrocarburos.

Bellizzia, A., 1967, Rocas ultramáficas en el sistema montañoso del Caribe y yacimientos minerales asociados: Caracas, Bol. Geología, v. VIII, no. 16, p. 159–214.

Bonis, S., 1967, Age of Guatemalan serpentinite: Geol. Soc. America, Abs. for 1967, Spec. Paper 115, p. 18.

Botero Arango, Gerardo, 1963, Contribución al conocimiento de la geología de Antioquia: Anales, Facultad de Minas, Medellín, no. 57, 101 p.

Bowin, Carl O, 1966, Geology of central Dominican Republic: Geol. Soc. America Mem. 98, p. 11–84.

Bowin, C. O., Nalwalk, A. J., and Hershey, J. B., 1966, Serpentinized peridotite from the north wall of the Puerto Rico trench: Geol. Soc. America Bull., v. 77, p. 257–270.

Bucher, W. H., 1952, Geologic structure and orogenic history of Venezuela: Geol. Soc. America Mem. 49, 113 p.

Burk, C. A., 1964, The AMSOC core hole in serpentinite near Mayagüez, Puerto Rico, in A study of serpentinite: Natl. Acad. Sci.–Natl. Research Council Pub. 1188, p. 1-6.

Butterlin, Jacques, 1956, La constitution géologique et la structure des Antilles: Paris, Centre Natl. de la Recherche Scientifique, 453 p.

—— 1960, Géologie générale et régionale de la République D'Haiti: Paris, Institut des Hautes Etudes de l'Amerique Latine, 194 p.

—— 1972, La position structurale des Andes de Colombie: Fourth Venezuelan Geological Congress Proc., Caracas, Ministerio de Minas e Hidrocarburos.

Case, J. E., Duran, L. G., Lopez, R. A., and Moore, W. R., 1968, Gravity anomalies and crustal structure, western Colombia: Geol. Soc. America, Abs. for 1968, Spec. Paper 121, p. 48–49.

Christensen, R. M., 1961, Geology of the Paria-Araya Peninsula, northeastern Venezuela [Ph. D. thesis]: Lincoln, Univ. Nebraska, 112 p.

Christman, Robert A., 1953, Geology of St. Bartholomew, St. Martin, and Anguilla, Lesser Antilles: Geol. Soc. America Bull., v. 64, p. 65–96.

del Giudice, D., and Recchi, G., 1969, Geología del área del proyecto minero de Azuero: Panamá: Adm. Recursos Minerales, 48 p.

Dengo, Gabriel, 1953, Geology of the Caracas region, Venezuela: Geol. Soc. America Bull., v. 64, p. 7–40.

—— 1962a, Estudio geológico de la región de Guanacaste, Costa Rica: San José, Costa Rica Inst. Geog. Nac. Informe Semestral, 112 p.

—— 1962b, Tectonic-igneous sequence in Costa Rica, in Engel, A.E.J., James, H. L., and Leonard, B. F., eds., Petrologic studies: A volume in honor of A. F. Buddington: Geol. Soc. America, p. 133–161.

—— 1967, Geological structure of Central America: Studies in tropical oceanography, No. 5: Miami, Univ. Miami, p. 56–73.

—— 1969a, Relación de las serpentinitas con la tectónica de América Central: Pan-American Symposium on the upper mantle, Mexico, p. 23–28.

—— 1969b, Problems of tectonic relations between Central America and the Caribbean: Gulf Coast Assoc. Geol. Socs. Trans., v. XIX, p. 311–320.

Dengo, Gabriel, and Bohnenberger, Otto, 1969, Structural development of northern Central America: Am. Assoc. Petroleum Geologists, Mem. 11, p. 203–220.

Dengo, G., Levy, E., Bohnenberger, O., and Caballeros, R., 1969, Metallogenic map of Central America, Guatemala, Inst. Centro-Americano Invest. Tec. Industr. (ICAITI), scale 1:2,000,000.

Donnelly, Thomas W., 1964, Evolution of eastern Greater Antilles island arc: Am. Assoc. Petroleum Geologists Bull., v. 48, No. 5, p. 680–696.

Ducloz, Ch., and Vaugnat, M., 1962, A propos de l'âge des serpentinites de Cuba: Génève Archives Sci., v. 15, fasc. 2, p. 309–332.

Ewing, J., Talwani, M., Ewing, M., and Edgar, T., 1967, Sediments of the Caribbean: Studies in tropical oceanography, No. 5: Miami, Univ. Miami, p. 88–102.

Ferencic, A., del Giudice, D., and Recchi, G., 1968, Tectomagmatic and metallogenic relationships of the region Central Panama–Costa Rica: Panama, Adm. Recursos Minerales (unpub. manuscript).

González de Juana, Clemente, and Muñoz, Nicolás, 1968, Rocas ultramáficas en la Península de Paria, Venezuela: Boletín Informativo, Asoc. Venezolana Geol., Minería y Petróleo, v. 11, no. 2, p. 28–43.

Guild, Philip W., 1947, Petrology and structure of the Moa chromite district, Oriente province, Cuba: Am. Geophys. Union Trans., v. 28, no. 2, p. 218–246.

Harrison, J. V., 1953, The geology of the Santa Elena Península in Costa Rica, Central America: Auckland-Christchurch, New Zealand, Seventh Pacific Sci. Congr. Proc., v. 2, p. 102–114.

Hatten, C. W., 1967, Principal features of Cuban geology: Discussion: Am. Assoc. Petroleum Geologists Bull., v. 51, no. 5, p. 780–789.

Hess, H. H., 1938, Gravity anomalies and island arc structure with particular reference to the West Indies: Am. Philos. Soc. Proc., v. 79, no. 1, p. 71–96.

—— 1955a Serpentines, orogeny and epeirogeny, in Poldervaart, A., Crust of the Earth: Geol. Soc. America Spec. Paper 62, p. 391–408.

—— 1955b, The oceanic crust: Sears Foundation Jour. Marine Research, v. 14, no. 4, p. 423–439.

—— 1960, Caribbean research project: Progress report: Geol. Soc. America Bull., v. 71, p. 235–240.

—— 1964, The oceanic crust, the upper mantle and the Mayagüez serpentinized peridotite, in A study of serpentinite: Natl. Acad. Sci.–Natl. Research Council Pub. 1188, p. 169–175.

—— 1966, Caribbean research project, 1965, and bathymetric chart: Geol. Soc. America Mem. 98, p. 1–10.

Hess, H. H., and Maxwell, J. C., 1949, Geological reconnaissance of the island of Margarita: Geol. Soc. America Bull., v. 60, p. 1857–1868.

—— 1953, Caribbean research project: Geol. Soc. America Bull., v. 64, p. 1–6.

Khudoley, K. M., 1967, Principal features of Cuban geology: Am. Assoc. Petroleum Geologists Bull., v. 51, no. 5, p. 668–677.

Khudoley, K. M., and Meyerhoff, A. A., 1971, Paleogeography and geological history of the Greater Antilles: Geol. Soc. America Mem. 129, 197 p.

King, P. B., 1969, Tectonic map of North America: U.S. Geol. Survey, Washington, D.C.

Knipper, A. L., and Puig, M., 1967, Protrusiones de las serpentinitas en el noroeste de Oriente: Havana, Acad. Ciencias Cuba, Revista de Geologiá, Año 1, no. 1, p. 122–137.

Kozary, Myron T., 1968, Ultramafic rocks in thrust zones of northwestern Oriente province, Cuba: Am. Assoc. Petroleum Geologists Bull., v. 52, no. 12, p. 2298–2317.

Lockwood, J. P., 1965, Geology of the Serranía de Jarara, Guajira Peninsula, Colombia [Ph. D. thesis]: Princeton, Princeton University, 237 p.

MacDonald, W. P., 1964, Geology of the Serranía de Macarena area, Guajira Peninsula, Colombia [Ph. D. thesis]: Princeton, Princeton University, 162 p.

MacGillavry, H. J., 1970, Geological history of the Caribbean: Amsterdam, Koninkl. Nederlandse Akad. Wetensch. Proc., ser. B, 73, no. 1, p. 64–96.

MacKenzie, D. B., 1960, High-temperature alpine-type peridotite from Venezuela: Geol. Soc. America Bull., v. 71, p. 303–318.

Matley, C. A., 1951, Geology and physiography of the Kingston district, Jamaica: Kingston, Jamaica, Inst. Jamaica, p. 1–75.

Matley, C. A., and Raw, F., 1951, Petrology of the Kingston district: Kingston, Jamaica, Inst. Jamaica, p. 92–102.

Mattson, Peter H., 1960, Geology of the Mayagüez area, Puerto Rico: Geol. Soc. America Bull., v. 71, p. 319–362.

—— 1964, Petrography and structure of serpentinite from Mayagüez, Puerto Rico, in A study of serpentinite: Natl. Acad. Sci.–Natl. Research Council Pub. 1188, p. 7–24.

—— 1968, Basement rocks, vulcanism and deformation in the Greater Antilles: Geol. Soc. America, Abs. for 1968, Spec. Paper 121, p. 191–192.

Maxwell, John C., 1948, Geology of Tobago, British West Indies: Geol. Soc. America Bull., v. 59, p. 801–854.

McBirney, A. R., 1963, Geology of a part of the Central Guatemalan Cordillera: Univ. California Pubs. Geol. Sci., v. 38, p. 177–242.

McBirney, Alexander R., and Bass, Manuel N., 1969, Structural relations of pre-Mesozoic rocks of northern Central America: Am. Assoc. Petroleum Geologists Mem. 11, p. 269–280.

Menéndez, Alfredo, 1966, Tectónica de la parte central de las montañas occidentales del Caribe, Venezuela: Caracas, Bol. de Geología, v. VIII, p. 116–139.

Meyerhoff, H. A., 1954, Antillean tectonics: New York Acad. Sci. Trans., ser. 2, v. 16, no. 3, p. 149–155.

Mitchell, R. C., 1955, The ages of serpentinized peridotites of the West Indies: Koninkl. Nederlandse Akad. Wetensch. Proc., ser. B., no. 3, p. 194–204, 205–212.

Molnar, Peter, and Sykes, Lynn R., 1969, Tectonics of the Caribbean and Middle America regions from focal mechanisms and seismicity: Geol. Soc. America Bull., v. 80, p. 1639–1684.

Nelson, H. W., 1962, Observaciones geológicas y estudio petrográfico sobre 54 muestras colectadas por el doctor Laureano Rincón en el departamento de Nariño en la vía Tambo, Peñol, Policarpa: Bogotá, Bol. de Geología, v. IX, nos. 1–3, p. 79–96.

Olsson, A. A., 1956, Colombia: Handbook of South American geology, Geol. Soc. America Mem. 65, p. 295–326.

Pardo, G., 1965, Stratigraphy and structure of central Cuba [abs.]: Miami, Internat. Conf. Tropical Oceanography (mimeographed), 2 p.

Radelli, Luigi, 1962, Epocas magmáticas y metalogenéticas en los Andes colombianos: Bogotá, Bol. de Geología, v. IX, nos. 1–3, p. 5–22.

—— 1967, Géologie des Andes colombiennes: Grenoble Univ. Fac. Sci. Lab. Géologie Travaux, 457 p.

Schubert, Carlos, 1971, Metamorphic rocks of the Araya Peninsula, eastern Venezuela: Geol. Rundschau, v. 60, no. 4, p. 1571–1600.

Seijas, F. J., 1972, Geología de la región de Carúpano, Estado Sucre, Venezuela: Fourth Venezuelan Geol. Congress Proc., Caracas Ministerio de Minas e Hidrocarburos.

Shagam, Reginald, 1960, Geology of central Aragua, Venezuela: Geol. Soc. America Bull., v. 71, p. 249–302.

Smith, F. D., 1962, Mapa geológico-tectónico del norte de Venezuela: Caracas, Primer Congreso Venezolano de Petróleo, scale 1:1,000,000.

Smith, Raymond J., 1953, Geology of the Los Teques–Cua region, Venezuela: Geol. Soc. America Bull., v. 64, p. 41–64.

Wassall, H., 1956, The relationship of oil and serpentine in Cuba: 20th Internat. Geol. Cong., Mexico, Sec. 3, Comptes Rendus, p. 67–77.

Weyl, R., 1966, Geologie der Antillen: Berlin, Gebrüder Borntraeger, 410 p.

Wokittel, Roberto, 1961, Geología económica del Chocó: Bogotá, Bol. de Geología, v. VII, nos. 1–3, p. 119–162.

Woodring, W. P., Brown, J. S., and Burbank, W. S., 1924, Geology of the Republic of Haiti: Port-au-Prince, Haiti, Dept. Travaux Publiques, 631 p.

Zans, V. A., 1961, Geology and mineral deposits of Jamaica: Kingston, Jamaica, Government Printer, 11 p.

Zans, V. A., Chubb, L. J., Versey, H. R., Williams, J. B., Robinson, E., and Cooke, D. L., 1963, Synopsis of the geology of Jamaica: Jamaica Geol. Survey Dept. Bull. 4, 72 p.

Zhiv, D. I., ed., 1966, Mapa tectónico de Cuba: Moscow, Academia de Ciencias de Cuba y Academia de Ciencias de la URSS, scale 1:1,250,000.

Manuscript Received by the Society March 29, 1971

PRINTED IN U.S.A.

THE GEOLOGICAL SOCIETY OF AMERICA, INC.
MEMOIR 132
© 1972

Zoned Ultramafic Complexes of the Alaskan Type: Feeder Pipes of Andesitic Volcanoes

C. G. MURRAY

Department of Geological and Geophysical Sciences,
Princeton University, Princeton, New Jersey 08540

ABSTRACT

It is proposed that zoned ultramafic complexes of the Alaskan type originate by fractional crystallization and flow differentiation of basaltic magma in the feeder pipes of volcanoes.

The parent magma of the complexes may be an olivine-rich tholeiitic basalt. Crystallization under hydrous conditions in the pressure range 2 to 11 kbars can produce the mineral assemblages of most zoned complexes. Segregation of crystals along the axis of the conduit by flow differentiation as the magma rises would produce the observed mineralogical and chemical zoning.

Separation of varying proportions of ultramafic materials from the parent magma could produce high-alumina tholeiitic basalt and andesite. Volcanic rocks of this type are spatially and temporally associated with ultramafic complexes in several areas.

INTRODUCTION

Concentrically zoned ultramafic complexes have been known for many years from the Ural Mountains of Russia (Wyssotsky, 1913; Duparc and Tikonowitch, 1920), southeastern Alaska and British Columbia (Buddington and Chapin, 1929; Kennedy and Walton, 1946; Walton, 1951; Aho, 1956; Ruckmick and Noble, 1959; Noble and Taylor, 1960; Irvine, 1963; Findlay, 1969), and may also be present in Japan (Onuki, 1963, 1965). Taylor (1967) and Taylor and Noble (1969) give admirable summaries of the geology and occurrence of these rocks, especially the Alaskan examples. Two zoned complexes of this type were recognized

313

by the writer during field work in northern Venezuela in 1969 and 1970. These are briefly described in the following section, since they display features typical of such intrusions.

In previous discussions of the origin of Alaskan zoned complexes, it has always been concluded that the ultramafic rocks crystallized from ultrabasic magmas and cannot be directly related genetically to associated gabbros (Ruckmick and Noble, 1959; Taylor and Noble, 1960; Irvine, 1963; Taylor and Noble, 1969). The main purpose of this paper is to propose an alternative origin for the zoned ultramafic bodies assuming a basaltic parent magma. This assumption is consistent with available experimental and field data, and suggests a simple genetic relation between the gabbroic and ultramafic rocks. It also leads to important conclusions regarding the origin of andesites.

ZONED ULTRAMAFIC COMPLEXES IN NORTHERN VENEZUELA

The larger of the two zoned ultramafic complexes in northern Venezuela occurs south of the Hacienda El Chacao about 10 km west of San Juan de los Morros in Guárico state (Fig. 1). The complex is an irregularly shaped body 8 km long and 6 km wide which intrudes the volcanic Villa de Cura Group of probable Lower Cretaceous age. It is divided into two parts by a narrow belt of contact metamorphosed rocks. In both parts of the intrusion, central cores of olivine pyroxenite consisting of about 75 percent clinopyroxene and 25 percent olivine with minor hornblende and opaque minerals are surrounded by hornblendite. The transition between these rock types is gradational through intermediate types such as pyroxenite, hornblende pyroxenite, and pyroxene hornblendite. Hornblende first becomes conspicuous as large poikilitic crystals enclosing and replacing clinopyroxene and olivine. The appearance of abundant hornblende is accompanied by the appearance of substantial quantities of magnetite, which forms a nearly constant proportion (about 15 percent) of all the hornblende-rich ultramafic rocks.

Wherever the contact with the enclosing Villa de Cura Group is exposed, either hornblendite or hornblende gabbro is the marginal rock type of the intrusion. Hornblende gabbro grades into feldspathic hornblendite with up to 80 percent

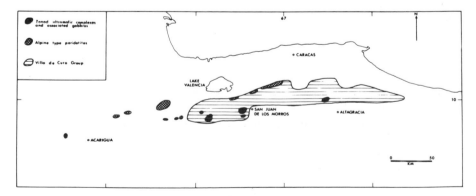

Figure 1. Map of north-central Venezuela showing the distribution of zoned ultramafic complexes and associated gabbros, alpine-type peridotites, and volcanic rocks of the Villa de Cura Group. The size of some ultramafic bodies is exaggerated.

hornblende, but no complete gradation to hornblendite was observed. Rather, the peripheral gabbro intrudes hornblendite and contains blocks of the latter. Dikes of hornblende gabbro cut the other rock types of the igneous complex and the surrounding metamorphic rocks. These relations suggest that the hornblende gabbro represents the last liquid fraction remaining from the magma which formed the zoned complex, and that the gabbro was still mobile after the solidification of the earlier pyroxenite and hornblendite.

The second zoned complex outcrops 25 km west of Acarigua in Portuguesa state (Fig. 1) and forms the conspicuous bare peak of Cerro Pelón. It is an elliptical body 2 km long and 1.5 km wide consisting of a dunite core (which makes up about half the area of the mass) surrounded by successive zones of pyroxenite and hornblende gabbro. Contacts between these rock types were not observed, but the transitions between them appear to be abrupt, since no intermediate rock types were found. The dunite is cut by veins and dikes of pyroxenite and by dikes of coarse grained hornblende gabbro. Gabbro dikes also cut pyroxenite. In contrast to the El Chacao intrusion, where hornblende-bearing rocks make up the bulk of the body, the hornblende gabbro at Cerro Pelón forms only a narrow outer rim to the complex.

Contact metamorphic rocks were found at the margins of the Cerro Pelon intrusion at a few localities. It intrudes low-grade regional metamorphic rocks characteristic of the Los Cristales Group of Lower Cretaceous age (Bellizzia and Rodriguez, 1968).

These two zoned ultramafic complexes occur in a belt of basic intrusions which stretches for at least 350 km along the southern edge of the Coast Ranges of Venezuela, from Altagracia in Guarico state to Acarigua (Fig. 1). This belt includes hornblende gabbros and hornblendites identical to those of the ultramafic complexes, and all the intrusions may be genetically related.

ORIGIN OF THE ZONED ULTRAMAFIC COMPLEXES

Introduction

Previous workers have argued that zoned complexes of the Alaskan type were produced from ultrabasic magmas, either (1) by a sequence of multiple intrusions produced by progressive fractional melting of the mantle (Ruckmick and Noble, 1959; Taylor and Noble, 1960, 1969); (2) by differentiation of an ultrabasic magma by vapor transfer (Walton, 1951); (3) by gravitational fractionation of a series of intrusions of ultrabasic magma of similar compositions (Irvine, 1963); or (4) by fractional crystallization of a single ultrabasic magma (Findlay, 1969).

An objection which can be raised against the last three hypotheses concerns the nature of the proposed parent magma, which is believed to be close to the average compositions of the Duke Island and Tulameen ultramafic complexes (Table 1, Nos. 1 and 2) or to the analysis of a hornblende pyroxenite from the Tulameen complex (Table 1, No. 3). Recent experimental work pertaining to partial melting of the mantle at high pressures (Green and Ringwood, 1967; O'Hara, 1968; Ito and Kennedy, 1968) gives no indication that a liquid of such unusual composition can be produced by this means. These workers have shown that olivine-normative basalts rather than ultrabasic liquids are the usual products of extensive partial melting of a lherzolite mantle. A further objection to an ultrabasic parent magma is provided by variation diagrams of the type devised by Pearce (1968). This is discussed in a later section.

The multiple intrusion hypothesis has been criticized by Presnall (1966), who found experimentally that the observed sequence of rock types in the Alaskan zoned complexes could be produced by low-pressure fractionation of a single ultra-basic magma. Presnall pointed out that it is unlikely that this sequence could also be derived by fractional melting of the mantle at high pressures for two reasons. Firstly, for systems exhibiting solid solutions or reaction relations involving the liquid phase, successive liquids produced by fractional melting are in general not the same as the sequence of crystal cumulate compositions formed during fractional crystallization. Secondly, these differences should be enhanced if compositions derived by fractional crystallization at low pressures are compared with those produced by melting at high pressures, when different phases are probably equilibrated.

For many years, petrologists have doubted the existence of ultramafic magmas because of the extremely high temperatures necessary for their formation and the lack of lavas of such compositions. To avoid this problem, a number of authors have suggested that specific ultramafic bodies are the residual material resulting from gravitational differentiation of basaltic magmas which accumulated in reservoirs beneath volcanoes (Brown, 1956; Lauder, 1965; Irvine and Smith, 1967; MacRae, 1969). Complementary liquid differentiates are believed to have been removed by periodic eruptions. This process has been proposed as a general mechanism of formation of ultramafic rocks by Challis (1965) and Challis and Lauder (1966). Osborn (1969a) suggested that alpine peridotites are cumulates resulting from fractional crystallization of basalt to produce andesite. Irvine (1967) noted

TABLE 1. COMPOSITIONS OF ROCKS FROM ZONED ULTRAMAFIC COMPLEXES
AND FROM EXPERIMENTAL RUNS OF GREEN AND RINGWOOD (1968)

	1	2	3	4	5	6
SiO_2	44.6	44.1	43.9	44.8	55.9	52.9
TiO_2	0.7	0.8	0.9	2.5	0.8	1.5
Al_2O_3	5.5	3.1	5.1	9.7	19.4	16.9
Fe_2O_3	4.3	7.4	7.6	—	0.4	0.3
FeO	7.9	7.0	7.2	9.1	7.2	7.9
MnO	0.1	0.2	0.2	—	0.3	0.2
MgO	20.2	22.4	14.1	14.6	4.4	7.0
CaO	15.8	14.3	17.4	16.6	7.8	10.0
Na_2O	0.5	0.4	0.5	1.0	3.3	2.7
K_2O	0.4	0.3	1.3	0.04	0.8	0.6
Total	100.0	100.0	100.0	98.34	100.3	100.0

1. Average composition of ultramafic rocks, Duke Island, Alaska (Irvine, 1963, weighted average from Table 1).

2. Average composition of ultramafic rocks, Tulameen, British Columbia (Findlay, 1969, Table VIII).

3. Hornblende clinopyroxenite, Tulameen (Findlay, 1969, Table VIII).

4. Analysis of early formed crystal residue separated from high-alumina quartz tholeiite (No. 6) to produce basaltic andesite (No. 5; Green and Ringwood, 1968, Table 26).

5. Basaltic andesite produced by separating 26 percent by volume of early formed crystals (No. 4) from high-alumina quartz tholeiite (No. 6; Green and Ringwood, 1968, Table 26).

6. High-alumina quartz tholeiite (Green and Ringwood, 1968, Table 2).

that the Duke Island zoned ultramafic complex might have formed in a reservoir beneath a volcano, although he did not specify the nature of the parent magma or the extrusive products.

It is the purpose of this paper to examine the possibility that zoned ultramafic complexes of the Alaskan type are differentiates of basaltic magma. It is postulated that such ultramafic complexes are formed by fractional crystallization and flow differentiation in the feeder pipes of volcanoes. The parent magma is assumed to be a hydrous olivine-rich tholeiitic basalt, and the eruptive products are dominated by andesitic compositions probably accompanied by high-alumina tholeiite. This mechanism overcomes many difficulties of the previous hypotheses of origin, and is compatible with the available experimental and field data discussed below.

Experimental Evidence

Presnall (1966) showed that the major rock types of the Alaskan zoned complexes can be produced by fractional crystallization of an ultrabasic magma under conditions of buffered oxygen fugacity. However, he noted that his experimental data are also consistent with the concept that the ultramafic rocks represent the crystal residue formed during the early stages of fractional crystallization of a basaltic magma.

Yoder and Tilley (1962) demonstrated that crystallization of various basalt compositions (olivine tholeiite, high-alumina basalt, and alkali basalt) under hydrous conditions at pressures in the range from 2 to 10 kbars commenced with separation of olivine, followed successively by clinopyroxene, hornblende, and plagioclase (Yoder and Tilley, Figs. 27, 28, 29). This is the sequence of crystallization of major silicate phases observed in zoned ultramafic complexes.

Watkinson and Irvine (1964) suggested that hornblende peridotite and hornblendite bodies near Quetico, Ontario, formed by fractional crystallization of olivine tholeiite at high water pressures, and discussed the limitations of applying Yoder and Tilley's experimental data to natural rock systems. Because the Quetico intrusions contain normative hypersthene and little if any primary magnetite, they considered them to be chemically distinct from the Alaskan zoned complexes. However, the amount of normative hypersthene might be expected to increase with the proportion of interstitial liquid remaining between the early formed crystals, and the relative abundance of magnetite must be controlled by the variation in oxygen fugacity during crystallization of individual intrusions.

Fractional crystallization of basaltic magma under constant oxygen fugacity, possibly achieved by high water pressure and diffusion of H_2, is the mechanism proposed by Osborn (1959, 1962, 1969b) to produce the calc-alkaline suite of igneous rocks. The zoned ultramafic complexes may represent the magnetite-rich crystalline residue resulting from this process. Green and Ringwood (1968) considered this possibility, but rejected it on the grounds that the complexes were believed to have crystallized from ultramafic magmas.

Green and Ringwood studied the crystallization of high-alumina quartz tholeiite under wet conditions (P_{H_2O} estimated at 2 to 5 kbars) at total pressures of 9 and 10 kbars, and obtained results similar to those of Yoder and Tilley (1962) except that olivine did not crystallize from the quartz normative basalt. Using microprobe analyses of the crystal phases, they calculated the compositions of successive liquids which would be derived by fractional crystallization of quartz tholeiite under these conditions. The composition of the crystal residue which would separate from the quartz tholeiite if a basaltic andesite were produced is shown in Table 1, No. 4. The similarity between this analysis and the average compositions of the Duke Island and Tulameen complexes (Table 1, Nos. 1 and 2) would be enhanced if an olivine normative basalt had been used in the experiments and if olivine had been present as the liquidus phase.

These analyses are low in Al_2O_3, and liquids derived by separation of this material must accordingly be enriched in this oxide, as shown by analyses of the basaltic andesite and the parent quartz tholeiite (Nos. 5 and 6, Table 1). It is to be expected that fractional crystallization of olivine and pyroxene (with or without hornblende and magnetite) from a basaltic parent magma would form a liquid differentiate with the composition of a high-alumina basalt.

Kuno (1968, Table 5) calculated compositions of solid materials which would be subtracted from high-alumina basalt and its derivatives to produce the successive magmas of the calc-alkali rock series. The compositions include high-alumina tholeiites. Many gabbros from zoned ultramafic complexes are high-alumina tholeiites (Taylor and Noble, 1969). The majority of analyses listed by Duparc and Tikonowitch (1920) for gabbros associated with the Ural ultramafic complexes have more than 20 percent Al_2O_3. Alumina-rich gabbros are also associated with the Union Bay complex in Alaska (Ruckmick and Noble, 1959) and the Tulameen complex in British Columbia (Findlay, 1969). These rocks have compositions similar to those calculated by Kuno and may represent cumulate material formed during the fractionation of high-alumina basalt to produce andesite and other members of the calc-alkali series. Findlay noted that the trend of the Tulameen gabbroic rocks plotted on an FeO-MgO-alkalies diagram is similar to Daly's average calc-alkaline trend.

Evidence from Variation Diagrams

Pearce (1968) devised a technique for plotting rock analyses which can provide more details of fractionation trends than conventional variation diagrams. He used this method (Pearce, 1969) to substantiate that the chemical variation of the Dundonald sill was the result of fractionation of olivine and pyroxene as suggested by Naldrett and Mason (1968). When applied to zoned complexes, this technique elucidates the mechanism of fractionation and the nature of the parent magma.

Fractionation of olivine from a liquid removes MgO, FeO, and SiO_2; absolute amounts of other components in the liquid (for example, Al_2O_3) remain constant. Thus the relation between absolute numbers of moles of MgO and SiO_2 during fractionation is the same as the relation between the molar ratios MgO/Al_2O_3 and SiO_2/Al_2O_3. If olivine is the only phase separating from the parent magma, a plot of molar ratios of MgO/Al_2O_3 versus SiO_2/Al_2O_3 for this magma, for the olivine cumulates, and for complementary liquids produced by olivine fractionation should lie on a straight line with the same slope as the MgO/SiO_2 ratio of the olivine. If olivine fractionation is replaced by pyroxene fractionation, the slope must change and become parallel to that of the MgO/SiO_2 ratio of the pyroxene. The positions of a series of rock analyses in such a plot can reveal whether they are related by olivine or pyroxene fractionation or some combination of these.

Data from the Tulameen ultramafic complex (Findlay, 1969) are plotted in Figures 2 and 3. Figure 2A shows the predicted variation in absolute amounts of MgO and SiO_2 during fractionation of olivine from an initial liquid (X) followed by separation of pyroxene from the derived liquid (Y). The compositions used for fractionating olivine and pyroxene are averages of the mineral analyses from Tables V and VI of Findlay (1969), and are assumed to remain constant. The fact that plots of rock analyses from Tulameen (Figs. 2B and 3B) fall along straight lines indicates that this assumption is justified. Figure 3A displays the equivalent variation of CaO and SiO_2 during fractionation under the same conditions as for Figure 2A. The axes of Figure 3A have the same relative values as in Figure 3B.

Figure 2B shows the MgO/Al_2O_3 versus SiO_2/Al_2O_3 variation of Tulameen dunites, pyroxenites, and gabbros, and Figure 3B shows their CaO/Al_2O_3 versus SiO_2/Al_2O_3 variation. A marked similarity between the slope and configuration of the predicted and observed trends is obvious. The slopes of the various lines

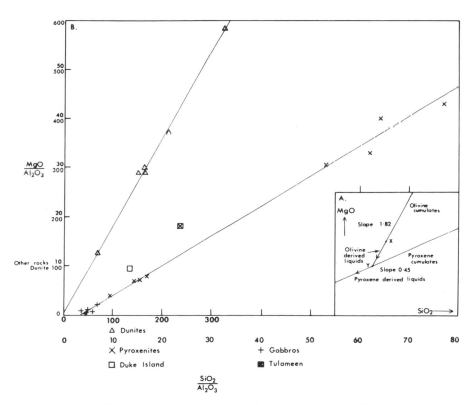

Figure 2. (A) Theoretical variation of absolute amounts of MgO and SiO_2 during fractionation of olivine from liquid X followed by fractionation of pyroxene from liquid Y. Arrows show the trends of derived liquids. The slopes are fixed by the theoretical fractionation mechanism and by the mineral analyses, which are averaged from Tables V and VI of Findlay (1969). (B) Observed relations of molar ratios MgO/Al_2O_3 and SiO_2/Al_2O_3 for Tulameen dunites, pyroxenites, and gabbros. Data from Tables II and IV of Findlay (1969). Compare with Figure 2A. Average compositions of the ultramafic rocks of Duke Island (Irvine, 1963) and Tulameen (Findlay, 1969, Table VIII) are also shown.

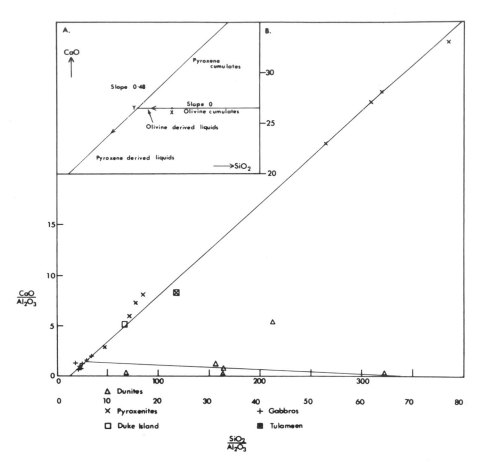

Figure 3. (A) Theoretical variation of absolute amounts of CaO and SiO₂ under the same conditions as for Figure 2A. (B) Observed relations of molar ratios CaO/Al₂O₃ and SiO₂/Al₂O₃ for Tulameen rocks. Compare with Figure 3A.

are listed in Table 2. The slopes of the observed olivine fractionation trends are close to the predicted values. However, the slope of the pyroxene fractionation line in Figure 2B differs from the expected results because the pyroxene contains an appreciable amount of Al₂O₃. When the observed value is corrected according to the method of Pearce (1969), good agreement is obtained (Table 2). If a similar correction is applied to the pyroxene slope in Figure 3B, the discrepancy between the theoretical and observed slopes is increased. Pearce encountered a similar problem in his treatment of the data from the Dundonald sill.

TABLE 2. SLOPES OF FRACTIONATION TRENDS IN FIGURES 2 AND 3

| | Olivine | | Pyroxene | | |
	Theoretical	Observed	Theoretical	Observed	Corrected observed
MgO/SiO₂	1.82	1.81	0.45	0.61	0.48
CaO/SiO₂	0.0	−0.005	0.48	0.45	0.40

Figures 2B and 3B provide strong support for the theory that the sequence of ultramafic rock types present in the Tulameen complex originated by fractionation of olivine (to produce dunite) and pyroxene (to produce pyroxenite).

The results also place certain restrictions on the composition of the parent magma. Experimental, chemical, and petrographic evidence (Duparc and Tikonowitch, 1920; Presnall, 1966; Irvine, 1967; Findlay, 1969) indicates that olivine is the first-formed silicate mineral in zoned ultramafic complexes. Thus the composition of the parent magma must plot on the olivine fractionation curve of Figures 2B and 3B. The estimated bulk compositions of the Duke Island and Tulameen complexes (Table 1, Nos. 1 and 2) plot between the pyroxene and olivine fractionation trends on Figures 2B and 3B, and are therefore considered to represent the products of accumulation of both these minerals. The hornblende pyroxenite analysis (Table 1, No. 3) which was proposed by Findlay (1969) as a possible parent magma for the Tulameen complex is one of the points which define the pyroxene trends in Figures 2B and 3B. It is concluded that none of these analyses can represent the parent magma of the Tulameen complex in particular or of all zoned ultramafic complexes in general.

Basalt analyses plot close to the origin in Figures 2B and 3B, in the vicinity of the intersection of the olivine and pyroxene trends. This intersection is not uniquely defined on the diagrams because of the inaccurate slopes of the pyroxene trends, especially in Figure 2B, the different scales used for different rock types, and the uncertainty over whether the Tulameen gabbros represent pyroxene cumulates or pyroxene-derived liquids. The data as presented do not exclude the possibility that the parent magma had basaltic chemistry, while not providing direct evidence to favor this idea. The results indirectly support a basaltic parent magma, however, insofar as they definitely exclude from consideration the ultrabasic parent compositions postulated by previous workers.

Plots of analyses from other zoned complexes give results similar to those shown in Figures 2 and 3.

Evidence from Hawaiian Volcanoes

Strong support for the concept that the ultramafic complexes are formed in the feeder pipes of volcanoes comes from detailed studies on the Hawaiian Islands.

Gravity and magnetic measurements have revealed the presence of high density plutonic rocks beneath the calderas of several Hawaiian volcanoes (Woollard, 1951; Eaton, 1963; Strange and others, 1965a; Strange and others, 1965b; Malahoff and Woollard, 1966). One of these bodies, beneath Koolau caldera on the island of Oahu, has also been investigated by a detailed seismic survey (Adams and Furumoto, 1965; Furumoto and others, 1965). The geophysical evidence suggests the presence of a 6-km-wide plug of material of density 3.2 gm/cm³ and seismic velocity 7.7 km/sec at a depth of 1,600 m below the caldera. Similar high density material extends laterally beneath the northwest rift zone of the volcano at a depth of about 4 km.

MacDonald (1965) suggested that these dense rocks could be olivine-rich cumulates formed in the feeder pipe and shallow magma reservoir below the volcano, and that fragments of dunite and wehrlite included in late stage lavas in other areas of Hawaii may be representatives of such rocks. Richter and Murata

(1961) described a remarkable occurrence of such inclusions on the island of Hawaii. More extensive studies of ultramafic inclusions in Hawaiian basalts have been made by White (1966) and Kuno (1969). Each of these workers concluded that inclusions of the wehrlite series (dunite, wehrlite, clinopyroxenite, clino-pyroxene gabbro) represent crystal cumulates formed from basaltic magma at comparatively shallow depths. All gradations exist between the members of this series.

Kuno (1969) showed that clinopyroxenes of the wehrlite series are chemically distinct from those in the lherzolite series of inclusions, being richer in CaO and poorer in Al_2O_3, Na_2O, and K_2O. The lherzolite nodules are believed to represent fragments of the upper mantle. Compositions of clinopyroxene phenocrysts in Hawaiian basalts are similar to those of the wehrlite series.

Olivine should be the main phase separating from liquids of basaltic composition as they rise through the mantle, due to increase in the crystallization field of olivine in such compositions with decreasing pressure (O'Hara, 1968; Kushiro, 1969a, 1969b). Separation of olivine during ascent results in tholeiitic chemistry at the surface (Ito and Kennedy, 1968). Thus olivine cumulates would be expected to form in the feeder pipes of the Hawaiian shield volcanoes, which are mainly tholeiitic basalt (Tilley, 1950; Powers, 1955). Olivine is probably joined by clinopyroxene and plagioclase at shallow depths as suggested by the results of Yoder and Tilley (1962) on anhydrous olivine tholeiite. The absence of horn-blende indicates a low water content in the magma.

The zoned ultramafic complexes may have the same form as and may have been produced by the same process as the high density plugs beneath the Hawaiian volcanoes. Steep walled intrusions showing concentric zoning, such as the Blashke Islands complex (Walton, 1951), formed in the feeder pipes of volcanoes. Rocks exhibiting gravitational layering, such as those on Duke Island (Irvine, 1963, 1967), may be cumulates produced in high-level magma reservoirs similar to the one existing at present beneath Kilauea volcano (Eaton and Murata, 1960).

Composition of the Parent Magma of the Zoned Ultramafic Complexes

Most zoned ultramafic complexes are associated with gabbroic rocks which in some cases are present in larger volumes than the ultramafic rocks themselves and completely surround the latter. This is especially true of occurrences in the Ural Mountains.

Because the bulk of these gabbros have tholeiitic chemistry and contain ortho-pyroxene, previous workers have concluded that they are not genetically related to the ultramafic intrusions (Ruckmick and Noble, 1959; Taylor and Noble, 1960, 1969; Irvine, 1967; Taylor, 1967). The results of Yoder and Tilley (1962), how-ever, indicate that it is possible to produce dunites, clinopyroxenites, and horn-blendites from basaltic liquids by fractional crystallization at high water pressures, and evidence outlined below suggests that any basaltic parent magma envisaged for zoned complexes must have tholeiitic chemistry. The writer considers an olivine-rich tholeiite to be the most likely parent magma of Alaskan type complexes. A genetic relation between ultramafics and associated gabbros can then be readily established.

The experimental data of Yoder and Tilley (1962) on the crystallization of various basalts under hydrous conditions gives some information on the chemistry of possible parent liquids. The sequence of crystallization of silicate phases observed in most zoned complexes is olivine, followed by clinopyroxene, amphibole, and plagioclase. Yoder and Tilley obtained this sequence in olivine tholeiite over the pressure range 2 to 11 kbars, in high-alumina basalt over the pressure range 2.5 to 8 kbars, and in alkali basalt over the range 1.5 to 5.5 kbars. Thus, the pressure range over which the required mineral sequence could be produced is larger for tholeiitic basalt than for the other varieties. Over most of this pressure range, the three ferromagnesian phases crystallize much earlier than the plagioclase, as shown in Figure 4. A large temperature interval before the appearance of plagioclase is necessary for the formation of ultramafic crystal residues by fractional crystallization.

Mica crystallized in the experimental runs on the alkali basalt, but not in the other experimental runs. Biotite is absent from the majority of zoned ultramafic complexes, or is present in minute quantities.

A number of workers (Wilkinson, 1956; Kushiro, 1960; Le Bas, 1962; Rothstein, 1962; Challis, 1965; Challis and Lauder, 1966) have shown that clinopyroxene compositions reflect the compositions of the magmas from which they crystallized. Irvine (1967) considered that the clinopyroxenes of the Duke Island complex indicated that any basaltic magma related to the formation of the ultramafic rocks must have been an alkalic type. This conclusion does not appear to be valid for zoned complexes in general.

Figure 5 is a portion of the pyroxene quadrilateral on which analyses of clinopyroxenes from zoned complexes, stratiform tholeiitic basalt intrusions, and alkali basalt sills have been plotted. It is readily apparent that the most Mg-rich (and presumably first formed) pyroxenes of the zoned complexes plot close to the early clinopyroxenes from tholeiitic intrusions such as Bushveld, Rhum, and Dawros. The trends of the pyroxenes from the zoned complexes are all similar, and differ markedly from the Bushveld trend defined by Atkins (1969), being much richer in Ca, so that the late Fe-rich pyroxenes plot in the same area as those from alkalic sills (Murray, 1954; Wilkinson, 1957).

The change in pyroxenes from zoned complexes from early tholeiitic types to those characteristic of alkalic rocks is supported by examination of other parameters. When plotted on Figures 2 and 5 (plots of SiO_2 versus Al_2O_3 and tetrahedral Al versus TiO_2, respectively) of Le Bas (1962), most analyses of pyroxenes from zoned ultramafic bodies fall in the tholeiitic fields defined on these diagrams. Only a few Fe-rich analyses extend into the alkalic fields.

Rothstein (1962), Challis (1965), and Challis and Lauder (1966) considered that the Na-content of clinopyroxenes depends on pressure and magma chemistry. According to the criteria defined by Challis and Challis and Lauder, pyroxenes from zoned complexes all fall in the same range as clinopyroxenes from the Bushveld, Rhum, and Dawros intrusions.

According to all the above criteria relating clinopyroxene chemistry to magma type (position in the pyroxene quadrilateral, and content of SiO_2, Al_2O_3, TiO_2, and Na_2O), the most Mg-rich and presumably first-formed pyroxenes from zoned ultramafic complexes indicate that if the parent magma of the complexes was a basalt, it must have been tholeiitic rather than alkalic.

Similar arguments were used by Deer and Abbott (1965), who proposed that the parent magma of the Kap Edvard Holm complex was a tholeiitic basalt, despite the facts that only clinopyroxene is present and that its trend is parallel to alkaline rather than tholeiitic clinopyroxenes (Fig. 5). They suggested that the Kap Edvard Holm trend was largely the result of relatively high water pressure during crystallization, which lowered the solidus temperature of the pyroxene so that it contained smaller quantities of $MgSiO_3$ in solid solution. This effect is believed to be partly responsible for the Ca-rich trend of the clinopyroxenes from zoned complexes. Probably more important is the fact that high water pressure would inhibit the crystallization of plagioclase (Fig. 4), and all Ca in the magma

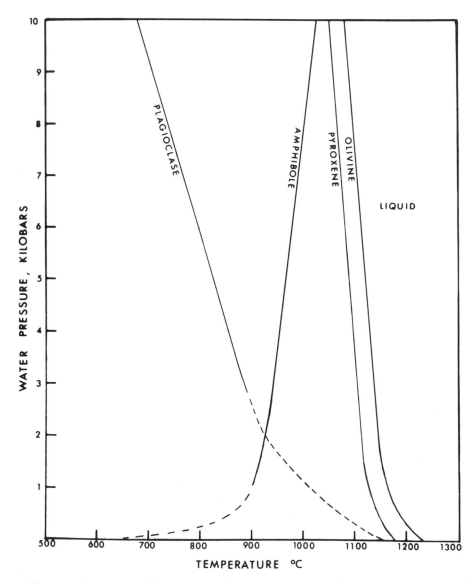

Figure 4. Projection of the olivine-tholeiite-water system, adapted from Figure 27 of Yoder and Tilley (1962). Each mineral is stable along that side of the curve where the name is shown. A chemical analysis of the olivine tholeiite is given by Yoder and Tilley.

was available to enter pyroxene. Enrichment of Al and Na in the later pyroxenes can also be explained by this argument.

The zoned ultramafic complex near Hope, British Columbia, is unique in that orthopyroxene is the dominant pyroxene, and appears earlier than clinopyroxene (Aho, 1956). The crystallization sequence of the early formed silicates (olivine followed by orthopyroxene and clinopyroxene) is the same as that observed by Green and Ringwood (1967) for the experimental crystallization of olivine tholeiite under dry conditions at pressures in the range of about 5 to 10 kbars. This suggests that the Hope body could have been derived from a magma similar to that which produced the other zoned complexes, but that the water content in this particular case was low. The presence of some water is indicated by the occurrence of hornblende in the peripheral zones of the intrusion. The fact that plagioclase appeared at the same time as, or slightly earlier than, hornblende in the Hope intrusion also indicates that the water pressure during crystallization was lower than in other zoned complexes.

If the parent magma were anhydrous, the mineral assemblage would be olivine, orthopyroxene, clinopyroxene, and plagioclase. This assemblage is present in the Red Hills intrusion, New Zealand (Challis, 1965; Walcott, 1969). Challis has proposed that the Red Hills body formed as subvolcanic differentiates produced from an olivine-rich tholeiitic parent magma, and that the liquid differentiates were removed by eruption, being represented by olivine-poor tholeiitic basalts associated with the ultramafic rocks.

The writer considers it possible that products as diverse as the Red Hills intrusion

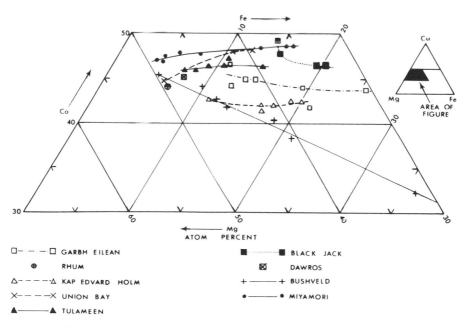

Figure 5. Portion of the pyroxene quadrilateral showing compositional trends of clinopyroxenes from zoned ultramafic complexes and other intrusions. Data from Murray (1954), Brown (1956), Wilkinson (1957), Rothstein (1958), Ruckmick and Noble (1959), Deer and Abbott (1965), Onuki (1965), Atkins (1969), and Findlay (1969). See text for discussion.

and the Alaskan zoned complexes can be derived by fractionation of olivine-rich tholeiitic magmas depending on the prevailing water pressure during crystallization. The Hope body may represent an intermediate type.

It is significant that the extrusive differentiates associated with the Red Hills intrusion are tholeiitic basalts. As will be shown in the next section, volcanic rocks associated with zoned complexes are members of the calc-alkaline suite, dominated by andesite. The evidence suggests that fractionation of olivine-rich tholeiite at crustal levels under anhydrous conditions produces olivine-poor tholeiite, and under hydrous conditions produces andesites.

This conclusion opposes Osborn's (1969a) hypothesis that andesites are the complementary differentiates of alpine-type peridotites, of which the Red Hills body is a typical example as regards mineralogy and composition. A further objection to his suggestion is provided by the fact that magnetite is absent from alpine ultramafics, and crystallization of magnetite is essential to produce the calc-alkaline trend.

Field Evidence

If the zoned ultramafic complexes represent feeder pipes for andesitic volcanoes, they should be spatially associated with extrusive andesites of the same age. This is indeed the case, and is believed to provide strong evidence in favor of the hypothesis proposed in this paper.

The major zoned complexes of southeastern Alaska are early middle Cretaceous in age (Lanphere and Eberlein, 1966) and outcrop in a linear belt 550 km long with a maximum width of 50 km (Taylor, 1967, Fig. 4.9). Paleogeologic reconstructions of Brew and others (1966, Figs. 8–11), and Brew (1968, Fig. 4B) show a narrow elongate belt of Lower Cretaceous fragmental volcanic rocks which coincides precisely with the distribution of the ultramafic complexes. This coincidence is shown in Figure 6. It is most readily explained by assuming a genetic connection between the intrusive and extrusive rocks. The Lower Cretaceous volcanic sequence is dominated by andesite, basaltic andesite, andesitic basalt, and basalt (Brew, 1968).

The Tulameen ultramafic-gabbro complex of British Columbia (Findlay, 1969) was emplaced in the Upper Triassic Nicola Group, which contains abundant basic and intermediate volcanic rocks, especially pyroxene andesite (Campbell, 1966). Radiometric dating suggests a Late Triassic age for the ultramafic rocks (Leech and others, 1963). The Tulameen complex might therefore represent a feeder for the volcanic material of the Nicola Group. Findlay suggested that the Nicola volcanics could be comagmatic with the Tulameen intrusion.

The zoned ultramafic complexes of the Ural Mountains are probably Middle to Upper Devonian in age (Smirnov, 1969). They outcrop along the site of the Tagil-Magnitogorsk trough (Plyusnin, 1963) which includes volcanic rocks of Silurian and Devonian age. The older volcanics (largely Silurian) consist of pillow lava basalts and spilites (Sharfman, 1968). By contrast, the Devonian volcanic rocks in this region are dominated by flows and tuffs of andesitic composition, with subordinate basalt and trachyte (Breivel and others, 1967). As in southeastern Alaska, andesitic volcanism is spatially and temporally associated with the intrusion of the zoned ultramafic complexes.

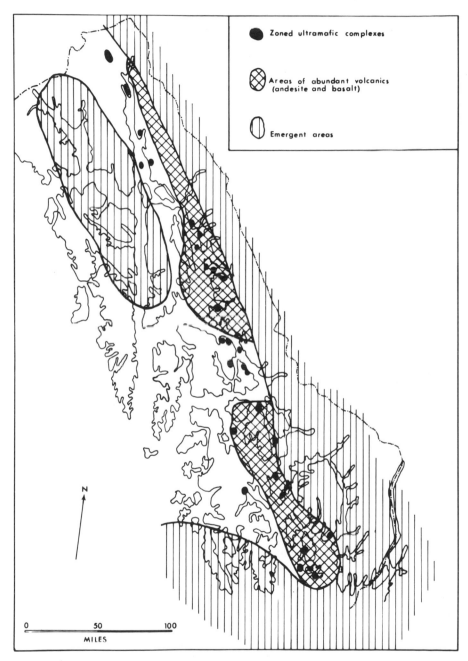

Figure 6. Paleogeologic map of southeastern Alaska for Lower Cretaceous time, showing the distribution of zoned ultramafic complexes and andesitic and basaltic volcanics.
Adapted from maps of Brew and others (1966), and Brew (1968).

No volcanic rocks are so obviously related to the Venezuelan zoned ultramafic bodies. Paleocene sediments near the Cerro Pelón intrusion contain basalt and andesite fragments (Von der Osten and Zozaya, 1957) which may represent the extrusive products of a volcano associated with the intrusion. The Villa de Cura Group cannot be related to the belt of ultramafic and gabbroic intrusions, since it was regionally metamorphosed prior to their emplacement. It consists of spilitic pillow basalts and keratophyres (Shagam, 1960) and may be equivalent to the "island arc tholeiitic series" of Jakes and Gill (1970), which precedes and forms a basement for calc-alkaline volcanics in many circum-Pacific island arcs.

Mechanism of Formation of the Zoned Complexes

The concentric zoning of the complexes can be explained by a process of fractional crystallization and flow differentiation.

Suspended solid particles in a flowing viscous liquid can be segregated as accumulates because of the flow properties of the mixture. Segregation takes place along the central axis of the conduit. This mechanism has been called flow differentiation (Bhattacharji and Smith, 1964; Bhattacharji, 1967; Simkin, 1967). In a vertical feeder pipe containing rising basaltic magma, early formed olivine will be segregated at the center, forming a solid core. The efficiency of flow differentiation in segregating olivine crystals at the centers of picritic dikes and sills and thereby altering the chemistry of the marginal liquid has been clearly demonstrated by Drever and Johnston (1966). As upward movement of the magma and core continues, further crystallization and flow differentiation in the marginal fluid would produce mineralogical and chemical zoning around the inner core (Bhattacharji, 1967). Early phases such as olivine would be restricted to the core, and late phases such as hornblende and plagioclase to peripheral zones. Also, the Mg content of a particular ferromagnesian phase such as clinopyroxene should decrease from the core outward. Both types of zoning are consistently observed in the ultramafic complexes (Taylor, 1967; Taylor and Noble, 1969).

If the process of flow differentiation operates continuously until the magma and suspended solids cease to rise and crystallization is complete, the various rock types should be gradational. This condition appears to be almost realized in the El Chacao complex of northern Venezuela. Fluctuating flow conditions, however, may cause the central column to break up and produce relative movement between the crystalline core and the enclosing fluid, resulting in apparently intrusive contacts or in a combination of intrusive and transitional relations. The intrusive contacts could indicate conflicting age relations. No chilled margins should form at contacts between the various rock types, and none are observed.

This mechanism seems particularly well suited to explain the features of the remarkable zoned body at Union Bay in southeastern Alaska (Ruckmick and Noble, 1959), which displays both intrusive and gradational contacts. There is marked similarity between the forms produced by flow differentiation in model experiments (Bhattacharji, 1967, Plate 1, Fig. A) and the cross section drawn through the Union Bay complex by Ruckmick and Noble (1959, Plate 4, Sec. A–A).

Figure 7 shows possible stages in the formation of a zoned complex. Separation of olivine and pyroxene from the parent liquid could produce high-alumina olivine-poor tholeiite which would precede the solid core, solidify at depth as gabbro, and

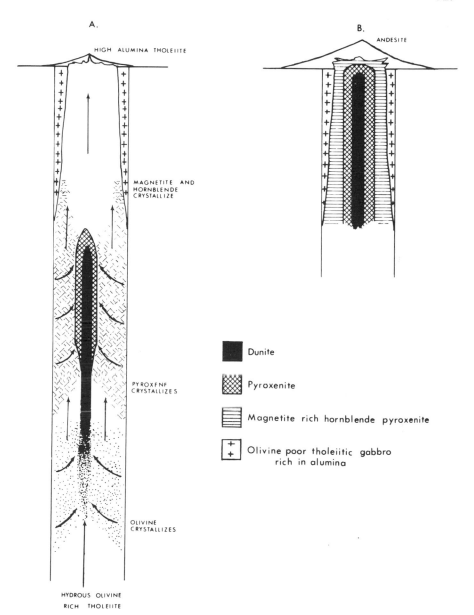

Figure 7. Schematic diagram showing possible stages in the formation of a zoned ultramafic complex by fractional crystallization and flow differentiation in the feeder pipe of a volcano. The width of the pipe is greatly exaggerated.

possibly be extruded at the surface (Fig. 7A). At a later stage, hornblende and magnetite crystallize, the derived liquids change to andesitic compositions, and the ultramafic core intrudes the earlier gabbros (Fig. 7B). This mechanism can explain most features of the zoned complexes, especially the common occurrence of high-alumina tholeiitic gabbros which enclose and appear to be intruded by the ultramafic rocks (Taylor, 1967; Taylor and Noble, 1969).

Flow differentiation as described above must be a general process operating in volcanic conduits, and would produce a variety of rock types depending on the composition of the initial magma. Many alkaline pyroxenites exhibit concentric zoning analogous to that of the Alaskan complexes (Upton, 1967). The pyroxene compositions, the relative abundance of biotite, and the presence of nepheline rather than feldspar in late stage differentiates all point to an alkalic, undersaturated parent magma. Wyllie (1967) suggested that flow differentiation might have been important in producing the alkalic associations.

SIGNIFICANCE OF ZONED ULTRAMAFIC COMPLEXES

Petrologic

The most significant aspect of the foregoing discussion is that andesites may be produced by fractional crystallization of basalt magma in the earth's crust under hydrous conditions. This is one of the mechanisms proposed by Green and Ringwood (1968) for the genesis of the calc-alkaline suite. The abundance of magnetite in the pyroxenites and hornblende pyroxenites suggests a high oxygen fugacity. Thus the field evidence also supports Osborn's contention (Osborn, 1959, 1962, 1969b) that andesites result from fractionation of basalt at high and constant P_{O_2}. The high P_{O_2} could be the result of diffusion of H_2 from the hydrous magma, as suggested by Osborn (1959).

If this scheme is a valid mechanism for the formation of andesitic magmas, then the means of producing a hydrous basaltic magma becomes the fundamental problem of the origin of calc-alkaline rocks.

Paleotectonic

The concept that zoned complexes represent feeder pipes of andesitic volcanoes is of considerable significance in reconstructing the tectonic history of areas where such intrusions occur.

Hess (1955) suggested that alpine-type serpentinites were intruded only during the first deformation of a mountain system. In terms of the modern hypothesis of plate tectonics (Morgan, 1968), the emplacement of alpine ultramafics in orogenic belts is usually envisioned as taking place in a subduction zone where oceanic crust descends into the mantle along a Benioff zone (Ernst, 1970; Moores, 1970). Andesitic volcanoes form above the downgoing slab at a distance of 100 to 200 km from the associated trench (Sugimura, 1967). Thus it might be expected that two parallel belts of ultramafic rocks would be exposed after uplift and erosion of an island arc, an older one consisting of alpine peridotites and ophiolites, and a younger one of zoned ultramafic complexes.

Such a situation exists in the Ural Mountains. An eastern belt of ultramafics consists of serpentinized harzburgite (Duparc and Tikonowitch, 1920) and is thought to be of early Paleozoic age (Ovchinnikov and others, 1968). These ultramafics are intimately related to high pressure metamorphic rocks such as glaucophane schist and eclogite (Sobolev and others, 1967; Hamilton, 1970). The Devonian zoned complexes form a parallel band about 80 km to the west (Taylor, 1967, Fig. 4.9). The eastern belt of peridotites and high-pressure metamorphic rocks may mark the original trench and subduction zone, while the zoned ultramafic bodies represent the eroded volcanic chain.

The existence of two parallel belts of ultramafic rocks, a northern one of alpine type and a southern belt of zoned complexes with associated gabbroic rocks, along the southern margin of the Coast Range of Venezuela (Fig. 1), supports the idea that a south-dipping Benioff zone extended beneath the present site of the Coast Range in Cretaceous time (Bell, 1972; Maresch, 1972).

Belts of alpine-type ultramafic rocks outcrop on both sides of the zoned complexes in southeastern Alaska (Taylor, 1967; Roddick and others, 1967). It seems premature to speculate on the significance of the two varieties of ultramafics in an area which has had a long and complex tectonic history that has continued to the present.

CONCLUSIONS

The following conclusions emerge from the arguments presented above:

1. Zoned ultramafic complexes of the Alaskan type could be produced by fractional crystallization and flow differentiation in the feeder pipes of volcanoes. They may be analogous to the high density plugs beneath Hawaiian calderas.

2. The parent magma of the complexes was probably an olivine-rich tholeiite. Fractional crystallization of such a magma under high water pressure produces the mineralogical sequence observed in most of the zoned complexes (olivine, clinopyroxene, hornblende, and plagioclase). Crystallization under anhydrous conditions forms orthopyroxene as well as clinopyroxene and retards the appearance of hornblende. In both cases, experimental evidence indicates that the assemblages formed at a total pressure in the range of about 3 to 10 kbars.

3. Flow differentiation in a vertical conduit can explain the mineralogical and chemical zoning displayed by the majority of the ultramafic intrusions.

4. Separation of varying proportions of early crystallized phases from the parent magma could produce high-alumina tholeiitic basalt and andesite as extrusive products.

5. Gabbros associated with the zoned complexes can be genetically related to the ultramafic rocks.

ACKNOWLEDGMENTS

The study of ultramafic complexes in Venezuela was undertaken as part of the requirements for the Ph.D. degree at Princeton University, and represents a continuation of the Princeton Caribbean Project initiated by H. H. Hess. The subject was suggested by H. H. Hess and R. Shagam of the University of Negev, Beer-Sheva, Israel. Field work was supported by the Ministerio de Minas e Hidrocarburos, Venezuela. I am grateful to the director, E. Araujo, and other members of the Direccion de Geologia for enabling me to carry out the field work. This manuscript was improved as a result of criticism by R. B. Hargraves and L. S. Hollister of Princeton University and T. N. Irvine of the Geological Survey of Canada, although all do not necessarily agree with the conclusions reached.

REFERENCES CITED

Adams, W. M., and Furumoto, A. S., 1965, A seismic refraction study of the Koolau volcanic plug: Pacific Sci., v. 19, p. 296–305.

Aho, A. E., 1956, Geology and genesis of ultrabasic nickel-copper-pyrrhotite deposits at the Pacific Nickel property, southwestern British Columbia: Econ. Geology, v. 51, p. 444–481.

Atkins, F. B., 1969, Pyroxenes of the Bushveld intrusion, South Africa: Jour. Petrology, v. 10, p. 222–249.

Bell, J. S., 1972, Global tectonics in the southern Caribbean area: Geol. Soc. America Mem. 132, p. 369–386.

Bellizzia, A., and Rodríguez, D., 1968, Consideraciones sobre la estratigrafia de los Estados Lara, Yaracuy, Cojedes y Carabobo: Bol. Geología, Min. de Minas e Hidrocarburos, Venezuela, v. 9, no. 18, p. 515–563.

Bhattacharji, S., 1967, Mechanics of flow differentiation in ultramafic and mafic sills: Jour. Geology, v. 75, p. 101–112.

Bhattacharji, S., and Smith, C. H., 1964, Flowage differentiation: Science, v. 145, p. 150–153.

Breivel, M. G., Eroshevskaya, R. I., Nestoyanova, O. A., and Khodalevich, A. N., 1967, Devonian of the eastern slope of the Urals, in Oswald, D. H., ed., International Symposium on the Devonian System, Vol. 1: Calgary, Alberta Soc. Petroleum Geologists, p. 421–432.

Brew, D. A., 1968, The role of volcanism in post-Carboniferous tectonics of southeastern Alaska and nearby regions, North America: Internat. Geol. Cong., 23d, Prague, Comptes Rendus, Sec. 2, p. 107–121.

Brew, D. A., Loney, R. A., and Muffler, L. J. P., 1966, Tectonic history of southeastern Alaska: Canadian Inst. Mining and Metallurgy Spec. Vol. no. 8, p. 149–170.

Brown, G. M., 1956, The layered ultrabasic rocks of Rhum, Inner Hebrides: Royal Soc. London Philos. Trans. B, v. 240, p. 1–53.

Buddington, A. F., and Chapin, T., 1929, Geology and mineral deposits of southeastern Alaska: U. S. Geol. Survey Bull. 800, 398 p.

Campbell, R. B., 1966, Tectonics of the south central Cordillera of British Columbia: Canadian Inst. Mining and Metallurgy Spec. Vol. no. 8, p. 61–71.

Challis, G. A., 1965, The origin of New Zealand ultramafic intrusions: Jour. Petrology, v. 6, p. 322–364.

Challis, G. A., and Lauder, W. R., 1966, The genetic position of "alpine" type ultramafic rocks: Bull. Volcanol., v. 29, p. 283–305.

Deer, W. A., and Abbott, D., 1965, Clinopyroxenes of the gabbro cumulates of the Kap Edvard Holm complex, east Greenland: Mineralog. Mag., v. 34, p. 177–193.

Drever, H. I., and Johnston, R., 1966, A natural high-lime silicate liquid more basic than basalt: Jour. Petrology, v. 7, p. 414–420.

Duparc, L., and Tikonowitch, M., 1920, Le platine et les gîtes platinifères de l'oural et du monde: Geneva, 542 p.

Eaton, J. P., 1963, Volcano geophysics: Am. Geophys. Union Trans., v. 44, p. 507–508.

Eaton, J. P., and Murata, K. J., 1960, How volcanoes grow: Science, v. 132, p. 925–938.

Ernst, W. G., 1970, Tectonic contact between the Franciscan melánge and the Great Valley sequence—crustal expression of a late Mesozoic Benioff zone: Jour. Geophys. Research, v. 75, p. 886–901.

Findlay, D. C., 1969, Origin of the Tulameen ultramafic-gabbro complex, southern British Columbia: Canadian Jour. Earth Sci., v. 6, p. 399–425.

Furumoto, A. S., Thompson, N. J., and Woollard, G. P., 1965, The structure of Koolau volcano from seismic refraction studies: Pacific Sci., v. 19, p. 306–314.

Green, D. H., and Ringwood, A. E., 1967, The genesis of basaltic magmas: Contr. Mineralogy and Petrology, v. 15, p. 103–190.

—— 1968, Genesis of the calc-alkaline igneous rock suite: Contr. Mineralogy and Petrology, v. 18, p. 105–162.

Hamilton, W., 1970, The Uralides and the motion of the Russian and Siberian platforms: Geol. Soc. America Bull., v. 81, p. 2553–2576.

Hess, H. H., 1955, Serpentines, orogeny and epeirogeny: Geol. Soc. America Spec. Paper 62, p. 391–408.

Irvine, T. N., 1963, Origin of the ultramafic complex at Duke Island, southeastern Alaska: Mineralog. Soc. America Spec. Paper 1, p. 36–45.

—— 1967, The Duke Island ultramafic complex, southeastern Alaska, *in* Wyllie, P. J., ed., Ultramafic and related rocks: New York, John Wiley & Sons, Inc., p. 84–97.

Irvine, T. N., and Smith, C. H., 1967, The ultramafic rocks of the Muskox intrusion, Northwest Territories, Canada, *in* Wyllie, P. J., ed., Ultramafic and related rocks: New York, John Wiley & Sons, Inc., p. 38–49.

Ito, K., and Kennedy, G. C., 1968, Melting and phase relations in the plane tholeiite-lherzolite-nepheline basanite to 40 kilobars and geological implications: Contr. Mineralogy and Petrology, v. 19, p. 177–211.

Jakes, P., and Gill, J., 1970, Rare earth elements and the island arc tholeiitic series: Earth and Planetary Sci. Letters, v. 9, p. 17–28.

Kennedy, G. C., and Walton, M. S., 1946, Geology and associated mineral deposits of some ultramafic rock bodies in southeastern Alaska: U.S. Geol. Survey Bull. 947–D, p. 65–84.

Kuno, H., 1968, Origin of andesite and its bearing on the island arc structure: Bull. Volcanol., v. 32, p. 141–176.

—— 1969, Mafic and ultramafic nodules in basaltic rocks of Hawaii: Geol. Soc. America Mem. 115, p. 189–234.

Kushiro, I., 1960, Si-Al relations in clinopyroxenes from igneous rocks: Am. Jour. Sci., v. 258, p. 548–554.

—— 1969a, Discussion of the paper "The origin of basaltic and nephelinitic magmas in the earth's mantle" by D. H. Green: Tectonophysics, v. 7, p. 427–436.

—— 1969b, The system forsterite-diopside-silica with and without water at high pressures: Am. Jour. Sci., v. 267A, p. 269–294.

Lanphere, M. A., and Eberlein, G. D., 1966, Potassium-argon ages of magnetite-bearing ultramafic complexes in southeastern Alaska: Geol. Soc. America Spec. Paper 87, p. 94.

Lauder, W. R., 1965, The geology of Dun Mountain, Nelson, New Zealand: Part 2—The petrology, structure and origin of the ultramafic rocks: New Zealand Jour. Geol. and Geophys., v. 8, p. 475–504.

Le Bas, M. J., 1962, The role of aluminum in igneous clinopyroxenes with relation to their parentage: Am. Jour. Sci., v. 260, p. 267–288.

Leech, G. B., Lowdon, J. A., Stockwell, C. H., and Wanless, R. K., 1963, Age determinations and geological studies: Canada Geol. Survey Paper 63–17.

MacDonald, G. A., 1965, Hawaiian calderas: Pacific Sci., v. 19, p. 320–334.

MacRae, N. D., 1969, Ultramafic intrusions of the Abitibi area, Ontario: Canadian Jour. Earth Sci., v. 6, p. 281–303.

Malahoff, A., and Woollard, G. P., 1966, Magnetic measurements over the Hawaiian Ridge and their vulcanological implications: Bull. Volcanol., v. 29, p. 735–760.

Maresch, W. V., 1972, Eclogitic-amphibolitic rocks on Isla Margarita, Venezuela: A preliminary account: Geol. Soc. America Mem. 132, p. 429–438.

Moores, E. M., 1970, Ultramafics and orogeny, with models of the U.S. Cordillera and the Tethys: Nature, v. 228, p. 837–842.

Morgan, W. J., 1968, Rises, trenches, great faults, and crustal blocks: Jour. Geophys. Research, v. 73, p. 1959–1982.

Murray, R. J., 1954, The clinopyroxenes of the Garbh Eilean sill, Shiant Isles: Geol. Mag., v. 91, p. 17–31.

Naldrett, A. J., and Mason, G. D., 1968, Contrasting Archean ultramafic igneous bodies in Dundonald and Clergue Townships, Ontario: Canadian Jour. Earth Sci., v. 5, p. 111–143.

Noble, J. A., and Taylor, H. P., Jr., 1960, Correlation of the ultramafic complexes of southeastern Alaska with those of other parts of North America and the world: Internat. Geol. Cong., 21st, Copenhagen, Comptes Rendus, Sec. 13, p. 188–197.

O'Hara, M. J., 1968, Bearing of phase equilibria studies in synthetic and natural systems on the origin and evolution of basic and ultrabasic rocks: Earth-Sci. Rev., v. 4, p. 69–133.

Onuki, H., 1963, Petrology of the Hayachine ultramafic complex in the Kitakami mountainland, northern Japan: Tohoku Univ. Sci. Repts., Ser. 3, v. 8, p. 241–295.

—— 1965, Petrochemical research on the Horoman and Miyamori ultramafic intrusives, northern Japan: Tohoku Univ. Sci. Repts., Ser. 3, v. 9, p. 217–276.

Osborn, E. F., 1959, Role of oxygen pressure in the crystallization and differentiation of basaltic magma: Am. Jour. Sci., v. 257, p. 609–647.

—— 1962, Reaction series for subalkaline igneous rocks based on different oxygen pressure conditions: Am. Mineralogist, v. 47, p. 211–226.

—— 1969a, The complementariness of orogenic andesite and alpine peridotite: Geochim. et Cosmochim. Acta, v. 33, p. 307–324.

—— 1969b, Experimental aspects of calc-alkaline differentiation: Oregon Dept. Geology and Mineral Industries Bull. 65, p. 33–42.

Ovchinnikov, L. N., Dunayev, V. A., Krasnobayev, A. A., and Stepanov, A. I., 1968, Age zones of the Urals based on radiologic data: Acad. Sci. USSR Doklady, Earth Sci. Sec., v. 180, p. 22–25 (A.G.I. transl.).

Pearce, T. H., 1968, A contribution to the theory of variation diagrams: Contr. Mineralogy and Petrology, v. 19, p. 142–157.

—— 1969, Some comments on the differentiation of the Dundonald sill, Ontario: Canadian Jour. Earth Sci., v. 6, p. 75–80.

Plyusnin, K. P., 1963, Tectonic zoning of the central and southern Urals: Acad. Sci. USSR Doklady, Earth Sci. Sec., v. 152, p. 92–94 (A.G.I. transl.).

Powers, H. A., 1955, Composition and origin of basaltic magma of the Hawaiian Islands: Geochim. et Cosmochim. Acta, v. 7, p. 77–107.

Presnall, D. C., 1966, The join forsterite-diopside-iron oxide and its bearing on the crystallization of basaltic and ultramafic magmas: Am. Jour. Sci., v. 264, p. 753–809.

Richter, D. H., and Murata, K. J., 1961, Xenolithic nodules in the 1800–1801 Kaupulehu flow of Hualalai volcano: U.S. Geol. Survey Prof. Paper 424–B, p. 215–217.

Roddick, J. A., Wheeler, J. O., Gabrielse, H., and Souther, J. G., 1967, Age and nature of the Canadian part of the circum-Pacific orogenic belt: Tectonophysics, v. 4, p. 319–337.

Rothstein, A. T. V., 1958, Pyroxenes from the Dawros peridotite and some comments on their nature: Geol. Mag., v. 95, p. 456–462.

—— 1962, Magmatic facies in ultrabasic igneous rocks of the tholeiitic series: Acad. Sci. USSR Moscow, 42 p. (Russian with English summary).

Ruckmick, J. C., and Noble, J. A., 1959, Origin of the ultramafic complex at Union Bay, southeastern Alaska: Geol. Soc. America Bull., v. 70, p. 981–1018.

Shagam, R., 1960, Geology of central Aragua, Venezuela: Geol. Soc. America Bull., v. 71, p. 249–302.

Sharfman, V. S., 1968, Average chemical properties of spilites: Acad. Sci. USSR Doklady, Earth Sci. Sec., v. 180, p. 164–165 (A.G.I. transl.).

Simkin, T., 1967, Flow differentiation in the picritic sills of north Skye, in Wyllie, P. J., ed., Ultramafic and related rocks: New York, John Wiley & Sons, Inc., p. 64–69.

Smirnov, Y. P., 1969, More information about Transural ultrabasites: Acad. Sci. USSR Doklady, Earth Sci. Sec., v. 180, p. 35–37 (A.G.I. transl.).

Sobolev, V. S., Dobretsov, N. L., Reverdatto, V. V., Sobolev, N. V., Ushrakova, E. N., and Khlestov, V. V., 1967, Metamorphic facies and series of facies in the USSR: Dansk Geol. Foren. Medd., v. 17, p. 458–472.

Strange, W. E., Machesky, L. F., and Woollard, G. P., 1965, A gravity survey of the island of Oahu, Hawaii: Pacific Sci., v. 19, p. 350–353.

Strange, W. E., Woollard, G. P., and Rose, J. C., 1965, An analysis of the gravity field over the Hawaiian Islands in terms of crustal structure: Pacific Sci., v. 19, p. 381–389.

Sugimura, A., 1967, Spatial relations of basaltic magmas in island arcs, in Hess, H. H., and Poldervaart, Arie, ed., Basalts: The Poldervaart treatise on rocks of basaltic composition: New York, Interscience Publishers, v. 2, p. 537–571.

Taylor, H. P., Jr., 1967, The zoned ultramafic complexes of southeastern Alaska, in Wyllie, P. J., ed., Ultramafic and related rocks: New York, John Wiley & Sons, Inc., p. 97–121.

Taylor, H. P., Jr., and Noble, J. A., 1960, Origin of the ultramafic complexes in southeastern Alaska: Internat. Geol. Cong., 21st, Copenhagen, Comptes Rendus, Sec. 13, p. 175–187.

—— 1969, Origin of magnetite in the zoned ultramafic complexes of southeastern Alaska, in Wilson, H. D. B., ed., Magmatic ore deposits: Econ. Geology Mon. 4, p. 209–230.

Tilley, C. E., 1950, Some aspects of magmatic evolution: Geol. Soc. London Quart. Jour., v. 106, p. 37–61.

Upton, B. G. J., 1967, Alkaline pyroxenites, *in* Wyllie, P. J., ed., Ultramafic and related rocks: New York, John Wiley & Sons, Inc., p. 281–288.

Von der Osten, E., and Zozaya, D., 1957, Geología de la parte suroeste del Estado Lara, region de Quíbor: Bol. Geología, Min. de Minas e Hidrocarburos, Venezuela, v. 4, no. 9, p. 3–52.

Walcott, R. I., 1969, Geology of the Red Hill complex, Nelson, New Zealand: Royal Soc. New Zealand Trans., Earth Sci., v. 7, p. 57–88.

Walton, M. S., 1951, The Blashke Islands ultrabasic complex: With notes on related areas in southeastern Alaska: New York Acad. Sci., Trans., v. 13, p. 320–323.

Watkinson, D. H., and Irvine, T. N., 1964, Peridotitic intrusions near Quetico and Shebandowan, northwestern Ontario: A contribution to the petrology and geochemistry of ultramafic rocks: Canadian Jour. Earth Sci., v. 1, p. 63–98.

White, R. W., 1966, Ultramafic inclusions in basaltic rocks from Hawaii: Contr. Mineralogy and Petrology, v. 12, p. 245–314.

Wilkinson, J. F. G., 1956, Clinopyroxenes of alkali olivine-basalt magma: Am. Mineralogist, v. 41, p. 724–743.

—— 1957, The clinopyroxenes of a differentiated teschenite sill near Gunnedah, New South Wales: Geol. Mag., v. 94, p. 123–134.

Woollard, G. P., 1951, A gravity reconnaissance of the island of Oahu: Am. Geophys. Union Trans., v. 32, p. 358–368.

Wyllie, P. J., 1967, Review, *in* Wyllie, P. J., ed., Ultramafic and related rocks: New York, John Wiley & Sons, Inc., p. 403–416.

Wyssotsky, N., 1913, Die Platinseifengebiete von Iss- und Nischny-Tagil im Ural: Comité Geol. Rus., Mem. N.S. no. 62, 694 p. (Russian with German summary).

Yoder, H. S., Jr., and Tilley, C. E., 1962, Origin of basalt magmas: An experimental study of natural and synthetic rock systems: Jour. Petrology, v. 3, p. 342–512.

MANUSCRIPT RECEIVED BY THE SOCIETY MARCH 29, 1971

Caribbean Geology

THE GEOLOGICAL SOCIETY OF AMERICA, INC.
MEMOIR 132
© 1972

Puerto Rico Trench Negative Gravity Anomaly Belt

CARL BOWIN
Woods Hole Oceanographic Institution
Woods Hole, Massachusetts 02543

ABSTRACT

The Puerto Rico trench negative free-air gravity anomaly belt extends from south of Barbados, around the Antillean arc, to eastern Cuba. The free-air minimum east of the Lesser Antilles is related to underthrusting of the Caribbean plate by the Atlantic Ocean plate. Here the axis of the free-air minimum lies very close to the eastern border of a zone of epicenters which apparently marks the commencement of crustal faulting along the underthrust. The line of trend of historic volcanoes of the Lesser Antillean arc is equidistant (160 km) from the axis of minimum free-air anomaly. It is postulated that differential shifts between the Caribbean plate and the underthrust Atlantic plate have occurred at least twice, once in late Eocene to early Oligocene, during which the outer island chain of the northern Lesser Antillean arc was formed, and again sometime since the late Miocene, when the Barbados ridge and associated uplifted topography were formed. This latter deformation caused the disappearance of a trench opposite the southern Lesser Antillean arc and a displacement of the Puerto Rico trench axis oceanward away from the axis of the negative free-air anomaly belt near the northeast corner of the Caribbean plate.

INTRODUCTION

Large negative gravity anomalies (free-air anomalies) were first discovered over the Puerto Rico trench on July 5, 1926 by Vening-Meinesz, who was en route to the East Indies from the Netherlands. He found another great negative gravity anomaly seaward of the East Indian archipelago. Vening-Meinesz (1930) developed the concept of a crustal downfold to explain the negative gravity anomalies associated with deep-sea trenches and attributed the ridges that occur in places within the negative gravity belt (offshore from Sumatra and the Island of Timor) to squeezing up of sedimentary rocks which are deformed as the main crust is downfolded. Kuenen (1936) gave the name "tectogene" to the concept wherein portions of the earth's crust are deeply downfolded. Hess's (1938) sketch of a tectogene (Fig. 1), in which he superimposed a general section of the Alps upon a downfolded crust some 25 km thick, has been a classic for several decades.

Hess (1957) summarized the history of the tectogene concept up to that date and reiterated his statement of 1938, "Vening-Meinesz's discovery of huge negative anomalies in the vicinity of island arcs is probably the most important contribution to knowledge of the nature of mountain building made in this century." Hess went on to say that a second great discovery of the twentieth century was the finding that the Mohorovicic discontinuity lies only about 5 km below the floor of the oceans. Hess recognized that, since the original tectogene hypothesis involved a crust of some 20 km, this finding would require a modification of the tectogene concept. That modification was never formally made. Instead, following closely the development of the concept of a youthful ocean floor (Hess, 1962), the concept of underthrusting of lithospheric plates was developed. The Vine-Matthews hypothesis (1963) and all the attendant discoveries, confirmations, and detailing of the new global tectonics are summarized in most geological articles today. Thus, I think that it is fair to state that Hess erred in attributing the most important contribution in this century to Vening-Meinesz's discovery of negative anomaly belts near island arcs, for in fact it has been his own vision of a youthful ocean floor that has inspired the modern revolution in our understanding of the Earth's features.

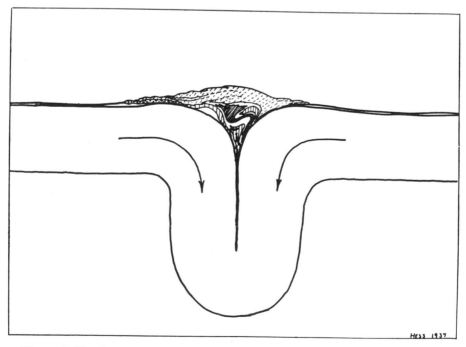

Figure 1. Sketch of a tectogene with a general section of the Alps superimposed. No vertical exaggeration. From Hess (1938).

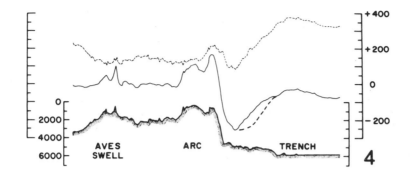

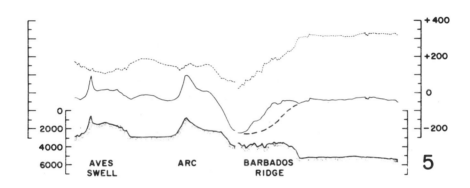

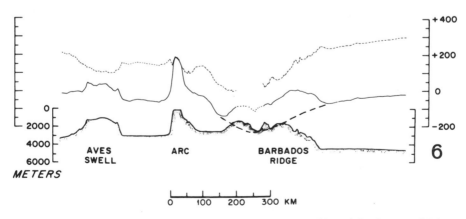

Figure 5. Gravity and bathymetric profiles across the Caribbean island arc. Solid line, free-air anomaly; dashed line, simple Bouguer anomaly; dotted line, inferred "normal" free-air anomaly profile (*see* text). Location of profiles is shown in Figure 4.

Sumatra in the East Indies. As stated earlier, these relations were attributed by Vening-Meinesz to the squeezing up of sedimentary material. Recent studies, to be discussed later, have continued to confirm that hypothesis.

The axis of the free-air minimum north of Puerto Rico is generally displaced slightly toward the southern wall of the trench. As the northeast corner of the Antilles arc is rounded, the axis of the trench diverges more and more from the axis of the free-air minimum. From north of Puerto Rico, the trench becomes shallower to the east and ceases to exist near 16° N., 59° W. An interpretation presented in this paper suggests that the divergence of the trench axis from the axis of the free-air minimum and the disappearance of a trench to the south along the Lesser Antillean arc have occurred largely since the late Miocene.

The axis of the free-air anomaly belt remains about equidistant from the central position of the root beneath the Antilles ridge in the northeast Antillean region (Fig. 7). The root position is inferred from the position of the minimum in Bouguer anomaly values. Opposite Guadaloupe Island and continuing to the south, the axis of the free-air anomaly minimum nearly coincides with the axis of the Bouguer anomaly low. Gravity observations on Tobago (Bowin, unpub. data) give low free-air and Bouguer anomalies (slightly more positive than those of Barbados) and suggest that the axes of minimum anomalies may continue southward, passing near Tobago.

UNDERTHRUSTING AND DEFORMATION
OF THE LESSER ANTILLEAN ARC

A classic example of underthrusting appears to be documented along the eastern side of the Lesser Antillean island arc. Here the mantle-crust interface dips downward as the arc is approached from the Atlantic Ocean side (Officer and others, 1959), earthquake slip mechanisms indicate thrust faulting to predominate in the Lesser Antillean arc (Molnar and Sykes, 1969), and continuous seismic profiling provides a glimpse of the trailing edge of the underthrust (Chase and Bunce, 1969). Chase and Bunce suggest that the thickness of sedimentary material above the oceanic basement increases threefold from lat 18° N. to 14° N., and is even thicker at lat 12° N. Bowin (1971) concluded from an examination of Bouguer gravity anomalies that the low-density material on the eastern flank of the Lesser Antillean ridge, probably sediments, thickens southward along the length of the arc. Bowin (1971) also suggested, on the basis of perturbations of the free-air anomaly profile, that there is a local mass excess within the mass deficient region indicated by the negative free-air anomaly belt. The locations of the perturbations are indicated in Figure 5 by dashed lines which delineate an inferred normal curve. The perturbations become greater toward the south along the length of the arc. This suggests that uplift of the sediments has been greater to the south, probably because in that direction a thicker pile of sediments is affected by compressive forces accompanying the underthrusting of the Lesser Antillean arc by the Atlantic Ocean crust. It further appears that the island of Barbados was raised above the sea floor during the present compressive regime.

The deviation of the free-air anomalies from the inferred normal curve (Fig. 5) is maximum at a point displaced to the east of the axis of the inferred minimum negative anomaly. This relation suggests that the over-all displacement of the sediments has been upward and eastward. This is compatible with the structures identi-

ceased in the outer island chain in the late Eocene to early Oligocene. This conclusion is based on the assumption that the hypabyssal basalt, andesite, and quartz diorite which intrude the Eocene volcanic rocks were reservoirs for surface volcanism which occurred before the erosion which exposed the hypabyssal intrusives that are the depositional surface for Oligo-Miocene limestone. On the inner island chain, on the other hand, the oldest rocks identified are Oligocene and consist of volcanic rocks. Volcanism there has continued with few interruptions from the Oligocene to the present. The products are predominantly andesite with some basalt and dacite (Maxwell, 1948). The shift in the center of volcanism was interpreted by Christman to represent the eastward movement of the crust toward the downbuckle (tectogene), which in the 1950s was thought to occur beneath the negative anomaly belt. This interpretation was never widely accepted because if the crust were continuously moving eastward into the downbuckle, why were there two chains of islands? Also, south of Dominica the islands of the Lesser Antilles have a history much the same as that of the inner arc, the difference being that Eocene rocks occur on Curacao and Grenada at the extreme south end (Martin-Kaye, 1960). Thus the southern half has remained a site of volcanism since the Eocene, which would be unlikely if the crust were continuously moving eastward into a downbuckle.

These relations can now be more easily explained by slight differential shifts of the Caribbean plate edge with respect to the site of the underthrusting. An eastward shift of the northern portion of the edge of the Caribbean plate in the latest Eocene to early Oligocene would have had the effect of stopping volcanic activity in what became the outer island chain, and the attendant crustal uplift would aid erosion, which would bring about the exposure of the hypabyssal rocks prior to late Oligocene deposition of limestone. The high positive free-air anomaly value on Barbuda may in part be due to such crustal uplift, which would be maintained as long as the underthrusting remained active. It is interesting to note that on Figure 7 the presently active volcanoes all lie about equidistant (160 km) from the axis of the free-air minimum. This implies that volcanism is linked closely to the geometry of the underthrusting. Minear and Toksöz (1970) and Oxburgh and Turcotte (1970) have analyzed the thermal structure of an underthrust slab of lithosphere beneath an island arc and have concluded that frictional dissipation along the underthrust plane (Benioff zone) is the major source of heat.

Following the eastward shift of the northern half of the Antillean arc, volcanism started to develop the inner island chain. The site of volcanism has remained in the same location since the Eocene in the southern half of the Antillean arc, which did not shift to the same extent as the northern half. Whether the differential motion between the northern and southern halves of the arc in the late Eocene to early Oligocene is caused by non-rigid deformation within the Caribbean plate edge, or by undetected fracturing of the plate, is unresolved. Why a shift might sometimes extinguish volcanism and other times not affect its relative location is beyond the scope of this paper.

FORMATION OF BARBADOS RIDGE AND ASSOCIATED UPLIFTED TOPOGRAPHY

On Barbados, the Paleocene to middle Eocene Scotland Group was apparently deposited as a chaotic mass by sheet dislocation (Baadsgaard, 1960). The sheet displacement may have been coincident in time with the postulated shift of the northern portion of the Lesser Antilles described above. Following emplacement of the Scotland Group, the Oceanic Group was deposited at great but not abyssal depths as an unbroken sequence from late Eocene to Miocene time (Baadsgaard, 1960). It is here inferred that sometime since late Miocene (within the last 12 m.y.) the eastern edge of the Caribbean plate again shifted (this time along the entire edge), squeezing the great thickness of sediments that had collected in the trench following the negative free-air anomaly belt. This squeezing resulted in the crumpled and contorted material seen on the eastern flank of the Antillean ridge as the Barbabos ridge, Tobago trough, Barbados basin, and other unnamed structures. The complexity of superimposed, folded, faulted, and overturned beds exposed on Barbados indicates the nature of the internal deformation this pile of sediments has undergone.

This deformation appears to continue to the south up to an inferred east-west transform fault along the Araya and Paria Peninsulas of Venezuela and the North Range of Trinidad, which defines the southern border of the Caribbean crustal plate (Fig. 7). Slip mechanisms (Molnar and Sykes, 1969) along this zone indicate dextral movements. The fault trace and sense of motion are indicated by the shift in location of bathymetric contours east of Trinidad on an unpublished Navy Oceanographic Office chart from which the contours of Figure 7 were selected. This fault zone separates the negative free-air anomaly minimum south of Barbados from the large negative anomaly (-200 mgal) of southern Trinidad and eastern Venezuela (Fig. 7).

With the formation of the Barbados ridge and its associated uplifted topography, the trench in which the sediments had accumulated ceased to exist. Farther north, the axis of the Puerto Rico trench became displaced oceanward from the axis of the negative free-air anomaly belt and became increasingly shallow to the south, ceasing to be identifiable near 16° N., 59° W.

The present historic volcanoes, with only one exception, lie close to the western edge of the ridge on which they occur. Also, the western slope of the ridge is much steeper than the eastern slope (see U.S. Naval Hydrographic Office BC charts 0703N, 0704N), although these relations are less pronounced for the northernmost islands (Montserrat to Saba). It is possible that these relations indicate that a slight westward shift in the site of volcanism did accompany the formation of the Barbados ridge.

ACKNOWLEDGMENTS

My years of association with Harry Hess will remain a source of wisdom, kindliness, and human warmth and wit. Albert Taylor of Massachusetts Institute of Technology kindly provided a magnetic tape copy of the U.S. Coast and Geodetic Survey earthquake compilation. Wilfred Bryan and Elizabeth Bunce suggested manuscript improvements. This study was supported by Grants GA-1209 and GA-12204 from the National Science Foundation and Contract Nonr-241 with the Office of Naval Research, Department of the Navy.

REFERENCES CITED

Baadsgaard, P. H., 1960, Barbados, W. I.: Exploration results 1950–1958: Internat. Geol. Cong., Part XVIII, Proc. of sed. 18, p. 21.

Bowin, Carl, 1971, Some aspects of gravity and tectonics in northern Caribbean: 5th Carib. Geol. Cong. Trans., St. Thomas, Virgin Islands, 1968, Queens College Press.

Bruins, G. J., 1960, Gravity expeditions 1948–1958; Neth. Geod. Comm. Pub., Delft, Waltman, 111 p.

Bunce, E. T., Phillips, J. D., Chase, R. L., and Bowin, C. O., 1971, The Lesser Antilles arc and the eastern margin of the Caribbean Sea, in The sea, V. 4: New York, Interscience.

Chase, R. L., and Bunce, E. T., 1969, Underthrusting of the eastern margin of the Antilles by the floor of the western North Atlantic Ocean, and origin of the Barbados ridge: Jour. Geophys. Research, v. 74, p. 1413–1420.

Christman, R. A., 1953, Geology of St. Bartholomew, St. Martin, and Anguilla, Lesser Antilles: Geol. Soc. America Bull., v. 64, p. 65–96.

Dickerson, R. E., 1940, Basic gravity-survey of Cuba: Am. Geophys. Union Trans., v. 21, p. 213–224.

Field, R. M., Brown, T. T., Collins, E. B., and Hess, H. H., 1933, The Navy-Princeton gravity expedition to the West Indies: U.S. Navy Hydrog. Office Tech. Rept., 55 p.

Hess, H. H., 1938, Gravity anomalies and island arc structure with particular reference to the West Indies: Am. Philos. Soc. Proc., v. 79, p. 71–96.

—— 1957, The Vening-Meinesz negative gravity anomaly belt of island arcs (1926–1956): Koninkl. Nederlandsch Geolog. Mijnbouwk. Genoot., Verhandelingen, Geolog. Serie, deel 18, p. 183–188.

—— 1962, History of ocean basins, in Engle, A.E.J., James, H. L., and Leonard, B. F., eds., Petrologic studies: A volume in honor of A. F. Buddington: Geol. Soc. America, p. 599–620.

Kuenen, P. H., 1936, The negative isostatic anomalies in the East Indies (with experiments): Leidse Geol. Meded., v. 8, p. 169–214.

Martin-Kaye, P., 1960, Discussion: 2d Caribbean Geol. Conf. Trans., Mayagüez, Puerto Rico, p. 155.

Maxwell, J. C., 1948, Geology of Tobago, B.W.I.: Geol. Soc. America Bull., v. 59, p. 801–854.

Minear, J. W., and Toksöz, M. N., 1970, Thermal regime of a down-going slab and new global tectonics: Jour. Geophys. Research, v. 75, p. 1397–1419.

Molnar, P., and Sykes, L. R., 1969, Tectonics of the Caribbean and Middle America regions from focal mechanisms and seismicity: Geol. Soc. America Bull., v. 80, p. 1639–1684.

Officer, C. B., Ewing, J., Hennion, J., Harkrider, D., and Miller, D., 1959, Geophysical investigations in the eastern Caribbean: Summary of the 1955 and 1956 cruises: Earth Phys. and Chem., v. 3, p. 17–109.

Oxburgh, E. R., and Turcotte, D. L., 1970, Thermal structure of island arcs: Geol. Soc. America Bull., v. 81, p. 1665–1688.

Robson, G. R., and Tomblin, J. F., 1966, Catalogue of the active volcanoes and solfatara fields of the West Indies: Part XX of catalogue of the active volcanoes of the world: Internat. Assoc. Volcanol.

Shurbet, G. L., and Ewing, M., 1956, Gravity reconnaissance survey of Puerto Rico: Geol. Soc. America Bull., v. 67, p. 511–534.

Shurbet, G. L., and Worzel, J. L., 1957, Gravity measurements in Oriente Province, Cuba: Geol. Soc. America Bull., v. 68, p. 119–124.

Shurbet, G. L., Worzel, J. L., and Ewing, M., 1956, Gravity measurements in the Virgin Islands: Geol. Soc. America Bull., v. 67, p. 1529–1536.

Sykes, L. R., and Ewing, M., 1965, The seismicity of the Caribbean region: Jour. Geophys. Research, v. 70, p. 5065–5074.

Vening-Meinesz, F. A., 1930, Maritime gravity survey in the Netherlands East Indies, tentative interpretation of the provisional results: Koninkl. Nederlandse Akad. Wetensch. Proc., Amsterdam, v. XXXIII, p. 566–577.

—— 1932, Gravity expeditions at sea, 1923–1930, I: Neth. Geod. Comm. Pub., Delft, Waltman, 109 p.

—— 1941, Gravity expeditions at sea, 1934–1939, III: Neth. Geod. Comm. Pub., Delft, Waltman, 97 p.

Vening-Meinesz, F. A., and Wright, F. E., 1930, The gravity measuring cruise of the U.S. Submarine S-21: U.S. Naval Obs. Publ. 13, Appendix I, 94 p.

350   C. Bowin

Vine, F. J., and Matthews, D. H., 1963, Magnetic anomalies over oceanic ridges: Nature, v. 199, p. 947–49.
Worzel, J. L., 1965, Pendulum gravity measurements at sea, 1936–1959: New York, John Wiley & Sons, Inc. 422 p.

MANUSCRIPT RECEIVED BY THE SOCIETY MARCH 29, 1971

CONTRIBUTION NUMBER 2611 OF THE WOODS HOLE OCEANOGRAPHIC INSTITUTION

THE GEOLOGICAL SOCIETY OF AMERICA, INC.
MEMOIR 132
© 1972

Continental Crust, Crustal Evolution, and the Caribbean

WILLIAM D. MACDONALD

State University of New York, Binghamton, New York 13901

ABSTRACT

Attention is focused on the genesis and tectonic behavior of the crust, especially the continental crust. A distinction is made between the rigid upper mantle, or *peridosphere,* and the crust which overlies it. Crust and peridosphere together make up the lithosphere.

Five major types of crust are recognized. *Continental* crust is distinguished especially by its great thickness and by the wide-spread distribution of Precambrian gneisses. *Oceanic* crust, overlying the Moho in the deep ocean basins, is thinner, younger, and more widespread over the Earth's surface. Undeformed crustal masses which contain great thicknesses of basaltic rocks overlying otherwise normal oceanic crust are called *platillo* crust. The undeformed crust of smaller ocean basins, being thicker than normal oceanic crust, is believed to represent a great thickness of sedimentary layers overlying normal oceanic crust. Such crust has been called *transitional* crust (Menard, 1967). Where Phanerozoic crustal rocks have accumulated into masses of tectonites of continental thickness, the crust is called *tectonitic.* A map of crustal types of the Caribbean area is presented as an illustration of classification.

Basalt is assumed to be the primary ultimate source of the continental crustal materials. Many processes, especially tectonic ones, have built up the continental crust to a relatively uniform equilibrium thickness. That thickness is determined by the interplay of tectonics, erosion, and isostasy.

Lateral additions probably dominate the process of continental growth. Sedimentary, igneous, and tectonic processes at the continental margins are important contributors. Away from the continental margins, some additions to the continental mass are possible by sedimentation on the surface, by limited intrusion near the surface, and perhaps by sill-like intrusions along the Moho. Intrusion of basic magma at the base of the continent is a mechanism which can explain epeirogenic uplift, the Conrad discontinuity, and the slight increase of density with depth in the continents.

The disappearance of widespread granulite metamorphic facies terrains at the end of the Precambrian is interpreted to be the result of cooling in the crust. It is suggested that the 550° C isotherm descended to levels below the continental crust near the end of Precambrian time. This would simultaneously permit the formation of serpentinite along the Moho and increase the rigidity of the crust. The possibility of slip, or free sliding, of the continents along the Moho is thus increased in Phanerozoic time.

Continental drift is seen to involve three possible processes: passive rafting, overriding, and free sliding. All three may operate simultaneously to produce the net effect known as drift.

CRUST AND LITHOSPHERE

Continental crust is the primary interest of this paper. In order to distinguish it from other crustal types from which it presumably has evolved by primarily tectonic processes, five major types of crust are recognized.

The Mohorovicic seismic discontinuity separates rock materials of the mantle below, with seismic P-wave velocity greater than 8 km/sec, from the rock materials of the crust above, with seismic P-wave velocities typically 7 km/sec or less. The Moho is typically 35 to 40 km below sea level under continents, and about 12 km below sea level in oceanic regions. Thus the Moho differentiates in a general way the thinner oceanic crust from the thicker continental crust. The relatively uniform depth to the Moho under continents, and the uniform but shallower depth under oceanic regions, suggest that the Moho is probably a compositional discontinuity between mantle and crust.

Before continuing with differences in crustal types, let us examine the subdivisions of the upper mantle briefly. According to Daly (1940), the outer shells of the Earth are the lithosphere, the asthenosphere, and the mesosphere. These are distinguished by their relative strengths. Morgan (1968) used the term *tectosphere* to include both the rigid upper mantle layer and the rigid crust, in a sense analagous to the *lithosphere* of Daly (1940), and of Isacks and others (1968). To distinguish between crust and mantle, the term *peridosphere* is introduced here for the uppermost layer of the mantle. The term derives from peridotite, a rock rich in the Mg-Fe silicates olivine and pyroxene, and one of the most likely constituents of the uppermost mantle. By definition, the peridosphere lies on the asthenosphere (Fig. 1) and its upper limit is the Moho.

The mantle is not everywhere easily distinguishable from the crust by seismic contrasts. For example, under the mid-Atlantic ridge, the depth to P-wave velocities of 8 km/sec is considerably greater than 12 km (Le Pichon and others, 1965). This does not necessarily indicate a thicker oceanic crust, nor even a different variety of crust. Instead, anomalously low seismic velocities in the upper peridosphere might be the explanation. These lowered velocities could be due to heating, fracturing, intrusion, and phase changes associated with the creation of new oceanic crust and peridosphere.

Types of Crust

Five types of crust are relatively easily recognized by criteria of thickness, composition, and structure. Most widespread, covering most of the surface of the Earth, is the oceanic crust. Next most abundant is continental crust. Next in

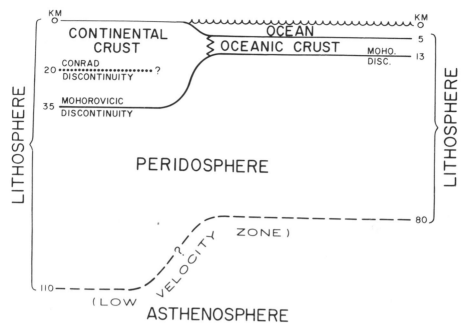

Figure 1. Crust, peridosphere, and lithosphere relations.

order of decreasing abundance are tectonitic, transitional, and platillo crust. The approximate distribution of these types in the Caribbean region (Fig. 2) illustrates the applicability of this classification.

Oceanic Crust

The oceanic crust typically lies beneath 4.5 to 5 km of water (Menard and Smith, 1966). Three principal layers are identifiable from seismic refraction studies (Raitt, 1963). The upper layer, variable from zero to a few kilometers in thickness, consists of sediments. The second layer, about 1.7 km thick, is possibly tholeiitic basalt or its slightly metamorphosed equivalent. The third layer, about 5 km thick, is of less certain composition. It may be chloritic-actinolitic greenstone, amphibolite, and some other metamorphic equivalent of basalt or gabbro (for example, Melson and van Andel, 1966). Alternatively, Hess (1960) suggested that this layer might be serpentinized peridotite representing hydrated upper mantle material. The total thickness of oceanic crust is about 6 to 8 km.

Continental Crust

The continental crust, about 35 to 40 km thick, has often been described as sialic, in reference to its notable content of silica and alumina. However, a primary distinguishing characteristic of continental crust is the presence of high-grade gneisses of Precambrian age. These metamorphic rocks indicate heating and re-crystallization of extensive regions of Precambrian rocks at temperatures of 600° C and above. Precambrian gneisses and related igneous rocks form the nuclei of the ontinents (Hurley and Rand, 1969). They are covered by extensive but relatively

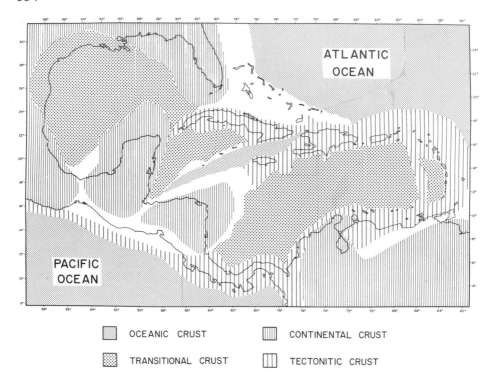

Figure 2. Crustal types of the Caribbean region.

thin veneers of younger, Phanerozoic deposits. Delimitation of the Precambrian nuclei of the continents is made difficult by the superficial cover, as well as by later remetamorphism and peripheral additions of tectonitic crust.

Tectonitic Crust

Mountain ranges and island arcs consist largely of deformed and metamorphosed sedimentary and volcanic rocks. Deformed rocks (tectonites) are accumulated by tectonic processes into masses reaching tens of kilometers in thickness. These crustal masses are referred to here as tectonitic crust. The initial materials of tectonitic crust are pre-existing crustal rocks, to which additions of igneous rocks such as andesite and diorite have been made during deformation. Examples of tectonitic crust are in the Greater Antilles, the Lesser Antilles, the Isthmus of Panama, and the western cordilleras of Colombia, and wherever deformed rocks are found lacking a base of Precambrian gneiss.

Metamorphic rocks are ultimately derived from pre-existing sedimentary and igneous rocks. Similarly, the *ultimate* source of most sedimentary rock materials is primarily igneous. It is therefore apparent that the large crustal plates of Precambrian metamorphic rocks have been derived ultimately from pre-existing igneous crustal sources. Because of its abundance in oceanic volcanoes and rift zones, and because of its abundance as an extrusive rock on the continents, basalt is thought to be the ultimate igneous source rock from which continental crust has been derived.

Apart from the widespread and very high-grade metamorphism of some Precambrian terrains, continental crust and tectonitic crust in this classification are distinguishable only by the age of deformation. The problem of distinguishing tectonitic crust from continental crust is most difficult in Paleozoic metamorphic terrains. An improvement in the present classification of crustal types might be achieved by mapping zones of tectonitic crust based on their age of deformation. This approach could ultimately eliminate the distinction between continental and tectonitic crust but requires a great effort in radiometric age determinations and field mapping.

Transitional Crust

The crust under some seas and small ocean basins is neither typical continental nor oceanic. It reaches 20 km or more in thickness, but has not been obviously deformed. Apparently it consists primarily of sedimentary rock, presumably resting on a base of (modified) oceanic crust. This type of crust has been termed transitional crust (Menard, 1967). The unusual thickness of this type of crust is possibly due to sedimentation in oceanic regions over abnormally long periods of time, or to abnormally greater sedimentation rates. But perhaps it results from abnormally slow spreading rates or from some as yet unknown process. The crusts of the Caribbean, Gulf of Mexico, and western Mediterranean basins are examples of transitional crust.

Platillo Crust

A thick pile of volcanic rock in an oceanic region might be regarded approximately as a volcanic analog of sedimentary transitional crust. Such anomalously thick volcanic crust is termed here platillo crust, from the Spanish word *platillo*, literally, a little plate. Platillo crust reaches tens of kilometers in thickness, and is produced by the extrusion of basaltic rock in large volume in an oceanic area, without deformation. Masses of platillo crust are typically discoidal in shape. Examples are Hawaii, Iceland, and Bermuda. The areal extent of platillo crust is much less than that of the other crustal types. Nevertheless, masses of platillo crust may become transformed into tectonitic crust and added to continental masses by tectonic processes at continental margins.

CARIBBEAN CRUST

The above classification has been applied to the Caribbean region (Fig. 2); this interpretation is modified from MacDonald (1967), and has been greatly influenced by the conclusions of Hess (1960).

One of the interesting observations is that the Cayman (Bartlett) trough is interpreted as normal oceanic crust. This suggests that the Cayman trough was formerly a spreading oceanic ridge or ridge section, or that it formed as a "leaky" transform fault. Its topographic expression and thin sedimentary layers suggest that it is a young Cenozoic feature. The magnetic patterns in the trough bottom should be very carefully studied for further clues on its origin and growth.

Platillo crust is of minor volume, as elsewhere on the Earth's surface.

Continental crust (Fig. 2) constitutes the Yucatan and Honduras microplates, as well as the Florida peninsula. Much of South America is similarly continental.

Early to mid-Paleozoic rocks, based on radiometric ages, are known from deep wells in Yucatan and Florida (Bass, 1969; Bass and Zartmann, 1969). Presumably those rocks rest on a Precambrian base. Precambrian to mid-Paleozoic rocks are also present in Guatemala near the tectonically disturbed zone between the Yucatan and Honduras microplates (Gomberg and others, 1968; McBirney and Bass, 1969).

Precambrian or Paleozoic rocks, however, are not known from the Greater Antilles, which are thought to consist mainly of deformed Mesozoic rocks. The oldest known fossiliferous strata in Cuba are of Jurassic age. The available radiometric age values from metamorphic rocks of Cuba, Hispaniola, and Puerto Rico are Cretaceous (Meyerhoff and others, 1969; MacDonald, 1968; Tobisch, 1968). Accordingly the Greater Antilles are interpreted as being of tectonitic crust, deformed during Cretaceous time. The Lesser Antilles are likewise thought to be of tectonitic crust, but the age of the principal deformation is early Cenozoic. The Aves Ridge is underlain by relatively dormant tectonitic crust, of probable Mesozoic age.

Tectonitic crust is also present in the northern South America region which has had a complex history of deformation. Very few radiometric ages of the metamorphic rocks of this region are available, but there are indications that some of the rocks of the Coast Range of Venezuela, for example, are as old as Ordovician (Hurley, 1969). The last major deformation of this area, however, was in Cretaceous to Paleocene time.

Tectonitic crust in southern Central America, Panama, and northwest Colombia is believed to have developed also in Cretaceous to Cenozoic time.

The anomalous crust of the Bahamas, left blank in Figure 2, has several kilometers of Cenozoic to Cretaceous limestone resting on an unknown base. According to the interpretation of Dietz and others (1970), this area probably is transitional crust.

The main basins (Yucatan, Venezuelan, and Colombian) of the Caribbean are underlain by transitional crust (Menard, 1967), a situation that bears significantly on the origin of the Caribbean. Two possibilities for the great thickness of Caribbean crust are, firstly, sedimentation over a very long period of time, extending perhaps back to earliest Mesozoic or pre-Mesozoic times (Ewing and others, 1967), and secondly, ponding of sediments combined with rapid sedimentation rates.

If the Caribbean antedates the disruption of the Atlantic basin, the Caribbean may represent a fragment of an older ocean basin, such as the Tethys. If so, the fragments of older ocean basins might be matched. For example, both the Mediterranean and the Caribbean may be fragments of a single pre-Atlantic basin, and the predisruption stratigraphic sections of those basins might be matched. Detailed deep seismic reflection studies could be useful for this purpose.

Alternatively, the great thickness of Caribbean crust could be due to high, or formerly high, sedimentation rates. It is interesting to note that the Andes and the coastal mountains of Venezuela have risen only relatively recently. In so doing, they have undoubtedly disrupted former drainages.

Perhaps the pre-Miocene or pre-Cenozoic major drainage of South America was northward into the Caribbean rather than eastward into the Atlantic. This could account for a greater thickness of sediment in a shorter period of time, and perhaps allow the Caribbean and Atlantic to have formed simultaneously. The presence of salt domes in both the Mediterranean and Gulf of Mexico is certainly enigmatic, and must be accounted for in the final interpretation of the origins of transitional crust.

Alternative explanations for the genesis of the Caribbean still might be required. For example, "oceanization" or basification of former continental crust has been proposed to account for the origin of the Caribbean (Skvor, 1969). Assuming the evolutionary concept of continental crust presented here, proof of the oceanization hypothesis requires demonstration that remnants of Precambrian gneisses will be found deep in the Caribbean crust. This writer regards this as unlikely.

CONTINENTAL CRUST AND TECTONIC PROCESSES

Continental Accretion and Decretion

If accretion is taken as the process of building up the mass of the continental crust, then the term decretion might reasonably be applied to the process of diminishing the mass of the continental crust. The most important mechanism of accretion is orogeny at the continental margin, by which new material is welded onto or intruded into the continental mass in tectonic deformations related to continental drift and peridospheric plate motions. Suturing together of continental plates is a related process by which the size of a continental mass is increased. Near-surface additions of material to the continental mass are also possible by volcanic extrusion, batholithic intrusion (Hamilton and Myers, 1967), and by some types of sedimentation which involve a net addition to the mass of the continent.

Intrusion of basic (tholeiitic) magma at the base of the continent along the Moho may also contribute to the mass of the continent. Such intrusion provides a mechanism of epeirogenic uplift, and also accounts for the deduced increase of density with depth in the continents. Basic intrusions of this sort probably recrystallize or metamorphose into amphibolite and hornblende-plagioclase gneiss. This mechanism may have contributed significantly to the growth of continents.

Continental drift implies that continental masses can be broken apart through some fissuring mechanism, and this is one of the important processes of decretion of continental mass. Other processes of decretion include mechanical and chemical weathering processes in erosion at the Earth's surface. Some of these losses are made up by sedimentation elsewhere on the continent. Subcrustal removal of mass from a continent, in ion migrations or in erosion by mantle convection currents, has also been suggested in the literature in various ways (Gidon, 1963).

Origin of the Continental Crust and Control of Its Thickness

Despite the detailed studies of many thousands of geologists, no unmodified 4-b.y.-old remnants of the earliest Earth's crust have been recognized; there are probably none remaining in the original form. These earliest crustal rocks were probably igneous. The present continental crustal rocks are construed here as having been derived ultimately from basalt modified by various sedimentary, igneous, metamorphic, and biologic differentiation processes, and thickened through tectonic processes.

The relatively uniform thickness to which the continental crust has been built up is remarkable. This is apparently the result of interaction of isostasy and erosion.

In the simplest case, platillo crust, for example, could be built up until sea level is reached. As the crust rises above sea level, erosional processes at and above sea level reduce the exposed landmass, and redistribute that mass around the seaward margins.

The thickness to which crust will build up, assuming effective erosion and constant volume of sea water, depends upon the relative densities and thicknesses of layers of water, crust, and upper mantle (Hess, 1955). If the volume or depth of sea water varies with time, then the equilibrium thickness of the crust also varies. The isostatic equilibrium of the crust relative to an oceanic section of equal thickness can be expressed in the form

$$d_1t_1 + d_2t_2 + d_3t_3 + \ldots = d_ct_c$$

where d_1 and t_1 are density and thickness, respectively, of layer 1, and so forth, of the oceanic section, and d_c and t_c similarly for the continental crustal section. Assume that an oceanic section of mantle ($d = 3.33$), overlaid by 5 km of water ($d = 1.01$) is in isostatic equilibrium with platillo crust ($d = 2.95$). Solution of the isostatic equilibrium equation indicates that platillo crust could pile up to an equilibrium thickness of 31 km.

The equilibrium thickness of continental crust can be obtained from an equation of the form

$$t_c = [d_m(t_w + t_{oc}) - (d_wt_w + d_{oc}t_{oc})]/(d_m - d_c)$$

(m = mantle, c = continental crust; w = water; oc = oceanic crust). From this it can be appreciated that the density difference ($d_m - d_c$) between mantle and crust is a very important factor in determining the isostasy-erosion equilibrium thickness of any crustal mass which rises to sea level.

Crustal Evolution

The evolutionary concept of continental crust embodied in the classification presented here also has nontectonic implications. For example, it indicates a primordial Precambrian Earth with no continental crust. Assuming the same volume of seawater in the past as at present, the primordial Earth was covered with an ocean of about 3 km depth, as opposed to the present 5 km depth. The concept also indicates the rate of accumulation of continental crust; estimating the present mass of continental crust including microcontinents at 19.8×10^{24} g, the rate of production of crustal materials over 4.5 b.y. is 4.4×10^{15} g/yr, equivalent to about 1.5 km^3/yr of crust. This is a rough estimate for the minimum rate of production of basalt at the Earth's surface over the same period.

Thermal History of Continental Crust

Widely exposed expanses of Precambrian rocks have been metamorphosed in the upper amphibolite and granulite facies, at temperatures of 600° C and above. Exposed younger metamorphic zones are generally much narrower in width and lower in grade of metamorphism. From this one might conclude that the 600° C isotherm descended to levels below the continent at about the end of Precambrian time. This assumes, among other things, that the high-grade metamorphism of Precambrian terrains took place within the continental crust rather than in the mantle. Morgan (1970, written commun.) has suggested that by extrapolating the present amount of radioactive material backward in time, the conclusion is reached that 3 b.y. ago there was roughly twice as much radiogenic heat produced as at present. Since the thermal conductivity of the crust would be about the same as now, the 600° C isotherm could have been at half its present depth.

Present temperatures of around 450° C have been estimated for the base of the tectonically quiescent continental crust (Roy and others, 1968). Sometime between Precambrian time and the present, the temperature at the base of the crust dropped below 550° C. This allowed the stable formation of serpentinite at the base of the crust and marked the cessation of granulite facies metamorphism in the crust. With lowered temperature gradients, the continental crust became more rigid at the same time that it acquired a weaker underfooting. Thus, with the passage of time, relative slip between continental crust and mantle became more probable during events of tectonic stress. This has possible significance in regard to the mechanisms of continental drift.

Drift Mechanics

The concept of continental drift implies that the continents have long-term rigidity, and that nondeformed margins can be fitted back together after separation (Carey, 1958; Bullard and others, 1965). However, the mechanisms by which relative motions of continents takes place is as yet little understood. The three potentially most significant processes of relative motions of continents can be characterized as free sliding, passive rafting, and overriding (Fig. 3), according to the relative motions of continental plate and underlying peridosphere.

Free Sliding

Free sliding implies a relative displacement between a continental plate and the entire underlying peridosphere. It was this concept that was greatly debated during the 1920s (Taylor, 1928; Chamberlin, 1928; Longwell, 1928). Although tidal, rotational, and gravitational forces have been invoked to explain such motion, no sufficiently compelling evidence in its favor has yet been presented. Nevertheless, assuming that free sliding can take place, three tectonic zones should be distinguishable (Fig. 3): (1) a frontal zone, where the leading continental margin slides over and depresses the adjacent peridosphere, (2) lateral zones of shear, where transcurrent displacement is found, and (3) a posterior zone of isostatically uplifted mantle. The frontal zone should remain out of isostatic adjustment during movement, and is a zone of great deformation including folding and thrusting. Pericontinental

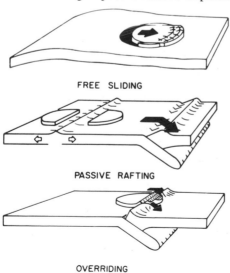

FREE SLIDING

PASSIVE RAFTING

OVERRIDING

Figure 3. Mechanisms of continental drift, showing relative motion of continental plates and underlying peridosphere.

sedimentary aprons might also be overridden in this region. The posterior zone of uplifted mantle would be characterized by a magnetic signature much different from that of ordinary oceanic crust, and would be overlain by an oceanic crust of quite different origin from that which forms in the spreading process. Perhaps the magnetically quiet zones of the eastern and western North Atlantic (Heirtzler and Hayes, 1967) are examples of the posterior zone of sliding. It might be possible to demonstrate that some structures of the ocean basin which antedate sliding are terminated at the edge of the posterior zone. It was noted earlier that serpentinization of the upper mantle under continents would form a weak zone at the base of a continent; this would facilitate the free sliding process.

Passive Rafting

The concept of passive rafting of continents (Hess, 1962) greatly promoted the acceptance of the drift concept. Passive rafting implies that a continent can be broken apart and the fragments dispersed by the relative motion of the peridosphere in which the fragments lie. This mechanism requires a zone between the separating fragments where new oceanic crust and peridosphere are created, and one elsewhere in which they are consumed. These have been shown to be, respectively, the oceanic ridge system (Vine, 1966) and the deep trenches (Isacks and others, 1968). Major deformation of a rafted continent does not take place except (1) by fissuring accompanying initial rifting of the lithosphere, and (2) by overriding when a rafted continent impinges upon a trench area or subduction zone. In rafting there is no relative motion of continental plate over the underlying peridosphere, except beyond a subduction zone where overriding can take place.

Overriding

In overriding, the effects observed will depend upon what mass overrides, and what mass is overridden. Continental crust may override adjacent peridosphere which either is contiguous with the main mass of peridosphere underlying the continent (general case of free sliding), or is separated, by a subduction zone, from the main mass of peridosphere underlying the continent (case of rafted continent reaching a trench).

The concept of overriding is not new to tectonics. Holmes (1965) indicated that the Himalayas are the isostatic uplift effect of overriding of the continental crusts of India and Asia. Also Wilson (1966) implied another variety of overriding, that of oceanic crust over oceanic crust, as an explanation for the very thick crust of the Caribbean. Note that, in the case of overriding across a subduction zone, the overriding crust is necessarily physically separated from the sinking peridosphere, and becomes supported by a different lithospheric plate.

Obviously free sliding and passive rafting can occur independently of one another. Overriding necessarily accompanies free sliding. Moreover, the possibility of free sliding during passive rafting is allowable, in which case free sliding, passive rafting, and overriding simultaneously operate to produce the drift effects.

REFERENCES CITED

Bass, M. N., 1969, Petrography and ages of crystalline basement rocks of Florida—some extrapolations: Am. Assoc. Petroleum Geologists Mem. 11, p. 283–310.

Bass, M. N., and Zartmann, R. E., 1969, The basement of the Yucatan Peninsula (abs.): Trans. Am. Geophys. Union, v. 50, p. 313.

Bullard, E. C., Everett, J. E., and Smith, A. G., 1965, The fit of the continents around the Atlantic: Royal Soc. London Philos. Trans., ser. A, v. 258, p. 41–51.

Carey, S. W., 1958, A tectonic approach to continental drift, *in* Continental drift, a symposium: Hobart, Geol. Dept., Univ. Tasmania, p. 177–355.

Chamberlin, R. T., 1928, Some of the objections to Wegener's theory, *in* Theory of continental drift: Tulsa, Am. Assoc. Petroleum Geologists, p. 83–87.

Daly, R. A., 1940, Strength and structure of the Earth: New York, Prentice-Hall, Inc., 434 p.

Dietz, R. S., Holden, J. C., and Sproll, W. P., 1970, Geotectonic evolution and subsidence of Bahama platform: Geol. Soc. America Bull., v. 81, p. 1915–1928.

Ewing, J., Talwani, M., Ewing, M., and Edgar, T., 1967, Sediments of the Caribbean: Studies of tropical oceanography: Univ. Miami, v. 5, p. 88–102.

Gidon, P., 1963, Courants magmatiques et evolution des continents: Paris, Masson et Cie., 155 p.

Gomberg, D. N., Banks, P. O., and McBirney, A. R., 1968, Guatemala: Preliminary zircon ages from the Central Cordillera: Science, v. 161, p. 121–122.

Hamilton, W., and Myers, W. B., 1967, The nature of batholiths: U.S. Geol. Survey Prof. Paper 554–C, 30 p.

Heirtzler, J. R., and Hayes, D. E., 1967, Magnetic boundaries in the north Atlantic Ocean: Science, v. 157, p. 185–187.

Hess, H. H., 1955, Serpentines, orogeny, and epeirogeny, *in* Poldervaart, A., ed., Crust of the Earth: Geol. Soc. America Spec. Paper 62, p. 391–408.

—— 1960, Caribbean research project: Progress report: Geol. Soc. America Bull, v. 71, p. 235–240.

—— 1962, History of ocean basins, *in* Engel, A. E. J., James, H. L., and Leonard, B. F., eds., Petrologic studies: A volume in honor of A. F. Buddington: Geol. Soc. America, p. 599–620.

Holmes, A., 1965, Principles of physical geology: New York, Ronald Press Co., 1288 p.

Hurley, P. M., 1969, Basement gneiss, Cordillera de la Costa, Venezuela, *in* Variations in isotopic abundances of strontium, calcium, and argon and related topics: Massachusetts Inst. Tech. Dept. Geol. and Geophys., M.I.T.–1381–16, Sixteenth Ann. Prog. Rept. for 1968, p. 81.

Hurley, P. M., and Rand, J. R., 1969, Pre-drift continental nuclei: Science, v. 164, p. 1229–1242.

Isacks, B., Oliver, J., and Sykes, L. R., 1968, Seismology and the new global tectonics: Jour. Geophys. Research, v. 73, p. 5855–5899.

LePichon, X., Houtz, R. E., Drake, C. L., and Nafe, J. E., 1965, Crustal structure of the mid-ocean ridges, 1: Seismic refraction measurements: Jour. Geophys. Research, v. 70, p. 319–339.

Longwell, C. R., 1928, Some physical tests of the displacement hypothesis, *in* Theory of continental drift: Tulsa, Am. Assoc. Petroleum Geologists, p. 145–157.

MacDonald, W. D., 1967, Continental drifting between the Americas: Paper presented at the UNESCO Symposium on continental drift emphasizing the history of the South Atlantic area, Montevideo, 1967: Am. Geophys. Union.

—— 1968, Continental drift and the Caribbean: Caracas, Boletin Informativo, v. 11, p. 17–24.

McBirney, A. R., and Bass, M. N., 1969, Structural relations of pre-Mesozoic rocks of northern Central America: Am. Assoc. Petroleum Geologists Mem. 11, p. 269–280.

Melson, W. G., and van Andel, Tj., 1966, Metamorphism in the mid-Atlantic ridge, 22° N latitude: Marine Geology, v. 4, p. 165–186.

Menard, H. W., 1967, Transitional types of crust under small ocean basins: Jour. Geophys. Research, v. 72, p. 3061–3073.

Menard, H. W., and Smith, S. M., 1966, Hypsometry of ocean basin provinces: Jour. Geophys. Research, v. 71, p. 4305–4325.

Meyerhoff, A. A., Khudoley, K. M., and Hatten, C. W., 1969, Geologic significance of radiometric dates from Cuba: Am. Assoc. Petroleum Geologists Bull., v. 53, p. 2494–2500.

Morgan, W. J., 1968, Rises, trenches, great faults, and crustal blocks: Jour. Geophys. Research, v. 73, p. 1959–1982.

Riatt, R. W., 1963, The crustal rocks, in Hill, Goldberg, E. D., Iselin, C. O'D., and Munk, W. H., eds., The sea: New York, Interscience Publ., v. 3, p. 85–102.

Roy, R. F., Blackwell, D. D., and Birch, F., 1968, Heat generation of plutonic rocks and continental heat flow provinces: Earth and Planetary Sci. Letters, v. 5, p. 1–12.

Skvor, V., 1969, The Caribbean area: A case of destruction and regeneration of a continent: Geol. Soc. America Bull., v. 80, p. 961–968.

Taylor, F. B., 1928, Sliding continents and tidal and rotational forces, in Theory of continental drift: Tulsa, Am. Assoc. Petroleum Geologists, p. 158–177.

Tobisch, O. T., 1968, Gneissic amphibolite at Las Palmas, Puerto Rico, and its significance in the early history of the Greater Antilles arc: Geol. Soc. America Bull., v. 79, p. 557–574.

Vine, F. J., 1966, Spreading of the ocean floor: New evidence: Science, v. 154, p. 1405–1415.

Wilson, J. T., 1966, Are the structures of the Caribbean and Scotia arc regions analogous to ice rafting?: Earth and Planetary Sci. Letters, v. 1, p. 335–338.

MANUSCRIPT RECEIVED BY THE SOCIETY MARCH 29, 1972

THE GEOLOGICAL SOCIETY OF AMERICA, INC.
MEMOIR 132
© 1972

Is the Entire
Caribbean Mountain Belt of
Northern Venezuela Allochthonous?

A. BELLIZZIA G.

*Dirección de Geología, Ministerio de Minas e Hidrocarburos,
Centro Bolívar, Caracas, Venezuela*

ABSTRACT

The Caribbean Mountains of northern Venezuela and the immediately adjacent
foothills and plains to the south may be described in terms of eight subparallel
tectonic belts (1 to 8 from south to north). Belts 5 through 8 consist dominantly of
metamorphic rocks and constitute the mountain belt proper; belts 1 through 4 to
the south involve sedimentary rock units. Early workers viewed all the tectonic
belts as autochthonous. Subsequently Hess and students concluded that belt 5, the
southernmost in the mountain range, was allochthonous and that its present location was the result of gravity sliding from a root zone located to the north in the
Caribbean Sea.

Extensive mapping by the writer and assistants has revealed that belt 4 (the
northernmost belt in the sedimentary foothills) is a flysch facies of Paleocene-
Eocene age which, when followed westward, is found to curve northward around
the western limits of the mountain belt to a point, in the Barquisimeto region,
where it is north of tectonic belt 8 (the northernmost in the ranges). A similar
flysch facies of similar age is found on Margarita Island north of the mainland. The
writer concludes that this flysch facies does not overlap onto the metamorphic
rocks of the mountain belt but rather that the mountain belt, consisting dominantly
of Jura-Cretaceous metamorphic rocks, is *embedded* in the Paleocene-Eocene flysch.

It is suggested that the allochthonous tectonic belt 5 rests on a mountain belt
which is itself allochthonous on the southern continent.

INTRODUCTION

The Caribbean Mountains located along the north coast of Venezuela have been
the subject of intensive study by H. H. Hess and students of Princeton University,

by geologists of the Geological Survey of Venezuela (Dirección de Geología, Ministerio de Minas e Hidrocarburos), and by the School of Geology, Central University of Venezuela.

Reference to virtually all previous work may be found in the tectonic summaries of Hess (1966), Menéndez (1965), and Bell (1971). Pertinent off-shore geophysical data are summarized by Ball and others (1971). The essential data is summarized on the regional map (Fig. 1).

PRINCIPAL GEOLOGICAL CHARACTERISTICS

As indicated on the map (Fig. 1), the geology of the mountain belt and adjacent areas of the foothills and plains to the south may be described conveniently in terms of eight subparallel tectonic belts numbered 1 through 8 from south to north. These belts were originally defined and characterized by Menéndez (1965) and Bell (1968, 1971). The principal features of these belts may be briefly summarized as follows.

Mountain Ranges Proper (Tectonic Belts 5 through 8)

The northernmost belt (Cordillera de la Costa) consists of a granitic basement complex (Sebastopol) overlain by schists of the Caracas Group. Rock units similar to the latter also occur in the Araya-Paria, Paraguaná and Goajira peninsulas, and the islands of Margarita and Orchila. In the central portion of the Caribbean Mountains the Caracas Group is at the epidote-amphibolite facies of regional metamorphism but contains conformable lenses of eclogite, eclogite-amphibolite, and garnet-amphibolite (Morgan, 1970). Tectonic belt 8 constitutes a large anticlinorium, the limbs of which are truncated by major longitudinal normal faults which define the boundaries of the belt. Postmetamorphic bodies of granite intrude the Caracas Group and one (Guaremal) has been dated at 79 ± 5 m.y. (Morgan, 1969). A few sparse faunas suggest a Late Jurassic–Early Cretaceous age range for the Caracas Group.

The Caucagua–El Tinaco tectonic belt (7) consists of the El Tinaco basement complex, which includes metamorphic rocks at the almandine-amphibolite facies of regional metamorphism unconformably overlain by Late Cretaceous metasedimentary and metaigneous rocks of the post-Caracas Group, which are at the greenschist facies. Most of the section overlying the Tinaco Complex was interpreted by Menéndez (1967) as of allochthonous character. K/Ar ages in the range 110 to 120 m.y. on minerals of the Tinaco Complex (see Piburn, 1968, p. 125) are interpreted to represent the thermal event corresponding to the rigonal metamorphism of the Caracas and post-Caracas Groups. As in the case of tectonic belt 8, regional structure is interpreted in terms of a broad anticlinorium, the limbs of which have been cut by major faults showing significant vertical movement.

Tectonic belt 6 comprises the Paracotos Formation from which a rich, well-preserved foraminiferal assemblage of Maestrichtian age has been recovered. The formation shows zeolite-pumpellyite-prehnite mineral assemblages. For the most part, the Paracotos Formation constitutes a south-dipping homocline. The northern limit of the belt is the Santa Rosa fault, believed of normal type; the southern limit is the Agua Fria fault which has been interpreted by Hess (*see* Menéndez, 1967) as a surface of bedding plane décollement.

The Villa de Cura tectonic belt (5) consists dominantly of metamorphosed basic volcanic rocks. Spilitic lavas and tuffs, and some quartz-albite schists, constitute the underlying blueschist sequence (Villa de Cura Group), which is overlain by basalts containing pumpellyite and prehnite of the Tiara Formation. Piburn (1968) outlines radiometric and stratigraphic evidence indicating that the Tiara Formation is of Late Cretaceous age with maximum age limit of 100 ± 10 m.y. Concerning the Villa de Cura Group, all that can be said is that it must be older than 100 ± 10 m.y. The rock units of tectonic belt 5 effectively constitute a south-dipping homocline. The southern boundary of the belt is believed to be a north-dipping thrust fault (Cantagallo), which has been interpreted to be continuous with the Agua Fria fault which constitutes the northern boundary of the belt (Menéndez, 1967; Bell, 1971). Hence the Villa de Cura belt is viewed as an allochthon overlying the Maestrichtian Paracotos Formation.

Region South of the Mountain Ranges (Tectonic Belts 1 Through 4)

These belts are composed entirely of unmetamorphosed sedimentary rocks. Pursuing our summary of the tectonic belts southward, the principal features of these belts are as follows.

The Foothills belt (4) is composed of marine sandstones of Late Cretaceous age which grade up into a turbidite section, of flysch-wildflysch aspect, of Paleocene-Eocene age. The turbidite section displays strong evidence (extensive mélange, olistostromes) indicative of gravitational tectonics in its deposition. In many aspects it resembles the "Argille Scagliose" of the Appenines. This tectonic belt is characterized by large recumbent folds and north-dipping thrust faults suggesting that this belt, too, is largely allochthonous and was deposited originally north of its present area of outcrop. The Foothills belt gives way to a narrow discontinuous belt of thrust faults (3) in which upper Eocene to Oligocene marine sedimentary rocks have been repeatedly shuffled by imbricate bedding plane thrusts. These rocks (Chacual Complex, Fig. 1) display the typical sedimentary structures of mélanges and olistostromes and, like those of the Foothills belt, resemble the "Argille Scagliose."

In turn the thrust fault belt gives way to the overturned tectonic belt (2) in which Oligocene-Miocene marine and continental deposits occur as a syncline overturned to the south. This structure is believed to have formed as a result of the southward sliding of the allochthon of the Foothills belt and of the Villa de Cura tectonic belt contained within the Foothills belt. The effects of gravitational tectonics peter out in the southernmost tectonic belt (1), the belt of gentle dips, characterized by dips of 2° to 5° to the south.

DEVELOPMENT OF IDEAS OF THE TECTONIC HISTORY

Early workers in the Caribbean Mountains interpreted the geologic history in terms of autochthonous faultbound blocks. Invariably, major obstacles were encountered in the attempt to explain satisfactorily the relations between the volcanic rocks of tectonic belt 5 and the dominantly metasedimentary sequences of belts 6 through 8 on the one hand, with the unmetamorphosed sedimentary rocks of belts 1 through 4 on the other. Syntheses in which belt 5 was viewed (a) as a block of upthrust basement, (b) as an in-situ section located stratigraphically between the Caracas and post-Caracas Groups, and (c) as an in-situ section stratigraphically above the post-Caracas Group eventually had to be abandoned for a variety of reasons (stratigraphic, structural, metamorphic, and so forth). In an attempt to resolve the impasse, Hess (see Menéndez, 1967) suggested that the Villa de Cura Group volcanic rocks of tectonic belt 5 were allochthonous, having reached their present setting by sliding from a root zone located to the north in the Caribbean Sea. The concept was developed and elaborated by Menéndez (1967), Seiders (1965), and Bell (1968, 1971, 1972). The sliding is considered to have begun in Late Cretaceous (late- or post-Maestrichtian) time and to have been renewed in the Oligocene. As elaborated, the theory views tectonic belts 3, 4, and 5 and the post-Caracas Group of belt 7 as allochthonous, the remaining belts as autochthonous. The basic concept of extensive gravitational tectonics has much to recommend it: for example, the structural and apparent stratigraphic relations of the Villa de Cura tectonic belt to the Foothills belt, and the recumbent folds, thrust faults, and extensive zones of mélange and olistostromes in the Foothills and thrust belts (4 and 3). An additional welcome dividend of the theory was to supply a mechanism for the metamorphism of the Caracas Group. As outlined by Piburn (1968) and Morgan (1970) the Villa de Cura Group in its southward progress would provide the necessary thickness of overburden to meet the estimated pressure-temperature conditions of formation of the eclogitic rocks in the Caracas Group.

DISCUSSION

The writer is in essential agreement with the concept of emplacement of allochthonous blocks of regional dimensions. On two points, however, his views are divergent with those enumerated above. Firstly, it appears unlikely, in the light of available structural and stratigraphic data, that the thickness of volcanics in the Villa de Cura tectonic belt (5) could have approached the required 20+ km estimated by Morgan (1970) as the minimum thickness of overburden to explain the formation of the eclogitic rocks in the Cordillera de la Costa belt (8). This opinion is supported by recent gravity surveys (Bonini, 1971, oral commun.) which suggest that the Villa de Cura allochthon may be as thin as 4 km. The writer believes that the regional metamorphism must be explained in terms of deformation and downbuckling at a subduction zone, followed by subsequent uplift, dismemberment, and gravity sliding into a deep flysch basin. Secondly, it is suggested that the concept of emplacement of allochthons requires further extension and that the entire Caribbean Mountain system of northern Venezuela is allochthonous. In support of this thesis the following evidence is offered:

1. Mapping by the writer and associates of the Geological Survey of Venezuela has revealed that the turbidites of tectonic belt 4 extend westward and appear to curve northward around the western termination of the metamorphic rocks of tectonic belts 6, 7, and 8 (see Fig. 1). The uniformity of sedimentary characters

throughout this belt, despite the fact that it shows a marked angular discordance to the other tectonic belts to the north, strongly suggests that the turbidites do not lap over belts 5, 6, 7, and 8 but rather that the latter are embedded in it.

2. In eastern and western Venezuela there are complete Cretaceous sections consisting of unmetamorphosed sedimentary rocks completely free of volcanic associations. In addition, there are marked parallels in lithologies between the two regions. As indicated above, the Cretaceous sections of the mountain belt are metamorphosed and associated with basic volcanic rocks. The autochthonous view of these spatial and temporal relations fails to explain the abrupt limits to the distribution of metamorphic and volcanic rocks, nor does it account for the uniformities of the unmetamorphosed Cretaceous sections in two distinct basins separated by hundreds of kilometers. It appears more reasonable to propose that the western and eastern sedimentary sections are continuous beneath the mountain belt and that both the metamorphism and volcanism associated with the Cretaceous belts of the mountain ranges occurred when those belts were located farther north in the Caribbean.

3. In the Goajira and Paraguaná peninsulas, metamorphic and igneous rocks bearing a strong resemblance to those of the Caribbean Mountains are known. In Figure 1 there appears to be some 200 km offset between the trends of the Caribbean Mountains and the Goajira-Paraguaná occurrences.

4. In the island chain north of Venezuela, outcropping metamorphic and igneous rocks (including granitic, basic, and ultramafic varieties) resemble those on the mainland. Sources of volcanism and sites for regional metamorphism were clearly available well to the north of the Caribbean Mountains.

5. In the overall view there appears to be a crude gradation in the character of regional metamorphism in the mountain belt, the grade apparently decreasing from north to south. But, as summarized above, metamorphic characteristics vary by tectonic belt, and in reality, distribution of facies is heterogeneous. Similarly, statements to the effect that basic volcanic rocks occur in all the tectonic belts obscures the fact that the detailed petrology of the basic rocks also varies by tectonic belt and is heterogeneous in the overall view. These facts accord better with a geologic history in which all belts are viewed as allochthonous.

Because the turbidite section of tectonic belt 4 is of Paleocene-Eocene age, the writer believes that the principal event of gravity sliding occurred in late- or post-Eocene time. From the evidence summarized above, it would appear that this event was preceded by décollement of the Villa de Cura tectonic belt over the Maestrichtian Paracotos Formation. Clearly the mechanics of emplacement of the allochthons is not likely to have been a simple single event. It still remains to check the feasibility of the allochthonous concept against the growing body of geophysical and other data for the Caribbean Sea.

Reasonable reconstructions of paleographic, paleofacies, and palinspastic type indicate that allochthonous blocks could only have come from the general Caribbean–Central American region. One possibility is to stipulate the existence of a subduction zone north of, and subparallel to, the coast line at a latitude not south of Margarita Island, inasmuch as the Paleocene-Eocene flysch facies outcrops on that island (see Fig. 1). Depending on the validity of the amount and direction of drift contained in various reconstructions of plate motions, one might postulate more distant locations for the root zones of the allochthons and the zone of subduction.

368   A. BELLIZZIA G.

REFERENCES CITED

Ball, M. M., Harrison, C.G.A., Supko, P. R., Bock, W., and Maloney, N. J., 1971, Marine geophysical measurements on the southern boundary of the Caribbean Sea: Geol. Soc. America Mem. 130, p. 1–33.

Bell, J. S., 1968, Geología de la region de Camatagua, Estado Aragua, Venezuela: Bol. Geología, Min. de Minas e Hidrocarburos, Venezuela, v. IX, p. 292–440.

——1971, Tectonic evolution of the central part of the Venezuelan Coast Ranges: Geol. Soc. America Mem. 130, p. 107–118.

——1972, Geotectonic evolution of the southern Caribbean area: Geol. Soc. America Mem. 132, p. 369–386.

Hess, H. H., 1966, Caribbean research project, 1965, and bathymetric chart: Geol. Soc. America Mem. 98, p. 1–11.

Menéndez, A., 1965, Geología del area de El Tinaco, centro norte del Estado Cojedes, Venezuela: Bol. Geología, Min. de Minas e Hidrocarburos, Venezuela, v. 6, no. 12, p. 417–543.

——1967, Tectonics of the central part of the western Caribbean Mountains, Venezuela: Internat. Conf., Tropical Oceanography Proc., Studies in Tropical Oceanography, v. 5, p. 103–130.

Morgan, B. A., 1969, Geología de la region de Valencia, Carabobo, Venezuela: Bol. Geología, Min. de Minas e Hidrocarburos, Venezuela, v. 10, no. 20, p. 3–136.

——1970, Petrology and mineralogy of eclogite and garnet amphibolite from Puerto Cabello, Venezuela: Jour. Petrology, v. 11, p. 101–145.

Piburn, M. D., 1968, Metamorfismo y estructura del Grupo Villa de Cura, Venezuela septentrional: Bol. Geología, Min. de Minas e Hidrocarburos, Venezuela, v. 9, no. 18, p. 183–289.

Seiders, V. M., 1965, Geología de Miranda central, Venezuela: Bol. Geología, Min. de Minas e Hidrocarburos, Venezuela, v. 6, no. 12, p. 289–416.

MANUSCRIPT RECEIVED BY THE SOCIETY MARCH 29, 1972

THE GEOLOGICAL SOCIETY OF AMERICA, INC.
MEMOIR 132
© 1972

Geotectonic Evolution of the Southern Caribbean Area

J. S. BELL

*Department of Geology, University of Alberta,
Edmonton, Alberta, Canada*

ABSTRACT

Thick marine sedimentary and volcanic sequences accumulated in northern Venezuela and Trinidad during ?Late Jurassic and Cretaceous time. By latest Cretaceous time, the northern part of this sequence had been regionally metamorphosed and rapidly uplifted. Flysch troughs developed to the south which received exotic masses and turbidite sands until early Eocene time. These events are interpreted as resulting from southward movement of the Caribbean crust and its subduction into a south-dipping Benioff zone located north of Venezuela. This postulated Benioff zone is believed to have become inactive in latest Cretaceous time, and the resulting isostatic rebound apparently caused overthrusting which initiated the flysch troughs.

Mid-Eocene orogeny caused major changes in basin geometry and initiated crustal shortening, overthrusting, uplift, and strike-slip faulting in Venezuela and Trinidad. These events are believed to be related to east-west right-lateral transcurrent movement between the Caribbean and Americas Plate, which appears to have begun in Eocene time. Initially, the Caribbean Plate was offset a minimum of 35 km along faults which extended into western Venezuela and Colombia, but later the Andes also became involved in the relative movements between the plates.

INTRODUCTION

In some of his earliest publications (1932, 1938), Hess postulated that the Caribbean area behaved as a rigid lithospheric plate which was moving eastward relative to surrounding areas. Left-lateral tear faulting occurred within and along strike of the Cayman trough, while right-lateral movements affected northern Venezuela and Trinidad, as the Caribbean crust was dragged toward, and then buckled down into a tectogene located east of the Antillean Island arc.

369

The general outlines of Hess's Caribbean Plate have been confirmed by the distribution of earthquakes along its margins (Gutenberg and Richter, 1954; Sykes and Ewing, 1965), which strongly suggest that the plate is acting as an approximately rigid body at depths of less than 200 km. However, epicentral distribution and focal mechanism solutions of earthquakes associated with the Antillean Island arc support the presence of a westward-dipping Benioff zone rather than Hess's hypothetical tectogene (Molnar and Sykes, 1969). An active tectogene was originally invoked to account for the Antillean negative gravity anomaly belt, but the gravity data can also be accommodated in a Benioff zone model involving westward underthrusting. Recent determinations of earthquake mechanisms do, however, confirm Hess's dynamic model with respect to strike-slip movements along the northern and southern margins of the Caribbean Plate (Molnar and Sykes, 1969). Thus, in the present global tectonic regime, the Caribbean Plate appears to be acting as a tongue of lithosphere moving towards the Americas Plate (Fig. 1), so that the latter plate is descending beneath the leading edge of the Caribbean Plate and is sliding past its sides.

Despite the essential role of seismic data in outlining present patterns of global tectonics, in many areas the data do not throw very much light on how these patterns have evolved or what patterns may have preceded them. For this, geological information is required. The aim of this paper is to assemble relevant evidence from northern Venezuela and Trinidad and reconstruct the Late Cretaceous and Tertiary geological history of this part of the southern Caribbean area in the context of global tectonics. Lack of data and definitive evidence makes many of the conclusions tentative, but it is hoped that the results will provide constructive restraints for subsequent models involving larger parts of the region.

Unless otherwise indicated, all localities mentioned are shown in Figure 2.

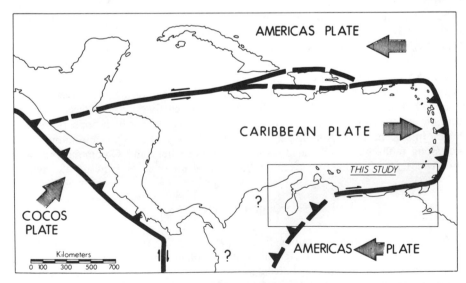

Figure 1. Present-day global tectonic framework of the Caribbean area (after Molnar and Sykes, 1969). Arrows indicate relative motions of lithospheric plates; thick lines mark the approximate position of plate margins.

Figure 2. Geological provinces of northern Venezuela and Trinidad.

PALEOGEOGRAPHY

General Statement

The southern Caribbean area is cut by many major linear fault systems, and discussion of their significance has dominated regional tectonic syntheses in recent years. Large lateral displacements have been postulated for many of these faults (Rod, 1956a; Alberding, 1957; and many others), although, with the exception of Stainforth (1969), few authors have given much attention to the chronology of movements. Large transcurrent offsets between crustal blocks are of major global tectonic significance, but the movements invoked must be compatible in timing and extent with a logical paleogeographic evolution of the area. Accordingly, the paleogeographic development of northern Venezuela and Trinidad during Jurassic, Cretaceous, and Tertiary time is outlined in this paper and is followed by a summary of relevant structural information. Finally, an attempt is made to weld all the data into a feasible evolutionary pattern.

Jurassic

So far as is known, most of Venezuela and Trinidad was above sea level during Jurassic time. Marine Jurassic sediments from the Guajira Peninsula in Colombia (Rollins, 1965), from the Paraguaná Peninsula (MacDonald, 1968b), from the Northern Range of Trinidad (Kugler, 1953), from the lowest part of the Caracas Group of the Venezuelan Coast Ranges (Dengo, 1953), and from the Uquire and Macuro Formations of northeastern Venezuela (González de Juana and others, 1965) probably record Late Jurassic southward marine transgression. Available evidence thus suggests that marine invasion of the northern edge of the region had begun prior to Cretaceous time.

Cretaceous

Lack of identifiable index fossils in the metamorphic rocks of northeastern Venezuela and Trinidad hinders detailed reconstructions of Early Cretaceous conditions, but it is clear from the record of unmetamorphosed sediments that a transgressing sea encroached on the area from the north (Salvador and Hotz, 1963). Thicknesses of metamorphic sequences of presumed Early Cretaceous age suggest that considerable subsidence was occurring near and parallel to the present northern coastline of Venezuela and Trinidad east of approximately 69° long (Morgan, 1969; Suter, 1960). In addition, some of the units appear to include turbidite deposits (Seiders, 1965; Bellizzia and Rodríguez, 1968).

These deeply buried sediments and volcanic rocks underwent regional metamorphism during Late Cretaceous time (Morgan, 1969; Piburn, 1968), while marine deposits apparently continued to accumulate above them (Fig. 3). At the same time, deposition to the southwest continued over wide areas. The most significant feature shown on the Late Cretaceous paleogeographic map is, however, the distribution of the metamorphic facies (Fig. 3). Morgan (1969) demonstrated that a belt of rocks in the epidote-amphibolite facies crops out along the northern part of the Venezuelan Coast Ranges and can be traced laterally between approximately 67° and 69° long. The belt probably extends farther eastward, since rocks of this facies are known from Margarita (Taylor, 1960; Maresch, 1970, oral commun.). All the autochthonous metamorphic rocks which crop out south of the epidote-amphibolite facies belt are assigned to the greenschist facies or lower grades, with the exception of almandine-amphibolite facies rocks from pre-Mesozoic basement terrains (Menéndez, 1966), which are presumed to have been metamorphosed considerably earlier than the events discussed in this paper. This southward passage into lower-temperature facies has been documented only in the central Coast Ranges, but a similar situation probably also exists in eastern Venezuela and Trinidad. The most perplexing metamorphic terrain in northern Venezuela is the very thick metavolcanic sequence contained within the Villa de Cura allochthon (Shagam, 1960). This terrain is clearly allochthonous and has been emplaced from a more northerly site (Menéndez, 1965; Seiders, 1965; Piburn, 1968; Bell, 1968). The lower part of the sequence contains mineral assemblages characteristic of the blueschist facies, whereas the highest exposed volcanics are assigned to the prehnite-pumpellyite metagraywacke facies (Piburn, 1968). Palinspastic reconstructions (Bell, 1971) involving overburden considerations (Morgan, 1970) suggest that the blueschist terrain was metamorphosed while it was above the epidote-amphibolite facies rocks in the northernmost part of the Coast Ranges, but the sequence may also have accumulated and been metamorphosed north of the present central Venezuelan coastline. The latter configuration is shown on the Late Cretaceous paleogeographic map (Fig. 3).

Figure 3 also shows the westward termination of this Late Cretaceous metamorphic belt, and the presence of lightly metamorphosed Mesozoic rocks on the Guajira and Paraguaná Peninsulas (MacDonald, 1968b). In constructing the map, the blueschist terrain has been placed in a possible original site, but no other palinspastic considerations have been taken into account. The Cretaceous deposits must have initially covered a wider area than is shown, since they have been involved in subsequent shortening.

Tertiary

During Paleocene and early Eocene time, a large part of the area previously covered by Cretaceous epicontinental seas was raised above sea level. Major subsidence along the northern coastline ceased (Fig. 3), but thick flysch and wildflysch

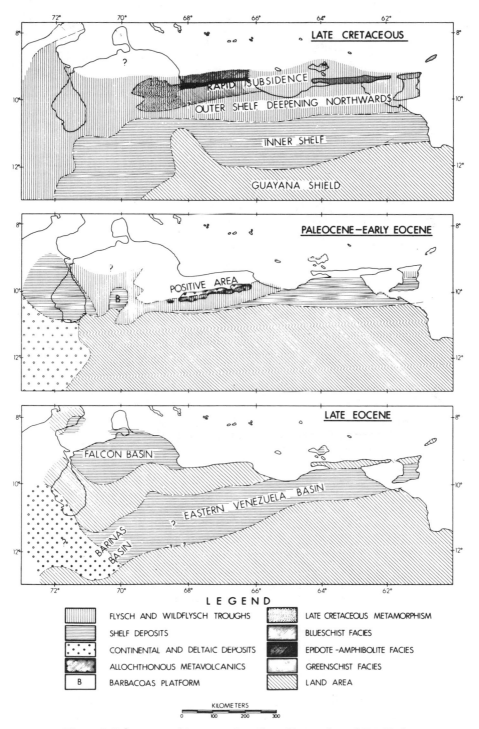

Figure 3. Paleogeographic maps of northern Venezuela and Trinidad.

Figure 4. Major late Tertiary structures of northern Venezuela and Trinidad. Thick lines represent faults; thin lines represent fold trends; closely spaced vertical lines cover areas which underwent compressive folding. Major faults are labelled as follows: B = Bocono; C = Cuisa; F = Frontal Thrust; G = Guarico; LB = Los Bajos; O = Oca; P = El Pilar; S = Soldado; SF = San Francisco; Sn = Sebastián; T = Tácata; U = Urica; and V = La Victoria.

deposits were laid down farther inland in the Barquisimeto area (for summary, *see* Bellizzia and Rodríguez, 1968), in north-central Venezuela (Peirson, 1965), and in the area of the Central Ranges of Trinidad (Kugler, 1953). It is probable that the areas of flysch sedimentation in Venezuela were connected, as has been shown on the paleogeographic map (Fig. 3), although there is no direct evidence of this. In addition, the last stages of the southward sliding of the Villa de Cura allochthon occurred in Paleocene time (Bell, 1968), ending a submarine sliding episode which had begun prior to early Maestrichtian time (Menéndez, 1966).

The most significant feature shown on the Paleocene-early Eocene paleogeographic map (Fig. 3) is the distribution of the marine deposits with the northward swing of the trough which extended westward from Trinidad. The eastern section of this trough has been involved in late Tertiary shortening and originally occupied a wider area than is shown.

Mid-Eocene orogenic events in northern Venezuela and Trinidad terminated the flysch troughs and caused widespread erosional unconformities. As Stainforth (1969) has emphasized, these events were extremely rapid, and in several areas, beds of middle Eocene age were deformed, yet the basal beds of the ensuing marine transgression are dated as late Eocene. In central Venezuela and Trinidad, basin axes were shifted southward (Hedberg, 1950; Renz and others, 1958; Kugler and Saunders, 1967; Bell, 1968), but instead of swinging northward in western Venezuela, the depositional axis continued southward as the Tertiary Barinas basin began to develop. Shallow marine sediments were deposited in the Falcón basin (Wheeler, 1963) and in the southern part of the Guajira Peninsula (MacDonald, 1964; Rollins, 1965). Eocene marine sediments are also known on Margarita (Taylor, 1960), but their regional significance is not clear.

TERTIARY STRUCTURAL DATA

General Statement

The importance of the mid-Eocene orogeny is that it delineated much of the structural configuration which exists to this day in the southern Caribbean area and ushered in the major late Tertiary tectonic events. These include: (1) the uplift of the Perijá and Venezuelan Andes; (2) downwarping and folding of the Falcon basin; (3) overthrusting in the northern Coast Ranges, accompanied by a southward shift of basin axes, and (4) the development of major linear fault zones. Considerable overthrusting, lateral shortening, and strike-slip faulting occurred, which I believe can be accounted for by the interaction of lithospheric plates. In addition, various areas such as the Maracaibo platform underwent minor folding and normal faulting, but these events are considered to be of less importance and are not discussed in detail in this paper.

Table 1 shows the timing and documentable effects of the development of the more important structures which involve a component of horizontal movement. In estimating the timing of these structures, the record of adjacent sediments has been invoked in several cases, so that listing the ages of deformed and undeformed rocks does not give a complete picture of the evidence. The table does show, however, where the data relevant for deducing periods of movement are scarce and gives some idea of the degree of uncertainty. The documentable effects listed are those which can be supported by conventional geologic evidence and which do not include unqualified speculations.

Trinidad and Eastern Venezuela

The earliest Tertiary event appears to be the southward thrusting of metamorphic rocks over unaltered sediments along the Chuparipal fault in pre-Eocene time (Metz, 1968). Although only documented in eastern Venezuela, this overthrusting presumably also affected northern Trinidad. Thrust plate remnants are found south of the El Pilar fault which, together with Cretaceous facies patterns, suggest that no more than 10 to 15 km of right-lateral movement has occurred along this fault zone (Metz, 1968). According to available data, the El Pilar fault was active only between approximately mid-Eocene and mid-Miocene time (Kugler, 1959; Stainforth, 1969).

Folding and thrusting along N. 75° E. axes had begun by Miocene time, and the resulting shortening initiated a later series of northwest-trending strike-slip faults which exhibit individual right-lateral offsets of up to 10 km. A parallel sequence of events affected both eastern Venezuela and Trinidad, but appears to have persisted for longer in Trinidad. The amount of north-south shortening may also be greater in eastern Venezuela.

Central Venezuela

The sequence of Tertiary events is far less well documented in central Venezuela. No thrust fault equivalent to the Chuparipal fault has been recognized, although the Santa Rosa fault (Menéndez, 1966) is possibly a similar structure, nor is there any evidence of mid-Tertiary transcurrent movements on east-trending faults. There is only evidence for offsets of the La Victoria and Sebastián faults in Miocene-Recent time (Smith, 1953; Picard and Pimentel, 1968; Morgan, 1969). North-south shortening through folding and thrusting in the southern part of the Coast Ranges may have begun as early as late Eocene time and clearly continued into post-Miocene time (Peirson, 1965; Bell, 1968). Northwest-trending right-

TABLE 1. TIMING AND EFFECTS OF TERTIARY DEFORMATIONAL EVENTS INVOLVING OVERTHRUSTING, LATERAL SHORTENING, OR STRIKE-SLIP MOVEMENTS

Structural features	Rocks involved = o; Rocks not involved = x							Probable timing												Comments[o]	Documentable effects[o]	Data source
	Cret.	Pal.	Eoc.	Olig.	Mioc.	Plioc.	Pleist./Rec.	Cret.	Pal.	E Eoc.	M Eoc.	L Eoc.	Olig.	E Mioc.	M Mioc.	L Mioc.	E Plioc.	L Plioc.	Pleist./Rec.			
TRINIDAD — El Pilar fault zone	o				o x	x	x							—						No data on SS	NS, nsu	Kugler, 1959; Stainforth, 1969
Central and Southern Ranges — Folding 1	o	o	o	o	o x	x	x							—							N-S shortening of 14 km +	Kugler, 1956, 1959; Suter, 1960
Central and Southern Ranges — Thrusting 1	o	o	o	o	o x	x	x							—								
Central and Southern Ranges — Folding 2	o	o	o	o	o	o x	x									— –	—					
Central and Southern Ranges — Thrusting 2	o	o	o	o	o	x	x										—					
Los Bajos fault					o	o x	x											–	—	SS 11 km rl	Kugler, 1959; Salvador and Stainforth, 1968; Suter, 1960; Wilson, 1958
EASTERN VENEZUELA — Chuparipal fault	o					x	x				—	? ?	? ?							pre-El Pilar	Southward overthrusting of 5 km +	Metz, 1968
El Pilar fault	o					x	x						—					?	Timing?	SS 5–15 km rl	Metz, 1968; Stainforth, 1969
S. de Interior — Folding	o	o	o	o	o x	x	x							—	—						N-S shortening of 20 km +	Hedberg, 1950; Lamb and Sulek, 1972; Salvador and Rosales, 1960
S. de Interior — Thrusting	o	o	o	o	o x	x	x						–	–		—						
Urica fault	o	o	o	o	o		x										—	—	post-El Pilar	SS 10 km rl	Rod, 1956a; Stainforth, 1969
San Francisco fault	o	o	o	o	o		x									–	—	—			SS few km rl	Salvador and Rosales, 1960

	Fault/Structure	Deformation	Symbols (time columns)	References
VENEZUELA CENTRAL	Sebastian fault	NS, ssu	o	Morgan, 1969
	La Victoria fault	NS, nsu; SS, ? rl	o o o x / o o o ? ?	Morgan, 1969; Picard and Pimentel, 1968; Seiders, 1965; Smith, 1953
	Foothills folds	N-S shortening of 20 km +	o o o x /	Bell, 1968; Peirson, 1965
	Foothills thrusts	Southward overthrusting of 15 km +	o o o x /	Bell, 1968; Peirson, 1963, 1965; Peirson and others, 1966
	Llanos folds/faults	N-S shortening of <5 km	o o o x	Bell, 1968; Peirson, 1963
	Guárico fault	SS 3–25 km rl	o o o x	Bell, 1968; Peirson, 1965; Shagam, 1960
	Tacata fault	? SS rl	o x x / ?	Smith, 1953
VENEZUELA WESTERN	Oca fault	SS 20 km rl	o o x x / ? ?	Bucher, 1950; Coronel, 1970; Feo-Codecido, 1972b; MacDonald, 1968a; Miller, 1962; Stainforth, 1969
	Eastern Andean marginal thrusts	E-W shortening	o o o / ?	Bucher, 1952; Feo-Codecido, 1972a; Hargraves and Shagam, 1969
	Perija Andes folds/thrusts	E-W shortening	o o o x	Miller, 1962
	Falcón basin folds	o o o o	Bucher, 1952; Wheeler, 1963
	Boconó fault	NS; SS c. 30 km rl	o o o x	Bushman, 1965; Grauch and Shagam, 1970; Rod, 1956a; Schubert and Sifontes, 1970

Abbreviations: NS, normal separation; SS, strike-slip separation; nsu, north side up; ssu, south side up; rl, right-lateral.

lateral strike-slip faults developed within the overthrust sequences toward the end of this period and, in the case of the Guárico fault, may be offset as much as 25 km (Bell, 1968). In the central part of the Coast Ranges, late Tertiary north-south shortening has probably exceeded 40 km. It may be less elsewhere in this region, but it is certainly greater than in eastern Venezuela and Trinidad.

Western Venezuela

Although large offsets have been proposed for the Oca fault (for summary, see Coronel, 1970), recent studies (MacDonald, 1968a; Feo-Codecido, 1972b) suggest that right-lateral offset does not exceed 20 km. According to Feo-Codecido's interpretation of surficial and subsurface data on both sides of the fault, structures involving Eocene rocks are offset approximately 20 km in the area south of the Guajira Peninsula. On the other hand, in the Falcón basin, where the eastward extension of the Oca fault should be found, Eocene units are folded but not noticeably offset (Bucher, 1950). Thus, most of the strike-slip movement along the Oca fault would appear to have occurred during a small part of Eocene time, although a small amount could be accommodated during the subsequent late Pliocene development of east-northeast-trending folds in the Falcón basin (Wheeler, 1963; Stainforth, 1969). The evolution of the late Tertiary basins on the Guajira Peninsula may be related to early movements of the Oca fault. These basins are bounded by northwest-trending faults exhibiting vertical separations, which probably have been active from Eocene time until the present (MacDonald, 1964).

Uplift of the Venezuelan and Perijá Andes involved the development of reverse faulted marginal zones (Dufour, 1955) which probably originated in Eocene time in the Venezuelan Andes (Hargraves and Shagam, 1969), but have been active from Miocene time until the present (Bucher, 1952; Feo-Codecido, 1972a). Post-Oligocene uplift is indicated in the Perijá Andes (Miller, 1962). The effect of these events is unknown in detail, but gravity profiles suggest that there is no root beneath the Venezuelan Andes and that there is an asymmetrical mass distribution. Hospers and Van Wijnen (1959) interpret this to indicate east-west-directed wedging, and they estimate a maximum crustal shortening of 50 km. The Boconó fault, which runs down the center of the Andes, is active at the present time (Rod, 1956b; Fiedler, 1961) and right-laterally offsets Pleistocene glacial terraces approximately 66 m (Schubert and Sifontes, 1970). Offset sedimentary and metamorphic terrains in the Barquisimeto area (Bushman, 1965; Zambrano and others, 1972) can be interpreted to indicate 30 km of post-Eocene right-lateral movement, but Grauch and Shagam (1970) caution that Tertiary displacements may be largely vertical, since identical basement terrains can be traced across the fault without deflections of features such as a sillimanite isograd.

TECTONIC EVOLUTION

The presumed configuration of metamorphic facies in northern Venezuela is believed to indicate that a southward-dipping Benioff zone existed in the southern Caribbean area in Late Cretaceous time. This would be a suitable tectonic setting for the eruption of the spilitic basalts and keratophyres of the Villa de Cura Group (Shagam, 1960) in an offshore island arc setting. By analogy with Ernst's (1970) model for the metamorphism of the Franciscan Group, a Benioff zone would also provide a suitable environment for forming blueschist assemblages in this sequence. Frictional heating associated with southward movement on this Benioff zone (Oxburgh and Turcotte, 1970) could have caused the continental metamorphism farther south. If such a feature existed, it may have been offset to the north in the western part of the area and may also have been responsible for the Cretaceous metamorphism of sediments on the Guajira and Paraguaná Peninsulas, assuming that the apparent displacement of metamorphic terrains is a primary feature and not the result of Tertiary movements (Fig. 5). Movement on this postulated Benioff zone is believed to have ceased in latest Cretaceous time, resulting in rapid local uplift due to isostatic rebound. This rapid uplift provided the structural environment which allowed the recently metamorphosed Villa de Cura Group volcanic rocks to slide southward, and also caused considerable overthrusting resulting in crustal thickening. As Price (1969) has shown, crustal thickening will cause the affected area to subside and will also cause subsidence in adjacent "foredeeps." This wedging mechanism is believed to have initiated the flysch basins which rim the southern margin of the uplifted metamorphic terrains in northern Venezuela (Fig. 6).

Major east-west right-lateral movements in the southern Caribbean area apparently occurred no earlier than Eocene time, suggesting that the Caribbean Plate began moving relatively eastward at this time. This view is reinforced by the fact that the earliest known andesitic volcanic rocks in the Lesser Antilles are of Eocene age (Tomblin, 1971), and it is not likely that they were extruded prior to the development of the Antillean Benioff zone at the eastern margin of the Caribbean Plate. Large right-lateral displacements have been suggested for the Oca-El Pilar

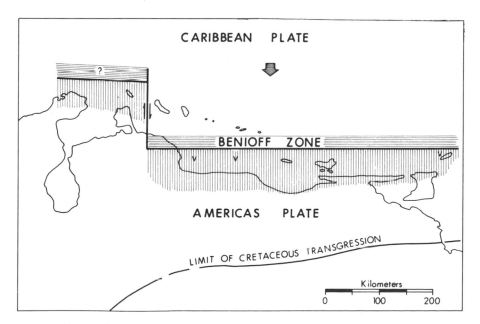

Figure 5. Postulated middle Cretaceous tectonic framework. At this time the Caribbean Plate is believed to have moved southward relative to the Americas Plate. Horizontal shading indicates location of Benioff zone; vertical shading indicates areas which were metamorphosed; and v denotes zone where submarine volcanics were erupted.

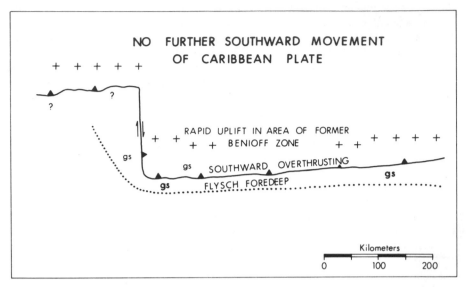

Figure 6. Paleocene tectonic events. Crosses denote regions which were uplifted; gs marks areas where gravity sliding of allochthonous masses occurred.

fault system (Rod, 1956a; Alberding, 1957; and many others) and, if the Caribbean Plate was directly connected to the East Pacific Plate at this time and no underthrusting of Central America was occurring, relative displacement of as much as 5.5 cm/yr would have been possible. At such a rate, the 400 km displacement of metamorphic terrains suggested by Alberding (1957) could have been accomplished in approximately 7 m.y. One cannot categorically deny that such movements occurred, since detailed evidence is lacking in many areas along the faults in question. For example, the pre-Eocene record of the Falcón basin is unknown, and this might shed considerable light on movements of the Oca fault. Available data support offsets of tens rather than hundreds of kilometers in northern Venezuela and also on the Guajira Peninsula, where the Cuisa fault shows approximately 15 km of right-lateral separation (Alvarez, 1967). Figure 7 accordingly presents a conservative view and also illustrates crustal thickening in the area of the former flysch basin and related subsidence to the south, which gave rise to the regional late Eocene transgression (Peirson, 1965). At that time the southern margin of the Caribbean Plate is believed to have extended eastward through northwestern Venezuela and Colombia. By Miocene time, the Americas Plate was no longer sliding past the Caribbean Plate in this manner, and westward movement of the former plate was being taken up by wedging mechanisms in the Andes in a pattern similar to that indicated for present day global tectonics (Fig. 1).

By Pliocene time, considerable folding and overthrusting as well as strike-slip faulting had occurred (Fig. 8). The pattern of overthrusting in central and eastern Venezuela and Trinidad suggests that the Americas and Caribbean Plates approached each other obliquely and that a minor north-south compressive component of motion existed. This type of oblique approach means that the eastern limit of the zone undergoing wedging and, presumably, uplift would shift eastward over a period of time. This may be the reason that the Miocene orogenic conglomerates in northern Venezuela become progressively younger eastward and that folding

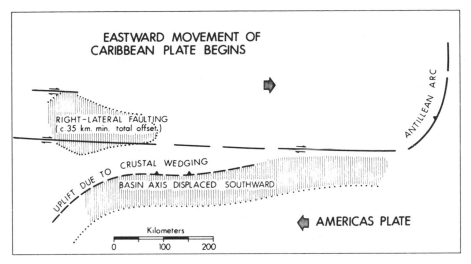

Figure 7. Late Eocene tectonic events. Eastward movement of the Caribbean Plate may have been greater than shown. Areas of marine sedimentation are indicated by vertical shading; arrows indicate relative plate motions.

and thrusting persisted for longer in Trinidad than in Venezuela (Salvador and Stainforth, 1968). It is puzzling that there is relatively little evidence of late Tertiary east-west-directed right-lateral faulting in the area, despite the fact that earthquake first motion records indicate that it is currently occurring and that reasonable extrapolations of late Tertiary plate motions require it. Tertiary strike-slip faulting may have occurred along the La Victoria fault and neighboring structures. The fault zone occupies a prominent valley along much of its length and, as with associated east-trending fault zones, mylonites crop out within it (Dengo, 1953), and drainage patterns suggest that a few hundred meters of right-lateral separation may have occurred recently (Smith, 1953). It is also not clear how much east-west movement can be accommodated by the network of northwest-trending faults, since

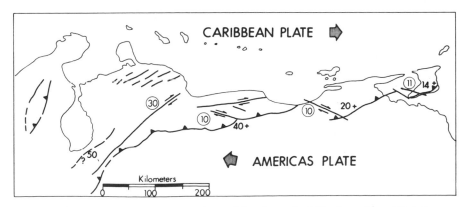

Figure 8. Tectonic configuration achieved by the end of Tertiary time. Circled numbers refer to number of kilometers of right-lateral strike-slip movement; uncircled numbers represent amounts of supra-crustal shortening in kilometers, taking into account folding and overthrusting; the structures involved are also shown on Figure 4. Arrows indicate relative plate motions.

many of these appear only to offset overthrust sequences, but the transcurrent movement must be less than 50 km. There remains the possibility that the offshore network of east-striking faults (Ball and others, 1969) are transcurrent features, as well as the likelihood that crustal shearing occurred beneath the overthrust terrains as they developed. Thrusting related to transcurrent movement is known in the San Andreas fault zone (Crowell, 1962) and has been reported in Puerto Rico (Glover and Mattson, 1960). In late Pliocene time, folding occurred within the Falcón basin and was probably related to a minor renewal of east-west right-lateral transcurrent shearing in that area. Later right-lateral movements may have offset what was formerly a continuous fault zone joining the Oca and Sebastián faults.

It is concluded that there is no well defined boundary between the Caribbean and Americas Plates at the present time and that one probably never existed in the past. Instead, this zone has been straddled by crustal wedges which have been deformed as a result of horizontal plate motions. No metamorphism appears to have accompanied any of the episodes of Tertiary deformation discussed above.

IMPLICATIONS

If an east-trending Late Cretaceous Benioff zone did exist, sea-floor spreading with a southward component to its motion may have been occurring in the southern Caribbean area at that time. This, in turn, suggests that an eastward-trending spreading ridge may have been active then and possibly earlier, as has been suggested by Dietz and Holden (1970). If such a ridge existed, it presumably branched off the Mid-Atlantic Ridge, so that a Late Cretaceous triple junction should still be present east of the Lesser Antilles. To my knowledge, no available magnetic profiles from either the Atlantic or the Caribbean unequivocally indicate the former existence of such a ridge and there is no obvious bathymetric expression of one (Hess, 1966). Alternatively, underthrusting in the southern Caribbean area may have been achieved by the convergence of a rigid plate of older oceanic crust against the South American craton. Either mechanism would account for the effects observed.

Relating subsidence to adjacent crustal thickening and relating this crustal thickening to horizontal motions of lithospheric plates has very broad implications. It provides a model which can be adapted to explain episodes of overthrusting and concurrent basin migration; it can account for successive emplacements of nappes; and it also provides a framework for explaining many unconformities. Downwarping of the central part of the east Venezuela basin in late Eocene time (Fig. 3) probably was caused by tectonic thickening through folding and thrusting of the Paleocene flysch sequence to the north, and this produced a northward transgression and unconformable deposition of the Peñas Blancas Formation (Bell, in prep.). Moreover, this unconformity is only one of many mid-Tertiary unconformities in northern South America, and the transgression is part of a clearly discernible pattern of advances and retreats of the seas in all the marine basins which ring the Brazilian Shield (Stainforth, 1968). It may be that such local transgressions and regressions can eventually be related to particular episodes of crustal thickening produced by deformation resulting from plate motions. Not all crustal thickening is due to wedging, since magmatic processes are frequently involved, but these too can be related to plate motions. Nor will local deformation account for worldwide advances and retreats of the sea; this is likely to be due to a larger scale process such as the rise and fall of mid-oceanic ridge systems (Moores, 1970). It is, therefore, reasonable to anticipate that much of the sedimentary record can be explained in detail in terms of crustal responses to the effects of global tectonics.

ACKNOWLEDGMENTS

Harry Hess introduced me to the Caribbean and supervised my Ph.D. thesis in Venezuela. This paper has grown from his interest and encouragement then and subsequently, and is thus dedicated to him in a very real way. It was a very great privilege to have known him and to have been his student in an area where he had worked so long and which he knew so well.

I am most grateful to E. Araujo, (Director), A. Vivas (Former Director), A. A. Bellizzia, C. Bellizzia Martín, A. Menéndez, C. Schubert, X. Picard, L. Gonzales, A. Sabater, and many other members of the staff of the Dirección de Geología of the Ministerio de Minas e Hidrocarburos of Venezuela for all the assistance and encouragement they gave me while I was engaged in field work in Venezuela between 1964 and 1970. Many of the ideas expressed in this paper also took shape in discussions with H. D. Hedberg, J. C. Maxwell, H. L. Metz, B. A. Morgan, M. D. Piburn, A. L. Pierson III, H. Rosales, R. Shagam, and R. M. Stainforth. I particularly acknowledge how much I have been influenced by the synthesis of the geological evolution of Venezuela recently presented by Stainforth (1969). The pattern of crustal events described in the previous pages differs very little from his reconstruction; our explanations, although reached from slightly different starting points (Stainforth discusses the development of possible convection cell systems), have much in common.

The Ministerio de Minas e Hidrocarburos of Venezuela supported my field work. Laboratory and compilation expenses have been paid by the Caribbean Research Fund at Princeton University, which has been sustained by contributions from the Creole, Mene Grande, Mobil, and Shell Oil companies in Venezuela; support was also obtained from intramural funds of the University of California at Riverside and the University of Alberta.

The paper has benefited from critical review by H. L. Metz, B. A. Morgan, R. Shagam, R. M. Stainforth, and F. B. Van Houten, for which I am most grateful.

REFERENCES CITED

Alberding, H., 1957, Application of principles of wrench-fault tectonics of Moody and Hill to northern South America: Geol. Soc. America Bull., v. 68, p. 785–790.

Alvarez, W. S., 1967, Geology of the Simarua and Carpintero areas, Guajira Peninsula, Colombia [Ph.D. thesis]: Princeton, New Jersey, Princeton Univ., 168 p.

Ball, M. M., Harrison, C. G. A., Supko, P. R., Bock, W. D., and Maloney, N. J., 1969, Fallamiento normal a lo largo de limite meridional del Mar Caribe, Bahía de Unare, Venezuela: Asoc. Venezolana Geología, Minería y Petróleo Bol. Inf., v. 12, no. 2, p. 23–39.

Bell, J. S., 1968, Geología de la región de Camatagua, Estado Aragua, Venezuela: Bol. Geología, Min. de Minas e Hidrocarburos, Venezuela, v. 9, no. 18, p. 292–440.

—— 1971, Tectonic evolution of the central part of the Venezuelan Coast Ranges, *in*

Donnelly, T. W., ed., Caribbean geologic, tectonic, and petrologic studies: Geol. Soc. America Mem. 130.

Bellizzia, A. A., and Rodríguez, G. D., 1968, Consideraciones sobre la estratigrafía de los Estados Lara, Yaracuy, Cojedes y Carabobo: Bol. Geología, Min. de Minas e Hidrocarburos, Venezuela, v. 9, no. 18, p. 515–565.

Bucher, W. H., 1950, Geologic-tectonic map of the United States of Venezuela: Geol. Soc. America, scale 1:1,000,000.

—— 1952, Geologic structure and orogenic history of Venezuela: Geol. Soc. America Mem. 49, 113 p.

Bushman, J. R., 1965, Geología del area de Barquisimeto, Venezuela: Bol. Geología, Min. de Minas e Hidrocarburos, Venezuela, v. 6, no. 11, p. 3–111.

Crowell, J. C., 1962, Displacement along the San Andreas fault, California: Geol. Soc. America Spec. Paper 71, 61 p.

Coronel, G., 1970, Why not a structural index of Venezuela?: Asoc. Venezolana Geología, Minería y Petróleo Bol. Inf., v. 13, no. 3, p. 109–121.

Dengo, G., 1953, Geology of the Caracas region, Venezuela: Geol. Soc. America Bull., v. 64, no. 1, p. 7–40.

Dietz, R. S., and Holden, J. C., 1970, The breakup of Pangea: Sci. American, v. 223, no. 4, p. 30–41.

Dufour, J., 1955, Some oil-geological characteristics of Venezuela: Rome, IV World Petrol. Cong. Proc., v. 1, p. 19–35.

Ernst, W. G., 1970, Tectonic contact between the Franciscan mélange and the Great Valley Sequence—Crustal expression of a late Mesozoic Benioff zone: Jour. Geophys. Research, v. 75, no. 5, p. 886–902.

Feo-Codecido, G., 1972a, Contribución a la estratigrafía de la cuenca Barinas-Apure: IV Cong. Geol. Venez. Proc., Caracas, Venezuela, 1969.

—— 1972b, Consideraciones estructurales sobre la Falla de Oca, Venezuela: Prim. Cong. Latinoamericano Geol., Lima, Peru, 1970.

Fiedler, G., 1961, Areas afectadas por terremotos en Venezuela: III Cong. Geol. Venez. Mem., v. 4, p. 1791–1810 (Pub. Esp. 3, Min. de Minas e Hidrocarburos, Venezuela).

Glover, L., III, and Mattson, P. H., 1960, Successive thrust and transcurrent faulting during early Tertiary in south-central Puerto Rico: U.S. Geol. Survey Prof. Paper 400–B, p. 363–365.

González de Juana, C., Munoz, N. G., and Vignali, M., 1965, Reconocimiento geológico de la parte oriental de Paria: Venezolana Geología, Minería y Petróleo Bol. Inf., v. 8, no. 9, p. 255–279.

Grauch, R. I., and Shagam, R., 1970, Distribution and tectonic significance of andalusite, kyanite, sillimanite and staurolite in the central Venezuelan Andes: Geol. Soc. America, Abs. with Programs (Ann. Mtg.), v. 2, no. 7, p. 560.

Gutenberg, B., and Richter, C. F., 1954, Seismicity of the Earth and associated phenomena: Princeton, New Jersey, Princeton Univ. Press, 310 p.

Hargraves, R. B., and Shagam, R., 1969, Paleomagnetic study of La Quinta Formation, Venezuela: Am. Assoc. Petroleum Geologists Bull., v. 53, p. 537–552.

Hedberg, H. D., 1950, Geology of the eastern Venezuela basin (Anzoátegui-Monagas-Sucre-eastern Guárico portion): Geol. Soc. America Bull., v. 61, p. 1173–1216.

Hess, H. H., 1932, Interpretation of geological and geophysical observations. Navy-Princeton Gravity Expedition to West Indies: U.S. Navy Hydrog. Office Tech. Rept., p. 27–54.

—— 1938, Gravity anomalies and island arc structures with particular reference to the West Indies: Am. Philos. Soc. Proc., v. 79, p. 71–96.

—— 1966, Caribbean research project, 1965, and bathymetric chart: Geol. Soc. America Mem. 98, p. 1–11.

Hospers, J., and Van Wijnen, J. C., 1959, The gravity field of the Venezuelan Andes and adjacent basins: Verh., Koninkl. Nederlandse Akad. Wetensch. Afd. Natuurk., eerste reeks, deel 23, no. 1, p. 5–95.

Kugler, H. G., 1953, Jurassic to Recent sedimentary environments in Trinidad: Assoc. Suisse Geol. Ing. Petr. Bull., v. 20, no. 59, p. 27–60.

—— 1956, Trinidad, in Jenks, W. F., ed., Handbook of South American geology: Geol. Soc. America Mem. 65, p. 351–365.

—— 1959, Geologic map of Trinidad and geologic sections through Trinidad: Zurich, Orell Fussli, S. A.

Kugler, H. G., and Saunders, J. B., 1967, On Tertiary turbidity-flow sediments in Trinidad, W.I.: Asoc. Venezolana Geología, Minería y Petróleo Bol. Inf., v. 10, no. 9, p. 243–259.

Lamb, J. L., and Sulek, J. A., 1968, Miocene turbidites in the Carapita Formation of eastern Venezuela: IV Caribbean Geol. Conf. Trans., Trindad, 1965.

MacDonald, W. D., 1964, Geology of the Serrania de Macuira area, Guajira Peninsula, Colombia [Ph.D. thesis]: Princeton, Princeton Univ., 167 p.

—— 1968a, Movement on the Oca fault, northern Colombia-Venezuela: V Caribbean Geol. Conf. Abstracts, St. Thomas, U.S. Virgin Is., 1968.

—— 1968b, Estratigrafía, estructura y metamorfismo, rocas del Jurasico Superior, Peninsula de Paraguaná, Venezuela: Bol. Geología, Min. de Minas e Hidrocarburos, Venezuela, v. 9, no. 18, p. 441–458.

Menéndez, V. de V. A., 1965, Geología del area de El Tinaco, centro norte del Estado Cojedes, Venezuela: Bol. Geología, Min. de Minas e Hidrocarburos, Venezuela, v. 6, no. 12, p. 417–543.

—— 1966, Tectónica de la parte central de las Montañas Occidentales del Caribe, Venezuela: Bol. Geología, Min. de Minas e Hidrocarburos, Venezuela, v. 8, no. 15, p. 116–139.

Metz, H. L., 1968, Geology of the El Pilar fault zone, state of Sucre, Venezuela: IV Caribbean Geol. Conf. Trans., Trindad, 1965, p. 293–298.

Miller, J. B., 1962, Tectonic trends in Sierra de Perijá and adjacent parts of Venezuela and Colombia: Am. Assoc. Petroleum Geologists Bull., v. 46, p. 1565–1595.

Molnar, P., and Sykes, L. R., 1969, Tectonics of the Caribbean and Middle America regions from focal mechanisms and seismicity: Geol. Soc. America Bull., v. 80, p. 1639–1684.

Moores, E. M., 1970, Patterns of continental fragmentation and reassembly; some implications: Geol. Soc. America, Abs. with Programs (Ann. Mtg.), v. 2, no. 7, p. 629.

Morgan, B. A., 1969, Geología de la región de Valencia, Carabobo, Venezuela: Bol. Geología, Min. de Minas e Hidrocarburos, Venezuela, v. 10, no. 20, p. 3–136.

—— 1970, Petrology and mineralogy of eclogite and garnet amphibolite from Puerto Cabello, Venezuela: Jour. Petrology, v. 11, p. 101–145.

Oxburgh, E. R., and Turcotte, D. L., 1970, Thermal structure of island arcs: Geol. Soc. America Bull., v. 81, p. 1665–1688.

Peirson, A. L., 1963, Galera member of the Quebradón Formation: Asoc. Venezolana Geología, Minería y Petróleo Bol. Inf., v. 6, no. 5, p. 141–150.

—— 1965, Geology of the Guárico mountain front: Asoc. Venezolana Geología, Minería y Petróleo. Bol. Inf., v. 8, no. 7, p. 183–212.

Peirson, A. L., Salvador, A., and Stainforth, R. M., 1966, The Guárico Formation of north-central Venezuela: Asoc. Venezolana Geología, Minería y Petróleo. Bol. Inf., v. 9, no. 7, p. 183–224.

Piburn, M. D., 1968, Metamorfismo y estructura del Grupo Villa de Cura, Venezuela septentrional: Bol. Geología, Min. de Minas e Hidrocarburos, Venezuela, v. 9, no. 18, p. 183–289.

Picard, X., and Pimentel, N., 1968, Geología de la cuenca de Santa Lucia-Ocumare del Tuy: Bol. Geología, Min. de Minas e Hidrocarburos, Venezuela, v. 9, no. 19, p. 263–296.

Price, R. A., 1969, The southern Canadian Rockies and the role of gravity in low-angle thrusting, foreland folding, and the evolution of migrating foredeeps: Geol. Soc. America, Abs. with Programs for 1969, Pt. 7 (Ann Mtg.), p. 284–286.

Renz, H. H., Alberding, H., Dallmus, K. F., Patterson, J. M., Robie, R. H., Weisbord, N. E., and Masvall, J., 1958, The eastern Venezuelan basin, in Weeks, L. G., ed., Habitat of oil: Am. Assoc. Petroleum Geologists, p. 551–600.

Rod, E., 1956a, Strike-slip faults of northern Venezuela: Am. Assoc. Petroleum Geologists Bull., v. 40, p. 457–476.

—— 1956b, Earthquakes of Venezuela related to strike-slip faults: Am. Assoc. Petroleum Geologists Bull., v. 40, p. 2509–2512.

Rollins, J. F., 1965, Stratigraphy and structure of the Goajira Peninsular, northwestern Venezuela and northeastern Colombia: Univ. Nebraska Studies, new ser., no. 30, 102 p.

Salvador, A., and Hotz, E., 1963, Petroleum occurrence in the Cretaceous of Venezuela: VI World Petrol. Cong. Proc., Frankfurt, 1963, v. 1, p. 115–140.

Salvador, A., and Rosales, H., 1960, Guía de la excursion A-3 Jusepin-Cumana: III Cong. Geol. Venez. Mem., v. 1, p. 63–74 (Pub. Esp. 3, Min. de Minas e Hidrocarburos, Venezuela).

Salvador, A., and Stainforth, R. M., 1968, Clues in Venezuela to the geology of Trinidad, and vice versa: IV Caribbean Geol. Conf. Trans., Trinidad, 1965, p. 31–40.

Schubert, C., and Sifontes, R. S., 1970, Boconó fault, Venezuelan Andes: Evidence of post-glacial movement: Science, v. 170, p. 66–69.

Seiders, V. M., 1965, Geología de Miranda central, Venezuela: Bol. Geología, Min. de Minas e Hidrocarburos, Venezuela, v. 6, no. 12, p. 289–416.

Shagam, R., 1960, Geology of central Aragua, Venezuela: Geol. Soc. America Bull., v. 71, p. 249–302.

Smith, R. J., 1953, Geology of the Los Teques-Cua region, Venezuela: Geol. Soc. America Bull., v. 64, p. 41–64.

Stainforth, R. M., 1968, Mid-Tertiary diastrophism in northern South America: IV Caribbean Geol. Conf. Trans., Trinidad, 1965, p. 159–174.

—— 1969, The concept of seafloor-spreading applied to Venezuela: Asoc. Venezolana Geología, Minería, y Petróleo. Bol. Inf., v. 12, no. 8, p. 257–274.

Suter, H. H., 1960, The general and economic geology of Trinidad, B.W.I. (2d ed.): London, H. M. Stationery Office, 145 p.

Sykes, L. R., and Ewing, M., 1965, The seismicity of the Caribbean region: Jour. Geophys. Research, v. 70, no. 10, p. 5065–5074.

Taylor, G. C., 1960, Geología de la Isla de Margarita, Venezuela: III Cong. Geol. Venez. Mem., v. 2, p. 838–893 (Pub. Esp. 3, Min. de Minas e Hidrocarburos, Venezuela).

Tomblin, J. F., 1971, Geochemistry and genesis of Lesser Antillean volcanic rocks: V Caribbean Geol. Conf. Trans., St. Thomas, U.S. Virgin Is., 1968.

Wheeler, C. B., 1963, Oligocene and lower Miocene stratigraphy of western and northeastern Falcón basin, Venezuela: Am. Assoc. Petroleum Geologists Bull., v. 47, p. 35–68.

Wilson, C. C., 1958, The Los Bajos fault and its relation to Trinidad's oilfield structures: Inst. Petroleum Jour., v. 44, no. 413, p. 124–136.

Zambrano, E., Vásquez, E., Duval, B., Latreille, M., and Coffiniers, B., 1972, Sintesis paleogeografica y petrolera del occidente de Venezuela: IV Cong. Geol. Venez. Proc., Caracas, Venezuela, 1969 (in press).

MANUSCRIPT RECEIVED BY THE SOCIETY MARCH 29, 1971
PRESENT ADDRESS: SHELL CANADA LTD., CALGARY, ALBERTA, CANADA.

THE GEOLOGICAL SOCIETY OF AMERICA, INC.
MEMOIR 132
© 1972

Origin of the Southern Caribbean Mountains

S.R.M. HARVEY

112 Fisher Place, Princeton, New Jersey 08540

ABSTRACT

The west section of the Southern Caribbean Mountains lies in north-central Venezuela and consists of three belts; a coastal range, an interior range, and a median zone. Previous workers devised a distinctive stratigraphic column for each belt, because each appeared to contain a unique succession of rocks, the rocks are too metamorphosed to be dated by paleontologic or radiometric means, and each belt is separated from the others by faults. In the more recent studies it has been suggested that the interior range originated far north of its present position and slid south in Maestrichtian-Paleocene times under the action of gravity.

The results of detailed mapping south of Caracas and of a reconnaissance of the entire mountain range indicate that the three belts may be essentially in place, representing a formerly continuous section shortened by thrusting.

The submergence, sedimentation, and volcanism, and the successive phases of metamorphism, faulting, pronounced uplift, and rapid erosion involved in the formation of the Southern Caribbean Mountains took place in overlapping regions. The superposition of these regions can be accounted for by assuming a linear, high-energy zone, localized as a depression at the surface of the mantle, being overridden progressively by thin continental crust which thickened inland. The broad llanos basin south of the Caribbean Mountains may represent the current site of the proposed depression, now covered by continental crust of average thickness. The major structural trends appear to have shifted progressively during formation of the mountains, suggesting that South America rotated anti-clockwise during the course of its inferred passage northward across the depression.

INTRODUCTION

The southeastern part of the Caribbean Sea is separated from the sedimentary basins of Trinidad and the Venezuelan llanos by fragmented parts of a mountain belt, extending eastward from the Andes for a distance of about 750 km. The western and central sections of the belt lie in Venezuela; the eastern section is represented by the Northern and Central Ranges of Trinidad.

The western section is the largest of the three and the only one to have been studied in detail. Aguerrevere and Zuloaga (1937) conducted a reconnaissance of the section. Between 1950 and 1967 workers from Princeton University completed thirteen areal studies covering about two-thirds of the section. These studies were initiated and directed by Harry Hess, in conjunction with the Venezuelan Ministry of Mines; they were undertaken as part of the Caribbean Research Project in order to examine the evolution of the mountain belt and its possible relation to the island arc represented by the Lesser and Greater Antilles (Hess, 1966).

The studies have defined three major provinces in the western section of the mountains; a coastal range composed largely of metasedimentary rocks covering basement, a median zone in which basement is covered by both metasedimentary and metavolcanic rocks and by unconsolidated sediments, and an interior range consisting of metavolcanic rocks. The mountains are bounded to the south by a piedmont zone containing flysch deposits and other folded sedimentary materials (Fig. 1.).

Rocks of the piedmont are fossiliferous and have been assigned to the Upper Cretaceous and Tertiary sections of the Eastern Venezuelan Basin, but in the three provinces to the north the ages and stratigraphic relations of units mapped are uncertain. The origin of metavolcanic rocks forming the interior range remains obscure. It has been suggested that they represent basement, or that they are inverted, constituting part of the lower limb of a nappe. Structural considerations have led Hess and others to suggest that the entire range, 280 km long, is an allochthonous block which slid over or off the present site of the coastal range at the close of the Cretaceous (Menéndez, 1967).

In 1961 and 1962 the writer investigated this proposal by mapping an area of 1,200 sq km lying south of Caracas and extending from the piedmont zone into the coastal range. The areas of other studies in the western, central, and eastern sections of the Southern Caribbean Mountains were examined at a reconnaissance level. Some of the conclusions emerging from this study are presented below.

INTRAMONTANE CORRELATION

Previous workers divided the metamorphic rocks of the western section of the mountains into four groups: the basement, the Caracas Group, the post-Caracas Group, and the Villa de Cura Group. The principal components of units assigned to the Caracas Group and the post-Caracas Group are shown in Figure 2. Columns in the upper half of the figure illustrate the sequences in which these units are reported to occur in different parts of the western section of the mountain belt.

Note that the Las Brisas Formation, the basal unit of the Caracas Group, is reportedly absent in the states of Guárico and in Southern Carabobo, though rocks resembling those of the Las Brisas are shown at the base of the metasedimentary section as representatives of the Las Mercedes Formation. However, rocks of the Las Mercedes Formation as originally mapped in the coastal range are akin to those of the Tucutunemo Formation of the median zone, being

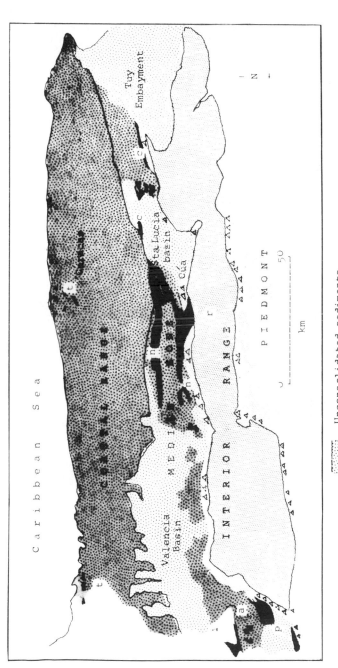

Figure 1. Major provinces, west section, Southern Caribbean Mountains.

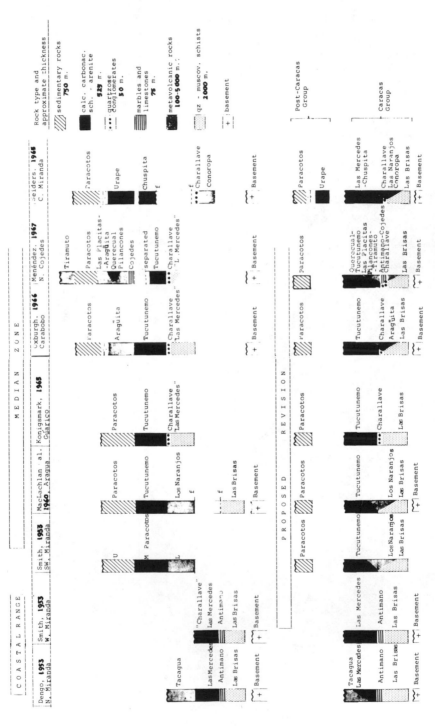

Figure 2. Correlation chart, west section, Southern Caribbean Mountains.

carbonaceous and calcareous in composition, not feldspathic. Rocks of the Tucu-tunemo Formation. The conglomerate extends farther west into northern Cojedes

The reason for the apparent duplication is as follows. Smith (1953) assigned the Charallave conglomerate to the upper part of the Las Mercedes Formation. Shagam (1960) mapped carbonaceous calcareous rocks similar to those of the Las Mercedes covering the conglomerate with apparent conformity and regarded them as a younger unit which he termed the Tucutunemo Formation. Later, the conglomerate was found in Guárico and in Carabobo where the rocks beneath it were referred to the Las Mercedes Formation, and those above it to the Tucu-tunemo Formation. The conglomerate extends farther west into northern Cojedes where, again, the overlying rocks have been assigned to the Tucutunemo Formation.

Smith based his observation on thin lenses of quartz conglomerate exposed in the coastal range, whereas the conglomerate extending east and west from Charallave along the southern edge of the median zone occurs in thick semicontinuous beds and is much coarser than the coastal range material. Furthermore, the Las Mercedes Formation of Smith terminates upward at the present erosion surface, or against faults, so that he could not establish the exact stratigraphic level of the conglom-erate with respect to the top of the formation.

Field relations observed during the course of the present study indicate that the Charallave conglomerate of the median zone is separated from basement by a thin section of metavolcanic rocks or quartzo-feldspathic schists and is overlain in turn by carbonaceous calcareous phyllites and arenites. In the terminology of the coastal range the underlying schistose materials belong to the Las Brisas Formation and the overlying materials are equivalent to those of the Las Mercedes Formation. The Charallave conglomerate marks the base of the Las Mercedes Formation, not the top, as previously asserted.

The changes resulting from this revision are illustrated in the lower part of Figure 2. In central Miranda the Tucutunemo Formation has not been identified, though the Chuspita Formation contains Tucutunemo-like materials and is found adjoining exposures of the Charallave conglomerate. Because of lithologic similari-ties and because of its proximity to the conglomerate, the Chuspita Formation is here considered correlative with both the Tucutunemo and the Las Mercedes Formations, rather than with the Tucutunemo alone, as suggested by Seiders (1965). On the basis of lithological correlation with two fossiliferous dated sedi-mentary units lying outside his map area, Oxburgh (1966) believes that the Aragüita Formation containing volcanic rocks may lie conformably above the Tucutunemo. However, Oxburgh mapped the Aragüita Formation as floored by basement and shows the formation to dip conformably *beneath* the Tucutunemo over the former formation's total exposed length of 10 km, thus duplicating the succession in Aragua where the Tucutunemo is underlain by the Los Naranjos volcanic member, and the succession in central Miranda where the Conoropa volcanic unit underlies the Chuspita Formation. Such correlations have been adopted in the revised part of Figure 2 and are extended to include the partly vol-canic Las Placitas Formation of Menéndez (1967), which he mapped as separating basement and the Tucutunemo Formation in northern Cojedes. Menéndez (1967) accepted the correlation and sequence proposed by Oxburgh and invoked a bedding plane thrust to account for the presence of the Tucutunemo Formation above the Las Placitas Formation. The thrusting is discounted in the proposed revision, the sequence mapped being taken as the depositional sequence.

In each area discussed above, conglomerate closely resembling the Charallave conglomerate occurs at or close to the gradational boundary between the bedded volcanic units and the conformably overlying calcareous carbonaceous schists and arenites. Whereas sequences shown in the upper part of Figure 2 do little to support the idea that the fault-separated median zone and coastal range might be linked, correlations indicated in the lower part suggest strongly that the two belts represent the fractured halves of a single province.

Figure 3 illustrates the interfingering of sedimentary volcanic rocks of the Caracas Group and volcanic rocks of the Villa de Cura Group, as observed in the Cúa region south of Caracas. The relation suggests that the median zone and the interior range, now separated by faults, formerly were continuous. The correlations of Figure 3 are based upon similarities in the mineralogy and texture, the lithology, and the observed sequences of various units, and upon structural and regional considerations. The Conoropa unit of Seiders (1965) may be equated with the Los Naranjos volcanic member of the Tucutunemo Formation described by MacLachlan and others (1960). The lower parts of these units, near Cúa, are glaucophanitic, and contain limestone, conglomerate, and lenses of quartzo-feldspathic material. A similar association occurs west of San Casimiro in glaucophanitic schists and granulites referred to the Santa Isabel Formation. The overlying volcanic units contain greater quantities of lava and of intercalated black chert and carbonaceous phyllite. Clastic components increase in volume northward, grading through the El Chino and El Caño Formations and, outside the map area, through the Aragüita and Las Placitas Formations into the Tucutunemo or Chuspita Formations. With increase in the proportion of volcanic materials southward, the pyroclastic rocks are observed to undergo a parallel change, becoming progressively coarser. Near San Casimiro, tuffs of the El Chino Formation can be followed continuously into breccias, here regarded as correlative with those of Menéndez's undated Tiramuto Formation.

Note that spatial relations of the Villa de Cura units as determined in the Cúa region and illustrated in Figure 3 differ from those suggested by other workers. Shagam (1960), who subdivided the Villa de Cura Group, considers the El Caño Formation the basal unit followed successively by the El Chino, El Carmen, and the Santa Isabel Formations. Seiders (1965) suggests the sequence may be overturned. Piburn (1968) investigated this possibility but concludes that Shagam's sequence is not inverted.

The sequences proposed in Figure 3 have been disrupted and shortened by faulting and interrupted by erosion. However, the only extensive developments of metavolcanic rocks of basaltic composition and submarine origin found within hundreds of kilometers of the occurrences forming the interior range occur along the northern edge of the range in the median zone, as components of the Caracas Group (Fig. 1). Similarly, substantial quantities of coarse detritus of metavolcanic composition do not occur anywhere in north-central Venezuela except in the piedmont flysch deposits, found adjoining the interior range (Fig. 1). It is difficult to believe that both these associations depend upon the fortuitous location of a volcanic block of allochthonous origin to which they are not related genetically. Furthermore, analysis of the faults found in the Cúa region and across the llanos suggests that the Villa de Cura materials have been displaced up to 50 km westnorthwest, rather than over 100 km southward as proposed by other workers (Fig. 4). Although the Paracotos Formation of the median zone visibly disappears

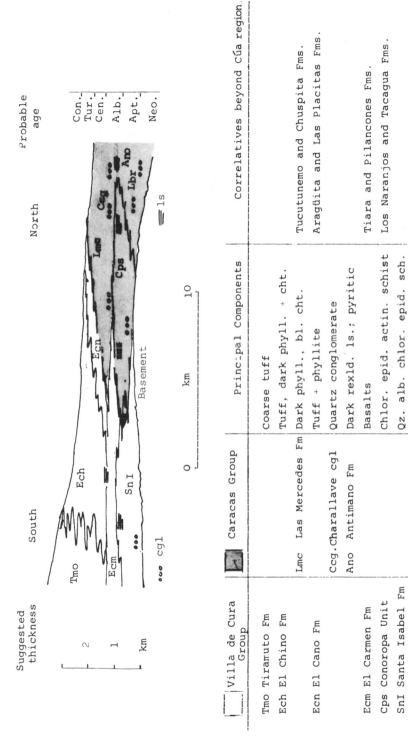

South   North

Suggested thickness

Probable age

Con.
Tur.
Cen.
Alb.
Apt.
Neo.

Tmo Ech Ecn Cps Ano Lbr Ccg Lmc

SnI Ecm

Basement

```
2

1

km
```

≡ ls

0   km   10

••• cgl

Villa de Cura Group	Caracas Group	Principal Components	Correlatives beyond Cúa region.
Tmo Tiraruto Fm		Coarse tuff	
Ech El Chino Fm		Tuff, dark phyll. + cht.	
	Lmc Las Mercedes Fm	Dark phyll., bl. cht.	Tucutunemo and Chuspita Fms.
Ecn El Cano Fm		Tuff + phyllite	Aragüita and Las Placitas Fms.
	Ccg Charallave cgl	Quartz conglomerate	
	Ano Antimano Fm	Dark rexld. ls.; pyritic	
Ecm El Carmen Fm		Basalts	Tiara and Pilancones Fms.
Cps Conoropa Unit		Chlor. epid. actin. schist	Los Naranjos and Tacagua Fms.
SnI Santa Isabel Fm	Lbr Las Brisas Fm	Qz. alb. chlor. epid. sch.	
		Qz. musc. sch., ss, gns.	

Figure 3. Structural relations, Villa de Cura and Caracas groups, Cúa region.

under Villa de Cura rocks and can be seen beneath them in windows found near the northern border of the interior range (Bell, 1968), the depositional relations are revealed west of Tácata where gently dipping unaltered Paracotos rocks are bedded upon markedly metamorphosed components of the Villa de Cura Group.

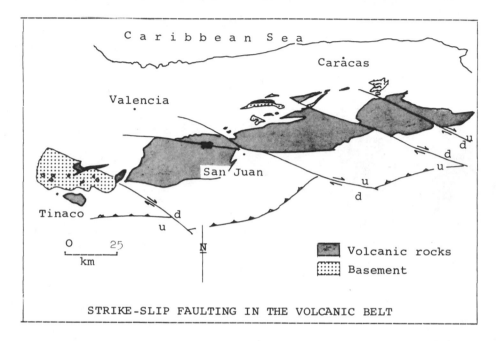

STRIKE-SLIP FAULTING IN THE VOLCANIC BELT

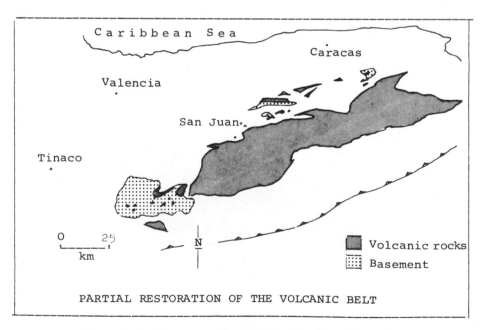

PARTIAL RESTORATION OF THE VOLCANIC BELT

Figure 4. Faulting, west section, Southern Caribbean Mountains.

REGIONAL CORRELATION

Stratigraphic columns for unmetamorphosed Cretaceous sections lying between eastern Colombia and Trinidad are illustrated in Figure 5. The average thicknesses and summary descriptions of major rock groups are generalized, but are correct in broad terms. The chart covers a time span of more than 60 m.y. and a region approximately 1,000 km long. Essentially the same succession of seven sedimentary rock types has been found throughout the region. Many of the units represented on the chart are fossiliferous and have been dated. They show that the corresponding lithological changes took place at similar times (Bucher, 1952; Mencher and others, 1953).

The western section of the Caribbean Mountains borders the central part of the 1,000-km belt represented on the chart. The sequence of metasedimentary units as revised in Figure 2 is shown in Figure 5, third column from the right, and appears to match the generalized succession devised for the unmetamorphosed Cretaceous sections of Venezuela, both in lithologic sequence and in terms of the relative thicknesses of units. Note that the metamorphic belt contains the lowermost six of the sequence of seven sedimentary rock types found beyond the belt. The chances of such correspondence existing between two juxtaposed sections of significantly different ages are extremely low. Consequently, the succession of the mountain belt is considered an extension of the unmetamorphosed Cretaceous sequence; individual units have been dated accordingly. As the metavolcanic section of the mountains has been tied to the metasedimentary section in Figure 3, the regional correlation above also serves to date materials of the interior range. Furthermore, the Pilancones and the Sans Souci volcanic rocks (Menéndez, 1967; Barr, 1962) which occur at the western and eastern ends of the Southern Caribbean Mountains are virtually unmetamorphosed, are regarded as probably Albian in part, and correspond in composition to the less-metamorphosed volcanic rocks of the inferior range and median zone.

Previous workers collected scant information on the ages of the metamorphic rocks. However, that which is available is in broad agreement with the ages arrived at through regional correlation. The flysch beds of the piedmont zone which contain fragments of metavolcanic rock probably derived from the interior range have been dated as Turonian–Santonian. Pyroxene diorite breccias have been reported from the piedmont cutting sedimentary rocks of Cenomanian–Santonian age. A K-Ar determination on a highly metamorphosed tuff representing the lower part of the Tiara Formation yielded an age of 100 ± 10 m.y., that is, approximately Albian–Cenomanian (Piburn, 1968). The Las Placitas Formation of Menéndez (1967) grades into the Cenomanian–Coniacian Querecual Formation. Smith (1953) describes a limestone pebble found north of Charallave by Hess which yielded three Foraminifera dated by Renz as Cretaceous, possibly Cenomanian. Other fossil fragments found by Smith appeared to Renz to resemble those of the El Cantíl limestone, which is approximately Upper Albian. Recently, a member of the Dirección de Geología, Venezuela, found ammonite remains north of Caucagua; the fossils are currently under investigation but appear to be indicative of a Late Albian age. At the western end of the mountain belt Bushman (1965) followed the Las Mercedes Formation into an unmetamorphosed section dated as Albian–Cenomanian and the Las Brisas Formation into a section of Neocomian–Aptian age. Limestones from the upper part of the lightly metamorphosed Para-

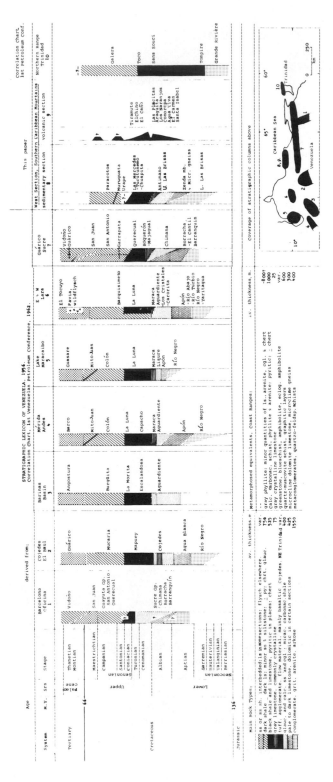

Figure 5. Regional correlation chart, northern Venezuela.

cotos Formation have been dated as Campanian–Maestrichtian (Shagam, 1960). In the piedmont zone, folding and thrusting affects beds of Eocene–Oligocene age (Bell, 1968). Little-disturbed sections of unconsolidated sediment dated as Oligocene–Miocene occur within the mountain belt.

GEOLOGIC HISTORY

The stratigraphy described in preceding paragraphs may be interpreted as follows. During the Early Cretaceous the Caribbean Sea invaded northern South America. Granitic basement remained exposed in a ridge extending along a line now marked by the axis of the median zone. Coarser detritus from the ridge was deposited at the coastline; limestones formed a short distance offshore. These materials are now represented by rocks of the Las Brisas, the Zenda, and Antimano units. Farther north a marginal deep developed in which the finer detritus accumulated, together with carbonaceous and calcareous matter, to form the components of the Las Mercedes Formation. The assemblage of both shallow- and deep-water deposits migrated southward through time as the sea encroached upon the basement ridge. By the Albian, most of the ridge was innundated, though islands remained. Volcanic vents opened along the axis of the ridge, erupting tuffs and lavas of basaltic composition which were deposited together with the limestones and coarse detritus collecting around the islands. The resultant materials are those of the Santa Isabel, Los Naranjos, and the Conoropa units. The products of isolated eruptions in deeper water were intercalated in the fine-grained carbonaceous sediments as the Tacagua Formation. The ridge continued to sink during the Cenomanian, pyroclastic materials from the axial vents being deposited along with dark silts and cherts as components of the El Chino, El Caño, Aragüita, and related formations. Volcanic activity then waned, though euxinic sediments continued to accumulate. Parts of the ridge began to re-emerge in the Coniacian–Santonian; uplift continued into the early Tertiary. The erosion resulting from these movements produced detritus forming the flysch beds of the Garrapata, Paracotos, and Pavia Formations. Parts of the ridge were covered by lakes and sea during mid-Tertiary and late Tertiary times, as indicated by the sedimentary sections of the Valencia and Santa Lucia basins and the Tuy embayment. Heating and recrystallization of the sediments and volcanic materials forming the Cretaceous section appears to have begun by Cenomanian time at the latest, and to have continued through to the Campanian–Maestrichtian. Marked deformation probably began in the Turonian and continued into the late Oligocene. Stratigraphic information coupled with data shown in Figures 1 and 4 indicates that during the Early Cretaceous the basement ridge, its coastline, and associated depositional boundaries and the main volcanic vents trended about N. 70° E. The principal fold axes of the mountains are superimposed upon the older trends, and run approximately N. 80° E. The current physiographic axis of the mountain belt runs N. 90° E., suggesting that the principal structural axis of the belt has rotated clockwise some 20° since the Early Cretaceous.

In summary, the subsidence, sedimentation, volcanism, recrystallization, intrusion, deformation, and rapid uplift and erosion leading to development of the west section of the Southern Caribbean Mountains took place in overlapping elongate belts, in the space of about 100 m.y. The succession of events may be

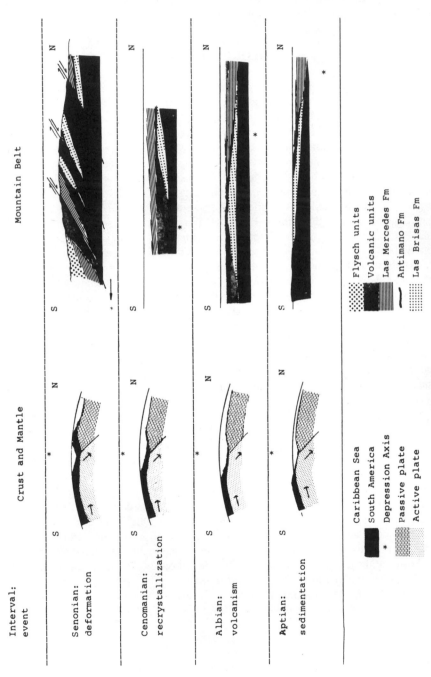

Figure 6. Evolution, Southern Caribbean Mountains.

accounted for by carrying continental crust across a linear depression localized at the surface of the mantle (Fig. 6). The drive is provided by subcrustal material moving into the depression. When the leading edge of a continental sheet approaches the depression and founders, sedimentation begins. Further advance of the crust causes the trough in which sedimentation takes place to migrate shoreward, producing an apparent transgression. Hot gases, liquids, and molten rock ascending near the axis of the depression break through the crust in a volcanic episode. The volatile materials and high temperatures associated with this activity cause extensive recrystallization within deeper layers of the covering materials. Faulting, gravity sliding, and thrusting, with much resultant folding of the cover rocks, occur as the continental margin is driven up the slope of the depression lying seaward of the axis by thicker crust being carried into and across the depression. The deformed materials ascending the outer slope emerge abruptly, yielding coarse detritus for the flysch deposits forming in the depositional basin now localized on the continental side of the emergent range. The final stages of igneous activity are marked by intrusions which enter some of the flysch beds.

The high-energy zone centered along the postulated depression at the mantle surface probably results from the interaction of the two sections of the upper mantle. One section may override the other, or, as the preferred explanation, one section moving horizontally may change direction, moving steeply downward on encountering a section which is inert or moving slowly.

Views in the last two paragraphs are corroborated in part by the work of Bell (1968) who shows development of the mountain belt to have been accompanied by southward migration of a marine basin.

Previous workers have suggested that northern developments of the Caracas Group which contain eclogite were metamorphosed while buried beneath a cover of Villa de Cura materials. In the present paper such metamorphism is related to abnormally high pressures and temperatures which may be expected to exist at comparatively shallow depths along the axes of depressions localized over interacting edges of mantle plates.

ACKNOWLEDGMENTS

H. H. Hess suggested the research topic and maintained great interest in the study. A. Vivas R., C. M. Bellizzia, A. Bellizzia G., J. Evanoff, and R. Shagam provided much assistance in Venezuela. R. Shagam, H. Metz, M. V. Maresch, and C. G. Murray have read the paper and contributed useful suggestions.

REFERENCES

Aguerrevere, S. E., and Zuloaga, G., 1937, Observaciones geológicas en la parte central de la Cordillera de la Costa, Venezuela; Bol. de Geol. y Min., v. 1, p. 3–22.

Barr, K. W., 1962, The geology of the Toco District, Trinidad, West Indies: Parts 1 and 2: Overseas Geol. and Min. Res., v. 8, p. 379–415 and v. 9, p. 1–29.

Bell, J. S., 1968, Geología de la región de Camatagua, Estado Aragua, Venezuela: Min. de Minas e Hidrocarb., Rep. de Venezuela, Bol. Geol., v. 9, p. 291–440.

Bucher, W. H., 1952, Geologic structure and orogenic history of Venezuela: Geol. Soc. America Mem. 49, 113 p.

Bushman, J. R., 1965, Geología del area de Barquisimeto, Venezuela: Min. de Minas e Hidrocarb., Rep. de Venezuela, Bol. Geol., v. 6, p. 3–111.

Dengo, G., 1953, Geology of the Caracas Region, Venezuela: Geol. Soc. America Bull., v. 64, p. 7–40.

Hess, H. H., 1966, Caribbean Research Project, 1965, and Bathymetric Chart: Geol. Soc. America Mem. 98, p. 1–10.

Konigsmark, T. A., 1965, Geología del area de Guárico Septentrional-Lago de Valencia, Venezuela: Min. de Minas e Hidrocarb., Rep. de Venezuela, Bol. Geol., v. 6, p. 209–285.

MacLachlan, J. C., Shagam, R., and Hess, H. H., 1960, Geology of the La Victoria area, Aragua, Venezuela: Geol. Soc. America Bull., v. 71, p. 241–248.

Mencher, E., Fichter, H. J., Renz, H. H., Wallis, W. E., Renz, H. H., Patterson, I. M., Robie, R. H., 1953, Geology of Venezuela and its oil fields: Amer. Assoc. Petroleum Geologists Bull., v. 37, p. 690–777.

Menéndez, A., 1967, Tectonics of the central part of the western Caribbean Mountains, Venezuela: Proc. Internat. Conf. Tropical Oceanography, Studies in Tropical Oceanography, v. 5, p. 103–130.

Oxburgh, E. R., 1966, Geology and metamorphism of Cretaceous rocks in Eastern Carabobo State, Venezuelan Coast Ranges: Geol. Soc. America Mem. 98, p. 241–310.

Piburn, M. D., 1968, Metamorfismo y estructura del Grupo Villa de Cura, Venezuela Septentrional: Min. de Minas e Hidrocarb., Rep. de Venezuela, Bol. Geol., v. 9, p. 183–290.

Seders, V. M., 1965, Geología de Miranda central, Venezuela: Min. de Minas e Hidrocarb., Rep. de Venezuela, Bol. Geol., v. 6, p. 289–416.

Shagam, R., 1960, Geology of Central Aragua, Venezuela: Geol. Soc. America Bull., v. 71, p. 249–302.

Smith, R. J., 1953, Geology of the Los Teques-Cúa region: Geol. Soc. America Bull., v. 64, p. 41–64.

Manuscript Received by the Society March 29, 1971

THE GEOLOGICAL SOCIETY OF AMERICA, INC.
MEMOIR 132
© 1972

Deep-Water, Shallow-Water, and Subaerial Island-Arc Volcanism: An Example from the Virgin Islands

THOMAS W. DONNELLY
Department of Geology,
State University of New York, Binghamton, New York 13901

ABSTRACT

The Water Island and Louisenhoj Formations of the U.S. Virgin Islands represent a mid-Cretaceous transition from deep-water to shallow-water and subaerial volcanic eruption of andesitic magmas. The upward increase in porphyritic texture, evidences of explosive eruption, segregation vesicles, and pumice and hyaloclastic fragments all suggest the dominating effect of ambient water pressure at the eruptive vent, which in turn controls the rapidity of separation of intrinsic volatiles during solidification.

INTRODUCTION

Island-arc volcanoes throughout the world are characterized by dominantly explosive volcanic activity which produces fragmental products ranging in composition from calc-alkaline basalt to dacite. These volcanic products are universally erupted today from subaerial or shallow submarine vents. However, the fact that most island arcs are developed on oceanic crust suggests the possibility that stratigraphically older deposits might represent eruption at greater water depth. Most of the earlier volcanic products of island arcs are buried beneath late Tertiary volcanic accumulations or metamorphosed beyond recognition. The Virgin Islands are nearly unique among the world's island arcs in affording an examination of volcanic activity prior to the existence of an emergent island platform (Donnelly, 1959, 1964, 1966a) and of the transition from apparently deep submarine to subaerial volcanic activity. Unfortunately, there appear to be few other examples for comparison (Donnelly, 1966b).

The present analysis stems from my earlier doctoral dissertation in St. Thomas and St. John (Donnelly, 1959), which was a part of the Princeton Caribbean Project. Subsequent work in the area, much in cooperation with other workers, has indicated the need for further development of the criteria by which I inferred the transition from deep-water to shallow-water to subaerial volcanism. I should emphasize that my arguments have been based on a variety of considerations, many of which (such as the chemical considerations) are amplified in an earlier paper (Donnelly, 1966a). The arguments presented here will be controversial, and the reader should be cautioned that these arguments are developed for one occurrence only. Also, it is tacitly assumed throughout that the Virgin Islands represent an orogenic magma sequence, and comparisons with nonorogenic magma sequences should be made only with great care. Especially relevant here is the probability that nonorogenic basalt magmas commonly have very low water contents and, thus, display consolidation characteristics which are considerably different from those of the Virgin Islands mafic volcanic rocks.

STRATIGRAPHY AND FIELD RELATIONS

The Water Island Formation of pre-Albian age and the overlying Louisenhoj Formation of probable Albian age[1] are the most widespread rock units of St. Thomas and St. John, Virgin Islands. The Water Island Formation consists predominantly of quartz-keratophyre flows with intercalated spilite flows. There are no terrigenous sediments, and pyroclastic units are scarce, except near the top of the formation. The Louisenhoj Formation, which overlies the Water Island Formation with a minor angular unconformity, consists of well-bedded porphyritic basaltic-andesitic pyroclastic beds with some intercalated conglomerates and submarine slump deposits.

I do not intend in this paper to develop arguments relevant to the "spilite" problem in general. My preference is to restrict the term spilite to rocks whose apparently low-grade mineralogy was achieved by essentially primary means. Mafic flows whose albite is distinctly secondary I refer to as "albitized" andesite or basalts. Consistent with this distinction the term spilite is restricted here to the Water Island Formation (which, however, does contain at least one basaltic andesite flow). The mafic igneous fragments of the Louisenhoj Formation are termed basaltic andesites.

The Louisenhoj Formation clearly represents the products of subaerial or shallow-water (with deposition in shallow water) volcanism; the beds are well graded and consist of angular fragments of basaltic andesite, which range from highly porphyritic and nearly nonvesicular to pumice and dense, hyaloclastic fragments. The field appearance of the unit bears a strong resemblance to that of the lower

---

[1] The ages of these units continue to be a problem. A trondhjemite on the island of Désirade has been dated at 142 m.y. (Fink, 1970); this plutonic rock appears to be associated with quartz keratophyres which are indistinguishable from those of the Water Island Formation. The radiometric age of 108 m.y. cited by Donnelly (1966a) suggests a younger age for the Virgin Island rocks but may represent later reheating. The Louisenhoj Formation has not been dated directly; the overlying Tutu Formation contains undoubted Albian rudists (Donnelly, 1966a) and ammonites (Young, 1971). My view is that there is not necessarily a major time break between the two units.

Tertiary Ohanapecosh Formation of the Mt. Rainier area (Fiske, 1963), except that the accumulation of pumice toward the tops of the beds is far less striking in the Virgin Islands. Abundant small-scale evidence of slumping, intercalated coarse submarine slump deposits, contemporaneously subaerially weathered beds, and coarse conglomerates all suggest further that the environment of deposition was the flank of an emergent submarine volcanic center.

An analogous environmental interpretation for the Water Island Formation is more difficult. Except in the uppermost part of the formation, virtually all of the eruptive products were massive flows. Intercalated terrigenous sediment is absent and pyroclastic beds are scarce. Slump structures are not seen. The flows themselves show only minor chilling, fragmentation, and explosion phenomena (except, again, near the top of the formation), pillows are virtually absent, and there are no columnar joints. For these reasons, reinforced with chemical and petrographic reasoning discussed previously (Donnelly, 1966a) and partly amplified below, I concluded that the formation was a very deep-water accumulation of flows.

Subsequent to my original field work, several new rock exposures near Charlotte Amalie, St. Thomas, have provided additional insights into the nature of the upper Water Island Formation. If the area were to be mapped today it would probably be feasible to recognize an additional upper member of the formation. Keratophyres of this upper unit are dominantly thin-bedded tuffs with angular crystal fragments (as at the Submarine Base and Flag Hill). Mafic extrusive units consist of partially pillowed, porphyritic augite-andesite (Careen Hill) and beds of bombs and blocks of a similar porphyritic andesite (Bluebeard Hill, northeast slope of Flag Hill). Transitional units consisting of relatively massive flows interlayered with more fragmental flows occur on Great St. James Island and in north-central St. John.

On the basis of simple field relations (nature of erupted material, character of bedding, presence or absence of terrigenous or biogenic sediment), I concluded (1966a) that Virgin Island stratigraphic sequence represents a transition from a deep-water to a shallow-water or emergent environment, with the relation between the ambient water pressure and the water pressure within the cooling melt being the most important factor.

CHEMICAL CRITERIA FOR ENVIRONMENT OF ERUPTION

The chemical characteristics of the siliceous and mafic erupted products of the two formations are strikingly similar (Donnelly, 1966a, and a more recent compilation of 359 Virgin Islands spilite analyses performed by R. Scott, R. Hekinian, and myself). Table 1 shows that spilite is chemically close to Louisenhoj mafic fragments. The MgO/FeO* ratio[2] is lower in the Louisenhoj Formation, though one of the younger Water Island flows (Flamingo Bay, Water Island) shows a similar degree of iron enrichment (average FeO* = 10.0 percent; MgO = 5.6 percent). Sodium is not enriched in the spilites compared to Antillean calc-alkaline basalts: the average of 359 analyses is 3.0 percent. Similarly, potassium is not depleted (average 0.62 percent), nor are the values widely scattered. Louisenhoj fragments have higher average sodium and lower potassium, and the individual values are

---

[2] FeO* means total iron computed as FeO.

more scattered in the 30 analyses of individual fragments considered here. I believe that the more extreme alkali ratio of the Louisenhoj represents some local alkali autometasomatism during metamorphism and indicates that the Louisenhoj fragments are approaching the so-called spilite, or soda-rich, composition of other authors. The chemical similarity of the Water Island spilites and the overlying Louisenhoj basaltic andesites precludes the necessity for a special spilite magma.

In contrast to these values, the alkali distribution of the keratophyres (Donnelly, 1966a) shows evidence for sea-water alkali-ion exchange. The intrusive keratophyres have a relatively low sodium/potassium ratio (Na_2O/K_2O about 3.5); however, the extrusive units are highly depleted in potassium and somewhat enriched in sodium. Further, several keratophyre flows have decreasing potassium toward the flow top. Alkali-exchange experiments (Orville, 1963) show that at higher temperatures silicates and aqueous fluids tend to approach equality in their sodium/potassium ratios. Because sea water has a sodium/potassium weight ratio of 28:1, we would expect a hot silicate with an initial weight ratio of 3:1 to lose potassium in exchange for sodium. At 700°C and 100 atmospheres (corresponding to sea-water depth of about 1 km) the density of pure water would be about 0.02 gm/cc (Kennedy and Holser, 1966). At the same temperature and 500 atmospheres pressure the density would be 0.13 gm/cc. In a higher pressure environment, therefore, we might expect the adjacent fluid phase to be a far more efficient medium for ion exchange. Water Island spilites show no evidence for alkali ion exchange, nor has such an exchange been invoked for abyssal basalts, in spite of the low potassium content of the basalt. Possible explanations are that the higher liquidus temperatures of mafic melts create a lower density surrounding aqueous phase or that divalent-ion-rich mafic melts do not undergo alkali ion exchange as readily.

PETROGRAPHY AND MINERALOGY

Previous accounts of petrography and mineralogy (Donnelly, 1962, 1963, 1966a) of these rocks have been augmented recently by Raymond E. Smith (unpub. data, 1969) and by Hekinian (1969, 1971). More precise mineralogical information must await microprobe analysis of some of the minerals; however, a few observations will help illuminate the general discussion.

TABLE 1. AVERAGE CHEMICAL COMPOSITIONS OF PARTIALLY ANALYZED SPILITES AND LOUISENHOJ BASALTIC ANDESITES (SiO_2 BETWEEN 45 AND 55 PERCENT) OF THE VIRGIN ISLANDS

	Spilite	Louisenhoj
SiO_2	49.5 (275)	50.6 (30)
TiO_2	0.63 (223)	0.61 (30)
Al_2O_3	15.0 (303)	16.5 (30)
FeO^*	9.2 (359)	10.3 (30)
MgO	7.4 (359)	4.8 (30)
CaO	8.0 (359)	8.1 (30)
Na_2O	3.0 (359)	3.7 (30)
K_2O	0.62 (359)	0.39 (30)

Number of samples in parentheses.
FeO^* is total iron as FeO.

Water Island spilites consist of an intersertal aggregate of plagioclase (generally low-T albite) in a groundmass of chlorite, with microphenocrysts of aluminous clinopyroxene ($Al_2O_3 = 6.0$, 4.3 percent in two samples). The rocks are commonly amygdular and contain irregularly distributed minerals (quartz, magnetite, calcite, epidote, and prehnite) generally associated with low-grade metamorphism. Because of the very small size of spilitic plagioclase microlites (commonly only a few microns wide—*see* Fig. 1), conventional optical determinative techniques have been difficult. Universal-stage determinations of larger grains generally give sodic compositions (Donnelly, 1963), but some smaller microlites have been found to be quite calcic[3].

Hekinian found two other minerals of great interest—green biotite and hydro-garnet. The latter is undoubtedly of metamorphic origin, but the biotite (which has a slightly lower birefringence than most biotites and whose basal x-ray peaks sharpen during heating, suggesting interlayered vermiculite) may be primary. Hekinian also noticed widespread very fine-grained actinolite.

Smith pointed out to me an example of brown hornblende adjacent to margins of vesicles in an apparently intrusive spilite body (Fig. 2). Although further work is indicated, I would note both its distinctly igneous character and the fact that comparable hornblende is unknown from stratigraphically higher basaltic andesites. Green hornblende microphenocrysts also occur in spilites from the deep core in St. John. Wairakite, whose occurrence is considered to be deuteric, was found rimming keratophyric xenoliths in a spilite flow (Donnelly, 1962).

Mineralogically the Louisenhoj Formation is similar to the Water Island Formation. Calcic plagioclase phenocrysts (labradorite to anorthite) are widespread, but about half of the plagioclase occurrences are now albite. Clinopyroxene is invari-

---

[3] The tedious U-stage search for calcic microlites was inspired by observations of R. E. Smith and R. Hekinian, who noted that some microlites appear to show positive relief against quartz.

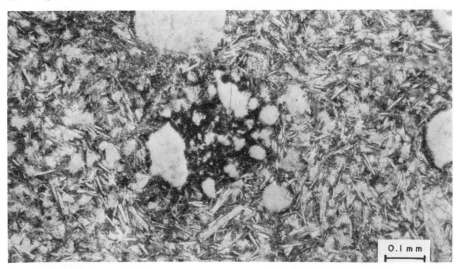

Figure 1. Photomicrograph of typical spilite from lower part of Water Island Formation (61–20a4, Concordia Bay, St. John), showing segregation vesicle and chlorite-filled amygdules.

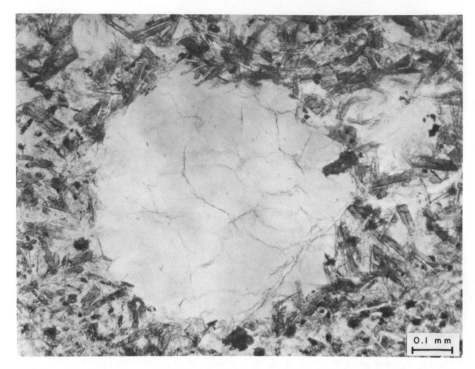

Figure 2. Photomicrograph of spilite showing abundant brown hornblende prisms rimming chlorite-filled amygdule. (61–57, Flamingo Point, Water Island.)

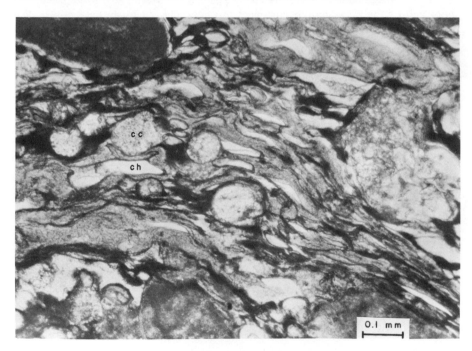

Figure 3. Photomicrograph showing pumiceous fragments in Louisenhoj basaltic andesite. Elongate vesicles filled with chlorite, (ch), more equant vesicles with calcite (cc). (ST–461, William Head, St. Thomas.)

ably fresh and not marginally altered to actinolite, as is the case with many spilites. The groundmass contains widespread epidote, prehnite, and calcite, as in the spilites, and also pumpellyite, which is uncommon in all but the stratigraphically higher Water Island spilites. Most of the occurrences of pumpellyite in the Louisenhoj Formation are secondary after glass, which occurs abundantly as pumiceous fragments (Fig. 3), which accumulate at the tops of some pyroclastic beds, and also as denser, angular glass fragments, which have recrystallized to almost pure, cryptocrystalline pumpellyite (Donnelly, 1966a). Some spilite flows (as on Great St. James Island), which are stratigraphically high in the sequence, have interlayered hyaloclastic fragmental beds; however, the scarcity of pumpellyite in the Water Island Formation might indicate a scarcity of original glass.

Keratophyric mineralogy (albite, quartz, muscovite, chlorite, and magnetite with some clearly secondary celadonite and stilpnomelane in stratigraphically higher tuffs) is not inconsistent with a metamorphic origin, with the most significant exception being the high-T albite and oligoclase in one sample (Donoe, St. Thomas; Donnelly, 1963). Siliceous fragments in the Louisenhoj Formation are scarce and aphanitic; however, I found high-T andesine, iron-rich clinopyroxene, iron-rich olivine, and brown hornblende from one locality (Donnelly, 1966a).

Seriate versus Porphyritic Texture

The variable extent of development of porphyritic texture of the mafic rocks has not been emphasized in past accounts of these rocks. Spilites low in the section are nonporphyritic, with seriate plagioclase (Fig. 2) and microphenocrysts of clinopyroxene (which are generally smaller than 0.2 mm). Stratigraphically higher spilites have a less seriate texture and distinctly larger plagioclase (Figs. 4, 5) and pyroxene (Fig. 6) crystals. The Careen Hill augite-andesite flow (Fig. 7) has a highly porphyritic texture with very large plagioclase and clinopyroxene phenocrysts, many of which are a few millimeters at their maximum dimension. Crystalline fragments from the Louisenhoj Formation (Figs. 8, 9) are commonly highly porphyritic.

Porphyritic texture reflects two stages of growth, conventionally taken as a slow intratelluric stage followed by a rapid hypabyssal or extrusive stage. The increase in porphyritic texture with the passage of time in the Virgin Islands could signify that successively later magmas were able to pause in their ascent to begin a slow crystallization prior to their final emplacement. However, the ambient water pressure at the volcanic vent might also be important. Wet magmas erupted into very high water pressures (500 atmospheres corresponds to the average ocean depth of 5 km) would tend to lose their volatiles slowly, and mainly subsequent to eruption, and their solidification would be a generally continuous process. The same wet magma, erupted either in shallow water or emplaced at shallow depth beneath a vent, might lose its water explosively within the vent prior to the conductive loss of much of its heat; therefore, such magmas might be expected to solidify in a two-stage, discontinuous fashion. According to this argument, the ubiquitous porphyritic texture of calc-alkaline andesites could be taken as an indication that the magma ascended to a low-pressure environment prior to its eruption. Separation of volatiles from the melt initiated intratelluric crystallization. Eruption of a porphyritic, less water-rich melt results in quenching, with forma-

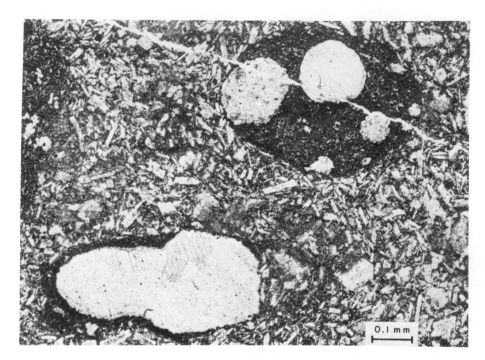

Figure 4. Photomicrograph showing stratigraphically high, more porphyritic spilite from the Water Island Formation. The segregation vesicle is filled with chlorite and epidote. Adjacent amygdule shows less evidence of shrinkage. (68–42a, West India Company Quarry, St. Thomas.)

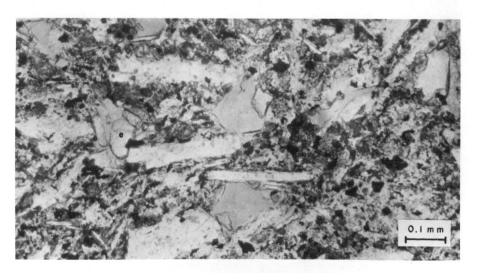

Figure 5. Photomicrograph of another stratigraphically high spilite showing relatively large sodic plagioclase crystals. Chlorite-filled amygdules have large epidote (e) crystals. (ST–201, West India Company Quarry, St. Thomas.)

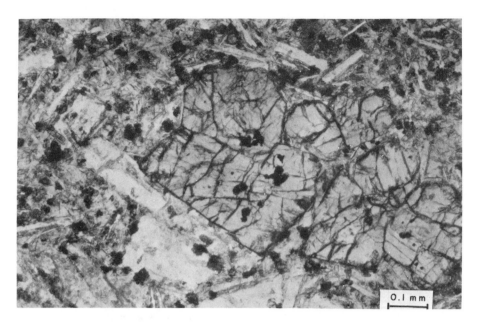

Figure 6. Photomicrograph of stratigraphically high Water Island spilite showing large microphenocryst of aluminous clinopyroxene intergrown with sodic plagioclase. (61–57, Flamingo Pt., Water Island.)

Figure 7. Photomicrograph of basaltic andesite flow from the uppermost Water Island Formation showing conspicuously porphyritic texture. P is a pseudomorphed mafic phenocryst; the remainder of the phenocrysts are labradorite replaced partially by albite. (ST–361, Careen Hill, St. Thomas.)

Figure 8. Photomicrograph showing fragments of porphyritic basaltic andesite from the Louisenhoj Formation. (ST–351d, Brewer's Bay, St. Thomas.)

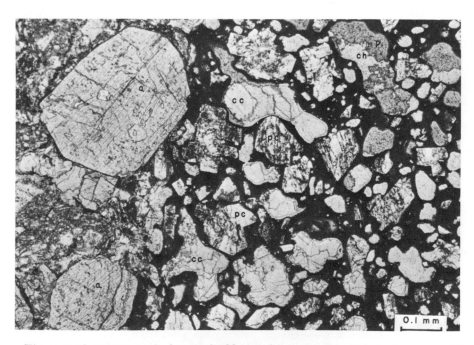

Figure 9. Photomicrograph showing highly porphyritic basaltic andesite fragment and crystal fragments from the Louisenhoj Formation. Amygdules filled with calcite (cc), chlorite (ch), and prehnite (p). Note large, altered plagioclase phenocrysts (pc) and augite fragments (a). (ST–452, Rosendal, St. Thomas.)

tion of a glassy or very finely crystalline groundmass. This argument is a corollary of a commonly expressed view that the partial intratelluric solidification of wet magmas increases the water pressure, which in turn triggers explosive eruption of the partially solidified melt.

Analogous arguments will not be developed for the keratophyres. The scarcity of siliceous fragments in the Louisenhoj Formation precludes any comparison of modes of crystallization between the two formations.

Segregation Vesicles

Smith's (1967) discussion of segregation vesicles, besides being a remarkably perceptive analysis of the phenomenon, has a particular relevance to the mafic lavas of the Virgin Islands. A segregation vesicle is one in which the original liquid filling (probably water) shrinks during solidification so that late-stage melt can enter the vesicle and occupy some of the former volume. If the original outline is preserved, an estimate can be made of the volume decrease during this portion of the cooling. Smith's examples showed shrinkages of about half; the Virgin

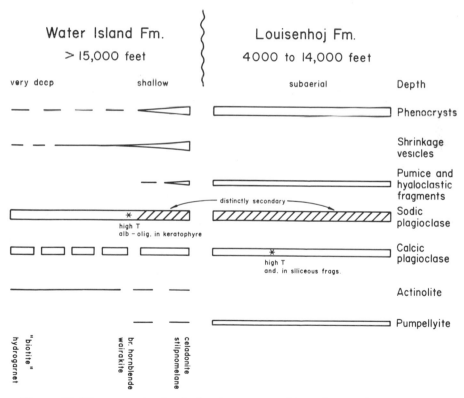

Figure 10. Diagrammatic sketch showing petrographic and mineralogic character of the transition through the Water Island Formation to the Louisenhoj Formation. The origin of the sodic plagioclase in the lower Water Island is neither distinctly secondary nor distinctly primary. Occurrences of some rare or scarce igneous and metamorphic minerals in the Water Island Formation are located stratigraphically along the bottom. Depth refers to position of the eruptive vent.

Islands spilites show many comparable examples (Figs. 1, 3). Smith's mechanism for obtaining these vesicles required a cooling basaltic lava to have flowed down a slope in sea water, such that the ambient pressure increased during cooling. Table 2 shows calculated shrinkage percentages for pure water (Kennedy and Holser, 1966) for various ranges of temperature and pressures. These values show that shrinkage, for a given temperature range (here 200°C) is enhanced by lower starting and finishing temperatures and by higher ambient pressure. Shrinkage is much more strongly enhanced by an increase in ambient pressure (here a range of 100 bars, corresponding to about 1 km depth of sea water) during this cooling. This effect is slightly greater at lower pressures.

A further important consideration is the specific volume of the fluid phase. If the magma is emplaced in a very low pressure environment, the expansion of even one percent of water will tear the magma apart by violent vesiculation, and no relatively orderly textural features can be expected to survive. The figures in parentheses in Table 2 show the specific volume of water for the final temperature in each case. Clearly higher ambient pressures will favor the orderly formation of vesicles.

The former requisite conditions for segregation vesicle formation are best met by eruption of magmas on the flanks of an emergent submarine volcanic edifice. The scarcity of segregation vesicles in early Water Island time implies flat topography (the flat sea floor), while their later prominence implies the formation of a submarine volcanic edifice.

METAMORPHIC VERSUS IGNEOUS ORIGIN

At first inspection, the Water Island spilites, as well as the Louisenhoj basaltic andesites, appear to have undergone low-grade metamorphism. In the case of the spilites (and the keratophyres), lack of obvious nonmetamorphosed analogues among younger West Indian rock sequences, as well as characters cited above, raise the possibility that these volcanic rocks are normal, wet calc-alkaline mafic melts that solidified in an unusual way and that, as a consequence of these conditions of solidification, they achieved a quasi-metamorphic mineralogy during their original cooling. Volcanic rocks retain characteristically metastable mineralogies only because the great majority solidify quickly and under essentially anhydrous conditions (those whose magmas were wet commonly lose their volatiles rapidly upon

TABLE 2. PERCENTAGE SHRINKAGE OF PURE WATER IN TWO 200°C TEMPERATURE INTERVALS AT DIFFERENT PRESSURES

	1,200 to 1,000°C	1,000 to 800°C
1 atm	13.6% (5875)	15.7% (4951)
100 atm	14.3% (58.39)	17.9% (48.54)
200 atm	15.1% (29.08)	18.1% (23.83)
500 atm	16.7% (11.59)	21.4% (9.104)
100 to 200 atm	57.3% (29.08)	59.2% (23.83)
200 to 300 atm	43.5% (23.24)	46.3% (15.61)

Data taken from or calculated from Kennedy and Holser (1966).

Figures in parentheses are specific volumes for water at the final temperature and pressure.

eruption). At high ambient water pressures, the retardation of the separation of intrinsic volatiles necessarily promotes successive subsolidus mineral reactions during cooling, such that mineral assemblages appropriate to "metamorphic" temperature and pressure ranges are produced.

In the Virgin Islands we are left with the dilemma that, if the above conclusion is accepted, there are few if any uniquely metamorphic features of the spilites. Smith (1968) proposed a metamorphic origin for similar rocks from Australia, partly on the basis of the inferred formation of epidote-rich patches subsequent to regional jointing. Similar patches are widespread in Virgin Island rocks and may also be metamorphic; however, I have seen no similar relations between patches and jointing in the Virgin Islands examples, and I prefer to reserve judgment on this question.

A doubtlessly controversial conclusion is that the Water Island Formation, although buried as deeply or deeper than the Louisenhoj Formation, may have actually been metamorphosed less than this unit. The higher formation, with abundant glass and phenocrysts of igneous minerals, was probably more susceptible than the lower unit to the temperature and pressure conditions promoting metamorphism. Still unanswered, but not really germane to the present argument, is the problem of establishing for each mineral in the spilite the exact time and circumstances of formation. This tangled skein of mineral paragenesis will require further analysis for its unraveling.

ACKNOWLEDGMENTS

I am grateful to Raymond E. Smith, a research associate in this department during 1968–1969, and to Roger Hekinian, who completed a doctoral thesis here in 1969 on the spilites of the St. John deep core, for many interesting discussions on the problem and for some new mineralogical observations. Clifford Hopson and Robert Hargraves provided careful critiques of the first draft of this paper, and I am extremely grateful to both of them for their views.

Though this volume was conceived as a tribute to Harry Hess, an acknowledgment to him would be necessary in any case. Hess's thoroughly nondogmatic attitudes, inquisitiveness, and completely fresh approach to geology provided a stimulating environment in which new and often heretical ideas began life with the expectancy of a proper evaluation.

My work has been supported by National Science Foundation grants G14407, GP–1352, GP–4695, and GA–588.

REFERENCES CITED

Donnelly, T. W., 1959, Geology of St. Thomas and St. John, U.S. Virgin Islands [Ph.D. thesis]: Princeton University, 179 p.

—— 1962, Wairakite in West Indian spilitic rocks: Am. Mineralogist, v. 47, p. 794–802.

—— 1963, Genesis of albite in early orogenic volcanic rocks: Am. Jour. Sci., v. 261, p. 957–972.

—— 1964, Evolution of eastern Antillean Island arc: Am. Assoc. Petroleum Geologists Bull., v. 48, p. 680–696.

—— 1966a, Geology of St. Thomas and St. John, Virgin Islands, in Hess, H. H., ed., Caribbean geological investigations: Geol. Soc. America Mem. 98, p. 85–176.

—— 1966b, Tectonic significance of the spilite-keratophyre association: Kingston, Jamaica, Trans. Third Caribbean Geol. Conf., Geol. Survey Dept., p. 13–26.

Fink, L. K., 1970, Evidence for the antiquity of the Lesser Antilles Island Arc [abs.], Trans. Am. Geophys. Union, v. 51, p. 326.

Fiske, R. S., 1963, Subaqueous pyroclastic flows in the Ohanapecosh Formation, Washington: Geol. Soc. America Bull., v. 74, p. 391–406.

Hekinian, Roger, 1969, Petrological and geochemical study of spilites and associated dike rocks from the Virgin Island core (Caribbean Island Arc) [Ph.D. thesis]: State University of New York at Binghamton, 204 p.

——, 1971, Petrological and geochemical study of spilites and associated rocks from St. John, U.S. Virgin Islands: Geol. Soc. America Bull., v. 82, p. 659–682.

Kennedy, G. C., and Holser, W. T., 1966, Pressure-volume-temperature and phase relations of water and carbon dioxide, in Clark, Sydney P., Jr., ed., Handbook of physical constants, revised edition: Geol. Soc. America Mem. 97, 587 p.

Orville, P. M., 1963, Alkali ion exchange between vapor and feldspar phase: Am. Jour. Sci., v. 261, p. 201–237.

Smith, R. E., 1967, Segregation vesicles in basaltic lava: Am. Jour. Sci., v. 265, p. 696–713.

—— 1968, Redistribution of major elements in the alteration of some basic lavas during burial metamorphism: Jour. Petrology, v. 9, p. 191–219.

Young, K., 1971, Ammonites from Puerto Rico and St. Thomas [Abs.]: 6th Caribbean Geological Conference, Margarita, Venezuela, p. 17–18.

Manuscript Received by the Society March 29, 1971

THE GEOLOGICAL SOCIETY OF AMERICA, INC.
MEMOIR 132
© 1972

Chaotic Sedimentation in North-Central Dominican Republic

FREDERICK NAGLE

*Department of Geology and Rosenstiel School of Marine
and Atmospheric Science
University of Miami, Miami, Florida 33149*

ABSTRACT

Chaotic, allochthonous, submarine gravity slide deposits (olistostromes) have been reported with increasing frequency during the past 15 yrs in stratigraphic columns throughout the world. One such unit, estimated to be several hundred meters thick, is distributed over 90 sq km in the Puerto Plata area of north-central Dominican Republic.

The gray, clay-sized, structureless matrix of this unit behaves as a quick-clay under shock, and is composed of kaolinite, quartz, montmorillonite, and illite. Pelagic Foraminifera date the matrix as Paleocene(?) or early Eocene(?).

Exotic blocks within the olistostrome include limestone, serpentinite, andesite, pillow volcanics, and tuffaceous rocks. The longest dimensions of the blocks range from 1 cm to 1.5 km, and the known ages range from pre-Paleocene to middle Eocene. Thus, the exotic blocks are older, the same age as, and younger than their enclosing matrix.

The matrix probably was once a marine tuffaceous unit of Paleocene–early Eocene age which was deposited rapidly and retained enough water to liquify spontaneously and to begin moving rapidly down a gentle submarine basin slope following an earthquake shock during the late-middle Eocene. Younger more competent overlying rocks were ruptured and incorporated into the moving mass, while older rocks were torn loose from the submarine slope. The entire unit was emplaced essentially instantaneously into a marine sedimentary sequence.

Although olistostromes record catastrophic events in the geologic record, they do not necessarily imply a major orogeny. However, most described olistostromes were formed during times of tectonic activity. The middle Eocene was such a time in the Puerto Plata area.

INTRODUCTION

During the course of field mapping of some 700 sq km in the vicinity of Puerto Plata, Dominican Republic (*see* Fig. 1 and Nagle, 1966, 1971), a major problem of geologic interpretation centered upon a mudstone unit incorporating a unique assemblage of boulders and blocks of various sizes and rock types. The San Marcos olistostrome was not at first recognized as a unit. During the initial stages of the investigation attention was centered on the blocks, and as more were examined, it became increasingly difficult to construct a reasonable map, no matter what sort of fold or fault patterns were invented. In the summer of 1961, H. H. Hess (oral commun.) suggested that this chaos was an allocthonous unit and that the blocks were exotic. In the following months, field work indicated that this was so.

The Puerto Plata area is located along the extension of the axial zone of the Puerto Rico trench which dies out 250 km east of Puerto Plata. Negative isostatic gravity anomalies extend beyond the topographic expression of the trench through the Puerto Plata area. The area is currently tectonically active (Fig. 2); two linear zones of epicenters intersect at roughly 70°10′ W., 19°30′ N., a point 50 km southeast of the mapped area. One of these zones continues along the south slope of the Puerto Rico trench, indicating the occurrence of a fault or fault zone along that line which may connect with known fault zones to the west in Hispaniola and Cuba. Faults of one type or another have been suggested for one or both slopes of the Puerto Rico trench by many authors, although past and present motions on these faults have been debated (*see,* for example, Bunce and Fahlquist, 1962; Glover, 1967; Chase and Hersey, 1968; Monroe, 1968; Molnar and Sykes, 1969; Bracey and Vogt, 1970; Nagle, 1970). Focal mechanism solutions from fault plane data (Molnar and Sykes, 1969) indicate current underthrusting nearly perpendicular to the Lesser Antillean portion of the West Indies arc but almost parallel to the Puerto Rico trench trend in the area north of the Virgin Islands to Hispaniola

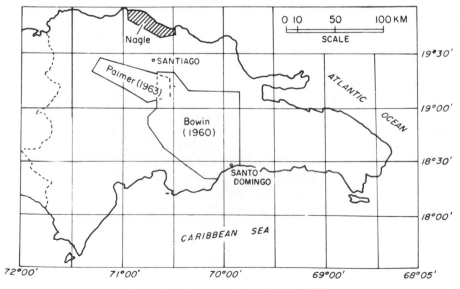

Figure 1. Index map, Dominican Republic.

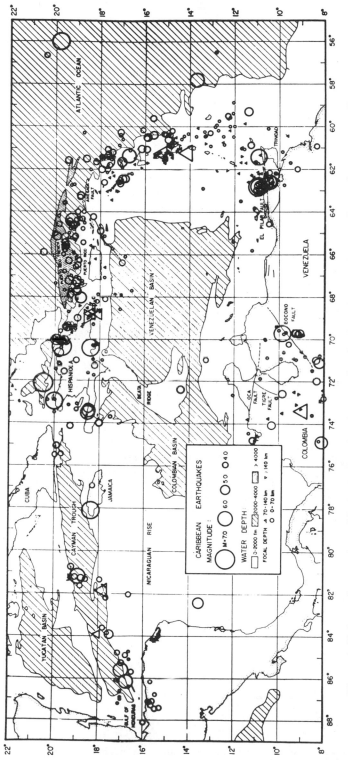

Figure 2. Epicentral map of Caribbean earthquakes for period 1950 through 1964 (*after* Sykes and Ewing, 1965).

(Fig. 3). Tectonic activity, especially vertical movements, can be documented from middle Eocene to the present in the Puerto Plata region, and I have suggested elsewhere (Nagle, 1971) that the rocks and tectonic history indicate that this region is an upfaulted sliver of the south slope of the Puerto Rico trench.

The San Marcos olistostrome is interpreted as a chaotic submarine gravity slide, containing blocks older, contemporaneous with, and younger than an unindurated clay-size matrix, the whole mass of which was emplaced during a single catastrophic event. There is no evidence for later tectonic remobilization.

Hsu (1968) made an important contribution to the subject by pointing out that two different sets of processes could lead to stratal disruption and mixing and suggests that the term *olistostrome* be restricted to sedimentary deposits as suggested by Flores (1955) and that the term *melange* be used to refer to those deposits which have resulted from tectonic mixing of consolidated rocks under an overburden after sedimentation. Hsu details several diagnostic characteristics of each type.

Olistostromes very similar to the San Marcos have been described in many parts of the world: Timor (Audley-Charles, 1965), Sicily (Marchetti, 1957), and New Zealand (Kear and Waterhouse, 1967).

Although neither the term olistostrome nor melange has been applied previously to describe gravity slide deposits in the Caribbean, one or the other of the terms might be applicable to certain units in Barbados (Senn, 1940; Saunders, 1968), Trinidad (Kugler, 1953; Kugler and Saunders, 1967; Barr and Saunders, 1968), Venezuela (Renz and others, 1955; Bushman, 1958; Bell, 1967), Cuba (Thayer and Guild, 1947; Kozary, 1968), and Puerto Rico (Glover, 1967). Other similar deposits around the world are listed in Horne (1969). Submarine slide and slump deposits on the present continental shelves are reviewed by Moore and others (1970).

STRATIGRAPHIC AND STRUCTURAL POSITION OF THE OLISTOSTROME

In the stratigraphic sequence for the Puerto Plata area (Fig. 4) the oldest rocks are serpentinites which represent basement upon which the Los Caños Formation

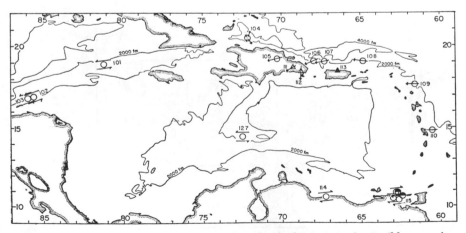

Figure 3. Azimuths of slip vectors for earthquake mechanisms in the Caribbean region. Earthquakes deeper than 100 km shown as triangles, those shallower than 100 km as circles (*after* Molnar and Sykes, 1969).

(pre-Paleocene) was deposited. The serpentinites occur as infaulted bodies in the Los Caños Formation and are probably detached fragments of the oceanic crust. The Los Caños Formation and serpentinites, as well as their gabbro, rodingite, and metamorphic associates, occur together near the core of the major anticlinal structure in the area.

The Los Caños Formation consists of andesite flows and tuffs, with spilites and keratophyres near the base. Relic glassy textures indicate the formation is submarine in origin. The exact age of the Los Caños Formation is unknown, but it is overlain unconformably by the Paleocene Imbert Formation. Its thickness is estimated at several kilometers.

Unconformably above the Los Caños Formation is a 1-km-thick succession of fine-grained graded calcareous tuffs which grade upward to vitric andesite and dacite tuffs, with rare interbedded green cherts and thin white aphanitic limestones.

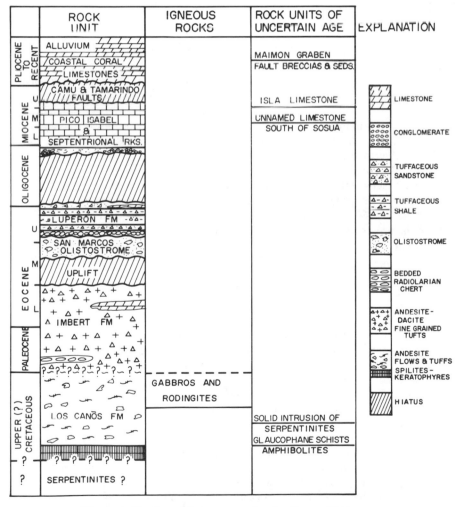

Figure 4. Stratigraphic sequence for the Puerto Plata area.

This marine succession, the Imbert Formation, has been dated as Paleocene to early Eocene.

The only known middle Eocene rocks, tuffaceous limestones, occur as exotic blocks in the San Marcos olistostrome. No middle Eocene rocks were found in situ. Upper Eocene rocks are present in the Luperon Formation, a 1-km-thick unit consisting of a coarse basal conglomerate and an alternating sequence of calcareous tuffaceous graded bedded sandstone and shales with rare interbedded bioclastic limestone. The Luperon Formation contains the first recognizable stream-worn detritus from the older formations in the area. It rests unconformably upon the Imbert Formation, but it is nowhere in contact with the San Marcos deposits.

The dominant structural feature of the area is a broad faulted anticline, 5 to 8 km wide and 25 km long, which trends N. 55° W., parallel to the north coast of the Dominican Republic, and plunges to the northwest. The San Marcos olistostrome is presently distributed over some 90 sq km within the core of this major anticline.

The contact relations of the San Marcos to adjoining units are puzzling. If the stratigraphic interpretation presented here is correct, the San Marcos unit should be or should have been enclosed within a middle Eocene section. I have been unable to document or disprove the occurrence of Middle Eocene rocks at either contact.

The age and nature of the rocks beneath the olistostrome are unknown. In the area west of Puerto Plata, the San Marcos rests unconformably(?) on Los Caños rocks, whereas to the south it is in fault contact with middle Miocene rocks of the Cordillera Septentrional. The San Marcos thins to the northwest and east of Puerto Plata where it is covered by recent alluvium.

There are no middle Eocene sediments now on top of the olistostrome. Either these sediments were removed by later slumping, faulting and erosion, or more likely the Puerto Plata area was generally rising during the middle Eocene and there was little or no post–San Marcos middle Eocene deposition.

Since contact relations of the San Marcos deposits with other units are unknown, nondiagnostic, or subject to alternative interpretations, I have interpreted the history of the San Marcos olistostrome primarily on the basis of known stratigraphy outside the area of olistostrome exposure and on the basis of fossils present in the olistostrome.

SAN MARCOS OLISTOSTROME

General Description

The hummocky topographic expression of the blocks within the San Marcos is a particularly outstanding feature of this deposit visible in the field and in aerial photographs (Fig. 5). A similar appearance in aerial photographs of other regions may indicate the presence of a similar unit.

The olistostrome has a minimum thickness of 300 m. However, this estimate is based only upon topographic exposures. There is no bedding preserved within the matrix, the lower contact was not found, and an upper contact with suspected late-middle Eocene or early-upper Eocene rocks was never proven. The oldest dated rocks now on top of the San Marcos are the late Oligocene–Miocene rocks of the Pico Isabel Formation, but these rocks are suspected to be an allochthonous block which slid over the olistostrome sometime in the Late Miocene from the south, where rocks of the same age and rock type were uplifted during that time. There

Figure 5. Aerial photograph of Puerto Plata area, D. R. San Marcos olistostrome in center portion of photograph; some olistoliths (exotic blocks) marked by arrows, many others visible. The city visible is Puerto Plata. Pico Isabel de Torres is the area outlined immediately south of Puerto Plata. Scale 1:60,000.

is some thinning of the San Marcos unit to the east as well as dilution of blocks in this direction, but no apparent thinning or dilution to the north or south.

Olistostrome Matrix

Fifty samples of the San Marcos matrix were collected over a distance of 20 km along the road running northwest about 4 km south of Pico Isabel. These samples represent the freshest matrix known in the area. The matrix is unconsolidated clay-sized sediment, sometimes streaked with red hematite or brown iron oxides, and it lacks bedding or directional features except for occasional swirling of layers in the immediate vicinity of a block. No blocks larger than 3 to 4 m were found surrounded entirely by the matrix, primarily because outcrops along this road and in stream valleys are limited to about 4 m in height.

The ratio of matrix to exotic blocks varies from about 1:1 to 5:1. Nowhere is the volume of blocks greater than that of the matrix.

Since there is no bedding or marker horizon in the San Marcos matrix, it is impossible to determine what part of the section was sampled. Samples in the west were collected from higher topographic levels than those in the east.

I believe all matrix samples collected are at least partially weathered. None were collected any deeper than 4 m from the present topographic surface; all samples were moist with ground water, and had the texture of wet clay. There is no evidence that the matrix was ever lithified. It is, however, undergoing weathering and probably represents the total product of source material, weathering in its present location, and possibly decomposition of the smaller enclosed blocks. Intense weathering has probably helped to eliminate any primary sedimentary features which survived initial churning.

X-ray diffraction traces of the less than 2 micron fraction indicate that the San Marcos matrix is composed of a mixture of quartz, kaolinite, montmorillonite and (or) chlorite and illite. The strongest peaks on all traces, other than those of quartz, are the 7 Å and 3.5 Å peaks of kaolinite. Three samples were heated to 550° C for several hours to separate kaolinite from chlorite. The 7Å and 37.5°2θ peaks disappeared, indicating kaolinite rather than chlorite. No samples were glycolated so that the 14 Å peak distinction of montmorillonite from chlorite is impossible, although the heat test mentioned above suggests montmorillonite rather than chlorite.

Portions of the coarse fractions (0.25 mm to 1 mm) of 20 samples viewed in thin section consisted mainly of turbid globs of clay(?) minerals and iron oxides, and rare unaltered broken crystals of plagioclase, probably volcanic ash remnants. No carbonate fragments were noted in the coarse fraction, nor did carbonate appear in the x-ray traces of the finer portion. This must indicate strong leaching or deep-sea deposition.

Thirty samples disaggregated in water in an ultrasonic vibrator and sieved through no. 60 (0.25 mm), no. 230 (0.62 mm) and no. 325 (0.044 mm) sieves were searched for organic remains. Only four of the samples yielded fossils. P. J. Bermudez examined this collection and found only members of the genus *Globigerina* and genus *Globorotalia* in addition to unidentified radiolaria. In his opinion (1962, written commun.) the fauna represents a marine deep-water environment and is of early Eocene–Paleocene age, although he notes that the collection "is insufficient to determine the age." Having no data to the contrary, I have tentatively assigned an early Eocene–Paleocene age to the matrix.

Olistostrome Blocks

The olistostrome blocks have sharp boundaries with the finer matrix, are invariably angular, show inconsistent strikes and dips and varied rock types, and range in size from 1 cm to 1.5 km in the longest dimension. Outcrops of blocks occur on the slopes of hills, on the tops of hills, or as boulders in stream valleys. Any recognizable pattern of outcrops or rock type is completely lacking. No geologic parameter is continuous from one locale to the next. Wise and Bird (1964) describe the blocks of the *argille scagliose* of the Apennines as a "chaotic littering of the landscape like so many popcorn boxes after a Saturday matinee." This analogy applies to the San Marcos blocks.

The most common rocks forming blocks of all sizes are gray to black tuffs and gray or buff tuffaceous limestones. Also rather common are blocks of andesite; pillow volcanics; veined, fractured, and recrystallized limestones; serpentinites; and rodingites. Approximately 30 different rock types have been recognized in thin sections. Some are definitely from the Los Caños and Imbert Formations and the older rocks (serpentinites and gabbros), whereas others have never been seen in place (for example, the recrystallized and veined limestones). The few paleontologic dates the blocks have yielded indicate a range in clasts from the Late Cretaceous(?) to the middle Eocene, excluding one doubtful Miocene age (FN-245-D) which I interpret as a fragment from a block which slumped over the olistostrome in late Miocene time forming Pico Isabel de Torres (*see* Fig. 4).

The olistolith (exotic block) dates are listed in Table 1. Except for sample *Bowin #202A* which is a gray limestone, all others listed are tuffaceous limestones. Textures of these rocks indicate that they were calcareous muds and calcarenites. None have grains coarser than 1 mm. In addition to organic and calcite fragments and irregular patches of calcite, volcanic rock, plagioclase, quartz, and pyroxene fragments are present in all samples.

Blocks or Basement?

Some rocks crop out continuously over distances of 1.5 km in the area of the San Marcos Formation. In all cases these rocks are serpentinites with associated rodingites or volcanic rocks, which in thin section look like Los Caños rocks.

These outcrops could be large blocks within the olistostrome. However, they cannot be seen completely in three dimensions so that they may also be bedrock poking through the olistostrome, representing basement irregularities that were

TABLE 1. PALEONTOLOGIC AGES OF SAN MARCOS OLISTOLITHS*

Sample No.	Age	Comments
Bowin 202-A (P. Bronniman)	Upper Cretaceous, Maestrichtian, or Eocene with reworked Upper Cretaceous	*Discocyclina* sp. (fragment?) *Rotalid* (large form nondescript) *Vaughanina cubensis* (fragments) *Orbitocyclina minima* *Calcisphaerula innominata* *Sulcoperculina* cf. *vermunti* *Kathina* sp.
FN 63CC (P. J. Bermúdez)	Probably Paleocene	*Actinosiphon barbadensis* (Vaughan) *Operculinoides* sp. (?) fragment *Ecology:* marine, shallow water (reef or near reef)
FN 245-D (W. S. Cole)	Possibly lower Miocene	Mostly planktonics—one or two larger Foraminifera suggestive of *Miogypsina*
FN 181-C (W. S. Cole)	Definitely Eocene	discocyclinid fragments *Eoconuloides* (?) . . . if so suggestive of middle Eocene
FN 121 (W. S. Cole)	Eocene	without question Eocene; could be middle or upper; would have to have more material to be certain
FN 111 (W. S. Cole)	Eocene	See comments under FN 121

*The name of the investigator who examined the fossils follows in parenthesis after the sample number. The comments and age determinations are those of the investigator listed.

never covered by the olistostrome or that have been stripped of their cover. Large areas around these outcrops are uncontaminated by blocks of other rock types, a feature not common to the olistostrome elsewhere, and they occur mainly in the east where the olistostrome appears to thin out.

On the other hand, there is no independent evidence that the base of the olistostrome has been discovered and there are no outcrops of the Imbert Formation (also older than the San Marcos) within the olistostrome which can be anything other than exotic blocks. On this basis, then, the large outcrops are considered to be blocks, rather than basement in place.

I can make no ordered pattern out of the distribution of the various types of blocks as Hsu (1968) suggests should be the case for melange units.

Origin of the Olistostrome

Any hypothesis of origin of the San Marcos olistostrome involves considerable speculation. The nonstratified and unconsolidated nature of the matrix, relative ages of the blocks and matrix, and lack of evidence for tectonic remobilization all indicate that some mechanism of subaqueous gravity sliding was most probably involved.

There is also no evidence for mud or salt diapirs or mud volcanism. Mud volcanism such as that described by Higgins and Saunders (1967) in Trinidad had been considered a likely alternative. However, in the Puerto Plata area there is no typical fault pattern usually associated with doming; the blocks in the San Marcos are larger than any described products of mud volcanism or mud diapirs; and, significantly, there are exotic blocks which are older than the matrix while those in mud volcanoes are invariably contemporaneous or younger than the matrix. Finally, folding usually associated with mud volcanoes is not present in the Puerto Plata area. All major rock deformation here can be explained by vertical tectonics; that is, tilting, normal faulting, slumping, and gravity sliding—not typical of mud volcano tectonics.

The few dates from the San Marcos matrix are tentatively accepted as indicating the time of deposition of the original unit which later became the matrix of the olistostrome. These Paleocene-Eocene dates are the time equivalent of Imbert tuffaceous deposition. Most likely the matrix represents a former tuffaceous unit within the Imbert Formation which has gone through a post-sliding alteration sequence of volcanic ash plus some montmorillonite and chlorite changing to kaolinite, a process which may still be continuing. That kaolinite can be formed from montmorillonite at low temperatures is suggested by the field work of Altschuler and others (1963) and the experimental work of Poncelet and Brindley (1967). Mackenzie and Mitchell (1966) point out that kaolinite or halloysite along with sesquioxides are generally the final products of rock alteration under humid tropical conditions irrespective of starting rock types. Hay (1959, 1960) reports that late Pleistocene andesitic subaerial ash deposits on St. Vincent, British West Indies, have weathered to halloysite, allophane, hydrated ferric oxide, and iddingsite.

One important feature of the San Marcos olistostrome was that it must have retained a large volume of pure water as other sediments rapidly buried it (Bredehoeft and Hanshaw, 1968), or it acquired excess water from a source layer at depth, perhaps its own montmorillonite (Hanshaw and Bredehoeft, 1968). That high water pressures may be maintained by unconsolidated sediments to burial depths of 2,100 to 5,000 m (more than twice the load necessary on top of the postulated tuffaceous layer in this case) eventually resulting in slumping has recently been documented by the work of Dickey and others (1968). These not unlikely conditions, in addition to probable random orientation of clay-sized particles (Liebling and Kerr, 1965) and location in an area of known tectonic activity set the stage for instability and mass flow, awaiting only a triggering mechanism such as earthquake shock or a series of shocks.

Shock-induced marine slumps in historical times are not uncommon; a good review is provided by Andresen and Bjerrum (1967) and by Morgenstern (1967), who lists several submarine slumps occurring on slopes of 3 to 20° following earthquake shocks of magnitude 6.75 to 8.5. Both figures are reasonable for the Puerto

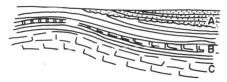

I. MIDDLE EOCENE TIME, PRIOR TO EARTHQUAKE SHOCK

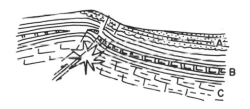

2. INSTANT OF EARTHQUAKE SHOCK.

3. POST EARTHQUAKE SHOCK SLUMPING.

EXPLANATION:

    A. OVERBURDEN

    B. TUFFACEOUS (?) UNIT IN IMBERT FM.

    C. LOWER IMBERT FORMATION AND OLDER ROCKS

    (THRUST FAULT IN 2,3 IS HYPOTHETICAL)

Figure 6. Sequence of events (from top to bottom) leading to production of olistostrome and olistoliths (adapted *from* Marchetti, 1957).

Plata area. It should be noted, however, that not all marine sediments on sloping surfaces are unstable and prone to liquid flow by shocking, a point made by Dill (1969) and reviewed by Morgenstern (1967).

The clay matrix of San Marcos Formation liquefies under uncontrolled laboratory conditions. Several of the matrix samples were placed in mason jars and varying amounts of water were added up to the saturation point of the sample. Liquefaction of the sample was caused by shaking or jarring. The degree of liquification brought about by this means increased with increasing water content.

The exotic blocks in the olistostrome are older than, contemporaneous with, and younger than their surrounding matrix; the youngest known are of middle Eocene age. This is taken as the probable time of shocking and consequent sliding, slipping taking place on a stratigraphically lower Paleocene–lower Eocene horizon. Flow was turbulent, presumably destroying the original stratification. Blocks younger than the matrix were incorporated as overlying layers ruptured and were swept into the moving mass. Older blocks from the Imbert Formation, the Los Caños Formation, and the peridotites were torn loose from a submarine slope where they had formed promontories in the sea floor (Fig. 6).

The San Marcos need not have traveled far and probably is near its original source. Where that is, is conjectural, although most likely it is only a few kilometers to the west where many of the rock types which form exotic blocks are in place.

Olistostromes do not necessarily indicate major orogenesis in progress. More likely they indicate shock-induced failure and catastrophic slumping of unstable sedimentary accumulations or perhaps sea-floor deposits scraped up at plate boundaries. However, many olistostromes are associated with tectonic movements, and the San Marcos deposit is no exception. The emplacement of the San Marcos probably was triggered by movement on the Camu fault, a major northwest-trending high-angle fault running across the entire mapped area, a fault which has been active since at least the middle Eocene. This fault strikes parallel to the south slope of the Puerto Rico trench.

ACKNOWLEDGMENTS

The writer gratefully acknowledges financial support from the National Science Foundation, Grant No. GA-14217, and the Woodrow Wilson Fellowship Foundation while at Princeton University; and current research support fron NSF Grant GA-13531 at the University of Miami Rosenstiel School of Marine and Atmospheric Science.

Helpful criticism of the manuscript was given by F. B. Van Houten and A. G. Fischer.

REFERENCES CITED

Altschuler, A. Z., Dwornik, E. J., and Kramer, H., 1963, Transformation of montmorillonite to kaolinite during weathering: Science, v. 141, p. 148–152.

Andresen, A., and Bjerrum, L., 1967, Slides in subaqueous slopes in loose sand and silt, *in* Richards, A. F., ed., Marine geotechnique: Urbana, Univ. Illinois Press, p. 221–239. p. 221–239.

Audley-Charles, M. G., 1965, A Miocene gravity slide from eastern Timor: Geol. Mag., v. 102, no. 3, p. 267–276.

Barr, K. W., and Saunders, J. B., 1968, An outline of the geology of Trinidad: Trans. Fourth Caribbean Geol. Conference, p. 1–10.

Bell, J. S., 1967, Geology of the Camatagua area, Estado Aragua, Venezuela [Ph.D. thesis]: Princeton, Department of Geology, Princeton Univ., 282 p.

Bracey, D. R., and Vogt, P. R., 1970, Plate tectonics in the Hispaniola area: Geol. Soc. America Bull., v. 81, p. 2855–2860.

Bredehoeft, J. D., and Hanshaw, B. B., 1968, On the maintenance of anomalous fluid pressures: I. Thick sedimentary sequences: Geol. Soc. America Bull., v. 79, p. 1097–1106.

Bunce, E. T., and Fahlquist, D. A., 1962, Geophysical investigation of the Puerto Rico trench and outer ridge: Jour. Geophys. Research, v. 67, p. 3955–3972.

Bushman, J. R., 1958, Geology of the Barquisimeto area, Venezuela [Ph.D. thesis]: Princeton, Department of Geology, Princeton Univ., 169 p.

Chase, R. L., and Hersey, J. B., 1968, Geology of the north slope of the Puerto Rico trench: Deep-Sea Research, v. 15, p. 297–317.

Dickey, P. A., Shriram, C. R., and Paine, W. R., 1968, Abnormal pressures in deep wells of southwestern Louisiana: Science, v. 160, p. 608–615.

Dill, R. F., 1969, Earthquake effects on fill of Scripps Submarine Canyon: Geol. Soc. America Bull., v. 80, p. 321–328.

Flores, G., 1955, Definition of olistostrome: Fourth World Petrol. Congress, Sec. 1, p. 122.

Glover, L., III, 1967, Geology of the Coamo area, Puerto Rico; with comments on Greater Antillean volcanic island arc-trench phenomena [Ph.D. thesis]: Princeton, Princeton Univ., 363 p.

Hanshaw, B. B., and Bredehoeft, J. D., 1968, On the maintenance of anomalous fluid pressures: II. Source layer at depth: Geol. Soc. America Bull., v. 79, p. 1107–1122.

Hay, R. L., 1959, Origin and weathering of Late Pleistocene ash deposits on St. Vincent, B.W.I.: Jour. Geology, v. 67, p. 65–87.

—— 1960, Rate of clay formation and mineral alteration in a 4000-year old volcanic ash soil on St. Vincent, D.W.I.: Am. Jour. Sci., v. 258, p. 354–368.

Higgins, G. E., and Saunders, J. B., 1967, Report on 1964 Chatham Mud Island, Erin Bay, Trinidad, West Indies: Am. Assoc. Petroleum Geologists Bull., v. 51, no. 1, p. 55–64.

Horne, G. S., 1969, Early Ordovician chaotic deposits in the central volcanic belt of northeastern Newfoundland: Geol. Soc. America Bull., v. 80, p. 2451–2464.

Hsu, K. J., 1968, Principles of melanges and their bearing on the Franciscan-Knoxville Paradox: Geol. Soc. America Bull., v. 79, p. 1063–1074.

Kear, D., and Waterhouse, B. C., 1967, Onerahi chaos-breccia of Northland: New Zealand Jour. Geology and Geophysics, v. 10, no. 3, p. 629–646.

Kozary, M. T., 1968, Ultramafic rocks in thrust zones of northwestern Oriente Province, Cuba: Am. Assoc. Petroleum Geologists Bull., v. 52, no. 12, p. 2298–2317.

Kugler, H. G., 1953, Jurassic to Recent sedimentary environments in Trinidad: Assoc. Suisse des Geol. et Ing. du Petrole Bull., v. 20, no. 59, p. 27–60.

Kugler, H. G., and Saunders, J. B., 1967, On Tertiary turbidity-flow sediments in Trinidad, B.W.I.: Asoc. Venezolana Geología, Minería y Petróleo Bol. Inf., v. 10, no. 9, p. 241–259.

Liebling, R. S., and Kerr, P. F., 1965, Observations on quick clay: Geol. Soc. America Bull., v. 76, p. 853–878.

Mackenzie, R. C., and Mitchell, B. D., 1966, Clay mineralogy: Earth-Sci. Rev., v. 2, p. 47–91.

Marchetti, M. P., 1957, The occurrence of slide and flowage materials (olistostromes) in the Tertiary series of Sicily: Internat. Geol. Cong., 20th, Mexico 1956, Comptes Rendus sec. 5, p. 209–225.

Molnar, P., and Sykes, L. R., 1969, Tectonics of the Caribbean Middle America regions from focal mechanisms and seismicity: Geol. Soc. America Bull., v. 80, p. 1639–1684.

Monroe, W. H., 1968, The age of the Puerto Rico trench: Geol. Soc. America Bull., v. 79, p. 487–494.

Moore, T. C., Jr., van Andel, Tj. H., Blow, W. H., and Heath, G. R., 1970, Large submarine slide off northeastern Continental Margin of Brazil: Am. Assoc. Petroleum Geologists Bull., v. 54, p. 125–128.

Morgenstern, N. R., 1967, Submarine slumping and the initiation of turbidity currents, in

Richards, A. F., ed., Marine geotechnique: Urbana, Univ. Illinois Press, p. 189–220.

Nagle, F., 1966, Geology of the Puerto Plata area, Dominican Republic [Ph.D. thesis]: Princeton, Princeton Univ., 171 p.

—— 1970, Serpentinites and metamorphic rocks at the Caribbean boundaries: Geol. Soc. America, Abs. with Programs (Ann. Mtg.), v. 2, no. 7, p. 633.

—— 1971, Geology of the Puerto Plata area, Dominican Republic, relative to the Puerto Rico Trench: Fifth Caribbean Geol. Conf. Trans., Geol. Bull. No. 5, Queens College Press, p. 79–84.

Poncelet, G. M., and Brindley, G. W., 1967, Experimental formation of kaolinite from montmorillonite at low temperatures: Am. Mineralogist, v. 52, p. 1161–1173.

Renz, O., Lakeman, R., and van der Meulen, E., 1955, Submarine sliding in western Venezuela: Am. Assoc. Petroleum Geologists Bull., v. 39, p. 2053–2067.

Saunders, J. B., 1968, Field trip guide, Barbados: Fourth Caribbean Geol. Conf. Trans., p. 443–450.

Senn, A., 1940, Paleogene of Barbados and its bearing on history and structure of Antillean-Caribbean region: Amer. Assoc. Petroleum Geologists Bull., v. 24, no. 9, p. 1548–1610.

Sykes, L. R., and Ewing, M., 1965, The seismicity of the Caribbean region: Jour. Geophys. Research, v. 70, no. 20, p. 5065–5074.

Thayer, T. P., and Guild, P. W., 1947, Thrust faults and related structures in eastern Cuba: Am. Geophys. Union Trans., v. 28, no. 6, p. 919–930.

Wise, D., and Bird, J. M., 1964, International Field Institute, Italy, 1964: Geotimes, v. 9, no. 5, p. 12–15.

MANUSCRIPT RECEIVED BY THE SOCIETY MARCH 29, 1971

CONTRIBUTION NO. 1471 FROM THE UNIVERSITY OF MIAMI ROSENSTIEL SCHOOL OF MARINE AND ATMOSPHERIC SCIENCE, MIAMI, FLORIDA 33149

THE GEOLOGICAL SOCIETY OF AMERICA, INC.
MEMOIR 132
© 1972

Eclogitic-Amphibolitic Rocks on Isla Margarita, Venezuela: A Preliminary Account

W. V. MARESCH

Princeton University, Princeton, New Jersey 08540

ABSTRACT

The Mesozoic medium-grade metamorphic terrain on the eastern part of Margarita Island, Venezuela, can be divided into a lower "greenstone division" and the conformably overlying Juan Griego Group. Highly varied rocks of eclogitic affinity, mostly found in situ, occur in both units but are most common in the greenstone division. The stratigraphic distribution of eclogitic and amphibolitic lenses within the Juan Griego Group resembles that of the lithologically equivalent Caracas Group near Puerto Cabello and Caracas on the Venezuelan mainland. The greenstone division and its eclogites on Margarita are interpreted to represent an original accumulation of basic volcanic material on ocean floor in the trough of a subduction zone off the margin of the continental crust. No lithologic unit corresponding to the greenstone division has been found elsewhere in the Coast Ranges.

INTRODUCTION

In 1968, some eight years after the completion of a previous Princeton doctorate thesis on Margarita Island (Taylor, 1960), H. H. Hess suggested a re-examination of the "Paraguachi Amphibolites,"[1] as defined by Taylor (1960). Hess felt that

---

[1] The term "Paraguachi Amphibolites" has now been declared invalid by the Stratigraphic Lexicon of Venezuela, 2nd Ed., 1970. Consequently, the informal designation "greenstone division," as defined and recommended in the above publication, will be used. Furthermore, in order to avoid confusion until a formal systematic definition of metamorphic units on Margarita Island can be made, all stratigraphic terms used in this note will be those designated by the Stratigraphic Lexicon of Venezuela, 2nd Ed., 1970, or, if no designation is given therein, the nomenclature in common usage in publications pertinent to Margarita Island.

in the light of the similarity of the amphibolitic gneisses of the "greenstone division"[1] to the pre-Mesozoic El Tinaco Complex of north-central Cojedes state (Menéndez, 1965), the presence of large volumes of ultramafic bodies, and the intricate petrological interrelations within the greenstone division, this unit might represent a Precambrian basement complex.

In light of present knowledge, although it is most probably of Mesozoic age, the greenstone division may nevertheless be interpreted as a type of basal complex, and clearly Hess's geological instinct again pointed the way to the study of an area that may well provide valuable information pertinent to the orogenic process in the Venezuelan Coast Ranges.

The writer has completed detailed mapping of the structurally complex greenstone division in the northern part of the mid to late Mesozoic metamorphic terrain of eastern Margarita. At present, mineralogic-petrologic studies are underway. The purpose of this note is to outline the general geology, to describe briefly the occurrence of eclogitic rocks, and to speculate upon their possible paleotectonic significance.

FIELD RELATIONS

The geology of Margarita Island is summarized by Taylor (1960), Jam and Méndez (1962), and González de Juana (1968). The metamorphic section, comprising the northern two-thirds of the eastern half of Margarita Island, has been folded into a tight, isoclinal, complex antiform plunging moderately to the southwest (*see* Fig. 1). The limbs of the antiform are essentially vertical. The greenstone division, representing the oldest rocks present, is exposed over approximately 120 sq km in the core of the antiform, It is conformably overlain to the southwest by the Juan Griego Group, comprising: (1) the chloritic unit, a locally chloritic, graphitic quartz-mica schist approximately 800 m thick; (2) the feldspathic unit, a quartz-feldspar-mica schist and (or) gneiss at least 1,200 m thick; and (3) the graphitic unit, a graphitic locally garnetiferous quartz-mica schist, at least 500 m thick, forming the top of the medium-grade section. The Los Robles Group, a series of low-grade phyllites, unconformably overlies the Juan Griego Group, but these rocks will not be considered here. Near the top of the feldspathic unit are prominent and stratigraphically restricted lenses of marble and minor occurrences of metaconglomerate. Voluminous ultramafic bodies and trondhjemitic intrusions are found throughout, but they occur mainly within the greenstone division.

OCCURRENCE OF ECLOGITIC ROCKS

Eclogitic rocks are volumetrically minor but widespread and varied. Only passing reference to the existence of such rock types on Margarita Island has been made in the literature. Rutten (1940, p. 831) briefly mentions an amphibole-eclogite sample from Punta Ausente; Weyl (1966, p. 285) lists eclogite as a rock type occurring on Margarita Island; Dengo (1950) notes three "amphibolite" locations on Margarita Island in connection with a discussion of eclogitic and glaucophanitic amphibolites in Venezuela; Delfino (1949) describes several sills of "schistose eclogite" containing "augite" near La Asuncion; and Jam and Méndez

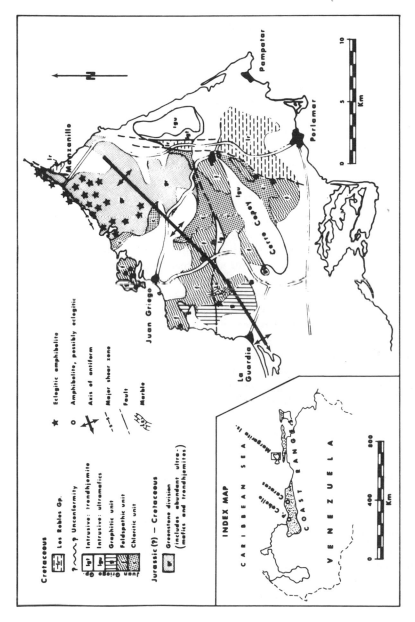

Figure 1. Generalized sketch map of the metamorphic terrain of the eastern part of Margarita Island (in part *after* Taylor, 1960).

(1962) in their petrological description of several "amphibolites" and "granulites" mention "augite" and "diallage," probably referring to the sodic pyroxene and relict sodic pyroxene symplektite respectively.

Within the greenstone division, sodic pyroxene-bearing assemblages are apparently restricted to the northern limb of the antiform (Fig. 1). They occur as abundant paraconformable sills and irregular bodies (some still exhibiting the intrusive character of the original basic sill) and as integral parts of the amphibolitic gneisses. Eclogitic rocks and garnet amphibolites are also found within ultramafic bodies and as irregular rotated blocks in a large northeast-trending shear zone along the north shore (Fig. 1).

Jam and Méndez (1962) have also mentioned the occurrence of small irregular bodies of dense green garnet amphibolite (the metadiorites of Taylor, 1960) within the quartz-mica schists and gneisses of the Juan Griego Group (Fig. 1). These occurrences are not common, and all examples studied contain sodic pyroxene relicts and display eclogitic texture.

PETROGRAPHY

Greenstone Division

The present study indicates that the following assemblage volumetrically represents the most important lithology of the greenstone division: (1) blue-green, subcalcic amphibole + sodic plagioclase + epidote (or clinozoisite) + garnet + quartz + rutile + sphene + opaques. White mica is also common. Chlorite and biotite occur as minor alteration products associated with amphibole. Garnet is also commonly partly replaced by chlorite.

Forming common but volumetrically minor interbands in the above gneisses is the following assemblage: (2) quartz + sodic plagioclase + white mica. Carbonate and graphite are widespread accessory minerals.

These schists and gneisses are extensively permeated by trondhjemitic material composed largely of quartz and sodic plagioclase.

Eclogitic Rocks

A summary of the eclogitic assemblages studied is given in Table 1. All assemblages listed contain amphibole, sodic pyroxene, and epidote or clinozoisite, and differ in containing various combinations of the essential phases sodic plagioclase, garnet, quartz, and sodic white mica. Assemblage (1) is the most common and contains the greatest number of mineral phases.

The commonly large number of phases suggests chemical disequilibrium; this is supported by textural evidence, the most striking being the partial to total replacement of sodic pyroxene laths by myrmekitelike intergrowths or symplektites of plagioclase and blue-green amphibole.

CORRELATION WITH THE VENEZUELAN MAINLAND

The stratigraphic distribution of eclogitic amphibolites (or amphibolites which *may* be eclogitic) within the Juan Griego Group is shown in Figure 1. Neglecting those occurrences south of the Cerro Copey peridotite, where the structure is complex, and mapping has been inconclusive, 8 of the 10 amphibolite bodies (including the two largest) occur near the contact between the graphitic and feldspathic units. This contact has been defined by Taylor (1960) as representing the first bed of graphitic schist above the highest lens of marble in the feldspathic unit (usually an interval of a few tens of meters). Clearly, these amphibolites are closely associated with the prominent marble-bearing horizon within the Juan Griego Group. Where lenses of marble are large enough to be mapped, as in the hills northeast of La Guardia (Fig. 1), this close association is very evident. However, since amphibolites also occur elsewhere, no direct genetic relation between the marble and amphibolite is indicated or implied here; only the stratigraphic association is emphasized.

At Puerto Cabello, 100 km west of Caracas (Fig. 1), Morgan (1970) found lenses and sills of eclogite, eclogite-amphibolite, and related mafic rocks occurring in a restricted stratigraphic interval close to the base of a section of graphitic schists which are underlain by prominent lenses of marble of the Antímano Formation. Furthermore, the Las Brisas Formation underlying the Antímano Formation is lithologically equivalent to the feldspathic unit of the Juan Griego Group on Margarita. In the Caracas region itself, Dengo (1953) and Smith (1953) have also described bodies of amphibolite and eclogitic amphibolite, most of which are found near the stratigraphic level of the Antímano marble (Dengo, 1953).

Thus, not only is the gross lithology of the Juan Griego Group of Margarita Island very similar to the Caracas Group of the Coast Ranges (Schubert, oral commun., 1969; Gonzalez de Silva, oral commun., 1970; Stainforth and others, 1970, p. 314), but the stratigraphic distribution of a variety of small bodies of mafic character in these two groups is extremely similar as well.

These similarities suggest that Caracas Group and Juan Griego Group sedimentation were probably related with respect to source area, were part of the same general region of deposition, and were probably closely related in time. Despite these obvious similarities, a fundamental difference exists: the Caracas Group unconformably overlies pre-Mesozoic granitic basement (Dengo, 1953; Morgan, 1967), whereas the Juan Griego Group conformably overlies and grades into the greenstone division.

TABLE 1. SUMMARY OF ECLOGITIC ASSEMBLAGES DESCRIBED

No.	No. of Examples	Blue-Green Amphibole	Sodic Pyroxene	Epidote or Clinozoisite	Sodic Plagioclase	Garnet	Quartz	Sodic White Mica
1	5	X	X	X	X	X	X	X
2	4	X	X	X	X	X		X
3	3	X	X	X	X	X		
4	3	X	X	X	X			
5	1	X	X	X	X		X	X
6	1	X	X	X		X	X	X
7	1	X	X	X		X		
8	1	X	X	X				

Common accessory minerals are rutile, sphene, opaque oxides, pyrite, and apatite. Chlorite and biotite occur as minor alteration products associated with amphibole.

TENTATIVE PALEOTECTONIC INTERPRETATION

To date, the evidence provided by the present study suggests that the green-stone division and its eclogites represent an original accumulation of basic volcanic material located on the ocean floor in a trough along the continental margin. This volcanic accumulation also contained intercalated nonvolcanic clastic sediments and large volumes of incorporated ultramafic material. Deformation was complex, and high-pressure–low-temperature conditions favorable for producing eclogites were apparently reached. An inferred subsequent change in the thermal gradient led to moderate heating which resulted in the intrusion of trondhjemitic plutons and recrystallization of the metamorphic rocks to mineral assemblages typical of the epidote-amphibolite facies of metamorphism (Eskola, 1939).

The existence of high-pressure–low-temperature metamorphic conditions associated with a long linear metamorphic belt (the Coast Ranges are 900 km long and approximately 100 km wide) leads logically to a comparison with subduction zone models as suggested by Miyashiro (1967), Ernst and others (1970), and as recently summarized by Dickinson (1970). The subduction zone model (Fig. 2) proposes the existence of paired belts of contrasting metamorphic conditions: a high-pressure–low-temperature belt associated with the ensimatic trench, and a lower-pressure–higher-temperature belt on the edge of the continental crust. These contrasting metamorphic belts are a direct consequence of thermal gradients associated with the downward motion of the slab of oceanic crust and mantle lithosphere along the Benioff zone (Oxburgh and Turcotte, 1970). Furthermore, the subsequent recrystallization at higher temperatures of many of the original high-pressure–low-temperature assemblages has been well documented in areas such as the Sanbagawa belt in Japan (Ernst and others, 1970). Ernst has attributed this heating to a return to a more normal geothermal gradient once rapid tectonic loading within the trench had passed. Oxburgh and Turcotte (1970), on the other hand, suggest that the inner edge of the trench (Fig. 2b) of an island arc grows oceanward due to the accretion of ocean-floor sediments scraped from the down-going slab. As the island arc, or in this case the continental plate, grows, they suggest that the isotherms will likewise migrate oceanward relative to the original inner edge of the trench. Parcel "A" (Fig. 2b) of the original high-pressure–low-temperature assemblage will, in effect, move along the lateral geothermal gradient to higher temperatures.

The metamorphic history and lithologic character of the greenstone division of Margarita, as they have been outlined in this note, suggest great similarity to geologic terrains which are inferred to have originated within the trenches of subduction zones under high-pressure–low-temperature conditions, and subsequently to have been recrystallized at higher temperatures. The contiguous lower-pressure–higher-temperature belt, in the case of Margarita, would be represented by the Caracas Group sediments of the Venezuelan Coast Ranges on the mainland to the south (Bell, 1972; insert, Fig. 1), although, as indicated by the presence of eclogites at Puerto Cabello and Caracas, the transition is not abrupt. Furthermore, the occurrence of Caracas-Group-type sediments (the Juan Griego Group) overlying the greenstone division indicates that the metamorphic section on Margarita originated close to this transition zone. From these relations it follows that the Benioff zone dipped to the south (Fig. 2a).

The cause of the subsequent recrystallization of this metamorphic section at

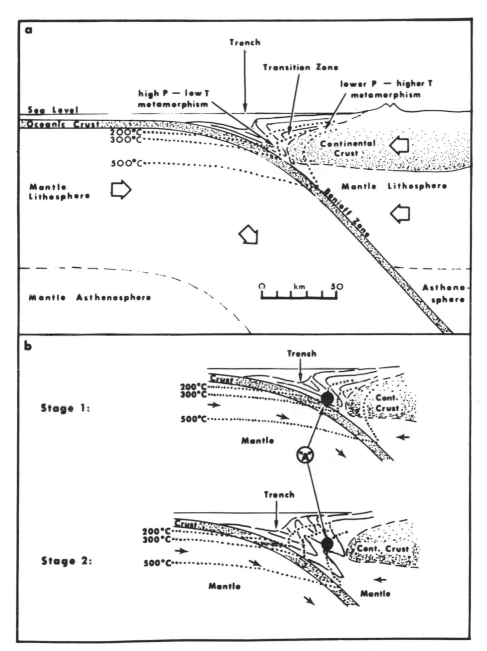

Figure 2. (a) Generalized representation of a subduction zone and the thermal gradients associated with the downward movement of the oceanic lithospheric plate (oceanic crust plus mantle lithosphere) beneath the continental lithospheric plate (continental crust plus mantle lithosphere) along the seismic Benioff zone. No vertical exaggeration implied. (Modified *after* Ernst, 1970; and Oxburgh and Turcotte, 1970.) (b) Schematic diagrams illustrating Oxburgh and Turcotte's (1970) model of migrating thermal gradients in subduction zones (*see* text). At stage 1, the parcel of sediments "A" is located on the 200°C isotherm and is being subjected to high pressures. At Stage 2, the trench and locus of subduction have shifted away from the original continental crust, and parcel "A" is now located on the 500°C isotherm. Scale is comparable to that in (a).

higher temperatures, in the light of the suggestions by Ernst and others (1970) and Oxburgh and Turcotte (1970) as outlined above, is not known. The distinction between the two suggestions could be useful, however, in judging the longevity of this particular subduction zone, since Oxburgh and Turcotte assume steady-state (that is, comparatively long-lived) subduction to have been attained, while the suggestion of Ernst and others carries no such constraints. Since abundant andesitic volcanic rocks are also normally associated with steady-state subduction zones (Oxburgh and Turcotte, 1970), their apparent lack in the Venezuelan Coast Ranges may be significant. In this context, any independent conclusion to the effect that a steady-state situation had nevertheless been achieved in this particular subduction zone might indicate that such volcanic rocks have been essentially eroded away and are now, as suggested by Murray (1972), represented only by scattered exposed feeder pipes.

In conclusion, Margarita Island may actually be the only locality in northern Venezuela where a now-modified high-pressure–low-temperature trench assemblage, a possible relic of a Mesozoic subduction zone, is now exposed. The detailed study of this area will reveal significant information on the Mesozoic history of the southern Caribbean.

ACKNOWLEDGMENTS

This note represents a preliminary account of field work completed for a dissertation at Princeton University. It is with great sadness and a sense of personal loss that I acknowledge my debt to the late H. H. Hess, who suggested this project to me and took great personal interest in its progress but was not able to see it to its completion. Thanks are also due R. B. Hargraves and L. S. Hollister for their subsequent encouragement and help, including critical appraisal of the manuscript.

Financial support in the field was provided by the Ministerio de Minas e Hidrocarburos of Venezuela and the Caribbean Research Fund, sponsored by Mobil, Shell, Mene Grande and Creole Oil Companies of Venezuela.

REFERENCES CITED

Bell, J. S., 1972, Geotectonic evolution of the southern Caribbean area: Geol. Soc. America Mem. 132.

Delfino, C., 1949, Reconocimiento geológico de la zona Manzanillo sur este (Edo. Nueva Esparta) [Undergraduate thesis]: Caracas, Univ. Central de Venezuela, Dpto. de Geología.

Dengo, G., 1950, Eclogitic and glaucophane amphibolites in Venezuela: Am. Geophys. Union Trans., v. 31, p. 873–878.

——1953, Geology of the Caracas region, Venezuela: Geol. Soc. America Bull., v. 64, p. 7–40.

Dickinson, W. R., 1970, Global tectonics: Science, v. 168, p. 1250–1259.

Ernst, W. G., 1970, Tectonic contact between the Franciscan melange and the Great Valley sequence—crustal expression of a Late Mesozoic Benioff zone: Jour. Geophys. Research, v. 75, p. 886–901.

Ernst, W. G., Seki, Y., Onuki, H., and Gilbert, M. C., 1970, Comparative study of low-grade metamorphism in the California Coast Ranges and the outer metamorphic belt of Japan: Geol. Soc. America Mem. 124, 276 p.

Eskola, P., 1939, Die Entstehung der Gesteine, in Barth, T.F.W., Correns, C. W., and Eskola, P., eds.: Berlin, J. Springer, 422 p.

Gonzáles de Juana, C., 1968, Guía de la excursión geológica a la parte oriental de la Isla de Margarita: Asoc. Venezolana Geología Minería, y Petróleo, 42 p.

Jam, L. P., and Méndez, A. M., 1962, Geología de las Islas Margarita, Coche, y Cubagua: Soc. Cien. Nat. La Salle Mem., v. 22, p. 51–93.

Menéndez, A., 1965, Geología del area de El Erinaco, centro-norte del Estado Cojedes, Venezuela: Bol. Geol. Venez., v. 6, p. 417–543.

Miyashiro, A., 1967, Orogeny, regional metamorphism, and magmatism in the Japanese islands: Geol. Fören Kobenhavn Medd. fra Dansk, v. 17, p. 390–446.

Morgan, B. A., 1967, Geology of the Valencia area, Carabobo, Venezuela [Ph.D. thesis]: Princeton, Princeton Univ., 220 p.

——1970, Petrology and mineralogy of eclogite and garnet amphibolite from Puerto Cabello, Venezuela: Jour. Petrology, v. 11, p. 101–145.

Murray, C. G., 1972, Zoned ultramafic complexes of the Alaskan-Ural type: Feeder pipes of andesitic volcanoes: Geol. Soc. America Mem. 132.

Oxburgh, E. R., and Turcotte, D. L., 1970, Thermal structure of island arcs: Geol. Soc. America Bull., v. 81, p. 1665–1688.

Rutten, L., 1940, On the geology of Margarita, Cubagua, and Coche (Venezuela): Amsterdam, Koninkl. Nederlandse Akad. Wetensch. Proc., v. 43, p. 828–841.

Smith, R. J., 1953, Geology of the Los Teques–Cua region, Venezuela: Geol. Soc. America Bull., v. 64, p. 41–64.

Stainforth, R. M.; de Juana, C. González; Bellizzia, C. Martin; de Rivero, F.; Petzall, C.; Bellizzia, G. A.; Menéndez, A.; de Ratmiroff, G.; Kiser, G. D.; Feo-Codecido, G.; Méndez, J.; de Graterol, M.; de Arozena, J.; de Gamero, Lourdes; Pimentel, N.; Mendoza, B.; de Gamboa, A.; Schwarck, A.; Picard, X.; and Bermúdez, P. J., eds., 1970, Lexico estratigrafico de Venezuela, 2nd ed.: Bol. Geol. Min. de Minas e Hidrocarburos Pub. Espec. No. 4, 756 p.

Taylor, G. C., 1960, Geology of the Island of Margarita, Venezuela [Ph.D. thesis]: Princeton, Princeton Univ., 121 p.

Weyl, R., 1966, Geologie der Antillen: Berlin, Gebrueder Borntraeger, 410 p.

MANUSCRIPT RECEIVED BY THE SOCIETY MARCH 29, 1971

THE GEOLOGICAL SOCIETY OF AMERICA, INC.
MEMOIR 132
© 1972

Volcanic Geology of Southwestern Antigua, B.W.I.

ROBERT A. CHRISTMAN

Department of Geology, Western Washington State College,
Bellingham, Washington 98225

ABSTRACT

The mountainous southwestern portion of Antigua consists mostly of calc-alkaline igneous rocks, ranging in composition from quartz basalt to dacite, with andesite being most common. These are associated with agglomerates, tuffs, and a few limestone lenses which have been dated as mid-Oligocene. A small area of quartz diorite suggests that the center of volcanic activity was just to the south of the island and that a considerable thickness of pyroclastics has been eroded from the southern part of the island to expose the roots of the volcanoes which were active in mid-Oligocene time. The relation of the geologic events in the Lesser Antilles to the present theories of ocean-floor spreading is discussed briefly.

INTRODUCTION

The island of Antigua in the Lesser Antilles is located northeast of the Caribbean Sea, east of Nevis and St. Kitts (Fig. 1). Antigua is part of the outer arc, called the "Limestone Caribbees," which is characterized by older rocks and the absence of volcanic activity in historic times, whereas Nevis and St. Kitts belong to the inner active volcanic arc. The island is 108 sq mi in area, being 14 mi from east to west and 11 mi from north to south. This report is concerned with the southwestern third of the island.

Antigua can be divided roughly into three zones: (1) a mountainous area in the southwest third which is underlain by igneous rocks and coarse- to fine-grained volcanic sedimentary rocks with minor limestones, (2) a low-lying central plain underlain by tuffs, argillites, and conglomerates with some beds of chert and limestone of mid-Oligocene age and, (3) an area of low hills in the northeast third which is underlain primarily by limestones and marls of late Oligocene age, termed

439

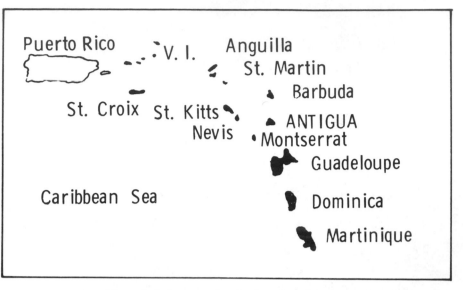

Figure 1. Index map showing location of Antigua.

the Antigua Formation by Spencer (1901). Except for the variations associated with the igneous rocks in the mountainous area, most of the sedimentary rocks strike northwest-southeast and dip northeast at angles ranging from about 20° adjacent to the mountainous area to about 5° in the northeastern part of the island. Thus, the sedimentary rocks in the mountainous area are the oldest, and they become progressively younger to the northeast. These rocks have been mapped and described most recently by Martin-Kaye (1959).

Because the geology of the mountainous area was poorly known, mapping was done during the summers of 1949 and 1960 to determine the nature and distribution of the volcanic rocks. The resulting contributions to the understanding of the geology of Antigua may be summarized as follows: (1) preparation of a detailed geologic map of the mountainous area; (2) recognition of lava flows dipping away from the mountainous area and forming dip slopes; (3) recognition of areas of complex mixtures of agglomerate, igneous rock, and diversely dipping volcanic sedimentary rocks which are postulated to represent the roots of volcanoes; (4) discovery near Old Road of a small area of phaneritic quartz diorite which indicates that a thick insulating layer of rocks has been removed and which suggests that the center of volcanic activity may have been situated just to the south and southwest of Antigua; (5) recognition that the tuffs and limestone in the volcanic area are similar to those in the central plain and are of similar age, as suggested by faunal evidence; and (6) identification of the igneous rocks as members of a calc-alkaline suite consisting mainly of andesite with lesser volumes of other igneous rocks ranging in composition from basalt to dacite; two chemical analyses are reported. Each of these six items will be discussed in the sections which follow.

GEOLOGIC MAP

The distribution of igneous rock, agglomerate, and tuff in the mountainous area (Fig. 2) reveals no orderly pattern except along the northeast flanks where the

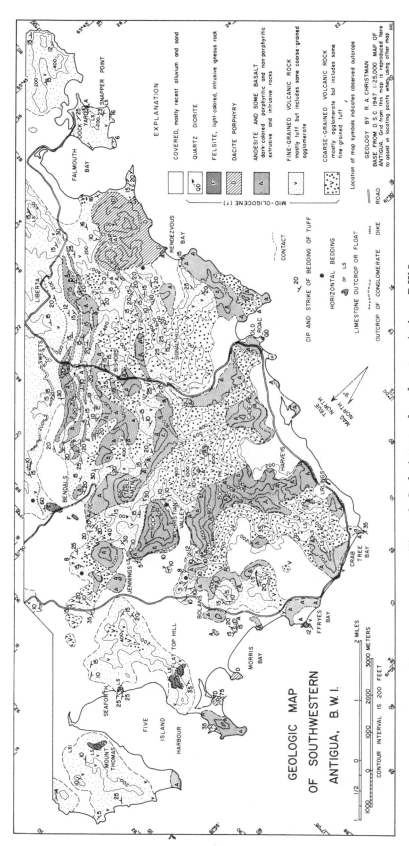

Figure 2. Geologic map of southwestern Antigua, B.W.I.

strikes and dips of tuffs and lavas are somewhat uniform. Precise identification of the rock units and detailed mapping of their contacts is difficult because of the scarcity and weathered condition of outcrops. For example, it was nearly impossible at some outcrops to determine whether the rock was highly weathered andesite, weathered andesite breccia, or weathered agglomerate composed of andesite fragments. Because decisions had to be based on poor outcrops and locally on float, the final map is somewhat interpretative. In general, complex relations between rock units are to be expected in a volcanic area: where outcrops are good, complex patterns can be mapped, and where outcrops are poor, simple patterns must be postulated. This must be done without preconceived ideas. Martin-Kaye's geologic map suggests that he believes that the northwest trend of the rock units of the central plain continues into the mountainous area where many of his map units of basalt, andesite, agglomerate, and tuff are interpreted as trending northwest. The author partially disagrees with this interpretation and was unable to substantiate many of these contacts in the field. However, because of the nature of the rock exposures, one should not be dogmatic; mapping by another geologist would probably result in a third interpretation.

Many small faults showing displacements of less than 10 ft and many narrow dikes were observed on the sea cliffs, but these could not be traced inland. No faults are shown on the geologic map and only a few of the dikes are mapped. However, the large numbers of faults and dikes exposed on the sea cliffs suggest that the inland areas are also riddled with faults and dikes.

LAVA FLOWS

A series of andesite lava flows was mapped in a band trending northwest between Bendals and a point about 1 mi south of Liberta. These flows dip 15° to 20° northeast and control the topography to form subdued escarpments or hogbacks. In places the lavas form dip slopes, but some of the gentle northeast slopes are developed on overlying conglomerates or breccias, and the outcrops of andesite are limited to the upper parts of the steeper southwest backslopes of the hogbacks. Four flows, 10 to 30 ft thick and interbedded with tuffs and breccias, were mapped. Except for several small isolated basic intrusions to the north of the mapped area, these flows represent the northeast limit of igneous rock. The next prominent overlying unit is a conglomerate which crops out rather continuously at the 200-ft elevation in a band running northeast of and parallel to the lavas.

The dips of the flows are such that, even without assuming a steepening of dip toward their presumed source, their projection to the southwest carries the rocks 1,000 to 2,000 ft over the present-day mountains. It is postulated that some of the andesite masses in the mountainous area may represent the roots of the volcanoes from which the lavas were derived.

AREAS WITH COMPLEX MIXTURES OF ROCKS

Complex interfingering of igneous rocks, agglomerate, and tuff occurs throughout much of the mountainous area. The irregular distribution and discordant contacts of andesite and basalt and their association with an overwhelming abundance of coarse agglomerate (pyroclastics which have not been deposited by water) in some areas, such as Boggy Peak, are interpreted as representing the roots of volcanoes. Some of the agglomerates may have formed in volcanic necks and some may have been deposited aerially in the development of the volcanic cones. Elsewhere, such as at Rock Peak, the mass of andesite is surrounded primarily by outward-dipping tuffs, which suggests the upward emplacement of the igneous rock in a discordant manner with minor tilting of adjacent sedimentary rocks.

Discordant relations between andesite, agglomerate, and tuff were observed in an area about ¾ mi east-northeast of Jennings (Fig. 3). The andesite (on the south) is bordered by a large amount of coarse-grained volcanic rock which lacks bedding and which, in part, may be intrusive agglomerate. Farther to the north the coarse-grained volcanic rocks are overlain by an andesite lava flow. Still farther north, gently dipping to horizontal, tuffs occur beneath the coarse-grained volcanics, and the discordant contact shows that the tuffs are older.

QUARTZ DIORITE

A very small outcrop of quartz diorite was discovered in the sea cliffs at the promontory south of Old Road (Fig. 2). The rock is clearly phaneritic and consists of labradorite (approx. 75 percent), quartz (approx. 15 percent), augite partly altered to chlorite (approx. 10 percent), and minor amounts of magnetite and pyrite. The coarse-grained nature of the rock suggests that several thousand feet of overlying rock, which served as insulation, have been removed since the rock crystallized. Also, because of its occurrence as one of the southernmost exposures on the island, one may speculate that additional quartz diorite may lie beneath the ocean to the south and southwest of the present shoreline. It is reasonable to assume that the quartz diorite is a deep-seated equivalent of part or all of the volcanic suite. Hence, its location suggests that the main center of volcanic activity may also have been to the south and southwest of the present island.

AGE OF THE VOLCANIC ROCKS

The igneous rocks, agglomerates, and tuffs of southwestern Antigua are assigned a mid-Oligocene age because they are, in part, conformably interbedded with tuffs and two limestone lenses which have been dated, occurring at Seaforth and about 1 mi east of Ffryes Bay. John W. Wells at Cornell University identified several specimens of coral, collected at the latter locality by the author, as *Antiguastrea*

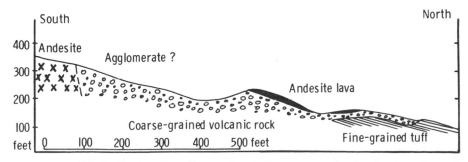

Figure 3. Sketch showing the relation of andesite, coarse-grained volcanic rocks, and fine-grained tuff in hills northwest of Jennings.

cellulose (Duncan) and assigned a mid-Oligocene age to them (1950, written commun.). As this locality lies well within the volcanic region, it may be used to date the surrounding volcanic rocks. Farther away from the volcanic centers, fossil corals at Seaforth, near Five Island Bay, have also been dated as middle Oligocene (Thomas, 1942). On the other hand, fossils from Snapper Point, south of Falmouth Bay, are dated as late Oligocene (Martin-Kaye, 1959).

It is likely that the major igneous rocks are only slightly younger than mid-Oligocene sedimentary rocks which they intrude. The tuffaceous sediments of the central plain contain progressively less coarse-grained pyroclastic materials in the younger stratigraphic units, and little evidence of volcanic activity is found in the overlying Antigua Formation, considered late Oligocene to early Miocene in age (Martin-Kaye, 1959). However, a small mass of basalt intruded into the Antigua Formation in the northern part of the island indicates that minor igneous activity continued after the deposition of the Antigua Formation.

COMPOSITION OF THE IGNEOUS ROCKS

Most of the igneous rocks in southwestern Antigua are mapped as andesite and basalt (Fig. 2), and only smaller areas, with the exception of Sugar Loaf Mountain, are mapped as dacite or felsite. Most of the identifications, which were made in the field on the basis of color and the presence or absence of quartz and plagioclase, were confirmed by the study of about 120 thin sections in the laboratory.

On the basis of their light gray color, with tints of blue or green, the majority of the rocks were identified as andesite or andesite porphyry. In places the rocks are darker gray in color and more appropriately may be called basalt or basalt porphyry, although the distinction between andesite and basalt is difficult to make. Even in thin sections, which show much evidence of weathering, it was difficult to differentiate between andesite and basalt because most of the rocks contain complexly twinned and zoned phenocrysts of labradorite in a fine-grained groundmass which appeared to be composed primarily of andesine microlites. Olivine was not found in any of the rocks and pyroxene was absent or constituted only a small percentage in some specimens.

The chemical analysis (Table 1) of one of the dark-colored porphyritic igneous rocks from Table Mountain, southwest of Bendals, shows it to be similar to other calc-alkaline rocks found in island arcs. Compared with other basalts it is high in SiO_2 and Al_2O_3 and low in K_2O, and on the basis of its norm, the rock is called quartz basalt. Chemically it is similar to a quartz basalt from St. Martin (Christman, 1953), and it resembles basalts from St. Kitts, Montserrat, and Guadeloupe (Table 2). In thin section the rock consists of phenocrysts of calcic labradorite (estimated by the Michel-Levy method of determination) and microlites of andesine with a few small phenocrysts of clinopyroxene. A few areas of "iddingsite" may indicate the former presence of a small amount of olivine.

The chemical analysis of the light-colored aphanitic rock which is mapped as felsite (quarried north of Bendals) is given in Table 1. This rock is probably a dacite in which the mafic components have been lost by alteration, as shown by the abnormally low FeO content and the minerals which appear in the norm. In thin section the rock consists of less than 10 percent highly altered andesine phenocrysts, maximum of 1 mm in length, with a glomeroporphyritic texture. A few grains of augite may be identified but generally only patches of chlorite, hematite, or calcite

Table 1. Chemical Data on Two Rocks from Antigua, B.W.I.

	Quartz Basalt, Table Mountain*	Felsite, Bendals Quarry*
SiO_2	50.98	66.60
Al_2O_3	22.29	15.39
Fe_2O_3	3.04	5.24
FeO	4.14	.27
MgO	3.60	1.24
CaO	11.73	2.36
Na_2O	1.90	5.32
K_2O	.33	1.08
H_2O^+	.52	.80
H_2O^-	.71	.61
TiO_2	.61	.68
P_2O_5	.06	.14
MnO	.18	.06
	100.09	99.79
Norms (calculated):		
Quartz	7.60	25.62
Orthoclase	2.22	6.39
Albite	16.24	44.54
Anorthite	51.29	10.84
Corundum		1.58
Hematite		5.24
Rutile		.28
Hypersthene	10.85	3.10
Diopside	5.07	
Magnetite	4.41	
Ilmenite	1.22	.76
Apatite	.17	.34
Feldspar	$Ab_{23}Or_3An_{74}$	$Ab_{72}Or_{10}An_{18}$

* Analyses by the laboratories at the University of Minnesota. Eileen K. Oslund, analyst, 1950.

suggest the former presence of other minerals. Most of the rock consists of a felted groundmass of plagioclase microlites (oligoclase?) and opaque minerals. The rock has indistinct flow banding and is intrusive into the surrounding tuffaceous sediments. Similar felsitic intrusions occur at Mount Thomas and Flat Top Hill.

The dacite porphyry at Sugar Loaf Mountain contains light-colored phenocrysts (35 percent) up to 5 mm in length, in a gray to pinkish-gray groundmass (65 percent). The rock contains about 10 percent quartz, which occurs as embayed crystals, and about 25 percent plagioclase of andesine composition, but which is highly zoned. A few squarish areas of chlorite and calcite suggest the former presence of a pyroxene. Similar rocks were also mapped along the shore of Morris Bay at two localities.

THEORETICAL CONSIDERATIONS

The present theories of ocean-floor spreading which have evolved from Hess's ideas (1962) hold that island arcs form where one oceanic lithospheric plate underthrusts another at the descending limb of a convection cell. From the direction of curvature of the West Indian arc and the occurrence of gravity anomalies and a

Table 2. Chemical Analyses of Basalts from the Lesser Antilles

	1	2	3	4	5
SiO_2	51.0	50.3	51.8	52.0	51.8
Al_2O_3	22.3	20.3	19.6	19.2	21.0
Fe_2O_3	3.0	2.6	3.5	2.7	4.0
FeO	4.1	6.4	5.5	5.6	5.1
MgO	3.6	5.2	4.0	5.5	3.1
CaO	11.7	11.3	10.1	10.6	9.7
Na_2O	1.9	2.0	3.6	2.5	2.6
K_2O	0.3	0.3	0.5	0.8	0.7
H_2O^+	0.5	0.2	—	0.2	0.5
H_2O^-	0.7	0.1	—	0.2	—
TiO	0.6	0.9	1.0	0.6	1.0
P_2O_5	0.1	0.1	0.1	0.1	0.1
MnO	0.2	0.2	0.2	0.1	—

1. Quartz basalt, this paper
2. Quartz basalt, St. Martin (Christman, 1953)
3. Basalt, Mt. Misery, St. Kitts (average of 4 analyses; Baker, 1968)
4. Olivine basalt, Montserrat (MacGregor, 1938)
5. Basalt, Guadeloupe (Lacroix, 1904; listed by Washington, 1917, p. 538–539)

buried trench on the Atlantic side of the islands, it has been postulated (LePichon, 1968; Bullard, 1969; Vine, 1969; and other authors) that a westward-moving Atlantic plate underthrusts a lobe of the Pacific plate in the eastern Caribbean area. A plot of foci of earthquakes suggest a westward-dipping plane of seismic activity beneath the islands (Sykes and Ewing, 1965; Molnar and Sykes, 1969). Moreover, the basement rock as determined by seismic reflection profiles appears to dip westward (Chase and Bunce, 1969).

The origin of the calc-alkaline igneous rocks characteristic of island arcs is still poorly understood. A comparison of the composition of the igneous rocks of the outer arc (Antigua, St. Martin, and so forth) with those of the inner arc (St. Kitts, Montserrat, and so forth) indicates that, whatever their origin, the rocks are remarkably similar despite their difference in age and location. However, more analytical data are needed for some of the common rocks on Antigua. Oxburgh and Turcotte (1970) have analyzed the thermal conditions along the zone between the stationary block and the descending block of oceanic crust and mantle (Fig. 4) and conclude that frictional heating may result in the production of the magmas from fusion of the oceanic crust and sedimentary materials which might be adhering to the descending block. An analysis of the same problem by Minear and Toksöz (1970) produced essentially similar results.

Dickinson and Hatherton (1967) postulated that the composition of the magma depends to some extent on its depth of origin, with higher contents of K_2O being characteristic of greater depths. The average K_2O content of the analyzed basalts and andesites from Antigua and St. Kitts (Baker, 1968) is very low, being less than one percent. According to the Dickinson-Hatherton curves this suggests a depth of about 80 km to the Benioff zone. As the trench lies some 250 km to the northeast, this would require that the Benioff zone have a shallow dip of about 18°. Because this seems unlikely, some factor other than depth of origin is probably responsible for the low K_2O values.

The age relations between the volcanic activity on the outer and inner arcs of the Lesser Antilles requires some speculation. On the inactive outer arc the oldest rocks are middle Eocene on St. Bartholomew, late Eocene on St. Martin (Christ-

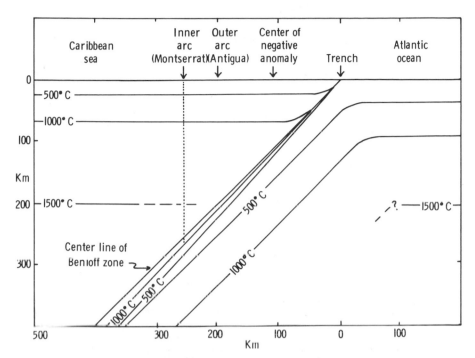

Figure 4. Relation of island arc features of Lesser Antilles to the general isothermal model proposed by Oxburgh and Turcotte (1970).

man, 1953), and middle Oligocene on Antigua. On the other hand, the inner arc is characterized by recent volcanic activity with the oldest rocks being late Miocene (Martin-Kaye, 1959). It is noteworthy that the date of cessation of volcanic activity on the outer arc is compatible with the mid-Cenozoic (25 to 30 m.y. ago) discontinuity in the tectonic history of the island arcs in the Pacific (Dott, 1969). Also, the earliest volcanic activity of the inner arc is compatible with evidence presented by Ewing and Ewing (1967) of an abrupt change in mechanics of ocean-floor spreading in the late Miocene.

The 55-km westward shift of volcanic activity from Antigua to Montserrat (Fig. 4) might be explained in several ways. If it is assumed that the magma is produced at a given depth, regardless of the mechanism, the shift might be explained by a relative movement of the overlying surface block. In Oligocene time, Antigua may have been directly over the source of magma, but subsequent eastward movement of the Pacific plate moved Antigua toward the trench so that Montserrat now lies over the area where magma is produced. This would require that the Pacific Plate move eastward over a stationary Atlantic Plate at the average rate of 0.25 cm/yr between mid-Oligocene and late Miocene time. As an alternate possibility the depth of magma generation may remain constant, but its location might shift westward if the dip of the Benioff zone became less steep. Such a change in dip could also be caused by the eastward movement of the overthrusting Pacific Plate. If it is assumed that the relations between the plates remained constant, the shift in site of volcanism might be caused by migration of the site of fusion down the Benioff zone. If friction is an important mechanism for producing heat (Oxburgh and Turcotte, 1970), a reduction of the speed of movement could cause this down-slope migration, other factors being equal.

ACKNOWLEDGMENTS

Field and laboratory expenses for the work conducted in 1949–1950 were provided under contract N6 onr-27008 between Princeton University and the Office of Naval Research for geological and geophysical investigations of the Caribbean region. Field expenses for the work conducted in 1960 were provided by the Geology Department, Princeton University. This paper has benefited from the suggestions and criticisms offered by Myrl Beck and Ada Swineford; their assistance is appreciated.

REFERENCES CITED

Baker, P. E., 1968, Petrology of Mt. Misery Volcano, St. Kitts, West Indies: Lithos, v. 1, p. 124–150.

Bullard, E., 1969, The origin of the oceans: Scientific American, v. 221, p. 66–75.

Chase, R. L., and Bunce, E. T., 1969, Underthrusting of the eastern margin of the Antilles by the floor of the western north Atlantic Ocean and origin of the Barbados Ridge: Jour. Geophys. Research, v. 74, p. 1413–1420.

Christman, R. A., 1953, Geology of St. Bartholomew, St. Martin and Anguilla, Lesser Antilles: Geol. Soc. American Bull., v. 64, p. 65–96.

Dickinson, W. R., and Hatherton, T., 1967, Andesitic volcanism and seismicity around the Pacific: Science, v. 157, p. 801–803.

Dott, R. H., Jr., 1969, Circum-Pacific late Cenozoic structural rejuvenation: Implications for sea-floor spreading: Science, v. 166, p. 874–876.

Ewing, J., and Ewing, M., 1967, Sediment distribution on the mid-ocean ridges with respect to spreading of the sea floor: Science, v. 156, p. 1590–1592.

Hess, H. H., 1962, History of ocean basins, in Engel, A.E.J., James, H. L., and Leonard, B. F., eds., Petrologic studies: A volume in honor of A. F. Buddington: Boulder, Colorado, Geol. Soc. America, 660 p.

LePichon, X., 1968, Sea-floor spreading and continental drift: Jour. Geophys. Research, v. 73, p. 3661–3697.

MacGregor, A. G., 1938, The Royal Society expedition to Montserrat, B.W.I.: The volcanic history and petrology of Montserrat with observations on Mt. Pele in Martinique: Royal Soc. of London Philos. Trans., B–229, p. 1–90.

Martin-Kaye, P.H.A., 1959, Reports on the geology of the Leeward and British Virgin Islands: Government report, St. Lucia, W.I., Voice Publishing Co. Ltd., 117 p.

Minear, J. W., and Toksöz, M. N., 1970, Thermal regime of a downgoing slab and new global tectonics: Jour. Geophys. Research, v. 75, p. 1397–1419.

Molnar, P., and Sykes, L. R., 1969, Tectonics of the Caribbean and Middle America regions from focal mechanisms and seismicity: Geol. Soc. America Bull., v. 80, p. 1639–1684.

Oxburgh, E. R., and Turcotte, D. L., 1970, Thermal structure of island arcs: Geol. Soc. America Bull., v. 81, p. 1665–1688.

Spencer, J.W.W., 1901, On the geological and physical development of Antigua: Geol. Soc. London Quart. Jour. 57, p. 490–505.

Sykes, L. R., and Ewing, M., 1965, The seismicity of the Caribbean region: Jour. Geophys. Research, v. 70, p. 5065–5074.

Thomas, H. D., 1942, On fossils from Antigua and the age of the Seaforth limestone: Geol. Mag., v. 79, p. 49–60.

Vine, F. J., 1969, Sea-floor spreading—new evidence: Jour. Geol. Education, v. 57, p. 6–15.

Washington, H. S., 1917, Chemical analyses of igneous rocks: U. S. Geol. Survey Prof. Paper 99, 1201 p.

MANUSCRIPT RECEIVED BY THE SOCIETY MARCH 29, 1971   PRINTED IN U.S.A.

THE GEOLOGICAL SOCIETY OF AMERICA, INC.
MEMOIR 132
© 1972

Andean Research Project, Venezuela: Principal Data and Tectonic Implications

R. SHAGAM

*Department of Geology, University of Negev
P. O. Box 2053, Beer-Sheva, Israel*

ABSTRACT

The principal conclusion drawn from an extensive mapping project in the Venezuelan Andes is that these ranges have had a long and complex orogenic history extending at least as far back as the early Paleozoic and probably the late Precambrian. The tectonic evolution of these ranges involved at least three and probably not less than four major orogenic pulses—late Precambrian?, Devono-Mississippian, latest Permian, and post-Paleozoic (principally at the end of the Eocene). Striking results of repeated orogeny include regional parallelism of structural trends, an apparent northwestward migration of the focus of granitic plutonism with time, and an event of high-grade regional metamorphism, only recently recognized, in late Permian time. It is pointed out that current concepts of sea-floor spreading will need elaboration or extension in order to make provision for superimposed orogenies.

INTRODUCTION

The geology of the continent has long been at the center of attention in the evolution of the concept of drift, and the tectonic setting of the Andes specifically has been integrated in modern hypotheses of sea-floor spreading (for example, Vine and Hess, 1968). The northwestern corner of the continent embracing Venezuela and Colombia is additionally of interest in that it would be expected to reveal evidence of the subsidiary effects of Caribbean tectonism.

Previous work in these ranges has been described by the writer (1969) and by Schubert (1969). Herein are reported the results of recent field and laboratory studies by the writer and students (L. Kovisars, R. I. Grauch), and by geologists of the Ministerio de Minas e Hidrocarburos (C. Schubert, C. Ramírez, R. Garcia, and V. Campos).

PRINCIPAL DATA

Figure 1 shows the regional geology in generalized form and the legend depicts the principal stratigraphic relations, thermal events, and inferred hiatuses. Significant aspects of structural deformation are shown in Figures 2 and 5. The most important addition to a recent summary of the tectonic evolution of these ranges (Shagam, 1969) is the stipulation of a major event of regional metaphorphism in hiatus H_7 (Late Permian—Early Triassic). Whereas the Tostós facies is now considered to represent, most probably, a metamorphosed portion of the Mucuchachí facies of the Upper Paleozoic section, it was formerly assigned to the Precambrian Iglesias Complex and considered coeval with the Bellavista facies (*see* legend to Fig. 1). This event of metamorphism was originally remarked by Liddle (1946) and later by Arnold (1966) and Kovisars (1971). Moreover, Bass and Shagam (1960) suggested the possibility of a major thermal event at the end of the Permian on the basis of Rb/Sr ages on micas from the basement Sierra Nevada facies. The writer discounted the significance of these opinions (because highly fossiliferous unmetamorphosed Permo-Carboniferous sections occur in many parts of these ranges) until conducted on field excursions by Garcia and Campos and by Grauch.

The tectonic evolution of the Andes may be summarized in terms of three principal "layers" of geology (Shagam, 1969) as presented in Table 1. From the viewpoint of geotectonics the following geologic characteristics are considered significant:

1. Locally there may be angular discordance between the structural trends of adjacent "layers" (note for example, the angular discordance between major structures of the Sierra Nevada and Mucuchachí facies in the area southwest of Mérida), but regionally there is a parallelism of structural trends through the three layers. The characteristics of structural style of the three layers are noted in Table 1. Noteworthy is the apparent superimposition of the San José synclinorium of the Mucuchachí facies over the Chacantá anticlinorium of the Sierra Nevada facies (Fig. 1, A-3 to B-4). Evidently much of the Paleozoic deformation took place by décollement over the Precambrian basement. Also significant is the marked change in structural style across the Permo-Triassic boundary—from one suggesting intense lateral compression to one of block uplift and tilting.

2. There was repeated development of unstable, commonly fault bound, sedimentary basins entirely or largely located within the confines of the present ranges. These are represented by the Lower Paleozoic section (± 2 km; south flanks of the ranges, Fig. 1, A-2 to C-3), Upper Paleozoic section (and Tostós facies (?); ± 5 km; mainly central Andes), Triassic red beds (± 2 km; southwest and northeast regions), marine Cretaceous and early Tertiary formations (up to ± 3 km; mainly in the southwest [Uribante trough]), and Plio(?)-Pleistocene terrestrial gravels (up to ± 300 m; largely restricted to a grabenlike trough corresponding approximately to the axis of the ranges). It is possible that the sedimentary accumulation represented by the Bellavista facies (± 2 km; south flanks) should be added to the list. Doubt arises because the southward extension of these rocks is not known. The Sierra Nevada facies of the basement Iglesias Complex is considered to represent sedimentary accumulations in excess of 6 km thick but is excluded from consideration because it is not known to be restricted to the area of the present ranges.

3. The earliest granitic event (G_1) refers to rootless pods of pegmatite believed to have formed in the regional metamorphism of hiatus H_1, and restricted to the Sierra Nevada facies of the Iglesias Complex. For the rest, granitic events (G_2 to G_7) refer to intrusive plutons which appear to be spatially differentiated with time, oldest (G_2) to the south. The northward sense of movement of the Paleozoic granitic events is roughly paralleled by the northward sense of movement of the sites of early and late Paleozoic sedimentary basins.

TABLE 1. MAJOR STRUCTURAL CHARACTERISTICS OF ANDEAN ROCK UNITS

"Layer"	Age	Rock Units	Structural Characteristics
3	Post-Paleozoic	Cretaceous and early Tertiary section Triass–Jura La Quinta Formation	Dominantly tilt blocks; dip ranges from 15 to 20° (flanks) to 40 to 55° (central), overall asymmetry (Fig. 5). Folds mainly restricted to intervening keystone grabens (Fig. 5). Some folding of flanking molasse by gravity sliding. Some gravity slide blocks into Rio Chama graben from north.
2	Late Paleozoic	Sabaneta and Palmarito facies	Tight concentric folds, minor flowage. Locally a spaced "fracture" cleavage with marked sigmoidal bending at bedding planes. Major folds plunge 0 to 10°.
		Mucuchachí, and Tostós facies	Penetrative slaty cleavage and schistosity//bedding. Later spaced slip-cleavage fans major synclinoria. Dominantly similar flow folds followed by brittle kink folding. Major synclinoria plunge 0 to 15°, involve major décollement over Sierra Nevada facies (Fig. 2).
	Early Paleozoic	lower Paleozoic section	Imbricate bedding-plane thrust slices and associated drag flow folds. Conglomeratic Wenlock-Ludlow section shows major concentric syncline, disharmonic to folding of underlying shaly section. Drag folds plunge 10 to 40°; poorly developed cleavage locally.
1	Precambrian(?)	Bellavista facies	Schistosity//bedding. Pronounced slip-cleavage. Similar flow folds followed by brittle kink folding. Refolding suggested by stereoplots of fold axes, cleavage lineations, and regional dip reversals (S_1) (Fig. 1).
		Sierra Nevada facies	Metamorphic foliation//(?) bedding. Overall moderate, locally tight, flexural flow folding; disharmonic flow folding and kinking of schist layers. Major folds plunge > 25°. Refolding indicated by folded boudins (abundant).

Note: Structures produced in older deformations do not necessarily bear direct geometric relations to those formed in younger deformations. For example, much of Permian folding took place by décollement over the Sierra Nevada facies. Also, some Cretaceous tilting involved draping over vertically rising Paleozoic blocks (compare Shagam and Hargraves, 1970, Fig. 5).

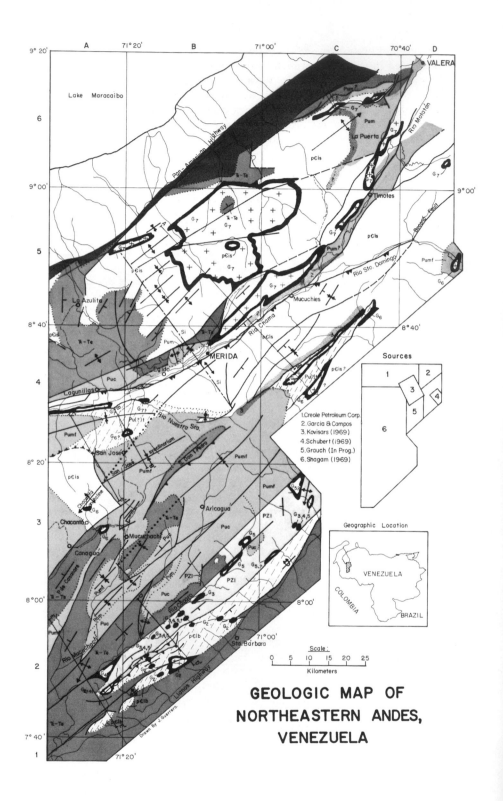

GEOLOGIC MAP OF
NORTHEASTERN ANDES,
VENEZUELA

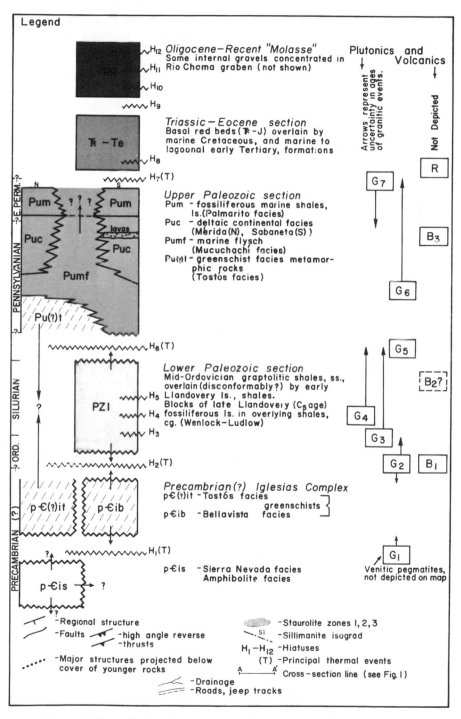

Figure 1. Generalized geologic map of the northeastern Andes, Venezuela.

4. Comparison of the Venezuelan Andes with the Appalachians reveals many gross parallels (*see* legend to Fig. 1; Table 1; and Fig. 2) in structural style and symmetry, sedimentary environments, granitic events, and orogenic phases. From the viewpoint of geotectonics the most significant difference is the absence, in the Andes, of ultramafic rocks.

Two important implications emerge from the geologic history as currently understood: (a) The evidence outlined above indicates repeated orogeny in space, at least since the early Paleozoic and possibly since Precambrian (Bellavista) time; and (b) the absence of ultramafics and the extreme paucity of basaltic volcanism (considerably less than 1 percent of the total section) imply a profound absence of direct mantle participation throughout the lengthy tectonic evolution of these ranges.

OUTSTANDING PROBLEMS

Tectonic interpretations are only as good as the knowledge of the stratigraphy and thermal events on which they are based. Major uncertainties yet remain in these areas of Andean geology.

Because of doubts concerning the stratigraphic position of the Tostós facies and the fact that the relations of the Bellavista facies to other basement facies are not known, the stratigraphic relations within the Iglesias Complex (*see* legend to

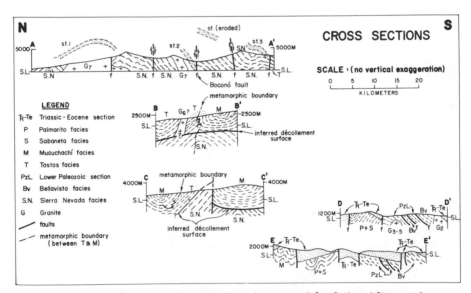

Figure 2. Simplified cross sections in approximate spatial relation (distance between C-C' and D-D', E-E' foreshortened; *see* Fig. 1 for locations). In A-A', staurolite zone 1 is projected diagrammatically into line of section from area of La Puerta (Fig. 1, C-6). Off-set of staurolite zones 2 and 3 suggests dominantly vertical movement on the Bocono fault. *See* Grauch (1971) for detailed sections across staurolite zone 3. Presumably G_6 granites in Tostos facies have been sheared off their roots (B-B'). For additional commentary on sections B-B' and C-C' see Figure 3. In sections D-D' and E-E' note concentric folding of Palmarito and Sabaneta facies to the north, and zone of imbricate thrusts to the south; also distinctive structural styles of Tr-Tpe section versus underlying sections.

Fig. 1) remain largely hypothetical. The suggestion that the Tostós facies represents a metamorphosed portion of the Mucuchachí facies solves a thorny structural problem (Fig. 3), on the other hand there is at least prima facie field evidence to suggest that the Tostós facies may grade down into the Sierra Nevada facies (the writer has inferred a décollement surface between these two facies as shown in Fig. 2).

The lower Paleozoic section, exposed in a narrow belt near the south flanks of the ranges, has been strongly affected by imbricate bedding-plane thrusting with the result that the stratigraphic section has been markedly shuffled. Consequently the very occurrence and (or) significance of hiatuses H_3–H_5 is not known beyond reasonable doubt. That at least part of the section has been obscured by deformation is suggested by the fact that, as reported by Boucot and others (1969), whereas brachiopod community assemblages representative of relatively deep and shallow water have been found, none of the community assemblages indicative of intermediate depths have been detected.

The writer's view of the upper Paleozoic section in terms of five lateral facies changes is not shared by Arnold (Compañia Shell and Creole, 1964; also Arnold, 1966) as shown in Figure 4. The relations favored by the writer are based on the occurrence northeast of Aricagua (Fig. 1, northeast quadrant of B-3) and immediately southwest of Lagunillas (Fig. 1, A-4) of Sabaneta-type continental conglomerates and sandstones interbedded with Mucuchachí-type marine slates, and the fact that all faunas recovered from the Mucuchachí facies appear to be of Permo-Carboniferous (probably mid-Pennsylvanian) age.

In the legend to Figure 1 the bulk of the Tertiary section is depicted as a flanking molasselike facies implying major uplift at the close of the Eocene. Such deposits are not always easily distinguishable from the upper Eocene section on the one hand and from Plio(?)-Pleistocene gravels on the other, and it is conceivable that continental deposits of Oligo-Mio-Pliocene age occur within the mountain belt. If found, such deposits would bear on the nature and number of the phases of uplift in "Andean" orogeny.

With respect to the thermal history, the most important uncertainty concerns the postulated regional metamorphism during hiatus H_7. Briefly stated, the data and

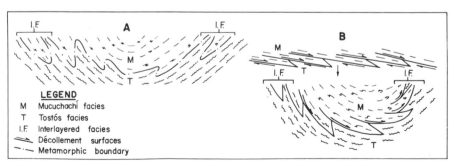

Figure 3. Stratigraphic and structural relations between Tostós and Mucuchachí facies. (A) (preferred): Tostós facies viewed as metamorphosed part of Mucuchachí facies, explaining structural conformity. Interlayering of schist and carbonaceous slate at contacts ascribed to compositional control over location of metamorphic boundary. (B) Tostós facies viewed as part of basement Iglesias Complex. Structural conformity accidental; tectonic mechanism to explain interlayered facies at contacts. Note: alternative (A) nonetheless requires décollement of Tostós over Sierra Nevada facies (see Fig. 1, sections B-B' and C-C').

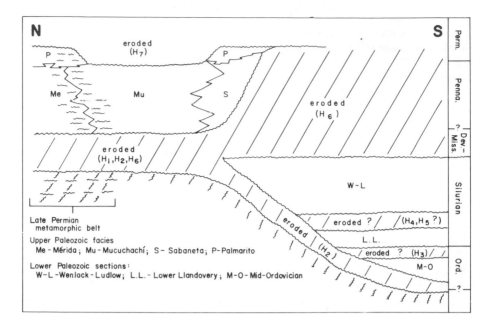

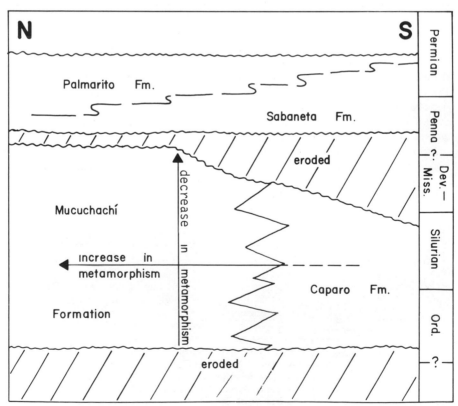

Figure 4. Contrasting views of Paleozoic sedimentary and stratigraphic relations: (A) This report (*see* also Shagam and Hargraves, 1970, Fig. 4). (B) After Arnold (*in* Compañia Shell and Creole, 1964; Arnold, 1966).

their implications are as follows: Garcia and Campos (1969, oral commun.) found that staurolite schists of the Las Torres Formation pass gradationally into a section of slates and limestones from which Permo-Carboniferous fossils have been recovered. Kovisars (1971) found staurolite in slates of the El Aguila Formation, a unit of upper Paleozoic lithologic aspect, though no fossils were found. The occurrence of staurolite in these formations coincides spatially (*see* Fig. 1, stippled pattern) with the occurrence of staurolite and alumino-silicates in the Sierra Nevada and Tostós facies. Grauch (1971, 1972) has shown that the staurolite zones enclose three isograds, closely associated in space, representing the first appearance of andalusite, kyanite, and sillimanite respectively. Furthermore, on the basis of textural criteria it appears that the alumino-silicates crystallized contemporaneously with or following staurolite formation. If these minerals were formed during H_7 what was the nature of the Iglesias Complex prior thereto? A tentative answer is suggested from the relations in the Chacantá–San José area (Fig. 1, A-3, 4). The effects of the H_7 regional metamorphism are relatively sharply demarcated, the transition to relatively unmetamorphosed slates occurring in less than 1 km laterally, such that the San José synclinorium of the Mucucachí facies was largely unaffected thereby. To the southwest of the San José synclinorium the spur of Sierra Nevada outcrop is at the amphibolite grade of regional metamorphism (although devoid of staurolite and alumino-silicates), suggesting that the Sierra Nevada facies had suffered high-grade regional metamorphism prior to the H_7 event. It remains to establish whether or not all the alumino-silicates now found in the Sierra Nevada facies were formed only during the H_7 event.

The metamorphism of the Bellavista facies clearly occurred in pre-Mid-Ordovician time inasmuch as the juxtaposed, highly fossiliferous, lower Paleozoic section is unmetamorphosed. The only isotopic age which bears on the minimum age of the metamorphism is a K/Ar measurement of 660 ± 30 m.y. reported by Arnold and Smith (*in* Compañia Shell and Creole, 1964, bottom of Fig. 5) on a G_2 pluton which clearly postdates the metamorphism (*see* Martín Bellizzia and others, 1968, for analytical details).

There are significant uncertainties concerning all plutonic and some volcanic events. Uncertainties concerning ages are expressed in the legend to Figure 1. The distinctions between granitic events G_3, G_4, and G_5 are based solely on the nature of contacts with fossiliferous portions of the lower Paleozoic section. Schubert (1969) reported Early Devonian K/Ar ages on biotite and hornblende from a specimen of granite collected from the Cerro Azul area, Barinas State, located along strike to the northeast from the G_3 to G_5 bodies shown on Figure 1.

The B_2 basalts occur in the lower Paleozoic section, but it is not clear whether they are flows or sills. Petrographically they closely resemble the B_3 volcanic rocks with which they may prove to be coeval. The rhyolitic tuffs (R) at the base of the La Quinta Formation in the type locality may be the surface equivalents of the G_7 plutons (*see* Hargraves and Shagam, 1969).

TECTONIC IMPLICATIONS

It is appropriate to view Andean tectonic evolution against the broad background of sea-floor spreading and plate tectonics. In so doing the writer is not merely trying to be *au courant*. The basic premise adopted here is that orogeny requires lateral compression operating at deep crustal and upper mantle depths

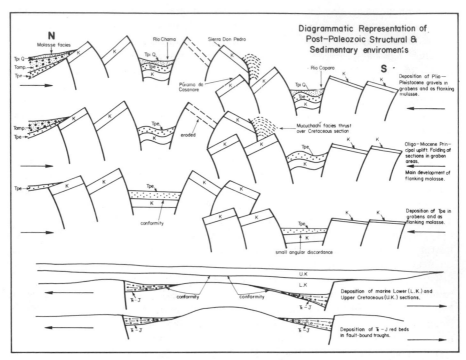

Figure 5. Post-Paleozoic relations. Tensional environment (Triassic through Early Cretaceous) followed by compressional environment (end of Cretaceous to present). Rios Chama and Caparo grabens arbitrarily used as data in constructing successive stages of uplift. Two principal axes of uplift defined by the two grabens. The two uplifts overlap (note opposed tilts) along the Páramo de Casanare–Sierra Don Pedro line (Fig. 1, A-3 to B-4). Gentle folding is restricted to the graben areas as a result of progressive shortening (shown diagrammatically); elsewhere shortening is absorbed by increased tilting. A minimum of three episodes in "Andean" uplift is indicated. The absence of Paleocene-Eocene section from tilt blocks is ascribed to restriction of Early Tertiary sedimentation to the graben areas. Molasse facies on south flank of ranges is not depicted.

even where the near-surface tectonic products indicate vertical uplift. The concept of drift expounded by Hess (1962) is particularly satisfying in that it supplies the mechanism for lateral compression.

For the most part the accumulated data is used in a negative sense, rather than as a basis for new geotectonic syntheses. Because the concept of plate tectonics (Morgan, 1968) is based almost entirely on the motions engendered by the latest cycle of sea-floor spreading, it is proposed to consider first the post-Paleozoic "layer" of geology.

The principal tectonic features of the post-Paleozoic section may be summarized as follows: (a) a complete absence of thermal events (see comment on rhyolitic tuffs (R) above); (b) a tectonic environment dominated by tensional conditions until approximately the end of the Cretaceous, at which time a compressional environment was instituted and maintained to the present (*see* Fig. 5); and (c) during the compressional phase release of stresses was accomplished dominantly by vertical uplift which was probably episodic rather than continuous judging from the structural and stratigraphic characteristics of the flanking molasse deposits.

One may reasonably infer that post-Paleozoic tectonism was a product of the mechanical reaction between crustal blocks. The responsible stress field may have been generated by deformation over the impaction zone along the western margin of the continent and (or) by the effects of shearing between the Caribbean and South American plates in the course of drift (Hargraves and Shagam, 1969, p. 544). Another possibility is suggested by the fact that in reconstructions of the distribution of the continents prior to the current drift cycle (Vine and Hess, Fig. 11, 1968) the Venezuelan Andes are in approximate linear alignment with the Appalachians, which also suffered significant vertical tectonism in post-Triassic time (Billings, 1960). On the basis of the age of the tectonism and the hypothesized geometry of the current drift cycle, it appears that uplift took place in the course of westward drift when the underlying mantle convective motion was horizontal. The orientation in space of the lineament defined by the mountain belts further suggests that uplift occurred diagonally across the westward-drifting plates. The possibility exists that the Venezuelan Andes—Appalachians uplift represents a gigantic disharmonic fold related to differential rates of movement in the underlying mantle convective motion, as shown in Figure 6. In any event, whatever the fundamental mechanism, there is the implication that significant geotectonism may take place in areas other than plate boundaries (subduction zones, ridges, transform faults). The concept in no way invalidates Morgan's (1968) concept of plate tectonics inasmuch as the orogenic effects are minuscule when compared to the size and amount of lateral drift of the plates. In all of the postulated environments the only available direction for release of accumulated stresses would be vertically upward, and one might expect such release to be episodic as accumulated stresses overcame crustal strength.

The significance of the apparent change of phase in tectonic environment from tensional to compressional across the Cretaceous-Tertiary boundary is not known. The effect is not restricted to the Venezuelan Andes but appears to extend westward into the Andes of Colombia and eastward into the Coast Ranges of Venezuela. The writer (1969) has suggested that the onset of compressional conditions may reflect the time of interaction of spreading from the East Pacific Rise and the Mid-Atlantic Ridge.

In marked contrast to post-Paleozoic events, Paleozoic tectonics involved significant thermal events and structures indicative of intense lateral compression. Interpreted in terms of plate tectonics, it is reasonable to infer that during much of that era the Venezuelan Andes were located at or close to a compressional boundary or boundaries. On the basis of the evidence of sedimentary environments, regional metamorphism, granitic events, and structure presented above, it may be said that the axes of orogenic belts were located approximately along the Rio Caparo (Devonian) and the Mérida-Valera line (Permian). Arnold's suggestion (in Compañia Shell and Creole, 1964) that the early Paleozoic basin of sedimentation migrated continuously northward into the Pennsylvanian Mucuchachí facies implies continuous orogeny in a relatively narrow belt (ca 60 to 100 km wide) for a long period of time (ca 250 m.y.). As indicated above, the writer believes there were two separate orogenies with a distinct pause through Devono-Mississippian time. The absence of serpentinites and the paucity of basaltic volcanic rocks suggest that no intervening island arc was involved in the collision

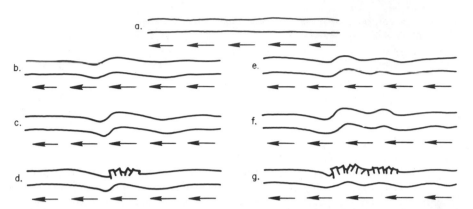

Figure 6. Diagrammatic representation of postulated orogenic mechanism related to differential rate of movement in horizontally moving mantle convection (vector arrows). (a) Initial steady state. (b), (c), (d) General sequence leading to block faulting. (e), (f), (g) Suggested double up-arching to explain block uplift pattern of Venezuelan Andes. In Appalachians, uplift appears to have ceased at stage corresponding to c or f. Horizontal motion of mantle requires that tectonism be directed vertically upward.

and that the principal thermal events (regional metamorphism, granitic intrusion) were related solely to crustal downbuckling. Keeping in mind the possible linear continuity of the Andes and the Appalachians prior to the Triassic, the parallels in orogenic events and overall tectonic setting suggest that the respective tectonism was related to the same drift cycle(s). The presence of abundant serpentinites in the Appalachians suggests that impaction there was related to a collision of ocean-to-continent type (compare Vine and Hess, 1968) involving an intervening island arc. The fact that granitic plutonism appears to become progressively younger as one moves northwest across the Andes (similar age/space relations occur in New England) suggests that the ranges were drifting from northwest to southeast (with reference to current compass directions).

Interpretation of the Iglesias Complex (layer 1) in terms of plate tectonics would be largely idle speculation; one pertinent comment is indicated: the association of granites, basaltic volcanic rocks, and greenschist metamorphic rocks in the Bella-vista belt along the southern flanks of the ranges suggests that an axis of an earlier orogenic event may have been located in that region. The true significance of the 660 m.y. measurement on the G_2 pluton (see above) and the time relations between the rock types mentioned are not known, but it is conceivable that the northward migration (continuous or discontinuous) of orogenic axes began in late Precambrian time, and is not strictly a Paleozoic phenomenon. Intense imbricate thrusting which affects the Bellavista facies and lower Paleozoic section suggests that the close proximity of the two orogenic belts may reflect significant lateral movement during the Devonian and (or) late Permian deformations (hiatuses H_6 and H_7).

In the light of the discussion of the principal data, the original and metamorphic ages of the Sierra Nevada facies are not known. Furthermore, it is not known that these rocks are specifically "Andean." Hence it is not possible to extend *Andean* orogenic history back to those times, even assuming that the deposition and metamorphism of the Sierra Nevada facies reflect the oldest events in the ranges.

CONCLUSIONS

Because this project is still very much in its infancy, it is not possible at this time to attempt a detailed integration of Andean orogenic history with the hypothesis of sea-floor spreading. In the abstract sea-floor spreading solves one of the long-standing problems of orogeny in that it provides the mechanism for lateral compression; in the specific there are signs that the theory of sea-floor spreading will require modification or extension to explain satisfactorily the factual data elicited in regional studies. The skeletal outlines of potential problems may be perceived if one assumes, following current concepts of plate tectonics, that the orogenic characteristics described above are best viewed in a setting of impacting plate boundaries.

1. Either there was remarkable coincidence in the location of the Andes with respect to past impacting boundaries or, if one assumes a single convection cycle of ca 600 m.y. duration, one has to explain how a narrow zone (ca 100 km) of crust was able to maintain its position relative to the zone of impact. The latter alternative implies that Gondwanaland was drifting westward until sometime in the Mesozoic when at least the western portion of Africa reversed its direction of drift relative to South America! The coincidences are all the more remarkable inasmuch as most mountain belts have a similar history of superimposed orogeny (for example, Appalachians, Rockies, Alps).

2. If two or more convective cycles are stipulated, it requires another remarkable coincidence to explain the parallelism of structural trends with time.

A portion of the objections in *1* may be overcome by stipulating a complex combination of external and internal motions of convection cells, but such a solution itself involves more remarkable coincidences. Similarly it may be argued in explanation of *2* that once impact occurs, a structural "grain" is established which controls subsequent deformations irrespective of drift geometry. This may be a satisfactory explanation where tectonism operates upward, as was suggested in explanation of the mechanics of deformation of the post-Paleozoic layer of Andean geology, but is difficult to apply where thermal events and structural characteristics suggest crustal downbuckling. In such cases active displacement of the upper mantle cannot be avoided and is difficult to reconcile simply with the influence of crustal grain.

Even though approached from a negative viewpoint, some speculation as to alternative explanations is not inappropriate in a volume honoring H. H. Hess. Convection may not be entirely random with respect to the distribution of continental masses. The continents may act as thermal or mechanical barriers which control, at least to some degree, the *surface* location of impact zones, specifically at continental margins. In this view, Andean orogeny merely reflects long continued location at or close to the continental margin, a coincidence far easier to accept than either listed in *1* above. It is noteworthy that by far the majority of currently hypothesized impacting plate boundaries are located at or close to continental margins.

Alternatively, it is suggested that certain types of orogenic belts (those with a history suggesting minimal direct mantle participation in their evolution) may develop *within* plates, thereby at least avoiding the problem of repeated location at impacting boundaries. That such may hold for orogeny involving block uplift has been mooted above. It is entirely another matter to hypothesize orogeny involving significant crustal thermal events within plates. The only possibility that the writer can visualize is that in extreme circumstances vertical uplift may pass

462   R. SHAGAM

through stages of asymmetry to form
crustal nappes, as shown in Figure 7.
Such a mechanism is attractive in that it
would explain crustal thermal events
without direct mantle participation, but
the geophysical objections are numerous.

In the last analysis, new geotectonic
syntheses must depend on reliable data
relative to stratigraphy and the ages of
thermal events. To this end P. O. Banks
and students, Case Western Reserve University, and the writer have instituted an
extensive study of the isotopic ages of
Andean plutonic and volcanic rocks. In
addition, important tectonic information
may come from a continuing series of
paleomagnetic studies by R. B. Hargraves, Princeton University, and the
writer.

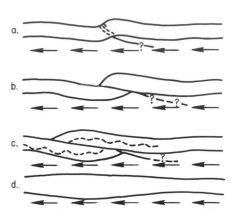

Figure 7. Possible mechanism (diagrammatic) to explain orogeny with crustal
thermal events *within* plates (over horizontally moving mantle). Initial vertical
uplift (compare Fig. 6b) gradually trans-
posed into crustal overthrust. Compressional environment of overthrust serves to
seal off mantle as source of thermal events.
Crustal thermal events related to doubling
of crustal thickness with concomitant rise
in geothermal gradient. As shown by reference plane in (c), folding need not be intense despite high grade metamorphism
(compare structure of Sierra Nevada
facies in cross sections of Fig. 2). Principal
objection concerns strength of crust and
ability to withstand overthrusting of postulated dimensions.

ACKNOWLEDGMENTS

Initial field studies in the Coast Range
of Venezuela under the supervision of
Hess (1960) provided the stimulus which
led to my present interests in the Andes.
Over and above this direct indebtedness
and that stemming from numerous discussions on Andean geology, I wish to
acknowledge a larger spiritual debt to
Harry Hess. To be under his guidance and supervision was at the same time to be
encouraged along the road of academic independence.

By far the major portion of the field and laboratory costs of this work have been
borne by the Ministerio de Minas e Hidrocarburos, Republic of Venezuela, to
whom I am deeply indebted.

REFERENCES CITED

Arnold, H. C., 1966, Upper Paleozoic Sabaneta-Palmarito sequence of Mérida Andes,
     Venezuela: Am. Assoc. Petroleum Geologists Bull., v. 50, no. 11, p. 2366–2387.
Bass, M., and Shagam, R., 1960, Edades Rb-Sr de las rocas cristalinas de los Andes
     Merideños, Venezuela: 3rd Venezuelan Geol. Cong., v. 1, p. 377–381: Caracas,
     Venezuela, Bol. Geol., Spec. Pub. no. 3.
Billings, M. P., 1960, Diastrophism and mountain building: Geol. Soc. America Bull.,
     v. 71, p. 363–398.
Boucot, A. J., Johnson, J. G., and Shagam, R., 1969, Braquiópodos Silúricos de los Andes
     Merideños, Venezuela: Caracas, 4th Venezuelan Geol. Cong.
Compañía Shell de Venezuela and Creole Petroleum Corporation, 1964, Paleozoic rocks
     of Mérida Andes, Venezuela: Am. Assoc. Petroleum Geologists Bull., v. 48, no. 1,
     p. 70–84.

Grauch, R. I., 1971, Geology of the Sierra Nevada south of Muchuchies, Venezuelan Andes: an aluminum-silicate-bearing metamorphic terrain [Ph.D. thesis]: Philadelphia, Pennsylvania Univ., 180 p.

——1972, Preliminary report of a late(?) Paleozoic metamorphic event in the Venezuelan Andes, in H. H. Hess Volume: Studies in earth and space science: Geol. Soc. America Mem. 132, p. 465–474.

Hargraves, R. B., and Shagam, R., 1969, Paleomagnetic study of La Quinta Formation, Venezuela: Am. Assoc. Petroleum Geologists Bull., v. 53, no. 3, p. 537–552.

Hess, H. H., 1960, Caribbean research project: Progress report: Geol. Soc. America Bull., v. 71, p. 235–240.

——1962, History of ocean basins, in Engel, A. E. J., James, H. L., and Leonard, B. F., eds., Petrologic studies: A volume in honor of A. F. Buddington: Geol. Soc. America, p. 599–620.

Kovisars, L., 1969, Geology of the eastern flank of the La Culata massif, Venezuelan Andes [Ph.D. thesis]: Philadelphia, Pennsylvania Univ., 236 p.

——1971, Geology of a portion of the north-central Venezuelan Andes: Geol. Soc. America Bull., v. 82, p. 3111–3138.

Liddle, R. A., 1946, The geology of Venezuela and Trinidad (2d ed.): Ithaca, New York, Palaeontographica Americana, 890 p.

Martín, Bellizzia, C., Ramírez, C., Menéndez, A., Ríos, J. H., and Benaím, N., 1968, Reseña geológica y descripción de las muestras de rocas Venezolanas sometidas a análisis de edades radiométricas: Bol. Geol. (Venezuela), v 10, no. 19, p. 339–355.

Morgan, W. J., 1968, Rises, trenches, great faults and crustal blocks: Jour. Geophys. Research, v. 73, no. 6, p. 1959–1982.

Schubert, C., 1969, Geologic structure of a part of the Barinas Mountain Front, Venezuelan Andes: Geol. Soc. America Bull., v. 80, no. 3, p. 443–458.

Shagam, R., 1969, Evolución tectónica de los Andes Venezolanos, in Simposio tectónico sobre Venezuela y regiones vecinas: Caracas, 4th Venezuelan Geol. Cong.

——1971, Geología de los Andes Centrales de Venezuela [Abs]: Caracas, 4th Venezuelan Geol. Cong., 1969.

Shagam, R., and Hargraves, R. B., 1970, Geologic and paleomagnetic study of Permo-Carboniferous redbeds (Sabaneta and Mérida facies), Venezuelan Andes: Am. Assoc. Petroleum Geologists Bull., v. 54, no. 12, p. 2336–2348.

Vine, F. J., and Hess, H. H., 1968, Sea-floor spreading: Princeton University pre-print, 60 p., prepared for v. IV, The Sea, A. E. Maxwell, E. C. Bullard, E. Goldberg, and J. L. Worzel, eds.

MANUSCRIPT RECEIVED BY THE SOCIETY MARCH 29, 1971

The Geological Society of America, Inc.
Memoir 132
© 1972

Preliminary Report of a Late(?) Paleozoic Metamorphic Event in the Venezuelan Andes

Richard I. Grauch
University of Pennsylvania
Philadelphia, Pennsylvania 19104

ABSTRACT

Recent studies in the central Venezuelan Andes suggest that a major episode of regional metamorphism took place in late Paleozoic time, affecting fossiliferous Permo-Carboniferous rocks and previously metamorphosed Precambrian(?) rocks.

Staurolite zones (±1 to 2 km wide) encompassing andalusite, kyanite, and sillimanite isograds spaced at approximately 100-m intervals have been mapped on a regional scale. One, possibly two, of these zones may truncate an earlier sillimanite isograd. On the basis of thin-section studies, mineral paragenesis within one of the staurolite zones appears to be as follows: (a) staurolite; (b) idioblastic andalusite; (c) andalusite → kyanite, staurolite; (d) andalusite → sillimanite, kyanite → sillimanite or fibrolite; (e) staurolite pseudomorphed by muscovite and sillimanite or fibrolite.

INTRODUCTION

Extensive areas of amphibolite grade (Sierra Nevada facies) and greenschist grade (Tostós facies) metamorphic rocks in the Venezuelan Andes are unconformably overlain by an upper Paleozoic section composed mainly of slates. Thus, the regional metamorphism has long been considered to be of pre-late Paleozoic age. Along the southern flank of the range a highly fossiliferous lower Paleozoic section (Caradoc-Ludlow age) shows no evidence of metamorphism and lies in fault contact with a metamorphic belt of greenschist grade (Bellavistá facies). A granitic pluton, postdating the regional metamorphism, intrudes the Bellavistá

465

schists. It has given an isotopic age of 660 m.y. (Rb/Sr, whole rock, Arnold, 1961), suggesting that the metamorphism occurred in Precambrian time. On the assumption that the Sierra Nevada facies is not likely to be younger than the Bellavistá facies, it has long been considered that the regional metamorphism of the former also occurred in Precambrian time.

Recent unpublished studies by Ramirez, and others, Kovisars (1969), Grauch and Shagam (1970), and Grauch (in prep.) indicate that several mineral phases characteristic of the Sierra Nevada metamorphic rocks are also present in restricted portions of the Late Paleozoic section in the central portion of the Andes. This evidence not only indicates that a regional metamorphic event occurred in latest Paleozoic time (probably Late Permian), but poses problems concerning the age and metamorphic history of the Sierra Nevada facies.

The purpose of this paper is to describe the distribution of metamorphic minerals in space and time in the central Andes and to outline possible models of metamorphic history for that region.

REGIONAL DISTRIBUTION OF STAUROLITE AND ALUMINOSILICATES

The distinctive phases characterizing the metamorphic rocks of the central Andes are andalusite, kyanite, sillimanite, and staurolite. The locations of the earliest sillimanite isograd and the staurolite zones are shown on the generalized geologic map of Figure 1. Two major features should be noted: (a) Three staurolite zones, separated by major faults, trend northeast sub-parallel to regional structures but cut formation contacts at small angles. Isograds marking the first appearance of andalusite, kyanite, and sillimanite have been mapped within the two southern zones and probably occur in zone 1. (b) A tentative sillimanite isograd (NW-trending) passes approximately through Merida: no sillimanite was found in ±60 thin sections of rocks from the region to the southwest, but it occurs in about 25 percent of ±200 specimens to the northeast.

Because the zones transect formation contacts and include rocks of widely varying composition, the first appearance of staurolite is considered to be a first approximation of the location of the staurolite isograd. Similarly, the disappearance of staurolite, which in some cases approximates the transition between the upper and lower sillimanite zones, is considered to represent an isograd.

The time and space relations between the NW-trending sillimanite isograd and the staurolite zones are not known. The former appears to intersect staurolite zone 2, suggesting two periods of metamorphism, but it is conceivable that the two trends are related to the same thermal event. The latter possibility would be indicated should further investigation reveal that the NE-trending aluminosilicate isograds do not terminate against the NW-trending isograd.

PETROGRAPHY OF THE STAUROLITE ZONES

Because zone 3 has been more extensively studied than the other two, most of the following discussion will be restricted to that zone. Several rock types occur within the zone; fine- to course-grained pelitic schists predominate over medium-grained gneisses, quartzites, and rare amphibolites. Bedding, defining layers 2 mm to 12 cm thick, is generally parallel to metamorphic foliation. With the exception of compositional variations and graded bedding, primary sedimentary structures are rare.

The spatial relations between the staurolite, andalusite, kyanite, and first sillimanite isograds as well as the boundary between the upper and lower sillimanite zones are shown in an idealized map of a small portion of zone 3 (Fig. 2). The parallel configuration of the isograds within this small area is apparently repeated along strike. However, it is possible that the isograds cross at small angles. Further mapping should resolve the problem. The typical mineral assemblages within the zones are shown in Table 1. The data are not sufficient to show significant modal variation with increasing grade.

The reaction(s) responsible for the first appearance of staurolite and andalusite are not known. However, the spatial distribution of andalusite (Fig. 3) suggests that its first appearance postdates that of staurolite. The kyanite isograd is marked by pseudomorphs after andalusite (Fig. 4a) and by rare xenoblastic kyanite in the matrix. A new generation(?) of staurolite is sometimes intergrown (epitaxially?) with the pseudomorphous kyanite. Sillimanite first appears as prismatic crystals pseudomorphous after andalusite and kyanite (Fig. 4b). Fibrolite is also pseudomorphous after kyanite. The transition from the lower to upper sillimanite zones is indicated by the disappearance of staurolite. Textural evidence similar to that recorded by Guidotti (1968) indicates that some staurolite has been pseudomorphically replaced by muscovite, sillimanite, or fibrolite (Fig 4c); garnet, biotite, and plagioclase may be minor products of the replacement reaction. However, fibrolite persists throughout the sillimanite terrain, both as pseudomorphs and as mats in the matrix, suggesting that increasing grade does not, in this case, lead to the formation of sillimanite at the expense of fibrolite.

The granite to the south of the zone (*see* Fig. 3) was probably emplaced tectonically with cold borders and is not directly responsible for any major metamorphic effects. The staurolite zone has been repeated by faulting both to the north and south of the granite (see the cross sections of Fig. 3). Post- and synmetamorphic deformation of the zone is demonstrated by rolled garnets, staurolites, and andalusites, swirled fibrolite mats, and folded sillimanite needles (Fig. 4d). Some staurolite crystals with clearly deformed interiors have undeformed marginal overgrowths. This, coupled with the rarity of deformed sillimanite, suggests that metamorphism continued after the major period of deformation.

Preliminary field work indicates that the mineral assemblages described above, the inferred sequence of mineral growth, and the suggested relation between deformation and metamorphism are repeated in zones 1 and 2 (Table 1). It is further suggested that a similar staurolite zone has been removed by erosion from an area approximately halfway between zones 2 and 3. A narrow zone of the Sierra Nevada Facies rocks containing rods of muscovite–sillimanite (± biotite, ± garnet), which appear to be pseudomorphous after andalusite and (or) staurolite, was traced parallel to zones 2 and 3. This zone of pseudomorphs may correspond to a metamorphic setting immediately upgrade from the second sillimanite isograd of zone 3 (*see* Fig. 1 and Fig. 5, a generalized cross section of the central Andes).

INTERPRETATION

The preceding observations suggest the following sequence of mineral development:

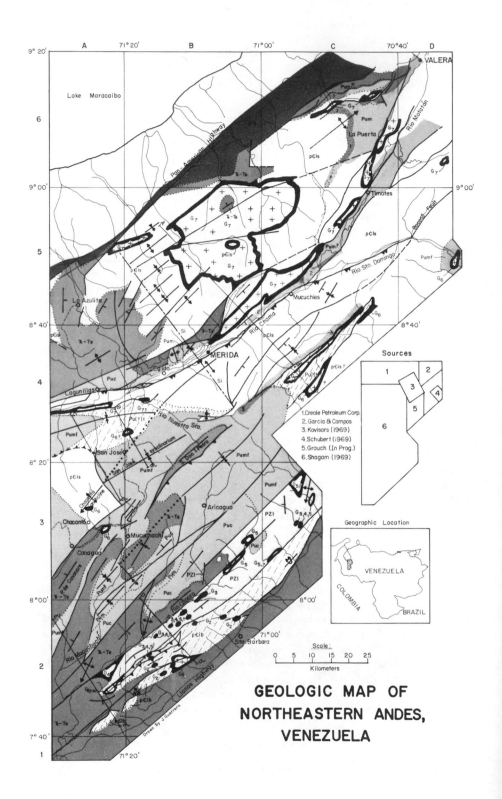

GEOLOGIC MAP OF
NORTHEASTERN ANDES,
VENEZUELA

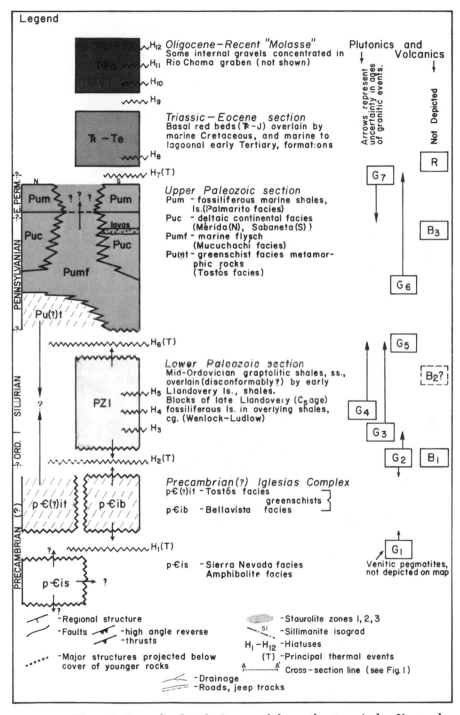

Figure 1. Generalized geologic map of the northeastern Andes, Venezuela.

TABLE 1. TYPICAL MINERAL ASSEMBLAGES OF THE STAUROLITE ZONE PELITES; STAUROLITE AND QUARTZ ARE INCLUDED IN ALL ASSEMBLAGES

Staurolite Zone	Formation (Facies)	Metamorphic Zone	Plagioclase	Muscovite	Biotite	Garnet	Andalusite	Kyanite	Sillimanite	Opaques
1	Upper Paleozoic	Staurolite			X	X				X
	Upper Paleozoic	Staurolite		X	X	X				X
	Sierra Nevada	Lower Sillimanite		X	X	X			X	X
	Sierra Nevada	Lower Sillimanite	X	X	X				X	X
	Upper Paleozoic	Lower Sillimanite	X	X	X	X				X
2	Sierra Nevada	Kyanite	X	X	X	X	X	X		X
	Sierra Nevada	Lower Sillimanite		X	X	X			X	X
3	Tostós	Staurolite	X	X	X	X				X
	Sierra Nevada	Andalusite	X	X	X	X	X			X
	Sierra Nevada	Kyanite	X	X	X	X		X		X
	Sierra Nevada	Kyanite	X	X	X	X	X	X	X	X
	Sierra Nevada	Lower Sillimanite	X	X	X	X	X	X	X	X
	Sierra Nevada	Lower Sillimanite	X	X	X	X	X	X	X	X
	Sierra Nevada	Lower Sillimanite	X	X	X	X		X	X	X

UPPER SILLIMANITE ZONE

LOWER SILLIMANITE ZONE
——————————SILLIMANITE ISOGRAD——
————————————KYANITE ISOGRAD————

STAUROLITE ZONE ——————————ANDALUSITE ISOGRAD——

SCHEMATIC SCALE

Figure 2. Idealized map view of a small portion of staurolite zone 3. *See* Figure 3 for location.

1. small prismatic staurolite crystals
2. idioblastic andalusite
3. andalusite → kyanite; oriented intergrowths of new(?) staurolite and kyanite
4. (a) andalusite → sillimanite (b) kyanite → sillimanite or fibrolite
5. staurolite pseudomorphically replaced by muscovite + sillimanite or fibrolite.

An alternate sequence in which an andalusite–sillimanite terrain was developed prior to the staurolite–kyanite–sillimanite terrain is also possible. This would reverse the positions of events 1 and 2. The apparent intersection of staurolite zone 2 with the northwest-trending sillimanite isograd suggests two periods of metamorphism. However, if the suggested parallelism of the isograds in zone 3 is accepted, it can be concluded that the staurolite-bearing assemblages are all the result of a single metamorphism. Further mapping of the staurolite zones and detailed analysis of the assemblages will be required to discriminate between these two hypotheses.

Regardless of which sequence of mineral formation is correct, the distribution of staurolite-bearing assemblages in both upper Paleozoic and Precambrian rocks indicates an intense late Paleozoic metamorphic event.

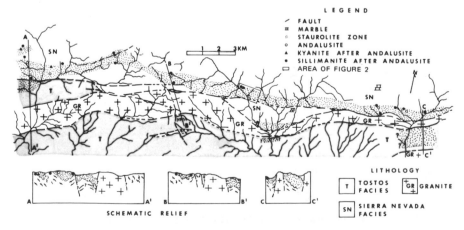

Figure 3. Geologic map of staurolite zone 3 (eastern portion).

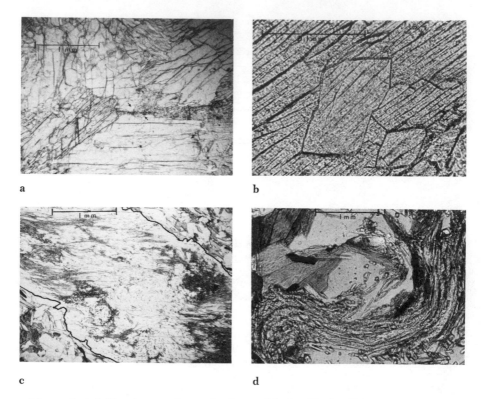

a   b

c   d

Figure 4. (a) Kyanite pseudomorph after andalusite (limbs of a partially preserved chiastolite cross are oriented parallel to the photomicrograph margins; the center of the cross is indicated by arrows); (b) sillimanite pseudomorph after kyanite; (c) single muscovite plate and sillimanite crystals; (d) folded sillimanite needles in Sierra Nevada schist.

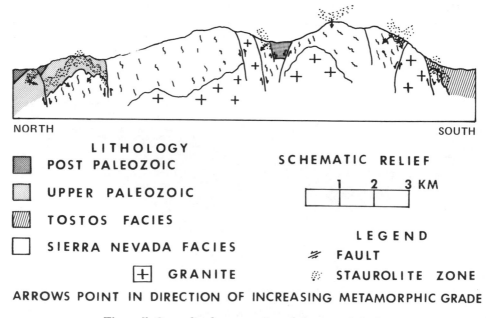

NORTH   SOUTH

LITHOLOGY

■ POST PALEOZOIC

□ UPPER PALEOZOIC

▩ TOSTOS FACIES

□ SIERRA NEVADA FACIES

⊞ GRANITE

SCHEMATIC RELIEF

| 1 | 2 | 3 KM |

LEGEND

∿ FAULT

⋯ STAUROLITE ZONE

ARROWS POINT IN DIRECTION OF INCREASING METAMORPHIC GRADE

Figure 5. Generalized cross section of the central Andes.

ACKNOWLEDGMENTS

I wish to thank the Ministerio de Minas e Hidrocarburos, Venezuela, for their generous support of field work in the Andes. In particular, I am deeply indebted to A. Vivas R., the former director of geology, and to his successor, E. M. Araujo. Many conversations with A. Bellizzia G., C. Ramirez C., R. Garcia J., and V. Campos C. were also of great assistance.

Reginald Shagam of the University of the Negev, Israel, and Alan Gaines and Robert Giegengack of the University of Pennsylvania have reviewed this manuscript and have made many useful suggestions.

Without the inspiration and guidance of Reginald Shagam, this study would not have been possible.

REFERENCES CITED

Arnold, H. C., 1961, The pre-Cretaceous geology of the Venezuelan Andes: Compania Shell de Venezuela, Exploration Department, Report #EPC 1807.

Grauch, R. I., and Shagam, R., 1970, Distribution and tectonic significance of andalusite, kyanite, sillimanite, and staurolite in the central Venezuelan Andes: Geol. Soc. America, Abs. with Programs (Ann. Mtg.), v. 2, no. 7, p. 560.

Guidotti, C. V., 1968, Prograde muscovite pseudomorphs after staurolite in the Rangeley–Oquossoc areas, Maine: Am. Mineralogist, v. 53, p 1368–1376.

Kovisars, L., 1969, Geology of the eastern flank of the La Culata Massif, Venezuelan Andes [Ph.D. thesis]: Philadelphia, University of Pennsylvania.

PRESENT ADDRESS: UNIVERSITY OF CALIFORNIA, LOS ANGELES, CALIFORNIA 90024.

MANUSCRIPT RECEIVED BY THE SOCIETY MARCH 29, 1971

Petrology

THE GEOLOGICAL SOCIETY OF AMERICA, INC.
MEMOIR 132
© 1972

Differentiation Trends and Parental Magmas for Anorthositic and Quartz Mangerite Series, Adirondacks, New York

A. F. BUDDINGTON
Department of Geological and Geophysical Sciences
Princeton University, Princeton, New Jersey 08540

ABSTRACT

Several geologists have drawn variation diagrams to demonstrate a continuous gradation in chemical composition between all members of the anorthositic and the quartz mangerite series and have inferred that this is a major support for the hypothesis of comagmatic origin. The alternative hypothesis elaborated here is that the plots of major oxides against differentiation indices are discontinuous and can be interpreted in terms of *two independent magma series,* each gradational throughout, but with contrasted trends of differentiation. A gabbroic anorthositic liquid may be generated in the *lower* crust and a quartz mangeritic liquid successively later in a relatively dry deep part of the *upper* crust, possibly by development of anatectic granitic liquid with concomitant reaction with and differential assimilation of appropriate constituents of associated undersaturated metagabbro.

INTRODUCTION

Interpretations of the parental magma for massif-type anorthosites range from high-alumina basalt through leucogabbro-gabbroic anorthosite to diorite or granodiorite (Isachsen, 1969). De Waard (1969) and others have plotted diagrams showing the variation of various major oxide abundances in the anorthositic and quartz mangerite series and conclude them to be consanguineous and comagmatic.

The opposite interpretation is elaborated in this paper. The term anorthositic series will be here used to include anorthosite, leucogabbro and leuconorite (gabbroic anorthosites), related gabbro and norite, and late-stage ferrogabbros, mela-

477

ferrogabbros, and Fe-Ti oxide mineral-rich rocks; the quartz mangerite series includes jotunite, mangerite, quartz mangerite (farsundite), and charnockite. The rocks of the stratiform Diana "syenite" complex (northwest Adirondacks) are included among the quartz mangerite series, although hypersthene is minor in much of these rocks and absent in hornblende quartz syenite which forms the bulk of the upper portion. On the whole, however, there is only a minor variation in the average chemical composition of the quartz-mangeritic series throughout the Adirondacks. All the rocks, with local exceptions, have undergone regional metamorphism in the granulite facies.

TRENDS OF DIFFERENTIATION

The succession of intrusion, decreasing volume relations, and the nature of oxide variations in the anorthositic series indicate that the trend of differentiation is anorthosite as a crystal cumulate by flow or gravity concentration, gabbroic anorthosite, feldspathic gabbro, and minor late stage ferrogabbros and feldspathic ferropyroxenites (melaferrogabbros) rich in Fe-Ti oxide minerals. The subordinate facies may occur either as conformable facies and sheets, or occasionally as dikes. Essentially all the rocks, except the late stage mafic facies with less than 40 percent normative feldspar, are slightly oversaturated or neutral; the mafic facies may in part show normative olivine. The only gabbroic rocks included in this study are those within the anorthositic massifs with composition, color, and texture characteristic and appropriate for the anorthositic series.

The mafic zones are in some places conformably layered, which suggests that they may have originated as crystal cumulates. However, ferrogabbros and other mafic bodies often occur as dikes, belying this origin. Davis (1969, Fig. 2) has shown that in mafic layers, the plagioclase is more sodic (in part primary, in part metamorphic) and the pyroxenes in part more FeO-rich than the equivalent minerals of the directly associated rocks, the reverse of what might be expected for cumulates.

All rocks of the anorthositic series, except those with more than about 85 percent normative feldspar, are interpreted as belonging to a liquid line of descent modified by some minor differential addition or subtraction of certain cumulus crystals and interdiffusion.

In variation diagrams, the compositions of the ferrogabbros of the Carthage, Marcy, Snowy Mountain, and Morin massifs all plot as continuous gradational extensions of the anorthositic series (Figs. 1 and 2). Papezik (1965) notes that the ferrogabbro dikes of the Morin massif show a relative scarcity of those elements characteristic of early differentiates, in keeping with the evidence of their origin as late stage residuals. He also gives a plot for Mg, Fe, and Na + K that shows an increase in Mg and Fe and decrease of Na + K with differentiation of the Morin massif. The origin of *ore* concentrations as residual magmas raises some petrogenetic problems, but a magmatic origin for ferrogabbros and Fe-Ti oxide mineral-rich feldspathic ferropyroxenites (melaferrogabbros) with no more than about one-quarter or one-third oxide minerals seems well supported. The anorthositic series from anorthosite through ferrogabbro is explicable essentially on the basis of early separation of plagioclase alone, followed by separation of plagioclase and pyroxene but with an exceptionally high ratio of the plagioclase. This results in an increase in MgO and a slight decrease in K_2O in contrast to the

Skaergaard trend, with decrease of MgO and increase of K_2O. The fO_2 is a factor affecting the trend of differentiation, and certain late-stage melanocratic ferrogabbros very rich in FeO and relatively poor in MgO may result from separation of a higher ratio of pyroxene under low fO_2. Ores very rich in ilmenite and titaniferous magnetite may have formed as a result of liquid immiscibility.

There is general agreement that the quartz-mangerite series usually differentiates from jotunite or mangerite through quartz-mangerite or farsundite to charnockite. In the Diana complex, the trend is from jotunite through pyroxene-syenite and quartz-syenite to hornblende-quartz-syenite and hornblende-granite (Buddington, 1939; Hargraves, 1969). The jotunites are in part the product of incorporation of material from border facies of the anorthositic massifs and from included mafic pyroxenic paragneiss, or are modified by such incorporation.

The trends of differentiation of the anorthositic and quartz mangerite series are thus in part reverse to each other (Buddington, 1969, p. 220–221).

IMPLICATIONS OF COMPOSITION GRAPHS

Curves for the variation of certain oxides with differentiation in the anorthositic and quartz mangerite series have been drawn by de Waard and Romey (1969) for the Snowy Mountain massif, by de Waard (1969) for the Carthage and Marcy complexes, and by Letteney (1969) for the Thirteenth Lake association.

De Waard and Romey concluded that there is a completely gradational compositional sequence that "favors an origin by differentiation from one magma or formation of part of the magma by anatexis rather than an origin involving two magmatic events."

Letteney (1969) however, found that the most appropriate plots for the Thirteenth Lake rocks yielded two discontinuous curves, essentially one for anorthosite-norite series, subjacent assimilation and differentiation during consoliterms of the emplacement first of a dioritic or leucogabbroic magma to yield the anorthosite-norite series, subjacent assimilation and differentiation during consolidation of the first series, and to the later intrusion and differentiation of a basic charnockite magma to yield the salic series. All are called consanguineous. Differential assimilation of salic material by an alumina-rich magma could result in the development of Na_2O- and K_2O-rich anorthosites, especially such as the Roseland massif (Herz, 1969).

Variation graphs for the Marcy and Carthage massifs based on those given by de Waard (1969) and interpreted by him as showing gradational differentiation from a common magma are given in Figure 1. I have superimposed a curve for the anorthositic series in accord with the alternative interpretation of differentiation from two parent magmas of contrasting composition. This distinction is emphasized by their reverse directions of differentiation. The plots for other anorthosite massifs of the Grenville province, given in Figure 2, are likewise interpreted in terms of two contrasting parental magmas with reverse differentiation trends. Data for the quartz mangerite series of the Morin massif[1] are not available. The data plotted

---

[1] The "Morin series" described by Philpotts (1966) are geographically separate from the "Morin massif" (compare de Waard, 1969). The analyzed mafic rocks described by Philpotts are a normal noritic gabbro group with but little anorthosite in one area and none in the other area described by him.

for the leucogabbro series of the Morin massif are the averages for chemical analyses of 7 anorthosites (91 to 96 percent normative feldspar), 10 gabbroic anorthosites (81 to 89 percent normative feldspar), 10 gabbroic anorthosites (70 to 80 percent normative feldspar), and 5 ferrogabbros (55 to 69 percent normative feldspar) given by Papezik (1965). The average composition of 6 ferrogabbros (36 percent normative feldspar) and 6 melaferrogabbros very rich in Fe-Ti oxide minerals (22 percent normative feldspar) taken from data of Rose (1969) for other bodies in anorthosites in the Grenville province have been added.

The quartz mangerite series (varying from jotunite to charnockite) is character-ized by K_2O being a critical oxide, and one might therefore expect the anorthositic series to show an increase in K_2O with progressive differentiation if both series were of comagmatic origin. In fact, however, the anorthositic series of the Marcy, Carthage, and Morin massifs all show a slight decrease in K_2O with differentiation (Figs. 1 and 2). In general, there is a clear discontinuity in K_2O content between the lowest values for the quartz-syenite or quartz-mangerite series and the maximum for the anorthositic series. Two examples follow.

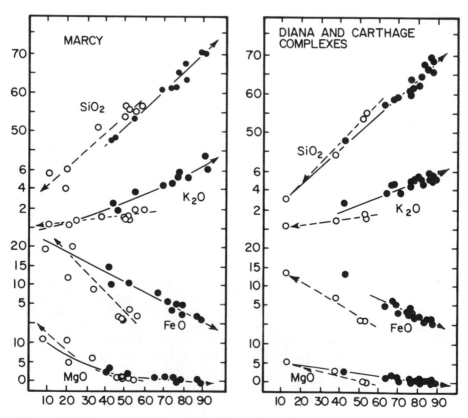

Figure 1. Differentiation-index (normative qtz+or+ab) variation graphs for Adiron-dack Marcy and Carthage anorthositic and associated quartz mangerite series. Modified after de Waard (1969). Circles, anorthositic series; dots, "syenitic" (quartz mangerite) series. Continuous lines as drawn by de Waard on basis of single cogenetic series; dashed lines drawn by Buddington on basis of two series of contrasted parentage. Arrows indi-cate trends of differentiation. Data in weight percent.

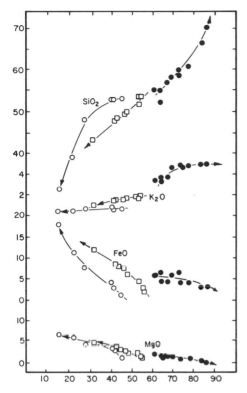

Figure 2. Differentiation-index—molecular (ionic percent) normative qtz + or + ab— variation graphs for the anorthositic series of the Morin massif and for the anorthositic and quartz mangerite series of the Snowy Mountain dome. Data for oxides in weight percent. Circles, leucogabbro series, Morin massif; squares, leucogabbro series; and dots, quartz mangerite series of Snowy Mountain dome. Arrows indicate trends of differentiation. Data for Morin from Papezik (1965) and Rose (1969); for Snowy Mountain from de Waard and Romey (1969) and Reynolds and others (1969).

Part of the Diana quartz-syenite complex overlies the Carthage anorthositic complex. Although field relations are obscured by intense shearing in the area, there appear to be relics of a screen of metasediment between the two. In addition, within the contact zone each contains incorporated fragments or material of the metasediment. The maximum percent K_2O in five analyses of the anorthositic series is 1.49 whereas the minimum in three jotunites of the Diana complex is 3.53. Zircons have been reported only from pegmatitic and ore facies of the anorthositic series. They are common in the jotunites and quartz-mangerite series.

Analyses (Buddington, 1939, 1953) of five rocks from the St. Regis dome of the main Adirondack anorthositic massif (Marcy) have a maximum percent K_2O of 1.09, whereas the minimum of two overlying jotunites is 3.06, again indicating a *discontinuity* in composition. Furthermore, local screens of metasediment occur between the anorthositic rocks and the overlying jotunites. The jotunites contain fragments of metasediment and of the anorthositic rocks. Where the jotunite has transgressed across the gabbroic anorthosite border facies into the anorthosite of the core, the latter may have 1.93 percent K_2O as a result of slight metasomatism. K-feldspar is common in marble at contact zones with members of the quartz mangerite series but effectively absent at contacts with leucogabbro where it might also be expected if quartz mangerite were a residual liquid in a leucogabbro crystalline portion.

Values for MgO, FeO, and $Na_2O + K_2O$ are plotted for the Marcy anorthositic and the associated quartz mangerite series in Figure 3. The trend for the successive liquids as estimated by Wager and Deer (1939, p. 314) for the differentiated Skaergaard gabbroic stratiform complex is also plotted for comparison. The trend for the Marcy rocks can be interpreted largely in terms of control by separation of plagioclase as far as rocks of ferrogabbroic composition. The melaferrogabbros (greater than 60 percent FeO) show, in addition, the effects of strong fractionation of pyroxenes with a trend parallel to that of the Skaergaard liquids. The Skaergaard liquid with maximum FeO comprises only about 5 percent by volume, and the Marcy melaferrogabbros similarly form not more than a few percent of the total at most. The last residual liquids of the Skaergaard trend are granophyric but form only about 1.5 percent by volume. A restudy of the Skaergaard data by Chayes (1970) suggests that there is "virtually no perceptible end-stage alkali enrichment."

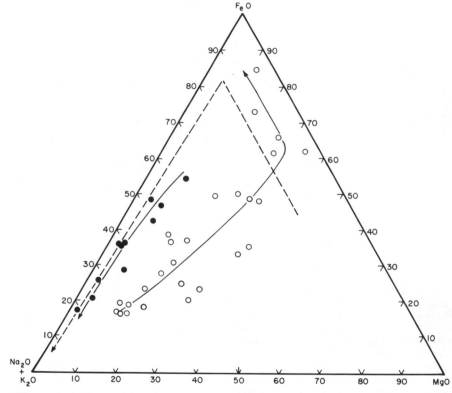

Figure 3. Analyses of anorthositic series, Marcy massif (circles) and bordering quartz mangerite series (dots) plotted on ternary diagram for FeO, MgO, and Na_2O+K_2O. Dashed line is for trend of composition of liquids of Skaergaard stratiform complex as estimated by Wager and Deer (1939, p. 314). Arrows indicate trend of differentiation. Data from Buddington (1939, 1952, 1953), Buddington and Leonard (1962, p. 57), and Papezik (1965, p. 686).

A plot of the ratios of FeO(+MnO) to FeO(+MnO) + MgO against a differentiation index for several anorthositic and quartz mangeritic series is given in Figure 4. All the quartz mangeritic series show a systematic increase with increase in salic minerals. The upper (stratigraphically) layers of the Diana complex are pink, hornblendic, and have 10 to 20 percent quartz in contrast to the lower layers that are green and pyroxenic with less than 10 percent quartz. The hornblendic facies shows greater oxidation, in part of primary and in part of secondary origin, and data for it is not plotted. Concentrations of pyroxenes and ore minerals occur in layers within the Diana pyroxenic facies and are interpreted as cumulative aggregates. Two analyses are shown in the plot. They differ from the melaferrogabbros of the anorthositic series in having a distinctly higher K_2O

and a relatively high zircon content (0.4 to 0.7 percent) and in occurring in mangerite.

The trend of differentiation for the Carthage anorthositic series is well documented in the field by successive intrusive phenomena and is characterized by an increase of ilmenite, magnetite, and pyroxenes, with an increase in ratio of FeO (+MnO)/FeO(+MnO) + MgO. The Snowy Mountain anorthositic series can be interpreted as showing a similar trend. The Marcy anorthositic series plot over a wide range. However, if only members showing a systematic increase in ilmenite and magnetite as in the Carthage and Snowy Mountain anorthositic series are considered, then a trend line of differentiation may be drawn as in the graph for part of the rocks. One gabbro shows concentration of ilmenite only. Interpretation of members with relatively low ratios of FeO(+MnO)/FeO + (MnO) +Mgo must await further studies. Locally, contamination with mafic paragneiss may affect a variation in composition and locally also there has been a slight secondary oxidation. Anorthosites (90 percent or more plagioclase) show a systematic variation such that a more calcic plagioclase is accompanied by a lower ratio of FeO/FeO + MgO.

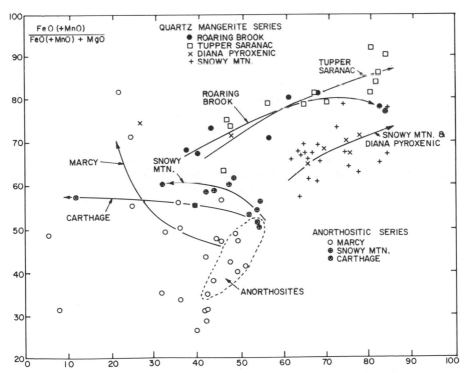

Figure 4. Molecular ratios FeO (+MnO)/Feo(+MnO)+MgO plotted against molecular norms (ionic percent) (qtz + ab + or) for some Adirondack anorthositic and quartz mangerite series. Curve for Marcy anorthositic series based only on members showing concentration of ilmenite and magnetite or ilmenite alone. Arrows indicate trend of differentiation. The Diana quartz mangerite series is adjacent to Carthage anorthositic series and the Roaring Brook and Tupper-Saranac quartz mangerite series border the Marcy massif. Data from Buddington (1939, 1952, 1953), Buddington and Leonard (1962, p. 57), de Waard (1970), de Waard and Romey (1969), and Reynolds and others (1969).

A diagram to support the hypothesis that the anorthositic and jotunite-char-nockite rocks in the Roaring Brook section of the Marcy massif show a continuous gradation has been plotted by de Waard (1970). He notes, however, that jotunite in the contact zone contains xenoliths of metasediment, anorthosite, and leuco-norite. This is consistent with screens of metasediment so commonly found in such zones.

The data as presented in Figures 1 to 4 for variation of chemical compositions are here interpreted as evidence that the anorthositic and quartz mangeritic series show quite contrasted trends of differentiation consistent with the geologic evidence for two independent series.

COMPOSITIONS OF POSSIBLE PARENTAL MAGMAS, ADIRONDACK ROCKS

An estimate (Buddington, 1939) of the volume ratio of the Adirondack anortho-sitic and the quartz mangeritic series to the total volume of the complexes gave 29 percent for the former and 71 percent for the latter. This estimate was based on the assumption that the volumes of the two series were proportional to their area of exposure. The bulk chemical composition of such a mixture is given as analysis 5, Table 1. It corresponds to that of a monzodiorite, monzonite, or mangerite with high alumina. However, in addition to other limitations this method ignores the high probability that there are several different quartz mangeritic complexes, each with widely differing volume relations to the anorthositic series with which they are associated. For example, the quartz mangeritic series of the Tupper-Saranac com-plex forms only about 20 percent of the total associated quartz mangeritic and anorthositic series in the Saranac and St. Regis quadrangles. This estimate is based on the interpretation that the anorthositic series forms a sheet about 4 km thick and the quartz mangeritic series a sheet about 1 km thick, the latter overlying the former. The average chemical composition of the Tupper-Saranac quartz mangeritic series is given as analysis 7, Table 1, the associated anorthositic series as analysis 3, and the average composition of the whole, consisting of 20 percent quartz mangerite and 80 percent anorthositic series, is given as analysis 4. By contrast, the Diana quartz mangeritic (quartz syenitic) complex and associated anorthositic series (Buddington, 1939; Hargraves, 1969) occur as folded sheets eroded and well exposed; the Diana quartz mangeritic series (analysis 6) can be seen to form more than 90 percent of the whole. Chemical analyses in the range between analyses Nos. 4 and 5, Table 1, are not adequately approximated by a mixing of simple gabbroic and anatectic salic magmas.

The evidence presented above is considered to indicate the essential magmatic independence of the anorthositic and quartz mangerite series. The chemical com-position here postulated for the parental magma of the Adirondack anorthositic series is given by analysis No. 3, Table 1. The magma is considered to have con-sisted of about 60 percent liquid of gabbroic anorthosite composition and 40 per-cent suspended andesine crystals. Its origin has been previously discussed (Budding-ton, 1969) and the possibility suggested that the liquid portion formed in the lower crust (or upper mantle) under conditions of relatively high P_{H_2O} and high load pressure as a fractional or residual melt and that the solid portion was derived from the cumulate plagioclase fraction of a differentiated or differentiating strati-form gabbroic sheet.

	1	2	3	4	5	6	7	8	9	10	11
SiO$_2$	54.27	53.40	54.05	55.74	59.61	63.0	62.5	62.85	58.7	69.4	46.62
Al$_2$O$_3$	25.94	23.96	25.44	23.47	18.70	16.0	15.6	16.80	15.6	14.0	17.48
Fe$_2$O$_3$	.70	.91	0.73	.98	1.82	2.7	2.0	2.96	2.32	1.75	2.10
FeO	1.36	3.02	1.8	2.46	3.65	3.1	5.4	2.89	5.83	2.03	10.60
MgO	1.07	1.88	1.3	1.22	1.11	1.0	.92	1.48	1.71	.40	8.10
CaO	10.32	9.85	10.2	8.88	5.34	3.3	3.60	3.24	4.00	1.60	8.76
Na$_2$O	4.67	4.17	4.54	4.38	4.42	4.5	3.76	4.09	3.90	3.30	2.90
K$_2$O	.93	.80	0.90	1.64	3.68	5.0	4.63	5.49	4.70	5.80	.68
TiO$_2$	.45	.77	.53	.59	.76	0.8	.83		1.50	.33	1.70
P$_2$O$_5$	.06	.18	.09	.14	.33	.33	.33		.68	.14	.30

C.I.P.W. Norm

	1	2	3	4	5	6	7	8	9	10	11
qz	0.2	2.49	0.6	4.0	6.7	10.0	13.1	8.58	7.0	23.9	—
or	5.56	4.73	5.5	10.0	21.7	29.5	27.8	32.80	27.8	34.3	3.89
ab	39.82	35.11	38.5	37.2	37.2	38.2	31.4	34.58	33.0	28.6	24.63
an	46.70	44.20	46.0	39.2	20.3	9.5	11.7	11.12	11.0	5.7	32.80
di	3.03	.80	3.4	3.2	3.5	3.6	2.8	3.43	2.7	—	7.30
hy	2.5	7.87	3.4	4.2	5.2	3.0	7.6	5.04	9.3	2.34	3.15
mt	1.0	1.28	1.1	1.4	2.6	4.0	2.8	4.41	3.3	2.50	3.02
ilm	.84	1.44	1.0	1.1	1.4	1.4	1.3		2.9	.60	1.06
ap	.27	.44	0.3	.4	0.8	0.9	0.8		1.7	.11	.67
ol											22.23

(1) Average composition anorthosite (Geol. Soc. Am. Mem. 7, p. 30, analyses A, 16, 17).

(2) Average composition 75 grab samples gabbroic anorthosite (leucogabbro) NE. of Reber (Geol. Soc. Am. Mem. 7, p. 30, No. 19).

(3) Estimated bulk composition of anorthositic series based on 3 average anorthosite plus one average gabbroic anorthosite.

(4) Estimated bulk composition of 80 parts average anorthositic series (analysis No. 3) plus 20 parts average Tupper-Saranac quartz mangeritic series (analysis No. 7); based on relative thickness of sheets.

(5) Estimated bulk composition of 29 parts gabbroic anorthosite (analysis No. 3) plus 71 parts average quartz mangerite (av. 47 analyses), based on relative exposed areas in mapped Adirondack Highlands.

(6) Estimated bulk composition of quartz mangerite (quartz syenite) series of Diana complex based on weighted average of 16 analyses.

(7) Estimated bulk composition of Tupper-Saranac quartz mangerite series based on average of 12 analyses (Buddington, 1939, 1952, 1953, 1957, 1962).

(8) Hornblende quartz-bearing monzonite, result of incorporation of metagabbro (analysis No. 11) by hornblende granite magma (analysis No. 10). (Geol. Soc. Am. Mem. 7, opp. p. 120).

(9) Quartz-bearing hornblende monzonite; result of incorporation of metagabbro (No. 11) by hornblende granite magma (No. 10). (Geol. Soc. Am. Mem. 7, p. 76).

(10) Average of 7 hornblende granite analyses, Adirondack Highlands (Bull. Geol. Soc. Am., 1957, p. 293).

(11) Average of 12 metagabbro analyses, Adirondack Highlands.

Emslie (1971) finds experimentally that the composition of the piercing point in the system plagioclase-($An_{60}Ab_{40}$)-diopside-enstatite at 15 kb is $75(An_{60}Ab_{40})$, 7 diopside and 18 enstatite at a temperature of $1,390° \pm 20°C$. The composition of the plagioclase in equilibrium with pyroxenes and liquid near the piercing point is about $An_{65}Ab_{35}$. This is persuasive evidence for the probability of development of a magma of leucogabbroic composition at the base of the crust at a time in the Precambrian, when the geothermal gradient was higher.

ORIGIN OF QUARTZ MANGERITIC MAGMA

A possible clue to the origin of the parental magma of the quartz mangeritic series in the Adirondacks is given by local facies of the younger batholithic hornblende granite where it has intruded undersaturated (commonly with olivine) metagabbros. In Table 1, analysis No. 11 is the average of 12 metagabbros and analysis 10 is of 7 hornblende granites, both of which rock types occur throughout the Adirondack Highlands. Analyses Nos. 8 and 9 are of facies of the granite contaminated with material from metagabbro. They are similar to the average quartz-bearing mangerite. At depths lower than that now exposed at the surface, in the lowest part of the upper crust, development of a granitic magma that assimilated material from associated garnet metagabbro sheets could yield a quartz mangeritic magma of appropriate composition. The granitic magma is partially desilicated and there is a differential lack of incorporation of magnesia. Lebedev (1937) has also proposed an origin for the quartz mangerite series by granitic magma assimilating metagabbro.

SUBCONFORMABLE RELATIONS OF TWO SERIES

The structural implications that the quartz mangerite series commonly subconformably overlies the anorthositic massifs is one appropriate to differentiation in place from a common magma. However, the concept of subconformable repetitive intrusion is well established in geology and the relation of the Bushveld granite to the Bushveld layered complex as described by Willemse (1969, p. 15–16) is quite apropos:

> The Bushveld granite by and large adapted itself to the stratiform character of the complex . . . The granite is generally separated from the mafic rocks by (screens) leptites, microgranite, partly feldspathized quartzite and granophyres. The granite is irregularly and markedly transgressive to its roof in places. . . . Several dikelike bodies of granite intrude the layered sequence . . . The Bushveld granite occupies obviously too large a volume to represent a differentiated product of the mafic magma and one has, perforce, to think in terms of anatexis of epicrustal rocks and of the sediments of the Transvaal System and older formations.

It may be noted that screens of country rock also preferentially occur between the quartz mangerite and anorthositic series of the Carthage and Marcy massifs (Buddington, 1969, p. 220) and of the Morin massif (Martignole and Schrijver, 1970, Plate 1). Composite sheets of leucogabbro and quartz mangerite are also explicable as subconformable successive intrusions. Each of the two rocks may also occur as independent sheets.

CONCLUSION

Plots of variation in composition of the major oxides and geologic relations of rocks of the associated anorthositic and quartz mangerite series of the Adirondacks can be interpreted to be consistent with the hypothesis of development from two independent magmas of gabbroic anorthosite and quartz mangerite composition, respectively.

ACKNOWLEDGMENTS

I am indebted to J. Martignole of the University of Montreal, Y. W. Isachsen and Philip Whitney of the New York State Geological Survey, and R. B. Hargraves for critical review of the manuscript.

REFERENCES CITED

Buddington, A. F., 1939, Adirondack igneous rocks and their metamorphism: Geol. Soc. America Mem. 7.
—— 1952, Chemical petrology of some metamorphosed Adirondack gabbroic, syenitic and quartz syenitic rocks: Am. Jour. Sci., Bowen Volume, p. 37–84.
—— 1953, Geology of the Saranac quadrangle, New York: New York State Museum Bull. No. 346.
—— 1957, Interrelated Precambrian granitic rocks, northwest Adirondacks, New York: Geol. Soc. America Bull., v. 68, p. 291–306.
—— 1969, Adirondack anorthositic series, in Isachsen, Y. W., ed., Origin of anorthosites and related rocks: New York State Museum and Science Service Mem. 18, p. 233–252.
Buddington, A. F., and Leonard, B. F., 1962, Regional geology of the St. Lawrence County magnetite district, northwest Adirondacks, New York: U. S. Geol. Survey Prof. Paper 376.
Chayes, F., 1970, On estimating the magnitude of the hidden zone and the composition of the residual liquids of the Skaergaard layered series: Jour. Petrology, v. 11, p. 1–14.
Davis, B. T. C., 1969, Anorthositic and quartz syenitic series of the St. Regis quadrangle, New York, in Isachsen, Y. W., ed., Origin of anorthosites and related rocks: New York State Museum and Science Service Mem. 18, p. 281–288.
de Waard, D., 1969, The anorthosite problem: The problem of the anorthosite-charnockite suite of rocks, in Isachsen, Y. W., ed., Origin of anorthosites and related rocks: New York State Museum and Science Service Mem. 18, p. 71–92.
—— 1970, The anorthosite-charnockite suite of rocks of Roaring Brook valley in the eastern Adirondacks (Marcy massif): Am. Mineralogist, v. 55, p. 2063–2070.
de Waard, D., and Romey, W. D., 1969, Chemical and petrologic trends in the anorthosite-charnockite series of the Snowy Mountain massif, Adirondack Highlands: Am. Mineralogist, v. 54, p. 529–537.
Emslie, R. F., 1971, Liquidus relations and subsolidus reactions in plagioclase-bearing systems: Ann. Rept. Director Geophys. Lab., Carnegie Institution, Washington, 1969–1970, p. 148–152.
Hargraves, R. B., 1969, A contribution to the geology of the Diana syenite complex, in Isachsen, Y. W., ed., Origin of anorthosite and related rocks: New York State Museum and Science Service Mem. 18, p. 343–356.
Herz, N., 1969, The Roseland alkalic anorthosite massif, Virginia, in Isachsen, Y. W., ed., Origin of anorthosite and related rocks: New York State Museum and Science Service Mem. 18, p. 357–368.
Isachsen, Y. W., 1969, Origin of anorthosite and related rocks—A summarization, in Isachsen, Y. W., ed., Origin of anorthosites and related rocks: New York State Museum and Science Service Mem. 18, p. 435–445.

Lebedev, P. I., 1937, Podolienne a czarnockites (Contributions a la petrographie du precambrian de l'Ukraine occidentale): Internat. Geol. Cong., 17th, Moscow 1937, Comptes Rendus, p. 71.

Letteney, C. D., 1969, The anorthosite-norite-charnockite series of the Thirteenth Lake dome, south-central Adirondacks, *in* Isachsen, Y. W., ed., Origin of anorthosite and related rocks: New York State Museum and Science Service Mem. 18, p. 329–342.

Martignole, J., and Schrijver, K., 1970, Tectonic setting and evolution of the Morin anorthosite, Grenville Province, Quebec: Bull. Geol. Soc. Finland, v. 42, p. 165–209.

Papezik, V. A., 1965, Geochemistry of some Canadian anorthosites: Geochim. et Cosmochim. Acta, v. 29, p. 673–709.

Philpotts, A. R., 1966, Origin of the anorthosite-mangerite rocks of southern Quebec: Jour. Petrology, v. 7, p. 1–64.

Reynolds, R. C., Whitney, P. R., and Isachsen, Y. W., 1969, K-Rb ratios in anorthositic and associated charnockitic rocks of the Adirondacks and their petrogenetic implications, *in* Isachsen, Y. W., ed., Origin of anorthosite and related rocks: New York State Museum and Science Service Mem. 18, p. 261–280.

Rose, E. R., 1969, Geology of titanium and titaniferous deposits of Canada: Canada Geol. Survey Econ. Geol. Rept. No. 25.

Wager, L. R., and Deer, W. A., 1939, The petrology of the Skaergaard intrusion, Kangerdluqssuaq, East Greenland: Medd. Grönland, Bs. 105, Nr. 4, 335 p.

Willemse, J., 1969, The geology of the Bushveld complex, the largest repository of magmatic ore deposits in the world, *in* Wilson, H. D. B., ed., Magmatic ore deposits: Econ. Geol. Publishing Co., p. 1–22.

MANUSCRIPT RECEIVED BY THE SOCIETY MARCH 29, 1971

THE GEOLOGICAL SOCIETY OF AMERICA, INC.
MEMOIR 132
© 1972

Petrology of Dikes Emplaced in the Ultramafic Rocks of Southeastern Quebec and Origin of the Rodingite

ANIRUDDHA DE
University of Calcutta,
Calcutta 19, India

ABSTRACT

The ultramafic rocks of the Appalachian region in the Eastern Townships of Quebec consist mainly of serpentinized harzburgite and dunite. They are intruded by dikes of dioritic rocks, quartz monzonites (adamellites), pegmatite, and albitite. Near contacts with the dikes the serpentinite with 7-Å serpentines has been transformed to a zone with 14-Å chlorites with higher Al/Si ratio; serpentinite may also be converted into diopside, talc, anthophyllite, and so forth, indicating dehydration and loss of water into the intruding dike. The granitic and dioritic rocks are modified especially at the borders to assemblages consisting of grossularite, diopside, vesuvianite, prehnite, zoisite, and calcite, forming a rodingite.

In general, the water-rich and silica-deficient environment of serpentinized ultramafic rock is responsible for affecting the normal sequence of crystallization of the dikes. This leads to the conversion of pyroxene to hornblende and finally to chlorite and (or) biotite; at the same time the crystallizing plagioclase becomes albitic. The lime and alumina that would normally form clinopyroxene or hornblende and plagioclase remain in the residual hydrous melt to form abundant lime-rich hydrothermal minerals. The lime in these rodingites is, therefore, not derived by serpentinization of the ultramafic wall rocks.

INTRODUCTION

Serpentinized ultramafic rocks of the Alpine type are commonly associated with intrusive dikes (white dikes) which show a wide variety of mineralogical and chemical compositions and include such special rock types as albitite and rodingite.

The large ultramafic bodies of the Eastern Townships of Quebec Province in Canada provide a major occurrence of this variety in the world. This paper is concerned with the original nature and the modifications of the "white dikes" resulting from their interaction with the ultramafic host rocks.

The ultramafic bodies trend parallel to the mountains forming the Appalachian range in Quebec (Fig. 1), and intrude the metasedimentary and metavolcanic rocks of the Caldwell group (Cambrian?) and the sedimentary and volcanic rocks of the Bauceville group (Ordovician). The ultramafic rocks consisting of more or less serpentinized harzburgite and dunite are intruded by a group of rocks ranging from gabbro to pyroxenite and dunite (Riordon, 1957). All these bodies intruded during the mid-Ordovician at the time of the Taconic orogeny (Hess, 1955). Finally a group of granitic and dioritic dikes and their modified equivalents, which are of interest here, were emplaced into the ultramafic rocks in the Devonian or later (Cooke, 1937).

Best exposures of these rocks are in the asbestos mines situated in the large ultramafic body of Thetford Mines–Black Lake. Thirty-five miles southeast of

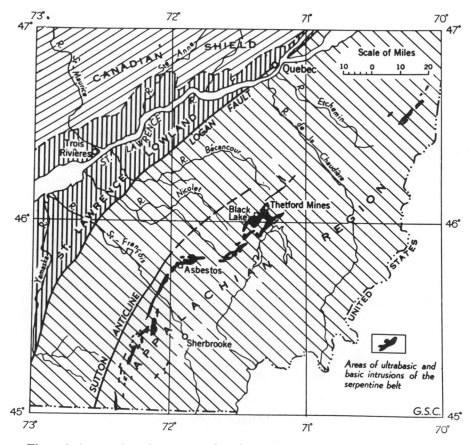

Figure 1. A map of southeastern Quebec showing the Appalachian region and the belt of ultramafic intrusions. The dikes investigated occur in the areas of Thetford Mines, Black Lake, and Asbestos (modified *after* Stockwell, 1957).

Thetford Mines on a large peridotite-dunite body is the Jeffrey mine, located in the mining town called Asbestos.

DIKE ROCKS

Granitic Rocks

The granitic rocks are the commonest dike rocks of the area. Granitic pegmatite dikes are also common. There are two main types of granitic rocks: (a) biotite-muscovite quartz monzonite, the most abundant granitic rock, particularly common in the Black Lake area; and (b) a leuco–quartz monzonite found in the Normandie mine, Black Lake, and hence termed the Normandie quartz monzonite. Several rock types have been produced by the desilication and subsequent albitization of the granitic rocks, out of which syenodiorite and albitite are noteworthy.

These dikes were emplaced in the epizone (Buddington, 1959, p. 677–690) by wedging along the shear zone, fractures, and joints in the serpentinized ultramafic rocks.

Both the above types of quartz monzonite consist of plagioclase An_{24-12}, potash feldspar Or_{77} $Ab(+An)_{23}$, and quartz. The micas are the important constituents, and a biotite of red-brown variety ($Ny = 1.643 \pm .001$; d (060) $= 1.543$ Å) is widely developed. A second generation of plagioclase, An_{0-10}, is commonly present.

The chemical, normative, and modal compositions of the granitic rocks show that the specimens of biotite-muscovite quartz monzonite and of the leuco-quartz monzonite are all typical quartz monzonites (adamellites); chemical analyses of two specimens (nos. 194 and 548) are given in Table 1. The normative quartz, orthoclase, and albite content of these two specimens (194 and 548), recalculated to 100, are 41.6, 33.0, 25.4; and 39.0, 26.8, 34.2, respectively. They fall close to

TABLE 1. CHEMICAL ANALYSES OF GRANITIC AND DIORITIC ROCKS

	194	548	9/144	570
SiO_2	71.98	76.78	41.31	47.16
TiO_2	0.34	0.09	0.81	0.88
Al_2O_3	14.09	13.37	16.36	20.82
Fe_2O_3	0.96	0.18	1.74	0.93
FeO	1.58	0.36	10.30	7.72
MnO	0.06	0.02	0.22	0.11
MgO	1.16	0.25	10.29	5.04
CaO	0.86	0.87	11.58	5.06
Na_2O	2.55	3.75	0.17	3.33
K_2O	4.75	4.25	1.72	3.12
P_2O_5	0.23	0.08	0.81	0.65
H_2O^+	1.49	0.79	4.32	4.94
H_2O^-	0.07	0.01	0.10	0.09
CO_2	0.00	0.00	0.04	0.00
Total	100.12	100.80	99.77	99.85

Analysts: H. B. Wiik (for 194, 548, and 570); J. Bouvier, Geol. Survey Canada (for 9/144).

194: Biotite-muscovite quartz monzonite, British Canadian mine, Black Lake.
548: Leuco-quartz monzonite, Normandie mine, Black Lake.
9/144: Hornblende-rich hornblende-biotite diorite, Vimy mine, Black Lake.
570: Hornblende-biotite diorite, Normandie mine, Black Lake.

the ternary minimum for 500 kg/cm² water pressure experimentally determined by Tuttle and Bowen (1958). This is consistent with the epizonal and magmatic nature of these granites.

None of these dikes shows any chilled contact against the serpentinized ultramafic rocks; instead a coarse-grained border facies is common for many granitic and pegmatitic rocks. In many cases large crystals of biotite are concentrated in this coarse-grained facies near the borders. These indicate that the dikes developed higher water pressure near the contacts during magmatic crystallization.

Hornblende-Biotite Diorite and Dioritic Biotite-Plagioclase Rock

The dioritic rocks are the oldest dike intrusions into the ultramafic rocks of the Black Lake, Thetford Mines, and Asbestos areas of Quebec. This group as a whole is characterized by saussuritized plagioclase and abundant brown biotite; a horn-blende-rich variety is also found in the Black Lake area.

The hornblende-biotite diorite contains porphyritic green hornblende, which poikilitically encloses small euhedral grains of plagioclase (normative composition basic andesine). The interstitial material consists of albite, quartz, clinozoisite, tremolite, prehnite, light greenish brown biotite, grossularite, sphene, chlorite, and microcline. Allanite, zircon, and apatite are common accessories.

The hornblende-biotite diorite and the hornblende-free dioritic biotite-plagioclase rocks of the Normandie and Vimy mines show every gradation in the field as well as in modal composition. The biotite-rich members of this gradational series are similar to the biotite-plagioclase rocks extensively developed in the Jeffrey mine at Asbestos.

CONTACT ZONES

The typical wall rock at a distance from a quartz monzonite or diorite dike is serpentinized peridotite or dunite in which chrysotile is the principal mineral. Antigorite, and in places picrolite, appear about 4 ft from the dike. Near the contact, commonly within 6 in. to 1 ft, the wall-rock serpentinite may be partly re-crystallized to antigorite and contain secondary asbestos fibers. At the contact with the quartz monzonite there is commonly a zone of dark-colored massive and fine-grained chlorite with some serpentine. This zone is generally 1 to 3 in. wide.

Next to the contact of this dark zone the dike shows a zone with abundant grossularite, diopside and (or) prehnite in places, with calcite and metahalloysite as well. These minerals have formed by replacement of normal minerals of the dike, and though both plagioclase and potash feldspar are present in this zone, quartz is commonly missing or subordinate.

A study of ten samples of dark contact zones, including one collected by H. H. Hess (oral commun., 1959) from the contact of a dikelet of grossularite in serpentinite occurring in Loma de Hierro, Central Aragua, Venezuela, showed that the major change in passing from the normal serpentinite into the dark contact zone is the first appearance of a common chlorite of penninite and clinochlore (Ny = 1.580 to 1.596 ± .001) variety with 14-Å basal spacing instead of the 7-Å serpentine, antigorite. The 14-Å peak of the x-ray diffraction pattern is totally absent in the serpentines beyond the dark zone. Concomitantly there is an increase in

A1 atoms replacing Si, as deduced from the data on the basal reflections (Brindley and Gillery, 1956). The chlorite forming at the contact of the grossularite-rich modified zone of the dikes is richer in aluminum, with an average of $Al_{0.65}Si_{3.35}$, $(d(004) = 3.574$ Å) in the tetrahedral position, while contacts against primary feldspar-rich borders of the dike and those against borders rich in diposide are much below $Al_{0.5}$ (corresponding to $d(004) = 3.620$ Å; *see* Fig. 2).

A consideration of the effect of granite or diorite magma intruding serpentinized peridotite shows that a temperature of 450° to 500°C could be reached at the contact principally by heat conduction (De, 1961). The pure magnesian serpentine,

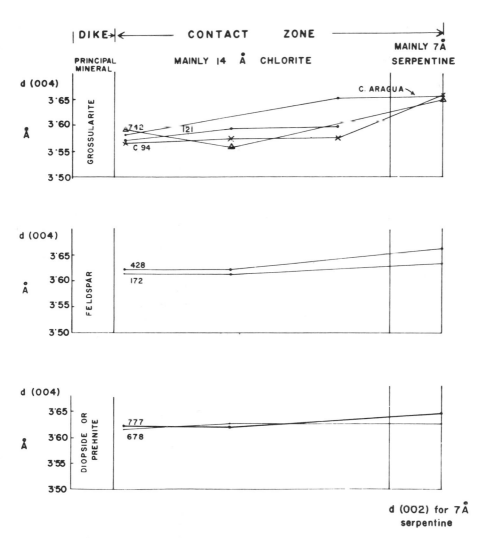

Figure 2. Variation in basal spacing of chlorite d(004) and serpentine d(002) in the dark contact zone formed against the rodingitic border of dikes. Specimens 121, 678, 742, and 777 are from contacts against granitic rocks, Black Lake. Specimens C94 and 172 are from contacts against dioritic rocks of Jeffrey mine and Thetford Mines respectively. Specimen from C. Aragua (Venezuela) is from contact against rodingite dike.

chrysotile, is unstable above 500°C (Bowen and Tuttle, 1949), but clinochlore has been synthesized by Yoder (1952) at temperatures between 520° and 680° at 2,000 to 30,000 psi; this upper limit of the stability of clinochlore is thus very close to the temperatures at which some granite magmas crystallize. The first appearance of the 14-Å chlorite out of septichlorites, as found in the present area at the contacts against the dike, may be significant in this respect. The width of the chlorite zone at the contact would also be dependent on the metasomatic introduction of alumina from the dike (*compare* Cooke, 1937), and the formation of these chlorites requires contact reactions in the presence of excess water (Yoder, 1952).

The abundance of water in these granitic and dioritic dikes has been evidenced by a profusion of hydrothermal minerals, such as kaolinite, metahalloysite, stilbite, pectolite, and montmorillonite, which replace large parts of these dikes. In some granitic dikes a porous texture indicates dissolution and removal of quartz and corrosion of plagioclase by the action of fluids. These alterations indicate that water was unable to escape from the cooling granite. Near the contact with the dense recrystallized chlorite zone, low temperature and high water-pressure conditions prevailed where the hydrothermal minerals became localized.

MINERAL ASSEMBLAGES AT THE CONTACTS, GARNETIZED DIKES AND RODINGITES

The contacts of the biotite-muscovite quartz monzonite and the Normandie quartz monzonite against the peridotite show profuse development of minerals rich in calcium (with or without aluminum) forming rodingitic rocks. The characteristic assemblages are as follows: (a) A grossularite-diopside assemblage is the commonest assemblage found at the contacts of all the granitic and pegmatitic rock types of the area; rarely the entire dike may be replaced by this assemblage. Chemical analyses of two specimens of such assemblages have been given in Table 2 (numbers 121A and A/31). The grossularite in specimen 121A shows $a_o = 11.847 \pm .003$ Å and refractive index $N = 1.733 \pm .002$. (b) A prehnite-calcite assemblage with minor diopside and grossularite is common in the biotite-muscovite quartz monzonite of the British Canadian mine and the adjacent Megantic mine. (c) Diopside aggregates also occur in the interstitial space between euhedral albite crystals in pegmatites in the Jeffrey mine. (d) Chrome diopside-calcite and chrome diopside-uvarovite ($a_o = 11.969 \pm .003$ Å) assemblages form 6- to 9-in. contact zones at the borders of the biotite-muscovite quartz monzonite in the British Canadian mine. (e) The serpentinite at the contacts of the Normandie quartz monzonite almost everywhere shows a 2- to 9-in.-wide zone in which the serpentinite has been metasomatically converted into nephrite-like fibrous aggregates of diopside with stringers of grossularite (Analysis 9/123C in Table 2).

Large crystals of quartz and fibrous diopside form veins along joints in Normandie quartz monzonite; the temperatures of formation of primary fluid inclusions in the quartz grains from two specimens occurring in separate locations in the dike are found to be 455° and 490°C (both corrected for a pressure of 800 bars). The same range of temperature also applies for the formation of diopside, as it shows petrographic evidence of simultaneous formation with quartz. It appears that the diopside in the calcium-rich rodingitic facies of the quartz monzonite occurring in

Table 2. Chemical Analyses of Lime-Rich Rodingitic Rocks
by Rapid Method

	121A	A/31	9/123 C
SiO_2	41.9	59.5	46.3
TiO_2	0.08	0.04	0.14
Al_2O_3	17.8	13.3	8.0
Fe_2O_3	<0.1	<0.1	0.2
FeO	1.6	0.8	3.2
MnO	0.31	0.28	0.22
MgO	4.0	2.1	16.2
CaO	28.1	16.5	19.3
Na_2O	0.1	5.5	0.4
K_2O	0.1	0.9	<0.1
P_2O_5	0.06	<0.02	0.27
H_2O_{Total}	3.5	0.6	3.5
CO_2	<0.1	<0.1	<0.1
Total	97.6	99.5	97.7

Analysts: Geol. Survey Canada, Rapid Method Group, Supervisor–S. Courville.

121 A: Rodingitic rock (grossularite, diopside, and methahalloysite) from border zone of biotite-muscovite quartz monzonite dike, British Canadian mine, Black Lake.

A / 31: Rodingitic rock from the border zone of andalusite granite, Jeffrey mine, Asbestos.

9/ 123 C: Rodingitic rock (nephritic diopside, grossularite, prehnite, and tremolite) from the border zone in serpentinite at contact with leuco-quartz monzonite, Normandie mine, Black Lake.

the neighborhood of these quartz-diopside veins may have formed in the same range of temperature.

Reactions which form anhydrous or less hydrous phases such as diopside, anthophyllite, talc, and similar minerals from serpentinite at the contacts release water which leads to higher water pressure inside the dike at the borders.

The hornblende-biotite diorite and the dioritic biotite-plagioclase rocks have been partially or very rarely completely replaced by hydrothermal minerals rich in lime. The lime-rich mineral assemblages show mainly four modes of occurrence as follows: (a) They are most commonly formed at the border of a dioritic or granitic dike at the contacts against the serpentinite wall rock, even when the dike itself may be completely unaffected in the core. (b) Commonly the mineral assemblages containing grossularite, diopside, prehnite, vesuvianite, zoisite, and chlorite occur as veins along primary joints in the dioritic biotite-plagioclase rocks in the Jeffrey mine and the Normandie mine. (c) The process of vein filling by grossularite, diopside, and other minerals commonly leads to some replacement of the vein walls. By intensification of this process the whole of the dike may be completely altered forming a rodingite. (d) Independent dikes consisting of grossularite and (or) diopside are also rarely found.

It is noteworthy that assemblages similar to those of the rodingites have been found at the contact between the ultramafic body and carbonaceous slates and phyllites in Asbestos. Their origin has been briefly discussed by the writer elsewhere (De, 1968); they appear to have formed in a manner different from that postulated in this paper for the origin of rodingitic rocks formed in the dikes.

ORIGIN OF THE CALCIUM-RICH ASSEMBLAGES

An increase in CaO in these contact zones between the dikes and the ultramafic rocks is indicated by the abundance of the lime-rich minerals. The source of lime here is a problem, because harzburgite or dunite contain very little CaO. Some peridotite xenoliths together with their anthophyllite reaction rim are completely enclosed by quartz monzonite in the Normandie mine, and also by pegmatite dikes in other mines. These xenoliths have been converted into diopside-rich assemblages, indicating that the lime-bearing solutions were available at the late stage of the crystallization of the granitic rocks. Though granitic rocks are normally poor in lime, several examples are known where they have formed lime-rich minerals in the late stages of crystallization; the formation of epidote-rich rocks like unakite (Jonas, 1935) and helsinkite (Laitakari, 1918), and lime-metasomatism of the wall rocks (Firman, 1957) by some granites are noteworthy in this respect.

Calcium occurs in the anorthite molecule of plagioclase in granitic and dioritic rocks, and is also a main constituent of hornblende present in the dioritic rocks. Owing to their high calcium content, mineralogical transformations involving plagioclase and hornblende are of considerable interest here.

The available experimental data indicate that anorthite is unstable under a wide range of hydrothermal conditions. Recent experiments by Adams (1966) show that anorthite is unstable in a hydrothermal environment from which SiO_2 and Al_2O_3 are being abstracted, in which case grossularite forms instead.

Ringwood (1959) made an examination of the role of high concentration of water, and particularly the behavior of OH^{-1} in silicate melts and its effect upon crystal fractionation.

> Ample experimental and theoretical evidence show that in crystals and silicate melts, ordering of OH^{-1} and O^{-2} anions between network forming Al^{+3} and Si^{+4} cations occurs. This results in Al^{+3} becoming preferentially co-ordinated by OH^{-1} ions, whilst Si^{+4} becomes co-ordinated by O^{-2}. . . . Aluminium is prevented from entering the anorthite component of plagioclase as AlO_4^{-5} groups, and consequently accumulates in the residual magma. Inhibition of anorthite crystallization causes Ca^{+2} and Si^{+4} to crystallize in the pyroxene or amphibole series. The net effect is to extend the field over which pyroxene or amphibole crystallize, and restrict the plagioclase field. At the same time, the plagioclase which finally crystallizes is soda rich. (Ringwood, 1959, p. 346.)

It is therefore concluded that anorthite crystallization may be inhibited under certain hydrothermal conditions leading to an enrichment of calcium and aluminum in the late-stage hydrous melt.

It is noteworthy that hornblende of dioritic rocks may be a source of calcium in cases of replacement of hornblende by biotite and (or) chlorite. The chemical analyses of the hornblende-rich diorite and hornblende-biotite diorite in Table 1 show that the gradation from the former to the latter is accompanied by a gain in SiO_2, Al_2O_3, K_2O, and Na_2O and a concomitant loss in total Fe, MnO, MgO, and CaO. The difference in CaO content is 6.5 percent. The transformation of hornblende to biotite is the main mineralogical change. Chemical analyses of co-existing hornblende and biotite of the hornblende-rich diorite give CaO contents of 11.5 and 1.5 percent respectively. Thus a transformation from hornblende to biotite or direct crystallization of biotite in place of hornblende would release a large amount of lime, giving rise to the diopside, prehnite, grossularite, and zoisite assemblages found in the interstitial spaces and also as veins in these rocks. The larger the potential hornblende content of the original magma the more lime would be released if biotite crystallized instead of hornblende. A pyroxene-bearing diorite or gabbro would also be suitable as an initial magma as clinopyroxene contains about 20 weight percent lime.

An important cause of the instability of hornblende might be high water pressure in the magma. The biotite-forming constituents appear to have been concentrated in the cooler and water-rich border zones of the dikes (*compare* Kennedy, 1955, p. 498–499), causing an inhibition of hornblende crystallization and the formation of biotite in place of hornblende with the release of lime. This also explains why the alteration is commonly confined to the borders of the dikes. A significant feature in some of the dioritic biotite-plagioclase rocks of the Jeffrey mine is the occurrence of a second generation of olive-brown-colored biotite (Ny = 1.636 ± .001), which entirely replaces the plagioclase in the manner of saussuritization, clearly showing that the biotite-forming constituents were mobile not only at the magmatic stage, as indicated earlier, but also in the deuteric stage of crystallization. It is noteworthy that the hornblende-bearing types contain only minor amounts of interstitial calcium-rich minerals, whereas the narrower dikes which are also hornblende-poor develop large amounts of the rodingitic assemblages.

ORIGIN OF THE RODINGITE

In the upper part of the Roding River valley in New Zealand, Marshall (*in* Bell and others, 1911) found numerous dikes of coarse-grained gabbro-like rocks penetrating the serpentinite. He called these *rodingite* because they differ from the gabbro in having a high percentage of lime as a result of the presence of a large amount of grossularite.

From a study of the rodingite or garnetized dikes of the type area in New Zealand, Grange (1927) found that they are comparable to the garnetized dikes described earlier from the present area in Black Lake, Quebec by Graham (1917).

Page (1967) studied serpentinized ultramafic rocks and found that CaO is lost during serpentinization. As pointed out by Condie and Madison (1969) there is still a "lack of quantitative geochemical and mineralogical data from a given ultramafic body relating the degree of serpentinization of a specific rock type to rock composition." In a systematic study of the Mayaguez serpentinized peridotite of the Alpine type, Hess and Otalora (1964) found that the maximum CaO content of serpentinized ultramafic rocks are as follows: lherzolite, 2.38; harzburgite, 1.65; and dunite 0.67 percent; and they concluded that with increasing serpentinization the Al_2O_3 and CaO show tendency to decrease (Hess and Otalora, 1964; Hess, 1964). They also observed that these constituents probably are deposited in fractured portions of the rock to form hydrogrossularite.

Dunite and harzburgite are the original ultramafic rock types in the ultramafic bodies of Thetford Mines–Black Lake and Asbestos; lherzolite has not been found. The CaO contents of two least-altered harzburgites are 0.9 and 0.5 percent and of dunite is 0.2 percent, whereas serpentinites have 0.2 percent CaO. These indicate that only a fraction of a percent CaO may have been lost during serpentinization of harzburgite; the loss of CaO by the dunite is not appreciable. It is also noteworthy that the calcium-rich pyroxene in these ultramafic rocks is comparatively resistant and persists until the advanced stages of serpentinization. Condie and Madison (1969) studied the progressive serpentinization of dunites of Webster-Addie and found that there is no systematic relation between the CaO and Al_2O_3 content of the rocks and the degree of serpentinization.

In view of the above and inasmuch as the dikes of the present area cut across the serpentinized harzburgite and dunite exposures without any change in the occurrence of rodingitic assemblages and alteration, the author (De, 1961, 1967) suggested that the lime and alumina were available in the dikes themselves.

That the lime needed to form the rodingitic assemblages originated elsewhere within the gabbroid dike rocks themselves has been recognized by Leech (1949), Grange (1927), Bloxam (1954), and Phemister (1964). Grange (1927) concluded that "the alteration of diallage will supply the lime and silica needed for the conversion of feldspar to garnet and prehnite. During this change albite in the feldspar will be set free."

More commonly however, the origin of the rodingitic rocks is ascribed to the metasomatism of the dikes of gabbro, dolerite, and other rocks by lime-rich solutions released by the ultramafic wall rocks at the time of serpentinization (Graham, 1917; Poitevin and Graham, 1918; Benson, 1918; Turner, 1933; Miles, 1950; Suzuki, 1954; Bilgrami and Howie, 1960; Thayer, 1966; and Coleman, 1966, 1967). Grange (1927) stated that the fine-grained spheroidal bodies of grossularite with minor amounts of diallage are formed by this process.

From a critical comparison of the garnetized dikes from a number of well-documented occurrences, the writer (De, 1961) found that in New Zealand (Grange, 1927), Gasquet quadrangle, California (Cater and Wells, 1953), Shulaps Range, British Columbia (Leech, 1949), and Hindubagh, West Pakistan (Bilgrami and Howie, 1960), the rodingites or the garnetized dikes originally consisted of a gabbro or dolerite containing clinopyroxene and plagioclase. Three separate stages of transformation of the primary mineral assemblages into the rodingitic assemblages could be distinguished, as given on Table 3, in which they are compared with the assemblages of the present area.

The nature of the above occurrences and the stages of alteration shown by the garnetized gabbros and the diorites indicate that the observed sequence is characterized by increasing pressure of water and (or) decreasing temperature. This ap-

TABLE 3. STAGES OF MINERALOGICAL TRANSFORMATIONS IN DIKES

	General sequence		Assemblages in the present area in Quebec	
Stage I	Augite/Diopside	Anorthite-rich plagioclase (An_{83-50})*	Not found	
Stage II	Hornblende	Andesine-Oligoclase (An_{50-20})*	Hornblende	Andesine
Stage III	Chlorite	Albite	Biotite Chlorite (=Amesite)	Albite

Stages II and III are accompanied by abundant grossularite, diopside, vesuvianite, prehnite, pectolite, and zoisite, while albite tends to form independent veins and patches. Abundant relict minerals are found in stages II and III. Hornblende has been recognized as a primary mineral in stage II of some occurrences.

* Indicates composition of plagioclase in Gasquet quadrangle, California (Cater and Wells, 1953).

pears to be related to the water-rich environment provided by the serpentinite in which the gabbro bodies crystallized.

Experiments on hydrothermal melting and crystallization of basalt show that the primary pyroxene becomes unstable at high water pressures and relatively low temperatures, being replaced by hornblende (Yoder and Tilley, 1956).

A necessary consequence of the conversion of primary diopsidic pyroxene to hornblende or chlorite and (or) biotite would be to enrich the residual fluid in lime. The anorthite molecule would also be unstable at the high water pressure, leading to the formation of prehnite, pectolite, vesuvianite, grossularite, zoisite, and to the releasing of the albite molecule. The lime-rich phases would be locally concentrated, depending upon the prevailing pressure-temperature conditions. This process therefore accounts for the common presence of albitite dikes in association with rodingites; this is because rodingite and albitite would be complementary products.

It is noteworthy that the presence of a water-rich environment provided by serpentinite, containing about 12 to 13 percent water by weight, would effectively control the nature of the phases crystallizing from a magma in the manner outlined above.

Podlike inclusions were found by Bowin (1966) in the Loma Caribe serpentinized peridotite of central Dominican Republic. These blocks are highly altered dioritic intrusive rocks consisting of altered andesine and hornblende (after pyroxene), prehnite, abundant grossularite, fibrous pyroxene (probably diopside) replacing amphibole, chlorite, and calcite. Bowin (1966) concluded that absorption of water by the dioritic magma from the serpentinized peridotite led to the modification of the dioritic intrusions according to the process previously suggested by the present writer. The mechanism suggested by the writer (De, 1961) is also consistent with the observations made by Phemister (1964) on the origin of rodingitic assemblages of Fetlar, Shetland Islands, Scotland. Phemister (1964) found that "the transformation which resulted in the rodingites of Fetlar was induced by residual solutions from the gabbroic rock which were effective in carrying off Si, Al and Na from the existing rock and possibly added Fe, Ca and Mg. Perhaps, and even probably these solutions were related in origin to those which were effective in the saussuritization of the gabbro." His conclusions are contrary to the view that the rodingitic transformation is genetically related to the serpentinization process.

The present investigation leads to the conclusion that rodingite is formed by the crystallization of gabbro, diorite, or less commonly granite, and by modification of the nature of the crystallizing minerals in the water-rich and silica-deficient environment of serpentinized ultramafic rocks.

ACKNOWLEDGMENTS

This paper is a part of the writer's Ph.D. thesis at Princeton University, 1961, carried out under the supervision of H. H. Hess and A. F. Buddington. Hess suggested the problem and supervised the investigation throughout. The writer is greatly indebted for his guidance in carrying out the petrological and x-ray diffraction work and for countless discussions on petrogenesis. Without his deep interest the present investigation would not have been possible. The writer is grateful to A. F. Buddington for his generous advice on petrological problems. Princeton University provided financial grants, including the Proctor Fellowship 1959–1960, to cover the entire expense of the writer's residence in Princeton and the field and laboratory investigations and chemical analyses.

The field work was possible through the co-operation of the asbestos mining companies of Quebec, for which the writer is thankful to P. H. Riordon, H. K. Conn, G. Rejohn, and M. Louzon. Thanks are due to C. H. Smith and T. N. Irvine for their interest in this investigation; the four additional chemical analyses reported here have been provided by the Geological Survey of Canada. Further work on the chemical petrology of the rocks of the area carried out by the writer during the tenure of a post-doctorate fellowship of the National Research Council of Canada at the Geological Survey of Canada during 1965–1967 will be reported elsewhere.

REFERENCES CITED

Adams, J. B., 1966, Compositional changes in plagioclase induced by hydrothermal leaching at high temperatures and pressure: Geol. Soc. America Abs. for 1965, Special Paper 87, p. 1.

Bell, J. M., Clarke, E. de C., and Marshall, P., 1911, The geology of the Dun Mountain subdivision, Nelson: New Zealand Geol. Survey Bull. 12, p. 29–40.

Benson, W. N., 1918, The origin of serpentinite, a historical and comparative study: Am. Jour. Sci., 4th Ser., v. 46, p. 693–731.

Bilgrami, S. A., and Howie, R. A., 1960, The mineralogy and petrology of a rodingite dike, Hindubagh, Pakistan: Am. Mineralogist, v. 45, p. 791–801.

Bloxam, T. W., 1954, Rodingite from the Girvan-Ballantrae complex, Ayrshire: Mineralog Mag., v. 30, p. 525–533.

Bowen, N. L., and Tuttle, O. F., 1949, The system MgO-SiO_2-H_2O: Geol. Soc. America Bull., v. 60, p. 439–460.

Bowin, C., 1966, Geology of central Dominican Republic (A case history of part of an island arc): Geol. Soc. America Mem. 98, p. 11–84.

Brindley, G. W., and Gillery, F. H., 1956, X-ray identification of chlorite species: Am. Mineralogist, v. 41, p. 169–186.

Buddington, A. F., 1959, Granite emplacement with special reference to North America: Geol. Soc. America Bull., v. 70, p. 671–747.

Cater, E. W., Jr., and Wells, F. G., 1953, Geology and mineral resources of the Gasquet quadrangle, California-Oregon: U.S. Geol. Survey Bull. 995–C, p. 92–104.

Coleman, R. G., 1966, New Zealand serpentinites and associated metasomatic rocks: New Zealand Geol. Survey Bull., v. 76, p. 1–97.

—— 1967, Low-temperature reaction zones and alpine ultramafic rocks of California, Oregon and Washington: U.S. Geol. Survey Bull. 1247, p. 1–49.

Condie, K. C., and Madison, J. A., 1969, Compositional and volume changes accompanying progressive serpentinization of dunites from the Webster-Addie ultramafic body, North Carolina: Am. Mineralogist, v. 54, p. 1173–1179.

Cooke, H. C., 1937, Thetford, Disraeli and eastern half of Warwick map-areas, Quebec: Geol. Survey Canada Mem. 211, p. 1–154.

De, Aniruddha, 1961, Petrology of dikes emplaced in the ultramafic rocks of southeastern Quebec [Ph.D. thesis]: Princeton, Princeton Univ., p. 1–201.

—— 1967, Origin of rodingitic assemblages in dikes emplaced in the ultramafic rocks of Quebec (Abs.): Am. Geophys. Union Trans., v. 48, p. 246.

—— 1968, Post-ultramafic dykes and associated rocks in the asbestos mining areas in the Eastern Townships, Quebec: Geol. Survey Canada Paper 68–1, pt. A, p. 108.

Firman, R. J., 1957, Fissure metasomatism in volcanic rocks adjacent to the Shap granite, Westmorland: Geol. Soc. London Quart. Jour., v. 113, p. 205–221.

Graham, R. P. D., 1917, Origin of massive serpentine and chrysotile asbestos, Black Lake–Thetford area, Quebec: Econ. Geology, v. 12, p. 154–202.

Grange, L. I., 1927, On the "Rodingite" of Nelson: New Zealand Inst. Trans. and Proc., v. 58, p. 160–166.

Hess, H. H., 1955, Serpentine, orogeny and epeirogeny, *in* Poldervaart, Arie, ed., Crust of the Earth: Geol. Soc. America Spec. Paper 62, p. 391–408.

—— 1964, The oceanic crust, the upper mantle and the Mayaguez serpentinized peridotite, in Burk, C. A., ed., A study of serpentinite, the AMSOC core hole near Mayaguez, Puerto Rico: Nat. Acad. Sci.–Nat. Res. Council Pub. No. 1188, p. 169–175.

Hess, H. H., and Otalora, G., 1964, Mineralogical and chemical composition of the Mayaguez serpentinite cores, *in* Burk, C. A., ed., A study of serpentinite, the AMSOC core hole near Mayaguez, Puerto Rico: Nat. Acad. Sci.–Nat. Res. Council Pub. No. 1188, p. 152–168.

Jonas, A. I., 1935, Hypersthene granodiorite in Virginia: Geol. Soc. America Bull., v. 46, p. 47–60.

Kennedy, G. C., 1955, Some aspects of the role of water in rock melts, *in* Poldervaart, Arie, ed., Crust of the Earth: Geol. Soc. America Spec. Paper 62, p. 489–503.

Laitakari, A., 1918, Einige Albitepidotegesteine von Südfinlande: Finlande Comm. Géol. Bull., no. 51, p. 1–13.

Leech, G. B., 1949, Petrology of the ultramafic and gabbroic intrusive rocks of the Shulaps Range, British Columbia [Ph.D. thesis]: Princeton, Princeton Univ.

Miles, K. R., 1950, Garnetised gabbros from the Eulamina district Mt. Margaret Goldfield: Western Australia Geol. Survey Bull. 103, p. 108–130.

Page, N., 1967, Serpentinization considered as a constant volume metasomatic process: A discussion: Am. Mineralogist, v. 52, p. 545–549.

Phemister, J., 1964, Rodingitic assemblages in Fetlar, Shetland Islands, Scotland, *in* Advancing frontiers in geology and geophysics, Krishnan Vol.: Hyderabad, Indian Geophys. Union, p. 279–295.

Poitevin, E., and Graham, R.P.D., 1918, Contributions to the mineralogy of Black Lake, Quebec: Geol. Survey Canada Museum Bull. No. 27, p. 1–103.

Ringwood, A. E., 1959, Genesis of the basalt-trachyte association: Beitr. Mineralogie u. Petrographie, Bd. 6, p. 346–351.

Riordon, P. H., 1957, The asbestos belt of southeastern Quebec, *in* The geology of Canadian industrial mineral deposits: 6th Commonwealth Mining and Metallurgical Cong., p. 3–8.

Stockwell, C. H., 1957, Geology and economic minerals of Canada: Canada Geol. Survey Econ. Geology Rept. no. 1, 4th ed.

Suzuki, J., 1954, On rodingitic rocks within the serpentinite masses of Hokkaido: Hokkaido Univ. Fac. Sci. Jour., Ser. IV, v. 8, p. 419–430.

Thayer, T. P., 1966, Serpentinization considered as a constant-volume metasomatic process: Am. Mineralogist, v. 51, p. 685–710.

Turner, F. J., 1933, The metamorphic and intrusive rocks of Southern Westland, Pt. II: New Zealand Inst. Trans. and Proc., v. 63, p. 256–284.

Tuttle, O. F., and Bowen, N. L., 1958, Origin of granite in the light of experimental studies in the system $NaAlSi_3O_8$–$KAlSi_3O_8$–SiO_2–H_2O: Geol. Soc. America Mem. 74, p. 1–142.

Yoder, H. S., Jr., 1952, The MgO-Al_2O_3-SiO_2-H_2O system and the related metamorphic facies: Am. Jour. Sci., Bowen Vol. 250A, p. 569–627.

Yoder, H. S., Jr., and Tilley, C. E., 1956, Natural tholeiite basalt–Water System: Ann. Rept. of the Director 1955–1956, Geophys. Lab. Paper no. 1265, p. 169–171.

Manuscript Received by the Society March 29, 1971

THE GEOLOGICAL SOCIETY OF AMERICA, INC.
MEMOIR 132
© 1972

Heat and Mass Transport During Crystallization of the Stillwater Igneous Complex

G. B. HESS

*Physics Department, University of
Virginia, Charlottesville, Virginia 22903*

ABSTRACT

During crystallization of a large intrusive igneous body such as the Stillwater complex, crystals formed in the magma settle and accumulate on the floor. The resulting layer of hot solid tends to block heat flow downward from the magma. In this paper, the rate of heat loss from the magma is calculated, taking into account the effect of the settled crystals. For likely values of thermal conductivity and other parameters, the heat flux into the floor is too small to crystallize the magma trapped in pore space between settled crystals before they are deeply buried. This conflicts with mineralogical evidence that significant exchange of ions between the pore liquid and the main body of magma occurred during crystallization of the former, and with field evidence that the unconsolidated layer was thin. On the relevant time scale, diffusion could be effective over distances no greater than a few centimeters. Possible convective transport in the pore space is discussed, but this conflict is not satisfactorily resolved. Some remarks are included on convection and crystal settling in the magma.

INTRODUCTION

Some years ago, Harry Hess asked me to collaborate in computing the rate of heat loss from the magma during crystallization of the Stillwater igneous complex in Montana. This work was not completed in time for inclusion in the Stillwater Memoir (Hess, 1960), and so an estimate was made there from published results for a somewhat different problem. In fact, most studies of thermal histories of intrusives (Lovering, 1935, 1955; Larson, 1945; Jaeger, 1957, 1959, 1964, 1968) are mainly concerned with external temperatures or with the period subsequent to solidification, and do not fully incorporate the effects of magma convection and

503

crystal settling. The Stillwater calculations are now completed, and some conclusions and their implications for the physical processes involved in Stillwater petrology are presented here.

Formulation of the appropriate thermal boundary value problem requires consideration of the effects of magma convection, crystal settling, and chemical diffusion. From a broader viewpoint, it is important to clarify the circumstances of these processes in petrogenesis. Therefore some remarks or speculations on these topics are included.

The Stillwater igneous complex was intruded as an extended sheet of basaltic magma about 8 km thick, under a roof of unknown thickness. After losing its superheat, if any, magma crystallized slowly as heat diffused into the overlying and underlying rocks. Although the greater part of the heat must have been lost to the roof, the crystals accumulated on the floor and formed a layer of mush, containing about 30 percent by volume of liquid-filled pore space (Hess, 1960; Jackson, 1961). The linear rate of crystallization was determined by the flux of heat out of the magma sheet. The heat flux into the floor determined the rate at which the mush was consolidated by crystallization of the pore liquid.

It is found (Hess, 1960, p. 109–113) that the final composition of the crystallized pore material is not that of the liquid which must have been trapped initially, allowing for possible reaction with the cumulus crystals. Rather, it is shifted strongly toward the composition of the first solid phases which would crystallize from such liquid. This implies that the pore liquid, as it crystallized, was able to exchange ions readily with the main body of magma. It is seen in episodes of slumping that the crystals were cemented (suggesting a little postcumulus crystallization) below a depth of 1 to 3 m from the top of the mush. Hess (1960) assumed that the total thickness of the mush was not much greater. Applying (essentially) the criterion that ionic diffusion would be effective if the diffusion coefficient exceeded the product of the mush layer thickness times its rate of advance by crystal accumulation, he concluded that diffusion might account for the observed composition shift, but only marginally. The principal task of this paper is to reexamine this conclusion.

Wager (Wager and Brown, 1968) draws a clear distinction between crystallization of pore liquid and enlargement in place (adcumulus growth) of the cumulus crystals forming the very top layer of the mush, resulting from direct contact with convecting supercooled magma. In extreme cases (adcumulus rocks) this process is supposed to proceed until practically no pore space is left, producing monomineralic rocks such as those found in the bronzitite and anorthosite layers of the Stillwater complex.

Such enlargement, if it occurs to a moderate degree, is not directly distinguishable from the original cumulus crystal. The pore space, as defined by Hess, is what remains after any such enlargement.

This paper is divided into sections dealing, respectively, with heat conduction from the magma, chemical transport in the pore space, adcumulus growth, convection in the bulk magma, and crystal settling.

HEAT CONDUCTION

Hess (1960) estimated that the magma temperature decreased from about 1,125°C at the start of crystallization to 1,100°C after solidification of 60 percent of the complex, which is the portion now exposed. More recent experiments by Yoder and Tilley (1962) on crystallization of natural basalts indicate that the initial liquidus temperature is more likely 1,200° or 1,235°C, and crystallization is nearly complete some 160° lower. If crystallization occurs at depth, the liquidus temperature is increased due to pressure by 2° or 3°/km.

For purposes of the conduction calculations, it is assumed that the bulk magma temperature had constant value $T_L = 1,225°C$ during crystallization, and that there was initially no superheat. The roof and floor rocks are assumed to have uniform thermal properties and to be initially at uniform temperature $T_o = 25°C$. No consideration is given to the normal geothermal gradient of perhaps 20°/km. The pore liquid is assumed to crystallize and to release its latent heat uniformly over a temperature range of 160°. Data of Wright and Weiblen (Shaw, 1969, Fig. 3) indicate that this linear approximation is reasonable. However, additional calculations have been made under the assumption that the intercumulus liquid crystallizes at the fixed temperature T_L. This latter model would be appropriate if an extremely rapid chemical transport mechanism were operative in the pore space, and it gives the thickness of mush required by purely thermal and mechanical considerations even in the absence of a finite temperature range of crystallization.

The parameter values used, which are derived primarily from Clark (1966), are given in Table 1. Subscripts refer to the various regions as follows: 0, magma; 1, roof; 2, original floor; 3, new rock; 4, mush. There is substantial uncertainty only in the values for conductivity, and this will be discussed later. Five dimensionless parameters (h, H, r, m, α) which appear in the conduction equations are also defined and evaluated in Table 1. Figure 1 shows the assumed geometry.

The Roof

The magma-roof contact is maintained at very nearly temperature T_L by con-

TABLE 1. Numerical Values Adopted for a Model of the Stillwater Intrusion

	(1) Roof	(2) Floor	(3) New Rock	(0) Magma
Density, $\rho(g/cm^3)$	2.65	2.65	3.05	2.60
Specific heat, c(cal/g-deg)	0.30	0.30	0.33	0.33
Conductivity, $k(10^{-3}$cal/cm-sec-deg)	5.0	5.0	6.5	6.5(?)
Thermal diffusivity, $\kappa(10^{-3}cm^2/sec)$ $\kappa = k/c\rho$	6.3	6.3	6.5	7.6(?)

Initial thickness of magma layer, $\ell_o = 8 \times 10^5$ cm
Initial liquidus temperature, $T_L = 1,225°C$
Temperature range of crystallization, $T_L - T_S = 160°$
Initial temperature of roof and floor, $T_o = 25°C$
Latent heat of fusion, $L = 100$ cal/g
Pore volume fraction, $p = 0.30$
Magma viscosity, $\eta = \nu\rho_o = 300$ poise
Magma volume expansivity, $\beta = 4 \times 10^{-5}$ deg^{-1}
Dimensionless parameters:
$H = c_3(T_L - T_o)/(\pi^{1/2}L) = 2.24$
$h = (k_1c_1\rho_1/k_3c_3\rho_3)^{1/2}H = 1.74$
$r = (k_2c_2\rho_2/k_1c_1\rho_1)^{1/2} = 1$
$m = (k_2c_2\rho_2/k_3c_3\rho_3)^{1/2} = 0.78$
$\alpha = (\kappa_3/\kappa_1)^{1/2} = 1.25$ (for $k_1 = k_3$)

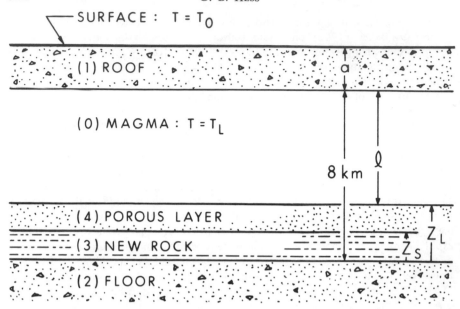

Figure 1. Vertical section of the intrusion during crystallization.

vection in the magma. Thus the heat flux into the roof does not depend on what is happening at the floor and may be obtained by an elementary calculation (Carslaw and Jaeger, 1959). If the heat flux from the bulk magma into the floor is neglected for the moment, the height $Z_L'(t)$ of the mush-liquid interface above the original floor as a function of time t after intrusion can be calculated with the assumptions that heat loss leads to rapid deposition of crystals on the floor and that these crystals pack to leave pore volume fraction p. For times $t < a^2/\pi\kappa_1$, where a is the roof thickness, the roof can be considered essentially infinite and the height of the mush-liquid interface is

$$Z_L'(t) = 2(1-p)^{-1} (k_1 c_1 \rho_1 t/\pi)^{1/2} (T_L - T_o)/L\rho_3 \qquad (1)$$
$$= 2(1-p)^{-1} h (\kappa_3 t)^{1/2}.$$

The dimensionless parameter h, defined implicitly here and also in Table 1, is a convenient measure of the heat flux to the roof, and the prime is a reminder that heat flux to the floor has been neglected.

For times $t > a^2/\pi\kappa_1$, assuming the upper surface is maintained at T_o, the roof has essentially steady-state temperature distribution, and Z_L' now increases linearly with t, rather than as $t^{1/2}$.

The Floor

The pore liquid is assumed to crystallize over the temperature range T_L to T_S (which we take to be 1,225° to 1,065°C). The latent heat can be treated as an additional specific heat over this temperature range, giving an effective specific heat for the mush region of

$$c_4 = c_3 + pL/(T_L - T_S). \qquad (2)$$

This expression neglects the possibility that the liquid phase might have a specific heat different from c_3. The boundary value problem for the floor is then a two-moving-interface problem, with changes in thermal properties but no latent heats at the two transitions. The interfaces are labelled $Z_S(t)$ between completely solidified rock and mush and $Z_L(t)$ between mush and free liquid. This problem is similar to a problem for which Carslaw and Jaeger (1959) give a solution, except that one boundary condition at Z_L must be modified to allow for the contribution due to heat loss to the roof, as given by equation (1). An analytic solution is possible because both interfaces and all isotherms advance from the original floor ($z = O$) as $t^{1/2}$, so long as the roof is effectively infinite so that equation (1) is valid. Thus one can write

$$Z_L(t) = 2 \, g(\kappa_3 t)^{1/2}, \; Z_S(t) = 2 \, f(\kappa_3 t)^{1/2} \tag{3}$$

and obtain coupled algebraic equations for the constants f and g, which are as follows:

$$g = h(1 - p)^{-1} + A \exp(-\alpha^2 g^2) \, [\text{erfc}(\alpha f) - \text{erfc}(\alpha g)] \tag{4}$$

with $A = c_4(T_L - T_S)/[\pi^{1/2} L(1 - p)\alpha]$ and $\alpha = (\kappa_3/\kappa_4)^{1/2}$; and

$$1 + m \, \text{erf}(f) = B \exp[(\alpha^2 - 1)f^2] \, [\text{erfc}(\alpha f) - \text{erfc}(\alpha g)] \tag{5}$$

with $B = (T_S - T_0)(T_L - T_S)^{-1} \, (mk_3/\alpha k_4)$. The error function is defined by

$$\text{erf}(x) = 1 - \text{erfc}(x) = 2\pi^{-1/2} \int_0^x \exp(-y^2) \, dy.$$

If the parameter αh is appreciably larger than unity, as it is for the values in Table 1, then the temperature gradient at the top of the mush becomes negligibly small, and correspondingly the second term on the right of equation (4) becomes negligible. In this case, $g \approx h(1 - p)^{-1}$ or $Z_L(t) \approx Z'_L(t)$, which is given by equation (1). The lower interface of the mush is obtained by solving equation (5) for f, which is easily done by successive approximations using Newton's method. With $\rho_4 = \rho_3$ and $k_4 = k_3$, the results are $g = 2.49$ and $f = 0.24g$, so that $Z_S(t) = 0.24 \, Z_L(t)$. Thus consolidation of the mush falls far behind its accumulation.

Measurements of high-temperature thermal conductivity in samples of basalt and ultramafic rocks (Kawada, 1966; Murase and McBirney, 1970) indicate that the conductivity is often nearly independent of temperature over the range 400° to 1,100°C, due apparently to compensating variations of the lattice and radiative contributions (Fukao and others, 1968). In the melting range, the conductivity increases rapidly, evidently because the radiative mean free path is larger in the liquid phase. To evaluate the effect this might have on the locations of the interfaces and on the temperature distribution in the mush, I have evaluated g and f also for the rather extreme case $k_4 = 2k_3$. (Of course, the actual conductivity is a continuous function of temperature.) In this case, $g = 2.50$ and $Z_S = 0.20Z_L$. The horizon z_N at which N percent of the intercumulus material has crystallized is given in the present model by

$$z_N = 2\zeta(\kappa_3 t)^{1/2} \tag{6}$$

with ζ given by

$$\frac{N}{100} = \frac{T_L - T}{T_L - T_S} = \frac{\text{erfc}(\alpha\zeta) - \text{erfc}(\alpha g)}{\text{erfc}(\alpha f) - \text{erfc}(\alpha g)}. \tag{7}$$

Figure 2 shows the fraction of intercumulus material which has crystallized (which is proportional to the temperature) as a function of height above the original floor. It is apparent that even a small degree of crystallization occurs only deep in the mush. The higher estimate of mush conductivity lowers slightly the horizon Z_S of complete crystallization, but raises the horizons for small percentages of crystallization.

If the roof reaches steady state, the interface Z_L advances more rapidly than given by equation (3), but the isotherms in the mush are virtually unaffected.

The rate of advance of the mush-liquid interface can be obtained by differentiating equation (3) [or, to a good approximation, equation (1)], so long as the roof is effectively infinite. This result is most usefully given as a function of height Z_L:

$$dZ_L/dt = 2g^2\kappa_3/Z_L = (25.4 \text{ cm/yr})/Z_L \text{ (in km)}. \tag{8}$$

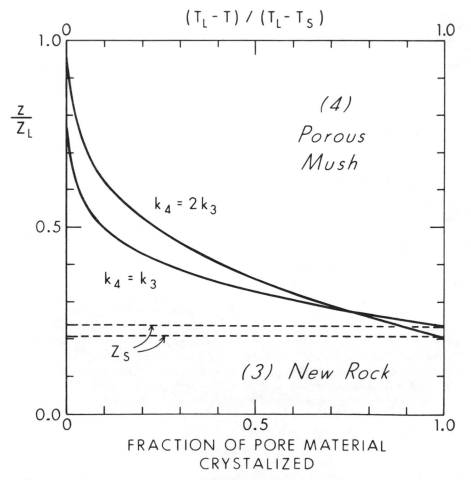

Figure 2. Height against fraction of the pore liquid crystallized, in the layer of crystals accumulated on the floor of the magma chamber. The two curves correspond to different assumed values of conductivity in the unconsolidated portion. The same curves with the upper scale give the temperature distribution. This model is not expected to be accurate in the region of nearly complete crystallization.

This is plotted in Figure 3, together with the results for several finite values of roof thickness. The rate of advance of the solid-mush interface as a function of Z_s is calculated the same way and is a factor $(f/g)^2$ smaller, or, for $k_4 = k_3$,

$$dZ_S/dt = (1.4 \text{ cm/yr})/Z_S(\text{in km}). \tag{9}$$

equation (9) will not hold in the upper portion of the complex. Z_L reaches the roof when $Z_S \approx 1.9$ km, and thereafter consolidation proceeds also from the top down.

Sharp Crystallization Temperature

The conditions under which thermal effects and the packing density of cumulus crystals alone lead to a thick mush layer can be determined by considering the case $T_L - T_S = 0$. In this case the mush-liquid interface $Z_L(t)$ is given by equation (1), and the solid-mush interface $Z_S(t)$ is determined by the rate at which heat flow into the floor can remove the latent heat of the intercumulus liquid. The solution of this boundary value problem is given by Carslaw and Jaeger (1959):

$$Z_S(t) = 2f(\kappa_3 t)^{1/2} \tag{10}$$

where f is the root of the equation

$$fp/rh = \exp(-f^2)/[1 + m \text{ } erf(f)] \tag{11}$$

with the constants defined previously. For the nominal values given in Table 1,

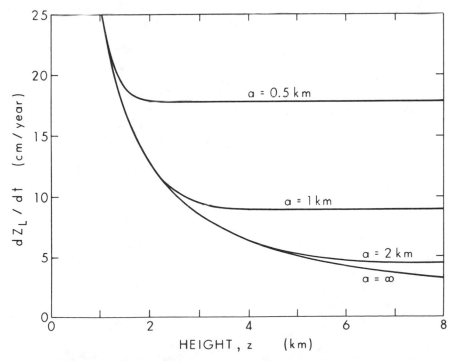

Figure 3. Rate of advance of the crystal mush-liquid interface as it passes height z above the base of the intrusion. Curves are given for several values of the thickness, a, of the roof over the magma chamber. Crystals are assumed to accumulate leaving 30 percent pore volume.

$f = 1.08$ and

$$Z_S(t)/Z_L(t) = f(1 - p)/h = 0.44.$$

This ratio can be forced to approach unity, corresponding to a very thin mush layer, by using a sufficiently large value for the conductivity k_3 of the new floor rock, relative to the conductivity k_1 of the roof. The required value is $k_3 > 8k_1$ if $k_2 = k_1$, or $k_3 > 6k_1$ if $k_2 = 2k_1$ (based on an assumed lower limit $H > 1.8$). Such large conductivity ratios are implausible.

CHEMICAL DIFFUSION

In view of the thermal results, the mineralogical evidence appears to require very effective chemical transport in the mush. Chemical transport by diffusion in the pore liquid can be estimated by extending the linear model of the crystallization range which has been used in the thermal calculation. Strictly speaking, one should consider chemical diffusion in the mush simultaneously with heat diffusion, as the former will alter the liquid-solid coexistence temperature and hence the effective value of T_S. However, chemical diffusion coefficients are so small compared to κ_4 that the effect is negligible. There is considerable uncertainty as to the appropriate value for the diffusion coefficient. Hess (1960) adopted the value $D = 2.9 \times 10^{-6}$ cm^2/sec measured by Bowen (1921) for mutual diffusion of diopside and laboradorite melts. This value is probably too large, as Bowen's measurements were made at 1,500°C. It might be a reasonable first approximation to assume that the activation energies for diffusion and viscosity are equal, in which case the product $D\eta/T$ would be independent of temperature. Measurements by Kani and Kozu (Clark, 1966) of the viscosity of $Di_2An_2Ab_1$ extrapolate to about 60 poise at 1,500°C and 2,400 poise at 1,200°C. The viscosity of basalt magma at 1,200°C is about 300 poise (Shaw, 1969). Thus an approximate allowance for 5 times larger viscosity yields the estimate $D = 5 \times 10^{-7}$ cm^2/sec (and $D/\kappa_4 = 1.2 \times 10^{-4}$, if $k_4 = k_3$). Data on diffusion of various ions in molten glass (*compare* Eitel, 1965) appear compatible with this estimate, although there are substantial differences for different ions and different glasses. New measurements on molten basalts would be useful. The effective bulk diffusion coefficient in the mush, D', is smaller than D by a tortuosity factor, which is probably on the order of 0.7 initially and decreases to zero near the bottom of the mush when interconnecting channels are pinched off.

The integrated change δX in intercumulus composition due to diffusion can be estimated, using a binary model and the thermal results given above. In the mush, the liquid must be in equilibrium with nearby crystal surfaces, and for a binary system this imposes a unique relation between liquid composition and temperature: $X = X(T)$. For simplicity it will be assumed that this relation is linear. The pore liquid initially has composition $X_L = X(T_L)$, and the last liquid has composition $X_S = X(T_S)$. The chemical flux is given by $-D'\,(dX/dT)\,(\partial T/\partial z)$, which may be evaluated using Equations (7) and (6). The integral of this flux with respect to time gives the total transport through each plane, and then differentiation with respect to z gives $-\delta X$. The result is

$$\delta X = (X_L - X_S)(D'/\kappa_4)F \tag{12}$$

with

$$1 + F = \pi^{-1/2}[(nf)^{-1}\exp(-n^2f^2) - (ng)^{-1}\exp(-n^2g^2)]/[\text{erfc}(nf) - \text{erfc}(ng). (13)$$

For nominal values and $k_4 = k_3$, $F = 0.50$; for $k_4 = 2k_3$, $F = 0.95$. Thus $\delta X/(X_L - X_S)$ is of order 10^{-4} rather than of order unity, as would be required to explain the mineralogical data. In fact, for the chemical diffusivity estimated above, effective chemical exchange requires a total mush thickness of no more than a few centimeters.

INTERCUMULUS CONVECTION

Natural convection may occur in the pore liquid, and may provide more effective chemical transport than diffusion alone. References to the theoretical and experimental literature on thermal convection in porous media are given by Elder (1967), whose numerical and experimental results will be cited here. Nield (1968) has calculated the onset of instability with simultaneous temperature and composition gradients.

In the present case the thermal stratification is stable, but comparable or larger density differences in the pore liquid may result from composition changes after partial crystallization. This density shift may have either sign, but at least in the ultramafic zone the pore liquid is progressively enriched in potential plagioclase. Thus the density decreases downward through the mush layer, and the pore liquid is potentially unstable. Results from thermal convection theory are easily transcribed to apply in this situation. The criterion for onset of convection is $R > R_c$, where the Rayleigh number for composition-driven convection is

$$R = \Delta\rho g \ell K / \eta D'. (14)$$

Here $\Delta\rho$ is the difference in liquid density across the layer, g is the acceleration due to gravity, is the layer thickness, K is the mush permeability, η is the liquid viscosity, and D' is the effective chemical diffusion coefficient. The critical value is $R_c = \pi^2$ in the case of a porous layer bounded below by a rigid surface and above by free liquid and with initially linear density profile. The actual boundary conditions may be somewhat different, but $R_c \approx 10$ is probably a good estimate. Note that the initial composition gradient is not established by the slow process of chemical diffusion but rather by heat diffusion and local fractional crystallization.

The permeability of the unconsolidated mush may be estimated from semi-empirical formulas (Carman, 1956). For pore fraction $p = 0.30$, a reasonable estimate seems to be $K = d^2/3,000$, where d is the mean particle diameter. Then, with $d = 0.2$ cm, $K = 1.3 \times 10^{-5}$ cm^2. Taking $\eta = 300$ poise (Shaw, 1969), $D' = 3.5 \times 10^{-7}$ cm^2/sec, and (arbitrarily) $l = 100$ cm, the requirement for the onset of convection becomes $\Delta\rho/\rho > 3 \times 10^{-4}$. By rough estimate, crystallization of enough olivine or pyroxene from the original magma to reduce the amount of liquid by 10 percent will produce an isothermal decrease in the density of the remaining liquid of 1 percent. The density change due to thermal contraction is of little consequence on account of the comparatively large diffusivity of heat (*compare* Nield, 1968); the medium will act as a counterflow heat exchanger.

Thus pore-space convection should occur whenever the course of fractional crystallization is reducing the liquid density (corrected to a fixed temperature) significantly. The remaining question is whether the rate of flow is appreciable. The average magnitude of the vertical component of velocity in the pore space can be estimated from Elder's (1967) results for steady convection as

$$v \approx (0.2) gK\Delta\rho/\eta p \qquad\qquad (15)$$

provided $R \gg R_c$. Note that this is independent of D' and l. (The factor $1/p$ appears because K is defined in terms of velocity averaged over the total volume.) The same expression up to a numerical factor can be derived from the balance of buoyant and drag forces on a discrete rising blob of liquid.

This velocity is to be compared with the rate of advance of the top of the mush, $v_L = dZ_L/dt$, or better, with the rate of advance of the horizon of 10 percent crystallization, $v_{10} = dz_{10}/dt$. As no effective mechanism of heat convection in the mush is predicted, the earlier heat conduction results should be valid. Thus intercumulus crystallization occurs long after emplacement of the cumulus crystals and is independent of the latter process. From Figure 2, $v_{10} = 0.5v_L$ (for $k_4 = k_3$). Then equation (8) gives $v_{10} = 4 \times 10^{-7}$ cm/sec when $z_{10} = 0.5$ km ($Z_L = 1$ km). If it turns out that $v \gg v_{10}$, the pore space would be well flushed with fresh magma until K is reduced by consolidation. If $v \ll v_{10}$, little upward displacement of differentiated pore liquid would occur.

On the basis of Hess's (1960) estimates of composition changes in the residual magma in the complex as a whole and measured liquid densities (Skinner, 1966), the magma density (corrected to fixed temperature) must have decreased by about 1 percent during crystallization of the first 9 percent of its volume, then remained nearly constant over the next 50 percent, and finally increased.

Consider a region at the middle of the ultramafic zone, 0.5 km above the base of the complex. Assuming the local pore liquid follows essentially the same course of differentiation as the bulk magma, the pore liquid density must decrease by about 0.5 percent as 5 percent or so of the pore liquid crystallizes. The above values in equation (15) give the estimate

$$v/v_{10} = 0.35 \text{ (at } z = 0.5/\text{km)}.$$

Higher in the complex, v_{10} is smaller, but $\Delta\rho$ is also smaller or of the wrong sign. Thus pore convection appears to provide significant chemical transport, but far less than thorough flushing of the pore space. This conclusion must be qualified somewhat because of uncertainty in the values of η and K.

It is also possible that larger density differences may occur due to failure of some solid phases to nucleate. For instance, if it is assumed in the preceding example that no plagioclase is nucleated, then after about 35 percent of the pore liquid has crystallized, the residual liquid will be nearly all potential plagioclase and will have a density about 5 percent less than the original magma. Then v/v_{35} may exceed unity.

As a second example, suppose no nucleation occurs in the initial stages of consolidation in a region in the anorthosite zone. The residual liquid will be enriched in potential pyroxene and will increase in density. At this epoch this region is underlain by a substantial thickness (more than 1 km) of incompletely consolidated mush, some containing pore liquid of lower density. Thus convection is likely to occur, producing some interchange of residual pore liquid between layers of differing cumulus composition.

Certain other mechanical processes deserve comment. Contraction of the intercumulus material on crystallization will draw additional magma down through the pore space, but this magma will have nearly the same composition as that originally trapped. Only displacement occurring after some differentiation will produce a shift in the final intercumulus composition. Figure 2 suggests that about 40 percent of the total intercumulus material in the region from the base of the complex to the horizon of 10 percent pore crystallization(z_{10}) remains uncrystallized at any epoch. If the average volume contraction on crystallization is 8 percent (*compare* Skinner, 1966) then residual pore liquid will be displaced downward. after significant differentiation by about 3 percent of the distance to the base of the complex. At the highest exposed portion of the complex, this amounts to 150 m. Thus this effect should shift the final intercumulus composition in regions of relatively small-scale layering. It could not explain, for instance, the deficiency of intercumulus pyroxene in the anorthosite subzones, which are 400 to 500 m thick.

On the other hand, the mush may be compressed by the nonhydrostatic stress due to overlying cumulus material, squeezing some of the pore liquid upward. If this occurs, it will have an effect opposite of the contraction just considered. Compression significantly above the horizon z_{10} will displace only undifferentiated magma. Compression much below z_{10} is unlikely because partial crystallization of pore liquid will rapidly strengthen the mush. It is unlikely that the strength of the mush would happen to be exceeded just near z_{10} except in a limited portion of the complex. Thus flow processes probably shifted the intercumulus composition in some layers of the complex, but the recognized processes appear inadequate to explain the observed composition changes.

SECONDARY ENLARGEMENT OF SETTLED CRYSTALS

The rate of crystal accumulation calculated above (for infinite roof) can be applied to study conditions for adcumulus growth. The mean diameter d of settled crystals is typically 2 mm (Hess, 1960; Jackson, 1961). The mean time τ that a crystal of this size remains in the top layer of the mush before being covered can be estimated from equation (8) as

$$\tau = d/(dZ_L/dt) = (2.6 \times 10^5 \text{ sec}) \times Z_L \text{ (in km)}. \tag{16}$$

The crystal will grow during this time if the adjacent magma is supersaturated. With supercooling of the order of 1°, ions must be exchanged with a volume of magma at least several hundred times the volume of crystal enlargement. The distance in the liquid over which diffusion is effective in time τ is approximately

$$\delta_D = (\pi D \tau)^{1/2} = (0.6 \text{ cm}) \times Z_L^{1/2} \ (Z_L \text{ in km}). \tag{17}$$

Thus secondary enlargement is possible in the top layer of crystals but generally not in buried layers. Whether enlargement proceeds to a significant extent depends on whether convection maintains supersaturated magma within a few millimeters of the mush-magma interface. (This condition becomes slightly less stringent higher in the complex.) Note that heat transport is no particular problem, as the analogous diffusion distance for heat is $(\kappa_4/D)^{1/2} \approx 90$ times longer than δ_D.

MAGMA CONVECTION

In order to clarify the process of crystal deposition, and especially to justify the boundary conditions assumed for the heat conduction calculation, it is desirable to establish the state of convection in the magma. Magma convection is complicated by the effects of crystal growth. Suspended crystals are probably present only in the lower portion of the magma chamber, so the theory of convection in a simple fluid should be applicable to the upper boundary layer. Hopefully it may also provide some insight into the more interesting problem of conditions near the lower boundary.

In the usual (Boussinesq) approximation, the nondimensional equations of viscous flow and heat transport in a horizontal sheet of fluid contain only two independent parameters, the Rayleigh number

$$R = 2\Delta T \, \beta g \, l^3 / \kappa_o \nu \tag{18}$$

and the Prandtl number $\sigma = \nu / \kappa_o$. Here $2\Delta T$ is the temperature difference between the boundaries of the sheet, β is the coefficient of thermal expansion, l is the thickness of the sheet, and $\nu = \eta / \rho$ is the kinematic viscosity. Convective instability occurs when R exceeds a critical value on the order of 10^3, depending on boundary conditions. For large R and $\sigma \gg 1$, ΔT is related to the average heat flux w by the semi-empirical equation

$$N \approx (0.1) R^{1/3} \tag{19}$$

where

$$N = wl / k_o 2\Delta T \tag{20}$$

is the Nusselt number. Equation (19) has the important property that l cancels out, which indicates that the dissipative processes limiting convection are predominantly localized in boundary layers.

In the present case, $\sigma = 1.5 \times 10^4$, $l = (8 \text{ km} - Z_L)$, and the heat flux into the roof, w, is given implicitly by equation (1). Then the product NR can be evaluated, and R and N can be found with equation (19). For $Z_L = 1$ km, $R \approx 10^{16}$ and $N \approx 2.3 \times 10^4$.

If $N \gg 1$, the horizontally averaged temperature is nearly constant in the interior of the fluid sheet, with the temperature difference ΔT appearing across a thin conduction layer at each boundary. The conduction layer thickness δ may be defined explicitly as $\delta = \Delta T / (dT/dz)_o$, where the denominator is the temperature gradient in the liquid approaching the boundary. By equation (20), $\delta = l/2N$.

For the parameters of the Stillwater magma,

$$\Delta T = (0.4°) Z_L^{-3/4} \ (Z_L \text{ in km}) \tag{21}$$

and

$$\delta = (15 \text{ cm}) Z_L^{1/4} \ (Z_L \text{ in km}). \tag{22}$$

Thus ΔT decreases from about 2.3° at the base of the ultramafic zone to 0.4° at the top of the ultramafic zone and 0.12° at the top of the exposed portion of the complex. The conduction layer thickness δ remains on the order of 15 cm. The fractional density difference required to drive convection is no greater than

$$(\Delta \rho / \rho)_{max} = \beta \Delta T \approx 2 \times 10^{-5} \text{ (for } Z_L = 1 \text{ km)}. \tag{23}$$

Estimation of velocities requires a more detailed theoretical model. Direct analytic or numerical methods are not practical at very large Rayleigh number. Experiments by Krishnamurti (1970) indicate that convection is both three dimensional and time dependent for $R > 6 \times 10^4$. Various model calculations (Kraichnan, 1962; Robinson, 1967) and experiments (Willis and Deardorff, 1967) suggest that convection will be fully turbulent when R exceeds, very roughly, $3 \times 10^4 \sigma^{3/2}$, which in the present case is about 6×10^{10}. Thus convection should be turbulent until Z_L becomes quite large, and a simple mixing-length model for turbulent convection proposed by Kraichnan (1962) may be applicable.

Kraichnan's model is based on the hypothesis that eddy transport at a distance z from a boundary is dominated by eddies of size of order z. He finds that outside the conduction layer there is a viscous boundary layer extending to $\delta_\nu \approx (10\sigma)^{1/2}\delta$, in which eddy conductivity but molecular viscosity are dominant. In an inertial zone between δ_ν and the center of the sheet, eddy conductivity and eddy viscosity dominate. The root-mean-square value v_z of the vertical component of velocity is proportional to the distance z from the boundary, within the viscous boundary layer. In the present application, the approximate magnitude is

$$v_z/z \approx (1 \times 10^{-4} \sec^{-1}) Z_L^{1/2} (\text{in km}) \quad (15 \text{ cm} < z < 60 \text{ m}). \quad (24)$$

The velocity increases more slowly as $z^{1/3}$ in the inertial zone, reaching about 1.8 cm/sec at the center. Several conclusions can be drawn:

1. The scale of eddy diameters is supposed to be comparable to z outside the conduction layer, so the eddy diffusivity for particles is of the order of zv_z, and hence is strongly dependent on distance from the boundary. Bartlett (1969) has calculated an equilibrium distribution for settling crystals, assuming a uniform eddy diffusivity.

2. The time to create a conduction layer from initially isothermal liquid is on the order of $\delta^2/\kappa_0 \approx 3 \times 10^4$ sec. The "overturn" time for the largest eddies, of diameter comparable to l, is very roughly $l/(v_z)_{max} \approx 4 \times 10^5$ sec. According to equation (8) the average thickness of crystals deposited in the latter time is 0.3 cm (for $Z_L = 1$ km). Thus igneous layering on any scale large compared to 1 cm is probably not correlated with simple overturn of a convection cell.

A more detailed picture of the boundary layer in turbulent convection has been proposed by Howard (1966). Howard suggested that a hot conduction layer forms in relatively quiescent fluid adjacent to the heated lower boundary, and, after a certain time, becomes unstable, collecting into blobs or "thermals" which break away and rise rapidly into the overlying fluid (and the analogous, inverted process occurs at the upper boundary). Each depleted region of conduction layer then reforms and repeats the cycle. Such instability is expected when a Rayleigh number based on the conduction layer $R_s = \Delta T \beta g \delta^3/\kappa_0$ exceeds a critical value of order 10^3. Foster (1965, 1968a, 1968b) has analyzed this initial instability problem in detail. With somewhat indeterminate numerical coefficients, his calculations lead to results similar to equations (19), (21), and (22) and also predict the separation λ of nearest neighbor thermals. In the present case, $\lambda \approx (200 \text{ cm}) Z_L^{1/4}$ (Z_L in km). Experimental results (Willis and Deardorff, 1967; Blair and Quinn, 1969; Foster, 1969; Sparrow and others, 1970) are consistent with Howard's model, and further indicate that the thermals, after rising a moderate distance, become unstable and mix with the surrounding fluid. Foster (1968b) has suggested that this may lead to subsequent instability of a larger layer of fluid, so that the process may cascade with increasing scales of length. If so, this would resemble Kraichnan's picture of the viscous boundary layer.

The principal implication of the foregoing is that convection near the boundaries is probably intermittent and localized, with peak velocities outward from the boundaries considerably exceeding the r.m.s. value given by equation (24).

None of this description of convection would be applicable if an appreciable density of crystals were present at the top of the magma chamber, as suggested by Hess (1960). A temperature difference of a few degrees between the interior of the magma and the upper boundary will drive convection with sufficient intensity to supply the heat removed at the roof. If the liquidus temperature increases 3° per km depth, then the magma in contact with the roof will be some 20° above its liquidus when crystallization is proceeding at the floor. Any crystals entrained by convection from the lower boundary region should be resorbed at a slightly higher level with little net effect on heat transport. However, suspended crystals could produce a stable density stratification, which might lead to progressive stagnation of the magma from the floor upward. Jackson (1961) has proposed such a mechanism of "variable-depth convection" to explain large-scale igneous layering. Only if this process continues nearly to the roof will two-phase convection occur in the upper boundary layer.

CRYSTAL SETTLING

The primary transport process in the lower boundary layer is crystal settling rather than heat conduction. In this section, the equations of crystal growth and settling are used to obtain a limited description of conditions in the boundary layer. The temperature difference across this layer is probably fixed by the supercooling required to produce sufficient nucleation to release latent heat at a rate equal to the heat flux into the roof. The thickness of the boundary layer is indeterminate until the complex interaction of convection with crystal growth and settling is understood. It is possible, for instance, that the zone of crystal nucleation and growth could be underlain by an arbitrary thickness of saturated magma, stabilized by the load of settling crystals.

No attempt will be made here to construct a model of the lower boundary layer (*compare* Jackson, 1961). However, several conclusions can be drawn simply from the observation that the typical diameter d of a settled crystal is about 2 mm:

1. The settling velocity of a pyroxene or olivine crystal of this size is about $v_c = 5 \times 10^{-3}$ cm/sec if $\eta = 300$ poise (Hess, 1960). This is just 3 times the r.m.s. convection velocity in the zone of maximum supercooling, as estimated by equation (24) with $z = \delta$.

2. The numerical density of settling crystals just above the floor can be estimated by comparing the settling velocity v_c with the rate of accumulation dZ_L/dt. The fractional increase in density of the magma due to these crystals is

$$\frac{\Delta\rho}{\rho} = \frac{(\rho_c - \rho_o)}{\rho_o}\frac{(1-p)}{v_c}\frac{dZ_L}{dt}$$

where ρ_c is the crystal density and ρ_o the magma density. With Stokes's law, this becomes

$$\frac{\Delta\rho}{\rho} = \frac{18\nu}{gd^2}(1-p)\frac{dZ_L}{dt}. \tag{25}$$

Substitution of equation (8), $d = 2$ mm, and kinematic viscosity $\nu = 115$ cm²/sec yields

$$\Delta\rho/\rho = (3 \times 10^{-5})Z_L^{-1} \ (Z_L \text{ in km}). \tag{26}$$

This is comparable in magnitude to the density difference required to drive convection, as given by equation (23), and so may be sufficient to stagnate convection in the region of crystal growth.

3. The growth of crystals during settling has been considered by Shaw (1965), but further results can be extracted. Frank (1950) has given a mathematical treatment of the growth of a (spherically symmetric) crystal, controlled by diffusion of heat and solvent into an initially uniform binary medium. Heat diffusion is sufficiently rapid to have negligible effect in the present case, and Frank's results reduce to an equation given also (in thermal notation), by Carslaw and Jaeger (1959). The diameter d as a function of time is given by

$$\frac{d^2}{8Dt} = \frac{\Delta X}{X_L - X_S},$$ (27)

provided that κ/D is sufficiently large and that the right-hand side is much less than unity. Here X_L and X_S are the equilibrium concentrations of coexisting liquid and solid, respectively, and ΔX is the supersaturation of the bulk liquid, stated as excess concentration. This treatment assumes that crystal growth is limited by bulk diffusion rather than surface kinetics, and so gives an upper limit for the growth rate.

One might suppose that a settling crystal would grow more rapidly than a stationary crystal as a result of moving into fresh supersaturated liquid. The magnitude of this effect can be evaluated using an empirical relation for the analogous heat transfer problem (Eckert and Drake, 1959): The dimensionless heat flux Nu from a heated sphere moving through a fluid of Prandtl number $Pr = \nu/\kappa$ at low Reynolds number $Re = vd/\nu$ is given by

$$Nu = 2 + 0.37(Re)^{0.6}(Pr)^{1/3}.$$

The preceding calculation for a stationary crystal corresponds to $Nu = 2$. The analog of Pr in the present case is $\nu/D \approx 2.3 \times 10^8$. The Reynolds number for a 2-mm diameter crystal is $Re \approx 9 \times 10^{-6}$. Thus $Nu \approx 2 + 0.21$, and the rate of crystal growth used above is low by about 10 percent at maximum velocity, due to neglect of settling.

An order of magnitude estimate of the growth time t_c follows from the somewhat arbitrary assignment $X_L - X_S \approx 0.5$, $\Delta X \approx \Delta T/400°$ in equation (27). The supercooling ΔT is probably not large compared to $1°$, judging from nucleation data (*compare* Winkler's data, discussed by Shaw, 1965), whereas a value much smaller than given by equation (21) would require an excessive convection velocity. For these values,

$$t_c \Delta T \approx (25°)d^2/D \approx 2 \times 10^6 \text{ sec deg.}$$ (28)

Stokes's law and equation (27) give for the settling velocity as a function of time

$$v = \frac{4gtD(\rho_c - \rho_o)}{9\nu} \cdot \frac{\Delta X}{\rho_o} \cdot \frac{1}{(X_L - X_S)}.$$

If the supersaturation ΔX is constant along the path, the settling distance over which a crystal grows to diameter d is then

$$l_c = \frac{gd^4}{288\,D\nu} \cdot \frac{(\rho_c - \rho_o)}{\rho_o} \cdot \frac{(X_L - X_S)}{\Delta X}$$ (29)

and for the values used previously,

$$l_c \; T \approx 5{,}000 \text{ cm deg.} \qquad\qquad (30)$$

If $\Delta T \approx 1°$, l_c is much larger than the thickness of a conduction layer. This suggests that a correspondingly thick stabilized layer of supercooled magma was present, or else crystals were convected through a significant portion of the magma layer before settling out.

CONCLUSIONS

1. The rate of crystal deposition on the floor of the magma chamber is evaluated and is given by equation (8) and Figure 3.

2. Heat conduction calculations show that, if the settled crystals form a mush with about 30 percent pore volume, then crystallization of the pore material occurs only long after deposition (*see* Fig. 2). Consequently, ionic diffusion between crystalizing pore material and the bulk magma is not possible. This conclusion is in conflict with mineralogical data described by Hess (1960). No satisfactory way to resolve this conflict has yet been found.

3. Concentration-driven convection in the porous layer is expected in some zones of the complex and would shift the final composition of the pore material, but was probably too slow to resolve the conflict cited in (2).

4. Occurrence of substantial secondary enlargement of crystals while on the top of the mush implies rather intense convection near the interface.

5. Parameters describing convection in the magma layer are estimated, including temperature and velocity distributions in the upper portion.

6. Several parameters relating to the lower boundary region of the magma are inferred from the observed sizes of settled crystals. In particular, it is shown that the crystals must have settled through many meters of supercooled magma.

After this manuscript was completed, I became aware of a recent paper by Irvine (1970) which deals with the same thermal conduction problem as the first part of this paper, together with a number of variants. It may be appropriate to discuss several additional effects treated by Irvine, in so far as they affect conclusions (1) and (2) above.

A. Most important is the gradually decreasing temperature of the bulk magma, for which Irvine gives an approximate treatment. For his model b and crystallization range $T_L - T_S = 160°$, the rate of crystal accumulation Z_L' is reduced by 41 percent (the effective value of the parameter h is reduced from 1.74 to about 1.03). This effect is partly due to reduced heat flux into the roof and in greater part because some of this flux comes from specific heat instead of latent heat. Then the coefficient in equation (8) should be reduced to about 9 cm/yr (but see below). The coefficient in equation (9) is reduced by only 4 percent. Conclusion (2) is weakened but not reversed; the conductivity ratio required, together with chemical convection, to prevent formation of a thick mush layer becomes $k_3/k_1 > 3$ (for $k_2 = k_1$). Also, there is a partially compensating effect not yet treated, in that a decreasing magma temperature will also reduce the downward heat flux in the mush.

B. Growth of an upper border zone would further reduce the accumulation rate. This is considered unlikely because of the probable magma temperature distribution discussed earlier and because of similarity to the Bushveld complex.

C. An endothermic metamorphic reaction or partial melting in the roof would increase the accumulation rate somewhat. This is probable, and will partially compensate for effect (A).

D. I have favored shallow emplacement. If the roof was thicker than 1 or 2 km, the bounding rocks would have significantly higher initial temperature T_0, and the accumulation rate would be reduced, according to equation (1).

ACKNOWLEDGMENTS

I have benefited from extended discussions with Harry Hess on matters relating to the Stillwater heat conduction problem. R. B. Hargraves kindly read a preliminary draft of this paper, and E. D. Jackson and H. R. Shaw provided valuable suggestions on the manuscript. This work was supported in part by the National Science Foundation through the Center for Advanced Studies, University of Virginia.

REFERENCES CITED

Bartlett, R. W., 1969, Magma convection, temperature distribution, and differentiation: Am. Jour. Sci., v. 267, p. 1067–1082.

Blair, L. M., and Quinn, J. A., 1969, The onset of cellular convection in a fluid layer with time-dependent density gradients: Jour. Fluid Mechanics, v. 36, p. 385–400.

Bowen, N. L., 1921, Diffusion in silicate melts: Jour. Geology, v. 29, p. 295–317.

Carman, P. C., 1956, Flow of gases through porous media: New York, Academic Press, 182 p.

Carslaw, H. S., and Jaeger, J. C., 1959, Conduction of heat in solids (2d ed.): Oxford, Clarendon Press, 510 p.

Clark, S. P., Jr., 1966, Handbook of physical constants (revised ed.): Geol. Soc. America Mem. 97, 587 p.

Eckert, E.R.G., and Drake, R. M., Jr., 1959, Heat and mass transfer (2d ed.): New York, McGraw-Hill Book Co., Inc., 530 p.

Eitel, W., 1965, Silicate science, Vol. 2: New York, Academic Press, 707 p.

Elder, J. W., 1967, Steady free convection in a porous medium heated from below: Jour. Fluid Mech., v. 27, p. 29–48.

Foster, T. D., 1965, Stability of a homogeneous fluid cooled uniformly from above: Physics Fluids, v. 8, p. 1249–1257.

—— 1968a, Effect of boundary conditions on the onset of convection: Physics Fluids, v. 11, p. 1257–1262.

—— 1968b, Haline convection induced by the freezing of sea water: Jour. Geophys. Research, v. 73, p. 1933–1938.

—— 1969, Onset of manifest convection in a layer of fluid with a time-dependent surface temperature: Physics Fluids, v. 12, p. 2482–2487.

Frank, F. C., 1950, Radially symmetric phase growth controlled by diffusion: Royal Soc. Proc. v. A 201, p. 586–599.

Fukao, Y., Muzutani, H., and Uyeda, S., 1968, Optical absorption spectra at high temperatures and radiative thermal conductivity of olivines: Physics Earth Planetary Interiors, v. 1, p. 57–62.

Hess, H. H., 1960, Stillwater igneous complex, Montana: Geol. Soc. America Mem. 80, 230 p.

Howard, R. H., 1966, Convection at large Rayleigh number, in Gortler, H., ed., Proc. 11th Internat. Cong. on Applied Mechanics: Berlin, Springer-Verlag, p. 1109.

Irvine, T. N., 1970, Heat transfer during solidification of layered intrusions: Pt. I: Sheets and sills: Canadian Jour. Earth Sci., v. 7, p. 1031–1061.

Jackson, E. D., 1961, Primary textures and mineral associations in the ultramafic zone of the Stillwater complex, Montana: U.S. Geol. Survey Prof. Paper 358, 106 p.

Jaeger, J. C., 1957, The temperature in the neighborhood of a cooling intrusive sheet: Am. Jour. Sci., v. 255, p. 306–316.

—— 1959, Temperatures outside a cooling intrusive sheet: Am. Jour. Sci., v. 257, p. 44–54.

—— 1964, Thermal effects of intrusions: Rev. Geophysics, v. 2, p. 443–466.

—— 1968, Cooling and solidification of igneous rocks: in Hess, H. H., and Poldervaart, Arie, eds., Basalts: The Poldervaart treatise on rocks of basaltic composition, Vol. 2: New York, John Wiley & Sons, Inc., p. 503–536.

Kawada, K., 1966, Studies of the thermal state of the earth, No. 17: Variation of thermal conductivity of rocks: Pt. 2: Tokyo Univ. Earthquake Research Inst. Bull., v. 44, p. 1071–1091.

Kraichnan, R. H., 1962, Turbulent thermal convection at arbitrary Prandtl number: Physics Fluids, v. 5, p. 1374–1389.

Krishnamurti, R., 1970, On the transition to turbulent convection: Pt. 2: The transition to time-dependent flow: Jour. Fluid Mechanics, v. 42, p. 309–320.

Larson, E. S., 1945, Time required for the crystallization of the great batholith of southern and lower California: Am. Jour. Sci., v. 243A, p. 399–416.

Lovering, T. S., 1935, Theory of heat conduction applied to geological problems: Geol. Soc. America Bull., v. 46, p. 69–94.

—— 1955, Temperatures in and near intrusions: Econ. Geology, Fiftieth Anniversary Vol., pt. 1, p. 249–288.

Murase, T., and McBirney, A. R., 1970, Thermal conductivity of lunar and terrestrial igneous rocks in their melting range: Science, v. 170, p. 165–167.

Nield, D. A., 1968, Onset of thermohaline convection in a porous medium: Water Resources Research, v. 4, p. 553–560.

Robinson, J. L., 1967, Finite amplitude convection cells: Jour. Fluid Mechanics, v. 30, p. 577–600.

Shaw, H. R., 1965, Comments on viscosity, crystal settling, and convection in granitic magmas: Am. Jour. Sci., v. 263, p. 120–152.

—— 1969, Rheology of basalt in the melting range: Jour. Petrology, v. 10, p. 510–535.

Skinner, B. J., 1966, Thermal expansion: Geol. Soc. America Mem. 97, p. 75.

Sparrow, E. M., Husar, R. B., and Goldstein, R. J., 1970, Observations and other characteristics of thermals: Jour. Fluid Mechanics, v. 41, p. 793–800.

Wager, L. R., and Brown, G. M., 1968, Layered igneous rocks: Edinburgh, Oliver and Boyd, 588 p.

Willis, G. E., and Deardorff, J. W., 1967, Development of short-period temperature fluctuations in thermal convection: Physics Fluids, v. 10, p. 931–937.

Yoder, H. S., Jr., and Tilley, C. E., 1962, Origin of basalt magmas: An experimental study of natural and synthetic rock systems: Jour. Petrology, v. 3, p. 342–532.

MANUSCRIPT RECEIVED BY THE SOCIETY MARCH 29, 1971

Mineralogy

THE GEOLOGICAL SOCIETY OF AMERICA, INC.
MEMOIR 132
© 1972

Pigeonitic Pyroxenes: A Review

G. MALCOLM BROWN
Department of Geology,
Durham University, Durham, England

ABSTRACT

The character and significance of pigeonite was first demonstrated by H. H. Hess in 1941. Extensive subsequent studies are reviewed in regard to crystallographic, chemical, and paragenetic relations among the calcium-poor members of the $CaSiO_3$-$MgSiO_3$-$FeSiO_3$ group of pyroxenes. Pigeonites of $P2_1/c$ and $C2/c$ symmetry comprise a link with $Pbca$ hypersthenes and $C2/c$ subcalcic augites, respectively, and constitute a vital key to our understanding of pyroxene relations. Recent structural and experimental thermal studies corroborate and refine Hess's views on the processes of subsolidus exsolution and inversion. The name "pigeonite" should be retained and extended to include new data on subcalcic, calcic, magnesian, and ferriferous variants.

INTRODUCTION

The vitally important contributions made by Harry Hess to an understanding of pyroxenes is well known to mineralogists. I began pyroxene studies under his guidance at Princeton (1954 to 1955) and am glad to have this opportunity to acknowledge the benefits gained from his ever-generous advice and encouragement.

Hess's work was on relations within the diopside-hedenbergite-enstatite-ferrosilite group, now familiar in terms of the "common pyroxene quadrilateral." He broke new ground in practically every aspect of this area, and in particular in establishing two-pyroxene relations, augite-series trends, subsolidus exsolution and inversion, and the chemical, optical, and unit-cell variations and interrelations. Before his classic paper in 1941, nobody had begun to show the sort of relations that now rank as primary indicators in igneous petrogenetic studies. Many workers have since added to or refined his data and observations. Hess was more than just one of a long line of pyroxene investigators, adding contemporary contributions of a quality to be expected of any competent mineralogist; he actually laid the foundations of modern pyroxene studies. Much valuable collaboration with the late Arie Poldervaart must be recognized, and studies by T.F.W. Barth, W. A. Deer, and the late Hisashi Kuno provided critical contemporary data, but Hess's vision was surely unique.

Hess coordinated a lot of diffuse data, and in part this characterizes his work on the augite trends and on the chemical and physical parameters. But the paragenetic relations *within* the quadrilateral were interpreted far in advance of experimental data which, even now, are only sparsely available.

The complex relations in the pyroxene quadrilateral hinge, to a very great extent, on the role played by pigeonite. Coexistence with augite (but not in alkali-basalt magmas), relations with more magnesian orthopyroxenes, exsolution and inversion to hypersthene, absence from the more ferriferous one-pyroxene field, and connection with subcalcic augites have since shown that pigeonites hold the key to adjacent phase-relations within the quadrilateral. The pigeonite crystal structure is still not fully understood, nor is the relation between its $P2_1/c$ space-group and the $C2/c$, $Pbca$, and $Pbcn$ space-groups of chemically similar pyroxenes.

Pigeonite (Winchell, 1900) was a rare subordinate member of the pyroxene group before Hess looked at it. He could refer to only two analyses of pure material (Hess, 1941), both characterized by a low optical angle and, more important, by about 9 per cent of Wo/Wo + En + Fs[1]. But he made a remarkable observation (Hess, 1941): that hypersthenes containing broad lamellae of augite, often arranged in herringbone pattern, should be viewed as "inverted pigeonites." He showed convincingly that their bulk wollastonite content indicated primary crystallization as monoclinic pigeonite, that subsolidus exsolution of augite took place along the (001) planes of the monoclinic phase, and that the orthorhombic hypersthene host was a result of subsolidus inversion. He had already postulated from optics (Hess and Phillips, 1938) that the (100) lamellae in orthopyroxene were due to exsolution of augite rather than to twinning, and this was confirmed by x-ray diffraction (Bown and Gay, 1960) and electron-probe analysis (Boyd and Brown, 1969). Hence, he emphasized subsolidus exsolution and inversion relations before similar relations in the feldspars were established by experimental and x-ray techniques. He could thus include inverted pigeonite data from gabbroic rocks, and progress to a consideration of orthopyroxene/pigeonite relations in cooling mafic magmas (Poldervaart and Hess, 1951). Many of the postulated pigeonite relations were demonstrated to hold true in detail in the single strongly fractionated Skaergaard intrusion (Brown, 1957). Atkins (1969) found the same to be true of the Bushveld pyroxenes, and the pattern was detected in several other basic intrusions (Wager and Brown, 1968). An excellent summary of Mg-Fe pyroxene relations has been provided by Smith (1969), the present review being aimed more directly at the additional problems introduced by a calcium component.

TEXTURAL RELATIONS AND OCCURRENCE

Monoclinic pigeonite in the uninverted state is often quoted as being present only as phenocrysts in lavas of basaltic or andesitic composition; however, it also occurs as rims to orthopyroxene phenocrysts in lavas (Lewis and White, 1967), and as primary crystals in sills (Hess, 1949) and slowly cooled plutonic complexes

---

[1] Wo, wollastonite; En, enstatite; Fs, ferrosilite.

(Brown, 1957). Hence, it may not always be a rapidly quenched product; but the plutonic occurrences are limited to ferriferous pigeonites that have a low inversion temperature (Fig. 2), and therefore the reaction rate is probably sluggish. Extra-terrestrial occurrences in meteorites are rare (Hess and Henderson, 1949; Mason, 1968), but a few Apollo 11 lunar rocks (for example, Weill and others, 1970) and many Apollo 12 lunar rocks (for example, Bence and others, 1970) contain pigeonite. It is rare in metamorphic rocks, but Bonnichsen (1969) has illustrated convincingly its presence in the inverted state in the Biwabik Iron Formation. Pigeonite has not so far been recorded from probable mantle-derived materials, although it appears to be stable to at least 20 kb (Brown, 1968, Fig. 10).

Inverted pigeonites are now accepted as such by most mineralogists, although only one reconnaissance-type experimental study of inversion relations has been attempted (Brown, 1968). The textural and chemical evidence for inverted pigeonites is the fact that, semiquantitatively (Brown, 1957; Atkins, 1969), analyses of hypersthenes with coarse augite lamellae give pigeonite bulk compositions (Wo_{8-10}). Probe analysis of host and lamellae should now be undertaken, supplemented by the method described by Ross and others (1969) for estimating the proportion of each phase. The main point is that all the analyses so far point toward a fairly uniform Ca content with more than a chance coincidence with the monoclinic pigeonite compositions.

The exsolution textures have been illustrated abundantly by Hess (1941, Fig. 3), Brown (1957, 1968), Wager and Brown (1968), and Bonnichsen (1969). The relict (100) twin composition planes of original monoclinic pigeonite crystals are preserved in the hypersthene crystals, in relation to which the augite lamellae mark out the original (001) planes of the pigeonites. Contrary to the conclusion reached by Poldervaart and Hess (1951), only rarely does the hypersthene retain the crystallographic axes orientation of the original pigeonite crystal. A case where the hypersthene has exsolved augite on (100) after the inversion episode shows direct illustration of a rotation of about 30° between this plane and the relict (100) plane of the pigeonite (Brown, 1957, Plate 17, Fig. 9). Studies involving a $C2/c$ phase are discussed later, but it is notable that Smyth (1970) found a 30° rotation of the silicate chain axis during inversion of heated orthopyroxene to $C2/c$ clinopyroxene. Morimoto and Tokonami (1969b) found both (001) and (100) exsolution of augite, termed augite (c) and augite (a), in uninverted pigeonite from the Moore County meteorite, and concluded that they were contemporaneous. Brown (1957, 1968) and Wager and Brown (1968) have shown cases of (010) and more complex exsolution patterns in inverted pigeonites, but the (010) exsolution is almost certainly from orthorhombic hypersthene, after inversion. Clearly, there is no single sequence of cooling events, although there are many cases that support Hess's concept of a sequence: (001) exsolution, inversion, (100) exsolution.

If Hess's view of the subsolidus phase relations is correct (1941, Fig. 13), augites coexisting with pigeonite should, on cooling, exsolve pigeonite while pigeonites are exsolving augite (c) on (001), and exsolve orthorhombic hypersthene while inverted pigeonites are exsolving augite (a) on (100). Pigeonite exsolution from augite is an acceptable concept, even though probe analysis generally shows the (001) lamellae in augite to be now clinohypersthene (subcalcic pigeonite), rather than pigeonite (Binns and others, 1963; Boyd and Brown, 1969). Boyd and Brown believe that pigeonite was exsolved initially, but that re-equilibration on cooling has led to the augite host becoming more calcic and the pigeonite guest less calcic. This is correlated precisely with the course followed by the coexisting pigeonite crystals, which became clinohypersthene hosts through Ca loss to the augite (exsolution lamellae) guests. The only difference thereafter was that the latter hosts inverted isochemically from clinohypersthene to orthorhombic hypersthene, whereas the clinohypersthene lamellae in augite remained uninverted. Since the (100) lamellae in augite are orthorhombic hypersthene (Bown and Gay, 1960) rather than clinohypersthene, we must conclude, as did Hess, that this stage of exsolution was at a lower temperature than the (001) exsolution.

CHEMICAL RELATIONS

The choice of a pigeonite field in the Di-Hed-En-Fs quadrilateral (Hess, 1941, Fig. 1) would seem to be a type example of "pigeonholing," if it were not for the crystallographic and phase relations that have subsequently confirmed the isolation of this compositional range (Fig. 1) from most of the rest of the quadrilateral species. Natural occurrences (for example, Muir, 1954; Bonnichsen, 1969) have extended the range to ferropigeonites, but only experimental studies at 20 kb pressure (Kushiro, 1969) have produced very magnesian pigeonites. There is a gradation, with Ca decrease, to what are usually termed clinohypersthenes (structurally, subcalcic pigeonites) at subsolidus temperatures, but the primary crystals are always of the more calcic pigeonite. With Ca increase, there might be a gradation to subcalcic augite, but natural terrestrial occurrences do not support this expectation (Kuno, 1955). We do not know enough about subcalcic augite

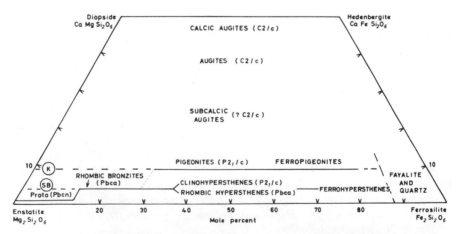

Figure 1. Common pyroxene quadrilateral showing compositional field of pigeonites. The adjacent pyroxene phases are shown with space-group symmetry in parenthesis. The proposed limitation of the protopyroxene field is after Brown (1968). The ferropigeonite field is after Lindsley and Munoz (1969, Fig. 7). The magnesiopigeonite extension (broken line to K) is related to synthesis by Kushiro (1969) at 20 kb only. Extension to the orthoenstatite (SB) from data by Sigurdsson and Brown (1970). Changes from *Pbcn* through *Pbca* to *P2₁/c* symmetry are shown in relation both to Ca and Fe²⁺ enrichment. Changes from *Pbca* through *P2₁/c* to *C2/c* may be a factor either of Ca enrichment or increasing temperature of quenching (for example, for hypersthene compositions around Fs50). Pressure effects must also influence the stability of the structural types.

(Wo_{15-30}), which is metastable below solidus temperatures. It may crystallize as either a calcic $P2_1/c$ pigeonite or a subcalcic $C2/c$ augite.

Pigeonites succeed orthopyroxenes in basaltic magma fractionation sequences, the change in primary crystalline phase generally occurring at an Mg:Fe ratio of about 7:3 in pyroxene composition. Hess (1941, Fig. 9) related this neatly to an intersection of crystallization and inversion curves, only modified slightly by Brown (1957, Fig. 5). Unfortunately this does not take into account the fact that at the temperature of crystallization, the Ca contents of the two phases are significantly different, although subsolidus exsolution prior to inversion leads to negligible differences. The fact remains that at the changeover, the monoclinic pigeonite is nearly twice as calcic as the orthorhombic bronzite, and the two may briefly coexist (Kuno and Nagashima, 1952). In contrast, Boyd and Schairer (1964) found, in the Ca-poor portion of the $MgSiO_3$-$CaMgSi_2O_6$ join, that the monoclinic phase was *less* calcic than the orthorhombic phase. This is due to the former being a quench product of protoenstatite; similar Ca contrasts are evident in a lava assemblage described by Dallwitz and others (1966). Chemical relations (Fig. 1) between proto-phases, ortho-phases, and clino-phases (pigeonite) were proposed by Brown (1968, Fig. 9). Unfortunately, it is at present impossible to decide whether the three-phase distribution is controlled primarily by Ca:Mg + Fe^{2+}, or by Mg:Fe^{2+} ratios. What is clear is that pigeonite ($P2_1/c$) stability is favored both by higher Fe and by higher Ca than found in either protopyroxene ($Pbcn$) or orthopyroxene ($Pbca$). In view of the findings of Perrotta and Stephenson (1965), Kushiro (1969), and Sigurdsson and Brown (1970), the effect of Ca (Fig. 1) seems to be the more important. This only applies to initial near-liquidus stability, since the conversion of pigeonite to clinohypersthene during exsolution of augite implies the temporary stability of a $P2_1/c$ structure with a very low Ca content at subsolidus temperatures. There seems little support for the stability of protopyroxene or orthopyroxene at temperatures higher than those for pigeonite (*compare* Yoder and others, 1963). Protohypersthene quenches to polysynthetically twinned clinopyroxene, but the latter could also have been derived, in the experiments, from the quenching of a high temperature C_2/c pyroxene, since diopside (Yoder and others, 1963) and hedenbergite (D. H. Lindsley, oral commun., 1967) also may quench to polysynthetically twinned aggregates.

CRYSTALLOGRAPHIC STUDIES

The problems of pigeonite crystal structure are related directly to those concerning solid solution and polymorphism within the quadrilateral. Structure refinements for the clinopyroxenes are now well advanced (Clark and others, 1969), as are the structural relations between orthoenstatite, clinoenstatite, and protoenstatite (Smith, 1969). Zussman (1968) has given a useful review of the crystal chemistry of pyroxenes.

Pigeonite has the $P2_1/c$ symmetry of clinoenstatite or clinohypersthene characteristic of two types of silicate-chain configuration (SiA and SiB), in contrast to the one type of chain for the $C2/c$ symmetry of diopside or augite. This, with the associated displacement of metal atoms in the M1 and M2 sites, gives rise to reflections of the $h + k =$ odd ($2n + 1$) category (Morimoto, 1956; Bown and Gay, 1957). If, as seems likely, this change in symmetry is a function of the amount of Ca relative to Mg or Fe^{2+} in the M2 site, we need to know how much Ca is tolerated in $P2_1/c$ pigeonites before they change to the $C2/c$ symmetry of subcalcic augites and augites.

Bown and Gay (1957) drew attention to the diffuseness of the characteristic pigeonite reflections ($h + k$ odd) in some specimens, and Morimoto and others (1960) postulated that the pigeonite structure is not the same as clinoenstatite in detail, but is intermediate between diopside and clinoenstatite. Morimoto and Güven (1968) viewed pigeonite as having a pseudo C lattice with a domain structure. The pigeonite examined (a lava phenocryst from Mull) showed appreciable cation-site ordering with Ca in M2, Mg in M1, and Fe^{2+} distribution between M2 and M1 of about 0.75:0.25. Morimoto and Tokonami (1969a) have now explained the pigeonite structure as a statistical averaging of domains, the domains having an anti-phase relation with $(a + b)/2$ shift, developed with the formation of crystallographically different SiA and SiB chains during transition from the C-centered lattice. Films with a diopsidic C-centered lattice between antiphase domains may have helped to suspend further growth of domains. Three important conclusions are: (a) the pigeonite domain structure grew from a high-temperature C-centered lattice, (b) the existence of a high-temperature orthorhombic phase is not so easily reconcilable with the structural interpretation, and (c) augite exsolution, involving extensive migration of metal atoms, would be more feasible in the C-lattice prior to the ordering of the SiA-SiB chains at lower temperatures. Smith (1969), Smyth (1970), and Prewitt and others (1970) also confirm that at high temperatures (perhaps as low as 700°C), transition from $P2_1/c$ to $C2/c$ symmetry occurs.

Augite and pigeonite have crystallized separately in several volcanic lavas, yet the diffuseness of the ($h + k$ odd) reflections in some of the pigeonite crystals suggests that the liquidus temperatures were high enough to have precipitated $C2/c$ pigeonite initially. Also, the sharpness of the reflections in certain plutonic pigeonites coexisting with separate augite may be related either to the higher Fe contents of the measured samples or to annealing-out of domains during slow cooling, rather than to the absence of a $C2/c$ stage in their crystallization history (Fig. 2). Hence, the existence of a $C2/c$ pigeonite at high temperatures has not, apparently, resulted in complete solid solution with $C2/c$ augite.

Virgo and Hafner (1969) have conducted a valuable study of order-disorder relations in heated Mg-Fe^{2+} orthopyroxenes. The low activation energies lead to a high degree of ordering through equilibration down to about 480°C. Their values for Fe^{2+} site occupancy numbers would suggest that the volcanic pigeonites referred to above (Morimoto and Güven) are relatively disordered, and in fact they comment on the probability of metastable distributions in volcanic pyroxenes. Even so, it would be important to have more data for order-disorder relations in Ca-bearing pigeonites from volcanic and plutonic (including meteoritic and lunar) environments.

Morimoto and Tokonami (1969b) have considered the mechanism of augite exsolution from pigeonite in terms of the lattice distortions and strain energies involved. They conclude that the (001) and (100) lamellae were exsolved simultaneously but that the slight strain energy differences would favor the former. Since strain energies for (010) or (110) exsolution are ten times greater, this mechanism should be precluded and therefore it seems that only orthorhombic hypersthene, *after* inversion from monoclinic pigeonite, has exolved augite on

(010) (Brown, 1957). One flaw in the strain-energy calculations is the assumption in regard to equivalent lamellae compositions, based on present compositions and unit-cell parameters. Boyd and Brown (1969) believe that the present compositional relations between host and lamellae is the result of successive re-equilibration with lowering temperatures, under subsolidus conditions. Hence the calculated strain energies should not necessarily correspond with those at the temperature of exsolution.

The structural history of pigeonites is clearly very complex, and a good range of natural samples is difficult to obtain. Experimental synthesis of compositions varying in Ca:Mg:Fe^{2+} across the pigeonite field is essential, together with studies on the structural and chemical changes associated with differing thermal regimes.

EXPERIMENTAL STUDIES

The conviction, held by petrologists, that monoclinic pigeonite would invert to orthorhombic hypersthene on cooling was momentarily shattered when a reverse thermal-structural relation was demonstrated for enstatite by Boyd and England (1965). Similarly, the hypothesis that the monoclinic phase was stable relative to the orthorhombic phase, because of its higher Ca content, was not apparently supported by Boyd and Schairer's work (1964) on the diopside-enstatite join. The natural and experimental observations, however, have been reconciled by the work of Munoz (1968) and others, who have explained the "low clinoenstatite" field according to the experimental conditions of shearing stress. Presumably, the absence of this phase in most metamorphic rocks is due to the persistence of temperatures above 600°C after shear-stress had ceased, thus permitting re-entry into the orthoenstatite stability field. In explaining Boyd and Schairer's data, Brown (1968) has proposed that when low-calcium clinopyroxene is the inversion product of protopyroxene, it is restricted in stability to the magnesian corner of the quadrilateral. Higher Ca would favor orthorhombic enstatite (Sigurdsson and Brown, 1970) and even more Ca would favor monoclinic magnesiopigeonite (Perrotta and Stephenson, 1965; Kushiro, 1969). However, at constant Ca but increasing Fe^{2+} (Fig. 1), the same sequence from proto → ortho → clino members of the hypersthene series apparently takes place, since clinohypersthene is the post-exsolution, pre-inversion product of cooled pigeonite (Boyd and Brown, 1969).

These complexities preclude any assumption that extrapolation from the side joins of the quadrilateral will suffice for an understanding of pigeonites. To the writer, the direct approach of studying inversion relations for natural pigeonitic compositions seemed the only way of eliminating problems in regard to minor-element control. The experiments (Brown, 1968, and Fig. 2) were designed to test Hess's hypotheses, so a volcanic monoclinic pigeonite and a plutonic "inverted pigeonite" of almost identical bulk composition were subjected to heating experiments. Difficulties at one atmosphere were overcome by achieving reaction in the presence of andesitic liquids, but the correspondence with data more easily obtained on separated crystals at high pressures suggests that future experiments could all be conducted in solid-media, piston-and-cylinder pressure apparatus and extrapolated below 5 kb. At one atmosphere, the clino ⇌ ortho inversion temperature for the pigeonites (Mg:Fe^{2+} about 0.45:0.55) is 980° ± 10°C. Examination of the formation and disappearance of augite blebs supported the hypothesis that either exsolution or homogenization took place above the inversion temperature and at ca. 1,020°C for maximum observable change. Pigeonites of this intermediate composition crystallize from andesitic (tholeiite-series) liquids, for which experiments showed entry of pigeonite on the liquidus at about 1,050°C (Fig. 2).

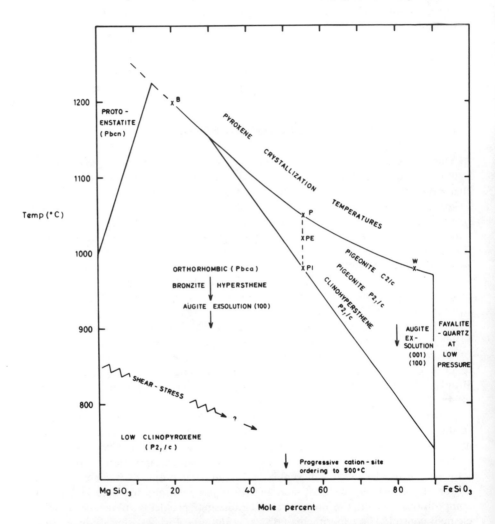

Figure 2. Hypothetical thermal relations, at atmospheric pressures, among the calcium-poor pyroxenes. See Figure 1 for effects of variable Ca content. Protoenstatite field probably limited, in natural magmas, at about Fe 15. Projection of low-clinopyroxene field to iron-bearing compositions not yet known. Data for bronzite (B) and pigeonite (P) from Brown (1968), including temperatures of notable exsolution (PE) and of inversion (PI). Crystallization temperatures for ferriferous pigeonites (W) assumed from data on ferroaugites (Lindsley and others, 1969). Subsolidus ordering to low temperatures is discussed in the text. The relations emphasize the narrow cooling range within which a C2/c pigeonite could crystallize, especially for magnesian compositions. This structural variant is more likely to occur amongst the ferropigeonite compositions (Fs 50-90). Depression of crystallization temperatures (as in some hydrous andesite magmas or in thermally metamorphosed rocks) would give primary crystallization of hypersthenes instead of pigeonites.

The P-T curve, although tentative, is clearly positive and shows the monoclinic phase to be lower in density than the orthorhombic phase (Brown, 1968). Smith (1969) accepts this critically different relation from the low clinopyroxene-orthopyroxene one, where the former is the higher density phase. Kuno (1966) introduced the term "high clinopyroxene" for what is really the pigeonite or clino-hypersthene that has been the subject of this paper. It was not a newly discussed phase or concept, but an attempt to refocus attention on the volcanic phase that could not be reconciled with Boyd and England's (1965) low-temperature clinopyroxene. If the latter is confined to shear-stress conditions then there is no question of a possible "high clinopyroxene-low clinopyroxene" interrelation in nature.

Intermediate pigeonites crystallize from andesitic magmas at about 1,050°C, exsolve augite around 1,020°C, and invert to hypersthene around 980°C. These experiments did not include a search for $C2/c$ phases, although the $P2_1/c$ reflections were sometimes weak on the diffractometer records and some inversion may well have taken place. It should be possible to restudy this system with more refinement, providing the high pressures necessary for reaction do not inhibit the formation of the $C2/c$ phase that could be stable in low-pressure lavas. If exsolution occurs from the $C2/c$ phase (Morimoto and Tokonami, 1969b) then inversion to $P2_1/c$ must occur in a narrow temperature interval between about 1,020 and 980°C (Fig. 2). Below the latter, further (100) exsolution from hypersthene and cation migration occur (Boyd and Brown, 1969), perhaps to about 500°C.

NOMENCLATURE

Natural terrestrial pyroxenes do not show complete miscibility between augites and pigeonites, despite the possibility of $C2/c$ symmetry throughout the quadrilateral above about 700°C (*compare* Prewitt and others, 1970). Coexistence of augite and pigeonite in both lavas and layered cumulates, or phenocrysts of one phase with subcalcic augite in the groundmass, indicate very limited miscibility at liquidus temperatures. However, pigeonites mantled by subcalcic augite need probe study to establish the extent of continuity. The lunar clinopyroxenes show, especially for some Apollo-12 specimens (for example, 12019, 12051, 12040, 12065), pigeonite zoned toward an augite mantle, but our studies show a compositional break, as also described by Bence and others (1970). Some of the lunar pigeonites may, especially in view of the metastability of the total mineral assemblage, eventually reveal a quenched $C2/c$ symmetry. If so, factors other than liquidus temperatures, such as the pyroxene minor-element chemistry or the magma bulk chemistry, may influence the crystal symmetry of the pyroxenes precipitating at liquidus temperatures. Intermediate pigeonites have high melting temperatures at one atmosphere of 1,330°C (Brown, 1968) to 1,380°C (Yoder and others, 1963), compared with about 1,300°C for augites with comparable Mg:Fe^{2+} ratios (Yoder and others, 1963). Hence zoning should generally be from pigeonite to augite, rather than the reverse. Ross and others (1970) find that the fine-scale pigeonite lamellae in lunar clinopyroxenes now have $P2_1/c$ symmetry but the bulk compositions (subcalcic augite field) suggest higher temperature, extensive miscibility with $C2/c$ augite.

It would appear that extensive solid solution between pigeonite and augite may occur in special environments, such that in bulk the compositions should be termed subcalcic augites. However, subsolidus division into $C2/c$ augite and a

$P2_1/c$ phase, with attendant cation migration between the phases, appears to give rise to convergence of $P2_1/c$ compositions on the low-calcium contents characteristic of "pigeonite." Hence terrestrial pigeonite, at least, seems to have a distinct chemical composition in regard to Ca:Ca + Mg + Fe^{2+} being close to a maximum of about 0.1. The $P2_1/c$ symmetry persists to lower Ca contents (clinohypersthene), so the term "pigeonitic" could apparently be extended to include "clinohypersthene." According to Morimoto and others (1960), the pigeonite structure is distinct from the clinoenstatite-type structure (*see* section on crystallographic studies, this paper) but increase in Fe/Mg may result in clinohypersthene approaching more closely the pigeonite structure. The existence of a stable $P2_1/c$ phase certainly justifies a contrast with subcalcic augite, which would probably have a $C2/c$ structure in its stable form.

The structural and phase relations between the Ca-poor pyroxenes of the common quadrilateral are much more complex than envisaged by Hess (1941). Even so, his concepts in regard to pigeonite were extremely well founded. The apparent absence of polymorphism amongst the commonest clinopyroxenes (augites) means that the terms "high clinopyroxene" and "low clinopyroxene" should not be used loosely. Nor should they be used to distinguish between either $P2_1/c$ pigeonites or subcalcic pigeonites (clinohypersthenes) formed under differing stress conditions, since the temperatures are only coincidental with the particular conditions required to promote or eliminate the effects of shearing stress (although "low clinopyroxene" is retained in Fig. 2 as an indicator of present concepts).

There is a case for revising the nomenclature based on the probability that $C2/c$ pigeonite is a high temperature form, and that some solid solution with $C2/c$ augite gives rise to the usually metastable, subcalcic augites. Equally, the $P2_1/c$ pigeonite is probably a slightly lower-temperature form that shows (from subsolidus exsolution data) complete solid solution with Ca-free, $P2_1/c$ clinohypersthene and thus accounts for the usually metastable subcalcic pigeonites (which invert to orthorhombic hypersthene at lower temperatures). This nomenclature is not shown on Figure 1 because of the need, first, for further studies on the stability and structure of this complex part of the quadrilateral and its relation to the magnesian members. Even so, it is probable that "C-pigeonite" and "P-pigeonite" comprise a critical link between the monoclinic augites and the orthorhombic hyperstheness, respectively, and that the name "pigeonite" should be retained and strengthened.

ACKNOWLEDGMENTS

I am grateful to M. G. Bown, J. Zussman, and L. Hollister for helpful and constructive comments during critical review of the manuscript.

REFERENCES CITED

Atkins, F. B., 1969, Pyroxenes of the Bushveld intrusion, South Africa: Jour. Petrology, v. 10, p. 222–249.
Bence, A. E., Papike, J. J., and Prewitt, C. T., 1970, Apollo 12 clinopyroxenes: chemical trends: Earth and Planetary Sci. Letters, v. 8, p. 393–399.
Binns, R. A., Long, J.V.P., and Reed, S.J.B., 1963, Some naturally occurring members of the clinoenstatite-clinoferrosilite mineral series: Nature, v. 198, p. 777–778.

Bonnichsen, B., 1969, Metamorphic pyroxenes and amphiboles in the Biwabik Iron Formation, Dunka River Area, Minnesota: Mineralog. Soc. America Spec. Paper 2, p. 217–239.

Bown, M. G., and Gay, P., 1957, Observations on pigeonite: Acta Cryst., v. 10, p. 440.

—— 1960, An x-ray study of exsolution phenomena in the Skaergaard pyroxenes: Mineralog. Mag., v. 32, p. 379–388.

Boyd, F. R., and Brown, G. M., 1969, Electron probe study of pyroxene exsolution: Mineralog. Soc. America Spec. Paper 2, p. 211–216.

Boyd, F. R., and England, J. L., 1965, The rhombic enstatite-clinoenstatite inversion: Carnegie Inst. Washington Year Book 64, p. 117–120.

Boyd, F. R., and Schairer, J. F., 1964, The system $MgSiO_3$-$CaMgSi_2O_6$: Jour. Petrology, v. 5, p. 275–309.

Brown, G. M., 1957, Pyroxenes from the early and middle stages of fractionation of the Skaergaard intrusion, East Greenland: Mineralog. Mag., v. 31, p. 511–543.

—— 1968, Experimental studies on inversion relations in natural pigeonitic pyroxenes: Carnegie Inst. Washington Year Book 66, p. 347–353.

Clark, J. R., Appleman, D. E., and Papike, J. J., 1969, Crystal-chemical characterization of clinopyroxenes based on eight new structure refinements: Mineralog. Soc. America Spec. Paper 2, p. 31–50.

Dallwitz, W. B., Green, D. H., and Thompson, J. E., 1966, Clinoenstatite in a volcanic rock from the Cape Vogel area, Papua: Jour. Petrology, v. 7, p. 375–403.

Hess, H. H., 1941, Pyroxenes of common mafic magmas: Am. Mineralogist, v. 26, p. 515–535, 573–594.

—— 1949, Chemical composition and optical properties of common clinopyroxenes: Am. Mineralogist, v. 34, p. 621–666.

Hess, H. H., and Henderson, E. P., 1949, The Moore County meteorite: a further study with comment on its primordial environment, Am. Mineralogist, v. 34, p. 494–507.

Hess, H. H., and Phillips, A. H., 1938, Orthopyroxenes of the Bushveld type: Am. Mineralogist, v. 23, p. 450–456.

Kuno, H., 1955, Ion substitution in the diopside-ferropigeonite series of clinopyroxenes: Am. Mineralogist, v. 40, p. 70–93.

—— 1966, Review of pyroxene relations in terrestrial rocks in the light of recent experimental works: Mineralog. Jour., v. 5, p. 21–43.

Kuno, H., and Nagashima, K., 1952, Chemical compositions of hypersthene and pigeonite in equilibrium in magma: Am. Mineralogist, v. 37, p. 1000–1006.

Kushiro, I., 1969, Synthesis and stability of iron-free pigeonite in the system $MgSiO_3$-$CaMgSi_2O_6$ at high pressures: Carnegie Inst. Washington Year Book 67, p. 80–83.

Lewis, J. F., and White, E. W., 1967, Pyroxene relations in basalts and basaltic andesites from the Soufriere volcano, St. Vincent, West Indies (abs.): Trans. Am. Geophys. Union, v. 48, p. 227.

Lindsley, D. H., and Munoz, J. L., 1969, Subsolidus relations along the join hedenbergite-ferrosilite: Am. Jour. Sci., v. 267–A, p. 295–324.

Lindsley, D. H., Brown, G. M., and Muir, I. D., 1969, Conditions of the ferrowollastonite-ferrohedenbergite inversion in the Skaergaard Intrusion, East Greenland; Mineralog. Soc. America Spec. Paper 2, p. 193–201.

Mason, B., 1968, Pyroxenes in meteorites: Lithos, v. 1, p. 1–11.

Morimoto, N., 1956, The existence of monoclinic pyroxenes with the space group $C\,^5_{2h}$ —$P2/c$: Japan Acad. Proc., v. 32, p. 750–752.

Morimoto, N., Appleman, D. E., and Evans, H. T., 1960, The crystal structures of clinoenstatite and pigeonite: Zeitschr. Kristallographie, v. 114, p. 120–147.

Morimoto, N., and Güven, N., 1968, Refinement of the crystal structure of pigeonite: Carnegie Inst. Washington Year Book 66, p. 494–497.

Morimoto, N., and Tokonami, M., 1969a, Domain structure of pigeonite and clinoenstatite: Am. Mineralogist, v. 54, p. 725–740.

—— 1969b, Oriented exsolution of augite in pigeonite: Am. Mineralogist, v. 54, p. 1101–1117.

Muir, I. D., 1954, Crystallization of pyroxenes in an iron-rich diabase from Minnesota: Mineralog. Mag., v. 30, p. 376–388.

Munoz, J. L., 1968, Effect of shearing on enstatite polymorphism: Carnegie Inst. Washington Year Book 66, p. 369–370.

Perrotta, A. J., and Stephenson, D. A., 1965, Clinoenstatite: high-low inversion: Science, v. 148, p. 1090–1091.

Poldervaart, A., and Hess, H. H., 1951, Pyroxenes in the crystallization of basaltic magma: Jour. Geology, v. 59, p. 472–489.

Prewitt, C. T., Papike, J. J., and Ross, M., 1970, Cummingtonite: a reversible, nonquenchable, transition from $P2_1/m$ to $C2/m$ symmetry (abs.): Am. Mineralogist, v. 55, p. 305–306.

Ross, M., Bence, A. E., Dwornik, E. J., Clark, J. R., and Papike, J. J., 1970, Mineralogy of the lunar clinopyroxenes, augite and pigeonite, in Levinson, A. A., ed., Proc. Apollo 11 Lunar Sci. Conf.: Elmsford, New York, Pergamon Press, Inc., v. 1, p. 839–848.

Ross, M., Papike, J. J., and Shaw, K. W., 1969, Exsolution textures in amphiboles as indicators of subsolidus thermal histories: Mineralog. Soc. America Spec. Paper 2, p. 275–299.

Sigurdsson, H., and Brown, G. M., 1970, An unusual enstatite-forsterite basalt from Kolbeinsey Island, north of Iceland: Jour. Petrology, v. 11, p. 205–220.

Smith, J. V., 1969, Crystal structure and stability of the $MgSiO_3$ polymorphs; physical properties and phase relations of Mg, Fe pyroxenes: Mineralog. Soc. America Spec. Paper 2, p. 3–29.

Smyth, J. R., 1970, High-temperature single-crystal x-ray studies of natural orthopyroxenes (abs.): Am. Mineralogist, v. 55, p. 312.

Virgo, D., and Hafner, S. S., 1969, Fe^{2+}, Mg order-disorder in heated orthopyroxenes: Mineralog. Soc. America Spec. Paper 2, p. 67–81.

Wager, L. R., and Brown, G. M., 1968, Layered igneous rocks: Edinburgh and London, Oliver and Boyd, 588 p.

Weill, D. F., McCallum, I. S., Bottinga, Y., Drake, M. J., and McKay, G. A., 1970, Petrology of a fine-grained igneous rock from the Sea of Tranquility: Science, v. 167, p. 635–638.

Winchell, A. N., 1900, Gabbroid rocks of Minnesota: Am. Geologist, v. 26, p. 203.

Yoder, H. S., Tilley, C. E., and Schairer, J. F., 1963, Pyroxenes and associated minerals in the crust and mantle: Carnegie Inst. Washington Year Book 62, p. 84–95.

Zussman, J., 1968, The crystal chemistry of pyroxenes and amphiboles, 1. Pyroxenes: Earth-Sci. Rev., v. 4, p. 39–67.

MANUSCRIPT RECEIVED BY THE SOCIETY MARCH 29, 1971

THE GEOLOGICAL SOCIETY OF AMERICA, INC.
MEMOIR 132
© 1972

Disordered Albite and Oligoclase from Veins in Metabasalt

MANUEL N. BASS
ARCH M. REID
AND
HIROSHI TAKEDA
Geochemistry Branch
NASA Manned Spacecraft Center,
Houston, Texas 77058

ABSTRACT

Metamorphosed pillow basalt from the actively spreading basin west of the Marianas Island Arc contains albite or oligoclase of intermediate to low structural state. A feldspathic vein occupying late fractures in one metabasalt was found to contain highly disordered plagioclase ranging in composition from An_3 to An_{31}. From the absence of thermal metamorphic effects in the chlorite-montmorillonite-actinolitic amphibole-epidote-sodic plagioclase assemblage of the host rock we infer formation of the vein at low temperature.

Individual feldspar grains consist of lamellae or irregular domains (compositionally distinct volumes, not the crystallographers' domains) of varied compositions. Electron microprobe analyses of these domains yield a bimodal distribution of compositions. The large sizes of the domains and the inferred low temperature of formation of the vein suggest to us that the domains may have formed as primary metastable growth features. The bimodality and other features of the compositional distribution suggest that the composition adjusted, but incompletely so, in accord with the peristerite solvus.

INTRODUCTION

The feldspar in dredged oceanic metabasalt and in the associated feldspathic veins and segregations is typically sodic plagioclase. The maximum possible anorthite content, corresponding to an assumed low-temperature structural state and a single feldspar composition, is generally in the range An_{22} to An_{35}, as estimated from peak separations on x-ray powder patterns using the methods of Smith and Yoder (1956), Bambauer and others (1967), and Smith and Gay (1958). Veins cutting a chloritic metabasalt from a dredge haul from the eastern Philippine Sea gave peak separations allowing any plagioclase from highly ordered labradorite to highly disordered albite. The possibilities are of interest because (1) labradorite veins are unknown in rocks of low metamorphic grade, and (2) disordered sodic plagioclase is uncommon in nature. This paper reports a detailed descriptive study of the vein feldspar, which is highly disordered albite and oligoclase.

MODE OF OCCURRENCE

Rock dredge scan-45D was recovered during scan Expedition (RV *Argo,* Scripps Institution of Oceanography) from an east-trending cross fracture in the actively spreading basin immediately west of the Marianas Island Arc between 17°33.4′ N., 144°52.5′ E. and 17°33.8′ N., 144°54.6′ E., starting at a depth of 2,430 fathoms and ending at a depth in excess of 2,110 fathoms (D. E. Karig, oral commun., 1969; Karig, 1971). The dredged samples are mostly low-grade metabasalt, including large and small fragments of pillows (Fig. 1), and a few fragments each of relatively fresh basalt and sedimentary rock.

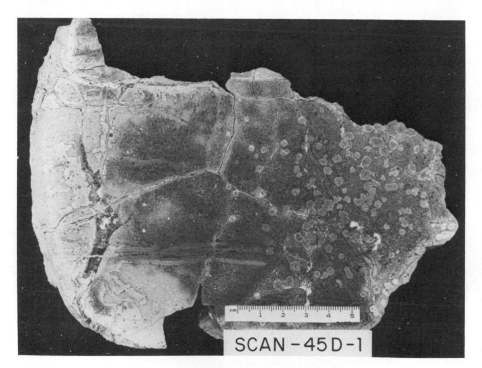

Figure 1. Section through metamorphosed basalt pillow. Outer, formerly glassy surface at left; color laminations indicate color-banded palagonite prior to metamorphism to chlorite. Note: relict columnar and other joints healed by veins; dark chloritic vein about 2 cm below outer surface; porphyroblastic plagioclase spherulites 4 to 8 mm in diameter; dark chloritic amygdules 1 mm in diameter or smaller; and irregular white veins in interior.

The mineral assemblage of the metabasalt is chiefly chlorite, plagioclase, quartz, montmorillonite, and actinolitic amphibole. Montmorillonite is considered to be a product of metamorphism, not sea-water alteration or weathering, because it is confined to the interior parts of pillows. This suggestion is supported by the observation that the 14 Å chlorite peak expands and broadens slightly on exposure to ethylene glycol vapor, indicating the presence of a minor amount of randomly interlayered montmorillonite.

Several types of veins are present in the metabasalts. Chloritic veins contain quartz and, rarely, large colorless grains of wairakite. Hard opaque white veins, in places showing agate-like banding in shades of white and gray along their outer margins, commonly consist of plagioclase, quartz, and a trace of either analcite or wairakite (x-ray peaks too weak for definite identification by criteria of Seki, 1968). The veins were deposited in late cracks that cut the metamorphic fabric;

Figure 2. Veins of highly disordered albite and oligoclase in late cracks in metabasalt (SCAN-45D-17). Porphyroblastic spherulites indicate specimen is from interior of a pillow (*compare* Fig. 1). Scale in cm.

deposition involved no visible alteration of the earlier metamorphic minerals. The maximum possible temperature of vein formation, therefore, is that of the lowest amphibolite facies, and the most likely temperature is that of the greenschist facies. Open vugs lined with terminated quartz indicate relatively low pressure. In no specimen is there any evidence of shear effects.

The white veins, which are the main object of this study, occur in specimen SCAN-45D-17 (Fig. 2) which contains porphyroblastic plagioclase spherulites similar to those in the interior of the pillow in Figure 1. Epidote is absent from the white veins but is present nearby in independent radial clusters of prisms that are probably of the same generation as the veins. The immediate wall rock of the veins is largely very fine-grained chlorite and minor plagioclase intergrown in groups with a radial or a combined radial and spiral structure that is at least partly a relict quench structure, but partly perhaps a metamorphic product. Abundant tiny patches of epidote dot the chlorite. The wall rock also includes quartz, sodic plagioclase, actinolitic amphibole, and minor amounts of montmorillonite. Texturally notable are skeletal or radial plagioclase clusters and chlorite pseudomorphs after euchedral, often glomeroporphyritic, olivine microphenocrysts.

Maximum anorthite contents of the plagioclase estimated from peak separations in x-ray powder patterns and assuming a single composition are given for various specimens in Table 1. It is noted that the plagioclases are generally sodic and have variable compositions or structural states or both, even within a single vein (SCAN-45D-17). This is true by either method of measurement, even though Γ systematically gives higher maximum anorthite values than $\Delta 2\theta$.

TABLE 1. COMPOSITION OF PLAGIOCLASE IN METABASALTS OF DREDGE HAUL SCAN-45D

Specimen number and type	$\Delta 2\theta^*$	Maximum AN Content (mol%)[†]	Γ[‡]	Maximum AN Content (mol%)[§]
1: porphyroblastic spherulite	1.69°	30	+0.33°	36
1: vein	1.57°	23	+0.14°	31
1: groundmass 25 cm below pillow surface	1.66°	28		
9: porphyroblastic spherulite	1.55°	23	−0.05°	25
17: porphyroblastic spherulite	1.64°	27	+0.19°	32
17: vein	1.88°	59	+0.69°	63
	1.84°	55	+0.63°	62
17: calculated by refinement of powder pattern	1.89°	60	+0.71°	64
18: porphyroblastic spherulite	1.60°	25	+0.16°	31

* $\Delta 2\theta = 2\theta(131) - 2\theta(1\bar{3}1)$.

† From Smith and Yoder (1956) and Baumbauer and others (1967), assuming a single composition is present.

‡ $\Gamma = 2\theta(131) + 2\theta(220) - 4\theta(1\bar{3}1)$; missing value due to interference from an amphibole peak.

§ From Smith and Gay (1958), assuming a single composition is present.

TEXTURE OF VEIN PLAGIOCLASE

The opaque white color of the veins is due in part to fine grain size, but stems mostly from abundant low-index, high-relief inclusions of unknown compositions in much of the plagioclase. Feldspar shapes in thin section vary from equant to elongate. The most common type is finely lamellar (Fig. 3) in a manner resembling that of chessboard albite. $2V_a$ in these grains is commonly in the range 60° to 90°, and uncommonly as low as 40° or 45° (estimated by separation of isogyres in a microscope field calibrated with 2V standards). The 40° or 45° $2V_a$ in a plagioclase is indicative of highly disordered AB-rich plagioclase near albite (Kano, 1955; Smith, 1958). Some of these grains contain a few lamellae that show no discernible contrast in refractive indices and are presumably twin lamellae, but most lamellae have indices contrasting distinctly with those of adjacent lamellae, indicating that they are due not to twinning but to exsolution or a regular intergrowth of compositionally contrasted growth domains ("domain" used solely in a compositional, not in a crystallographic sense). These finely lamellar grains commonly contain inclusions of high relief, either throughout or only peripherally. Other, commonly elongate, grains rich in peripheral inclusions show wavy extinction, simple rather than polysynthetic twinning, and no lamellar structure. These grains are probably sectioned parallel to twin or compositional lamellae similar to those in Figure 3.

The second most common type of grain (Fig. 4) is an aggregate of complexly twinned, crudely diamond-shaped areas, the spaces between which are filled with

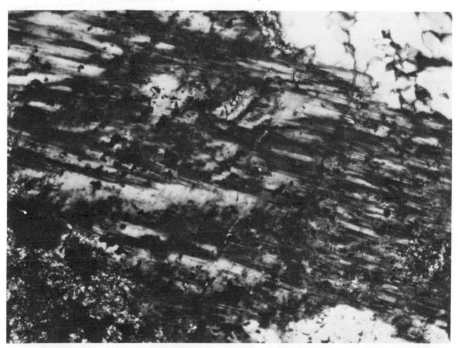

Figure 3. Most common type of plagioclase grain in veins of Figure 2. Lamellae due both to twinning and to compositional differences. Fine-grained chlorite of wall rock in lower left. Crossed polarizers. Field size 0.85 × 0.65 mm.

Figure 4. Grain 2. Second most common type of plagioclase grain, located on edge of polished section adjacent to epoxy resin. Clear, twinned, crudely diamond-shaped areas separated by untwinned dusty plagioclase of distinctly lower refractive indices. Crossed polarizers. Several twinned areas with strong Becke lines along edges outlined in white; intervening twinned areas are below the surface of the polished section and edges are vague. Field size 1.4 × 1.1 mm.

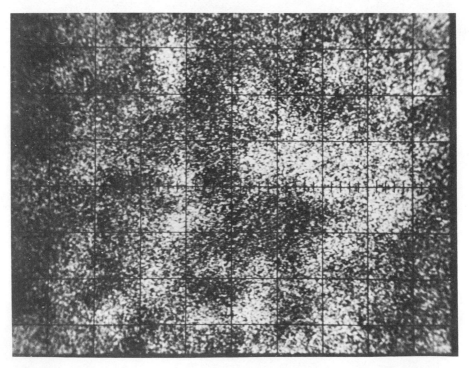

Figure 5. Ca x-ray image of grain similar to that in Figure 4, but with smaller diamond-shaped areas. Grid squares are 2 × 2 μ.

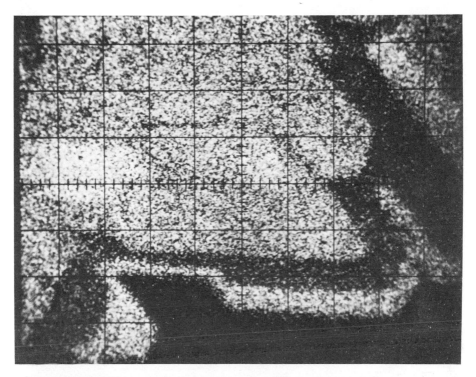

Figure 6. Grain 11. Ca x ray image showing euhedral zones bordering on vug partly filled with quartz, and interior AN-rich domains. Grid squares are $10 \times 10 \ \mu$.

Figure 7. Grain 3. Unusual grain type with relatively AN-rich core (outlined by white dots). Vague contact twins in core, interpenetration twins in rim. Crossed polarizers. Field size 0.80×0.60 mm.

untwinned, inclusion-rich plagioclase of distinctly lower refractive indices. The compositional contrast in a grain similar to that in Figure 4 is seen in a Ca x-ray image (Fig. 5). These grains commonly border on vugs, against which they show euhedrally bounded zones of sharply contrasting compositions (Fig. 6), and indeed the diamond-shaped areas may once have been euhedra growing unobstructed into an open space, and the low-index plagioclase between them may be late deposits similar to the euhedral border zones in Figure 6. There is no consistent sequence of compositional changes toward the vugs, which are partially or wholly filled with quartz. Figure 7 illustrates a single grain shaped similar to the diamond-shaped areas of Figure 5. The relatively AN-rich core contains vague contact twins, and the relatively AB-rich rim contains interpenetration twins. Only the prominent composition plane roughly bisecting the grain is shared by both core and rim. None of the twin laws has been identified.

COMPOSITION OF VEIN PLAGIOCLASE

Examination in oils of a crushed sample of the broadly exposed vein in Figure 2 established that the refractive indices of the feldspar are generally lower than those of quartz, but the abundant inclusions, fine twinning, lamellar structure, compositional variation, and a notable failure to cleave well prevented accurate determination of X' and Z' of cleavage plates. These results, taken alone, allow us to state only that the compositions are mostly An_{26} and more sodic.

Electron microprobe analyses (accelerating voltage 15 kilovolts, sample current 0.02 microamperes, spot size 1μ) were carried out for Ca, Na, and K at intervals of 2, 4, or 10μ along traverses across several grains. The analyses reveal highly variable compositions over short distances without any apparent order. Calculation of the compositions on the assumption of ideal feldspar stoichiometry yields a wide range of analytic totals, which is predictable in view of the fact that the measured elements comprise less than 10 percent of the total. Twenty-eight analyses were rejected because, as a result of holes or cracks in the polished surface, they failed to total as much as 89 percent. The totals of the remaining 234 analyses range from 89 to 105 percent, 30 percent of them within ±1 percent and 60 percent within ±3 percent of 100 percent. The molar proportions of OR ($KAl Si_3O_8$), AB ($NaAlSi_3O_8$), and AN ($CaAl_2Si_2O_8$) in the 234 analyses were normalized to a total of 100 and plotted on the feldspar ternary for each grain separately (Fig. 8). Some of the grains are illustrated in Figures 3 through 7.

Immediately notable are the wide ranges of composition. The bimodal distribution in grain 3 reflects the core and rim shown in Figure 7. Grain 11, which may in fact be part of grain 2, was measured across the area shown in Figure 6. The cluster of AB-rich points is from the AN-poor zone second removed from the vug; the points near An_{25} are from the euhedral AN-rich zone bordering the vug; and the point near An_{29} is from the smaller of the two interior AN-rich areas (immediately to the right of center of Fig. 6).

Also notable are the low contents of K. Although two grains (1 and 11) show a slightly higher OR content in AN-richer domains (as found by Crawford, 1966), grain 7 shows the opposite and most grains show no trend at all. Only two analyses (one each in grains 1 and 7) show significantly more than 1 mole percent OR.

The finely lamellar, chessboardlike grains (1 and 9) have wide ranges of composition, supporting the suggestion that some of the apparent twinning is in fact due to an intergrowth of layers of different compositions. Grain 7 was sectioned almost parallel to the lamellae and as a result yields a restricted range of composition.

Only one cleavage flake suitable for x-ray precession measurements was found. It is an (010) flake composed dominantly of a single crystal with a minor proportion of material twinned on the albite law. Like grain 7, and for the same reasons, the (010) flake shows a restricted range of composition. The compositions of four points analyzed with a beam spot 50 μ in diameter are given in Table 2 and plotted as X's in Figure 8. Note the very low Fe contents.

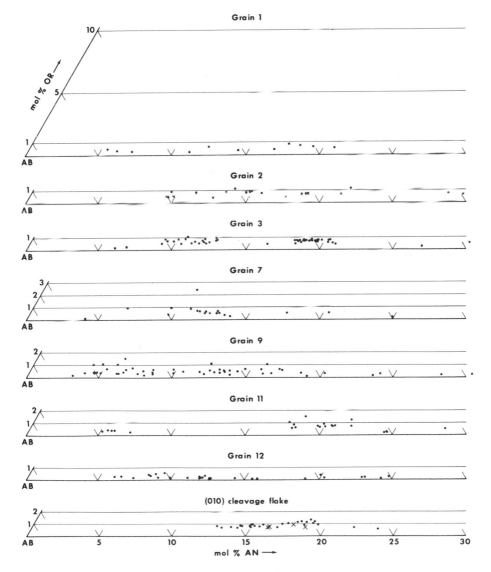

Figure 8. Electron microprobe analyses of vein plagioclase plotted separately in the feldspar ternary for each grain.

TABLE 2. ELECTRON MICROPROBE ANALYSES (Weight %) OF (010) CLEAVAGE FLAKE FROM VEIN IN SCAN-45D-17*

SiO_2	62.49	62.02	62.89	62.45
Al_2O_3	23.78	23.78	23.40	23.38
Fe_2O_3	0.00	0.01	0.01	0.01
CaO	3.69	3.89	3.35	3.35
K_2O	0.17	0.13	0.13	0.14
Na_2O	9.38	9.39	9.61	9.52
Total	99.51	99.12	99.39	98.85
mole percent AB	81.34	80.65	83.18	83.10
mole percent AN	17.70	18.66	16.06	16.14
mole percent OR	0.96	0.69	0.75	0.76

* Electron beam spot 50μ in diameter.

All points from Figure 8 are plotted in a single histogram in Figure 9. Most of the points lie in the range An_4 to An_{22} (average about An_{14}) and the distribution is bimodal.

The sizes and shapes of the compositional domains can be estimated from Ca x-ray profiles across several of the analyzed grains (Fig. 10). The profiles superficially resemble the Na profiles measured by Vogel (1970) in orthoclase. Figure 10B shows much more pronounced compositional variation across than along the

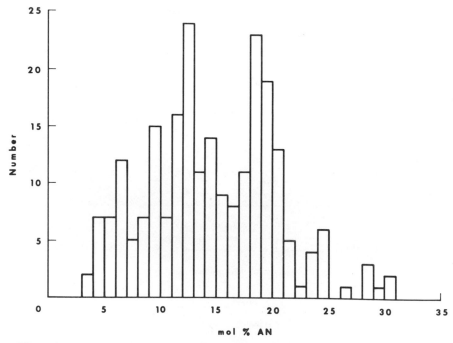

Figure 9. Histogram of AN contents for all analyses shown in Figure 8. Note bimodal distribution.

traces of interlamellar surfaces in grain 9. Traverses roughly parallel to [001] across (010) sections (Figs. 10A, D, E) show an apparent broadening of the compositional lamellae. These observations confirm indications from strong Becke lines, reflecting refractive index contrasts, that the compositional domains are lamellae nearly parallel to (010). The traverses across the (010) cleavage flake (Figs. 10D, E) show long stretches with relatively restricted compositional variation, corresponding to the limited variation seen in Figure 8. The widths of AN-rich lamellae in traverses across the traces of interlamellar surfaces in grains cut at a high angle to (010) (Figs. 10B, C) vary from about 2.5 to more than 30μ, most of them being less than 12μ. These widths are such that some of the compositional lamellae may coincide with apparent twin lamellae in finely lamellar grains (Fig. 3), whereas others may encompass two or more twin lamellae with little change of composition.

STRUCTURAL STATE OF VEIN PLAGIOCLASE

The compositions of the plagioclase indicate that the x-ray peak separations (Table 1) reflect a high degree of Si-Al disorder in a sodic plagioclase rather than an ordered intermediate plagioclase (Smith and Yoder, 1956; Bambauer and others, 1967; Smith and Gay, 1958). The fact that no 2V smaller than 40° was observed suggests that disorder is not complete (Schneider, 1957; Ribbe, 1960) An independent measure of disorder was obtained from the reciprocal lattice angle γ^*, determined by least-squares refinement of a powder pattern[1] using the computer program of Evans and others (1963) as modified by Appleman and Handwerker, and by direct measurement by x-ray precession photography. The results are shown in Table 3. The lattice parameters from the precession photography are based on strong sharp reflections. Weak reflections indicate albite twinning and slightly misoriented crystals whose compositions and structural states could be slightly different. The strong reflections are consistent with the $C\bar{1}$ space group previously established for oligoclase (Colville and Ribbe, 1968); Borg and Smith, 1968; Smith and Ribbe, 1969), but the quality of the photograph is inadequate for eliminating the presence of very weak e, s, or f reflections.

The γ^* values and ranges of composition from Figures 8 and 9 are plotted in Figure 11, from which it is clear that the plagioclases are highly disordered. It may be further inferred that the γ^* values, especially that from the powder pattern refinement, are probably averages of a range of values representing different degrees of Si-Al disorder.

DISCUSSION

The disordered plagioclase fills late cracks in a chloritic metabasalt without any detectable alteration of the host rock. The plagioclase is evidently contemporaneous with relatively coarse prismatic epidote and apparently grew, therefore, at low

---

[1] Norelco diffractometer, monochromatic Cu Kα radiation, 40 kilovolts, 15 milliamperes, scanning rate 1/4°/minute, time constant 1 second, counting rate 200 to 1,000 cps, depending on peak height; diffraction angles relative to internal spinel standard ($a_o = 8.0833$ Å).

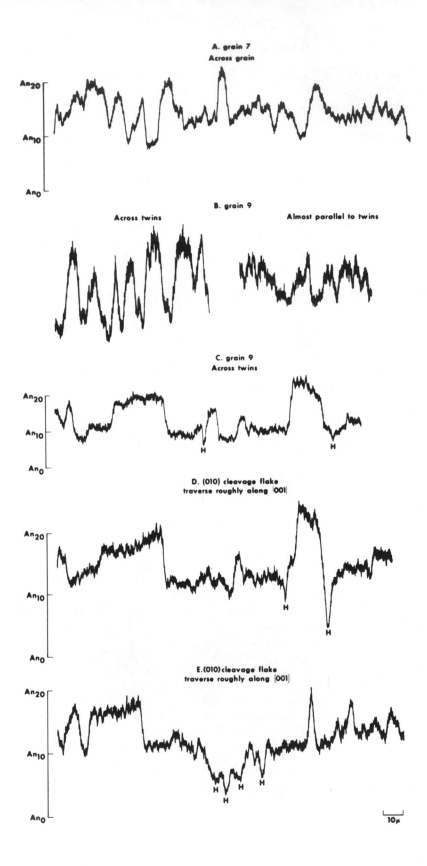

A. grain 7
Across grain

B. grain 9

Across twins   Almost parallel to twins

C. grain 9
Across twins

D. (010) cleavage flake
traverse roughly along [001]

E. (010) cleavage flake
traverse roughly along [001]

10µ

temperature in the epidote-plagioclase stability field (Christie, 1962).[2] The Si-Al disorder is most simply ascribed, as in albites from alpine veins (MacKenzie, 1957), to growth disorder at such low temperatures that subsequent ordering was kinetically inhibited. The relation of disorder to crystal morphology is not known, so we cannot speculate at this time on a possible relation between disorder and compositional sector zoning as found in staurolite and other minerals (Hollister, 1970).

The older plagioclase of the host rock, both in the groundmass and in porphyroblastic spherulites, can be explained as having grown at higher temperatures or for longer periods of time so that thermal activation permitted a significant and possibly variable degree of secondary Si-Al ordering. The variability, as suggested by variable x-ray peak separations (Table 1), could be confirmed only with information on the compositions of the host rock plagioclases. Compositions could not be measured on groundmass plagioclases because of their fine grain size and intimate admixture with the greatly dominant chlorite, but preliminary microprobe analyses of 28 points in a porphyroblastic spherulite gave a compositional range from An_5 to An_{55} (average An_{23}). Evidently the cause of variable plagioclase composition did not operate solely during deposition of the veins. These results indicate that the peak separations for porphyroblastic spherulites in Table 1 ($1.55°$ to $1.69°$ for $\Delta 2\theta$), measured on powdered bulk samples, must be average values for a wide compositional range and cannot be interpreted unequivocally. At the AB-rich end of the range, the separations necessitate intermediate degrees of disorder. At the AN-rich end, they would indicate improbably low degrees of disorder; the OR content, ranging from 0 to 1.5 mole percent (average 0.75 mole percent), is too low to account for such small $\Delta 2\theta$ values in the range An_{50} to An_{55} (Bambauer and others, 1967). At the average composition (An_{23}) the peak separations indicate low degrees of disorder comparable to those of most natural plutonic plagioclases. Thus there is a suggestion that the porphyroblastic plagioclase is, on the average, more ordered than the vein plagioclase, but more detailed data on the correlation of compositions and peak separations are obviously needed.

Experimental evidence suggests that several factors may influence Si-Al ordering. The rate of ordering in albite at low temperature is known to be accelerated by high Fe content (Raase and Kern, 1969) and high Na^+/H^+ (Martin, 1969b, 1970). Similarly, high K^+/H^+ accelerates ordering in K-feldspar (Martin, 1969a). On the other hand, even slightly peraluminous solutions strongly inhibit ordering

---

[2] Waldbaum and Kastner (1968) suggest lack of cathodoluminescence as a criterion for a low-temperature origin of feldspar. In scan-45D-17 the plagioclases in both the veins and porphyroblastic spherulites luminesce equally intensely in shades of light blue and dark, dull red that are patchily intermixed. The groundmass plagioclase, in those rare areas where its luminescence is not obscured by chlorite, luminesces essentially identically. Thus, cathodoluminescence provides no definite evidence regarding the relative temperatures of formation of the various plagioclases.

◄———————————————————————————————————————

Figure 10. Ca x-ray profiles across several grains showing sizes and shapes of compositional domains. "Across twins" and "Almost parallel to twins" indicate traverses essentially normal and parallel, respectively, to traces of interlamellar surfaces in plane of polished surface. Note relatively restricted ranges of composition along large parts of traverses A and E, and greater variation across than parallel to interlamellar traces in B. H = hole or crack in polished section; horizontal scale holds for all profiles; vertical scales, where shown, are based on oligoclase or labradorite standard.

(Martin, 1969a, 1969b). Martin suggests that the structurally intermediate states of many reported metastable albites which grew at low temperature are not to be ascribed simply to kinetic inhibition of ordering at low temperature subsequent to growth, but that they are consequences primarily of a Na-poor or Al-rich chemical environment at the time of growth. Under otherwise similar conditions a Na-rich or Al-poor environment would have promoted a high degree of ordering. Thus, he suggests, structurally intermediate authigenic albites (Baskin, 1956; Kastner and Waldbaum, 1968) would be formed in the presumed Na-poor environments of limestones and dolomites, and structurally intermediate metasomatic albites (*see* Donnelly, 1963; Rusinov, 1965; and Callegari and De Pieri, 1967) would be formed in local peraluminous environments resulting from the replacement of intermediate or calcic plagioclase by sodic plagioclase during the metasomatism which produced spilites and keratophyres. Martin's (1969b) presumption of Na-poor solutions in limestones and dolomites at the time of authigenic albite formation is debatable because authigenic feldspars appear to form early in diagenesis when the sedimentary section is presumably below sea level and connate sea water is still present.

The disordered vein plagioclase in SCAN-45D-17 is almost devoid of Fe (Table 2). Fe was not present, therefore, in amounts sufficient to catalyze Si-Al ordering.

Assuming that the host rock prior to metamorphism was basalt, the original calcic plagioclase in the rock has been replaced by more sodic plagioclase (Table 1), and we might appeal to local peraluminous solutions due to breakdown of intermediate or calcic plagioclase to explain the establishment and persistence of a high state of Si-Al disorder. However, disorder from this cause is expected mainly in the plagioclase of the groundmass and porphyroblastic spherulites, not in the plagioclase of veins which filled open cracks (Fig. 2) at a time when intermediate and calcic plagioclase no longer remained in the host rock. Contrary to this expectation, the compositions of spherulite plagioclases and the evidence from Table 1 and Figure 11 imply that groundmass and spherulitic plagioclases are less disordered than those of the vein. Therefore, a strained appeal to Al from distant calcic plagioclase sources is needed to maintain this explanation for the highly disordered state of the vein plagioclase.

If the higher temperature boundary of the epidote-plagioclase field can lie above all or part of the peristerite solvus, and if, during cooling, the rate of reaction of

TABLE 3. LATTICE PARAMETERS OF VEIN PLAGIOCLASE FROM SCAN-45D-17

	Refinement of Powder Pattern†	X-Ray Precession Photography‡
a	8.154 ± 0.001 Å	8.12 ± 0.02 Å
b	12.860 ± 0.002 Å	12.84 ± 0.04 Å
c	7.119 ± 0.001 Å	7.097 ± 0.005 Å
α	$93°37.5' \pm 0.7'$	$93°23.5' \pm 2'$
β	$116°27.6' \pm 0.4'$	$116°00' \pm 5'$
γ	$89°55.1' \pm 0.7'$	$90°0.7' \pm 5'$
Volume	666.67 ± 0.09 Å³	663.51 Å³
γ*	$88°17.1' \pm 0.7'$	$88.20' \pm 5'$

† Starting parameters from calcic oligoclase (Colville and Ribbe, 1966; Borg and Smith, 1968); uncertainties are standard errors.

‡ (010) cleavage flake mounted along b^*; a^*, b^*, c^*, α^*, and γ^* measured directly on film; β measured on dial of precession camera; all other parameters calculated.

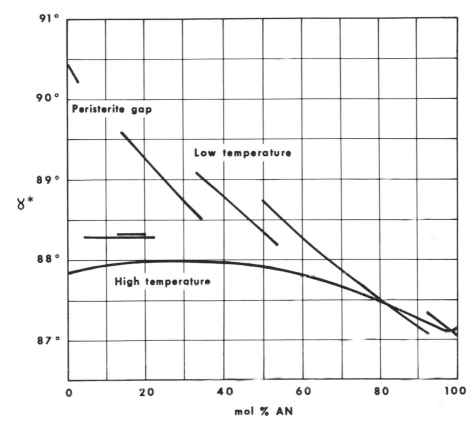

Figure 11. γ^* versus composition plot. Low-temperature data from Doman and others (1965); high-temperature data from Smith and Ribbe (1969). Longer horizontal line from main compositional range in Figure 9 and γ^* from refinement of x-ray powder pattern (Table 3); shorter horizontal line from compositional range for (010) cleavage flake (Fig. 8) and γ^* by x-ray precession photography (Table 3).

plagioclase to more sodic plagioclase plus epidote is slow compared to the rate of exsolution, then the bimodal distribution of Figure 9 could reflect a partial equilibration in accord with the solvus at relatively high temperatures. The shape and position of the solvus appropriate to these highly disordered plagioclases is not known (Viswanathan and Eberhard, 1968). The failure to exsolve completely to almost pure albite and calcic oligoclase would be ascribable to kinetic inhibition due to low temperature, limited time, or both (MacKenzie, 1957). A problem is posed by the fact that the highly ordered state so common in natural pure albite is not attained in the vein plagioclases. MacKenzie (1957) suggests that in the alkali feldspar system albite attains a given degree of ordering at higher temperature (that is, conditions of faster equilibration) than does K-feldspar. The disorder of the vein plagioclases would be explained if even slightly anorthitic plagioclases behave relative to pure albite in the same relatively sluggish manner as K-feldspar. This explanation is in fact supported by studies on the kinetics of disordering (Schneider, 1957) and by the subsolidus relations proposed by Ribbe (1962). This general interpretation, centering on the peristerite solvus, conflicts with assertions that ordering precedes exsolution (Brown, 1960; Viswanathan and Eberhard, 1968), but can, if the degrees of disorder and exsolution are both variable, be reconciled with the opinions that ordering and exsolution either proceed concurrently (Ribbe, 1962; Smith and Ribbe, 1969) or are not directly related (Crawford, 1966). The contrast in degree of ordering between the plagioclases of the veins and the porphyroblastic spherulites for similar compositional ranges is consistent with Crawford's suggestion.

Several features do not appear to be in accord with an origin by exsolution. Peristerite exsolution blebs are normally, at most, 0.5μ in diameter (Ribbe and Van Cott, 1960, 1962; Ribbe, 1960) and exsolution lamellae rarely exceed the $1\ \mu$ thickness reported by Raith (1969; the 3-μ-thick lamellae reported by Brown, 1960, are of uncertain nature). Exsolution lamellae follow $(0\overline{8}1)$, which is at a $14°$ angle to (010) (Raith, 1969, and references therein). In contrast, the compositional lamellae in the vein plagioclase of scan-45D-17 are up to $30\ \mu$ wide and are almost or quite parallel to (010) or to free-growth surfaces bounding vugs (Fig. 6). Had these thick lamellae formed by exsolution, the plagioclase would presumably be more ordered than it was observed to be, as are most other low-temperature plagioclases subject to conditions of high internal ion mobility. The lamellae may have formed, however, not by exsolution, but by independent subsolvus growth (Crawford, 1966) coupled with epitaxial orientation of new lamellae on earlier ones. This mechanism of formation would not exclude secondary submicroscopic peristerite exsolution within the thick (010) lamellae.

A unique explanation of the origin of the vein plagioclase is not obvious. Metastable crystallization at low temperatures is the simplest and most likely cause of the origin and preservation of Si-Al disorder, but the exact nature of the crystallization processes is not clear. The variation and frequency distribution of composition (Fig. 9) may reflect changes both in the composition of the vein-forming fluid and in the metastable crystallization process. Control of that metastable process by the peristerite solvus is conceivable if partial Si-Al ordering and adjustment to the solvus occurred with progressively decreasing efficiency as temperature fell. Alternatively, disordered plagioclase initially grew stably in the epidote-plagioclase field above the peristerite solvus. Then due to falling temperature, the plagioclase lay in the two-phase field under the solvus and underwent partial exsolution with further decrease of temperature, under conditions such that epidote did not continue to form. The greater part of the exsolution occurred at a relatively high temperature, at which the observed compositional modes (Fig. 9) approximate the equilibrium compositions appropriate to that temperature. MacKenzie's (1957) findings on extended hydrothermal treatment of albite, if they represent equilibrium, would indicate vein deposition near 750°C if the feldspars were pure albite, but the host rock petrology and the somewhat anorthitic compositions of the vein plagioclases indicate considerably lower temperatures.

As a final note, there is no evidence for post-crystallization deformation. One may thus conclude that the chessboardlike twinning in lamellar grains (Fig. 3), if it is such, is not mechanical twinning. The twinning is most likely a primary growth phenomenon.

SUMMARY

The conditions of crystallization of plagioclase in the metabasalt sequence in the tectonically active basin west of the Marianas Island Arc varied with time. Early metamorphic plagioclase appears to be relatively ordered, whereas late vein plagioclase is highly disordered. Both types are highly variable in composition. Whatever the detailed mechanism of deposition of the vein plagioclase, its texture, bimodally variable compositional distribution, and Si-Al disorder may represent primary growth features. These features can also be reconciled with variable degrees of Si-Al ordering and exsolution, in accord with the peristerite solvus, of what may originally have been a structurally homogeneous, disordered plagioclase.

ACKNOWLEDGMENTS

The dredged samples were donated by D. E. Karig. F. K. Aitken kindly instructed M. N. Bass in the application of the computer program for refining powder patterns. Discussions with S. W. Bailey were most helpful. A brief encounter with D. B. Stewart, who can bid fairly as "the fastest bibliographer in the East," afforded an entree into the literature. The authors held NRC Resident Research Associateship at the NASA Manned Spacecraft Center during the course of this work.

REFERENCES CITED

Bambauer, H. U., Corlett, M., Eberhard, E., and Viswanathan, K., 1967, Diagrams for the determination of plagioclases using x-ray powder methods (Part III of laboratory investigations on plagioclases); Schweizer. Mineralog. u. Petrog. Mitt., v. 47, p. 333–349.

Baskin, Y., 1956, A study of authigenic feldspars: Jour. Geology, v. 64, p. 132–155.

Borg, I. Y., and Smith, D. K., 1968, Calculated powder patterns: I. Five plagioclases: Am. Mineralogist, v. 53, p. 1709–1723.

Brown, W. L., 1960, The crystallographic and petrologic significance of peristerite unmixing in the acid plagioclases: Zeitschr. Kristallographie, v. 113, p. 330–344.

Callegari, E., and De Pieri, R., 1967, Crystallographical observations on some chessboard albites: Schweizer. Mineralog. u. Petrog. Mitt., v. 47, p. 99–110.

Christie, O.H.J., 1962, Discussion: Feldspar structure and the equilibrium between plagioclase and epidote: Am. Jour. Sci., v. 260, p. 149–153.

Colville, A. A., and Ribbe, P. H., 1968, The crystal structure of oligoclase: Geol. Soc. America, Abs. for 1966, Spec. Paper 101, p. 41–42.

Crawford, M. L., 1966, Composition of plagioclase and associated minerals in some schists from Vermont, U.S.A., and South Westland, New Zealand, with inferences about the peristerite solvus: Contr. Mineralogy and Petrology, v. 13, p. 269–294.

Doman, R. C., Cinnamon, C. G., and Bailey, S. W., 1965, Structural discontinuities in the plagioclase feldspar series: Am. Mineralogist, v. 50, p. 724–740.

Donnelly, T. W., 1963, Genesis of albite in early orogenic volcanic rocks: Am. Jour. Sci., v. 261, p. 957–972.

Evans, H. T., Jr., Appleman, D. E., and Handwerker, D. S., 1963, The least squares refinement of crystal unit cells with powder diffraction data by an automatic computer indexing method [abs.]: Am. Crystallog. Assoc. Ann. Mtg. Program, Cambridge, Mass., p. 42–43.

Hollister, L. S., 1970, Origin, mechanism, and consequences of compositional sector-zoning in staurolite: Am. Mineralogist, v. 55, p. 742–766.

Kano, H., 1955, High-temperature optics of natural sodic plagioclases: Mineralog. Jour., v. 1, p. 255–277.

Karig, D. E., 1971, Structural history of the Mariana Island Arc system: Geol. Soc. America Bull., v. 82, p. 323–344.

Kastner, M., and Waldbaum, D. R., 1968, Authigenic albite from Rhodes: Am. Mineralogist, v. 53, p. 1579–1602.

MacKenzie, W. S., 1957, The crystalline modifications of $NaAlSi_3O_8$: Am. Jour. Sci., v. 255, p. 481–516.

Martin, R. F., 1969a, Effect of fluid composition on structural state of alkali feldspars [abs.]: Am. Geophys. Union Trans., v. 50, p. 350.
—— 1969b, The hydrothermal synthesis of low albite: Contr. Mineralogy and Petrology, v. 21, p. 323–339.
—— 1970, Cell parameters and infrared absorption of synthetic high to low albites: Contr. Mineralogy and Petrology, v. 26, p. 62–74.
Raase, P., and Kern, H., 1969, Ueber die Synthese von Albiten bei Temperaturen von 250 bis 700°C: Contr. Mineralogy and Petrology, v. 21, p. 225–237.
Raith, M., 1969, Peristerite aus alpidisch metamorphen Gneisen der Zillertaler Alpen (Tirol, Oesterreich): Contr. Mineralogy and Petrology, v. 21, p. 357–364.
Ribbe, P. H., 1960, An x-ray and optical investigation of the peristerite plagioclases: Am. Mineralogist, v. 45, p. 626–644.
—— 1962, Observations on the nature of unmixing in peristerite plagioclases: Norsk Geol. Tidsskr., v. 42, p. 138–151.
Ribbe, P. H., and Van Cott, H. C., 1960, Unmixing in peristerite plagioclases observed by x-ray and microscopic techniques [abs.]: Canadian Mining and Metall. Bull., v. 53, p. 200.
—— 1962, Unmixing in peristerite plagioclases observed by dark-field and phase-contrast microscopy: Canadian Mineralogist, v. 7, p. 278–290.
Rusinov, V. L., 1965, Disordered hydrothermal albite and its petrologic implication (in Russian): Akad. Nauk SSSR Doklady, v. 164, p. 410–413.
Schneider, T. R., 1957, Roentgenographische und optische Untersuchung der Umwandlung Albit-Analbit-Monalbit: Zeitschr Kristallographie, v. 109, p. 245–271.
Seki, Y., 1968, Synthesized wairakites: Their difference from natural wairakites: Geol. Soc. Japan Jour., v. 74, p. 457–458.
Smith, J. R., 1958, The optical properties of heated plagioclases: Am. Mineralogist, v. 43, p. 1179–1194.
Smith, J. R., and Yoder, H. S., Jr., 1956, Variations in x-ray powder diffractions patterns of plagioclase feldspars: Am. Mineralogist, v. 41, p. 632–647.
Smith, J. V., and Gay, P., 1958, The powder patterns and lattice parameters of plagioclase feldspars. II.: Mineralog. Mag., v. 31, p. 744–762.
Smith, J. V., and Ribbe, P. H., 1969, Atomic movements in plagioclase feldspars: Kinetic interpretation: Contr. Mineralogy and Petrology, v. 21, p. 157–202.
Viswanathan, K., and Eberhard, E., 1968, The peristerite problem: Schweizer Mineralog. u. Petrog. Mitt., v. 48, p. 803–814.
Vogel, T. A., 1970, Albite-rich domains in potash feldspar: Contr. Mineralogy and Petrology, v. 25, p. 138–143.
Waldbaum, D. R., and Kastner, M., 1968, Authigenic albite from the Island of Rhodes: Geol. Soc. America, Abs. for 1967, Spec. Paper 115, p. 230–231.

MANUSCRIPT RECEIVED BY THE SOCIETY MARCH 29, 1971

THE GEOLOGICAL SOCIETY OF AMERICA, INC.
MEMOIR 132
© 1972

Al^{IV}-Si^{IV} Disorder in Sillimanite and its Effect on Phase Relations of the Aluminum Silicate Minerals

H. J. GREENWOOD
Department of Geology
University of British Columbia
Vancouver 8, Canada

ABSTRACT

An amount of disorder among tetrahedral silicon and aluminum in sillimanite too small to measure readily by means of x-rays may have significant effects on phase relations of the Al$_2$SiO$_5$ minerals. The effect is nearly as great as that due to solid solution of minor constituents and cannot be ignored.

A statistical thermodynamic treatment permits evaluation of the Gibbs' free energy of sillimanite as a function of disorder at any temperature on the assumption of a most stable degree of order at a temperature. Reasonable permutations of the assumed order and its associated temperature indicate that the enthalpy change associated with complete disorder is approximately 4 kilocalories and that sillimanite may become completely disordered at temperatures of 1,700°C or greater. The processes of ordering and disordering are likely to occur by homogeneous diffusion rather than by nucleation of domains.

INTRODUCTION

The possibility that tetrahedrally coordinated aluminum and silicon in sillimanite may not be perfectly ordered has been suggested by Burnham (1963), Anderson and Kleppa (1969), Beger and others (1970), Chinner and others (1969), and by Zen (1969). Zen points out that if such disorder is possible "the petrological implications are considerable."

This paper is an attempt to explore in some detail the consequences of such an hypothesis. It must be stressed that the approach taken here is one of hypothesis, assumption, and testing numerically the consequences of these assumptions. As will become clear, if disorder is present in sillimanite, it is too slight to detect by conventional x-ray crystallographic techniques except, perhaps, by very careful single-crystal measurements and structural refinements. Nevertheless it will also develop that even such a slight amount of disorder as this can have substantial effects on the stability relations of the aluminum silicate minerals, and that the possibility of disorder cannot lightly be dismissed.

The arguments that follow are based on the assumption that some small amount of disorder is likely to exist, and that for any temperature there is a most stable degree of order. The coupled assumptions of a most stable degree of order and of a temperature at which this degree of order is stable permit calculation of the degree of order that is stable at any other temperature, the enthalpy and free energy of sillimanite at any degree of order and any temperature, and the effect of such assumed disorder on the phase reactions involving sillimanite. It is further possible to test the sensitivity of the conclusions reached to variation in the assumptions used, permitting one to evaluate the degree of confidence that may be placed in the conclusions.

MODELS

Sillimanite, Al_2SiO_5, can be regarded as being $Al^{VI}Al^{IV}Si^{IV}O_5$ in which the Roman superscripts refer to the first coordination of the cations. The Al^{IV} and Si^{IV} in course-grained natural sillimanite studied by Burnham (1963) are perfectly ordered, at least as far as can be determined by single-crystal x-ray analysis and structural refinement. In this condition (Fig. 1) there are no 2-coordinated oxygens shared between aluminum atoms. All such bridging oxygens, O_c, are located between Si-tetrahedra and Al-tetrahedra which share a common apex. Each IV-site is surrounded by four oxygen ions, three of which are coordinated to three cations and the fourth of which is the bridging oxygen, O_c, described above. Of the 3-coordinated oxygens, one, $O_a(Si)$ or $O_b(Al)$, is coordinated to the IV-site under consideration and to two Al^{VI} ions. The other two 3-coordinated oxygens, O_d, are coordinated to the IV-site under consideration and to one Al^{VI} and one neighboring IV-site. This coordination scheme applies to both the aluminum and silicon atoms.

Two different possible schemes of disordering come to mind. The first, here called model A (suggested by Zen, 1969) involves the complete randomization of Al^{IV} and Si^{IV} over the structure. This raises the likelihood that there will be some bridging oxygens coordinated only to two Al^{IV} atoms. This linkage is not common, but it is not unknown (Zen, 1969—$CaAl_2O_4$, $KAlO_2$, mizzonite) and thus cannot be rejected out of hand. Model B is a modification of model A formed by adding the constraint that any randomization is possible that does not result in linkages of the form Al^{IV}-O-Al^{IV}. This amounts to regarding the Al^{IV}-O-Si^{IV} linkage as a vector and to then considering the number of ways that a linear string of such vectors can be randomized over the structure. There are obviously half as many of these vectors as there are IV-sites in the structure so that the entropy and energy effects will be correspondingly less. There follows an outline theoretical

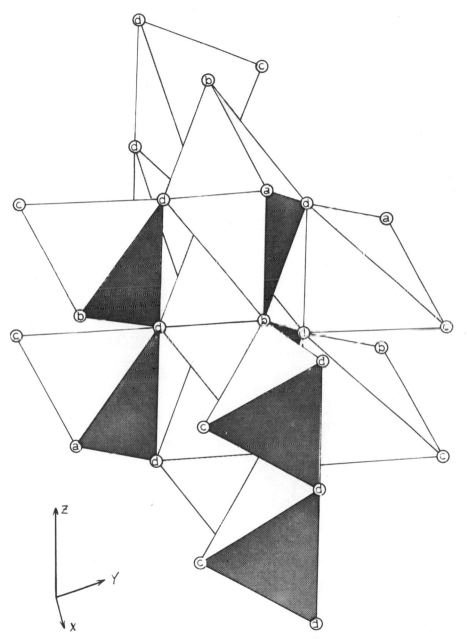

Figure 1. Polyhedral diagram of the sillimanite structure (Burnham, 1963). Chains of Al^{VI} octahedra are arranged along crystallographic c, with 0_a-0_b the common edge. The corners of these octahedra are connected to Al^{IV} and Si^{IV} tetrahedra which tie the octahedral chains together. The oxygen atoms are lettered in the same way as those in Burnham (1963, Fig. 4).

development of model A and the energetic relationships between model A and model B. After these formalities the numerical consequences of the models are examined.

FORMALITIES

A quantitative assessment of disorder in minerals requires the adoption of a parameter describing the state of the system and a notation describing the locations of the atoms or ions involved in the disordering. We will first define the atomic locations with reference to a perfectly ordered sillimanite. In such a crystal the Si^{IV} and Al^{IV} are perfectly ordered (Fig. 1) with Si- and Al-tetrahedra regularly alternating along crystallographic c and across the bridging oxygen O_c. We will refer to the locations of Al^{IV} and Si^{IV} in this perfectly ordered structure as α-sites and β-sites respectively.

A parameter in common use for describing the state of order in the system is the Bragg-Williams order parameter s (Moelwyn-Hughes, 1964) and it will serve our present purpose adequately. This parameter is not readily measured but it does afford a convenient description of the system. In a slightly disordered system there is never certainty about which of the ions involved will be encountered at any site, but there is a well-defined probability of which atom will be there. In the long run of examining a very large number of IV-sites the probability p of finding a silicon atom in a silicon-site and of finding an aluminum atom in an aluminum site is just the ratio of the total number of aluminum atoms that are on Al-sites to the total number of Al^{IV} atoms in the structure, that is,

$$pAl,\alpha = pSi,\beta = \frac{\text{number of } Al^{IV} \text{ on } \alpha\text{-sites}}{\text{total number of } Al^{IV}}. \tag{1}$$

The two probabilities must be identical if the mineral stays on the sillimanite composition. The probability that any particular site will contain the wrong cation is of course $(1 - p)$. The probability defined above is less convenient as an order parameter than s which is defined as

$$s = \frac{p - p_{complete\ disorder}}{p_{perfect\ order} - p_{complete\ disorder}}, \tag{2}$$

or alternatively as

$$p = \tfrac{1}{2}(1 + s), \text{ and } (1 - p) = \tfrac{1}{2}(1 - s). \tag{3}$$

The configurational entropy of disordered systems has received thorough treatment in many textbooks among which can be recommended the presentations of Kittel (1963) and Moelwyn-Hughes (1964). These are essentially identical, and their method of approach is followed closely here. The configurational entropy can be expressed as

$$\Delta S_{config} = R\ ln\ (w), \tag{4}$$

where w is the number of distinguishable arrangements of the system. The assumption is made that individual aluminum atoms are indistinguishable from one another as are silicon atoms. Thus, w must be the product of two combinatorial terms, explicitly,

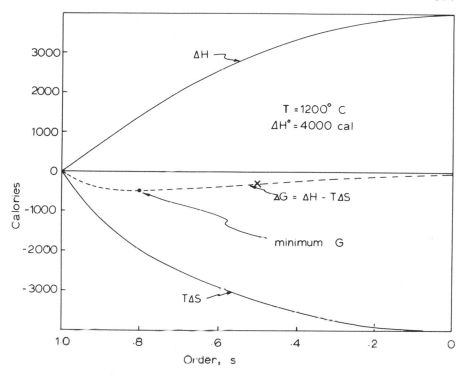

Figure 2. Enthalpy, entropy, and free energy of sillimanite disordered according to model A with $\Delta H° = 4{,}000$ calories at a temperature of $1{,}200°$C. Note the position and broad shallow shape of the minimum in the free-energy curve.

$$w = \left(c^{N Al\alpha}_{N Al}\right) \times \left(c^{N Al\beta}_{N Al}\right), \tag{5}$$

where $N Al\alpha$ is the number of aluminum atoms situated on α-sites and $N Al\beta$ is the number of aluminum atoms situated on β-sites. Using equations (3), (4), and (5), and applying Stirling's approximation for the factorials of large numbers one may arrive at the expression

$$\Delta S = R[2 ln2 - \{(1+s) ln(1+s) + (1-s) ln(1-s)\}], \quad \text{(model A)}. \tag{6}$$

It is of some interest to note that under conditions of complete disorder $(s = 0)$ the slope of this function is zero, and that where order is perfect the slope is infinite. The function, multiplied by a constant temperature of $1{,}200°$C $(1{,}473.1°$K) is illustrated in Figure 2. On the scale of the figure the infinite slope is not apparent, but its significance is that from an energetic standpoint it should be physically impossible to have a *perfectly* ordered sillimanite at any temperature except $0°$K. The entropy effect for complete disorder in model A is 2.7549 cal deg^{-1}mole^{-1}, which when multiplied by geologically reasonable temperatures produces a substantial energy effect, particularly in minerals like the aluminum silicates, which have small changes in free energy and entropy associated with their phase transitions.

The configurational entropy associated with model B is exactly half that associated with model A. This can be demonstrated by noting that in N moles of silli-manite there are $N/2$ moles of α-vectors and $N/2$ moles of β-vectors, and that having placed the α-vectors in position in the structure there is no remaining freedom to place the β-vectors. Consequently for model B

$$w = C \frac{\mathrm{N}_a}{\mathrm{N}_{aa}}, \tag{7}$$

where N_a is the number of α-vectors in an ordered sillimanite and N_{aa} is the number of α-vectors actually found in their correct place in the structure. The probabilities and the order parameter are defined in the same way in model B as in model A, but of course in model B the same numerical value of the parameter s corresponds to a less-disturbed structure than it does in model A.

The disordering of atoms that are normally ordered produces a change in the internal energy of the crystal as well as in its configurational entropy. This part of the effect is represented in the heat content or enthalpy of the mineral as a function of the order parameter. This is more difficult to evaluate than the configurational entropy because it properly needs calorimetric data on a sillimanite of known degree of order. Such data are not presently available. In the section that follows it will be shown that even though we have no measured information on the enthalpy of disordering we can nevertheless deduce the functional form of the equation that describes the variation of enthalpy with order. This deduction is necessarily based on assumptions about the energy associated with a particular atom on a particular site. Combination of the functional form of the $\Delta H(s)$ curve with the equation for entropy of disorder permits calculation of the Gibbs free energy associated with disorder, which further permits an estimate of the most stable degree of order at any temperature.

The enthalpy of disorder is assumed here to be entirely composed of a temperature-independent potential energy summation. The contribution to the enthalpy of a 4-coordinated cation is assumed to be the sum of its energies of interaction with all its first and second coordination neighbors. The actual values of these terms are unknown, but upon treating them algebraically and taking the sum it turns out that the difference in enthalpy on these assumptions, between ordered and partially disordered sillimanite, contains an unknown energy constant characteristic of sillimanite modified by the factor $(1 - s^2)$ with no terms in s. Because of this simple result it is possible to calculate this energy constant (enthalpy of complete disorder) such that it is consistent with any assumed values of temperature and most stable degree of order. There follows an outline account of the model.

In model A there are eight distinct coordination hook-ups and it is assumed that there are thus eight distinct interaction energies. These unknown energy terms, ϕ_1, ϕ_2, and so forth, may be labelled:

ϕ_1 $\mathrm{Al^{IV}\text{-}O_c\text{-}Si^{IV}}$

ϕ_2 $\mathrm{Al^{IV}\text{-}O_c\text{-}Al^{IV}}$

ϕ_3 $\mathrm{Si^{IV}\text{-}O_c\text{-}Si^{IV}}$

ϕ_4 $\mathrm{Al^{IV}\text{-}O_d} \Big\langle {}^{Si^{IV}}_{Al^{VI}}$

ϕ_5 $\mathrm{Al^{IV}\text{-}O_d} \Big\langle {}^{Al^{VI}}_{Al^{IV}}$

ϕ_6 $\mathrm{Si^{IV}\text{-}O_d} \Big\langle {}^{Al^{VI}}_{Si^{IV}}$

ϕ_7 $\mathrm{Al^{IV}\text{-}O_{b,a}} \Big\langle {}^{Al^{VI}}_{Al^{VI}}$

ϕ_8 $\mathrm{Si^{IV}\text{-}O_{a,b}} \Big\langle {}^{Al^{VI}}_{Al^{VI}}$

With the bookkeeping convention already adopted, the energy contribution of an aluminum atom on an α-site is equal to the interaction energy of Al on α multiplied by the probability p that the site will actually be occupied by an aluminum atom. Each energy term, ϕ_1, and so forth, must also be multiplied by the appropriate probability that we are in fact considering the correct stereochemical arrangement. If $U_{Al, \alpha}$ is the total energy contribution made by aluminum atoms on α-sites it follows that

$$U_{Al}\alpha = \tfrac{1}{2}p[p\phi_1 + (1 - p)\phi_2 + p\phi_4 + (1 - p)\phi_5 + p\phi_4 + (1 - p)\phi_5 + \phi_7]. \quad (8)$$

The pre-multiplier of $\tfrac{1}{2}$ is necessary because in simply adding up all the interactions we have counted every interaction twice. The factor p before the parentheses is used to account for the probability that the α-site we are considering is actually occupied by an aluminum atom. Within the main parentheses the factors p and $(1 - p)$ account for the probability that the interaction energy noted is actually in effect at the site under consideration. Collecting terms and rewriting,

$$U_{Al, a} = \tfrac{1}{2}p[p(\phi_1 + 2\phi_4) + (1 - p)(\phi_2 + 2\phi_5) + \phi_7], \quad (9)$$

and similarly for the energies $U_{Al, \beta}$, $U_{Si, a}$, $U_{Si, \beta}$, giving

$$U_{Al, \beta} = \tfrac{1}{2}(1 - p)[p(\phi_2 + 2\phi_5) + (1 - p)(\phi_1 + 2\phi_4) + \phi_7], \quad (10)$$

$$U_{Si, \beta} = \tfrac{1}{2}p[(1 - p)(\phi_3 + 2\phi_6) + p(\phi_1 + 2\phi_4) + \phi_8], \quad (11)$$

$$U_{Si, a} = \tfrac{1}{2}(1 - p)[p(\phi_3 + 2\phi_6) + (1 - p)(\phi_1 + 2\phi_4) + \phi_8]. \quad (12)$$

Addition of these four energy terms yields

$$\begin{aligned} U_{Total} = \tfrac{1}{2}[\phi_1 &+ \tfrac{1}{2}(\phi_2 + \phi_3) + 2\phi_4 + \phi_5 + \phi_6 + \phi_7 \\ &+ \phi_8 + s^2(\phi_1 - \tfrac{1}{2}(\phi_2 + \phi_3) + 2\phi_4 - \phi_5 - \phi_6)], \end{aligned} \quad (13)$$

which can be simplified to

$$U_{Total} = \Phi_1 + S^2\Phi_2, \quad (14)$$

where

$$\Phi_1 = \tfrac{1}{2}[\phi_1 + \tfrac{1}{2}(\phi_2 + \phi_3) + 2\phi_4 + \phi_5 + \phi_6 + \phi_7 + \phi_8] \quad (15)$$

and

$$\Phi_2 = \tfrac{1}{2}[\phi_1 - \tfrac{1}{2}(\phi_2 + \phi_3) + 2\phi_4 - \phi_5 - \phi_6]. \quad (16)$$

Note that Φ_1 and Φ_2 are energy (enthalpy) constants characteristic of sillimanite and are, at this point, unknown. Evaluating equation (14) under conditions of perfect order ($s = 1$) results in

$$U_{Total, s=1} = \Phi_1 + \Phi_2, \quad (17)$$

and taking the difference between equations (16) and (17) gives

$$(U_{Total, s=1} - U_{Total}) = \Phi_2(1 - s^2) \quad (18)$$

as the difference in energy between perfectly ordered and partially disordered sillimanite. If we now regard this energy term Φ_2 as enthalpy we can write[1]

---

[1] This is equivalent to ignoring the PV term in the thermodynamic identity $H = U + PV$, a reasonable assumption at one atmosphere and an untested but possibly less reasonable one at pressures of tens of kilobars.

$$\Delta H_{\text{disorder}} = \Delta H^\circ (1 - s^2) \tag{19}$$

Equation (19) provides the justification for using a simple function to describe the heat effect associated with Al^{IV}-Si^{IV} disorder in sillimanite, according to the constraints of model A. It should be noted at this point that for crystals of identical structure and possibilities for disorder, but of different chemistry, the ΔH of disorder will depend on the chemistry alone and the ΔS of disorder on the structure alone. Consequently, for a particular temperature the minimum in free energy will occur at different degrees of order if the compositions are different. However, if two isostructural phases of different composition should fortuitously have free-energy minima at the same degree of order at a particular temperature it would be necessary that they have the same ΔH° also, granting the assumptions of ideal disordering followed here.

It is a simple matter to deduce the form of the enthalpy versus order function required by model B if we note that for model B equations (8) to (16) inclusive apply exactly except that in model B, ϕ_1 always applies and ϕ_2 and ϕ_3 never apply. Accordingly, we replace ϕ_2 by ϕ_1 and ϕ_3 by ϕ_1 in those equations yielding

$$U_{\text{Total}} = \Phi_3 + s^2 \Phi_4 \tag{20}$$

where

$$\Phi_3 = \tfrac{1}{2} (2\phi_1 + 2\phi_4 + \phi_5 + \phi_6 + \phi_7 + \phi_8), \tag{21}$$

and

$$\Phi_4 = \tfrac{1}{2} (2\phi_4 - \phi_5 - \phi_6). \tag{22}$$

Again, taking differences between ordered and disordered states gives

$$\Delta H_{\text{disorder}} = \Delta H^\circ (1 - s^2), \tag{23}$$

which is identical with equation (19) for model A.

The Gibbs' free energy associated with the change from ordered to disordered sillimanite can now be found by combining equations (6) and (19) according to the expression

$$\Delta G = \Delta H - T\Delta S, \tag{24}$$

to give for model A

$$\Delta G_A = \Delta H^\circ_A (1 - s^2) + RT[(1 + s)ln(1 + s) + (1 - s)ln(1 - s) - 2ln2]. \tag{25}$$

Both model A and model B are of the "convergent" type of disordering (Thompson, 1969) as sites α and β are topologically indistinguishable when their occupancies are the same.

It can be seen by inspection of equation (25) that if one knew or assumed a value for ΔH°_A it would be possible to calculate the Gibbs' free energy of disorder at any arbitrary combination of temperature and the order parameter s. ΔH°_A is not available from calorimetric measurements, but if one is willing to assume a most stable degree of order and its associated temperature, it is possible to find the ΔH°_A that is consistent with these assumptions.

The most stable degree of order occurs at that value of s where the ΔG versus s

curve (eq. 25) passes through a minimum. Accordingly, we find the first derivative of equation (25) with respect to s, set it equal to zero and, rearranging terms, we find

$$\Delta H_A^\circ = RT\left[\frac{ln(1+s) - ln(1-s)}{2s}\right] \qquad (G\text{min}). \qquad (26)$$

Substitution of paired values of T and s in equation (26) permits calculation of ΔH_A°, or conversely, a knowledge of ΔH_A° permits calculation of the most stable degree of order at any arbitrary temperature.

Before evaluating the consequences of the models, it will be useful to derive one further expression. The needed expression is for the location of the point of inflection in the G versus s curve, separating the parts of the curve that are convex upward from the part that is convex downward. This is found by setting equal to zero the second derivative of ΔG with respect to s and results in

$$s = \sqrt{1 - \frac{RT}{\Delta H_A^\circ}} \qquad \text{(inflection point)}. \qquad (27)$$

EVALUATION AND CONSEQUENCES OF THE MODEL

We have now arrived at the point of evaluating equation (26) and from it the rest of the free-energy function, equation (25). As we do not *know* either ΔH_A° or an appropriate set of (T, s) the approach has been to permute assumed values of (T, s) through what seems to be a reasonable range and from this to find the range of ΔH_A° that results. The basis of the assumptions made is as follows: the infinite slope of equation (25) where $s = 1$ ensures that some degree of disorder will be present, however slight; the methods of single-crystal structural analysis do not permit recognition of less than about 5 percent of disorder (Charles W. Burnham, 1968, oral commun.); the most ordered natural sillimanite must have formed at the lowest temperature possible, which is the temperature of the triple point between kyanite, andalusite, and sillimanite. I have therefore evaluated equation (26) over the range of s from $s = 0.9$ to $s = 1.0$, at temperatures of 500°C, 550°C, 600°C, 650°C, and 700°C. The results are presented in Figure 3. It can be seen that the entire range of reasonable assumptions regarding the most stable degree of order at specific temperatures results in a range of ΔH_A° from about 3,000 calories per mole to about 5,500 calories per mole of sillimanite. If natural sillimanites disordering according to model A are more ordered than $s = 0.995$, the consistent ΔH_A° rapidly becomes very large and the estimate of this quantity is correspondingly weak. It is of interest to compare the values of ΔH_A° from Figure 3 with the enthalpy of disordering Al^{IV} and Si^{IV} in albite. Holm and Kleppa (1968) present results from their study of albite to indicate that the heat effect is 2,400 calories for the first step of disordering and 1,000 for the second step, giving a total heat of disorder of 3,400 calories. These values are sufficiently close to the values deduced here for sillimanite as to provide some measure of confidence in the results.

Any assumed value of ΔH_A° implies a particular position of the point of minimum free energy at any temperature, hence it is possible to calculate the degree of order that is most stable at any temperature for a particular ΔH_A°. The results of

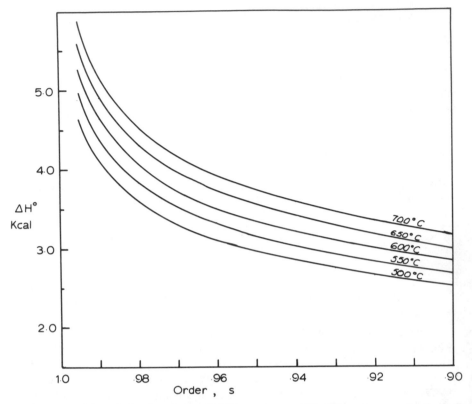

Figure 3. The variation in calculated $\Delta H°$ consistent with different assumed values of equilibrium disorder s and its associated temperature. Note that the range of most reasonable s (that amount that would be difficult or impossible to detect by x-rays) coupled with experimentally and geologically reasonable temperatures results in values of $\Delta H°$ ranging only from 3 kilocalories to 5 kilocalories.

these calculations are presented in Figures 4 and 5. The latter is an enlarged view of the extreme left-hand side of Figure 4. From these diagrams can be read the most stable degree of order s for any temperature under any of the assumed values of the enthalpy of disorder. It will be seen that under all reasonable assumptions of ΔH_A° the degree of disorder is very slight at temperatures below 500°K and that models having higher enthalpy of disorder are more ordered at any temperature.

Figures 4 and 5 are for model A, but they also apply to model B if the energy labels on the individual curves are divided by 2.0. It can also be seen from Figure 5 that the disorder would be virtually undetectable below 500°C if the ΔH_A° were 3,340 calories, but that for the same enthalpy change it should be detectable in the neighborhood of 1,000°C. Below 500°C all curves require that s be more than 0.97, and if $\Delta H_A^{\circ} = 4,000$ calories, s must be more than 0.987.

The effect of pressure on the order parameter is unknown at present because of the impossibility of estimating the volume change, if any, that results from disorder. It seems likely that the volume change is small, and on this assumption the degree of order has been assumed to be independent of pressure although Beger

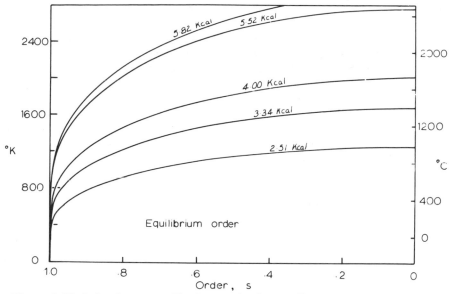

Figure 4. Variation in most stable degree of order in sillimanite required by a range of temperatures and enthalpies of disorder. If $\Delta H° = 4,000$ calories, disorder is complete at 1,740°C.

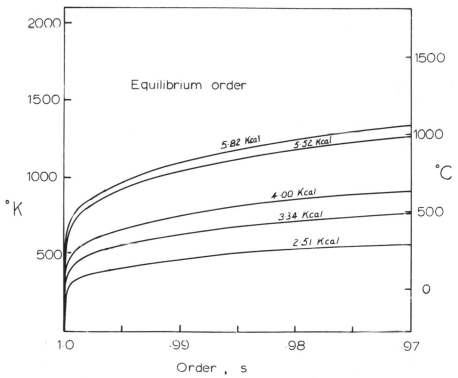

Figure 5. Variation in most stable degree of order in sillimanite required by a range in temperatures and enthalpies of disorder. Note how the order rapidly becomes almost perfect as absolute zero is approached, and that only at 0°K is order perfect.

and others (1970) indicate that there may be a pressure effect. Figure 6 has been
prepared for the condition that $\Delta H_A^\circ = 4,000$ calories per mole with the degree
of order independent of pressure. In this figure the positions of the kyanite-silliman-
ite-andalusite boundaries are taken from Richardson and others (1969) and the
curves involving mullite from thermochemical data in Robie and Waldbaum
(1968). In this figure a simple Clapeyron calculation was used, and no modification of the boundaries by the disorder in sillimanite was introduced. The following
points may be noted from the diagram. Any temperature requires a particular
degree of Al^{IV}-Si^{IV} disorder. Any point on a univariant equilibrium involving sillimanite requires a particular degree of order if perfect equilibrium is obtained. The
possibility of complete disorder of the Al^{IV} and Si^{IV} in sillimanite over about
1,700°C permits the possibility of complete solid solution between sillimanite and

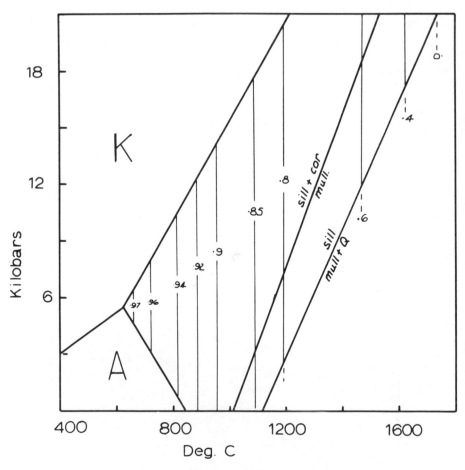

Figure 6. Stability relations of the minerals andalusite, kyanite, and sillimanite after
Richardson and others (1969) and of mullite from Clapeyron calculation on thermodynamic data in Robie and Waldbaum (1968). None of the phase relations have been
modified from their experimental or calculated positions by means of the disorder effects
considered here. The sillimanite field is contoured by lines of constant degree of order
on the assumptions that the degree of order is independent of pressure and that the
$\Delta H^\circ = 4,000$ calories.

mullite, which has a structure corresponding to a completely disordered silliman-ite. In particular, if at some pressure and temperature mullite and sillimanite should reach the same composition by such solid solution, the reaction sillimanite = mullite + quartz would end at an upper critical end point. The evidence recently presented by Hariya and others (1969) in support of continuous solid solution between sillimanite and mullite may in fact be a manifestation of a convergence in structure made possible by Al^{IV}-Si^{IV} disorder in sillimanite.

Isotherms of the order parameter s versus Gibbs' free energy have been calcu-lated for several different assumptions of ΔH_A° and are illustrated in Figures 2, 7, 8, 9, and 10. All the curves have the same general aspect, and although calculated specifically for model A, they apply equally well to model B if the numbers on the energy scale are halved and the ΔH° label is also halved. The following features of all the isotherms may be noted. Every isotherm has a minimum and an inflection point. The inflection point is always at a more disordered position than the point of minimum free energy. The minimum is broad and shallow with particularly low slopes on the low-s side of the minimum. A consequence of this is that the energy difference driving the system to reach the state of lowest energy is small, and the process of ordering will probably be slow. A final point to note is that the difference

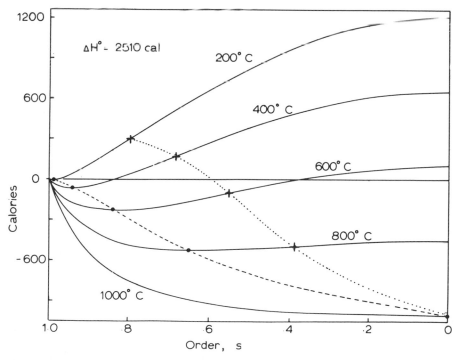

Figure 7. Gibbs' free energy of disordered sillimanite at different temperatures under the assumption that $\Delta H^\circ = 2,510$ calories, model A. The dashed curve connecting the solid dots gives the locus of the minima in Gibbs' free energy, and the dotted line con-necting the crosses gives the locus of the points of inflection in the $G(s)$ curves. Points to the right of the dotted curve are unstable with respect to diffusion and domain forma-tion, and points on the dashed curve correspond to sillimanites with the most stable degree of order.

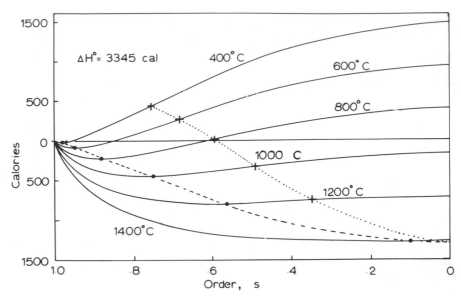

Figure 8. Gibbs' free energy of disordered sillimanite at different temperatures under the assumption that $\Delta H^\circ = 3{,}345$ calories, model A. *See* caption for Figure 7.

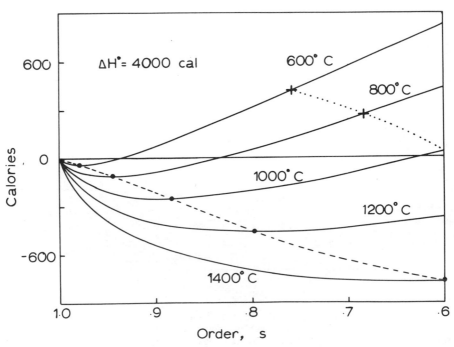

Figure 9. Gibbs' free energy of disordered sillimanite at different temperatures under the assumption that $\Delta H^\circ = 4{,}000$ calories, model A. *See* caption for Figure 7.

TABLE 1. FREE-ENERGIES OF SOLID SOLUTION AND AL-SI DISORDER
ACCORDING TO MODEL A

ΔH°_A Calories	T °C	order s at minimum ΔG	Minimum ΔG Calories	$RT \ln s$ at minimum ΔG Calories
5520	600	.9947	− 6.03	− 9.22
5520	800	.9880	− 24.75	− 25.74
5520	1200	.9448	−146.94	−166.22
4000	600	.9782	− 36.03	− 38.24
4000	800	.9436	−109.25	−123.80
4000	1000	.8852	−246.22	−308.51
4000	1200	.7960	−463.34	−667.91
3345	600	.9500	− 79.52	− 89.90
3345	800	.8816	−213.39	−268.73
3345	1000	.7680	−443.84	−667.83

in free energy between sillimanites at different temperatures and the same degree of order depends on the model and on the ΔH°.

Figures 7, 8, 9, and 10 together with Table 1 illustrate the magnitudes of the energy effects involved in the different models, and permit comparison with an ideal solution model. From this table it can be seen that an amount of ideal solid solution numerically equal in terms of mole fraction to the order parameter produces energy effects comparable with the disorder model. For mineral equilibria within this system we must therefore give as much weight to the effects of Al^{IV}-Si^{IV} disorder as to the possible effects of contamination and solid solution.

A detailed examination of Figures 8, 9, and 10 can be aided by reference to Table 2, which presents numerical values illustrating the excess free energies that become available upon changing the temperature of a sillimanite without changing the degree of order. Table 2 lists the excess free energy that results from taking a stably disordered sillimanite from either 800°C or 1,000°C to 600°C without changing the degree of order to that which is most stable. For example, consider Figure 8, for model A with $\Delta H^\circ_A = 3,345$ calories. At 1,000°C the most stable state of order has $s = 0.7680$. Reduction of the temperature to 600°C without change of order causes the curve labelled 600°C to apply, and at this degree of order at 600°C sillimanite has a free energy that is 206.9 calories higher than the minimum of −79.5 calories at $s = 0.9500$. Other values of excess free energy are presented in Table 2 and may be similarly compared with the curves of Figures 7, 8, 9, and 10.

The amounts of energy involved appear small, perhaps trivial, and would be dismissed lightly for most reactions. However, the reactions in the Al$_2$SiO$_5$ system are unusual in that they have small entropy changes and thus have equilibrium temperatures that are very sensitive to small changes in the free energy of any of their participants. To appreciate this, consider a definition of the entropy change of reaction ΔS_r of

$$\frac{d \Delta G}{dT} = - \Delta S_r. \tag{28}$$

If the entropy change is essentially independent of temperature over a small range

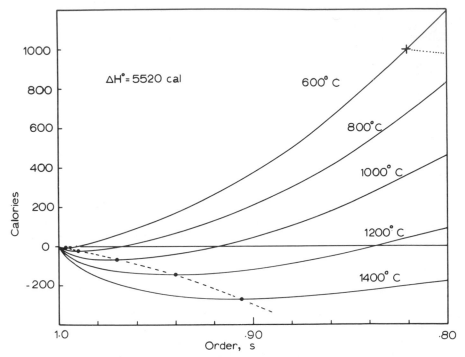

Figure 10. Gibbs' free energy of disordered sillimanite at different temperatures under the assumption that $\Delta H° = 5{,}520$ calories, model A. *See* caption for Figure 7. In comparing Figures 7, 8, 9, and 10 note that as $\Delta H°$ increases, the magnitude of the free energy effect at minimum G is decreased. The figures also apply to model B if the figures on the energy scale are halved.

near the equilibrium temperature, we can write

$$\frac{d\,T_e}{d\,\Delta G_r} = \frac{-1}{\Delta S_r},\qquad (29)$$

where T_e is the equilibrium temperature, indicating that perturbation of the equilibrium temperature by perturbation in the free energy of any reactant or product is equal to the reciprocal of the entropy change. Obviously a small entropy change will have a large reciprocal and the equilibrium temperature will be sensitive to small changes in G of any of the involved phases. Thompson (1955, p. 73) has also remarked on the sensitivity of equilibrium temperature to perturbations of free energy at low ΔS. Equation (29) has been applied to the excess free energies of Table 2 and the results shown in the same table. It will be seen that the effect of metastable persistence of disorder in sillimanite may have appreciable effect on the equilibrium temperatures of reactions involving stillimanite, and these may be important in both nature and the laboratory.

In experiments the use of material formed at one temperature to test "equilibrium" at another temperature may produce quite misleading results, with metastable sillimanite transforming to kyanite or andalusite well above the stable transformation temperature. Such transformations may well be reversible, even though they are not taking place between the *most stable* phases. It seems possible, in particular, that some of the differences between the results of Richardson and others (1969) and Althaus (1969) may as well be attributed to differing states of disorder in the sillimanites used as to differences in composition. Certainly both should be influential. It seems unlikely in view of the shape of the shallow minimum in free energy with disorder that reordering of sillimanite could occur in times convenient for laboratory study, and thus, if disorder occurs, it should persist.

In nature similar complexities are presumably possible and may provide a partial explanation of naturally coexisting Al_2SiO_5 polymorphs. For example, in progressive metamorphism with increasing temperature, the Al_2SiO_5 polymorph produced by reaction might be sillimanite if the sillimanite nuclei were of the stable state of order but kyanite or andalusite if the sillimanite nuclei were somewhat more disordered than permitted by equilibrium. Perhaps, for example, fibrolite may be a disordered variety of sillimanite (Chinner and others, 1969) and thus might behave in this way. If there were a range in the degree of order of the sillimanite nuclei, then both kyanite (stable with respect to disordered sillimanite and metastable with respect to ordered sillimanite) and stably ordered sillimanite would form simultaneously. Perhaps mixed associations should be more the rule than the exception, and this view is not at variance with the natural occurrences.

In Figures 7, 8, 9, and 10 are marked dotted curves passing through + marks. These curves trace out the loci of the points of inflection as temperature is changed. They are analogous to the spinodal curves of binary solid solutions and may be interpreted in much the same way. At any temperature, the curve to the left of the + mark is concave upward and any mixture of two sillimanites of differing s in this range has a higher free energy than a sillimanite of intermediate s lying on the curve. Hence there is no potential driving the homogeneous sillimanite in this range to separate into two or more domains of higher and lower degree of

TABLE 2. Excess Free Energies and Changes in P-T Boundaries Due to Metastable Persistence of Disorder Upon Change of Temperature

	ΔG excess at 600°C Calories	ΔT°C Ky-Sill	ΔP bars Ky-Sill	ΔT°C And-Sill	ΔP, bars And-Sill
$\Delta H_A^\circ = 5,520$ cal Stable at 800°C	8.0	2.7°	57.4	11.6°	205.
$\Delta H_A^\circ = 5,520$ cal Stable at 1,000°C	58.0	19.7°	419.	84.1°	1488.
$\Delta H_A^\circ = 4,000$ cal Stable at 800°C	31.5	10.7°	227.	45.6°	807.
$\Delta H_A^\circ = 4,000$ cal Stable at 1,000°C	141.2	47.9°	1018.	205.°	3630.
$\Delta H_A^\circ = 3,345$ cal Stable at 800°C	47.2	16.0°	340.	68.5°	1212.
$\Delta H_A^\circ = 3,345$ cal Stable at 1,000°C	206.9	70.2°	1490.	300.°	5310.

order. To the right of the + mark, however, the free-energy curve is concave downward, and sillimanites in this range would tend to spontaneously separate into two or more different sillimanites having higher and lower degrees of order. From the practical standpoint, however, except for models of very low $\Delta H°$ (for example, 2,500 cal), no reasonable reduction in temperature could render a sillimanite unstable with respect to domain formation, so it seems likely that the processes of ordering and disordering will occur by means of homogeneous diffusion rather than by the heterogeneous processs of domain formation, even ignoring the matter of the size required of nuclei for continued growth.

SUMMARY

Assumption of the possibility of Al^{IV}-Si^{IV} disorder in sillimanite leads to the conclusions that sillimanite should co-exist metastably with kyanite and andalusite over a range of temperature, that use of sillimanite formed at one temperature to test equilibrium at another temperature may be misleading, that discrepant experimental results may be due in part to different degrees of order in the sillimanite used, that Al^{IV}-Si^{IV} disorder is as important as solid solution in displacing Al_2SiO_5 equilibria, and that in general any solid-solid reaction having a small entropy change will be highly sensitive to small perturbations in free energy whatever the source, be it disorder, solid solution, or even grain size and surface energy.

ACKNOWLEDGMENTS

The ideas presented here have benefitted greatly from discussions with more scientists than I can correctly recall, among whom I am most grateful to note E-an Zen, Charles W. Burnham, G. V. Gibbs, E. P. Meagher, Egon Althaus, F. Seifert, W. G. Ernst, Reolof Schuiling, and P. M. Orville. David Waldbaum and E-an Zen contributed much by their careful and perceptive reviews. I would hasten to add that none of these helpful and interested people should be held responsible for ideas expressed here. Financial support of computations done at the University of British Columbia Computing Center and of manuscript preparation were borne by grant 67-A-4222 of the National Research Council of Canada.

REFERENCES CITED

Althaus, Egon, 1969, Experimental evidence that the reaction of kyanite to form sillimanite is at least bivariant: Am. Jour. Sci., v. 267, p. 273–277.

Anderson, P.A.M., and Kleppa, O. J., 1969, The thermochemistry of the kyanite-sillimanite equilibrium: Am. Jour. Sci., v. 267, p. 285–290.

Beger, R. M., Burnham, C. W., and Hays, J. F., 1970, Structural changes in sillimanite at high temperature: Geol. Soc. America, Abs. with Programs (Ann. Mtg.), v. 2, no. 7, p. 490–491.

Burnham, C. W., 1963, Refinement of the crystal structure of sillimanite: Zeitschr. Kristallographie, v. 118, p. 127–148.

Chinner, G. A., Smith, J. V., and Knowles, C. R., 1969, Transition-metal contents of Al_2SiO_5 polymorphs: Am. Jour. Sci., v. 267–A (Schairer Volume), p. 96–113.

Hariya, Y., Dollase, W. A., and Kennedy, G. C., 1969, An experimental investigation of the relationship of mullite to sillimanite: Am. Mineralogist, v. 54, p. 1419–1441.

Holm, J. L., and Kleppa, O. J., 1968, Thermodynamics of the disordering process in albite: Am. Mineralogist, v. 53, p. 123–133.

Kittel, Charles, 1963, Introduction to solid state physics: John Wiley & Sons, Inc., New York, 617 p.

Moelwyn-Hughes, E. A., 1964, Physical chemistry: Oxford, England, Pergamon Press Ltd., 1334 p.

Richardson, S. W., Gilbert, M. C., and Bell, P. M., 1969, Experimental determination of kyanite-andalusite and andalusite-sillimanite equilibria; the aluminum silicate triple point: Am. Jour. Sci., v. 267, p. 259–272.

Robie, R. A., and Waldbaum, D. R., 1968, Thermodynamic properties of minerals and related substances at 298.15°K (25.0°C) and one atmosphere (1.013 bars) pressure and at higher temperatures: U.S. Geol. Survey Bull. 1259, 256 p.

Thompson, J. B., Jr., 1955, The thermodynamic basis for the mineral facies concept: Am. Jour. Sci., v. 253, p. 65–103.

——1969, Chemical reactions in crystals: Am. Mineralogist, v. 54, p. 341–375.

Zen, E-an, 1969, The stability relations of the polymorphs of aluminum silicate; a survey and some comments: Am. Jour. Sci., v. 267, p. 297–309.

Manuscript Received by the Society March 29, 1971

THE GEOLOGICAL SOCIETY OF AMERICA, INC.
MEMOIR 132
© 1972

Matrix-Rich Pleistocene Sediments from Western Washington: Incipient Graywacke-Type Sedimentary Rocks?

JOHN T. WHETTEN
University of Washington, Department of Geological Sciences
Seattle, Washington 98195

ABSTRACT

Pleistocene gravels with abundant interstitial matrix of silt and clay are common in western Washington. At one locality, the matrix appears to result from chemical alteration of detrital clasts. In an adjacent sand bed, hypersthene is altering to clay. A similar process may have formed matrix in graywackes. Some matrix in the gravels may have been introduced by other mechanisms.

INTRODUCTION

Many Pleistocene outwash gravels in western Washington have a distinctive bimodal size distribution similar to that of graywacke-type sedimentary rocks. The interstitial silt and clay matrix coats and surrounds coarse detrital clasts. This is noteworthy, because unconsolidated sediments with abundant matrix are uncommon (Cummins, 1962; Hollister and Heezen, 1964).

The purpose of this paper is to call attention to these matrix-rich sediments, for they may represent an important early stage in the formation of graywacke. Sediments from three gravel pits near Sumner, Dungeness, and Kalaloch (Fig. 1) are described. There are still some unresolved questions regarding the origin of the matrix.

These and other Pleistocene sediments in western Washington were deposited in a complex pattern of depositional environments ranging from glacial to lacustrine to marine. Sediment sources were also varied; clasts from plutonic, volcanic, and metamorphic rocks characterize the deposits originating in the Cascade Range, whereas Olympic Peninsula sources contributed mostly clastic sedimentary rocks.

573

STUCK DRIFT NEAR SUMNER, WASHINGTON

Fluvial and lacustrine sediments (Stuck Drift and Puyallup Formation, respectively; Table 1) are exposed in a gravel pit south of Sumner, Washington (Fig. 1), on the south side of State Highway 162 1 km west of the junction with Highway 167. The sediments in this area were described most recently by Crandell (1963), although this gravel pit had not been excavated at the time of Crandell's study.

Figure 1. Index map showing localities mentioned in text.

Approximately 12 m of the upper Stuck Drift and the lowermost 4 m of the Puyallup Formation are exposed. The drift consists of cross-bedded gravels with two interbedded sand lenses up to 2 m thick. Till is not present in this outcrop but presumably lies below it. The sediments are probably recessional outwash deposited during retreat of the Stuck Glacier. About 150 m of younger Pleistocene sediments overlie the gravel pit.

Of particular interest is the gravel between approximately 1.5 and 3 m above the quarry floor (Fig. 2A). Moderately sorted pebbles and cobbles up to 15 cm in diameter, composed of a variety of volcanic, plutonic, and metamorphic rocks (derived from the Cascades), are surrounded by a matrix of silt and clay which fills the voids between clasts (Fig. 2B). The difference in particle size between large framework clasts and interstitial matrix is striking. Sizes in the range of fine gravel, sand, and coarse silt are virtually absent. X-ray analysis shows that the matrix clay is composed of montmorillonite, illite, and kaolinite, in decreasing order of abundance.

Many pebbles and cobbles in the matrix-rich gravel are decomposing (Fig. 2C); some cannot be removed intact without crumbling. Many are altering to clay, and some course-grained rocks are disaggregated. Meaningful size analyses of these sediments cannot be made because of the disintegrating clasts.

Underlying the matrix-rich gravel is gravel in which the voids between clasts are filled with coarse sand, and the pebbles are not noticeably altered (Fig. 2A). A lens of sand lies on top.

TABLE 1. MAJOR PLEISTOCENE GLACIAL AND INTERGLACIAL EPISODES IN THE SOUTHERN PUGET SOUND LOWLAND (IN ORDER OF INCREASING AGE)

Fraser Glaciation

Olympia Interglaciation

Salmon Springs Glaciation

Puyallup Interglaciation

Stuck Glaciation

Alderton Interglaciation

Orting Glaciation

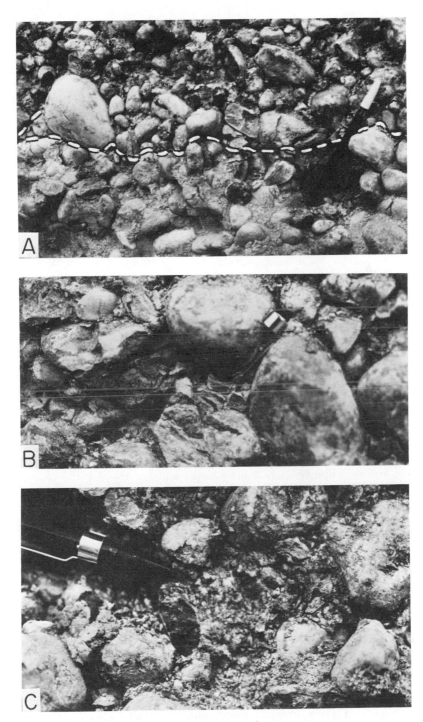

Figure 2. Stuck Drift near Sumner, Washington. (A) Matrix-rich gravels
and altered pebbles (above dashed line) and sandy gravel with fresh pebbles
(below). (B) Silty clay matrix coats pebbles and fills voids between pebbles
(note void-filling matrix at tip of pen clip). (C) Decomposed plutonic(?) pebble (at tip of pen point).

DUNGENESS KAME

A kame (identified by D. Biederman, oral commun.) formed during the Fraser Glaciation (Table 1) approximately 14,000 yrs B.P. has been excavated for gravel 2 km south of Dungeness (Fig. 1) on the Sequim Highway. Judging from the age of trees on the gravel pit floor the excavation is probably about 15 to 20 yrs old.

Although the walls of the pit are partly covered by sliderock, the internal structure of the kame appears to consist mostly of cross-bedded gravels and sandy gravels, with local interbedded sands and silts.

A large amount of silt and clay matrix is present in the gravels in the lower part of the exposure (Fig. 3B). Virtually none occurs near the top. This contrast is apparent in size analyses of representative samples from the top and the bottom of the kame (Fig. 4). As at Sumner, the clay matrix fills voids and surrounds the pebbles and cobbles. X-ray analysis of the clay fraction shows that montmorillonite, chlorite, and illite are present, in decreasing order of abundance.

The pebbles and cobbles are composed of plutonic, volcanic, and metamorphic rocks, probably from the Cascade Range. Some appear to be altering in outcrop, but others are fresh. A number of cobbles of volcanic and plutonic rocks have rolled to the base of taluses and have decomposed and disintegrated by weathering since the gravel pit was worked (Fig. 3C).

No younger sediments overlie the kame, and the soil on its surface is thin and weakly structured.

KALALOCH SEDIMENTS

A gravel pit 4.5 km south of Kalaloch (Fig. 1) on the west side of Highway 101 has exposed approximately 5 m of sands and gravels of Pleistocene(?) age (Fig. 5). A silty clay bed 1 m thick overlies the coarse sediments.

There is a very large amount of matrix clay in the gravels in the lowermost 1.5 m of the exposure but very little above this level. The matrix-rich sediments show up on the vertical wall of the quarry as a "wet zone" underlying the "dry zone" of gravels without matrix. Size analyses of samples representing these two sediment types are shown in Figure 6. Chlorite and illite are present in the clay fraction.

A discontinuity may occur between the matrix-rich and matrix-poor sediments, as the latter have large-scale cross-beds which are not apparent in the sediments below (Fig. 5).

The composition and mean size of the pebbles appears to be uniform throughout the gravelly portion of this exposure. Virtually all the clasts are graywackes derived from the interior of the Olympic Peninsula.

The large volume of silt and clay matrix in some places completely obscures the pebbles. Irregular stringers of clay-rich sediment appear to cut across stratigraphic boundaries. In addition, black and dark red oxides (as yet unidentified) have precipitated along irregular discontinuous surfaces apparently cut across bedding planes.

Although the silt and clay matrix is virtually absent in most of the dry zone, locally a reddish-colored matrix (composed of chlorite and illite) has developed in irregular oxidized zones (Fig. 5). The reddish matrix does not occur in the wet zone.

Figure 3. Altered rock and mineral, and matrix-rich gravel. (A) Hypersthene grain with strongly etched surface from sand bed in Stuck Drift near Sumner, Washington. (B) Silty clay matrix (arrow) separating pebbles in kame near Dungeness, Washington. (C) Decomposed gabbro boulder at base of talus, Dungeness kame.

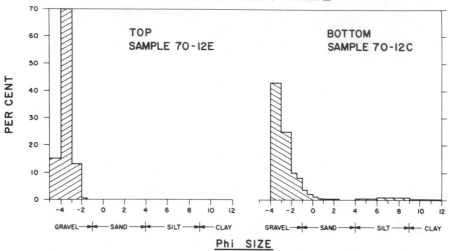

Figure 4. Size analyses (wet sieve and pipette) of sediments from near top (left) and bottom (right) of Dungeness kame exposure.

ORIGIN OF MATRIX

Three processes could result in matrix-rich outwash sediments: (1) contemporaneous deposition of clay, silt, pebbles, and cobbles; (2) mechanical introduction of matrix following deposition; and (3) postdepositional chemical alteration of detrital material.

It is difficult to postulate conditions by which rivers or streams could deposit fine silt and clay contemporaneously with pebbles and cobbles. In the exposures described above, cross-beds and scour features suggest that currents were too vigorous to deposit silt and clay. Furthermore, if the matrix were primary, one might expect a *gradation* in particle sizes from pebbles to clay, rather than a distinctly bimodal distribution. Sand is not a component of matrix. It undoubtedly was present in the sediment load and was commonly deposited in beds adjacent to the matrix-rich sediments (Fig. 5). Contemporaneous deposition, therefore, does not appear to be a reasonable explanation.

The matrix could have been formed by reworking detrital silt and clay and redepositing them in voids of open-work gravel. Such an explanation requires a source of the fine sediment and sufficiently high ground-water flow velocities to cause erosion and transportation of silt- and clay-size material. Perhaps silt and clay deposited on top of open-work gravel was later percolated downward into the voids between pebbles and cobbles.

In the exposures described here there is no very good evidence that this has happened, and some evidence that it has not: (1) Fine sediments do not, at present, directly overlie matrix-rich sediments. If once present, they must have been totally removed. (Fine sediments apparently do not underlie the matrix-rich sediments, either.) (2) At the Sumner and Kalaloch exposures (Fig. 5) sandy beds do overlie the matrix-rich sediments, but it is highly unlikely that silt and clay

Figure 5. Gravel pit near Kalaloch, Washington. Matrix-rich gravel (below dashed line), matrix-poor gravel (above line), reddish oxidized zones (dotted lines), silty clay bed (SC). Note hammer for scale (arrow).

KALALOCH GRAVEL PIT

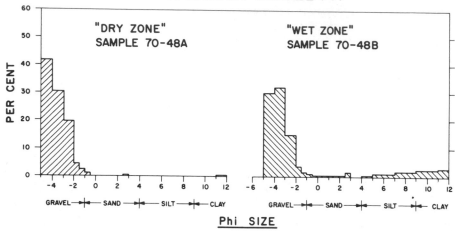

Figure 6. Size analyses (wet sieve and pipette) of sediments from matrix-poor gravels (left) and matrix-rich gravels (right) of Kalaloch exposure.

could be transported through sand to the underlying gravel. (3) No obvious channels occur along which matrix may have migrated. (4) Matrix commonly coats pebbles and cobbles on all sides uniformly. If the fines came from a single direction, they might have been deposited preferentially on the lee side of clasts. (5) The 1-m-thick silty clay bed at the top of the Kalaloch exposure (Fig. 5) does not appear to have contributed to the open-work gravel immediately below.

The matrix at the Sumner quarry clearly was formed, at least in part, by postdepositional chemical alteration of detrital material, because: (1) A large number of pebbles and cobbles are clearly altering in situ. Almost certainly some clasts have been completely altered to matrix and are no longer recognizable. (2) The sand bed overlying the matrix-rich gravels also shows incipient alteration. Hypersthene and hornblende are the most common mafic minerals in the sand, and virtually every hypersthene grain is badly corroded at both ends of the long axis of the mineral (Fig. 3A). A clay envelope, which is probably a product of the alteration, surrounds many altered grains. (3) No zones of weathering were found within the Stuck Drift. Deposition is considered to have been relatively rapid (Crandell, 1963, p. 23–24). (4) The Stuck Drift is well within the Pleistocene sequence (Middle Pleistocene(?); Crandell, 1963, p. 25), and is probably sufficiently old for alteration to have taken place in the mild and moist conditions prevailing in the region.

It seems reasonable to postulate that the matrix-rich gravel at Sumner was once an open-work gravel which served as a permeable bed for movement of ground water. The ground water, well below the zone of weathering, altered geochemically unstable pebbles and cobbles to matrix. Other gravels, which had voids filled with sand, were less permeable and, consequently, remain relatively unaltered.

The young age of the Dungeness kame and its low percentage of pebbles altering in situ argue against postdepositional alteration of detrital material at that locality. However, cobbles that must have been intact and presumably unaltered at the time the gravel pit was operated have been thoroughly decomposed by weathering within a relatively few years (Fig. 3C). Presumably matrix may have similarly been derived from sediments that altered after burial. The origin of the Dungeness matrix minerals is still unresolved, but an origin by chemical processes is not ruled out.

Chemical alteration at Kalaloch is virtually precluded because graywacke pebbles and cobbles are practically the only constituents in both the matrix-rich and matrix-poor sediments. The clasts in the matrix-rich sediment show slight surface alteration, but not nearly enough to have produced the extremely large volume of matrix.

Clearly, not all questions relating to the origin of matrix have been resolved. Additional work with new tools, such as the scanning electron microscope, may bring to light additional evidence on the problem.

DISCUSSION

Graywackes are sandstones in which the sand grains are set in a muddy matrix. The origin of matrix has been debated at length. Cummins (1962) postulated that graywacke matrix must be the result of diagenetic alteration of normal sand. Whetten and Hawkins (1970) present experimental evidence supporting Cummins's views.

It may well be, therefore, that environment of deposition has relatively little to do with whether a sediment will eventually become a graywacke. Certainly most graywackes are neither products of outwash sedimentation nor conglomeratic, and many graywackes have more matrix than the sediments described here. However, if the matrix in the Pleistocene gravels has been introduced postdepositionally, either chemically or mechanically, then the sediments should prove useful in determining the course and pattern of "graywackization."

There is good evidence that some or all of the matrix at the Sumner locality is of postdepositional origin resulting from chemical processes affecting detrital sediments below the zone of weathering. The origin of matrix at the other localities cannot be as easily determined. Further study of these and other Pleistocene sedimentary deposits composed of thermodynamically unstable clasts should reveal interesting information on the early stages of diagenesis.

ACKNOWLEDGMENTS

This research was sponsored by the National Science Foundation (Grant GA-15706). T. R. Walker offered much helpful advice, and S. C. Porter kindly read the manuscript and made suggestions for improving it.

REFERENCES CITED

Crandall, D. R., 1963, Surficial geology and geomorphology of the Lake Tapps quadrangle, Washington: U.S. Geol. Survey Prof. Paper 388-A, 84 p.
Cummins, W. A., 1962, The greywacke problem: Geol. Jour., v. 3, p. 51–72.
Hollister, C. D., and Heezen, B. C., 1964, Modern graywacke-type sands: Science, v. 146, p. 1573–1574.
Whetten, J. T., and Hawkins, J. W., Jr., 1970, Diagenetic origin of graywacke matrix minerals: Sedimentology, v. 15, p. 347–361.

MANUSCRIPT RECEIVED BY THE SOCIETY MARCH 29, 1972

Deformation of Materials

THE GEOLOGICAL SOCIETY OF AMERICA, INC.
MEMOIR 132
© 1972

Mechanical Twinning in Sphene at 8 Kbar, 25° to 500°C[1]

I. Y. BORG

AND

H. C. HEARD

Lawrence Livermore Laboratory, University of California
Livermore, California 94550

ABSTRACT

Mechanical twinning in sphene ($CaTiSiO_5$) has been produced in laboratory experiments at 8 kbar confining pressure, 25° to 500°C, $\epsilon = \sim 10^{-5}/\text{sec}$. The slip system is $K_1 = \sim \{221\}$, $K_2 = \{131\}$, $N_1 = <110>$, $N_2 =$ irrational, and $s = 0.60$. The conjugate twin law ($K_1 = \{131\}$, $K_2 = \sim \{221\}$, $N_1 =$ irrational) was not observed. Increasing the temperature of the experiment by 500°C lowers the critical resolved shear stress for twinning (τ_c) by about 50 percent. At 500°C, τ_c is about 1.3 kbar.

INTRODUCTION

During routine microscopic examination of granodiorite taken from the immediate vicinity of several nuclear explosions, it has been noted that accessory grains of sphene have been highly twinned compared with pre-event material (Borg, 1970). From the known relation between peak shock pressure and distance from the shot point, this twinning was estimated to occur first at ~ 14 to 18 kbar peak pressure and to become increasingly conspicuous with proximity to the shock source and increased pressure (Borg, 1972). Under such conditions of dynamic deformation, the temperature remains essentially ambient, strain rates are of the

---

[1] Work performed under the auspices of the U.S. Atomic Energy Commission.

order of 10^4 to 10^5/sec, and differential stresses are thought to be in the 8 to 9 kbar range (Cherry and Rapp, 1968).

Discovery of these deformation features prompted us to attempt to reproduce the mechanical <110> twinning in the laboratory under more carefully controlled conditions (that is, in static experiments at moderate confining pressures and strain rates of 10^{-3} to 10^{-5}/sec). If such twinning could be experimentally produced at relatively low stress levels (compared to those required for glide in quartz or the plagioclase series), it might be a useful tool for deducing the principal stress directions in some granitic rocks at the time of their deformation. However, the strain rates commonly observed in naturally deformed rocks are usually very much lower ($\sim 10^{-14}$/sec). Thus, results of these static laboratory tests are not strictly comparable to those of either field (nuclear) experiments or to natural deformation. However, the static tests probably do set lower limits on the stress differences at which twinning can be initiated under shock conditions. They probably also set upper limits for natural deformation occurring in the earth's crust.

ORIENTATION OF STARTING MATERIAL

Mügge (1889) concluded that polysynthetic twins quasi-parallel to {221} in sphene were probably mechanical in origin. The slip system deduced was only recently confirmed by Borg (1970) and is shown in Table 1.

There are two potential twin glide planes and slip lines ~54° apart corresponding to ~(221) and ~(2$\bar{1}$1) and to [1$\bar{1}$0] and [110], respectively. It is thus impossible to prepare a test crystal cylinder in which both twin glide systems have the maximum coefficient of resolved shear stress (that is, $S_o = 0.5$). Four orientations of samples from the large single crystal of starting material were required for testing. In each case, one of the two planes of the {221} form was oriented approximately 45° to the maximum principal compressive stress (σ_1). In two instances, the [1$\bar{1}$0] axis contained by the ~(221) plane lay near the circular section of the cylinder (approximately parallel to σ_3); in the other two it was coplanar with the cylindrical sample axis and the normal to (221). Individual cylinders within each of the two sets differed by the sense of movement possible along the slip direction; we have arbitrarily designated these "+" and "−."

In Figure 1, for the <110> slip on K_1 to be considered +, σ_1 must fall within the stippled area. Conversely, it is considered − if σ_1 falls in the unstippled field. For simplicity, + and − senses for twinning (or slip) are shown for the (2$\bar{2}$1), [1$\bar{1}$0] system only; comparable fields can be drawn for the ($\bar{2}\bar{2}$1), [1$\bar{1}$0] system.

TABLE 1. TWIN ELEMENTS IN SPHENE

K_1 = Irrational, near {221}	Slip plane
K_2 = {$\bar{1}$31}	Other plane of no distortion
N_1 = <110>	Slip line
N_2 = Irrational	Trace of K_2 on S
S = Irrational, near {$\bar{1}\bar{1}$2}	Plane of deformation
s = 0.60	Shear

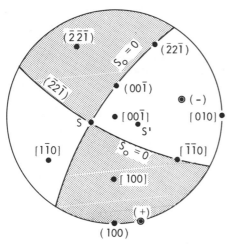

Figure 1. Stereogram of twin elements in [110] twinning. s and s' = poles to deformation planes. If σ_1 falls within stippled area, $S_o > 0$ and sense of movement on (221) is +. If σ_1 occurs within open area, sense is —. Circled points correspond to orientation of σ_1 when $S_o = 0.5$. Lower hemisphere projection.

In each case, these regions are bounded by the slip plane and the plane normal to the slip line. The sense of the potentail slip and value of S_o are given in Table 2 for each of the two {221} planes in the cylinders. The S_o can be calculated by $S_o = \sin \chi_o \cos \lambda_o$ where χ_o and λ_o are the angles between σ_1 and the silp plane and the slip line, respectively, before deformation.

STARTING MATERIAL AND EXPERIMENTAL PROCEDURE

The sphene samples used in our deformation experiments were prepared with a diamond-impregnated coring bit from a large (15 cm × 10 cm × 5 cm), brown, single crystal from Arkarolla Station, Australia. The cell dimensions as determined by least-square reduction of powder x-ray data are $a = 6.558$, $b = 8.718$, $c = 7.450 \pm 0.002$, $\beta = 120.00 \pm 0.01°$, C 2/c Z = 4. The crystal was locally vuggy and contained minor actinolite (<1 percent). Bounding surfaces were well-formed {221} partings which paralleled pre-existing, regularly spaced, through-going twins.

Right cylinders, 4.5 to 5.0 mm in diameter by 9 to 10 mm in length, were cut from a rough core in each of the four orientations noted above. The finished

TABLE 2. RESOLVED SHEAR STRESS COEFFICIENTS

Experiment no.	Temperature (°C)	Plane	S_o	Sense of Twin Glide	Amount of Twinning
848	25	221	0.41	—	None
		2̄21	0.09	+	None
816	25	221	0.12	+	Some
		2̄21	0.21	—	None (possibly some translation)
849	25	221	0.40	+	Moderate, local
		2̄21	0.41	+	Moderate, local
815	25	221	0.12	+	Some, local
		2̄21	0.19–0.28	+	Some, local
850	300	221	0.44	+	Highly twinned, 30% of central sectors
		2̄21	0.34	+	Little or none
851	500	221	0.41	+	Highly twinned, 40 to 50% of central sectors
		2̄21	0.31	+	Little or none

sample-cylinders were mechanically jacketed in annealed copper and compressed parallel to their axes between tungsten carbide anvils in a triaxial deformation apparatus described by Heard and Carter (1968). Confining pressure (σ_3) remained constant at 8.0 kbar during all experiments. The temperature ranged from 25° to 500°C, and the strain rates (inelastic) ranged from 2 to 4×10^{-5}/sec. All stress-strain data have been corrected for the loads supported by the copper jackets.

RESULTS

Figure 2 summarizes the differential stress-strain behavior of all compression experiments reported here. In tests where the cylinder orientations favored <110> slip in the negative sense, the sphene was very strong (σ_1-$\sigma_3 > 15$ kbar) and minimal inelastic deformation occurred. When the crystal cylinders were oriented to favor <110> slip in the positive sense, however, the yield point was much lower (the amount depended upon temperature and S_o), and permanent strains of up to ~7 percent were achieved. In no case was an experiment terminated by violent fracture of a sample cylinder. Some faulting was noted, however, in thin-section examination of test 849. From the shape and anomalously low level of the stress-strain curve for 849, we regard it as marginal for strength correlation purposes. Such behavior could result from either the vuggy nature of the sample or a partial leak in the jacket. It may be also noted (Fig. 2) that an increase in temperature from 300° to 500°C markedly lowers the stress-strain curve. The few experiments shown in Figure 2 do not allow a quantitative determination of the change of τ_c with temperature. Qualitatively, it seems to decrease by a factor of about two in the 25° to 500°C interval. The τ_c at 500°C, calculated from the well-defined yield point of 3.1 kbar in experiment 851, is about 1.3 kbar.

Localized areas of prolific <110> twin development occur in all deformed crystals except 848 (compare especially Figs. 3a and 3b). The twins tend to be short and closely spaced, and thus depart markedly in appearance from the through-going, widely spaced, pre-existing twins noted in the starting material. Furthermore, their consistently greater abundance in the central areas of all cylinders suggests production during, rather

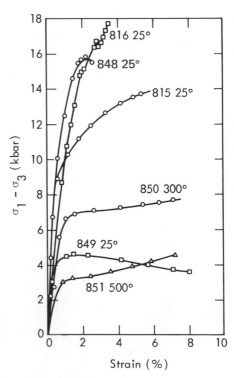

Figure 2. Differential stress-strain curves (compression) for sphene single crystals of several orientations. Confining pressure 8 kbar, strain rate 2 to 4×10^{-5}/sec. Experiment number and temperature indicated.

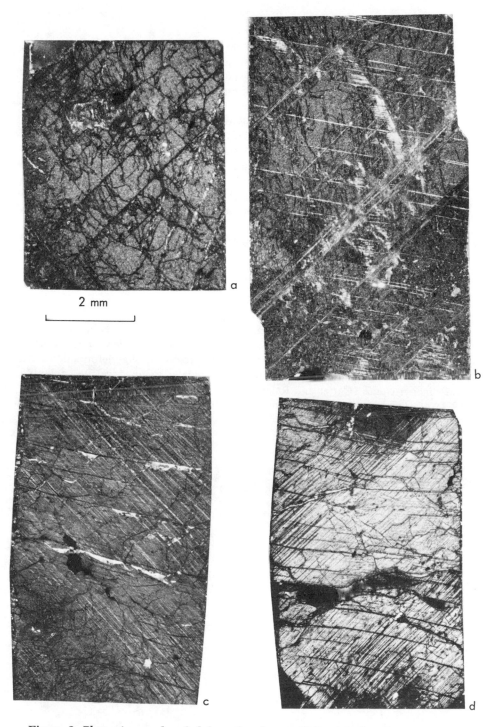

Figure 3. Photomicrographs of deformed sphene cylinders. Scale = 2 mm, crossed polaroids: (a) 848, 25°C, no twinning; (b) 815, 25°C, minor twinning; (c) 850, 300°C, highly twinned; (d) 851, 500°C, highly twinned.

than before, the induced deformation. Where twinning is better developed (that is, in experiments at 300°C and 500°C), twins traverse the specimen and occupy > 30 percent of central areas (Figs. 3c and 3d). The pervasiveness of twinning shows some variability in the deformed samples depending on the magnitude of positive S_0 and temperature of the experiment (Table 1). Specimens 848, 816, 815, 849, 850, and 851 are listed in order of observed increase in twinning. From Figure 2, we see that this is also generally the order of decreasing yield point. The sense of the movement along <110> is + in all instances where twinning is considered to have been produced to some degree by the deformation. This corresponds to the shear geometry in Figure 4. On an atomic level, the mechanical twin procss so described involves minor shuffles and rotation of Ti-octahedra and Si-tetrahedra and movements associated with a homogeneous shear (Borg, 1970). The theoretically possible conjugate twin law [$K_1 = \{\bar{1}31\}$, $K_2 =$ irrational near $\{221\}$, $N_1 =$ irrational slip line, and $N_2 = <110>$] was not observed in any of the deformed cylinders examined here.

DISCUSSION

The demonstration of laboratory induced <110> mechanical twinning in only one sense in sphene makes this accessory mineral a potentially valuable tool for deducing principal stress directions in granitic rocks during deformation. The method is analogous to those used for mechanically twinned calcite, dolomite, and clinopyroxenes (Carter and Raleigh, 1969) and plagioclase (Borg and Heard, 1970). In the case of sphene, limitations on the technique are determined by the

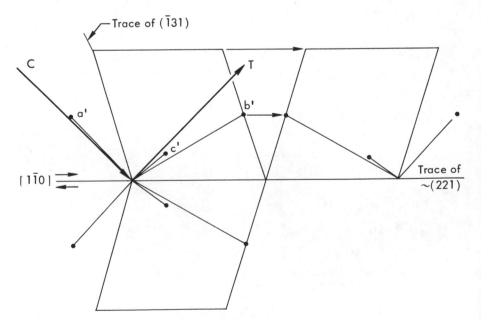

Figure 4. Geometry of twin gliding in sphene. Rhombs are defined by traces of $K_1 = \sim(221)$ and $K_2 = (\bar{1}31)$ on deformation plane (plane of drawing). $N_1 = [1\bar{1}0]$. Traces of crystallographic axes (a', b', c') are also shown. C and T are orientation of compression or extension axis associated with a + sense of gliding on $\{221\}$.

facts that (1) it is generally a minor constituent of rocks in which it occurs, thereby posing sampling problems, and (2) it is difficult to measure using conventional U-stage techniques because of customary small grain size, high refractive indices, and almost complete reflection of incident light. It appears metallic in reflected light. These same points account in part for the fact that <110> twinning has infrequently been recognized in naturally deformed rocks. However, Mügge (1889) cites numerous examples in association with other twinned minerals such as calcite and augite in defense of his thesis that polysynthetic <110> twinning in sphene is deformational in origin. Except for a play of abnormal interference colors along the lamellae when viewed obliquely to the composition plane, <110> twinning, when poorly developed, is inconspicuous. It is not likely confused with {100} growth twinning, inasmuch as the latter typically consists of simple individuals (contact or penetration twins).

In granitic rocks, accessory sphene can be expected to be a more sensitive structural element than the more ubiquitous quartz, since the τ_c to activate <110> twinning at the experimental conditions reported here (25° to 500°C at strain rates of $\sim 10^{-5}$/sec) is less than that required for sub-basal slip in quartz by at least an order of magnitude (Heard and Carter, 1968). This, of course, assumes that the τ_c for slip in each mineral is lessened to the same degree by the much lower "geologic" strain rates. Other primary minerals (for example, biotite) or some secondary ones (calcite) can also serve in a similar fashion and are probably even more sensitive.

REFERENCES CITED

Borg, I. Y., 1970, Mechanical <110> twinning in shocked sphene: Am. Mineralogist, v. 55, p. 1876–1888.
—— 1972, Shock effects in granodiorite to 270 kbar, in Heard, H. C., Borg, I. Y., Carter, N., and Raleigh, C. B., eds., Flow and fracture of rocks: Am. Geophys. Union Geophys. Mon. 16.
Borg, I. Y. and Heard, H. C., 1970, Experimental deformation of plagioclases, in Paulitsch, P., ed., Experimental and natural rock deformation: Berlin, Heidelberg, Springer-Verlag, p. 375–403.
Carter, N. L., and Raleigh, C. B., 1969, Principal stress directions from plastic flow in crystals: Geol. Soc. America Bull., v. 80, p. 1231–1264.
Cherry, J. T., and Rapp, E. G., 1968, Calculation of free-field motion for the Piledriver event: Lawrence Radiation Laboratory, Rept. UCRL-50373, Livermore, California.
Heard, H., and Carter, N., 1968, Experimentally induced "natural" intragranular flow in quartzite: Am. Jour. Sci., v. 266, p. 1–42.
Mügge, O., 1889, Ueber durch Druck enstandene Zwillinge von Titanit nach den Kanten [110] and [1̄10]: Neues Jahrb. Min. Geologie u. Paläontologie II, s. 98–115.

MANUSCRIPT RECEIVED BY THE SOCIETY MARCH 29, 1971

THE GEOLOGICAL SOCIETY OF AMERICA, INC.
MEMOIR 132
© 1972

Peridotite Fabrics and Velocity Anisotropy in the Earth's Mantle

DAVID B. MACKENZIE

Marathon Oil Company, Littleton, Colorado 80120

ABSTRACT

In deformed peridotites and other olivine-rich rocks, the olivine grains commonly show preferential crystallographic orientation. Of six or more fabric models reported in naturally occurring peridotite tectonites, two are consistent with experimentally determined slip systems; one is consistent with experimentally verified recrystallization under a known stress system. Because of a strong anisotropy in compressional wave velocity in olivine crystals, seismic velocities in deformed peridotites are significantly higher in the direction of [100] maxima and lower in the direction of [010] maxima.

Recent refraction seismic work in the northeast Pacific (Raitt and others, 1969; Morris and others, 1969) has demonstrated anisotropies ranging from 0.3 to 0.6 km/sec. Small anisotropies can be explained by numerous combinations of fabrics and fabric orientations. Consider, for example, a common model having a [100] maximum horizontal, and [010] and [001] evenly distributed in a girdle normal to it. Using this model, an anisotropy of 0.3 km/sec is consistent with deformation of a peridotitic mantle which is about 17 percent serpentinized and in which 21 percent of the olivine grains are preferentially oriented. The model would require [100] to be subparallel to the direction of sea-floor spreading.

An anisotropy as great as 0.6 km/sec imposes more severe constraints. The model described above would require 42 percent of the olivine crystals to be oriented. Many other models have insufficient anisotropy and would not therefore be acceptable. Among them is a commonly encountered one having mutually perpendicular [100], [010], and [001] maxima.

INTRODUCTION

The hypothesis advanced by Hess (1964) of seismic anisotropy of the uppermost mantle has subsequently been more strongly confirmed by a Scripps Institution of Oceanography group (Raitt and others, 1969; Morris and others, 1969) for three areas of the northeast Pacific between California and Hawaii. Hess's concept resulted initially from considering some long-established petrofabric data in the context of sea-floor spreading by convection in a peridotitic mantle. These data had to do with the preferred orientation of olivine crystals within plastically deformed or recrystallied peridotites (Turner, 1942). An essential key was provided by the laboratory demonstration of a pronounced dependency of sonic velocity in olivine on the crystallographic direction of propagation (Verma, 1960). Although the confirmation of seismic anisotropy in three small areas does not demonstrate that the upper mantle is composed of deformed peridotite as hypothesized by Hess, and subsequently by Kumazawa and Shimazu (1967) and Francis (1969), it remains the most probable explanation.

Not for the first time, Hess was on the right road in the wrong vehicle. He called for [010] of olivine, the normal to the principal cleavage and glide plane, to tend to be aligned normal to the plane of the Mendocino fracture zone. A more probable model calls for [100] of olivine to have a preferred orientation horizontal and subparallel to the Mendocino fracture zone, with [010] either vertical or distributed radially and normally to [100] (Francis, 1969).

Of course, no unique interpretation of anisotropy is possible; various orientations of other fabric models may be consistent with the velocity anisotropies reported from uppermost mantle of the Pacific. But an anisotropy as large as 0.6 km/sec (Morris and others, 1969) imposes severe constraints. Probably only olivine and pyroxene have sufficient differences between crystallographically dependent fastest and slowest sonic velocities to be eligible constituent minerals. In the case of olivine-rich rocks, many of the fabrics reported in naturally occurring peridotites are too weakly developed to have sufficient anisotropy.

The purpose of this paper is to show how velocity anisotrophy, in conjunction with mean velocity, provides constraints on acceptable models. Velocity anisotropy of three areas of the northeast Pacific will be interpreted in terms of two common peridotite fabric models, with particular attention to perfection of crystallographic orientation of the olivine grains and degree of serpentinization.

OLIVINE AND ITS EXPERIMENTAL DEFORMATON

The crystallographic directions of olivine, with related optical refractive indices and sonic velocities (Verma, 1960), are shown in Figure 1. The directions of fastest and slowest sonic velocities coincide with those of the slowest (Z) and fastest (X) vibration directions for light, respectively. (Data on the elastic moduli for olivine have also been recently presented by Kumazawa and Anderson, 1969.)

In crystal habit, olivine tends to be flattened along the *b* crystallographic axis. The fabic of some peridotites believed to exhibit magmatic flow structures reflects this crystal shape of olivine.

The mechanisms of plastic flow of peridotite have been determined experimentally (Raleigh, 1967, 1968; Carter and Ave'Lallemant, 1970). Experiments

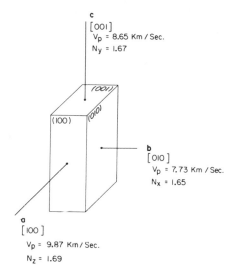

c
[001]
V_p = 8.65 Km / Sec.
N_y = 1.67

(100)

b
[010]
V_p = 7.73 Km / Sec.
N_x = 1.65

a
[100]
V_p = 9.87 Km/ Sec.
N_z = 1.69

Figure 1. Crystallographic directions, seismic velocities (V_p), and optical refractive indices (N_x, N_y, N_z) for magnesian olivine (after Hess, 1964).

have now been conducted at confining pressures up to 30 kb, at temperatures up to 1,400°C, and at strain rates of 10^{-3} to 10^{-8}/sec. From a microscopic study of deformation lamellae and crystal rotations in the deformed specimens, the predominant flow mechanisms of olivine were deduced by Carter and Ave'Lallemant to be those shown in Figure 2. The two lower temperature slip systems (B and C) confirm the results of Raleigh (1968). He likened the system {$0kl$} [100][1] to a pencil glide in which the plane in the zone [100] having the highest shearing stress is the active slip plane.

An increase in pressure of 1 kb appears to lower the transition temperature between slip systems by about 7°C. As shown in Figure 2, a decrease in strain rate by a factor of 10 lowers the temperature of transition between slip mechanisms by 55° to 65°C. A linear extrapolation of the experiments to a naturally occurring strain rate of 10^{-14}/sec places the lower and upper boundaries of pencil glide (B) at about 350° and 550°C at 15 kb, and each about 100°C higher at 2 kb.

The deformation fabrics are further complicated by recrystallization which takes place concurrently with plastic flow. On the basis of laboratory evidence (Ave'Lallemant and Carter, 1970) and field evidence (Ave'Lallemant, 1967), syntectonic recrystallization is an important, possibly the principal, mode of flow at temperatures greater than 500°C. New intergranular crystals formed during deformation have [010] maxima parallel to the principal stress, σ_1, and [001] and [100] girdles in the $\sigma_2 = \sigma_3$ plane. What happens in nature where $\sigma_2 \neq \sigma_3$ is a matter of some debate as indicated in the next section.

FABRICS OF PERIDOTITES

Beginning with Andreatta (1934), several tens of petrofabric studies of olivine-rich rocks have been made. The study by Turner (1942) on dunites of Dun Mountain, New Zealand, is particularly noteworthy. Reviews of the literature have been made by Collée (1963), Den Tex (1969), and Ave'Lallemant and Carter (1970), among others.

At least six different fabric models have been reported. Included among the common ones are two which are consistent with the higher temperature experimentally determined slip systems of Figure 2: (010) and [100], and {$0kl$} [100].

---

[1] The first part of each couplet designates the glide plane; the second part designates the glide direction in the plane.

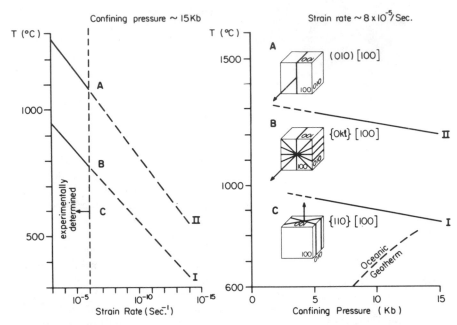

Figure 2. Slip mechanisms in experimentally deformed olivine (Modified from Carter and Ave'Lallemant, 1970).

These will be referred to as models A and B.[2] Fabrics resulting from slip on the low temperature system, {110} [001], are rare in nature.

Since many olivine nodules in basalts may represent fragments of the upper mantle, their fabrics are of particular interest. Many have concentrations of X, Y, and Z maxima at right angles to one another, without any visible foliation (Turner, 1942; Talbot and others, 1963). Where a foliation is present, it is commonly normal to the X maximum, just as in many alpine-type peridotites (Ave'Lallemant and Carter, 1970). In tectonite fabrics of all kinds of olivine-rich rocks, rarely, if ever, do more than 50 percent of the grains observed conform to any particular deformation model; usually only 20 to 40 percent do so.

The most controversial point is whether Z maxima within foliation planes tend to be parallel to tectonic *b* or to tectonic *a* (normal or parallel to the direction of tectonic transport). Z parallel to tectonic *b* has been reported at Dun Mountain (Battey, 1960) and in the French Pyrenees (Ave'Lallemant, 1967); Z parallel to tectonic *a* has been recorded in Japan (Yoshino, 1961). Evidence in support of Z parallel to tectonic *a* is also recorded at Tinaquillo, Venezuela, in a high-temperature alpine-type peridotite. There, a Z maximum for olivine is parallel to elongated laths of orthopyroxene flattened in the foliation plane of the peridotite. As discussed elsewhere (MacKenzie, 1960), the form and orientation of these laths is explicable only if they are subparallel to the direction of tectonic transport. Moreover, experimental work (Turner and others, 1960) indicates that the glide system for enstatite is (100) [001], consistent with the observations at Tinaquillo.

Perhaps the fabric orientation depends to some degree on whether syntectonic

---

[2] Fabric models A and B may also result entirely from syntectonic recrystallization rather than from plastic flowage.

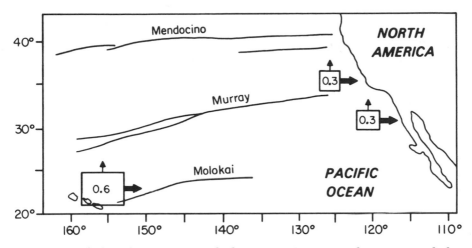

Figure 3. Blocks indicate areas in which seismic anisotropy in the upper mantle has been established, with maximum P-wave velocities east-west and minimum velocities north-south. Numbers within blocks indicate magnitude of anisotrophy in km/sec. Data from Raitt and others (1969) and Morris and others (1969).

recrystallization or plastic flow is the dominant mechanism. Where deformation is intense, the least principal stress, σ_3, may flip from normal to parallel to the axis of folding, as in an isoclinal fold (Christie and Raleigh, 1959). If new crystal growth is dominant and has a preferred orientation of Z parallel to σ_3, Z would be parallel to tectoic b. This is the situation inferred by Ave'Lallemant (1967) for the axial zones of folds in lherzolites of the French Pyrenees. In deformations of less intensity, σ_2 would more likely be parallel to tectonic b (Dieterich and Carter, 1969) and Z would be parallel to tectonic a by either deformation mechanism.

IMPLICATIONS TO OBSERVATIONS
OF VELOCITY IN UPPERMOST MANTLE

Velocity anisotropy has been observed in three different areas of the northeast Pacific (Fig. 3). In an area northeast of Hawaii (where Jackson, 1968, has interpreted many olivine nodules as metamorphic fragments of the underlying mantle), maximum compressional wave velocity occurs in an east-west direction and is 0.6 km/sec greater than the minimum which occurs at right angles to it (Morris and others, 1969). The mean P-wave velocity in the uppermost mantle is 8.16 km/sec. In two areas off the coast of California (Raitt and others, 1969), the maximum and minimum velocities have the same relative orientations (east-west and north-south, respectively), but the difference between them is only 0.3 km/sec. The mean P-wave velocity is 7.98 km/sec in one area and 8.14 km/sec in the other. (The standard error is 0.02 km/sec or less.)

Shown in Figure 4 are the velocity implications of fabric models A and B. Whichever one is applicable, the observation in the northeast Pacific that the velocity maxima are oriented approximately east-west and subparallel to the fracture zones suggests that the Z maxima are east-west, horizontal, and subparallel to the direction of tectonic transport or sea-floor spreading.

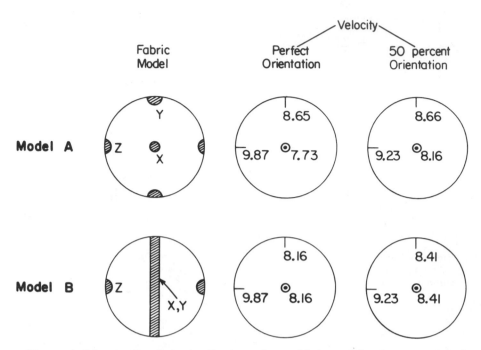

Figure 4. Directional velocity implications of two fabric models of unserpentinized peridotite. Velocities in km/sec are based on those of olivine, using $1/V_{xy} = 1/2(1/V_x + 1/V_y)$ for averaging velocities.

Model B is exactly analogous to the transverse isotropy model of Christensen and Crosson (1968) and Crosson and Christensen (1969), but [100] rather than [010] is the axis of symmetry. The comparison calls for [100] to be tilted 90° from vertical in an east-west direction. (Transversely isotropic media possess an axis of symmetry such that all plans containing the axis are equivalent.)

Assuming the uppermost mantle does have a peridotite composition, its mean P-wave velocity and velocity anisotropy depend on a number of factors. Among them are the model for the deformation fabric as discussed above, the proportion of grains that conform to the model, the composition of olivine, the proportion and mineralogy of the non-olivine fraction, the degree of serpentinization, and the confining pressure. In the following discussion, it is assumed that the olivine is Fo_{90} or more magnesian, and that the non-olivine fraction makes up less than about 10 percent.

Laboratory data on the velocity and anisotropy of compressional waves in peridotite have been presented by Birch (1960, 1961) and Christensen (1966). At 2 kb (the approximate confining pressure in the uppermost mantle beneath the oceans), compressional velocities (V_p) of the Twin Sisters dunite, among the least serpentinized of the samples measured, range from 8.22 to 8.96 km/sec (Christensen, 1966). (The specimen has a model A fabric; maximum concentrations of olivine X, Y, and Z axes correlate with directions of propagation having minimum, intermediate, and maximum velocities, respectively.) The mean velocity for this dunite is 8.52 km/sec, which is 0.1 km/sec greater than for the sample of Tinaquillo peridotite cited by Hess (1960). Averaging the elastic constants of olivine for aggregates of

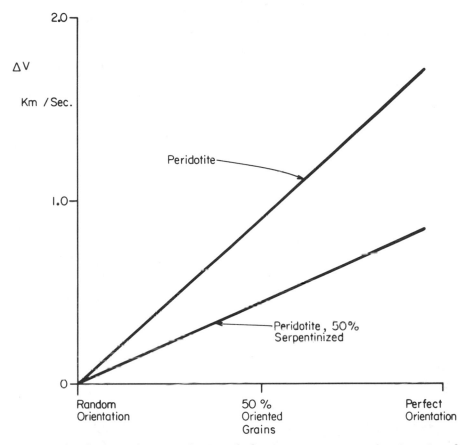

Figure 5. Velocity anisotropy in horizontal plane versus percentage of grains oriented according to model B fabric.

randomly oriented olivine grains yields a theoretical V_p for dunite of 8.4 or 8.56 km/sec, depending on the method used (Birch, 1961, p. 2210). Averaging the directional velocities for olivine by the equation $V_p = 1/3(1/V_x + 1/V_y + 1/V_z)$ gives $V_p = 8.67$ km/sec.

The value adopted in this paper is $V_p = 8.6$ km/sec.

Figure 5 is shown velocity anisotropy as a function of the proportion of olivine grains in a peridotite oriented according to model B. (It is assumed that the remainder are randomly oriented.) In few or none of the peridotites in the literature do more than 50 percent of the grains have a preferred orientation. So, from this factor alone, deformation of an unserpentinized peridotite according to model B would result in a velocity anisotropy in the horizontal plane of no more than about 0.9 km/sec. (Note that the difference between the laboratory-measured maximum and minimum velocities of the Twin Sisters dunite is about 0.7 km/sec.)

As discussed by Hess (1959), serpentinization of the uppermost mantle would result in a dramatic reduction of velocity. The curve at the left of Figure 6 is adapted from data of Christensen (1966). Raitt and others (1969) have observed that the uppermost mantle velocity ranged from 7.8 to nearly 9 km/sec world-wide.

This would suggest a maximum serpentinization of about 25 percent. But this boundary is artificial, since velocities lower than 7.8 km/sec would be arbitrarily assigned to the crust rather than to the mantle.

Figure 6 shows the interrelations of mean velocity, velocity anisotropy, percent serpentization, and percent of oriented olivine crystals for a peridotite with a model B fabric. The curve at the left assumes perfect (100 percent) orientation, with the mean velocity for unserpentinized peridotite being approximately $1/2(1/9.87 + 1/8.16) = 8.94$ km/sec.

Figure 6 can be used in the following way. For the observed mean P-wave velocity, carry a horizontal line over to the graph in the right until it intersects the interpolated sloping line which represents the velocity anisotropy. The interpolated vertical line through the intersection then gives the percentage of grains oriented according to model B. By interpolation between the curves at the left, the approximate percent serpentinization can then be read.

Thus, for an area northeast of Hawaii (Morris and others, 1969), the mean velocity in the horizontal plane is 8.16 km/sec and the velocity anisotropy is 0.6 km/sec. The intersection at the chart to the right shows that 42 percent of the olivine grains would be oriented. Using this value of 42 percent for interpolation on the left, the peridotite of this area would be about 15 percent serpentinized.

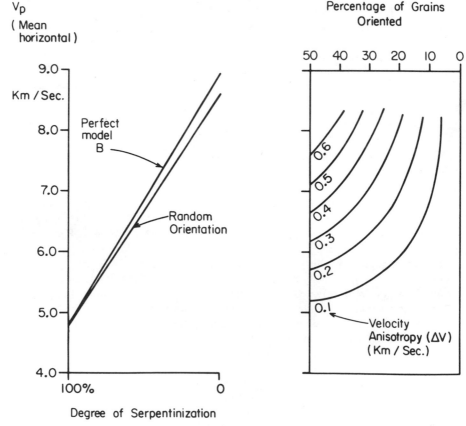

Figure 6. Interrelations between mean horizontal velocity (V_p), velocity anisotropy, percent serpentinization, and percent olivine grains oriented according to model B fabric.

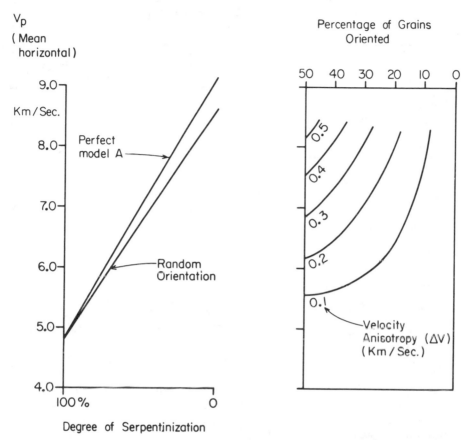

Figure 7. Interrelations between mean horizontal velocity (V_p), velocity anisotropy, percent serpentinization, and percent olivine grains oriented according to model A fabric.

For the two areas studied by the Scripps group in offshore California (Raitt and others, 1969), the mean velocity is 8.05 km/sec and the anisotropy is 0.3 km/sec. Following a similar type of analysis, the peridotite of the uppermost mantle in this area would be about 17 percent serpentinized and would have only 21 percent of its olivine grains oriented according to model B.

The above values calculated for percent serpentinization and percent of olivine crystal lattices oriented are not unreasonable. A similar type of analysis for a model A fabric (Fig. 7) leads to the values shown in the right-hand part of Table 1.

As will be seen from Figure 7, the combination of mean velocity and velocity anisotropy for northeast Hawaii is consistent with a model A fabric only if substantially more than 50 percent of the grains are oriented according to the model. Since such a high perfection of orientation has rarely, if ever, been encountered in exposed peridotites, model A would not be appropriate for this area.

CONCLUDING REMARKS

Velocity anistropy is an important additional parameter for characterizing the uppermost mantle. The degree to which it is consistent with mean P-wave velocity,

TABLE 1. OLIVINE ORIENTATION AND SERPENTINIZATION OF
PACIFIC UPPERMOST MANTLE

	Model B		Model A	
	Northeast Hawaii	Offshore California	Northeast Hawaii	Offshore California
Mean Velocity (km/sec)	8.16	8.05	8.16	8.05
Velocity Anisotropy	0.6	0.3	0.6	0.3
Percent Olivine Grains Oriented	42	21	—	31
Percent Serpentinization	15	17	—	19

in the context of peridotite fabrics, may provide added insight to the composition and variability of the upper mantle. Anisotropy measurements would be still more useful if more laboratory measurements of directional compressional wave velocities were available on peridotites of known petrofabric (for example, Christensen, 1966).

Where does the fabric originate? Either it is frozen in at least by the time the uppermost mantle begins to move away from the rise where it surfaces, or it develops and is continuously modified as spreading continues. At the conditions prevailing in the uppermost mantle of the northeast Pacific (depth 11 km, pressure 2 kb, and temperature 200°C), temperatures are too low for any deformation except on the rarely observed slip system {110} [001]. For this reason, I concur with Francis (1969) and Ave'Lallemant and Carter (1970) that the observed anisotropy is fossilized; almost certainly it was developed during or prior to the birth of the mantle containing it.

Among many of the possibilities, the difference in magnitude of anisotropy by a factor of 2 between northeast Hawaii and offshore California could be related to a difference between model B and model A fabrics. If so, it might suggest an increase in heat flow along the East Pacific Rise during the 40+ m.y. which separate the two areas. In any case, as anisotropy measurements become available from other areas, it will be useful to look for correlations between inferred fabrics and rate of spreading, distance from spreading centers, geothermal gradients, and so forth.

In conclusion, the setting forth and testing by Hess of the hypothesis of seismic anisotropy in the upper mantle is a good example of his perceptiveness and his approach. He had a remarkable facility for utilizing seemingly unrelated observations in imaginative invention. Our understanding of this aspect of the mantle, as well as many other earth characteristics and processes, is greatly enhanced as a result.

ACKNOWLEDGMENTS

This paper has benefited from critical comments by J. B. Hayes, C. H. Hewitt, and C. B. Raleigh.

REFERENCES CITED

Andreatta, C., 1934, Analisi strutturale di rocce metamorfiche; V Olivinite: Period. Mineralogia, Anno V, p. 237–253.

Ave'Lallemant, H. G., 1967, Structural and petrofabric analysis of an "alpine-type" periodotite: The lherzolite of the French Pyrenees: Leidse Geol. Meded., v. 42, p. 1–57.

Ave'Lallemant, H. G., and Carter, N. L., 1970, Syntectonic recrystallization of olivine and modes of flow in the upper mantle: Geol. Soc. America Bull., v. 81, p. 2203–2220.

Battey, M. H., 1960, The relationship between preferred orientation of olivine in dunite and the tectonic environment: Am. Jour. Sci., v. 258, p. 716–727.

Birch, F., 1960, The velocity of compressional waves in rocks to 10 kilobars, part 1: Jour. Geophys. Research, v. 65, p. 1083–1102.

——1961, The velocity of compressional waves in rocks to 10 kilobars, part 2: Jour. Geophys. Research, v. 66, p. 2199–2224.

Carter, N. L., and Ave'Lallemant, H. G., 1970, High temperature flow of dunite and peridotite: Geol. Soc. America Bull., v. 81, p. 2181–2202.

Christensen, N. I., 1966, Elasticity of ultrabasic rocks: Jour. Geophys. Research, v. 71, p. 5921–5931.

Christensen, N. I., and Crosson, R. S., 1968, Seismic anisotropy in the upper mantle: Tectonophysics, v. 6, p. 93–107.

Christie, J. M., and Raleigh, C. B., 1959, The origin of deformation lamellae in quartz: Am. Jour. Sci., v. 257, p. 385–407.

Collée, A. L. G., 1963, A fabric study of lherzolites, with special reference to ultrabasic nodular inclusions in the lavas of Auvergne (France): Leidse Geol. Meded., v. 28, p. 3–102.

Crosson, R. S., and Christensen, N. I., 1969, Transverse isotrophy of the upper mantle in the vicinity of Pacific fracture zones: Seismol. Soc. America Bull., v. 59, p. 59–72.

Den Tex, E., 1969, Origin of ultramafic rocks, their tectonic setting and history: A contribution to the discussion of the paper "The origin of ultramafic and ultrabasic rocks" by P. J. Wyllie: Tectonophysics, v. 7, p. 457–488.

Dieterich, J. H., and Carter, N. L., 1969, Stress-history of folding: Am. Jour. Sci., v. 267, p. 129–154.

Francis, T. J. G., 1969, Generation of seismic anisotropy in the upper mantle along the mid-oceanic ridges: Nature, v. 221, p. 162–165.

Hess, H. H., 1959, The AMSOC hole to the earth's mantle: Am. Geophys. Union Trans., v. 40, p. 340–345.

——1960, Caribbean Research Project progress report: Geol. Soc. America Bull., v. 71, p. 235–240.

——1964, Seismic anisotropy of the uppermost mantle under oceans: Nature, v. 203, p. 629–631.

Jackson, E. D., 1968, The character of the lower crust and upper mantle beneath the Hawaiian Islands: Internat. Geol. Cong., 23d, Prague 1968, Comptes Rendus, pt. 1, p. 135–150.

Kumazawa, M., and Anderson, O. L., 1969, Elastic moduli, pressure derivatives, and temperature derivatives of single-crystal olivine and single-crystal forsterite: Jour. Geophys. Research, v. 74, p. 5961–5972.

Kumazawa, M., and Shimazu, Y., 1967, Preferred mineral orientation by recrystallization under non-hydrostatic stress and geotectonic stress in the upper mantle: Royal Astron. Soc. Geophys. Jour., v. 14, p. 57–60.

MacKenzie, D. B., 1960, High-temperature alpine-type peridotite from Venezuela: Geol. Soc. America Bull., v. 71, p. 303–318.

Morris, G. B., Raitt, R. W., and Shor, G. G., Jr., 1969, Velocity anisotropy and delay-time maps of the mantle near Hawaii: Jour. Geophys. Research, v. 74, p. 4300–4316.

Raitt, R. W., Shor, G. G., Jr., Francis, T. J. G., and Morris, G. B., 1969, Anisotropy of the Pacific upper mantle: Jour. Geophys. Research, v. 74, p. 3095–3109.

Raleigh, C. B., 1967, Experimental deformation of ultramafic rocks and minerals, in Wyllie, P. J., Ultramafic and related rocks: New York, John Wiley & Sons, Inc., p. 191–199.

——1968, Mechanisms of clastic deformation of olivines: Jour. Geophys. Research, v. 73, p. 5391–5406.

Talbot, J. L., Hobbs, B. E., Wilshire, H. G., and Sweatman, T. R., 1963, Xenoliths and xenocrysts from lavas of the Kerguelen Archipelago: Am. Mineralogist, v. 48, p. 159–179.

Turner, F. J., 1942, Preferred orientation of olivine crystals in peridotites with special
    reference to New Zealand examples: Royal Soc. New Zealand Trans., v. 72, p. 280–
    300.
Turner, F. J., Heard, H. C., and Griggs, D. T., 1960, Experimental deformation of en-
    statite and accompanying inversion to clinoenstatite: Internat. Geol. Cong., 21st,
    Copenhagen 1960, Comptes Rendus, pt. 18, p. 399–408.
Verma, R. K., 1960, Elasticity of some high-density crystals: Jour. Geophys. Research,
    v. 65, p. 757–766.
Yoshino, G., 1961, Structural-petrological studies of peridotite and associated rocks of the
    Higashi-Akaishi-Yama district, Shikoku, Japan: Hiroshima Univ. Jour Sci., Ser. C3,
    p. 343–402.

MANUSCRIPT RECEIVED BY THE SOCIETY MARCH 29, 1971

THE GEOLOGICAL SOCIETY OF AMERICA, INC.
MEMOIR 132
© 1972

North American Cryptoexplosion Structures: Interpreted as Diapirs which Obtain Release from Strong Lateral Confinement

L. O. NICOLAYSEN

Bernard Price Institute of Geophysical Research, University of the Witwatersrand, Johannesburg, South Africa

ABSTRACT

Several students of cryptoexplosion structures have emphasized the importance of very high fluid pressures. Data on the behavior of mud and sand under high pore-fluid pressures support this concept. Extending the analysis of Goguel and Bucher, it is proposed that such pressures in a specific rock formation can initiate a diapir and drive it to the cryptoexplosion stage. As this formation moves laterally and radially inward to supply the diapir core, viscous coupling with overlying strata will cause these strata to move inward too. Both in the focal zone where the diapir forms and in the collar region through which the central uplift must be forced, there are strong inward horizontal confining stresses.

To understand the subsequent behavior of this highly compressed vertical column of rock, data on the behavior of compressed metals and rocks are used. When fluid under high pressure surrounds a metal billet in an extrusion container, increased hydrostatic stress induced in the billet makes the metal capable of an increasingly ductile behavior. Extrusion under these conditions is termed hydrostatic extrusion. The movement of the central uplift can be compared to a hydrostatic extrusion process. If a sufficient compressive strain energy is stored in the upward-moving rock column, Lüders band phenomena are expected where the compressed column approaches the earth's surface and encounters a zone of sudden, marked relief from the horizontal confining stresses. The large deviatoric stress acts within a very narrow Lüders front, which moves downward relative to the rising extruded rock.

Lüders fronts are a very special category of shock front, and this mechanism can

account for the so-called shock deformation textures, characteristic of central uplifts of cryptoexplosion structures. Shatter cones are also explained through penetration of tensile joints into the plastic zone where Lüders bands form. Certain
other workers have deduced that the shock front must be associated with meteorite
impact, but this deduction is thought to be in error. Violently formed breccias and
pseudotachylites are also accounted for in this category of diapir; it forms a consistent internal mechanism for generating cryptoexplosion structures.

INTRODUCTION

The cryptoexplosion structures analyzed in this paper are domical features of
nearly circular to polygonal outline, in which central uplifts of older units contain
intensely deformed and brecciated rocks. The uplift is usually surrounded by a well-
marked ring syncline of younger units, exhibiting much less intense deformation
although always faulted. Sometimes the ring syncline may be subdivided into concentric synclines, separated by a circular horst.

The intense deformations affecting the central uplift may include localized fusion
of rock and production of closely spaced cleavages, lamellae, and "planar features"
in quartz (which does not normally exhibit such textures). In certain structures,
other minerals (notably plagioclase) also exhibit intense cleavages and distinctive
microstructures. There is no preferred orientation or fabric of the mineral grains.
At the same time a very rare and distinctive mode of fracture termed "shatter cone"
is present at about 16 of the cryptoexplosion sites; parallel orientation of the axes
of these "shatter cones," at any single locality, does impart a marked vector regular-
ity to the deformed rock.

The origin of cryptoexplosion structures has been in dispute for several decades.
A massive energy source is clearly required, to account for the work done in uplifting the deep-lying strata and for the work expended in shearing, fracturing, fus-
ing, and brecciating the uplifted rocks. In certain structures there is evidence that
the intensity of deformation and brecciation dies out with increasing depth.

A meteorite impact origin for the North American cryptoexplosion structures was
first proposed by Boon and Albritton (1936); the kinetic energy of the meteorite
was seen as a suitable energy source. Downward diminution of the violent and
chaotic brecciation was accounted for neatly. This theory has rapidly gained favor
among North American geologists during the past decade. There were two mile-
stones in this development: the claim (by Dietz, 1959) that shatter cones uniquely fingerprint meteorite impact sites, and the discovery of close similarities in
the quartz cleavages, basal lamellae, and planar features present at these sites and
at known meteorite impact craters (Carter, 1965). The ultimate accolade was be-
stowed on impact theory in the Tectonic Map of North America, prepared and
published by the United States Geological Survey (King, 1969). In the legend of
this map the following descriptions of astroblemes occur: "Probably produced by
extra-terrestrial impact. Includes so-called meteorite craters, cryptoexplosions and
cryptovolcanic structures." The sites on the map are reproduced in Figure 1, and
they include almost all of the structures previously classified as cryptoexplosive.

Another portent of the change in opinion came in a conference on shock metamorphism of natural materials held in Washington, D.C. in 1966. A large amount
of valuable new information demonstrated that many microscopic features were

Figure 1. Stars—meteorite impact craters; open circles—structures which are also classified as Astroblemes in the Tectonic Map of North America (King, 1969); and closed circles—additional structures classified as probable cryptoexplosion structures in French and Short (1968, p. 256, 293).

shared by rocks deformed in the following environments: (*a*) laboratory and nuclear explosion shock wave experiments; (*b*) known meteorite impact craters; (*c*) certain cryptoexplosion structures; and (*d*) static laboratory experiments at elevated pressures (15 to 30 kbars) and temperatures (300 to 1,500°C).

Great significance was assigned to the first three of these environments, and the massive Conference Volume, edited by French and Short (1968) presented an account of almost unanimous consent among those present that the genetic problems of cryptoexplosion structures were solved by a meteorite impact origin. In the introduction, French stated that

"The passage of time since the Conference has brought about a realization that "cryptoexplosion" structures, regardless of whether they occur in bedded limestones or in granitic gneisses, are the result of meteorite impacts. The increasingly common association of shatter cones with petrographic shock-metamorphic features also supports the theory that shatter cones themselves are unique indicators of meteorite impacts . . ."

Data from previous investigations which support an internal origin for these structures were given short shrift (French and Short, 1968, p. 8–9).

Nevertheless, careful and penetrating reviews of the geologic evidence relating to some of these same cryptoexplosion structures were published by Bucher (1963) and Snyder and Gerdemann (1965). The crucial point of their analyses—and surely a rational point—is that the distribution of impact structures should be essentially random. This is especially true of structures caused by impact at different times. In Figure 1, note the prominence of cryptoexplosion structures in a part of the east-central United States. The literature on individual structures shows that they have different ages. There are several in eastern Canada. They are extremely rare elsewhere, even in terranes very similar to the east-central United States. Worldwide, outside of North America (Fig. 1), only about 13 cryptoexplosion structures are known.

The second point made by Bucher, and by Snyder and Gerdermann, is that certain of the cryptoexplosion structures are closely associated, in space and time, with structures of deep-seated magmatic origin. This point is documented in great detail for the Crooked Creek, Decaturville, and Wells Creek structures (of the east-central United States), and no further review is needed here.

Bucher's third criticism is that "Not one of the 'bona fide' meteorite craters shows the least sign of a central uplift whereas every one of the so-called cryptoexplosion structures has a central uplift and marginal collapse."

INTERNAL MECHANISM

Proponents of an impact origin have often stated that an externally induced origin is inescapable, since there is no plausible internal mechanism for generating the specific features of cryptoexplosion structures, such as the chaotic breccias, "shock" textures, and shatter cones. This paper deals with an internal mechanism, which is based on the gross geology of cryptoexplosion structures; examples are illustrated in Figure 2 for (a) Wells Creek (Tennessee) and (b) Sierra Madera (Texas). We note at the outset that, where the structures affect purely sedimentary successions, carbonate or sulphate beds play an important part in the stratigraphic sequence affected. The mechanism is proposed (in the first instance) for such structures, which have a typical and frequent development in the Palaeozoic strata of the east-central stable platform of the United States, the area within the dashed boundary outlined in Figure 1. While it may be relevant to other North American structures, those whose yield of rock samples is restricted to drill core specimens are likely to remain enigmatic for a long time.

1. The ring syncline and the central core have a strongly complementary appearance. The gross form is similar to that of experimentally produced diapirs (Ramberg, 1967), in which the syncline sinks because of a radial inward abstraction of rocks lying below the syncline, toward the central uplift (Fig. 3). The rocks which form the fundamental source of the central uplift may be termed source-beds; Amstutz (1964) was the first to emphasize these specifically diapiric features.

2. The polygonal form shown in plan by specific beds in the uplift exhibits a unique feature: The total outcrop length, summed over many slightly curved segments, is considerably greater than the perimeter of a circle transcribed on the median position of that particular bed. Each bed exhibiting this feature must have

moved inward from its original position, toward the center of the structure (Manton, 1965; Wilshire and Howard, 1968). The simplest explanation of such movement is viscous coupling of these strata with the underlying source-beds. The first postulate can therefore be extended: When the uplift was initiated, the central sector must have been subjected to higher-than-normal horizontal confining stresses. These stresses must have been approximately equal in all azimuths (from the gross symmetry).

3. Ramberg's photographs of model diapirs allow us to estimate the height of the zone in which intense confining stresses act on the upward-moving rock column of the central sector. It will not be a narrow ring, but more likely a wide collar, with vertical dimension a substantial fraction of the central uplift radius (Fig. 4a). This stress regime invites comparison with the innovation in die pressing termed hydrostatic extrusion (Kronberger, 1969), illustrated in Figure 5. If an otherwise brittle metal is subject to high hydrostatic stresses from a confining fluid, it takes on a markedly greater capacity for ductile behavior. For example, it may be forced through a much narrower die than was previously possible; there is a zone of intense plastic yielding around the die.

4. After the unusually compressed, mechanically weakened, rock column has been driven upward to a level near the earth's surface, it will encounter a zone of marked relief from the lateral compressive stresses (Fig. 4b)[1]. The rock will fail by lateral expansion; the expansion is accomplished by passage of Lüders bands. This is a mode not often seen by structural geologists, but very familiar in experimental deformation of rocks in the brittle-ductile transition (Heard, 1960). As a given point in the central rock column moves upward, it remains in a state of hydrostatic compression until it suddenly meets a band of plastic deformation. Each elementary volume of rock therefore experiences passage through a sharply defined wave of plastic deformation which has a downward sense of motion, relative to this elementary volume. This point is clarified in Figure 6, drawn from an experimental deformation of a limestone in the brittle-ductile transition region.

The model explains distinctive features of cryptoexplosion structures, as detailed below.

Production of "Shock Deformation Textures"

The production of intense cleavages, lamellae, and planar features in quartz has hitherto been identified with passage of shock waves. Carter (1965, 1968) placed the greatest importance on the lamellae parallel to the basal plane, formed by basal slip. These features are rare in tectonites, but they occur in specimens from many cryptoexplosion structures. Results were quoted from static experimental studies (in which Carter played a leading role) where quartz deformed by slip on several systems. Under most of the experimental conditions (15 to 30 kb confining pressure, 300 to 1,500°C, strain rate $\dot{\epsilon} = 10^{-3}$ to 10^{-6}/sec) basal slip parallel to the a-axis had the lowest critical resolved shear stress. The quartz deformation lamellae in specimens from cryptoexplosion structures and impact structures were found to

---

[1] Bucher (1963, p. 637) noted that ". . . the major stage of the disturbance was beginning, when the center column was raised bodily from below and the surrounding ring of rock was settling downward, producing lateral tension while pressure from below imposed a maximum compression in the vertical direction."

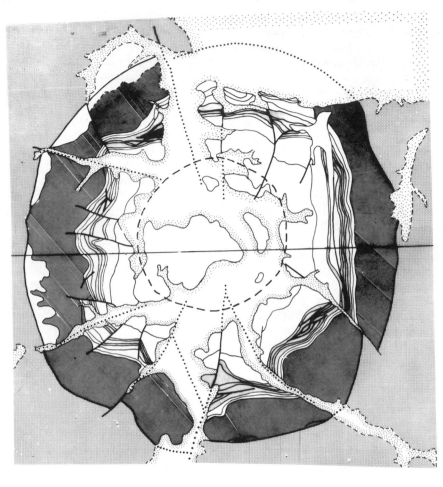

0   1   2 Km

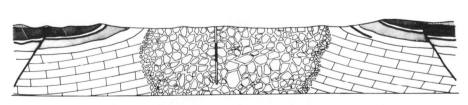

Figure 2a. Simplified geological map and section for the Wells Creek (Tennessee) cryptoexplosion structure. *After* **Wilson and Stearns (1968).**

have identical optical properties to those produced both in static experiments and naturally under static conditions (Carter, 1968, p. 463). Why then did Carter state that the cryptoexplosion quartz features could only have formed during the intense shock deformation around impact sites? Their frequent development and random spatial orientation, in specimens from Vredefort and Middlesboro struc-

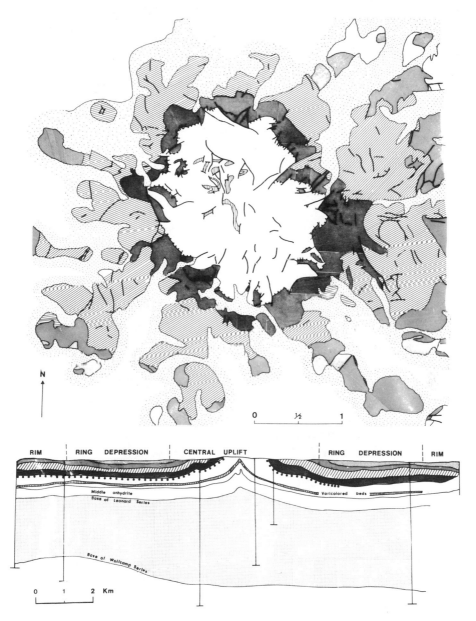

Figure 2b. Simplified geological map and section for the Sierra Madera (Texas) cryptoexplosion structure. *After* Wilshire and Howard (1968).

tures, were the crucial features which led him to state (1968, p. 469): "I can conceive of no reasonable loading condition other than shock, with its accompanying complex reflections and rarefactions, that could account for these patterns which show that virtually every grain in a random aggregate must have been stressed sufficiently highly to promote basal slip to the exclusion of slip on other systems."

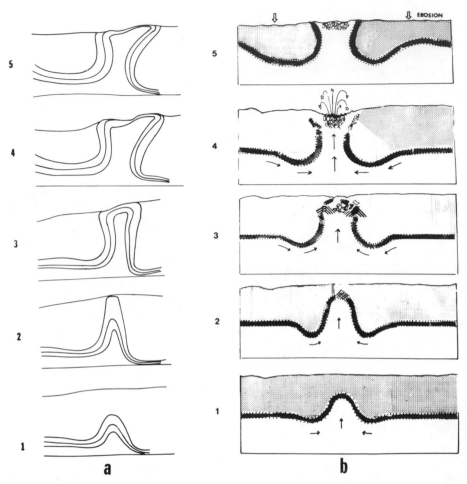

Figure 3. (a) Tracing of "bed" positions at successive stages of an experimentally produced diapir (using laboratory materials) *from* Ramberg (1967). (b) Schematic interpretation of cryptoexplosion structures as diapirs. Stage 3 is interpreted as the time when the upward-moving rocks of the central core encounter release from confining stresses (at the top of the diapir).

In this paper, the crucial feature (intense basal lamellar development coupled with random grain orientation) is assigned to passage of a plastic wave of Lüders band character. Cottrell (1964) shows how such plastic waves form a special category of shock front because the large strain change (which defines the transition from the unstrained to the strained state) occurs only in a very thin band.

The strain rate localized in the front itself is

$$\dot{\epsilon} = \frac{v\Delta\epsilon}{S}$$

where v = velocity of propagation of the Lüders front,
   $\Delta\epsilon$ = the discontinuity in strain across the Lüders front,
   S = thickness of the front, of the order of the typical grain size of the material.

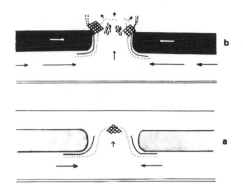

Figure 4. Highly schematic diagram to indicate how the laterally compressed central uplift in a diapir resembles a material under hydrostatic extrusion. As the diapiric structure develops from stage (a) to stage (b), viscous coupling between "source bed" and covering strata causes increasing confinement of the uplifted rocks in the central zone. The portion of the more rigid layer (of covering strata) which is closest to the central uplift is shaded to denote the mechanical resemblance between this zone of constriction and a die. The central uplift rocks which escape from the lateral constraints of the die zone are thought to experience Lüders band development (diagonal hachuring) as soon as they emerge.

Cottrell cites a typical laboratory experiment in which a steel bar is stressed in tension; v is 1 cm/sec, $\Delta\epsilon = 0.03$, and $S = 0.01$ cm, so that $\dot{\epsilon} = 3$/sec. Thus, an exceptionally high strain rate can be brought about by conditions far removed from meteorite impact.

Referring back to Figure 4, and to Figure 6: a large amount of elastic strain energy per unit volume in maintained in the compressed rock rising vertically within the central uplift; conversion of this energy into the work of shear deformation is delayed until the thin, sharply defined Lüders front moves through the material. After it has passed, the shear deformation is complete. There is no time available for grain orientation processes to take place at such high strain rates.

Production of Shatter Cones

The main difference between the laboratory experiment (Fig. 6) and the expanding central uplift (Fig. 4) lies in the asymmetry of Figure 4. As the rock is released from lateral constraint, and expands laterally, cracks form. However, as the tensile joint penetrates downward, it will dilate a material in an unusual rheologic condition, that is, the plastically deforming rock in the Lüders band (Fig. 7). Here the crack becomes plastically blunted. The distinctive stepped form of shatter cones is due to rapid alternation of blunting and further penetration of a crack tip in this material. This leads to an en echelon array of cracks (when viewed in a vertical section); the geometry of coalescence of these cracks to form shatter cones has been demonstrated (Nicolaysen, in prep.).

Explanation for the Geometry of the Cone Axes

Decompression and jointing of a diapir top can account for the geometry of the cone axes. Two simple concepts are required: (a) When the diapir top undergoes decompression near the free surface, the axis of greatest compressive stress is vertical, and the axis of least compressive stress is horizontal. Thus shatter cone axes are vertical when the cones are formed. (b) As the diapir top first begins to undergo decompression, this portion of the structure is anticlinal.

These concepts are combined in Figure 8a. Figure 8b illustrates the geometry of cone axes after overturning of the strata; their attitudes closely resemble the observed orientations. If this reconstruction is correct, then Figure 8c indicates (in

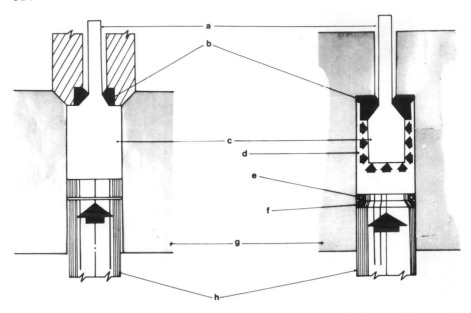

Figure 5. (a) extrusion; (b) die; (c) billet; (d) high-pressure fluid; (e) "O" ring; (f) miter seal ring; (g) high-pressure vessel; and (h) ram or plunger. In conventional extrusion (left), a heated metal billet is compressed by a ram and forced through a die in the direction of ram movement. In hydrostatic extrusion (right), the compressive stress is applied to a fluid surrounding the billet. Hydrostatic stress exists throughout the metal except at the die aperture. The metal is now more prone to ductile behavior and is more readily extruded. Increased capacity for plastic deformation around the die brings about this ready extrusion. *After* Kronberger (1969).

a schematic way) how the cone axes might give a *fictional* focus at a central explosion when the strata are restored to a horizontal attitude.

Production of Intensely Brecciated Rocks, "Shock Melts," and Pseudotachylites

The delicate stepped features preserved on shatter cone surfaces demonstrate that there has been no gross shearing motion between opposite faces. The breccias and so-called shock melts (which occur in close proximity) are sharply contrasted in their evidence for chaotic movement, violent comminution, and melting. We return to the rapid conversion of compressional elastic strain energy into the work of shear deformation during the passage of the thin Lüders front. Plastic work is accomplished by the deviatoric stresses, constrained to act in an intense manner over an exceedingly narrow zone. This system is vulnerable to instabilities. First consider a cold environment: an increase in strain rate, or a perturbation in the geometry of the maximum compressive stress and the confining stresses can cause the material to behave in a more brittle manner. The formation of through-going faults would take place violently; if there are dilatancy phenomena, they would be accompanied by a large stress drop. Next, consider a metamorphic environment: where the velocity of the central uplift relative to the front of decompression exceeds a limiting value, the conversion of elastic strain energy into the work of plastic deformation, and then into heat, would occur at a rate which is too high for this

Figure 6. The theory of plastic deformation of continuum solids defines particular planes (Lüders bands) which divide yielded and unyielded regions and which advance through the solid as a group of parallel shear bands as deformation progresses. This figure illustrates the development of Lüders bands in a limestone specimen undergoing deformation in the brittle-ductile transition region (*after* Heard, 1960). In this experiment, the Lüders front (shown schematically in shaded form) is moving downward through the specimen. In a cryptoexplosion structure, a given point in the central uplift is envisaged as moving upward relative to a Lüders front in the expansion zone of Figure 4b. The maximum compressive stress is vertical.

Figure 7. Here the development of shatter cones is linked to the time when central uplift rocks move up relative to the Lüders front. The zone of lateral expansion is identical to the zone of lateral expansion in Figure 4b. When the expansive lateral strain exceeds a certain level, joints open. (This process is assisted by pore fluid pressures existing in these rocks.) These joints penetrate downward into the zone where Lüders bands are forming. Intense plastic blunting of the fracture tip results, leading to en echelon arrays of coalescing cracks, with crack planes orthogonal to this diagram. These are shatter-cone arrays, with apices pointing up.

heat to be effectively dissipated in the environment by conduction or material transfer. This leads to partial melting, which would tend to begin in highly stressed zones, and would violently accelerate any tendency to faulting. This process is similar to the Griggs and Handin (1960, p. 362) mechanism for violent propagation of a fault by shear melting; it is proposed for the category of pseudotachylites which show mylonite-like features in the field.

DRIVING FORCE FOR DIAPIRISM IN CRYPTOEXPLOSION STRUCTURES

Quantitative physical studies of instability in layered rocks leading to diapir formation, for example, those of Ramberg (1967) and Odé (1966), are based on gravitational instability. Field studies of salt and shale diapirs usually offer strong evidence of a density contrast; geologists are therefore well acquainted with the mechanics of diapirs "driven" by the geostatic loading of a buried layer of lower density than its overburden. Obvious density contrasts must have assisted in driving certain cryptoexplosion diapirs, the most likely example being the Vredefort structure. However, there are many North American cryptoexplosion structures in which there is no obvious density contrast between the (dry) rocks of the central uplift and the (dry) rocks of its surrounding; another agency for the release of gravitational energy must be sought.

Many previous investigators and interpreters of regions with cryptoexplosion structures have emphasized the role of fluid under high pressure. Their deductions are summarized below: (a) There was an exceptionally high volatile content in

alkalic and peridotitic intrusives which
occur on the same deep-going fault system as the Crooked Creek and Decaturville structures. Nearly all these intrusives are carbonatized as well (Zartman
and others, 1967; Snyder and Gerdermann, 1965). (b) A peculiar gas-rich
medium (characteristic of the kimber-
lite-alnoite rock suite?) is responsible
(Bucher, 1963; Holmes, 1965). (c)
Alkaline carbonatite fluids—probably
related to alkaline ultramafics—have a
critical importance in certain Canadian
cryptoexplosion structures (Currie and
Shafiquallah, 1968). (d) Eruption of a
volatile-rich alkaline ultramafic yielded
an incipient carbonatite volcano below
the Gross Brukkaros breccia vent—
which has certain striking similarities
to cryptoexplosion structures (Janse,
1969). (e) There is a long seismic an-
orogenic zone, cut by certain well-defined deep-going fault zones, traversing
eastern North America from the St.
Lawrence to the Mississippi embayment.
It has important analogs with the East
Africa rift valley system, and the cryptoexplosion structures have some relation

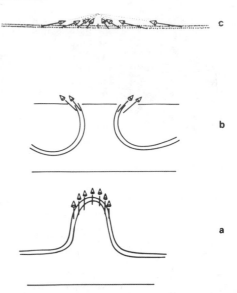

Figure 8. The geometry of shatter cones
and their cone axes formed in cryptoexplosion structures. (a) Approximate vertical orientation of cone axes shown at the
time of shatter cone formation. (b) Orientation of cone axes after folding is complete (schematic). (c) "Astrobleme" reconstruction of the geometry of cone axes
(when the strata are horizontal) yields a
fictional focus at a central point.

to explosively emplaced carbonatites (independently suggested by McCall, 1964;
Kumarapeli and Saull, 1966).

MECHANICS OF DIAPIRS DRIVEN BY FLUID PRESSURE

Goguel (1963) gives a closely reasoned analysis of the way in which high fluid
pressures can lead to diapirism and a cryptoexplosion structure. (Although the
specific word "diapirism" is not used, it is quite clear that a similar physical process
is involved.) The physics used by Goguel is close to that developed independently
by engineers for the behavior of a deep sand layer in which "liquefaction" of the
sand occurs during an earthquake. During a large earthquake such as the Quetta
event (West, 1936), dilatancy phenomena may cause continued increases in the
pore fluid pressure in one deep sand layer, until it is equal to the external confining
pressure on that layer. Seed (1968) and co-workers showed rigorously how the
layer then becomes liquefied and cohesionless. If the fluid pressure exceeds the confining pressure, the layer may rise at one point in dome form and may extrude
violently from the surface as a "sand blow" (Housner, 1958). The inherent instability of these structures may be gauged by the difference between the fluid pressure (of the critical sand layer) and the lithostatic pressure at various points on its
ascent. If the pore fluid pressure is maintained, this difference increases at successively shallower depths. Dike-like intrusions of sand into surficial strata may also
occur. Further examples of diapirs driven by fluid pressure in weak geological

materials are mud diapirs, mud volcanoes, and certain Persian salt domes. Releases of fluid under abnormal pressures can take place when all these reach a near-surface zone (Gansser, 1960).

A fluid-pressure-driven rock diapir, which culminates in a cryptoexplosion structure, is forced up by the same gross mechanics causing these well-understood sand blows and mud diapirs. At first, such structures appear unlikely because of the high cohesion exhibited by rocks (compared to sands and muds). I draw attention to the large body of experimental work which shows that compressed jacketed rock specimens have increased cohesive strength under pressure, but if the jacket leaks, admitting the pressure fluid to the specimen, there is a substantial reduction in strength. Silicate rocks can also exhibit a marked loss of strength from an effect termed hydrolytic weakening. (Both these processes are reviewed by Griggs, 1967.) It is envisaged that the accession of high-pressure fluid to a particular rock can result in sufficient weakening and sufficient reduction in density (over a well-defined region) to allow that rock to form a diapir.

In those sedimentary successions of the United States affected by cryptoexplosion structures, the development of a mechanically weak character in limestone or anhydrite seems feasible, as it accords with the experimental behavior of these rocks under high pressure. Heard (1960) demonstrated that when the pore fluid pressure approaches the confining pressure in Solenhofen limestone, and the rock is subsequently stressed to failure, the mode of failure is characteristic of the brittle-ductile transition, provided that the confining pressure exceeds a certain limiting value. (At 150°C, this limit is $\sim$ 1.5 kb, and at 300°C, it is estimated to be $\sim$ 1 kb.) The strength of this rock is unaffected by fluid pressure if the sample remains in the ductile field but falls when it enters the brittle-ductile transition and brittle regions. In silicate rocks, the capacity for plastic behavior is less well known, and hydrolytic weakening of the silicates may be essential.

SUMMARY OF THE MECHANISM FOR NORTH AMERICAN CRYPTOEXPLOSION STRUCTURES

When fluid pressure is slowly increased within a particular rock layer, at a certain critical pressure the roof will bulge slightly, and the fluid moves radially inward to this focus. The rock exhibits successively lower cohesive strength and moves inward as well. In this focal region, it effectively senses the high ambient fluid pressures and also the highest horizontal confining pressures. At sufficient values of these pressures, one particular restricted volume of rock takes on a mechanically weak character and initiates a central uplift.

Now consider the region, nearer the earth's surface, where the compressed column obtains relief from the confining stresses and can undergo expansion in the horizontal plane. Hydrostatic extrusion is coupled to development of a Lüders front in this relief region. This front promotes the vertical movement of the rock in the central uplift and also accounts for the intensely developed deformation lamellae in random orientations. Certain zones suffer a sharply defined drop in fluid pressure, so that a strong and abrupt increase in the effective stress moves as a front through the rock mass.

Shatter cones are also accounted for, as a form of jointing which affects rock in the brittle-ductile transition. As the compressed diapir with high pore fluid pressure moves toward the earth's surface and is affected by the front of unloading, it fails. Cracking and crack coalescence are accompanied by an orderly loss of pore fluid pressure from the compressed rock to fluid-inflated microcracks[2] and then out of the system. Due to simultaneous loss of pore-fluid pressure and confining pressure, the material is constrained to remain in the brittle-ductile transition while shatter coning passes through a large volume of rock. The inflation of cracks by high-pressure fluid enables fracturing to occur at confining pressures greater than those normally penetrated by tensile joints. This model of shatter cone development explains their association with deformation lamellae formed in static laboratory experiments at elevated pressures and the formation of folds after shatter-cone development.

ENERGY BALANCE IN CRYPTOEXPLOSION STRUCTURES

When a buried rock mass undergoes uplift, the work required to open up a tensile joint comes from the release of the compressive strain energy in the rock. The shatter-cone mode, in which the rock yields and cracks simultaneously, absorbs this compressive strain energy (per unit area of crack advance) several times more effectively than does an ordinary tensile joint (Nicolaysen, in prep.). The upward movement of the central column in a cryptoexplosion structure differs from normal domical uplift primarily in the high strain energy stored and the high strain energy released during jointing.

At the beginning of this paper, it was recognized that a massive energy source is needed to explain many of the features of cryptoexplosion structures. Consider the many tests of pressure vessels and pressurized aircraft cabins which have been carried out to the stage of failure. The failure pressure level is little affected by substituting air for water as a pressurizing medium. However, there is immensely greater damage by violent fracture and shattering when air is used, due to the greater stored energy, or compliance, of the test system. A rising diapir with tight confinement of high pore fluid pressures would be characterized by a higher compliance than is expected from the corresponding "dry" structure. Elastic energy stored in compressed rock and compressed pore fluid is therefore thought to be the essential energy source for these North American cryptoexplosion structures.

In concluding, note one major difference between cryptoexplosion structures and the diapirs developed in metamorphic terranes and in Ramberg's experiments: cryptoexplosion structures lack thinning of strata along the sides of the uplift. While the sand blow analogy is believed to yield insights concerning the initiation and driving force for these structures, it is not an appropriate model for some other aspects of their development.

ACKNOWLEDGMENTS

H. H. Hess and J. C. Maxwell made it possible for the author to visit the United States in 1968, and to acquaint himself with the problems presented by these struc-

---

[2] Bucher (1963, p. 635) first emphasized such fracturing. He referred to a rise in gas pressure in the pores of the rock leading "to the formation of shatter cone fractures that are small compared to the dimensions of the whole deforming rock body."

tures. Important geology was learned from B. M. French (NASA) and from K. A. Howard, D. Milton, T. W. Offield, E. M. Shoemaker, and H. G. Wilshire, all of the Astrogeology Division of the U.S. Geological Survey. J. D. Byerlee and N.G.W. Cook provided very stimulating discussions on the possible role of Lüders fronts in geologic structures. This period of research was supported by Princeton University.

Sir Edward Bullard enabled the author to visit the Department of Geodesy and Geophysics, Cambridge University, during the Lent term, 1969; much of the paper was written during this period.

J. Goguel, J. Ferguson, and N. C. Gay gave valuable criticisms of the manuscript.

REFERENCES CITED

Amstutz, G. C., 1964, Impact, cryptoexplosion, or diapiric movements?: Kansas Acad. Sci. Trans., v. 67, no. 2, p. 343–356.

Boon, J. D., and Albritton, C. C., Jr., 1936, Meteorite craters and their possible relationship to "Cryptovolcanic Structures": Field and Laboratory, v. V, p. 1–9.

Bucher, W. H., 1963, Cryptoexplosion structures caused from without or within the earth? ("Astroblemes" or "Geoblemes?"): Am. Jour. Sci., v. 261, p. 597–649.

Carter, N. L., 1965, Basal quartz deformation lamellae, a criterion for recognition of impactites: Am. Jour. Sci., v. 263, p. 786–806.

Carter, N. L., 1968, Dynamic deformation of quartz, in Shock metamorphism of natural materials: Baltimore, Mono Book Corp., p. 453–474.

Cottrell, A. H., 1964, The mechanical properties of matter: New York, John Wiley & Sons, Inc., p. 168–170.

Currie, K. L., and Shafiqullah, M., 1968, Geochemistry of some large Canadian craters: Nature, v. 218, p. 457–459.

Dietz, R. S., 1959, Shatter cones in cryptoexplosion structures (meteorite impact?): Jour. Geology, v. 67, p. 496–505.

French, B. M., and Short, N. M., eds., 1968, Shock metamorphism of natural materials: Proc. of the First Conference held in Greenbelt, Maryland, April, 1966: Baltimore, Mono Book Corp., 644 p.

Gansser, A., 1960, Uber Schlammvulkane und Salzdome: Naturf. Gesell. Zürich Verh., v. 105, p. 1–46.

Goguel, J., 1963, A hypothesis on the origin of the "cryptovolcanic structures" of the central platform of North America: Am. Jour. Sci., v. 261, p. 665–667.

Griggs, D., 1967, Hydrolytic weakening of quartz and other silicates: Royal Astron. Soc. Geophys. Jour., v. 14, p. 19–32.

Griggs, D., and Handin, J., 1960, Observations on fracture and a hypothesis of earthquakes: Rock deformation: Geol. Soc. America Mem. 79, p. 362–363.

Heard, H. C., 1960, Transition from brittle to ductile flow in Solenhofen limestone as a function of temperature, confining pressure and interstitial fluid pressure: Rock deformation: Geol. Soc. America Mem. 79, p. 347.

Holmes, A., 1965, Physical geology (Revised edition): London, Thos. Nelson, p. 1051–1053.

Housner, G. W., 1958, The mechanism of sand blows: Bull. Seismol. Soc. America, v. 48, p. 155–161.

Janse, A.J.A., 1969, Gross Brukkaros, a probable carbonatite volcano in the Nama Plateau of Southwest Africa: Geol. Soc. America Bull., v. 80, p. 573–586.

King, 1969, Tectonic map of North America: U. S. Geol. Survey Map.

Kronberger, H., 1969, Hydrostatic extrusion: Royal Soc. [London] Proc., A 311, p. 331–347.

Kumarapeli, P. S., and Saull, V. A., 1966, The St. Lawrence Valley system: A North American equivalent of the East African Rift Valley System: Canadian Jour. Earth Sci., v. 3, p. 639–658.

Manton, W. I., 1965, The orientation and origin of shatter-cones in the Vredefort Ring: New York Acad. Sci. Annals, v. 123, p. 1017–1048.

620   L. O. NICOLAYSEN

McCall, G.J.H., 1964, Are cryptovolcanic structures due to meteorite impact?: Nature, v. 201, p. 251–254.

Odé, H., 1966, Gravitational instability of a multilayered system of high viscosity: Koninkl. Nederlandse Akad. Wetensch. Verh., Afd. Natuurk. Eerste Reeks, Deel XXIV, no. 1, 96 p.

Ramberg, H., 1967, Gravity, deformation and the earth's crust: New York, Academic Press, Inc., 214 p.

Seed, H. B., 1968, Landslides during earthquakes due to soil liquefaction: The Fourth Terzaghi Lecture: Am. Soc. Civil Engineers Proc., Jour. Soil Mechanics and Found. Div., SM 5, p. 1053–1122.

Snyder, F. G., and Gerdemann, P. E., 1965, Explosive igneous activity along an Illinois-Missouri-Kansas axis: Am. Jour. Sci., v. 263, p. 465–493.

West, W. D., 1936, Geological account of the Quetta Earthquake: Mining Inst. India Trans., v. 30.

Wilshire, H. G., and Howard, K. A., 1968, Structural pattern in central uplifts of crypto-explosion structures as typified by Sierra Madera: Science, v. 162, p. 258–260.

Wilson, C. W., Jr., and Stearns, R. G., 1968, Geology of the Wells Creek structure, Tennessee: Tennessee Div. Geology Bull. 68, 236 p.

Zartman, R. E., Brock, M. R., Heyl, A. V., and Thomas, H. H., 1967, K–Ar and Rb–Sr ages of some alkalic intrusive rocks from Central and Eastern United States: Am. Jour. Sci., v. 265, p. 848–870.

MANUSCRIPT RECEIVED BY THE SOCIETY MARCH 29, 1971

Environmental Geology

THE GEOLOGICAL SOCIETY OF AMERICA, INC.
MEMOIR 132
© 1972

Waste Discharges as Sedimentological Experiments

M. GRANT GROSS
Marine Sciences Research Center
State University of New York
Stony Brook, New York 11790

ABSTRACT

Large-volume waste discharges provide opportunities for sediment studies and experimentation on a regional scale during short periods of time. Studies of sedimentary dispersal systems have used (1) large-scale discharges of waste solids, (2) changed erosion rates, and (3) natural sources "labelled" by distinctive substances such as radionuclides. Physical characteristics (including grain size) and chemical and mineral composition aid identification of certain deposits of waste solids. Combustion products, including mullite from coal and soot from petroleum, are potentially useful as tracers. Industrial materials and ceramics containing minerals alien to most marine sediments may also serve as tracers. Dumping of large volumes of wastes on small areas of the continental shelf provides a useful means for experiments. Greatly increased erosion resulting from agricultural or construction activities may also afford opportunities for studies of sedimentary processes. Radionuclides have been used to study sediment movement in rivers, on continental shelves, and in ocean water.

DISCHARGE OF WASTE SOLIDS

Large-volume discharges of waste solids afford unique opportunities for the study of sediments and sedimentary processes involving large areas and various time scales. These deposits yield valuable information about rates and routes of movement through rivers, the coastal ocean, and the atmosphere. In effect they can serve as large-scale field experiments.

Waste discharges offer advantages not commonly encountered in natural sediment systems. Sources of waste solids usually are known or can be determined from public records (Gross, 1970a). Disposal site locations are known and vary little with time. Furthermore, data on the volume of discharged materials commonly are available. In contrast, chemical and mineral composition and physical properties of the wastes must be determined, in most instances, as part of the study (Gross, 1970b).

Waste discharges may be used in several ways to study sedimentary processes:

1. Discharges of materials rarely present in the disposal areas may be used to observe dispersal of solids under different conditions. Changing disposal areas or types of material discharged provide opportunity for experimentation.

2. Natural discharge rates may be changed for varying periods of time by construction or agricultural activities, thus permitting investigation of transient phenomena.

3. Uniquely identifiable tracers, such as radionuclides or distinctive suites of minor elements, mixed with the natural sediment load or water discharged by rivers, aid the identification of deposits far from their sources.

Each of these methods is illustrated by selected examples of the application, and by showing its potential uses.

The simplest type of experimental field study involves minerals or other materials not normally present in sediments unaffected by waste discharges. Usually present in small quantity, such materials require sensitive and selective analytical procedures to determine their presence and concentrations. Modern analytical techniques provide substantial opportunities, among them x-ray diffraction. A promising application of x-ray diffraction technique is the study of the distribution of high-temperature minerals or industrial ceramics found in sediments as a result of waste disposal contamination. For example, mullite, a typical mineral in coal ash (Gross, 1970b), does not occur naturally in sediments. When present, it points to waste disposal operations. Because of the large amount of coal burned throughout the world and the wide dispersal of ash, mullite should be quite common in sediments accumulating near urban areas. For example, Muller (1966) detected both coal (0.07 to 1.43 percent) and ash (3.14 to 7.13 percent) in sediments from Lake Constance on the Swiss-German-Austrian border.

Industrial minerals and ceramics are potentially useful as tracers. Among them are silicon carbide and alumina, widely used as abrasives. Although produced in large quantities, these materials may be too rare in sediments for ready detection without preliminary concentrations, such as by density separations. More common materials could also be used; for example, glass fragments are probably widespread and might be useful tracers.

Studies may also be made of sediment discharges controlled or affected by human activities. In most coastal urban areas, large volumes of sediment and other wastes are removed by dredging navigable waterways; these wastes are deposited elsewhere in a harbor or nearby coastal area. For example, millions of tons of dredged materials are removed each year from New York Harbor and deposited on the adjacent continental shelf (Gross, 1970a). These dredged materials have physical and chemical characteristics quite different from the natural sediments occurring in the waste-disposal area (Figs. 1 and 2). Harbor deposits are dominantly silts (median diameter 30 μ), apparently supplied by the Hudson River.

Sand (median diameter 100 to 200 μ) predominates on most continental

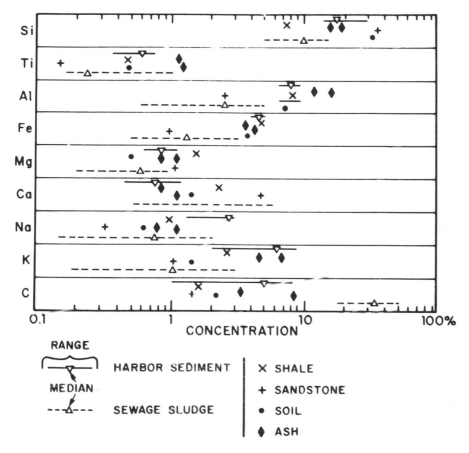

Figure 1. Abundance of major elements in deposits from the continental shelf near New York Harbor and from disposal areas used for dredged wastes and sewage sludges. Except for the abnormally high carbon contents and low calcium contents, the waste deposits appear to be mixtures of sand and shale-like materials (Gross, 1970b).

shelves. Therefore, it is easy to detect fine-grained sediments rich in carbon from wastes that are mixed with the shelf sand. Studies of the behavior of these materials provide an opportunity to learn more about sedimentary processes in an area otherwise nearly devoid of recently deposited sediment (Emery, 1968).

CHANGED EROSION RATES

On land, construction of highways and subdivisions (Anderson and McCall, 1968; Vice and others, 1969), supply large volumes of sediments eroded over short periods of time (Table 1). These pulses of sediment affect river channels and deposit downstream from the sources, providing information about sediment movement in streams (Wolman and Schick, 1967). Such sediment sources are common in most coastal areas.

Sediment eroding from agricultural land also provides opportunities for study of sediment movement and deposition in streams. It is estimated that sediment

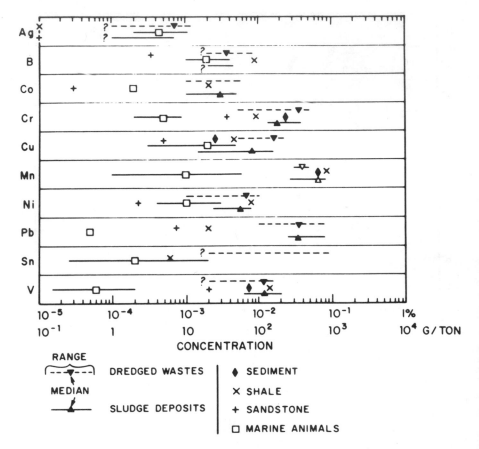

Figure 2. Abundance of minor elements in deposits from the continental shelf near New York Harbor and from designated disposal areas used for dredged wastes and sewage sludges. Note the abnormally high concentrations of copper and lead in the waste deposits relative to those in nearby sediments or in typical sands and shales (Gross, 1970b).

production in the United States is about 4 billion tons per year (Wadleigh, 1968, p. 36), much of which comes from agricultural land. Anomalously high sediment discharges seem to prevail over large areas (Strakhov, 1967, p. 17). Judson (1968) showed that human activity near Rome caused a ten-fold increase in erosion rates. Meade (1969) estimated that U.S. rivers draining into the Atlantic Ocean carry sediment loads four to five times greater than if they had not been disturbed by man.

In the past, most attention has been given to water transport of sediment, although wind transport does occur. Studies of mineral composition of deep-ocean sediment indicate that wind-borne material may be far more important than previously recognized (Rex and Goldberg, 1962). Detection of talc in wind-blown dust in rain and snow demonstrates the effect of agricultural activities (Windom and others, 1967). Talc apparently is injected into the atmosphere through its use as a diluent and carrier for agricultural pesticides.

Other industrial minerals may be useful tracers for wind-borne materials. These include mullite (from coal ash), asbestos, and soot (Gorham, 1957; Delaney and

TABLE 1. SOLIDS DISCHARGED PER UNIT AREA OF
DRAINAGE BASIN BY VARIOUS SOURCES

Source of Solids	Average (Range) (Tons/yr/km² of drainage basin)	
Rural drainage area, Maryland	126	(6–350)
Strip mine, Kentucky	440	
Housing subdivisions, Maryland	1,500	(390–46,000)
Construction sites and		
Industrial parks, Maryland	31,000	(28,000–54,000)

After Wolman and Schick, 1967.

others, 1967) from combustion of petroleum. Iron oxide particles from industrial wastes and paper fibers from sewage solids have also been detected in coastal waters (Manheim and others, 1970). Because of the minute quantities involved, detection of such tracers in sediments will require sensitive techniques using electron microscope, electron microprobe, and x-ray diffraction.

LABELLING BY DISTINCTIVE SUBSTANCES

Sediment-associated radionuclides have been used to study the dispersal and deposition of modern sediments in rivers and the coastal ocean. The interest in using radionuclides results, in part, from the sensitive instrumentation and analytical techniques available.

Low-level radioactive wastes were discharged by the Oak Ridge National Laboratory into tributaries of the Tennessee River for at least 20 yrs beginning in 1943 (Carrigan, 1969). These radionuclides, primarily fission products, were detectable in sediments more than 30 km downstream from the discharge points (Pickering, 1969). Variations of radionuclide concentrations with depth in sediment cores corresponded closely to known discharge rates from the laboratory, thus providing a quantitative measure of rates of sediment accumulation over a 30-km portion of the lower Clinch River, a tributary of the Tennessee River.

Detailed studies of mineral and chemical composition of sediment from the same river system showed that the uptake of radionuclides by sediment particles was determined by specific reactions of several types (Jenne and Wahlberg, 1968). Cesium-137 was sorbed primarily by clay minerals. Cobalt-60 was taken up by manganese and iron oxides, occurring primarily as coatings on mineral grains. Strontium-90 was controlled by in-situ precipitation of carbonates.

The larger continued discharges of radionuclides by the plutonium-producing reactors on the Columbia River, at Hanford, Washington, were used to study the river and its interactions with the northeast Pacific Ocean (Barnes and Gross, 1966). Unlike the relatively long-lived fission products discharged by the Oak Ridge facilities, the Hanford reactors discharged large quantities of relatively short-lived neutron-activated radionuclides, including zinc-65 (half-life 245 days), chromium-51 (28 days), cobalt-60 (5.3 yrs), which proved useful for sediment studies (Nelson and others, 1966). Studies of radionuclides in the river showed that sediment deposition occurred in reservoirs impounded by dams on the river. About one-third of the annual sediment accumulation was scoured and moved downstream during the annual spring–early summer flood (Nelson and others, 1966). Studies of radionuclide uptake showed that chromium-51, present primarily in anionic form, was removed by relatively slow reactions with sediment particles. Radionuclides occurring as cations were removed more rapidly (Osterberg and others, 1966).

Radionuclides sorbed by sediment served as tracers for Columbia River sediment recently deposited on the continental shelf near the river mouth. Distribution of radionuclides indicated that Columbia River sediment moved primarily northward along the continental shelf off the state of Washington (Gross, 1966). Using changes in activity of zinc-65 and cobalt-60 in sediment, it was possible to show that the radioactive portion of the sediment apparently moved approximately 30 km/yr parallel to the coast, but only about 10 km/yr perpendicular to the coast (Gross and Nelson, 1966). Such studies require detailed knowledge of the minerals and chemical properties of the sediment systems involved.

Distribution of radionuclides in surface waters of the northeast Pacific Ocean also provided useful information, such as the seasonally variable surface circulation (Gross and others, 1965). For example, the results confirmed that Columbia River water moves northward along the Washington coast during winter (Osterberg and others, 1966), and generally southwestward during summer (Barnes and Gross, 1966).

Radionuclides from atmospheric fallout have also been used to study sedimentary processes (Gross, 1967). The most abundant fission products have moderately short half-lives (typically three to four months). Furthermore, the source is more dispersed, and the input is more variable in time. The largest releases of fallout occurred during 1961 to 1963, a time of active testing of nuclear weapon systems in the atmosphere by the Soviet Union and the United States (Glasstone, 1964).

One study made during this period involved the behavior of estuarine sediments and their associated radionuclides in San Francisco Bay (Klingeman and Kaufman, 1965). During 1961 to 1963, zirconium-95 and niobium-95 (half-life 65 days) were the dominant fallout nuclides associated with the sediment. Because of the relatively short half-lives and relatively high levels of activity, it was possible to trace movements of sediment brought to the bay by individual storms. The study showed a continual exchange of sedimentary materials between recently deposited sediments and particles still suspended in the water.

Surveys of radioactivity in sediment on the continental rise indicated that radioactive particles sank at rates of 10 m or more per day (Gross, 1967). Apparently silt-sized particles carried most of the activity. Since atmospheric tests of nuclear devices by the United States and the Soviet Union ceased, the amount of radioactive fallout has decreased markedly. Some of the longer-lived fission products could still be useful for certain types of studies. Techniques capable of detecting low levels of cesium-137 (half-life 30 yrs) might be applied to studies of short-term sediment accumulation rates. Furthermore, the well-known spring peak of atmospheric fallout (Holland, 1963) might be used to study short-term sediment transport.

Other sources of potentially useful radionuclides include nuclear-powered electrical generation plants, nuclear-powered ships, and plants processing reactor fuel elements (Templeton and Preston, 1966). These sources release relatively small amounts of activity to the environment so that the experiments must be carefully planned and capable of detecting low levels of activity. Another source of widely dispersed release of radionuclides is the weathering of mining debris, which can be used for studies of sediment movement in and around areas of uranium mining and ore processing.

In summary, there appear to be many opportunities to use discharges of various wastes in studies of sedimentary processes on a regional basis over comparatively short periods of time. In some instances, it may even be possible to conduct experiments by changing the nature or location of waste disposal operations.

REFERENCES CITED

Anderson, P. W., and McCall, J. E., 1968, Urbanization's effect on sediment yield in New Jersey: Jour. Soil and Water Conserv., v. 23, no. 4, p. 142–144.

Barnes, C. A., and Gross, M. G., 1966, Distribution at sea of Columbia River water and its load of radionuclides, in Disposal of radioactive wastes into seas, oceans and surface waters: Vienna, Internat. Atomic Energy Agency, p. 291–301.

Carrigan, P. H., Jr., 1969, Inventory of radionuclides in bottom sediment of the Clinch River, Eastern Tennessee: U.S. Geol. Survey Prof. Paper 433-I, 18 p.

Delaney, A. C., Delaney, Audrey Claire, Parkin, D. W., Griffin, J. J., Goldberg, E. D., and Reimann, B.E.F., 1967, Airborne dust collected at Barbados: Geochim. et Cosmochim. Acta, v. 31, no. 5, p. 885–909.

Emery, K. O., 1968, Relict sediment on the continental shelves of the world: Am. Assoc. Petroleum Geologists Bull., v. 52, p. 445–464.

Glasstone, S., 1964, The effects of nuclear weapons: Washington, D.C., U.S. Government Printing Office.

Gorham, Evile, 1957, On the acidity and salinity of rain. Geochim. et Cosmochim. Acta, v. 7, p. 231–239.

Gross, M. G., 1966, Distribution of radioactive marine sediment derived from the Columbia River: Jour. Geophys. Research, v. 71, no. 8, p. 2017–2021.

——1967, Sinking rates of radioactive fallout particles in the northeast Pacific Ocean, 1961–62: Nature, v. 216, p. 670–672.

——1970a, New York Metrolopitan region—a major sediment source: Water Resources Research, v. 6, no. 3, p. 927–931.

——1970b, Analyses of dredged wastes, fly ash, and waste chemicals—New York Metropolitan region: State University of New York, Stony Brook, New York, Marine Sciences Research Center, Technical Report no. 7, 33 p.

Gross, M. G., and Nelson, J. L., 1966, Sediment movement on the continental shelf near Washington and Oregon: Science, v. 154, no. 3750, p. 879–885.

Gross, M. G., Barnes, C. A., and Riel, G. K., 1965, Radioactivity of the Columbia River effluent: Science, v. 149, no. 3688, p. 1088–1090.

Holland, J. Z., 1963, Distribution and physical-chemical nature of fallout: Federation Proceedings, Federation American Society of Experimental Biology, v. 22, no. 6, p. 1390–1399.

Jenne, E. A., and Wahlberg, J. S., 1968, Role of certain stream-sediment components in radioion sorption: U.S. Geol. Survey Prof. Paper 433-F, 16 p.

Judson, S., 1968, Erosion rates near Rome, Italy: Science, v. 160, p. 1443–1446.

Klingeman, P. C., and Kaufman, W. J., 1965, Transport of radionuclides with suspended sediment in estuarine systems: California Univ. Sanitary Engineering Research Laboratory Report no. 1, 65–15, 222 p.

Manheim, F. T., Meade, R. H., and Bond, G. C., 1970, Suspended matter in surface waters of the Atlantic continental margin from Cape Cod to the Florida Keys: Science, v. 167, p. 371–376.

Meade, R. H., 1969, Errors in using modern stream-load data to estimate natural rates of denudation: Geol. Soc. America Bull., v. 80, p. 1265–1274.

Muller, G., 1966, Die Sedimentbildung in Bodensee: Naturwissenschaften, v. 53, no. 10, p. 237–247.

Nelson, J. L., Perkins, R. W., Nielsen, J. M., and Haushild, W. L., 1966, Reactions of radionuclides from the Hanford reactors with Columbia River sediments, *in* Disposal of radioactive wastes into seas, oceans, and surface waters: Vienna, Internat. Atomic Energy Agency, p. 139–162.

Osterberg, C. L., Cutshall, N., Johnson, V., Cronin, J., Jennings, D., and Frederick, L., 1966, Some non-biological aspects of Columbia River radioactivity, *in* Disposal of radioactive wastes into seas, oceans, surface waters: Vienna, Internat. Atomic Energy Agency, p. 321–336.

Pickering, R. J., 1969, Distribution of radioactivity in Clinch River bottom sediment: U.S. Geol. Survey Prof. Paper 433-H, 25 p.

Rex, R. W., and Goldberg, E. D., 1962, Isolubles, *in* Hill, M. N., ed., The sea: Vol. 1, Physical oceanography: New York, Interscience Publishers, p. 295–304.

Strakhov, M. N., 1967, Principles of lithogenesis: New York, Consultants Bureau, v. 1, 245 p.

Templeton, W. L., and Preston, A., 1966, Transport and distribution of radioactive effluents in coastal and estuarine waters of the United Kingdom, *in* Disposal of radioactive wastes into seas, oceans, and surface waters: Vienna, Internat. Atomic Energy Agency, p. 267–288.

Vice, R. B., Guy, H. P., and Ferguson, G. E., 1969, Sediment movement in an area of suburban highway construction, Scott Run Basin, Fairfax County, Virginia 1961–1964: U.S. Geol. Survey Water-Supply Paper 1591-E, 41 p.

Wadleigh, C. H., 1968, Wastes in relation to agriculture and forestry: U.S. Dept. Agriculture Misc. Pub. 1065, 112 p.

Windom, H., Griffin, J., and Goldberg, E. D., 1967, Talc in atmospheric dust: Environmental Sci. and Technology, v. 1, no. 11, p. 923–926.

Wolman, M. G., and Schick, A. P., 1967, Effects of construction on fluvial sediments, urban and suburban areas of Maryland: Water Resources Research, v. 5, no. 2, p. 451–464.

MANUSCRIPT RECEIVED BY THE SOCIETY MARCH 29, 1971

THE GEOLOGICAL SOCIETY OF AMERICA, INC.
MEMOIR 132
© 1972

Prediction of Object Penetration into Seafloor Sediments

R. J. SMITH

U. S. Navy, Monterey, California 93940

ABSTRACT

It has become necessary to predict with some accuracy the depth of penetration into softer sea-floor sediments of various size objects that fall through the water column. A provisional method for doing this has been developed utilizing data from tests on core samples. The method is based on the equivalence of the kinetic and potential energy of the falling object to the work done during progressive shearing of the sediment in the penetration event.

INTRODUCTION

A surprisingly large array of objects fall to the sea floor which are sought for retrieval or subsequent observation. These objects range in scale, for example, from full-size ships to small and costly components related to offshore drilling.

Efforts to locate these objects are usually time consuming, expensive, and often unproductive, even in those instances in which the drop point is well known. This difficulty usually results from the combined effects of the general inability to relocate points very precisely both on the surface of and within, the sea, and also from the limited visibility afforded within deeper water. Other factors contribute to each particular circumstance.

The search procedures are usually carried out primarily by visual means, either through submerged camera arrays or with submersibles, and are sometimes augmented by use of various types of sonar, magnetometers, and related sensing equipment. The search mode is, however, in large part dependent on the object being exposed at the interface between the sediment and the water column.

At some point in a search the possibility is almost always raised that perhaps the object cannot be found by use of the above procedures because it has penetrated into soft sediments to a depth such that it is not exposed at the interface and hence is effectively masked. This is sometimes a very real possibility, particularly with dense loads in areas floored by an extremely soft bottom. The strength of the sea-floor materials has been found to vary widely, and exhibits a greater range of variation in areas closer to land-derived detrital sources. These locations are usually where the greatest use of the sea is taking place. Development of a fairly accurate means of estimating the possible depth of penetration has therefore become necessary.

It should not be interpreted that all potential applications of sea-floor penetration prediction are related to search and recovery efforts. Numerous objects are now purposefully placed using a free-fall mode through the water column, and the relation of penetration to the newer types of anchors and related equipment is evident. There is a growing need for a valid method to predict the penetrability of sea-floor sediments, and one would therefore have wide applications.

CONTRIBUTION OF THE U.S.S. *THRESHER*

In consideration of the intent of this summary, it is well that some mention be made of the search experience associated with the loss of the submarine *Thresher* in April of 1963. This incident forcibly brought attention to the need for a capability for making penetration predictions.

The vessel disappeared while conducting a series of test dives in 8,000 ft deep waters on the continental slope off the northeast coast of the United States. The location of the last contact with the submarine was well, though not precisely, defined by its surface escort vessel. Search procedures were instituted and followed various stages as outlined, for example, by Hersey (1964), Keach (1964), Hurley (1964), and Andrews (1965). In that the submarine was nuclear powered, the Joint Committee on Atomic Energy conducted a series of hearings that resulted in a comprehensive U.S. Congressional Report (1965).

For a long time the search yielded no trace, and numerous speculations arose as to the fate of the vessel. One of these considered the possibility that its velocity of fall resulted in deep penetration into the bottom and subsequent covering by the soft, unconsolidated, glacially derived sediments present in the area. No backlog of experience existed to permit an evaluation of this possibility. It was subsequently found that the vessel broke into fragments prior to impact with the bottom.

In 1963 the Secretary of the Navy established the Deep Submergence Systems Review Group for the purpose of examining the existing deep-water search and recovery procedures, with particular reference to the *Thresher* incident. The scientific advisor to this review group was designated as Rear Admiral Harry H. Hess. It was he who critically recognized the widespread implications of being able to make accurate penetration predictions and aided in initiating efforts toward solution of the problem.

PREVIOUS PENETRATION EXPERIENCE

The penetration problem is found to be closely related to a host of other subject fields including marine geology, soil mechanics, hydrodynamics, rheology, pile driving, and ballistics. Its study has achieved the greatest degree of sophistication in connection with the piercement behavior of projectiles. The projectile situation is actually handled in a very similar manner to penetrations into marine sediments, with the basic equations of Prandtl (1920) applying to both circumstances.

The first attempts to analyze earth penetrations include those of Euler (1745) and of Poncelet (1829). Robertson (1941) presents a summary of experiments to determine the depths of penetration of delayed-action bombs. Refinements to the above thinking are offered by Brooks and Reis (1963), Thompson and Colp (1964), Caudle and others (1967), and Young (1969). These more recent studies are primarily concerned with the development of high-velocity penetration devices, and generally indicate that higher-velocity impact phenomena differ considerably from the low-velocity case. Some specific aspects of the features of penetration into earth materials have been summarized by Schmid (1970). The extensive investigations of the actions involved in pile driving are too empirical to be generally applicable to sea-floor penetration studies, though there is a great similarity between the processes.

Extremely little is known about the mechanics of low-velocity penetration of objects into earth materials, on the land surface as well as at sea. Penetration into soft marine sediments appears to be a complex process not amenable to an accurate and simple solution.

GENERAL APPROACH

Estimating the depth of object penetration usually requires taking a series of cores in the area of concern, subjecting these to a variety of physical tests, and using the results to calculate a prediction. None of the steps used in the process can be considered to be free from error.

The possible disturbance of, or introduction of other detrimental factors to, deep-sea samples by using normal oceanographic coring tools for their procurement is a subject area that is presently almost completely unexplored. Errors may be unknowingly developed at any stage in the coring, handling, or storage procedures. Some recent work by A. Inderbitzen (unpub. report), using the submersible *Deep Quest,* does suggest, however, that test values obtained from short cores may yield approximately correct numbers to use in prediction. In most penetration prediction instances the length of valid cores is far from sufficient to satisfy needs, and thus unsupportable extrapolations of properties must almost always be made. A penetration analysis therefore commonly begins by using samples of indeterminate validity.

The physical tests to which the samples are subjected are slightly more reliable. Most of the index, classification, strength, and consolidation measurements made in the laboratory are considered as somewhat crude approximations of in-situ circumstances. Some of these tests are known to be grossly in error. An accurate penetration prediction does not permit utilization of safety factors such as are commonly employed with these test procedures for engineering use. The proper interpretation of these laboratory tests, when applied in the light of the compositional information available, does give some numerical quantities for predictive manipulation.

The previous information is then used to make the estimate of possible penetration. This final step is undoubtedly less well-understood than the earlier phases, but in the past it has proven necessary to supply the best available estimates. An almost complete unawareness of the validity of the theoretical equations used in the sediment prediction process exists as a result of the present lack of confirmation by sea-floor tests.

The following sections outline in summary fashion the predictive procedure as is currently in use.

PENETRATION PREDICTION PROCEDURE

We assume that a falling body possesses a certain amount of energy, and that during the process of penetrating, this energy is consumed by work done to the sediment. The assumption is then made that essentially all of the energy involved is dissipated in a progressive shearing of the material into which it passes. More specifically, the energy is translated to a continual bearing failure applied over the penetrated distance.

The kinetic energy available to penetrate can be determined from suitable modifications of the basic relation:

$$KE = \frac{1}{2}mv^2,$$

so as to involve the applicable mass of the object and its velocity of fall at the instant of impact.

It may be noted that the velocity V is squared in obtaining the quantity of energy available. It is therefore necessary that the fall velocity be accurately known if a valid estimate is to be made. In spite of an apparent ease of determining the velocity of fall of objects through water, no unanimity exists among hydrodynamicists as to the way that this velocity should be calculated. For the same object of larger size, velocity estimates can range from 10 to 100 ft/sec. The consequences of such a variation on the squared term of kinetic energy value used are evident. The results of penetration studies must therefore not be related to a single velocity, but must be presented so as to permit the effects of various velocities to be apparent.

The initial penetration predictions using the above energy relation indicated that an appreciable error existed in instances of very low impact velocity and extremely soft sediment. In such caces, the effect of the change in the potential energy serves to predominate over the influence of the kinetic energy. It is therefore appropriate to numerically convert the kinetic energy to a potential energy value and then to use the sum of the two energies for the prediction. In accordance with Figure 1,

the energy available for conversion to work is hence considered as:

Energy available $= W (h + d_t)$,

in which W equals weight of object, h equals computed height of fall, and d_t

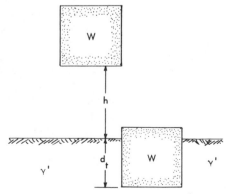

Figure 1. Height relations used in penetration prediction.

equals total depth of penetration. The term h in the above corresponds to the computed free-fall height (in air) necessary to produce the assumed velocity at the instant of impact v, as determined in accordance with the relation:

$$v^2 = 2gh,$$

where v equals assumed velocity of fall, g equals acceleration of gravity, and h equals height of free fall.

The work done in absorbing this energy is the product of the load bearing capability of the sediment and the depth of penetration as expressed by:

$$\text{work done} = q_d A d_t,$$

where q_d equals ultimate unit bearing capacity, A equals area of base of object, and d_t equals total depth of penetration. The bearing capacity requires further explanation. Various semi-empirical relations have been devised over the years to account for the load-supporting ability of earth materials. The most appropriate of these relative to the sea-floor penetration circumstance is that offered by Terzaghi and Peck (1948), in which, for a square loaded area:

$$q_{ds} = 1.3cN_c + \gamma d N_q + 0.4\gamma B N_\gamma,$$

where q_{ds} equals ultimate unit bearing capacity for a square base, c equals cohesion of the sediment, γ equals unit weight of the sediment, d equals depth of load base below sediment surface, B equals width of load, and N_c, N_q, and N_γ equal bearing capacity factors.

It is currently generally accepted that the values of shear strength and cohesion for fine-grained sea-floor materials are equivalent. It is also believed that the friction angle is close to zero, although subsequent experimentation may well prove this assumption to be in error. Nonetheless, with a friction angle of zero, N_c is 5.7, N_q is 1, and N_γ is 0. The submerged unit weight of the sediment γ' is also applicable. The ultimate unit bearing capacity for a square based load on marine sediments is therefore summarily:

$$q_{ds} = 7.4c + \gamma'd.$$

For loads having a circular base, it has been found that the bearing capacity is 85 percent of that for a square base:

$$q_{dc} = 0.85\,(7.4c + \gamma'd),$$

where q_{dc} equals ultimate unit bearing capacity for a circular base, c equals cohesion of the sediment, γ' equals submerged unit weight of the sediment, and d equals depth of load base below sediment surface.

For objects of rectangular base the relation is:

$$q_{dr} = 5.7c\,(1 + 0.3\frac{B}{L}) + \gamma'd,$$

where q_{dr} equals ultimate unit bearing capacity for a rectangular base, B equals width of base, and L equals length of base.

APPLICATION TO A SIMPLE CASE

If it is desired to predict the depth of penetration of a square-based object under the circumstances shown on Figure 1, the following relation is applied:

$$W \left(h + d_t \right) = A d_t \left(7.4c + \gamma \frac{d_t}{2} \right).$$

The weight of the object and its basal area are assumed known. The term h can be determined for several different velocities so as to indicate the range of possible penetration depths. In the simplified circumstance, both the cohesion and the submerged unit weight can be taken as averaged representative quantities. The total depth of penetration term in the bearing capacity portion of the above relation has been divided by two in order to account for its variation from zero to full depth during the penetration event.

The solution is most readily accomplished by the graphical means shown by Figure 2. Increasing values of depth of penetration are used to compute the quantities for the two curves. The point of intersection of the curves gives the predicted depth of total penetration.

APPLICATION TO MORE COMPLEX CASES

The basic approach previously outlined can also be applied to more complex situations where the variations with depth of the cohesion and submerged unit weight can be taken into account. Here the load is penetrated by increments that are small enough to allow using more correct values of the sediment properties. The cohesion and unit weight quantities applied are those averaged over the increment involved.

Objects having more irregularly shaped bases, as, for example, approximations to spheres, cylinders, or cones, can be simulated by a series of thin horizontal plates of proper dimensions. These plates are then incrementally penetrated in the same

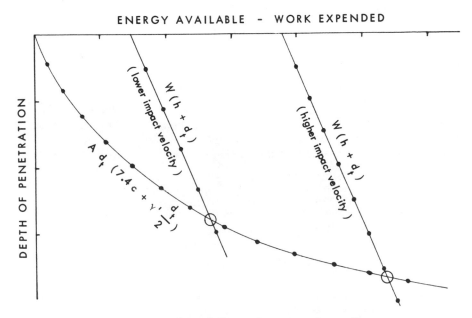

Figure 2. Graphic solution to the penetration problem.

fashion as the above. The use of a computer can be of material assistance in the handling of the quantities of numbers produced in the process.

A recent series of air-drop penetration tests have permitted more detailed study of at least some of the parameters that are involved (J. Carlmark, in prep.). Eventually a series of large object drops at sea will be necessary to better define all of the empirical quantities that act to control this particular penetration situation.

ACKNOWLEDGMENTS

This somewhat novel, from a strictly geological aspect, summary has been presented here for two reasons. First it serves to emphasize another of the broad range of participation and interests of Professor Harry Hess. Second, the subject material represents my own last professional contact with his thinking, with numerous hours spent discussing the factors that act to control the strength parameters of sea-floor sediments. The more than several years of discussion, controversy, and solution of varied aspects of Caribbean tectonics are additionally most vivid remembrances.

Credit is also due Fred Knoop, another member of the Deep Submergence Systems Review Group, and the Naval Facilities Engineering Command for additionally providing initial impetus toward study of the sea-floor penetration problem and for support throughout the years. It must be noted that the penetration prediction thinking presented here is not solely that of the author. Countless hours of consideration of the problem were had with many different persons affiliated with numerous institutions and firms throughout the years, and their contributions should be realized. Appreciation is expressed to A. Inderbitzen, M. C. Hironaka, and R. J. Malloy for critically reading and offering comments regarding the manuscript.

REFERENCES CITED

Andrews, F. A., 1965, Search operations in the *Thresher* area: Naval Engrs. Jour., v. 77, p. 549–561.

Brooks, W. B., and Reis, G. E., 1963, Soil penetration theory: Albuquerque, Sandia Corp. Rept. SC-4950, 46 p.

Caudle, W. N., Pope, A. Y., McNeill, R. L., and Margason, B. E., 1967, The feasibility of rapid soil investigations using high-speed, earth-penetrating projectiles: Albuquerque, Internat. Symposium on Wave Propagation and Dynamic Properties of Earth Materials, 23 p.

Euler, L., 1745, Neue grundsatze der artillerie: Berlin, Haude, 409 p.

Hersey, J. B., 1964, Search for *Thresher* by photomosaic: U. S. Navy Jour. Underwater Acoustics, v. 14, p. 311–318.

Hurley, R. J., 1964, Bathymetric data from the search for U.S.S. *Thresher:* Internat. Hydrog. Rev., v. 41, no. 2, p. 43–52.

Keach, D. L., 1964, Down to *Thresher* by bathyscaph: Natl. Geog. Mag., v. 125, p. 765–777.

Poncelet, J. V., 1829, Cours de mechanique industrielle: Paris, Clouet, 158 p.

Prandtl, L., 1920, Uber die harte plastischer korper: Nachr. Ges. Wiss. zu Gottingen, p. 74–85.

Robertson, H. P., 1941, Terminal ballistics: Natl. Research Council, Prelim. Rept. for Comm. on Passive Resistance Against Bombing, 83 p.

Schmid, W. E., 1970, Penetration of objects into the ocean bottom: U. S. Naval Civil Engr. Lab. Rept. CR-69.030, 118 p.

Terzaghi, K., and Peck, R. B., 1948, Soil mechanics in engineering practice: New York, John Wiley and Sons, Inc., 566 p.

Thompson, L. J., and Colp, J. L., 1964, Preliminary evaluation of earth targets for use in impact effects studies: Albuquerque, Sandia Corp. Rept. SC-DR-316-63, 65 p.

U. S. Congress, 1965, Loss of the U.S.S. *Thresher:* Washington, D.C., Hearings before Joint Committee on Atomic Energy, 192 p.

Young, C. W., 1969, Depth prediction for earth-penetrating projectiles: Am. Soc. Civil Engineers Proc., v. 95, no. SM3, p. 803–817.

MANUSCRIPT RECEIVED BY THE SOCIETY MARCH 29, 1971

Space Science

THE GEOLOGICAL SOCIETY OF AMERICA, INC.
MEMOIR 132
© 1972

Crystallization Histories of Two Apollo 12 Basalts

L. S. HOLLISTER
W. E. TRZCIENSKI, JR.
R. B. HARGRAVES
AND
C. G. KULICK
Department of Geological and Geophysical Sciences
Princeton University
Princeton, New Jersey 08540

ABSTRACT

The spherulitic and variolitic textures of Apollo 12 samples 12063 and 12065, the order of crystallization of all phases, the compositional trends and sector-zoning of the pyroxenes, the chrome partition between phases, the occurrences of metastable minerals, and the wide variations of edge compositions of the olivine and pyroxene suggest that essentially all the crystallization of both rocks took place under highly supercooled conditions at or near the lunar surface. The composition of the original magmas after nucleation of pyroxene and olivine approached those of the present rocks.

INTRODUCTION

This paper summarizes work done at Princeton on two crystalline rocks returned by the Apollo 12 lunar mission and collected by a Princeton alumnus (Pete Conrad). We feel it is doubly appropriate that a paper on the lunar rocks be included in this volume; not only was the study of geologic processes on the moon a major interest of Harry Hess, but he was the original Principal Investigator on the NASA contract which supported our work.

Additional data on and description of the pyroxenes in the two samples, 12063 and 12065, discussed in this paper have been published in the Proceedings of the Second Lunar Science Conference (Hollister and others, 1971).

DESCRIPTION OF TWO APOLLO 12 BASALTS

Samples 12065.6 and 12063.15 are similar in bulk composition (Smales and others, 1971; Willis and others, 1971) but are markedly dissimilar in texture and mineralogy. Our approach has been to study the compositional zoning of pyroxene and olivine in these samples in an attempt to determine the crystallization sequence and the physical conditions of crystal growth. These data are essential to any interpretation of the history of the lunar basalts.

Sample 12065 is a medium- to fine-grained microporphyritic melabasalt, with randomly oriented, elongate (0.3 by 4 mm) pyroxene microphenocrysts (31 percent by volume) and equant olivine (2 percent) set in a variolitic groundmass consisting of sheaves of plagioclase crystals (22 percent, less than 1 mm), granular pyroxene (36 percent), and thin, lathlike (wispy) opaques (9 percent, < 0.5 mm).

The pyroxene phenocrysts are conspicuously and complexly zoned, generally from colorless pigeonite ($2V = 0$) to pinkish augite ($2V_z = 47°$) with peripheral iron enrichment ($2V_z = 30°$, analysis 6, Table 1). These crystals grew initially with a "hollow" (001) sector. As a result, many sections through the phenocrysts appear to enclose cores containing mineral aggregates typical of the matrix.

The olivine grains in sample 12065 are inclusion-free and equant. The matrix plagioclase crystals are elongate, are in some sections skeletal, and are intergrown with matrix pyroxene. The predominant opaque mineral is ilmenite, which occurs as thin and elongate crystals. Neither the plagioclase nor the ilmenite was observed to be in physical contact with the pigeonite or the pinkish augite. In addition, there are minor amounts of equant, zoned spinel grains; these consist of a bluish aluminous chrome spinel core with a brownish ulvospinel border. One aggregate of chrome spinel with native iron (containing small amounts of Ni and Co), but without ulvospinel, was enclosed in a pigeonite phenocryst. Globules of native iron are also present in accessory troilite grains. Cristobalite is present in trace amounts.

In contrast to the porphyritic texture of 12065, sample 12063 is a relatively massive, fine-grained, equigranular olivine basalt consisting of pyroxene (57 percent), olivine (early 6 percent, late 2 percent), plagioclase (25 percent), opaques (9 percent), and cristobalite (1 percent). The texture has a spherulitic aspect, with what appear to be kernels of crystallization 5 mm across separated by variolitic sheaves. Conspicuous residual aggregates of cristobalite, inclusion-charged fayalitic olivine, and opaques occur in the variolitic areas.

Olivine grains are prominent and distinctly zoned ($2V_x = 80°$ to $70°$); they are in part enclosed in or overgrown by augite. In contrast to sample 12065, the pyroxenes (< 1 mm) have only rare, very small pigeonite cores; they are compositionally zoned from augite (commonly twinned) toward ferroaugite ($2V_z = 46°$ to $30°$). Plagioclase is lathlike and is most abundant in the variolitic "matrix"; some plagioclase grains have grown in skeletal or "swallow-tail" form and appear to enclose pyroxene aggregates.

TABLE 1

Sample 12065

	Pyroxene						Olivine	
Oxide	Hyp. 1	Pig. 2	Pig. 3	Aug. 4	Aug. 5	Pxf. 6	Center 7	Edge 8
Na$_2$O	0.0	.01	.03	.03	.05	.03	0.0	0.2?
MgO	24.26	22.63	20.85	14.57	12.40	.34	36.26	29.69
Al$_2$O$_3$	0.66	1.69	2.04	3.80	5.41	.54	0.0	0.0
SiO$_2$	53.51	51.63	51.55	48.98	47.10	45.23	37.59	36.18
CaO	2.61	3.77	5.00	14.09	15.76	6.43	.29	.33
TiO$_2$	.35	.70	.79	1.61	2.32	.73	.05	.04
Cr$_2$O$_3$	.74	.99	1.06	1.28	.94	.05	.46	.24
MnO	.29	.31	.33	.27	.28	.56	.29	.38
FeO	16.69	17.90	18.25	13.57	14.85	46.54	25.76	33.67
Total	99.11	99.63	99.90	98.20	99.11	100.45	100.70	100.73

Sample 12063

	Pyroxene			Olivine				Celsian
Oxide	Pig. 9	Aug. 10	Aug. 11	Center 12	Edge-px 13	Edge-pl 14	Fayalite 15	Celsian 16
Na$_2$O	.02	.02	.03	.01?	0.0	0.0	0.0	0.13
MgO	20.49	14.71	11.46	33.75	30.72	24.00	1.81	0.0
Al$_2$O$_3$	.98	3.97	1.79	.06?	.07?	.05?	.15?	21.35
SiO$_2$	52.64	49.20	49.31	36.38	36.06	34.44	30.08	54.45
K$_2$O	—	—	—	—	—	—	—	8.83
CaO	4.51	14.77	11.16	.30	.27	.45	.71	.55
TiO$_2$	.72	2.21	1.44	.07	.06	.10	.13	n.d.
Cr$_2$O$_3$	.50	.64	.31	.23	.16	.06	.05	n.d.
MnO	.31	.30	.35	.29	.33	.36	.71	n.d.
FeO	20.40	14.13	24.11	28.60	32.11	40.22	65.70	1.00
BaO	—	—	—	—	—	—	—	13.45
Total	100.57	99.95	99.96	99.69	99.78	99.68	99.34	98.46

Analyses 1 through 5 are representative analyses from one crystal. Analyses 1 and 2, and 3 and 4 were taken at compositional discontinuities.

Analyses 9 through 11 are from a single crystal. Analysis 10 is representative of high-Al core augite, and analysis 11 is representative of the low-Al rim pyroxene. Compositional discontinuities occur between analyses 9 and 10 and between 10 and 11. Analyses 13 and 14 are edge compositions of olivine against pyroxene and plagioclase, respectively.

The opaque minerals include bluish chrome spinel enclosing native iron, overgrown by ulvospinel and, at the rim, by ilmenite. Equant spinel grains are most common in the crystallization kernels, ilmenite laths in the variolitic matrix.

The latest minerals to crystallize include fayalite ($2V_x = 40°$, analysis 15, Table 1), cristobalite, aggregates of troilite and native iron, ulvospinel with ilmenite(?) lamellae in octahedral planes, and discrete ilmenite. Adjacent high-iron clinopyroxene is markedly pinkish brown ($2V_z \approx 30°$). Microprobe studies also revealed a barium-bearing high-potassium, high-silica phase included in and associated with fayalite and a potassium-bearing high-barium phase (celsian?, analysis 16, Table 1) in the interstices between fayalite grains.

ANALYTICAL RESULTS

The electron microprobe results on the pyroxenes of samples 12065 and 12063 are summarized in Figures 1 and 2. Table 1 lists a few representative analyses. Figure 1 illustrates the Ca, Fe + Mn, and Mg data on the pyroxene quadrilateral diagram. Figure 2 illustrates the relative percent of Al, Ti, and Cr in the pyroxenes.

In general (with growth) the pyroxenes of samples 12065 and 12063 become strongly enriched in Fe relative to Mg; in some cases the composition of later pyroxene lies directly on the $CaFeSi_2O_6$–$Fe_2Si_2O_6$ side of the pyroxene quadrilateral (Fig. 1). The composition of this pyroxene suggests it is pyroxferroite (analysis 6, Table 1; Chao and others, 1970). Ca and Al, on the other hand, first increase with growth, then drop at a discontinuity, and decrease with further growth. The pyroxene that grew after the discontinuity is termed "low-A1 rim pyroxene" in sample 12065 and "group 3 pyroxene" in sample 12063. These compositional zoning trends differ significantly from those reported for most Apollo 11 pyroxenes in that Ca and Al *increase* with growth in the central parts of the Apollo 12 pyroxene crystals.

It is clear from the compositional data of Figure 1a and b that in the (110) sectors of the pyroxene phenocrysts of sample 12065 there is a crystallization sequence in single phenocrysts from hypersthene (*chemically* defined) to pigeonite to augite (analyses 1 through 5, Table 1) and finally to rim compositions in the ferroaugite, ferropigeonite, and pyroxferroite compositional fields. In the (010) sectors, pigeonite continues to grow contemporaneously, and presumably metastably, with the augite. The data for the pyroxenes of sample 12063 (Fig. 1c) are similar, but (1) there are fewer points (only 1) in the extremely iron-enriched region (the "forbidden zone" of Lindsley and Munoz, 1969), (2) at the growth centers the hypersthene composition is absent, and (3) the pigeonite composition (analysis 9, Table 1) plots near the end of the (110) pigeonite trend of sample 12065 (Fig. 1b; analysis 3, Table 1). Also, the region of low-A1 rim pyroxene (group 3) in the pyroxene of sample 12063 is about half the area (as viewed in thin section) of the entire pyroxene crystals, much wider than that rimming the phenocrysts of sample 12065, which is less than about a tenth the diameter of the phenocrysts.

DISCUSSION

It is clear from the data shown in Figure 2 that the pigeonite of samples 12063 and 12065 probably nucleated without containing a significant amount of the

⸻⸻⸻⸻⸻⸻⸻⸻⸻⸻⸻⸻⸻⸻⸻⸻⸻⸻⸻⸻⸻⸻➤

Figure 1. (a) Pyroxene compositional zoning trends in sample 12065, based on analyses interpolated from electron microprobe step scans at 2-μm intervals with a beam diameter less than 1 μm. Solid lines represent regions of continuous zoning; dashed lines connect analysis points across discontinuities. (b) Schematic summary of data in (a), showing distinction between trends in the (110) and (010) sectors. Other trends in (010) sectors and along the [001] direction span the gap between pigeonite and augite and are shown in (a). For comparison, analyses of Apollo 11 samples 10058 and 10062 are plotted from the data in Hollister and Hargraves (1970). (c) Pyroxene analyses in sample 12063. Most points represent analyses on either side of a discontinuity between high-Al pyroxene (Group 2) and low-Al pyroxene (Group 3). Dashed and solid lines with arrows show the compositional trend of one crystal.

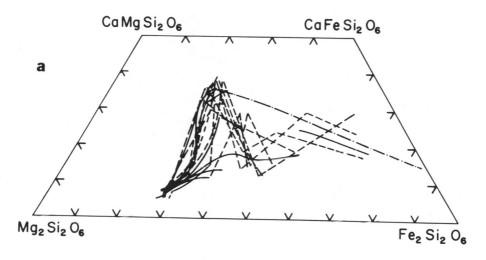

a

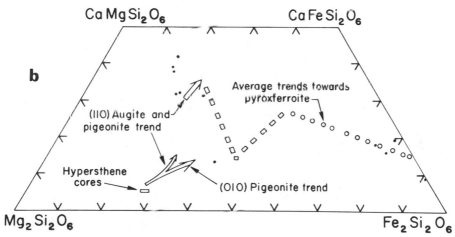

b

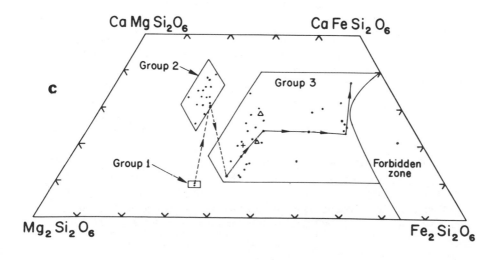

c

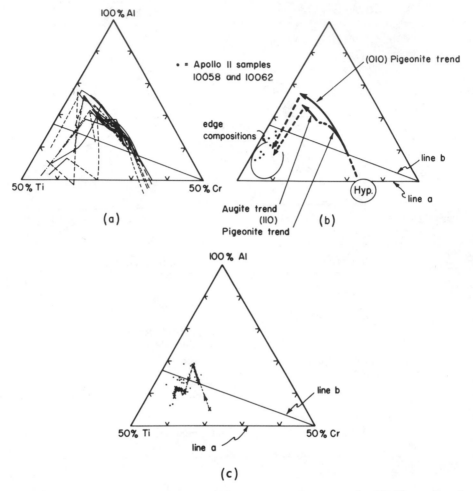

Figure 2. (a) Pyroxene compositional zoning trends in sample 12065, expressed in terms of relative Al, Ti, and Cr contents. Solid lines represent regions of continuous zoning; dashed lines connect analysis points across discontinuities. (b) Schematic summary of data in (a), showing distinction between trends in the (110) and (010) sectors. A point lying on line a implies Al substitution for Si is balanced by Ti^{+3} and Cr^{+3} (assuming no Cr^{+2}) substitution for divalent cations. A point lying on line b implies Al substitution for Si is balanced by Ti^{+4} (assuming no Ti^{+3}) and Cr^{+3} for divalent cations. A point above line b implies Al substitution for divalent cations. Apollo 11 analyses of pyroxenes in samples 10058 and 10062 (Hollister and Hargraves, 1970) all plot below line b. (c) Al, Ti, and Cr pyroxene compositions in sample 12063. Group 2 analyses plot above line b, Groups 1 and 3 below. Dashed and solid lines with arrows show compositional trend of one crystal.

component $CaAl_2SiO_6$. If it should be argued that the pigeonite nucleated with this component, then, in order to maintain charge balance, it must also be demonstrated that a substantial part of the Ti and Cr was trivalent and divalent, respectively. The component $CaAl_2SiO_6$ progressively increases until the point at which the low-Al rim pyroxene abruptly commenced to grow. It is here that it is reasonable to presume that plagioclase began to crystallize; this is consistent with the fact that in both samples the only pyroxene that encloses plagioclase is the low-Al rim pyroxene.

Other evidence favoring this interpretation comes from the results on the Apollo 11 samples. The pyroxene analyses of samples 10058 and 10062 (Fig. 2b) plot in the low-Al pyroxene field, and Hollister and Hargraves (1970) concluded for these samples as did Ross and others (1970) for sample 10050, that plagioclase and pyroxene crystallized together during most if not all the pyroxene growth history. For the only sample (10022) from the Apollo 11 landing site similar to many from the Apollo 12 landing site, Weill and others (1970, p. 948) correlated a discontinuity of Ca and Al, similar to that found in sample 12063 and 12065, with the beginning of crystallization of plagioclase.

The Al, Ti, and Cr compositional trends shown in Figure 2, prior to the abrupt decrease in Al, do not lie on a line that intersects the Al corner as would be expected if the pyroxene growth resulted simply in enrichment of Al in melts with constant Cr/Ti ratios. Instead, the trends are essentially on a line which intersects the Cr corner (not shown) of the triangles. This implies that, during most of the growth of the pyroxene phenocrysts of sample 12065 and during the early part of the growth of the pyroxene of sample 12063, Cr was simultaneously being depleted from the melt by the cocrystallization of a chrome-bearing phase, possibly the early chrome-bearing spinel; it may have been depleted by the pyroxene itself.

The fact that the early trends in the Al-Ti-Cr diagrams are significantly different for the pyroxenes of samples 12063 and 12065 suggests that these pyroxenes grew from melts which had initially different compositions, at least with respect to the ratio of Ti/Cr. There is a difference in bulk chrome content between the two samples (Smales and others, 1971; Willis and others, 1971): 0.50 percent Cr_2O_3 in sample 12065 and 0.44 percent Cr_2O_3 in sample 12063. This difference is, however, too small to account for the fact that the chrome contents of the centers of the olivines and pigeonites of sample 12065 are twice those of sample 12063 (Table 1). The empirical partition of chrome between olivine and melt of sample 12065 is 0.9, the same as determined experimentally for lunar rocks by Ringwood (1970, p. 6458). However, the empirical chrome portion between olivine and melt of sample 12063 is 0.5, suggesting that the Cr_2O_3 content of the melt when olivine nucleated was 0.25 percent, rather than the 0.44 percent now present in the rock. To account for the observed content of chrome in rock 12063, it is concluded that a substantial amount of chrome-bearing spinel must have crystallized *prior* to the nucleation of the olivine; and, of course, it did not separate from the melt, because it is still present in the rock.

The partition factors for chrome determined between the centers of olivine and centers of pigeonite of the two samples are identical: 2.2. This coincidence, in spite of the large differences in chrome content of the olivines and the pigeonites of the two specimens, suggests that, for both rocks, pigeonite nucleated in the same environment as olivine. This conclusion is not consistent with any model involving substantial addition or subtraction of olivine or pigeonite to the rocks by gravity settling or flow differentiation.

There is no direct evidence, such as cumulate textures, to suggest that the olivines or pyroxenes nucleated in melts with compositions any different from those of the bulk rocks in which they now occur. In fact, the lack of cumulate textures coupled with the fact that the olivines and pyroxenes are highly zoned and that the compositional zoning trends of each pyroxene grain are nearly identical (Figs. 1 and 2) suggest crystallization in a closed system.

The zoning patterns in the pyroxenes of samples 12063 and 12065, the composition of olivine (both early and late) where it is in contact with pyroxene in sample 12063, the arguments based on the empirical chrome partition coefficients, and the textural relations we described above suggest the sequence of crystallization for the two rocks shown in Figure 3. The sequence of crystallization of Apollo 11 sample 10058 (Hargraves and others, 1970) is included for comparison. The relative positions for the end and beginning of the arrows are meant to imply reaction relations with the melt.

The initial nucleus of hypersthene in the pyroxene phenocryst of sample 12065 appears to have grown with a hollow (001) sector. It is certain that pigeonite grew with a hollow (001) sector. These hollow sectors were subsequently closed off from the matrix by growth of augite. Both the pigeonite and the augite show compositional sector-zoning effects, albeit minor. Skeletal crystals and crystals with compositional sector zoning have been interpreted to have grown under highly supersaturated conditions (Drever and Johnston, 1957; Hollister, 1970); on similar grounds, we conclude that the pyroxene of sample 12065 grew rapidly under highly supersaturated (supercooled) conditions. We find it difficult to believe that a sufficient degree of supercooling would have been achieved in the lunar interior; this situation is most likely to prevail at the time of extrusion of a magma onto the lunar surface.

There are several other lines of evidence for the rapid crystallization of both rocks: (1) the compositional trends near the edge of the pyroxene and olivine crystals depend on the contiguous phase assemblage (see, for example, differences in olivine composition where it is in contact with low-Al rim pyroxene and with plagioclase, analyses 13 and 14, Table 1), implying growth rates in excess of diffusion rates in the melt; and (2) oscillation of low-Al rim compositions between calcic and subcalcic pyroxene (Fig. 2a) suggests alternating local supersaturation and undersaturation of calcic pyroxene component in the adjacent melt.

Sample 12063 does not appear to have crystallized as rapidly as sample 12065. The pyroxenes did not grow with hollow sectors, they do not have pronounced sectoral differences in composition, and they rarely reach compositions in the "forbidden zone." Instead, iron-rich compositions are typically represented by the fayalite-crystobalite-ferroaugite assemblage. However the plagioclase typically has a "swallow-tail" form, as does the plagioclase in sample 12065, suggesting a relatively rapid cooling rate. The difference in cooling rates between the two rocks may be reflected in their marked difference in texture. The difference in mineralogy can be accounted for essentially by the difference in silica content of the two rocks.

Conditions of extreme supersaturation can lead to nucleation of metastable phases and to metastably prolonged growth of phases already present, thereby significantly deviating the crystallization path of the melt. It is tempting to suggest that much of the diversity of the rock types (thought to represent original liquids) found on the moon may result from such metastable nucleation and crystallization. Perhaps the maria basalts were remelted residua formed by the prolonged metastable crystallization of plagioclase in the earliest stages of the moon's history.

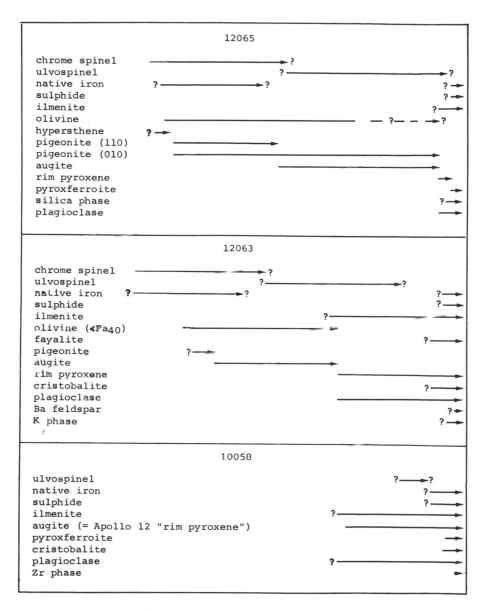

Figure 3. Inferred order of crystallization of most minerals in samples 12065, 12063, and 10058. Within a single sample group, beginning of crystallization of a phase at the same position where another phase disappears implies a reaction relation with the melt. Arrows are queried (?) where the data are less conclusive, but positions of arrows reflect our best judgment. The Zr phase (sample 10058) is mentioned in Hargraves and others (1970). Ba feldspar is the celsian of analysis 16, Table 1, sample 12063. In sample 12063, olivine zones to about Fa₅₀ where it is in contact with plagioclase only.

ACKNOWLEDGMENTS

This paper benefitted from many discussions with colleagues at the Second Lunar Science Conference in Houston, January, 1971, where it was presented orally. We thank H. D. Holland for critically reviewing the manuscript. The work was supported by NASA contract NAS 9–7897.

REFERENCES

Chao, E.C.T., Minkin, J., Frondel, C., Klein, C., Drake, J., Fuchs, L., Tani, B., Smith, J. V., Anderson, A., Moore, P., Zechman, G., Trail, R., Plant, A., Douglas, J., and Dence, M., 1970, Pyroxferroite, a new calcium-bearing iron silicate from Tranquility Base: Apollo 11 Lunar Sci. Conf. Proc.: Pergamon Press, v 1, p. 65–79.

Drever, H. I., and Johnston, R., 1957, Crystal growth of forsteritic olivine in magmas and melts: Royal Soc. Edinburgh Trans., v. 63, p. 289–315.

Hargraves, R. B., Hollister, L., and Otalora, G., 1970, Compositional zoning and its significance in pyroxenes from three coarse-grained lunar samples: Science, v. 167, p. 631–633.

Hollister, L. S., 1970, Origin, mechanism, and consequences of compositional sector zoning in staurolite: Am. Mineralogist, v. 55, p. 742–766.

Hollister, L. S., and Hargraves, R. B., 1970, Compositional zoning and its significance in pyroxenes from two coarse-grained Apollo 11 samples: Geochim. et Cosmochim. Acta, Suppl. 1, Proceedings of the Apollo 11 Lunar Science Conference, v. 1, p. 541–550.

Hollister, L. S., Trzcienski, W. E., Jr., Hargraves, R. B., and Kulick, C. G., 1971, Petrogenetic significance of pyroxenes in two Apollo 12 samples: Proceedings of the Second Lunar Science Conference: MIT Press, v. 1, p. 529–557.

Lindsley, D. H., and Munoz, J. L., 1969, Subsolidus relations along the join hedenbergite-ferrosilite: Am. Jour. Sci., Schairer Vol., 267-A, p. 295–324.

Ringwood, A. E., 1970, Petrogenesis of Apollo 11 basalts and implications for lunar origin: Jour. Geophys. Research, v. 75, p. 6453–6479.

Ross, M., Bence, A. E., Dwornik, E. J., Clark, J. R., and Papike, J. J., 1970, Mineralogy of the lunar clinopyroxenes, augite and pigeonite: Geochim. et Cosmochim. Acta, Suppl. 1, Proceedings of the Apollo 11 Lunar Science Conference, v. 1, p. 839–848.

Smales, A. A., Mapper, D., Webb, M. S. W., Webster, R. K., Wilson, J. D., and Hislop, J. S., 1971, Elemental composition of lunar surface material (pt. 2): Proceedings of the Second Lunar Science Conference: MIT Press, v. 2, p. 1253–1258.

Weill, D. F., McCallum, I., Bottinga, Y., Drake, M., and McKay, G., 1970, Mineralogy and petrology of some Apollo 11 igneous rocks: Proceedings of the Apollo 11 Lunar Science Conference: New York, Pergamon Press, v. 1, p. 937–955.

Willis, J. P., Ahrens, L. H., Danchin, R. V., Erlank, A. J., Gurney, J. J., Hofmeyer, P. K., McCarthy, T. S., and Orren, M. J., 1971, Some interelement relationships between lunar rocks and fines, and stony meteorites: Proceedings of the Second Lunar Science Conference: MIT Press, v. 2, p. 1123–1138.

MANUSCRIPT RECEIVED BY THE SOCIETY MARCH 29, 1971

THE GEOLOGICAL SOCIETY OF AMERICA, INC.
MEMOIR 132
© 1972

Calcium in Chondritic Olivine

R. T. DODD
Department of Earth and Space Sciences
State University of New York at Stony Brook,
Stony Brook, New York 11790

ABSTRACT

The abundance of calcium in olivine from 14 equilibrated low-iron chondrites was determined by microprobe analysis. With two exceptions (Shaw and Kunashak), median CaO abundances fall between 0.03 and 0.06 wt percent, substantially lower than those encountered in type 3 chondrites (for example, 0.15 percent in Sharps). It appears that much of the calcium in type 3 olivine was expelled during the metamorphic transition from type 3 to type 4. By contrast, changes from type 4 to type 6 were modest, though the data suggest slight calcium depletion in this interval.

High CaO contents (0.09 and 0.07 percent) in the Kunashak and Shaw chondrites, both type 6, suggest unusual thermal histories for these meteorites: shock reheating in the former case, and unusually high metamorphic temperatures in the latter.

INTRODUCTION

The need to use mineralogical data to infer the thermal histories of rocks is nowhere more acute than in the study of chondritic meteorites, whose field relations are unknown and whose textures are difficult to interpret unambiguously. Although it now seems clear that most or all such meteorites have experienced thermal metamorphism, the metamorphic conditions and, in particular, the direction of temperature change remain obscure. We are not yet certain whether metamorphism of the chondrites took place during cooling of a hot agglomerate or during heating of initially cold material, though the available data pertinent to this question (reviewed in detail by Dodd, 1969) favor the latter interpretation.

The calcium content of olivine is a potentially useful geothermometer for the chondrites. Olivine is abundant in most such meteorites, and extensive chemical

651

data on terrestrial olivine (Smith, 1966; Simkin and Smith, 1970) indicate that its calcium content varies systematically with the environment of crystallization. Thus olivines from plutonic igneous rocks and ultrabasic inclusions in basalt typically contain less than 0.1 wt percent Ca, while those from volcanic and hypabyssal rocks normally contain more than twice this amount. The few available data for metamorphic olivines suggest that they contain less calcium than those in igneous rocks: Smith (1966) reports values of 0.04, 0.02, 0.02, and 0.01 wt percent for four occurrences. Early data on lunar olivines (Brown and others, 1970; Chao and others, 1970) suggest that most of them resemble those in terrestrial volcanic rocks, though Chao and others report one lunar olivine (from rock no. 10084, "feldspar-rich poikilitic") whose low-calcium content (0.04 percent) seems more consistent with a plutonic origin.

Although a relation between composition and environment is well established for terrestrial olivines, it is not yet clear which factors in the environment are responsible for it. Possibilities include (1) the availability of calcium; (2) the temperature and (or) pressure of crystallization; and (3) the rate of crystallization and (or) subsequent cooling. Simkin and Smith (1970) describe several occurrences of olivine, each of which suggests a different control. They suggest that pressure is the most important determinant of calcium content, but stress the need for further experimental study to check this conclusion.

The complicated bulk chemical and pressure variations which impede interpretation of terrestrial olivines are largely absent from the ordinary chondrites. Recent superior analyses (Ahrens and others, 1969) indicate that each chondrite group (H, L, and LL) is chemically homogeneous, and all pressure indicators examined to date suggest low-pressure histories (approx. 1 to 2 kb; Wood, 1967; Dodd, 1969). For all practical purposes, the calcium content of olivines in metamorphosed chondrites should therefore be directly and simply related to the thermal histories of these meteorites.

The genetic implications of high calcium abundances in olivines of the unequilibrated ordinary chondrites (Dodd and Van Schmus, 1965; Dodd and others, 1967) are considered in detail elsewhere (Dodd, 1968, 1969, 1971) and only briefly reviewed here. The purposes of this paper are to report and interpret new analyses of olivines in 14 equilibrated low-iron chondrites of petrologic types 4 to 6 (Van Schmus and Wood, 1967).

ANALYTICAL PROCEDURES

Partial analyses for Fe, Ca, and Mg were obtained on an ARL EMX-SM microprobe. Accelerating potential and specimen current were maintained at 15 kv and 0.015 μa, respectively, and beam integration 10^{-7} coulombs; integration time 7.5 sec) was used to minimize drift. Prime and working standards for the study were Elba hematite and synthetic fayalite (Fe), pseudowollastonite and synthetic wollastonite glass (Ca), and synthetic MgO and enstatite glass (Mg).

Analyses were made at 10-μ intervals on traverses 100 to 200 μ apart. Such random sampling permits accurate determination of the degree of Fe-Mg and Ca inhomogeneity in the olivines of each meteorite (Dodd and others, 1967), and this advantage outweighs the risk of occasional spurious analyses caused by proximity to grain boundaries. It was found that an adequate number of olivine measurements (25 or more) resulted from total sample populations of 100 to 150 points.

The microprobe data were reduced by computer, using a program described elsewhere (Dodd, 1968) and modified to utilize the Bence-Albee correction procedures (Bence and Albee, 1968). This program identifies each analyzed point by adding sufficient Si and O to form orthosilicate molecules. These are summed, and the point is reported as olivine if the molecular sum of Fe_2SiO_4, Ca_2SiO_4, Mg_2SiO_4 falls between 98 and 102 percent.[1]

As the concentrations and hence intensities of Fe and Mg are high in chondritic olivines (approx. 5,000 counts for Fe and 9,500 counts for Mg, both for 7.5 sec of integration), the precision of FeO and MgO for each data point is estimated to be ± 2 to 5 percent relative. The precision for each chondrite, based on 25 or more point analyses, is thought to be ± 2 percent relative.

Because calcium is a minor constituent in chondritic olivines, its accurate determination requires very careful measurement of the background. This was done separately for each chondrite, using long integrations (75 sec) on several points. The calcium background was found to vary little both within and among the samples considered here.

The detection limit for calcium in olivine and the precision of the calcium analyses were estimated with formulas developed by Ziebold (1967). If 25 point analyses are made, CaO can be detected with 95 percent confidence at levels above 0.024 percent. This limit falls to 0.019 percent CaO for a sample population of 40 points. The precision of a mean CaO value of 0.06 percent is estimated as ± 0.025 percent for 25 points and ± 0.020 percent for 40 points, both at the 95 percent confidence level. The observed precision of two analyses of the Honolulu chondrite (Table 2, Fig. 2) agrees very well with these estimates.

RESULTS

Unequilibrated Chondrites

That the ferromagnesian silicates in carbonaceous and type 3 ordinary chondrites are grossly inhomogeneous (Dodd and Van Schmus, 1965; Dodd and others, 1967; Fredriksson and Keil, 1964; Keil and others, 1964; Wood, 1967) signifies that these meteorites have not achieved chemical equilibrium. Detailed study of individual chondrules (for example, by Keil and others, 1964; Van Schmus, 1967; Kurat, 1967a, 1967b; Dodd, 1968) reveals wide mineralogical and textural variations, which in turn imply complex and varied preaccumulation histories. These diverse histories are also reflected in variable CaO abundances in olivine. Although Keil and others (1964) found olivine in most chondrules of the Chainpur (type LL-3) chondrite to contain less than 0.14 wt percent CaO (their detection limit), they report some with up to 0.5 wt percent CaO. A larger range was found in Murray (C-2) carbonaceous chondrite (Fredriksson and Keil, 1964), and data for 96 chondrules in the Sharps (type H-3) chondrite (Fig. 1; Dodd, 1971) indicate wide variations of the olivine composition within and among chondrules and systematic variations with chondrule type.

---

[1] The only other component present in any appreciable abundance in chondritic olivine is Mn_2SiO_4, whose level was established by Van Schmus and Koffman (1967) as 0.5 mol percent.

TABLE 1. SAMPLES STUDIED

Meteorite	Type*	Source	Remarks†
Beenham, New Mexico	L-5	Arizona State University, No. 414.100	Find; moderate shock
Bjurböle, Finland	L-4	American Museum of Natural History	Fall; light to no shock
Colby, Wisconsin	L-6	Arizona State University, No. 21a	Fall; moderate to heavy shock
Crumlin, Ireland	L-5	British Museum, No. 86115	Fall; light to moderate shock
Goodland, Kansas	L-4	Arizona State University, No. 425.2x	Find; moderate to heavy shock
Homestead, Iowa	L-5	Yale University, No. P-56	Fall; light to moderate shock
Honolulu, Hawaii	L-5	Yale University, No. P-188	Fall; moderate to heavy shock
Jhung, Pakistan	L-5	British Museum, No. 51190	Fall; light to moderate shock
Kunashak, USSR	L-6	Arizona State University	Fall; moderate to heavy shock
Modoc, Kansas (1905)	L-6	Arizona State University, No. 415a	Fall; moderate shock
New Concord, Ohio	L-6	Harvard University	Fall; moderate shock
Saratov, USSR	L-4	Arizona State University	Fall; light shock
Sharps, Virginia	H-3	University of Virginia	Fall; light (?) shock
Shaw, Colorado	L-6	Princeton University	Find; light shock
Slobodka, USSR	L-4	Vienna Museum of Natural History, No. VA 223	Fall; light to moderate shock

* *After* Van Schmus and Wood (1967).
† Based on the criteria of Carter and others (1968).

Equilibrated Ordinary Chrondrites

In contrast to the mafic silicates in type 3 ordinary chondrites, those in the higher petrologic types show little or no Fe-Mg variation. According to Van Schmus and Wood (1967) the percent mean deviation (PMD)[2] of iron is less than 5.0 in type 4 olivines and negligible in those of types 5 and 6. It is inferred from mineralogical data that the silicate phases in type 4 to type 6 chondrites reached chemical equilibrium at metamorphic temperatures of roughly 600° to more than 850°C (Van Schmus and Koffman, 1967; Dodd, 1969; Binns, 1970).

Keil and Fredriksson (1964) determined the Fe, Ca, and Mg contents of olivine and low-Ca pyroxene in 95 ordinary chondrites, most of which are of types 4 to 6. They found olivine in all of these to contain less than 0.14 percent CaO (their detection limit). The only apparent exception among type 4 to type 6 chondrites is Shaw, an unusually highly recrystallized type L-6 chondrite, whose olivine is reported to contain 0.18 percent CaO (Fredriksson and Mason, 1967).

New olivine data for 14 type L4 to L6 chondrites are presented in Table 2 and Figures 1 and 2. Because the olivine in some meteorites shows a skewed

---

[2] Defined by Dodd and others (1967) as the mean deviation of the oxide abundance divided by mean abundance and multiplied by 100.

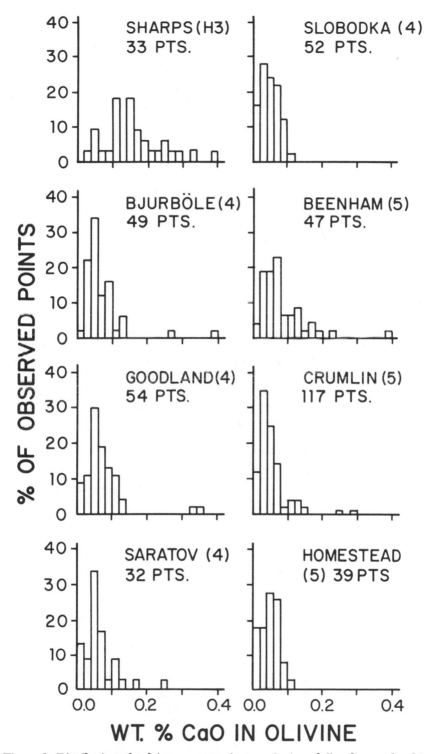

Figure 1. Distribution of calcium contents in type 3, 4, and 5 ordinary chondrites. Designations of petrologic type (brackets) follow Van Schmus and Wood (1967). All but Sharps are L-group chondrites.

CaO distribution, both mean and median values of CaO are reported in Table 2. The latter are thought to be more representative of the equilibrium abundances than the former. Rare high values in Figures 1 and 2 are probably spurious and due to interference from adjacent pyroxene crystals.

DISCUSSION

The data presented here confirm the conclusion of Keil and Fredriksson (1964) that most chondritic olivine contains very little calcium, and the observation of Fredriksson and Mason (1967) that the olivine in Shaw is somewhat more calcic than is normal for type 6 chondrites. They also suggest a systematic relation between olivine composition and petrologic type, and some apparent exceptions to this relation. The data show that the inhomogeneous, commonly calcic olivines found in type 3 chondrites are largely absent from types 4 to 6. It appears that chondritic olivine lost most of its calcium during the metamorphic transition from type 3 to type 4, that is, at temperatures below roughly 600°C (Dodd, 1969).

By contrast, variations among the higher types are minor. If data for the Shaw and Kunashak chondrites are omitted (for reasons discussed below), it appears that the median calcium content of olivine decreases slightly from type 4 (0.05 to 0.06 percent CaO) to type 6 (0.03 percent). The data are too few to be more than suggestive, but they are consistent with the view that types 4 to 6 chondrites represent metamorphism at progressively higher temperatures.

Three meteorites—Kunashak, Beenham, and Shaw—are apparent exceptions to the main trend. The histograms for Kunashak and Shaw (Fig. 2) indicate that the olivine in these meteorites is less homogeneous and more calcic than that in other type 6 (and most type 5) chondrites. Beenham differs from other type 5 chondrites in the same way, though to a smaller degree.

TABLE 2. FeO, CaO, AND MgO CONTENTS OF CHONDRITIC OLIVINES

	FeO		CaO			MgO		FeO/(FeO + MgO)	
								This work	Mason '63
	Mean	PMD	Mean	Median	PMD	Mean	PMD		
Type 4:									
Bjurböle	23.71	1.63	0.07	.05	52.8	38.01	1.25	25.9	26
Slobodka	23.13	1.84	0.04	.05	50.1	38.57	1.14	25.2	23
Goodland	22.35	1.71	0.07	.06	52.8	39.30	1.46	24.2	25
Saratov	21.96	1.61	0.09	.06	77.0	39.88	1.20	23.6	24
Type 5:									
Beenham	22.07	1.34	.08	.06	59.1	39.04	1.02	24.1	23
Crumlin	23.09	1.48	.06	.04	74.3	38.67	1.03	25.1	24
Homestead	21.97	1.45	.04	.05	53.1	39.59	1.20	23.7	24
Honolulu (1)	23.29	1.56	.05	.04	65.6	38.80	1.23	25.2	24
Honolulu (2)	23.29	1.00	.03	.03	64.2	38.96	0.62	25.1	
Jhung	22.65	1.58	.05	.05	41.1	38.82	0.90	24.7	25
Type 6:									
Colby	22.88	1.78	0.04	.03	55.7	38.83	1.30	24.8	25
Kunashak	22.90	1.75	.10	.09	37.8	38.56	1.19	25.0	24
Modoc	23.16	1.53	.04	.03	43.6	38.88	1.06	25.0	23
New Concord	22.80	2.16	0.03	.03	56.9	38.57	1.57	24.9	24
Shaw	21.05	1.22	0.07	.07	45.9	39.85	0.78	22.9	23

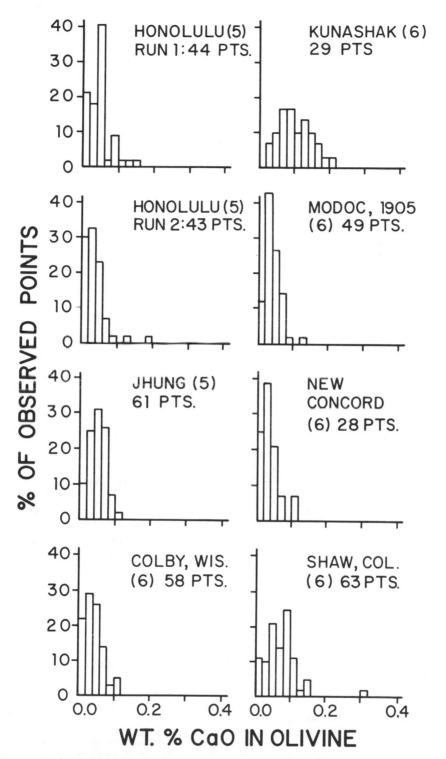

Figure 2. Distribution of calcium contents in type 5 and 6 ordinary chondrites. Conventions conform with Figure 2.

The high and variable CaO content of olivine in Kunashak may be secondary and due to moderate-to-heavy shock metamorphism of this meteorite. Such metamorphism is suggested by the presence of olivine crystals which display mosaic extinction and contain profuse black inclusions (Carter and others, 1968). That shock *per se* or (more probably) transient high temperature associated with it can facilitate entry of calcium into olivine is suggested by the fact that calcic olivine is present in the intensely shocked Ramsdorf chondrite (Begemann and Wlotzka, 1969) and in highly shocked individual chondrules of the Sharps meteorite (Dodd, 1971). This explanation for Kunashak is not entirely convincing, however, for other chondrites studied appear petrographically to be nearly as highly shocked, yet they show no anomaly in the calcium abundance (Tables 1, 2). A more detailed study of the relation between shock and olivine composition, including several of the heavily shocked black chondrites, is therefore planned as an extension of the present work.

This interpretation is less satisfactory for Beenham, which has apparently experienced only moderate shock on the basis of the criteria of Carter and others (1968).[3] Beenham is more readily explained as transitional in mineralogy and thermal history to the type 4 chondrites.

The shock interpretation is wholly inadequate for Shaw, whose texture resembles that found by Carter and others (1968) to be typical of lightly shocked material. The writer concurs with Fredricksson and Mason (1967), who suggest that the extremely coarse texture of Shaw and the presence in it of calcic orthopyroxene and subcalcic diopside signify that it experienced much higher temperatures ($\geqq$1,100° to 1,200° C) than those indicated by other type 6 chrondrites ($\geqq$850° C, Van Schmus and Koffman, 1967).[4] The deficiency of Fe, Ni, and S which these writers noted in Shaw is also consistent with this interpretation, for the inferred metamorphic temperatures are high enough to cause extensive melting of metal and troilite (Kullerud, 1963).

That Shaw contains a more calcic olivine than is found in other type 6 chondrites suggests that the tendency for olivine to lose calcium during metamorphism is reversed at near-liquidus temperatures. This is in accord with experimental evidence that the forsterite-monticellite solvus becomes narrow at such temperatures (Ricker and Osborn, 1965). It would be interesting to determine where the reversal occurs in the chondrites, but thus far no meteorites intermediate in character between Shaw and the more orthodox type 6 chondrites have been reported.

CONCLUSIONS

The data presented here, and those published elsewhere for type 3 chondrites, indicate that calcium was expelled from chondritic olivine during the transformation of type 3 to type 4 material (that is, at metamorphic temperatures below

---

[3] Shock features in Beenham are discussed in detail by Buseck (1967).

[4] Its chemistry, mineralogy, and texture set Shaw apart from the other type 6 chondrites. It deserves separate classification (perhaps as petrologic type 7), for its inclusion in petrologic type 6 obscures its significance as the most intensely metamorphosed chondrite known and as a possible intermediate between the chondrites and achondrites.

about 600° C—Dodd, 1969). They suggest, with less certainty, that olivine continued to lose calcium during the type 4 to type 6 transition (temperatures of or higher than 820° C—Van Schmus and Koffman, 1967). This trend was apparently reversed at the very high temperatures experienced by Shaw (Fredriksson and Mason, 1967). It may also have been reversed in some stones (for example, Kunashak) by intense heating due to shock.

Because it varies subtly if at all from type 4 to type 6, the calcium content of olivine is of limited value for chondrite classification. However, it provides a useful means of identifying recrystallized xenoliths in type 3 chondrites (Dodd, 1968), and it may prove to be a useful criterion for shock.

ACKNOWLEDGMENTS

The writer is grateful to Sara Jacobson for assistance in data reduction, to Penelope Campbell for typing the manuscript, and to Christopher Miele for drafting illustrations. A. E. Bence, C. B. Moore, Gordon Brown, and Patrick Butler kindly read and criticized the manuscript, to its considerable benefit.

The work reported here is part of a program of meteorite research supported by the National Science Foundation Grant No. GA-1674.

REFERENCES CITED

Ahrens, L. H., Von Michaelis, H., Erlank, A. J., and Willis, J. P., 1969, Fractionation of some abundant lithophile element ratios in chondrites, *in* Millman, P. M., ed., Meteorite research: Dordrecht-Holland, D. Reidel, p. 166–173.

Begemann, F., and Wlotzka, F., 1969, Shock induced thermal metamorphism and mechanical deformations in the Ramsdorf chondrites: Geochim. et Cosmochim. Acta, v. 33, p. 1351–1370.

Bence, A. E., and Albee, A. L., 1968, Empirical correction factors for the electron microanalysis of silicates and oxides: Jour. Geology, v. 76, p. 382–403.

Binns, R. A., 1970, Pyroxenes from non-carbonaceous chondritic meteorites: Mineralog. Mag., v. 37, p. 649–669.

Brown, G. M., Emeleus, C. H., Holland, J. G., and Phillips, R., 1970, Petrographic, mineralogic, and x-ray fluorescence analysis of lunar igneous-type rocks and spherules: Science, v. 167, p. 599–601.

Buseck, P. R., 1967, The post-formational history of a hypersthene chondrite-Beenham: Geochim. et Cosmochim. Acta, v. 31, p. 1583–1587.

Carter, N. L., Raleigh, C. B., and De Carli, P., 1968, Deformation of olivine in stony meteorites: Jour. Geophys. Research, v. 73, p. 5439–5461.

Chao, E.C.T., James, O. B., Minkin, J. A., Boreman, J. A., Jackson, E. D., and Raleigh, C. B., 1970, Petrology of unshocked crystalline rocks and shock effects in lunar rocks and minerals: Science, v. 167, p. 644–647.

Dodd, R. T., 1968, Recrystallized chondrules in the Sharps (H-3) chondrite: Geochim. et Cosmochim. Acta, v. 32, p. 1111–1120.

—— 1969, Metamorphism of the ordinary chondrites: A review: Geochim. et Cosmochim. Acta, v. 33, p. 161–203.

—— 1971, The petrology of chondrules in the Sharps meteorite: Contr. Mineralogy and Petrology, v. 31, p. 201–227.

Dodd, R. T., and Van Schmus, W. R., 1965, Significance of the unequilibrated ordinary chondrites: Jour. Geophys. Research, v. 70, p. 3801–3811.

Dodd, R. T., Van Schmus, W. R., and Koffman, D. M., 1967, A survey of the unequilibrated ordinary chondrites: Geochim. et Cosmochim. Acta, v. 31, p. 921–951.

Fredriksson, K., and Keil, K., 1964, The iron, magnesium, calcium and nickel distribution in the Murray carbonaceous meteorite: Meteorites, v. 2, p. 201–217.

Fredriksson, K., and Mason, B., 1967, The Shaw meteorite: Geochim. et Cosmochim. Acta, v. 31, p. 1705–1709.

Keil, K., and Fredriksson, K., 1964, The iron, magnesium and calcium distribution in coexisting olivines and rhombic pyroxenes of chondrites: Jour. Geophys. Research, v. 69, p. 3487–3515.

Keil, K., Mason, B., Wiik, H. B., and Fredriksson, K., 1964, The Chainpur meteorite: Am. Mus. Novitates No. 2173.

Kullerud, G., 1963, The Fe-Ni-S system: Carnegie Inst. Washington Year Book 62, p. 175–189.

mechanical deformations in the Ramsdorf chondrite: Geochim. et Cosmochim. et Cosmochim. Acta, v. 31, p. 1843–1857.

——1967b, Zur Entstehung der Chondren: Geochim. et Cosmochim. Acta, v. 31, p. 491–502.

Ricker, R. W., and Osborn, E. F., 1954, Additional phase equilibrium data for the system CaO-MgO-SiO$_2$: Am. Ceramic Soc. Jour., v. 37, p. 133–136.

Simkin, T., and Smith, J. V., 1970, Minor-element distribution in olivine: Jour. Geology, v. 78, p. 304–325.

Smith, J. V., 1966, X-ray-emission microanalysis of rock-forming minerals: Pt. II: Olivines: Jour. Geology, v. 74, p. 1–16.

Van Schmus, W. R., 1967, Polymict structure of the Mezö-Madaras chondrite: Geochim. et Cosmochim. Acta, v. 31, p. 2027–2042.

Van Schmus, W. R., and Koffman, D. M., 1967, Equilibrium temperatures of iron and magnesium in chondritic meteorites: Science, v. 155, p. 1009–1011.

Van Schmus, W. R., and Wood, J. A., 1967, A chemical-petrologic classification for the chondritic meteorites: Geochim. et Cosmochim. Acta, v. 31, p. 747–765.

Wood, J. A., 1967, Chondrites: Their metallic minerals, thermal histories, and parent planets: Icarus, v. 6, p. 1–49.

Ziebold, T. O., 1967, Precision and sensitivity in electron microprobe analysis: Anal. Chemistry, v. 39, p. 858–861.

MANUSCRIPT RECEIVED BY THE SOCIETY MARCH 29, 1971

THE GEOLOGICAL SOCIETY OF AMERICA, INC.
MEMOIR 132
© 1972

Are Lunar Rilles
Inverted Eskers?

CHARLES E. HELSLEY
Geosciences Division
The University of Texas at Dallas
P.O. Box 30365
Dallas, Texas 75230

ABSTRACT

Sinuous rilles on the lunar surface may have been formed by the movement of volatiles ($H_2O + CO_2$) in a channel between the basement surface and a surficial permafrost layer. Later sublimation of the ice formed in the vicinity of this channel would produce the features observed today.

INTRODUCTION

In April 1969 Harry Hess, in his characteristic way, suddenly confronted me with this statement: "I think I know how the lunar rilles were formed—they're inverted eskers." During the ensuing discussion regarding the behavior of water in the near vacuum of the lunar surface, many of the features of the sinuous rilles seemed to be explained by this idea. In light of the current interest in the origin of lunar rilles, it seems appropriate to discuss this hypothesis, as formulated by Harry Hess, and in which ideas contributed by the two of us have become somewhat indiscriminately mixed.

PREVIOUS HYPOTHESES

Lunar rilles have been the object of much speculation since their early description by Schroeter (1788). Their essential features have recently been summarized

661

very succinctly by Mutch (1970, p. 189 to 196). In general, they can be classified as either linear or sinuous. The sinuous rilles are perhaps the most interesting, for they have many features that remind one of river channels. Yet, when considered in detail, many of the most prominent features of rivers are missing. For example, tributaries are very rare, there are no pronounced levees or cut-off meanders, and, perhaps most important of all, there is no evidence of deltas at their ends even though the volume of material that must have been moved exceeds several hundred cubic kilometers for many of the rilles. This perplexing assemblage of features has led to many hypotheses regarding their origin, varying from running water on the surface (for example, Firstoff, 1961; Gilvarry, 1969), collapsed lava tubes (Kuiper and others, 1966), channels scoured by ignimbrites (Cameron, 1964), erosion by flowing water beneath an ice sheet (Lingenfelter and others, 1968; Schubert and others, 1970) and, most recently, coalescence of craters produced by the fluidization of surficial material due to gas venting from beneath the regolith (Mills, 1969; Schumm, 1970).

Of these hypotheses, the one closest to the hypothesis suggested here is that of Lingenfelter and others (1968), expanded upon by Schubert and others (1970). These authors conclude that the sinuous rilles are formed as the result of water flowing on the lunar surface under a 0.4-m-thick protective layer of ice which prevents the water from boiling explosively. According to their hypothesis, the channel is eroded to the depth of the permafrost, a depth estimated to be some 200 m, and has a more or less constant depth and probably also a uniformly decreasing gradient. On most sinuous rilles the depth does not appear to be constant (*see* Figs. 1 and 2 or Figs. 7 through 15 of Schumm, 1970). Furthermore, the Lunar Orbiter photographs of the lunar rilles often show discontinuities in gradient. Moreover, there is no evidence for hydrated minerals in the Apollo 11 and Apollo 12 samples, and therefore it is doubtful that there was ever an appreciable quantity of liquid water in contact with the lunar surface.

COMMON FEATURES OF SINUOUS RILLES

Schubert and others (1970) point out that there is a close association between mare areas and rilles. Rilles are absent in highland areas and the Lunar Orbiter photography of the backside of the moon shows no sinuous rilles. Rilles are most abundant in volcanic areas, for example the Marius Hills, and at the margins of mare basins.

Most rilles are seen to originate (that is, have their apparent source) in a breached crater or in source regions obscured because of an overabundance of craters. A few rilles apparently have no source, for example NASA Lunar Orbiter photo IV 158H (Mutch, 1970, p. 195). However, in this case one could argue that there are two rilles, both beginning at a deep crater near the upper end of the deep portion of the rille as figured by Mutch. For those rilles beginning in clearly defined craters, many have a smaller rille within the much larger main rille (*see* Fig. 1).

Features common to the central segment of most rilles are slumped walls, an irregular floor, a general disregard for which way is "downhill," an absence of features associated with rivers (such as levees, braiding, or tributaries), and the presence of numerous small craters, a crater chain, or a smaller rille within the

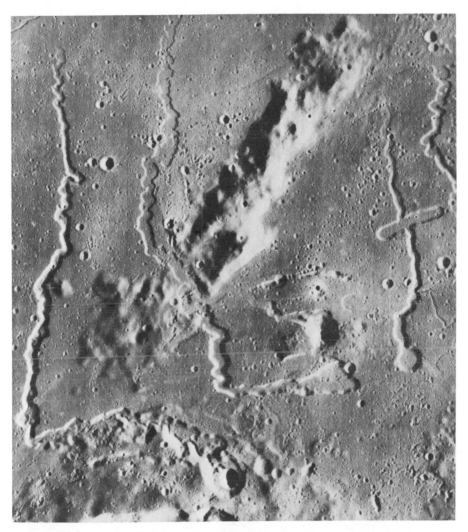

Figure 1. Part of the Harbinger Mountains Plateau north of the crater Prinz showing several sinuous rilles including Rima Prinz I (left) and Rima Prinz II (center). Lunar orbiter 5 medium-resolution frame 191.

rille. In general, the sinuous rilles do not cross "bed rock" highs and major topographic features, although a few are observed to cross low ridges with virtually no change in width or other characteristics (*see* Fig. 1). Some rilles are locally discontinuous as a result of burial by secondary events or perhaps because their mode of origin allows the creation of discontinuous pattern.

A feature common to what appears to be the terminal end of most sinuous rilles is a general decrease in channel size, both in width and depth, as one approaches the end (*see* Fig. 1). There is no delta or distributary system in evidence on any of the rilles.

These features generally appear to be similar to features produced by running water as has been suggested by many authors, for example Urey (1967), Lingen-

Figure 2. High resolution photograph of a portion of a rille in the marius Hills region. Lunar Orbiter 5 high-resolution frame 213.

felter and others (1968), and Schubert (1970). However, as has been noted above, when examined closely, many of the features expected for a river are not present. The fluidization origin advocated by Schumm (1970) offers many intriguing possibilities but has the shortcoming that it requires a uniform source of volatiles along the length of the rille (but virtually zero source at right angles to the rille) and a very complicated fracture pattern. Thus, an alternative explanation seems to be required.

HYPOTHESIS

An esker is a ridge of gravel and sand deposited by a subglacial river flowing in a channel that has been eroded upward into the ice sheet. When the enclosing ice melts, the channel deposits are left as a pronounced ridge. An inverted esker can therefore be thought of as a sinuous valley resulting either from the erosion of material from beneath the ice or from the melting of a sinuous ridge of ice enclosed within outwash gravels. Such features are not observed on the earth since the ground beneath the ice is also frozen, and the water doing the erosion comes from the surface melting of the ice sheet. On the moon, however, surface melt waters are not possible due to the high vacuum. The volatiles are coming from below and thus would tend to erode downward into the unfrozen regolith rather than upward into the permafrost layer.

In all models for the formation of lunar rilles by the action of water, one is required to assume that a degassing of the moon took place in order to derive the necessary fluids to do the erosion at the surface. This hypothesis is no different than the others in this regard. The timing of the degassing events is unknown but is required to be later than the mare-filling events in the other hypotheses. This model is also compatible with late degassing but is easier to visualize as being related to the late stages of the mare-filling events as discussed below. In all hypotheses involving the action of water, including this one, the mare material is considered to be fragmental, that is, to be easily erodable by water, to a depth of several hundred meters. This does not preclude the existence of lava flows within the mare but does require that they not exist near the surface in the sinuous rille areas.

The temperature within the moon (Gold, 1966) is such that water vapor will condense and freeze as it approaches the surface to form a layer of permafrost which would effectively prevent further escape of volatiles. Gold argues that the temperature regime is now such that the permafrost would form to a depth of over 1 km. However, during earlier stages of the Moon's history, the permafrost would be thinner or perhaps absent except for a thin zone near the surface produced as a consequence of the rapid loss of heat due to evaporation (or sublimation). Thus it seems very likely that an effectively impermeable cap would be formed above any region emitting water vapor. Should the emission of the volatiles be concentrated at series of local sources comparable to volcanoes on earth, one would expect the permafrost to be formed first in these regions. Subsequent emissions would be confined beneath this layer of permafrost and would collect and flow or percolate in the zone beneath the permafrost and above the impermeable bedrock either as a liquid (H_2O) or a gas (CO_2). When these volatiles reached the margin of the permafrost they would again be able to break through to the surface and dissipate into space. Since this amounts to a fairly sudden release in pressure, the rapid egress of the confined volatiles would erode a channel beneath the permafrost and might form a series of craterlets at the permafrost boundary until such time as enough ice had been formed to seal the vent and prevent further escape of the volatiles to the surface. The volatiles would then slowly accumulate and continue migrating until they again broke through to the surface, and the violent erosive process would be repeated.

The mechanism is similar to that observed in a geyser except that the gaseous phase would be CO_2 rather than steam. The CO_2 dissolved in the water would be released whenever the confining pressure dropped, that is, whenever new marginal venting occurred, and would assist in removing material from beneath the permafrost by increasing the velocity of the transporting fluid. The eroded and redeposited material would be a mixture of lunar regolith and particles of ice produced during the venting of the water and CO_2. This process would continue as long as volatiles were emitted from the source and, due to the self-sealing nature of the permafrost developed, these volatiles would be channeled along whatever more or less random path happened to develop. This channeling would provide a mechanism for subpermafrost erosion whenever a new surge of volatiles occurred as a result of a new vent at the surface. This surface venting would be very sudden and would help excavate the rille by throwing material to the side as well as venting material eroded from a channel beneath the permafrost.

There are several interesting consequences of such a mechanism. First, there is little need for the rille to develop in a "downhill" direction since the primary controls are pressure gradient and bedrock topography rather than gravity, and the pressure gradient is primarily controlled by the permeability of the material through which the volatiles must move. A further consequence is that the channel will be controlled by permeability and porosity contrasts and thus will be confined to area of high porosity and permeability and will avoid "basement" areas much as the burrow of a mole avoids rocks or large roots in the soil. However, the channel will still be able to cross ridges of loose material such as is observed for Rima Prinz II (Fig. 1) so long as there is sufficient permeability. Branching would only occur very rarely, since the channel would be completely enclosed by permafrost both above and to the sides. There is no need for the development of a delta, since the products of the craterlets would be deposited along the margins of the rille as it grows in length.

The question of what happens to the material that was in the channel should perhaps be considered further. The degassing of the moon will occur shortly after, or perhaps concurrently with, the final stages of mare filling, since this appears to be a period of major volcanism. Thus the material being removed from the channel would be mixed with other material being added to the basin. This material would be much like the present regolith and could include lava flows as well as rock debris, breccia, and other fragments produced during small impact events. Lava flows would tend to inhibit the development of rilles by this method, since they would be competent, impermeable units, and may be the reason that some rilles are discontinuous. At the time that the degassing ceases, or becomes inconsequential, the surface in the vicinity of the rille would be more or less smooth with a number of small craters and virtually no expression of the rille itself being visible (*see* Fig. 3). The future rille at this time would be represented by a channel in the subsurface, either open or virtually filled with ice, and a zone of ice, rock debris, and the original permafrost at the surface. Most of the ice in the future rille is formed during the marginal venting process as a result of the "explosive boiling" of the water as it comes in contact with the vacuum at the surface. Thus a wide zone of ice-rich material would be formed above and beside the subsurface channel. After the period of active rille formation comes to an end, the ice near the surface will begin to sublime away more rapidly than it is formed, and as it does so, the rille itself develops by compaction and slumping as long as there is ice to sublime away. This would in the end be limited by the amount of heat available to sublimate the ice. Schubert and others (1970) calculate that sublimation could have proceeded to a depth of several hundred meters in 10^9 yrs if the surface were to remain near 243° K. As the permafrost becomes very thin, the channel that has been transporting volatiles and debris to the margin may collapse if it has not previously been filled with ice and, since this would be a well-defined sinuous channel, the possibility exists that a more or less continuous rille would be developed within the broader collapse zone. It is possible that some liquid water would be present at this stage, and this might contribute to the rapid venting of gas to the surface with the production of small craterlets in the bottom of the rille using the fluidization mechanism suggested by Mills (1969) or Schumm (1970). The crater at the head of the rille could also be accounted for by the collapse of material back into the vent as the ice in the vent sublimed away.

DISCUSSION

The hypothesis thus accounts for many of the features observed in sinuous rilles (Figs. 1 and 2). One of the major criticisms of such an hypothesis is that the thickness of the permafrost as postulated by Gold (1966) is about 1 km, and this is too thick to be sublimed away within a reasonable time. However, Gold's calculations are based upon the thermal regime being that of today. At the time of mare filling the heat flow and thus the temperature gradient would have been much higher due to local heating events during mare formation. Consequently, the permafrost would have been much thinner, so that volatiles would have had a good chance of moving between the thin permafrost layer and the top of the basement. At least some of the rilles are as old as the rest of the mare surface, since the crater density appears to be identical on the flat floor of the rille and on the mare surface outside of the rille. The only regions of low crater density are the walls themselves, and here the surface is quite steep (up to 30°) and has

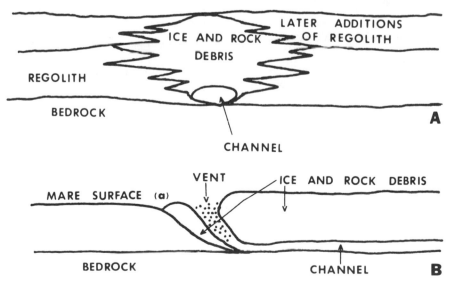

Figure 3. Longitudinal section of terminal vent. Next vent will develop in area near (A).

been extensively modified by slumping. The volume of water required will be not much greater than the volume of the rille and only 3 percent of that required by Lingenfelter and others (1968) or Schubert and others (1970).

It is concluded that escaping volatiles (CO_2 and H_2O) could form a sinuous channel beneath the developing permafrost and above the impermeable basement. Concurrent with the development of the subsurface channel would be the production of a large amount of fragmented ice mixed with rock debris at the surface above and along the sides of the sinuous subsurface channel. The eventual sublimation of the ice in the surficial material down to and perhaps including the subsurface would develop a feature at the surface that would be morphologically very similar to a valley produced by running water.

ACKNOWLEDGMENTS

I wish to acknowledge the late Henry Hess for his stimulating ideas that led to this paper. Discussions with A. L. Hales were most helpful as was his thoughtful criticism of the manuscript. This work was made possible by funds from the University of Texas at Dallas and NASA contract no. NAS 9–8767.

REFERENCES CITED

Cameron, W. S., 1964, An interpretation of Schroeter's valley and other lunar sinuous rilles: Jour. Geophys. Research, v. 69, p. 2423–2430.
Firstoff, V. A., 1961, Surface of the moon: London, Hutchinson, 128 p.
Gilvarry, J. J., 1969, What are the Mascons? Saturday Review, v. 52, p. 54–57.
Gold, T. A., 1966, The moon's surface, in Hess, W. N., Menzel, D. H., and O'Keefe, J. A., eds., The nature of the lunar surface: Baltimore, Johns Hopkins, p. 107–129.
Kuiper, G. P., Strom, R. G., and Le Poole, R. S., 1966, Interpretation of Ranger records: Part I, sinuous rilles, section 3 of Ranger VIII and IX, Part 2: Experimenters' analysis and interpretations: Pasadena, Jet Propulsion Lab. Tech. Rept. 32–800, California Inst. Technology, p. 2423–2430.
Lingenfelter, R. E., Peale, S. J., and Schubert, G., 1968, Lunar rivers: Science, v. 161, p. 226–269.
Mills, A. A., 1969, Fluidization phenomena and possible implications for the origin of lunar craters: Nature, v. 224, p. 863–866.
Mutch, T. A., 1970, Geology of the moon—a stratigraphic view: Princeton, New Jersey, Princeton Univ. Press, 324 p.
Schroeter, J. H., 1788, Astron. Jarbuch für das Jahr 1791: Astronomisches Recheninstitut, p. 201.
Schubert, G., Lingenfelter, R. E., and Peale, S. J., 1970, The morphology, distribution and origin of lunar sinuous rilles: Rev. Geophysics, v. 8, p. 199–224.
Schumm, S. A., 1970, Experimental studies on the formation of lunar surface features by fluidization: Geol. Soc. America Bull., v. 81, p. 2539–2552.
Urey, H. C., 1967, Water on the moon: Nature, v. 216, p. 1094–1095.

MANUSCRIPT RECEIVED BY THE SOCIETY MARCH 29, 1971

CONTRIBUTION NO. 169, GEOSCIENCES DIVISION, THE UNIVERSITY OF TEXAS AT DALLAS, P.O. BOX 30365, DALLAS, TEXAS

THE GEOLOGICAL SOCIETY OF AMERICA, INC.
MEMOIR 132
© 1972

Hess Lunar Crater

DONALD U. WISE
University of Massachusetts, Department of Geology
Amherst, Massachusetts 01002

Harry Hess was a key figure in the scientific guidance leading to the first manned landing on the Moon, just two months before his death. He served as chairman of the Space Science Board of the National Academy of Sciences. He also served as a member of many advisory groups including those that helped screen astronaut candidates, decide on the experiments to be carried on the first mission, and judge the types of analyses to be performed on the returned samples. Indeed, at the time of his death he was chairing a National Academy group advising on plans for further lunar exploration.

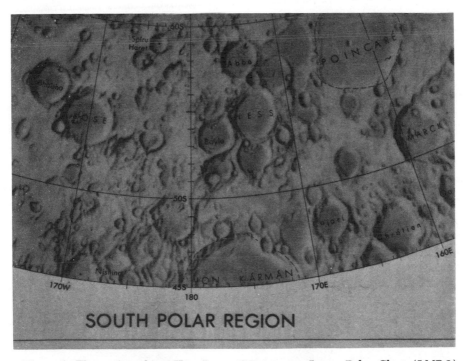

Figure 1. The region about Hess Lunar Crater. NASA Lunar Polar Chart (LMP-3) Second edition, October 1970. Note that map is printed with south at top.

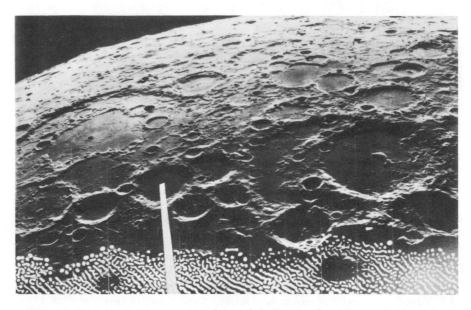

Figure 2. Hess Crater Region viewed from left (east) side of Figure 1. Von Karmen is large crater in right middle foreground. Poincaré Basin is in the left distance. NASA Lunar Orbiter V, Photo 65 M.

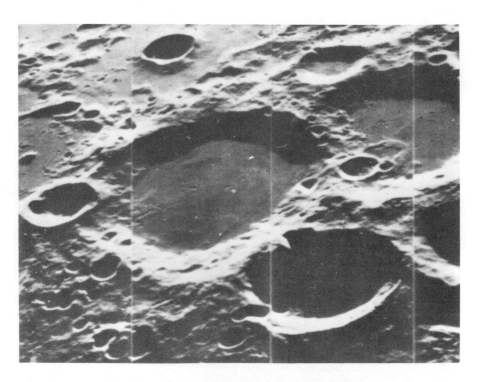

Figure 3. Enlargement of Hess Crater from Figure 2.

Accordingly, it was appropriate that the International Astronomical Union, meeting in 1970 at Brighton, England, approved the naming of a lunar crater in honor of Harry Hess and of Victor Hess of cosmic ray fame. The crater is located in the highlands of the lunar farside toward the south polar regions at 54° S., 174° E., as indicated on Figure 1. The 85-km diameter crater is part of a cluster with a hint of a north-south alignment or chain. These craters are largely younger than, and overlap onto, the projection of the outermost scarp of the 250-km diameter Poincaré Basin (Figs. 1 and 2).

Unfortunately, the only photographs of the crater presently available are quite oblique, low resolution, Lunar Orbiter imagery. These photos (Figs. 2 and 3) suggest that Hess Crater has ancient landslides on its walls with their toes buried by a younger filling of the floor. The crater was further modified by younger satellite craters on the rim and walls. A series of small rilles and crater chains trends northerly across the floor.

Future lunar explorations, for which Hess helped lay the groundwork, may return adequate imagery to do detailed geology of his crater.

Manuscript Received by the Society March 29, 1971

Index